Carbon Nanomaterials for Agri-food and Environmental Applications

Micro and Nano Technologies

Carbon Nanomaterials for Agri-food and Environmental Applications

Edited by

Kamel A. Abd-Elsalam

ELSEVIER

Elsevier
Radarweg 29, PO Box 211, 1000 AE Amsterdam, Netherlands
The Boulevard, Langford Lane, Kidlington, Oxford OX5 1GB, United Kingdom
50 Hampshire Street, 5th Floor, Cambridge, MA 02139, United States

Library of Congress Cataloging-in-Publication Data
A catalog record for this book is available from the Library of Congress

British Library Cataloguing-in-Publication Data
A catalogue record for this book is available from the British Library

ISBN: 978-0-12-819786-8

For information on all Elsevier publications
visit our website at https://www.elsevier.com/books-and-journals

Publisher: Matthew Deans
Acquisition Editor: Simon Holt
Editorial Project Manager: Mariana Pizzolatto Henriques
Production Project Manager: Nirmala Arumugam
Cover Designer: Greg Harris

Typeset by SPi Global, India

Contents

Contributors

Ruchi Aacharya Department of Botany, Mohanlal Sukhadia University, Udaipur, India

Abuelmagd M. Abdelmonem Special Food and Nutrition Research, Food Technology Research Institute, Agricultural Research Center, Oula, Egypt; Functional Nanostructures, Institute of Physical Chemistry and Electrochemistry, Leibniz Universität Hannover, Hannover, Germany

Kamel A. Abd-Elsalam Plant Pathology Research Institute, Agricultural Research Center (ARC); Unit of Excellence in Nano-Molecular Plant Pathology Research Center—Plant Pathology Research Institute, Giza, Egypt; CIHEAM IAMB, Mediterranean Agronomic Institute of Bari, Valenzano, Italy

Farah K. Ahmed Biotechnology English Program, Faculty of Agriculture, Ain Shams University, Shoubra El-Kheima, Cairo, Egypt

Mousa Alghuthaymi Department of Biology, Science and Humanities College, Shaqra University, Alquwayiyah, Saudi Arabia

Ramez A. Al-Mansob Faculty of Engineering, Technology and Built Environment, UCSI University, Kuala Lumpur, Malaysia

Hassan Almoammar National Centre for Biotechnology, King Abdulaziz City for Science and Technology (KACST), Riyadh, Saudi Arabia; ETH Zürich, Department of Biology, Institute of Microbiology, Zürich, Switzerland

Jamal M.A. Alsharef Department of Civil Engineering, Faculty of Engineering Science and Technology, Sebha University, Sebha, Libya; Department of Civil and Structural Engineering, Universiti Kebangsaan Malaysia (UKM), Bangi, Malaysia

Asran-Amal Plant Pathology Research Institute, Agricultural Research Center (ARC); Unit of Excellence in Nano-Molecular Plant Pathology Research Center—Plant Pathology Research Institute, Giza, Egypt

Katerina Atkovska Faculty of Technology and Metallurgy, University Ss. Cyril and Methodius, Skopje, Macedonia

Sadia Zafar Bajwa Nanobiotechnology Group, Industrial Biotechnology Division, National Institute for Biotechnology and Genetic Engineering (NIBGE), Faisalabad, Pakistan

Altaf H. Basta Cellulose and Paper Department, National Research Centre, Giza, Egypt

Suman V. Budihal Department of Physiology, Kasturba Medical College, Mangalore, Manipal Academy of Higher Education, Manipal, India

Luigi Calabrese Department of Engineering, University of Messina, Messina, Italy

Rajkuberan Chandrasekaran Department of Biotechnology, Karpagam Academy of Higher Education, Coimbatore, India

Somenath Chatterjee Centre for Materials Science and Nanotechnology, Sikkim Manipal Institute of Technology, Sikkim Manipal University, Gangtok, India

Hemraj Chhipa College of Horticulture and Forestry, Agriculture University, Kota, India

Aleksandar Dimitrov Faculty of Technology and Metallurgy, University Ss. Cyril and Methodius, Skopje, Macedonia

Ahmed M.A. El Hamaky Department of Mycology and Mycotoxins, Animal Health Research Institute, Agriculture Research Center, Dokki, Egypt

Ahmed I. EL-Seesy Institute for Energy Research, Jiangsu University, Zhenjiang, China; Department of Mechanical Engineering, Benha Faculty of Engineering, Benha University, Benha, Egypt

Nady A. Fathy Physical Chemistry Department, National Research Centre, Giza, Egypt

Ali Akbar Firoozi Department of Civil and Structural Engineering, Universiti Kebangsaan Malaysia (UKM), Bangi, Malaysia

Esraa Gabal Agriculture Science and Resource Management in the Tropics and Subtropics, Faculty of Agriculture, Bonn University, Bonn, Germany

Mohamed Amine Gacem Laboratory of Ecosystems Protection in Arid and Semi-Arid Area, University of Kasdi Merbah, Ouargla; Department of Biology, Faculty of Science, University of Amar Telidji, Laghouat, Algeria

Hiba Gacem Epidemiology Service and Preventive Medicine, Hassani Abdelkader University Hospital Center, Faculty of Medicine, University of Djillali Liabes, Sidi-Bel-Abbes, Algeria

Gumani Gangashe Nanotechnology and Water Sustainability Research Unit, College of Science, Engineering and Technology, University of South Africa, Roodepoort, South Africa

Panbarasi Govindasamy Civil Engineering Dept, Politeknik Ungku Omar, Ipoh, Malaysia

Anita Grozdanov Faculty of Technology and Metallurgy, University Ss. Cyril and Methodius, Skopje, Macedonia

Said Fatouh Hamed Fats and Oils Department, National Research Centre, Egypt

Ayat F. Hashim Fats and Oils Department, National Research Centre, Egypt

Hamdy Hassan Department of Energy Resources Engineering, Egypt-Japan University of Science and Technology (E-JUST), Alexandria, Egypt

Atef A. Hassan Department of Mycology and Mycotoxins, Animal Health Research Institute, Agriculture Research Center, Dokki, Egypt

Mona Kamal Hassanien Department of Microbiology, Faculty of Science, Ain Shams University, Cairo, Egypt

Zhixia He Institute for Energy Research, Jiangsu University, Zhenjiang, China

Asma Irshad National Centre of Excellence in Molecular Biology, University of the Punjab, Lahore, Pakistan

Irfan Irshad Department of Pathology, University of Veterinary and Animal Sciences, Pattoki Campus, Pakistan

Josef Jampílek Department of Analytical Chemistry, Faculty of Natural Sciences, Comenius University, Bratislava, Slovakia; Regional Centre of Advanced Technologies and Materials, Faculty of Science, Palacky University, Olomouc, Czech Republic

Khaled Kasem Department of Plant Protection, Faculty of Agriculture, Hama University, Hama, Syria

Waheed S. Khan Nanobiotechnology Group, Industrial Biotechnology Division, National Institute for Biotechnology and Genetic Engineering (NIBGE), Faisalabad, Pakistan

Katarína Kráľová Institute of Chemistry, Faculty of Natural Sciences, Comenius University, Bratislava, Slovakia

George Z. Kyzas Department of Chemistry, International Hellenic University, Kavala, Greece

Utkarsha M. Lekhak Department of Biochemistry, The Institute of Science, Mumbai, Mumbai, India

Kiril Lisichkov Faculty of Technology and Metallurgy, University Ss. Cyril and Methodius, Skopje, Macedonia

Vivian F. Lotfy Cellulose and Paper Department, National Research Centre, Giza, Egypt

Gcina Mamba Nanotechnology and Water Sustainability Research Unit, College of Science, Engineering and Technology, University of South Africa, Roodepoort, South Africa

Mogda K. Mansour Department of Biochemistry, Animal Health Research Institute, Agriculture Research Center, Dokki, Egypt

Emanuela Mastronardo Thermochemical Processes Unit, IMDEA Energy Institute, Madrid, Spain; Materials Science and Engineering, Northwestern University, Evanston, IL, United States

Kostas A. Matis Division of Chemical Technology and Industrial Chemistry, Department of Chemistry, Aristotle University of Thessaloniki, Thessaloniki, Greece

Candida Milone Department of Engineering, University of Messina, Messina, Italy

Abd El-Moez A. Mohamed School of Metallurgy and Materials, University of Birmingham, Birmingham, United Kingdom; Department of Physics, Faculty of Science, University of Sohag, Sohag, Egypt

Mohamed A. Mohamed Plant Pathology Research Institute, Agricultural Research Center (ARC), Giza, Egypt

Lerato Moss Nanotechnology and Water Sustainability Research Unit, College of Science, Engineering and Technology, University of South Africa, Roodepoort, South Africa

Manal Mostafa Plant Pathology Research Institute, Agricultural Research Center (ARC), Giza, Egypt

Anam Munawar Nanobiotechnology Group, Industrial Biotechnology Division, National Institute for Biotechnology and Genetic Engineering (NIBGE), Faisalabad, Pakistan

Velluchamy Muthuraj School of Chemistry, Shoolini University, Solan, India

Thabo T.I. Nkambule Nanotechnology and Water Sustainability Research Unit, College of Science, Engineering and Technology, University of South Africa, Roodepoort, South Africa

S. Ookawara Department of Chemical Science and Engineering, School of Materials and Chemical Technology, Tokyo Institute of Technology, Tokyo, Japan

Noha H. Oraby Department of Mycology and Mycotoxins, Animal Health Research Institute, Agriculture Research Center, Dokki, Egypt

Aminata Ould-El-Hadj-Khelil Laboratory of Ecosystems Protection in Arid and Semi-Arid Area, University of Kasdi Merbah, Ouargla, Algeria

Khalid Parwez Department of Clinical Laboratory Science, Faculty of Applied Medical Science, Al-Dawadmi, Shaqra University, Al-Dawadmi, Saudi Arabia

Sayali S. Patil Department of Environmental Science, Savitribai Phule Pune University, Pune, India

Perica Paunovik Faculty of Technology and Metallurgy, University Ss. Cyril and Methodius, Skopje, Macedonia

Elpida Piperopoulos Department of Engineering, University of Messina, Messina, Italy

Edoardo Proverbio Department of Engineering, University of Messina, Messina, Italy

P. Rajiv Department of Biotechnology, Karpagam Academy of Higher Education, Coimbatore, India

Mohamed M. Ramadan Plant Pathology Research Institute, Agricultural Research Center (ARC); Unit of Excellence in Nano-Molecular Plant Pathology Research Center—Plant Pathology Research Institute, Giza, Egypt

Sergio Ruffo Roberto Agricultural Research Center, Department of Agronomy, Londrina State University, Londrina, Brazil

Nadeem Sarwar Department of Computer Science, Bahria University Lahore Campus, Lahore, Pakistan

Rasha M.H. Sayed El Ahl Department of Mycology and Mycotoxins, Animal Health Research Institute, Agriculture Research Center, Dokki, Egypt

Tahsin Shoala Department of Environmental Biotechnology, College of Biotechnology, Misr University for Science and Technology, 6th of October City, Egypt

Mohd Raihan Taha Department of Civil and Structural Engineering, Universiti Kebangsaan Malaysia (UKM), Bangi, Malaysia

Sourbh Thakur Department of Chemistry, PSG College of Technology, Coimbatore, India; Institute of Materials Science of Kaunas University of Technology, Kaunas, Lithuania

Sethumathavan Vadivel Department of Chemistry, VHNSN College, Virudhunagar, India

Gcina D. Vilakati Department of Chemistry, University of Swaziland, Kwaluseni, Swaziland

Khamis Youssef Plant Pathology Research Institute, Agricultural Research Center (ARC), Giza, Egypt

Maira Zahid National Centre of Excellence in Molecular Biology, University of the Punjab, Lahore, Pakistan

About the editor

Kamel A. Abd-Elsalam, Ph.D., is a head researcher at the Plant Pathology Research Institute, Agricultural Research Center, Giza, Egypt. Kamel's research interests include developing, improving, and deploying plant biosecurity diagnostic tools, understanding and exploiting fungal pathogen genomes, and developing ecofriendly hybrid nanomaterials for controlling toxicogenic fungi and plant diseases. He has published 6 book chapters, 6 review articles, 1 translated book, and more than 120 research articles in international peer-reviewed journals, including *Fungal Diversity*, *Fungal Biology*, *International Journal of Biological Macromolecules*, *FEMS Microbiology Reviews*, *PLOS One*, and *PLOS Genetics*. He is an associate editor for *Mycosphere*, a review editor for *Frontiers in Genomic Assay Technology*, and a referee for journals, including *Plant Pathology*, *Journal of Phytopathology*, *Crop Protection*, *IET Nanotechnology*, *Fungal Diversity*, *BMC Genomics*, and *Foodborne Pathogens and Diseases*. Kamel pursued his Ph.D. in molecular plant pathology from Christian Alberchts University of Kiel (Germany) and Suez Canal University (Egypt); he was then awarded a postdoctoral fellowship from Christian Alberchts University of Kiel in 2008. Kamel served as visiting associate professor at the Institute of Excellence in Fungal Research, Mae Fah Luang University, Thailand; the Institute of Microbiology, TUM, Germany; the Laboratory of Phytopathology, Wageningen University, The Netherlands; and the Plant Protection Department, Sassari University, Italy.

Preface

Since the first record of three types of carbon nanomaterials–fullerenes in 1985, carbon nanotubes in 1991, and graphene in 2004–the specific characters of carbon nanomaterials have attracted great attention, which has promoted the production techniques for extensive commercial production. However, to the best of our knowledge, the applications of carbon nanomaterials (CNMs) in the agrifood and environmental sectors are still not well described. Also, there is no such book available that deals with agrifood applications for carbon nanostructures. This book provides in-depth knowledge about the properties and applications of advanced carbon nanomaterials for agroecosystem applications. This book contains 26 chapters prepared by outstanding authors from Algeria, Brazil, the Czech Republic, Egypt, England, Germany, Greece, India, Italy, Japan, Kingdom of Saudi Arabia, Libya, Lithuania, Macedonia, Malaysia, Pakistan, Slovakia, South Africa, Spain, Swaziland, Syria, and the United States. The book's authors have collected knowledge, discoveries, and the fruitful findings of carbon nanostructures and their applications in agriculture, food, and the environment. Considering this, the book has been divided into three parts: Part I deals with the synthesis and characterization of various kinds of carbon nanomaterials, Part II reviews the appropriate technological developments in environmental applications, and Part III is targeted at agrifood applications. Unique carbon nanomaterials may be applied for fast detection and separation of heavy metal ions, pesticides, and herbicides from environmental and industrial waste materials as well as from air, oil, and water solutions. The probable applications of carbon nanomaterials in agrifood sectors are manifold: the development of sensors, monitoring the environment, the treatment of wastewater, the remediation of contaminated soils, increasing crop yield (e.g., nanopesticides or nanofertilizers), biosecurity (e.g., sensors for detecting fungal plant pathogens and their mycotoxins), gene editing and delivery, antimicrobial effects in additions to soil-microbe interactions. The hazardous impact of constructed carbon nanoparticles on the environment, plants, and animals will be reviewed. Finally, the quality control and risk control of carbon nanomaterials will be discussed to provide an assessment of the health and environmental risks associated with manufactured carbon nanomaterials.

The book is of an interdisciplinary nature and will be very useful for students, teachers, researchers, and agrifood environmental scientists working in nanotechnology, materials science, biology, chemistry, physics, plant science, chemical technology, microbiology, plant physiology, biotechnology, and further groups of interest such as industry. Additionally, this will be a useful tool for industrial scientists investigating technology to update their nanotoxicology and nanosafety understanding. We hope to offer a balanced, interesting, and innovative perspective in the area, not only for advanced readers but also for industrial decision makers and those approaching the field with limited knowledge. The expected audience for the current book includes researchers, graduate students, postgraduate students, and carbon nanomaterial engineers from various fields of science and technology. We hope that this book will be useful for all researchers who also are working in this field of study.

I am really grateful to all authors who contributed chapters and provided their valuable ideas and knowledge to this edited book. Without their enthusiasm and support, the compilation of this book might never have been feasible. We also thank all the reviewers who devoted their valuable time to make critical comments on each chapter. I thank the Elsevier team for recognizing the demand for such a book and for their good assistance at every single stage of production. Finally, I would personally say thanks to my family members for their ongoing support and assistance. Any suggestions toward its further improvement will be thankfully acknowledged and incorporated in the next editions.

Kamel A. Abd-Elsalam
Agricultural Research Center, Giza, Egypt

Carbon nanomaterials: 30 years of research in agroecosystems

1

Kamel A. Abd-Elsalam
Plant Pathology Research Institute, Agricultural Research Center (ARC), Giza, Egypt

1 Introduction

The first discovery of carbon nanotubes (CNTs) must be attributed to Radushkevich and Lukyanovich in 1952. Employing a vapor growth method, Oberlin in 1976 noticed hollow carbon materials with nanometer-scale diameters (Monthioux and Kuznetsov, 2006). Carbon-based nanomaterials detected in 1985 when the fullerene C_{60} was discovered; the discovery of C_{60} was released in nature at the end of 1985 by Harry Kroto a chemist at the University of Sussex Richard Smalley at the University of Sussex and Rice University, synthesized fullerene-based carbon nanotube. Fullerenes and buckyballs are constructed exclusively of carbon of differing sizes and molecules, and are similar to a hollow sphere or tube (Kroto et al., 1985). Nevertheless, CNTs were really described in 1991 when Japanese researcher Sumio Iijima determined the presence of multiwalled CNTs (MWCNTs) as a derivative in the production of phularen (Iijima, 1991). The extensive application of CNTs in consumer supplements, compound materials, and biomedical utilization has resulted in their quickly rising development (De Volder et al., 2013). The worldwide CNT market is estimated to reach \$8.1 billion by 2024 (Cano et al., 2016; Das et al., 2019). CNTs can be employed in agriculture to improve the plant growth of different crops and enhance the performance of pesticides and fertilizers. For example, the application of CNTs in soils increased the flower, fruit production, and crop yield of tomatoes (Khodakovskaya et al., 2012). The function of carbon nanomaterials may be valuable for the gene delivery of nucleic acid, pesticides, and fertilizers to plants. It may also be valuable for wastewater purification, management of plant diseases caused by various pathogens, and sensing of essential plant molecules with a large level of sensitivity (Mohamed et al., 2018). There is not a specific book available called "Carbon Nanostructures: Agri-food and Environmental Applications"; hence, the current book will be a unique contribution in the field with almost no competition. Most of the contents of Part I deal with the synthesis and characterization of carbon nanotubes, Part II reviews new appropriate technological developments in environmental applications. Fifty percent of the book's content is in Part III, focused on agri-food applications. There are no books available on the application of carbon nanomaterials in the mentioned sectors. In the introductory chapter, the impact of CNMs is described in terms of sustainable agriculture as well as environmental and green technologies such as wastewater treatment, air

Carbon Nanomaterials for Agri-food and Environmental Applications. https://doi.org/10.1016/B978-0-12-819786-8.00001-3

pollution monitoring, pesticide sensing and degradation, gene delivery, antimicrobial activity, plant protection, etc.

2 Types of carbon nanomaterials

Soccer ball-shaped CNPs called fullerenes were discovered less than 30 years by Kroto et al. (1985), launching a new development in carbon research. The rediscovery of carbon nanotubes and the unpredicted finding of graphene provided a strong impulse for the additional production of carbon nanomaterials. Previously, these nanocarbons, along with nanodiamonds and nanofibers, were viewed as "traditional" carbon nanostructures (Kharisov and Kharissova, 2019). There are various different classifications of common carbon allotropes. Based on their particular dimensionality, low-dimensional carbon nanomaterials are usually separated into groups of zero-dimensional (0D) to one-dimensional (1D) and two-dimensional (2D), depending on their nanoscale range (Mostofizadeh et al., 2011; Tiwari et al., 2012). These kinds of nanomaterials are allotropes of carbon made of graphite, and are generally produced in cylindrical tubes with a diameter in the nanometer scale and a length of several millimeters (Hirlekar et al., 2009; Singh et al., 2012). The classic carbon nanostructures include the following nanomaterials (Fig. 1).

2.1 Carbon nanotubes

CNTs may well be among the most mentioned nanomaterials in recent publications due to their very specific characteristics, such as being highly stable and offering promising applications. For this reason, they are attracting research and industry. The researchers divided CNTs into three types: single, double, and multiwalled nanotubes. CNTs are commonly arranged depending on the number of sidewalls and the atomic structure in the sidewalls (Sinnott and Andrews, 2001). CNTs are possibly important in a large number of applications, including solar cells, nanomedicine, actuators, capacitors, sensors, nanoelectronics, optics, etc. Also, they may have various environmental applications such as wastewater purification to eradicate oil pollutants such as oils, fertilizers, heavy metals, pesticides, pharmaceuticals, etc. The utilization of CNTs as an adsorbent is common because of their unique cylindrical form with exclusive properties. Carbon nanotubes are very sensitive and selective when it comes to the detection of environmental contaminants that are shown by carbon nanotube-based sensors. Sensor manufacturing is among the most critical applications of functionalized CNTs. DNA sequences in the body are often recognized by CNT-based nanobiosensors (Qiu et al., 2015). For example, CNT-based sensors are generally integrated into circuitry in smart packaging that enables real-time information on the state of food using smartphones or other connected devices (Zhu et al., 2017) as well as the detection of fruit ripening (Esser et al., 2012), food spoilage (Liu et al., 2015a,b), and pesticides (Krishnan et al., 2019). The sensing features of CNT-based sensors and the use of CNT-based devices present in chemical realizing was investigated, also the

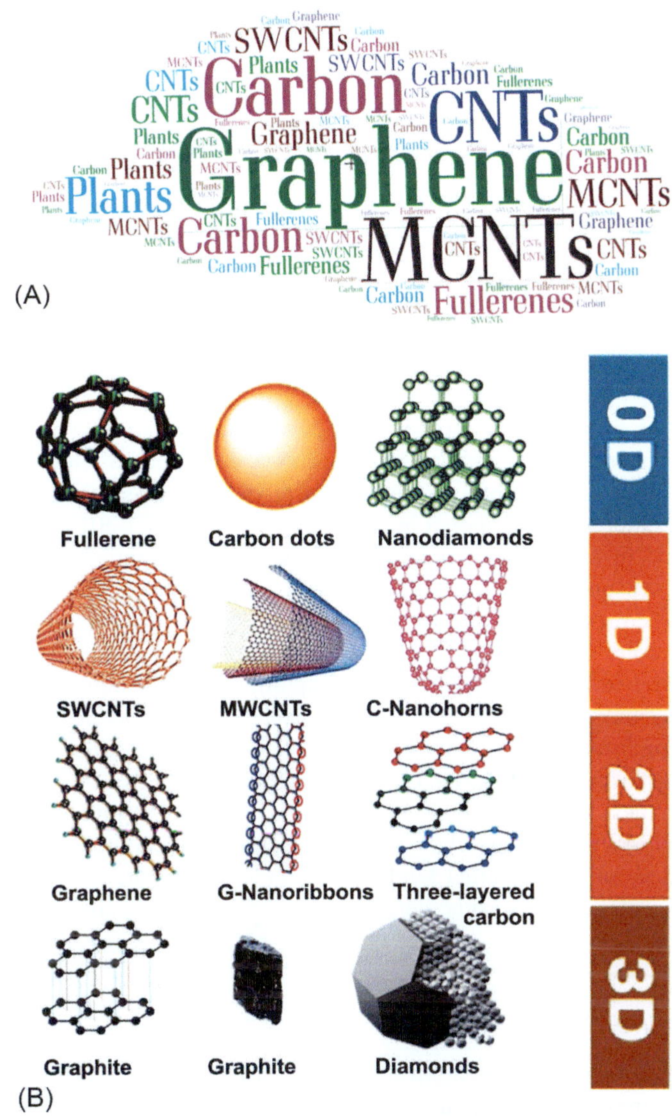

Fig. 1 A. Carbon is one of the most flexible elements in the periodic table and forms a large number of allotropes. Carbon nanomaterials present exceptional variations in the chemical structure, which consist of carbon nanotubes (CNTs), graphene, carbon dots, fullerene (C_{60}, C_{36}…), graphene quantum dots, carbon nanohorns, onion-like carbon, nanodiamondoids, and carbon nanoribbons regarding the nanoscale range, <100 nm. B. Regarding their dimensionalities, carbon nanomaterials may be separated into several categories: (i) zero-dimensional (0D) including fullerene, carbon quantum dots, and nanodiamonds, (ii) one-dimensional (1D) such as carbon nanotubes and carbon nanohorns, and (iii) two-dimensional (2D) that include graphene, carbon (graphene) nanoribbons, and few-layered graphene (graphite and diamonds are classified as 3D carbon allotropes).

considerable applications of CNT-based sensors to diverse aspects including ecosystem monitoring, farming and foodstuff applications, biological devices, and nationwide security was first described in details simply by Schroeder et al. (2019).

2.2 Graphene

An additional important type of CNMs are graphene and graphene oxide is a great group of graphite-like products. Graphene is the sole membrane of carbon atoms closely loaded into a 2D honeycomb crystal lattice (Debgupta et al., 2014) with zero-energy band gaps and unique characteristics that include large conductivity, significant surface area, and high mechanical and thermal stability. Graphene is extremely hydrophobic in nature, is not soluble in the hydrophilic solvent, and forms large aggregates via pep interaction/van der Waals interaction. These characteristics give graphene beneficial materials in numerous applicable studies. Lately, many review articles have focused on graphene-based composites with nanoparticles and polymers, highlighting mainly the methods and properties while even modest focus is paid to probable applications (Chen et al., 2012; Fei et al., 2014; Upadhyay et al., 2014). Graphene has been employed in a number of applications, incorporating anticorrosion films and paints, active sensors, microelectronics batteries, supercapacitor electrode materials, quicker DNA sequencing, fluorescence bioimaging, photocatalysis, gene and bioactive delivery, water treatment, and biosensing (Huang et al., 2012; Upadhyay et al., 2014; Xiaoqiang and Jimmy, 2011). The hydrophilicity of GO improves as a result of the presence of oxygenated molecules, which usually permits its utilization in water and natural conditions. A high amount of graphene and its manual composites are applied for wastewater purification. Contaminants and toxins that include small-scale harmful molecules, heavy metals, dyes, pesticides, aromatic pollutants, and antibiotics are eliminated by using adsorbent applying the graphene composites (Yang et al., 2013). Activated carbon and other kinds of carbon allotropes are perfect examples of adsorbents of water contaminants; and graphene owing a great specific surface area for pollutants remediations.

2.3 Fullerenes

Fullerenes are a type of carbon allotropes, and consist of nanotubes and buckyballs, likewise Fullerenes of C_{60}, C_{70}, and C_{84} well known. The functions of fullerene with a functionalized byproduct are elevated in diverse biomedical applications including pharmaceutical drugs, medicines, photovoltaics, etc. (Mudshinge et al., 2011). Fullerenes can also be screened for the adsorption of organic ingredients. Fullerenes are not great sorbents intended for organic chemical substances, but they are effective for removing organometallic pollutants (Ballesteros et al., 2000). Fullerenes are employed for the delivery of medication and image resolution instruments in some practical modules (Bolskar, 2012). Fullerenes and fullerols have received much attention from plant researchers on probable agri-food applications such as managing plant growth and crop improvement. An increase in the correct quantity of fullerenes to the reformatory plant system can easily boost the performance of the plants for nitrogen fixation and reduction of organic contaminants such as trichloroethylene. This implies that fullerenes may be beneficial as a positive alternative in a plant remediation system (Verma et al., 2019).

3 Carbon nanomaterial applications

CNTs have been the main research theme for more than 20 years. Reflecting this unique academic effort is the global industrial interest, resulting in the development capacity of some thousand tons of CNTs per year (De Volder et al., 2013). Graphene has various applications in different fields as well, including electronics, various biochemical sensors, and solar cells (Choi et al., 2010). These developments have provided approaches to an extensive array of applications (Schnorr and Swager, 2010) in microelectronics (Park et al., 2013), computing (Shulaker et al., 2013), medicinal therapy (Hong et al., 2015), and electrochemical biosensors (Wang, 2005). In addition, a more significant effort may be utilized in the production of functional CNT-based chemical sensors (Meyyappan, 2016; Schroeder et al., 2019). Multiple functions in food administration and agriculture production might be examined and improved by using chemical sensing. Particularly, these functions involve monitoring of food ripeness, elimination of food spoilage, providing high-quality food storage (Fig. 2), and diagnosis of pesticides and plant pathogens. Economically practical

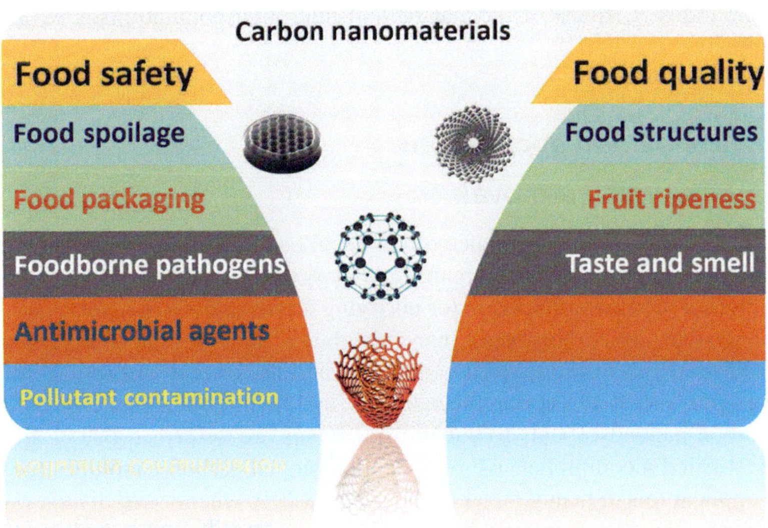

Fig. 2 Carbon-based nanomaterial applications in the food industry include (1) food safety, food spoilage, food packaging, food-borne pathogens, antimicrobial agents, and pollutant contamination while (2) food quality includes the detection of food structures, fruit ripeness, and screening the food taste and smell.

sensing devices can advise both producers and customers, strengthen the skills of harvesting and shipping, screen storage in both commercial and purchaser settings, and determine the pesticide residue in polluted food. For example, CNT-based sensors are generally integrated into circuitry that allows for real-time facts on the status of food through mobile phones or different connected devices. This can be applied in pesticide detection (Chen et al., 2008), fruit ripening (Esser et al., 2012), smart packaging (Liu et al., 2015a,b), and food spoilage (Zhu et al., 2017). Recently, quantum-dot CD have several applications such as food area, regarding the screening object, discovering environment, analysis of nutrients, limited or prohibited substances, food-borne pathogenic bacteria and their toxins and the mode of action used in the sensors were reviewed (Qu et al., 2018).

A comprehensive understanding of the mechanistic interactions of carbon nanomaterials (CNMs) with the plant rhizosphere and associated microbial communities is required to assess the problems linked with common use in agriculture. Mukherjee et al. (2016) provided a complete analysis of the existing research materials on CNM's fate in the soil, introducing the impact on plants correlated with soil microorganisms. Carbon nanomaterials are available to be developed on a significant scale for an extensive range of utility fields incorporating the environmental and agri-food sectors (Zaytseva and Neumann, 2016). Also, special interest was offered to the biotic applications of graphene (antimicrobial packaging material for food) and in sensing food quality (detection of chemical pollutants, recognition of food structure, detection of pesticides). In the current book, we targeted the applications of micro/nano biochar in the decontamination of wastewater, the bioremediation of pesticides, antimicrobials, fertilizers, slow-release fertilizers, plant improvement, silage agents, plant disease control, the indirect effects of biochar on soil microbial populations, etc., as shown below (Fig. 3).

3.1 Environmental applications

3.1.1 Heavy metal removal

A great surface area, the convenience of chemical or physical changes, and the capability of eliminating each of those organic and inorganic pollutants make carbon-based nanomaterials a promising selection for purifying wastewater (Rodrigues et al., 2015). Different designs of carbon-based nanoadsorbents and their adsorbent activities, widespread strategies to improve carbon nanomaterials, several adsorption isotherms, kinetic designs, varied adsorption mechanisms, and functional applications of the improved carbon nanoadsorbents to aquatic specimens are described (Yu et al., 2018). They investigated a comprehensive review of employed carbon nanotubes for the adsorption applications of heavy metals. Abbas et al. (2016) mentioned diverse variables that can impact the adsorption behavior of CNTs during pollutant remediation stages. CNTs may have a high absorbent level and a large adsorption proficiency while the convenience of recovery and recycling has triggered common attention to remedy heavy metals and water contaminants (Mallakpour and Khadem, 2019; Zhang et al., 2015).

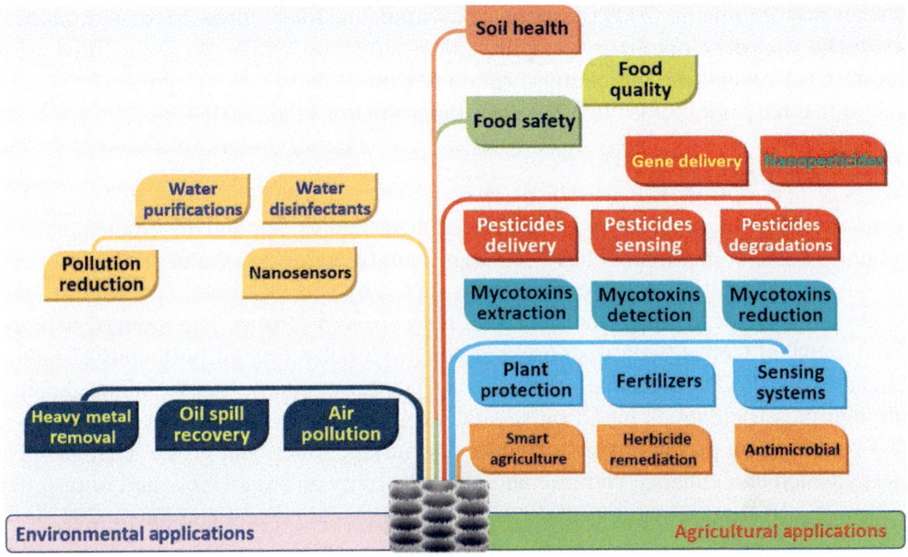

Fig. 3 Carbon-based nanomaterial applications in the environmental and agri-food sectors, environmental applications of carbon nanomaterials including reducing pollution such as air pollution, heavy metals, oil spill, and water contaminations. For instance, carbon nanosensors can be used in monitoring of various pollutant air, soil and wastewater pollutant such as heavy metals, toxic materials and pesticides. Another application of carbon nanosorbents in agroecosystem treatment is capturing heavy metal ions and organic contaminants from different media. While the CNMs applications in agri-food sectors including, smart agriculture, the low release of agrochemicals, pesticides detection, pesticides degradations, antimicrobial, gene delivery, plant diseases management, detection and reductions different mycotoxins. For example, CNMs enhance seed germination in several plant species, improve nutrients and water uptake in plants, increase in vegetative growth and enhanced crop yield, also carbon dots able to activate the plant photosystem, finally, MWCNTs facilitate water channels and increase nitrogenase activity. As nanotransporters for different bioactive materials in plant, soil and animal systems such as DNA delivery, pesticides and fertilizers; smart carbon core-shell nanomaterials can be target specific parts in plant, in addition to carbon nanofibers in gene therapy of plants. Different applications of carbon-based nanomaterial may apply for improved agricultural production and in environmental protection.

3.1.2 Oil spill recovery

Numerous forms of carbon nanomaterials have always been frequently explored for water filtration, water/oil separation, oil-spill cleanup, wastewater remedies, and gas separation and purification (Gupta and Tai, 2016). Multishell nanocarboscavengers (NCS) with good treatment proficiency via a mix of absorption and dispersion characteristics were used for petroleum spill remediation through booms and fine mesh equipment (Daza et al., 2017). Because NCS is produced from organic carbohydrate origins, the calibrating particle structure and passivating providers with huge nitrogen and phosphorous components can increase NCS ability as a bioremediation product to hydrocarbon-degrading microorganisms in oil spill management

(Joseph and Lehmann, 2009). Silicone polyurethane foam formulated with carbon nanotubes was developed for oil spills and purification while the foam filled with pristine CNT reveals the maximum sorption potential (800 wt% in virgin naphtha) and exceptional recycling (close to 10 times) (Piperopoulos et al., 2018).

3.1.3 Air pollution

Most recently, the production of carbon nanostructures applied to eradicate large amounts of gas contaminants has been significantly reviewed (Wang et al., 2013). A carbon nanotube filtration system was constructed by adding carbon nanotubes to the membrane layer of distinct materials with pore sizes of 10 mm. The reduced activity of the prepared CNT filters was observed whenever collecting air pollutants of motor vehicle exhaust. CNTs can be applied to reduce gases dangerous to human health, although the efficiency is based on several variables (Romero-Guzmán et al., 2018). CNT composition permits hybrids of over one surface functional group (OH, COOH, C=O), which can enhance carbon nanotube selectivity and stableness and impact the maximum adsorbent capability of the remediation model (Gupta and Saleh, 2013).

3.1.4 Water purification

Among the promising aspects of CNT and graphene-based materials is their application for water purification, which is used a low quantity of material required to achieve great absorbent or filtration potential (Sui et al., 2012). All these nanomaterials are structured on a graphite-like carbon platform and generally employ an adsorption procedure for cleaning out diverse pollutants. These types of carbon nanomaterials have also revealed promise as size exclusion membrane filters that provide water movement when preventing the flow of pollutants and toxins (Sui et al., 2012). Specific chemical improvements or modified composites of the described nanomaterials are also shown as practical for neutralization of hazardous water-borne pathogens (Karumuri et al., 2013). The current fee for producing such nanoscale graphene or CNT water purification materials makes them as yet improper for the industrial scale, although most researchers find these materials as the new strategy for the next era in water treatment with large efficiency and flexibility features (Smith and Rodrigues, 2015). The results of recently formulated CNMs, CNTs, and graphene materials and their particular composites for water purification were discussed in detail (Sweetman et al., 2017).

3.2 Agricultural applications

3.2.1 Smart agriculture

Recently, researchers developed nanosensor-based monitoring devices for enhancing plant health that include detection and management of diseases, host-parasite relationships. At the molecular biology level employing nanosensors to detect plant disease, reduction pollutants from soil and water, postharvest control of vegetables and flowers, and reclamation of salt-affected soils, etc. Nanobiosensors can also be applied for sensing a diverse number of plant pathogens and fertilizers, decreasing the damage of nutrients, and improving crop yields by using an integrated plant nutrient management system (Kaushal and Wani, 2017). In this framework, a digital device for real-time sensing of harmful gases based on CNTs, which can be inserted into insects or plants, may be used in distant sensing systems for plant diseases (Lee et al., 2014). An acetylcholinesterase biosensor depending on the assembly of multiwalled carbon nanotubes onto liposome bioreactors was designed for the diagnosis of organophosphate pesticides (Yan et al., 2013). SWCNT-based nanosensor hybrids with metal/metal oxide nanoparticles or nanotubes for gases, namely, ammonia, nitrogen oxide, hydrogen sulfide, sulfur dioxide, and volatile organics, have potential applications in measuring agroecosystem contaminants, assessing the effects on living matter or health, and improving crop yields (Sekhon, 2014).

3.2.2 Antimicrobial agents

Carbon nanostructures (CNSs) including fullerene, CNTs, graphene, and diamond-like carbon (DLC) are generally shown to acquire strong broad-spectrum antimicrobial efficiency against diverse food, water-borne, and plant pathogens (Al-Jumaili et al., 2017; Dizaj et al., 2015; Jung et al., 2011). CNSs may possibly offer essential applications for plant protection, but the fungicidal properties of MWCNTs with unique surface areas against important phytopathogenic fungi such as *Fusarium graminearum* were tested (Wang et al., 2014). The antimicrobial activities of CNTs, the analysis of their varied modes of action against different pathogens, and their particular use for water treatment, microbial control, and water purification procedures are additionally presented by Liu et al. (2018b).

3.2.3 Fertilizers

Utilizing diverse low-cost carbon nanomaterials such as graphene oxide helps the extensive production of encapsulated fertilizers with a controlled release strategy to be available at a more economical price (Zhang et al., 2014). Several of these types of fertilizers are based on amendments of nutrient and biofertilizers with carbon nanomaterials, which in most conditions serves as a fertilizer synergist with the purpose of enhancing plant nutrient supply while minimizing beneficial nutrient deficits and simulative plant health (Zaytseva and Neumann, 2016). Carbon nanotubes can be utilized as a nutrient transporter for both macro- and microelements and even as slow-release fertilizers, which may lessen the excessive concentrations that are commonly applied (Hasaneen et al., 2017). The growth and yield structure differ from plant to plant and are influenced by the number of CNMs implemented. Scientific reports observed a boost in plant vigor growth and the yield of fruits/seeds at a lower concentration of CNMs, but a decline in all these parameters was also documented when bigger concentrations of CNMs were employed (Verma et al., 2019).

3.2.4 Nanopesticides

Because of their antifungal activities, carbon nanomaterials are encouraging substances for the production of different types of nanopesticides. The impact of industrial MWCNTs on fungal conidia of the entomopathogenic *Paecilomyces fumosoroseus* was explored, and the CNTs highly reduced the spore formation of mycelium in employed concentrations (Gorczyca et al., 2009). Sarlak et al. (2014) have also applied the MWCNT cover with poly (citric acid) as a transporter for the water-dissolvable dithiocarbamate fungicides. Nanobiopesticides depict the next generation of common synthetic pesticides that might offer better features, including elevated efficiency and longevity, with fewer quantities of active chemicals needed below the field or greenhouse conditions. In fact, CNMs may apply as innovative approaches for eco-friendly antimicrobial agents against plant pathogenic and toxigenic fungi (Abd-Elsalam et al., 2019).

3.2.5 Pesticides sensing

Soil pollution by pesticide residues is a severe issue in the agricultural sector. A few researchers have examined the feasibility of carbon dots in pesticide detection in various agricultural and food samples (Wang et al., 2016). CNT-based optical and electrochemical sensors possess a competent potential to examine very tiny quantities of insecticide residues specifically on the plant surface, which can certainly offer vital benefits in foodstuff production and industry (Wong et al., 2017). Zhao et al. (2018) evaluated modern innovations and unique developments in carbon nanomaterial-enabled biosensors for the detection of pesticides, incorporating the application of graphene, carbon nanotubes, carbon dots, and various other types of carbon nanomaterials with a couple of major biosensing techniques such as enzyme and enzyme-free biosensors. Additionally, a comprehensive review reference for researchers in biosensors, graphene, materials science, nanotechnology, chemistry, and electrochemistry was explored by Krishnan et al. (2019).

3.2.6 Pesticide remediation

Carbon nanomaterials, which include biochar and activated carbon, described by substantial specific surface areas are generally employed as successful adsorbents to reduce a diverse collection of pollutants over the long term. Recently more research interest is dedicated to nanosize carbon-based adsorbents that can be ready to remediate different pesticides or may also significantly contribute to the improved degradation of pesticides in water and soil media. The sensible adsorbent metal-organic platform magnetic multiwalled carbon nanotubes were effectively developed to reduce some organophosphorus pesticides from environmental water and soil samples (Dehghani et al., 2019; Liu et al., 2018a).

3.2.7 Plant protection

Carbon nanomaterials may be applied effectively to plant diseases to manage several plant pathogens and detection methods and minimize fungal aflatoxins. MWCNTs were layered with distinct metallic nanomaterials such as Ag NPs or Zn NPs in a good approach to inhibit the fungal growth of the toxigenic fungi *Aspergillus fumigatus* and *Aspergillus ochraceus* (Fosso-Kankeu et al., 2016). Zaytseva and Neumann recorded interesting results that showed the broad-spectrum antifungal efficacy of those nanomaterials against some fungal pathogens, namely *F. graminearum* and *Fusarium poae*, compared to other carbon-based nanomaterials such as fullerenes and graphene oxide (Zaytseva and Neumann, 2016).

3.2.8 Mycotoxin extraction

Recently, MWCNTs, a unique type of solid phase extraction absorbent, have generally been applied to purify mycotoxins from various samples because of their particular adsorption ability, specific composition, and efficiency of being functionally changed as carboxylic MWCNT (MWCNT–COOH) and hydroxyl MWCNT (MWCNT–OH) (Moazzen et al., 2013). Activated carbon (AC) was extracted from the seed shells of *Jatropha curcas* and was employed to extract the zearalenone (ZEA) mycotoxin (Kalagatur et al., 2017). A new and quick one-step washing method using MWCNT–COOH $+C_{18}$ as the SPE sorbent identified the 21 particular mycotoxins, which was confirmed in corn and wheat, recommending an application of the suggested approach in the investigation of the distinct mycotoxins in several crops (Jiang et al., 2018).

3.2.9 Mycotoxins detection

Some types of carbon nanomaterials such as SWCNTs, MWCNTs, graphene, and nanodiamonds are generally used in the production of optical biosensors to get rapid and efficient detection of mycotoxins (Guo et al., 2011). Singh et al. (2013) developed a unique electrophoretically deposited C-MWCNT sensor for the covalent bond of anti-AFB1 for delicate detectors of aflatoxin B1 using an electrochemical procedure. A new electrochemical immunosensor was developed for the detection of AFB1 by combining SWNTs/CS with an enzymatic signal monitor. The design methodology showed an appropriate fabrication reproducibility and a substantial level of specificity and sensitivity for the AFB1 assay (Zhang et al., 2016). Wang et al. (2016) employed AuNPs as a quenching element of the fluorescent nitrogen-doped carbon dots (N, C-dots) constructed on an AuNP-modified aptamer. In the existence of AFB1, the fluorescence of the N, the C-dots retrieved, and the measurement of Fl intensity allowed the diagnosis of AFB1 with a LOD of 0.005 μg/L. The test applicability has been confirmed in corn trials with a great level of sensitivity, considerably better than that provided by the classic HPLC technique.

3.2.10 Mycotoxin reduction

Activated charcoal continues to be utilized for mycotoxin reduction for the long term. Chemically, the carbon-carbon covalent bonds and crystalline composition offer particular characteristics that include durability, flexibility, and good conductivity. Different types of carbon nanomaterials such as graphene, graphene oxide, nanodiamonds, fullerenes, fiber, and nanotubes possess a great capability to be innovative adsorbents for mycotoxin reduction (Horky et al., 2018). Preliminary reports of the impact of fullerenol $C_{60}(OH)_{24}$ nanoparticles on the fungal development, aflatoxin formation, and oxidative stress modulation in *Aspergillus flavus* demonstrated that mycelial biomass weight fairly decreased but aflatoxin accumulation in media was considerably reduced. Using a concentration of 10 ng/mL of fullerenol reduced construction when compared with fungus control, which usually resulted from the inhibition of oxidative stress in the fungal culture (Kovac et al., 2017).

3.2.11 Gene delivery

However, some experts mentioned that both MWCNTs and SWCNHs were recognized via Raman spectroscopy and transmission electron microscopy (TEM) (Lahiani et al., 2015). Serag and his coworkers noticed that single-stranded DNA molecules were warped surrounding SWCNTs to target the cytoplasms of the plant cell wall. These features could be used to insert RNA pieces into the nucleus to stimulate or silence the targeted genes. Likewise, protoplast can be an objective for supplying bigger DNA molecules, which includes the transport of plasmids inside the plant cell genome. Therefore, it is necessary to consider all practical impacts of carbon-based materials applied as the DNA delivery system on the plant genome and proteome (Serag et al., 2015). Recently, Kwak et al. (2019) attempted a novel strategy to deliver the transgene into the chloroplast specifically by nanotubes. During his investigation, he produced chitosan-complexed single-walled carbon nanotubes that selectively deliver plasmid DNA to the chloroplasts of a few adult plants such as *Eruca sativa*, *Nasturtium officinale*, *Nicotiana tabacum*, and *Spinacia oleracea* as well as in the isolated mesophyll protoplasts of *Arabidopsis thaliana*.

4 Conclusion and future outlook

Presently, carbon-allotrope forms are one of the most effective growing areas of science and nanobiotechnology, leading to unique horizons of advanced research. This chapter offers an extensive literature review on carbon nanomaterials with specialized focus on plant reactions as well as environmental and agricultural applications. CNMs possess great potential in the agri-food sector, offering exclusive novel economic strategies for plant promotions and crop protection with ecofriendly and negligible constraints. Various carbon nanostructures such as fullerene, carbon nanotubes, graphene, and diamond-like carbon have exhibited their antimicrobial activities against an extensive range of food or plant pathogens. CNT-based sensors have already been produced to discover food spoilage by using biomarkers involved with food spoilage directly or by verifying the integrity of the packaging. The application of CNMs for large-scale commercial production will lead to the continued accumulation of CNMs in soils, which can reach a concentration that is regarding plant uptake and human vulnerability. However, many investigations have confirmed the confident impact of CNMs on plant growth, with superb potential for agri-food applications. Controlling the environmental threats and application of CNMs requires more reports regarding their concentration in environmental media, such as agroecosystems. For environmental problem examination, the probable influence of CNMs on nontarget microorganisms also needs detailed analysis prior to the commercialization of new nanoagrochemicals. Trusted information concerning the outcome of nanocarbons, which includes CNTs on plant physiology and plant production at the species level, is limited. However, there are various issues that still must be resolved to understand most agricultural and environmental applications. The current book regarding the impact of a large variety of carbon nanomaterials also addresses the current most relevant knowledge gaps.

References

Abbas, A., Al-Amer, A.M., Laoui, T., Al-Marri, M.J., Nasser, M.S., Khraisheh, M., Atieh, M.A., 2016. Heavy metal removal from aqueous solution by advanced carbon nanotubes: critical review of adsorption applications. Sep. Purif. Technol. 157, 141–161.

Abd-Elsalam, K.A., Al-Dhabaan, F.A., Alghuthaymi, M., Njobeh, P.B., Almoammar, H., 2019. Nanobiofungicides: present concept and future perspectives in fungal control. In: Nano-Biopesticides Today and Future Perspectives. Academic Press, pp. 315–351.

Al-Jumaili, A., Alancherry, S., Bazaka, K., Jacob, M., 2017. Review on the antimicrobial properties of carbon nanostructures. Materials 10 (9), 1066.

Ballesteros, E., Gallego, M., Valcarcel, M., 2000. Analytical potential of fullerene as adsorbent for organic and organometallic compounds from aqueous solutions. J. Chromatogr. A 869 (1–2), 101–110.

Bolskar, R.D., 2012. Fullerenes for drug delivery. In: Bhushan, B. (Ed.), Encyclopedia of Nanotechnology. Springer, Amsterdam, pp. 898–911.

Cano, A.M., Kohl, K., Deleon, S., Payton, P., Irin, F., Saed, M., Shah, S.A., Green, M.J., Cañas-Carrell, J.E., 2016. Determination of uptake, accumulation, and stress effects in corn (*Zea mays* L.) grown in single–wall carbon nanotube contaminated soil. Chemosphere 152, 117–122.

Chen, H., Zuo, X., Su, S., Tang, Z., Wu, A., Song, S., Zhang, D., Fan, C., 2008. An electrochemical sensor for pesticide assays based on carbon nanotube–enhanced acetycholinesterase activity. Analyst 133 (9), 1182–1186.

Chen, D., Feng, H., Li, J., 2012. Graphene oxide: preparation, functionalization, and electrochemical applications. Chem. Rev. 112, 6027–6053.

Choi, W., Lahiri, I., Seelaboyina, R., Kang, Y.S., 2010. Synthesis of graphene and its applications: a review. Crit. Rev. Solid State Mater. Sci. 35 (1), 52–71.

Das, K.K., Nava, V., Chang, C.W., Chan, J.W., Xing, B., Yang, Y., 2019. Emerging investigator series: quantification of multiwall carbon nanotubes in plant tissues with spectroscopic analysis. Environ. Sci. Nano 6 (2), 380–387.

Daza, E.A., Misra, S.K., Scott, J., Tripathi, I., Promisel, C., Sharma, B.K., Topczewski, J., Chaudhuri, S., Pan, D., 2017. Multi–shell nano–carboscavengers for petroleum spill remediation. Sci. Rep. 7, 41880.

De Volder, M.F., Tawfick, S.H., Baughman, R.H., Hart, A.J., 2013. Carbon nanotubes: present and future commercial applications. Science 339 (6119), 535–539.

Debgupta, J., Mandal, S., Kalita, H., Aslam, M., Patra, A., Pillai, V., 2014. Photophysical and photoconductivity properties of thiol–functionalized graphene–CdSe QD composites. RSC Adv. 4, 13788–13795.

Dehghani, M.H., Kamalian, S., Shayeghi, M., Yousefi, M., Heidarinejad, Z., Agarwal, S., Gupta, V.K., 2019. High–performance removal of diazinon pesticide from water using multi–walled carbon nanotubes. Microchem. J. 145, 486–491.

Dizaj, S.M., Mennati, A., Jafari, S., Khezri, K., Adibkia, K., 2015. Antimicrobial activity of carbon–based nanoparticles. Adv. Pharm. Bull. 5 (1), 19.

Esser, B., Schnorr, J.M., Swager, T.M., 2012. Selective detection of ethylene gas using carbon nanotube-based devices: utility in determination of fruit ripeness. Angew. Chem. Int. Ed. 51, 5752–5756.

Fei, H., Peng, Z., Li, L., Yang, Y., Lu, W., Samuel, E.L., Fan, X., Tour, J.M., 2014. Preparation of carbon–coated iron oxide nanoparticles dispersed on graphene sheets and applications as advanced anode materials for lithium–ion batteries. Nano Res. 7, 502–507.

Fosso-Kankeu, E., De Klerk, C.M., Botha, T.A., Waanders, F., Phoku, J., Pandey, S., 2016. The antifungal activities of multi-walled carbon nanotubes decorated with silver, copper and zinc oxide particles. In: International Conference on Advances in Science, Engineering, Technology and Natural Resources (ICASETNR-16), Parys, South Africa, pp. 24–25.

Gorczyca, A., Kasprowicz, M.J., Lemek, T., 2009. Physiological effect of multi-walled carbon nanotubes (MWCNTs) on conidia of the entomopathogenic fungus, *Paecilomyces fumosoroseus* (*Deuteromycotina: Hyphomycetes*). J. Environ. Sci. Health Part A 44 (14), 1592–1597.

Guo, Z., Ren, J., Wang, J., Wang, E., 2011. Single–walled carbon nanotubes based quenching of free FAM–aptamer for selective determination of ochratoxin A. Talanta 85 (5), 2517–2521.

Gupta, V.K., Saleh, T.A., 2013. Sorption of pollutants by porous carbon, carbon nanotubes and fullerene–an overview. Environ. Sci. Pollut. Res. 20 (5), 2828–2843.

Gupta, S., Tai, N.H., 2016. Carbon materials as oil sorbents: a review on the synthesis and performance. J. Mater. Chem. A 4 (5), 1550–1565.

Hasaneen, N., Abdel-Aziz, H., Omer, A., 2017. Characterization of carbon nanotubes loaded with nitrogen, phosphorus and potassium fertilizers. Am. J. Nano Res. Appl. 5 (2), 12–18.

Hirlekar, R., Yamagar, M., Garse, H., Vij, M., Kadam, V., 2009. Carbon nanotubes and its applications: a review. Asian J. Pharm. Clin. Res. 2 (4), 17–27.

Hong, G., Diao, S., Antaris, A.L., Dai, H., 2015. Carbon nanomaterials for biological imaging and nanomedicinal therapy. Chem. Rev. 115 (19), 10816–10906.

Horky, P., Skalickova, S., Baholet, D., Skladanka, J., 2018. Nanoparticles as a solution for eliminating the risk of mycotoxins. Nanomaterials 8 (9), 727.

Huang, C., Li, C., Shi, G., 2012. Graphene based catalysts. Energy Environ. Sci. 5, 8848–8868.

Iijima, S., 1991. Helical microtubules of graphitic carbon. Nature 354 (6348), 56.

Jiang, D., Wei, D., Wang, L., Ma, S., Du, Y., Wang, M., 2018. Multiwalled carbon nanotube for one–step cleanup of 21 mycotoxins in corn and wheat prior to ultraperformance liquid chromatography–tandem mass spectrometry analysis. Toxins 10 (10), 409.

Joseph, S., Lehmann, J., 2009. Biochar for environmental management: Science and technology (No. 631.422 B615bi). Earthscan, London.

Jung, J.H., Hwang, G.B., Lee, J.E., Bae, G.N., 2011. Preparation of airborne Ag/CNT hybrid nanoparticles using an aerosol process and their application to antimicrobial air filtration. Langmuir 27 (16), 10256–10264.

Kalagatur, N.K., Karthick, K., Allen, J.A., Ghosh, N., Sivaraman, O., Chandranayaka, S., Gupta, V.K., Krishna, K., Mudili, V., 2017. Application of activated carbon derived from seed shells of *Jatropha curcas* for decontamination of zearalenone mycotoxin. Front. Pharmacol. 8, 760.

Karumuri, A.K., Oswal, D.P., Hostetler, H.A., Mukhopadhyay, S.M., 2013. Silver nanoparticles attached to porous carbon substrates: robust materials for chemical–free water disinfection. Mater. Lett. 109, 83–87.

Kaushal, M., Wani, S.P., 2017. Nanosensors: frontiers in precision agriculture. In: Nanotechnology. Springer, Singapore, pp. 279–291.

Kharisov, B.I., Kharissova, O.V., 2019. Carbon Allotropes: Metal–Complex Chemistry, Properties and Applications. Springer.

Khodakovskaya, M.V., De Silva, K., Biris, A.S., Dervishi, E., Villagarcia, H., 2012. Carbon nanotubes induce growth enhancement of tobacco cells. ACS Nano 6 (3), 2128–2135.

Kovac, T., Sarkanj, B., Klapec, T., Borisev, I., Kovac, M., Nevistic, A., Strelec, I., 2017. Fullerol C60(OH)24 nanoparticles and mycotoxigenic fungi: a preliminary investigation into modulation of mycotoxin production. Environ. Sci. Pollut. Res. 24, 16673–16681.

Krishnan, S.K., Singh, E., Singh, P., Meyyappan, M., Nalwa, H.S., 2019. A review on graphene–based nanocomposites for electrochemical and fluorescent biosensors. RSC Adv. 9 (16), 8778–8881.

Kroto, H.W., Heath, J.R., O'Brien, S.C., Curl, R.F., Smalley, R.E., 1985. C60: buckminsterfullerene. Nature 318, 162–163.

Kwak, S.Y., Lew, T.T.S., Sweeney, C.J., Koman, V.B., Wong, M.H., Bohmert-Tatarev, K., Snell, K.D., Seo, J.S., Chua, N.H., Strano, M.S., 2019. Chloroplast–selective gene delivery and expression in planta using chitosan–complexed single–walled carbon nanotube carriers. Nat. Nanotechol. 14 (5), 447.

Lahiani, M.H., Chen, J., Irin, F., Puretzky, A.A., Green, M.J., Khodakovskaya, M.V., 2015. Interaction of carbon nanohorns with plants: uptake and biological effects. Carbon 81, 607–619.

Lee, K., Park, J., Lee, M.S., Kim, J., Hyun, B.G., Kang, D.J., Na, K., Lee, C.Y., Bien, F., Park, J.U., 2014. In–situ synthesis of carbon nanotube–graphite electronic devices and their integrations onto surfaces of live plants and insects. Nano Lett. 14 (5), 2647–2654.

Liu, S.F., Moh, L.C., Swager, T.M., 2015a. Single–walled carbon nanotube–metalloporphyrin chemiresistive gas sensor arrays for volatile organic compounds. Chem. Mater. 27 (10), 3560–3563.

Liu, S.F., Petty, A.R., Sazama, G.T., Swager, T.M., 2015b. Single-walled carbon nanotube/metalloporphyrin composites for the chemiresistive detection of amines and meat spoilage. Angew. Chem. 54 (22), 6554–6557.

Liu, G., Li, L., Huang, X., Zheng, S., Xu, X., Liu, Z., Zhang, Y., Wang, J., Lin, H., Xu, D., 2018a. Adsorption and removal of organophosphorus pesticides from environmental water and soil samples by using magnetic multi–walled carbon nanotubes@ organic framework ZIF–8. J. Mater. Sci. 53, 10772–10783.

Liu, D., Mao, Y., Ding, L., 2018b. Carbon nanotubes as antimicrobial agents for water disinfection and pathogen control. J. Water Health 16 (2), 171–180.

Mallakpour, S., Khadem, E., 2019. Carbon nanotubes for heavy metals removal. In: Composite Nanoadsorbents. Elsevier, pp. 181–210.

Meyyappan, M., 2016. Carbon nanotube-based chemical sensors. Small 12 (16), 2118–2129.

Moazzen, M., Ahmadkhaniha, R., Gorji, M.E.H., Yunesian, M., Rastkari, N., 2013. Magnetic solid–phase extraction based on magnetic multi–walled carbon nanotubes for the determination of polycyclic aromatic hydrocarbons in grilled meat samples. Talanta 115, 957–965.

Mohamed, M.A., Hashim, A.F., Alghuthaymi, M.A., Abd-Elsalam, K.A., 2018. Nano-carbon: plant growth promotion and protection. In: Nanobiotechnology Applications in Plant Protection. Springer, Cham, pp. 155–188.

Monthioux, M., Kuznetsov, V.L., 2006. Who should be given the credit for the discovery of carbon nanotubes? Carbon 44 (9), 1621–1623.

Mostofizadeh, A., Li, Y., Song, B., Huang, Y., 2011. Synthesis, properties, and applications of low–dimensional carbon–related nanomaterials. J. Nanomater. 2011, 685081. 21 pages.

Mudshinge, S.R., Deore, A.B., Patil, S., Bhalgat, C.M., 2011. Nanoparticles: emerging carriers for drug delivery. Saudi Pharm. J. 19 (3), 129–141.

Mukherjee, A., Majumdar, S., Servin, A.D., Pagano, L., Dhankher, O.P., White, J.C., 2016. Carbon nanomaterials in agriculture: a critical review. Front. Plant Sci. 7, 172.

Park, S., Vosguerichian, M., Bao, Z., 2013. A review of fabrication and applications of carbon nanotube film–based flexible electronics. Nanoscale 5 (5), 1727–1752.

Piperopoulos, E., Calabrese, L., Mastronardo, E., Proverbio, E., Milone, C., 2018. Synthesis of reusable silicone foam containing carbon nanotubes for oil spill remediation. J. Appl. Polym. Sci. 135 (14), 46067.

Qiu, W., Xu, H., Takalkar, S., Gurung, A.S., Liu, B., Zheng, Y., Guo, Z., Baloda, M., Baryeh, K., Liu, G., 2015. Carbon nanotube–based lateral flow biosensor for sensitive and rapid detection of DNA sequence. Biosens. Bioelectron. 64, 367–372.

Qu, J.H., Wei, Q., Sun, D.W., 2018. Carbon dots: principles and their applications in food quality and safety detection. Crit. Rev. Food Sci. Nutr. 58 (14), 2466–2475.

Rodrigues, D.F., Advincula, R.C., Claydon, F., Santos, C.M., Tria, M.C.R., 2015. Nanocomposite Polymer-Carbon Based Nanomaterial Filters for the Simultaneous Removal of Bacteria and Heavy Metals. U.S. Patent 8,925,736. University of Houston.

Romero-Guzmán, L., Reyes-Gutiérrez, L.R., Romero-Guzmán, E.T., Savedra-Labastida, E., 2018. Carbon nanotube filters for removal of air pollutants from mobile sources. J. Miner. Mater. Charact. Eng. 6, 105–118.

Sarlak, N., Taherifar, A., Salehi, F., 2014. Synthesis of pesticide nanopesticides encapsulating nanoparticles by using functionalized carbon nanotubes and application of new nanocomposite for plant disease treatment. J Agric. Food Chem. 62 (21), 4833–4838.

Schnorr, J.M., Swager, T.M., 2010. Emerging applications of carbon nanotubes. Chem. Mater. 23 (3), 646–657.

Schroeder, V., Savagatrup, S., He, M., Lin, S., Swager, T.M., 2019. Carbon nanotube chemical sensors. Chem. Rev. 119 (1), 599–663.

Sekhon, B.S., 2014. Nanotechnology in agri–food production: an overview. Nanotechnol. Sci. Appl. 7, 31.

Serag, M.F., Kaji, N., Tokeshi, M., Baba, Y., 2015. Carbon nanotubes and modern nanoagriculture. In: Nanotechnology and Plant Sciences. Springer, Cham, pp. 183–201.

Shulaker, M.M., Hills, G., Patil, N., Wei, H., Chen, H.Y., Wong, H.S.P., Mitra, S., 2013. Carbon nanotube computer. Nature 501 (7468), 526.

Singh, B.G.P., Baburao, C., Pispati, V., Pathipati, H., Muthy, N., Prassana, S.R.V., Rathode, B.G., 2012. Carbon nanotubes. A novel drug delivery system. Int. J. Res. Pharm. Chem. 2 (2), 523–532.

Singh, C., Srivastava, S., Ali, M.A., Gupta, T.K., Sumana, G., Srivastava, A., Mathur, R.B., Malhotra, B.D., 2013. Carboxylated multiwalled carbon nanotubes–based biosensor for aflatoxin detection. Sensors Actuators B Chem. 185, 258–264.

Sinnott, S.B., Andrews, R., 2001. Carbon nanotubes: synthesis, properties, and applications. Crit. Rev. Solid State Mater. Sci. 26 (3), 145–249.

Smith, S.C., Rodrigues, D.F., 2015. Carbon–based nanomaterials for removal of chemical and biological contaminants from water: a review of mechanisms and applications. Carbon 91, 122–143.

Sui, Z., Meng, Q., Zhang, X., Ma, R., Cao, B., 2012. Green synthesis of carbon nanotube–graphene hybrid aerogels and their use as versatile agents for water purification. J. Mater. Chem. 22 (18), 8767–8771.

Sweetman, M., May, S., Mebberson, N., Pendleton, P., Vasilev, K., Plush, S., Hayball, J., 2017. Activated carbon, carbon nanotubes and graphene: materials and composites for advanced water purification. J. Carbon Res. 3 (2), 18.

Tiwari, J.N., Tiwari, R.N., Kim, K.S., 2012. Zero–dimensional, one–dimensional, two–dimensional and three–dimensional nanostructured materials for advanced electrochemical energy devices. Prog. Mater. Sci. 57, 724–803.

Upadhyay, R.K., Soin, N., Roy, S.S., 2014. Role of graphene/metal oxide composites as photocatalysts, adsorbents and disinfectants in water treatment: a review. RSC Adv. 4, 3823–3851.

Verma, S.K., Das, A.K., Gantait, S., Kumar, V., Gurel, E., 2019. Applications of carbon nanomaterials in the plant system: a perspective view on the pros and cons. Sci. Total Environ. 667, 485–499.

Wang, J., 2005. Carbon-nanotube based electrochemical biosensors: a review. Electroynalysis 17 (1), 7–14.

Wang, S., Sun, H., Ang, H.M., Tadé, M.O., 2013. Adsorptive remediation of environmental pollutants using novel graphene–based nanomaterials. Chem. Eng. J. 226, 336–347.

Wang, X., Liu, X., Chen, J., Han, H., Yuan, Z., 2014. Evaluation and mechanism of antifungal effects of carbon nanomaterials in controlling plant fungal pathogen. Carbon 68, 798–806.

Wang, B., Chen, Y., Wu, Y., Weng, B., Liu, Y., Lu, Z., Li, C.M., Yu, C., 2016. Aptamer induced assembly of fluorescent nitrogen–doped carbon dots on gold nanoparticles for sensitive detection of AFB1. Biosens. Bioelectron. 78, 23–30.

Wong, A., Silva, T., Caetano, F., Bergamini, M., Marcolino-Junior, L., Fatibello-Filho, O., Janegitz, B., 2017. An overview of pesticide monitoring at environmental samples using carbon nanotubes–based electrochemical sensor. Chem. Eng. J. 255, 107–113.

Xiaoqiang, A., Jimmy, C.Y., 2011. Graphene–based photocatalytic composites. RSC Adv. 1, 1426–1434.

Yan, J., Guan, H., Yu, J., Chi, D., 2013. Acetylcholinesterase biosensor based on assembly of multiwall carbon nanotubes onto liposome bioreactors for detection of organophosphates pesticides. Pestic. Biochem. Physiol. 105 (3), 197–202.

Yang, M.Q., Zhang, N., Xu, Y.J., 2013. Synthesis of fullerene–, carbon nanotube–, and graphene–TiO2 nanocomposite photo–catalysts for selective oxidation: a comparative study. ACS Appl. Mater. Interfaces 5, 1156–1164.

Yu, G., Lu, Y., Guo, J., Patel, M., Bafana, A., Wang, X., Qiu, B., Jeffryes, C., Wei, S., Guo, Z., Wujcik, E.K., 2018. Carbon nanotubes, graphene, and their derivatives for heavy metal removal. Adv. Compos. Hybrid Mater. 1 (1), 56–78.

Zaytseva, O., Neumann, G., 2016. Carbon nanomaterials: production, impact on plant development, agricultural and environmental applications. Chem. Biol. Technol. Agric. 3, 17.

Zhang, H., Huang, H., Lin, Z., Su, X., 2014. A turn-on fluorescence-sensing technique for glucose determination based on graphene oxide–DNA interaction. Anal. Bioanal. Chem. 406 (27), 6925–6932.

Zhang, X., Huang, Q., Liu, M., Tian, J., Zeng, G., Li, Z., Wang, K., Zhang, Q., Wan, Q., Deng, F., Wei, Y., 2015. Preparation of amine functionalized carbon nanotubes via a bioinspired strategy and their application in Cu2+ removal. Appl. Surf. Sci. 343, 19–27.

Zhang, X., Li, C.R., Wang, W.C., Xue, J., Huang, Y.L., Yang, X.X., Tan, B., Zhou, X.P., Shao, C., Ding, S.J., Qiu, J.F., 2016. A novel electrochemical immunosensor for highly sensitive detection of aflatoxin B1 in corn using single–walled carbon nanotubes/chitosan. Food Chem. 192, 197–202.

Zhao, F., Wu, J., Ying, Y., She, Y., Wang, J., Ping, J., 2018. Carbon nanomaterial–enabled pesticide biosensors: design strategy, biosensing mechanism, and practical application. TrAC Trends Anal. Chem. https://doi.org/10.1016/j.trac.2018.06.017.

Zhu, R., Desroches, M., Yoon, B., Swager, T.M., 2017. Wireless oxygen sensors enabled by Fe (II)–polymer wrapped carbon nanotubes. ACS Sens. 2 (7), 1044–1050.

Further reading

Han, Y., Xu, Z., Gao, C., 2013. Ultrathin graphene nanofiltration membrane for water purification. Adv. Funct. Mater. 23 (29), 3693–3700.

Part One

Synthesis and characterization

Carbon nanotubes: Synthesis, characterization, and applications

2

Abd El-Moez A. Mohamed[a,b], *Mohamed A. Mohamed*[c]
[a]School of Metallurgy and Materials, University of Birmingham, Birmingham, United Kingdom, [b]Department of Physics, Faculty of Science, University of Sohag, Sohag, Egypt, [c]Plant Pathology Research Institute, Agricultural Research Center (ARC), Giza, Egypt

1 Introduction

The carbon atom has six electrons distributed among 1s, 2s, and 2p orbitals. The 1s electrons are tightly bound, forming the core while the 2s and 2p electrons are the valence electrons. The wave function of 2s and 2p electrons overlaps due to the small energy difference between these orbitals than the chemical bond. This is known as hybridization (Saito et al., 1998). The type of hybridization is determined according to the number of overlapped p orbitals, which are sp^1, sp^2, and sp^3. In the sp^1 type, the 2s orbital hybrids with one p orbital with 180 degree. In sp^2, the 2s is hybridized with two of p orbitals in the same plane with 120^0. In sp^3, the 2s is hybridized with three p orbitals separated by 109.5^0 forming a tetrahedral bond shape. Carbon atom shows different allotropes diamond 3D structure and carbon nanotubes (1D structure). Carbon nanotubes (CNTs) are a hexagonal rolled-up layer of graphene, forming a hollow seamless cylindrical tubular structure; the carbon atoms are connected via sp^2 bonds. According to the number of these rolled graphene sheets, CNTs can be classified into single-walled carbon nanotubes or multiwalled carbon nanotubes (SWCNTs and MWCNTs) (see Fig. 1). In MWCNTs, layers are separated by a small distance due to the van der Waals force. CNTs show superior properties such as a specific strength that is higher than steel in addition to being lightweight with high elasticity. According to their various properties, CNTs are good candidates for different applications. For example, the large surface area and aspect ratio are ideal for biomedical applications such as drug and enzyme delivery. Thermal stability, high thermal conductivity, and large current capacity properties are good for interconnect applications. In this chapter, we focus on the physical and chemical synthesis methods of carbon CNTs, their characterizations, and their applications.

1.1 Electrical and thermal properties

The graphene unit cell contains two carbon atoms with an even number of electrons. Therefore, it can be a semiconductor or a metal. However, the conductivity nature depends on the degree of graphene sheet twist. The graphene sheet conductivity is calculated via the $\sigma = (2e^2/h) \cdot (2\pi l_e E/h\nu)$ equation where l_e is the scattering length and h is the Plank constant. Experimentally, the one-dimensional nature of

Carbon Nanomaterials for Agri-food and Environmental Applications. https://doi.org/10.1016/B978-0-12-819786-8.00002-5

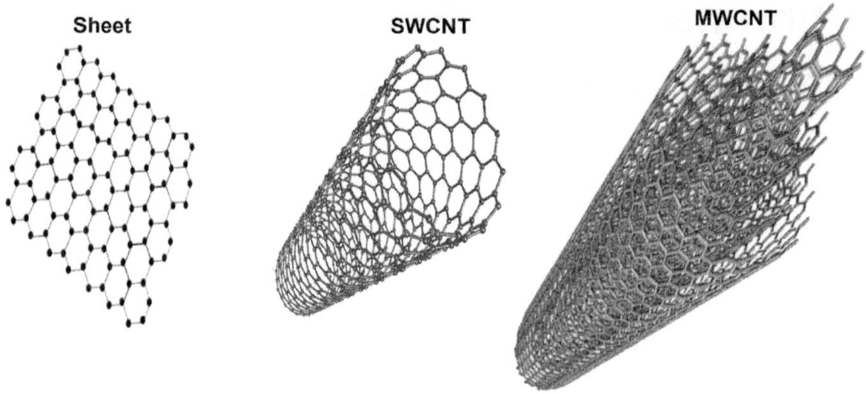

Fig. 1 Graphene sheet, single-walled CNT, and multiwalled CNT.

an SWCNT makes it hard to measure the electrical resistance due to the additional contact because of the mismatch between the external wire contact and the tube (Datta, 1995). According to the Landauer formula and considering the perfect contact, the intrinsic resistance of SCNT is 10^{-4}Ohm^{-1} cm^{-1} at 27°C. On the other hand, MWCNT conductivity is complex due to the interwall interactions. When a magnetic field is applied, the band gap oscillates with the increase in magnetic field intensity, so the metallic CNT will be a semiconductor first, then go back to metallic behavior (Kittel, 1976).

Diamond and graphene show high thermal conductivity, which is controlled via phonons (Dresselhaus et al., 2008). The MWCNTs show the same thermal conductivity behavior of bulk graphene, in contrast with SWCNTs. This difference in SWCNTs is attributed to the optical and acoustic phonons. The expected thermal conductivity by theoretical works is 6600 W mK^{-1} (Berber et al., 2000). However, the measured value was >200 W mK^{-1}, which is very close to the good metal value. Unlike several materials, the coherence length in graphene is <1000 A, and the phonon free path is controlled via boundary scattering above 140 K. Also, the defect-free CNTs are suggested to be in long lengths that exceed several microns, which is longer than the crystallite diameter. In addition, the thermal expansion of CNTs is a little bit different from graphene. It has been shown that the thermal radial expansion coefficient of MWCNTs will be identical to the on-axis thermal expansion coefficient, despite the fact that they are separated by similar distances to the interplanar separation in graphene (Ruoff, 1992). This is because the nanotube sheet is wrapped onto itself, therefore the radial expansion is governed entirely by the carbon covalent bonding network; the van der Waals interaction between nested cylinders is, therefore, incidental to the radial thermal expansion. We, therefore, expect that the thermal coefficient of expansion will be isotropic in a defect-free SWNT or MWNT.

1.2 *Mechanical properties*

The strain energy of nanotubes can be predicted using the elastic properties of graphene sheets (Tersoff, 1992). The elastic strain energy that results from simple calculations based on continuum elastic deformation of a planar sheet compares very favorably with the more sophisticated ab initio results. The results have been confirmed by the ab initio calculations of Mintmire et al. (1994). This suggests that the mechanical properties of nanotubes can be predicted with some confidence from the known properties of single crystal graphene. In defect-free CNTs, the stiffness constant of SWCNTs can be calculated using the elastic modulus of graphite. Moreover, in spite of theoretical studies that predict a high tensile strength and a Young's modulus that reaches 1 TPa for one CNT, the experimental results have shown results far from this predicted value, which refers to the existing defects in CNTs (Hyung et al., 2017). Hyung et al. (2017) designed an accurate technique of tensile testing for MWCNTs, where they found a tensile strength and Young's modulus value of 0.85 and 34.65 GPa, respectively.

2 Synthesis methods

2.1 *Arc discharge*

It is a traditional and common method in CNT production where carbon rods are put between the cathode and anode at a very close distance surrounded by an inert gas. Then, a high dc electric current is applied, leading to a high temperature resulting from the discharge. This evaporates the rod surface, forming small rods; however, the plasma arc uniformity controls the CNT formation (see Fig. 2).

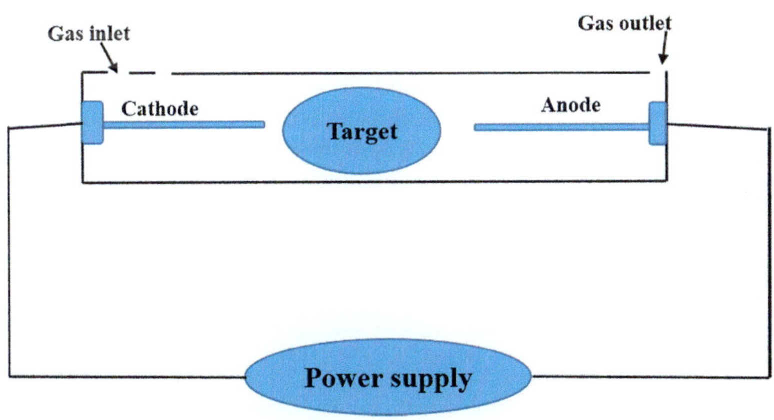

Fig. 2 Schematic design for the arc discharge technique.

2.2 Laser ablation

This is a similar way of arc discharge, although here the laser is used in carbon evaporation (see Fig. 3), where the pulsed and contentious laser is used to heat the graphene surface. A graphene rod with nickel and cobalt (as a catalyst) is put in an oven at 1200°C in an inert gas atmosphere. Two pulses are used to break down the larger particles and produce CNTs. The dimension of the produced CNTs can vary according to the catalyst composition and temperature. In addition, it was found that a carbon dioxide laser could be used in producing SWCNTs, where it was found that the increase in laser power increases the CNT diameter. Also, there is a thermal synthesis process method that is similar to laser ablation and arc discharge. However, the only difference is that in this method, only heat is used without exceeding 1200°C.

There are disadvantages to laser ablation and arc discharge. In spite of the high CNT quality produced by these two methods, they have some disadvantages that limit their use:

1. The large bulk graphene target needed for the process.
2. The huge energy needed in these processes to produce CNTs.
3. The necessary purification process for the CNTs to get rid of unwanted carbon and catalysts.

3 Ball milling

An annealing process should follow this method to ensure the CNT formation. Graphene powder is put in an evacuated container with stainless steel balls filled with an inert gas for a long time, then this powder is annealed at 1400°C for 6h. This method is effective in MWCNTs more than SWCNTs.

3.1 Chemical vapor deposition

The advantage of this method is that it can produce a large number of CNTs with controlling the growth direction on a substrate (Monthioux and Kuznetsov, 2006). A hot substrate of 700–900°C is needed with hydrocarbon gas and process gas such as nitrogen. Due to the hydrocarbon gas decomposition, CNTs grow on the substrate that works as a catalyst. In fact, the substrate type is an important factor in the preparation processes, as it determines the type of the obtained CNTs. The porous silicon substrate

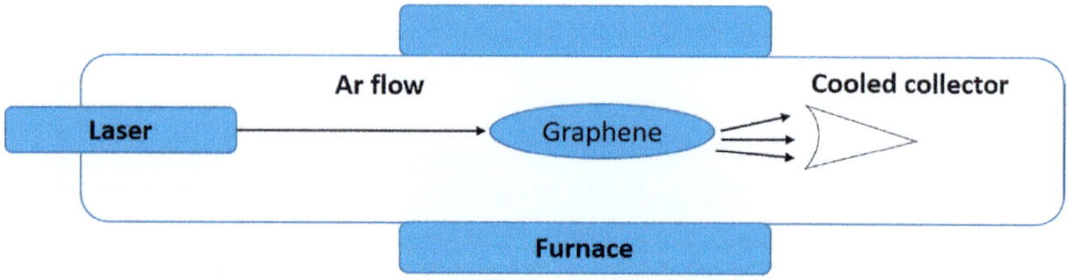

Fig. 3 Schematic plot of laser ablation system.

is the best choice for growing self-oriented CNTs. In more details, CNTs are grown on the substrate and implanted with nanoparticles such as Fe, Ni, or Co. Afterward, the hydrocarbon will decompose and deposit on the substrate and carbon will connect to the metal particles in the holes, creating CNTs.

4 Flame method

This method has a great potential to produce large amounts of CNTs at a low cost. Three main components are needed for this method: a metallic catalyst particle, a heat source, and a source of carbon. The flame parameters can be tuned to obtain different sizes of CNTs, where Lee et al. (2004) found that flame temperature has a great effect on CNT fabrication and alignment. SWCNT flame fabrication is not easy because it requires a gas phase catalyst. However, they can be formed in special conditions using a premixed acetylene, oxygen, and argon flame at 50Torr with Fe vapor pentacarbonyl as the catalyst.

4.1 Saline solution method

A carbon paper substrate containing electric current for heating is submerged in a Co:Ni saline solution catalyst at a 1:1 ratio, then a carbon gas source is passed over the substrate. This leads to an interaction between the gas and the catalyst, forming CNTs on the substrate (Bhaskar et al., 2013).

4.2 Spray pyrolysis method

This is a modified technique from the CVD method to fabricate high-purity MWCNTs (Gul et al., 2019). The samples are mounted on a stainless substrate in a tube furnace, then an inert gas is introduced. When the furnace temperature reaches 170°C, the inert gas flow rate is increased. The mixture of ferrocene dissolved dinxylene is rapidly injected until the solution reaches the injector head, producing CNTs that are ready for collection and purification from the catalyst using nitric acid. The advantage of this method in comparison with CVD is that we can still contentiously introduce hydrocarbons and catalysts, decreasing the expense of CNT production (Wasel et al., 2007).

4.2.1 Characterization

Optical characterization of CNTs is a good means to explore their internal properties. Several techniques use the optical response such as infrared spectroscopy, Fourier transform infrared, and Raman spectroscopy. In this part, we will present some of these techniques.

4.3 X-ray photoelectron spectroscopy

When CNTs are exposed to X-rays, they release photoelectrons. The binding energy estimation of these electrons enables us to determine the surface composition. This refers to the inelastic photoelectron mean free path that allows the recognition of the elemental composition of the surface, except the hydrogen (Cano et al., 2017). In addition, X-ray photoelectron spectroscopy (XPS) introduces information about other existing elements on the surface because the binding energy is easily affected by neighbor atoms. The only problem with XPS is the necessary large mass of the sample.

4.4 Infrared spectroscopy

Infrared spectroscopy is a familiar technique in detecting the vibrational motion of the functional groups attached on the CNT surface. However, one disadvantage related to this technique is that it cannot give a quantitative analysis for the detected groups.

4.5 Raman spectroscopy

Raman spectroscopy is a vibrational spectroscopic technique that provides information about molecular vibrations and crystal structures. When a material is exposed to light, almost all the photons are elastically scattered with the same energy. But a small number of photons are scattered with different energy and have inelastic collisions with the material; that is known as the Raman effect. The normal Raman spectra for an ordered sidewall CNT is the G-band, in contrast with the nonordered CNT side wall that shows a D-band. The defect density in CNTs can be easily determined from the ratio of these two bands (IG:ID), which introduces information about the structure change. In some cases, the radial breathing mode is used to estimate the CNT diameter according to their peaks (Rao et al., 2001). The advantages of the Raman spectroscopy characterization technique are a fast analysis tool, no sample preparation, nondestructive analysis, and all allotropic forms of carbon are active for this spectroscopy (Arepalli et al., 2004).

4.6 Transmission electron microscope

Microscopy techniques play an important role in the easy detection of impurities with CNTs, which affects the structure and other properties. This is because most of the synthesis methods result in amorphous carbon that is adsorbed in the CNT wall side and residual metals, which are embedded between CNT layers. Using transmission electron microscopy (TEM) has several advantages, as it can distinguish these impurities based on the difference in transmission properties (Hyung et al., 2017), determine the length of CNTs using low magnification, measure the inner and outer diameter, and sometimes count the number of CNTs. Using high-magnification TEM, Kiang et al. found that the intershell spacing varies from 0.34 to 0.39 nm according to the CNT diameter (Belin and Epron, 2004), which is greater than the interplaner distance of graphite. This difference refers to the graphene sheet curvature, leading to an increase in the repulsive force and the size effect. However, in some cases, the electron used in TEM itself damages the CNT walls, so it was suggested to expose only the target area for short durations.

4.7 X-ray diffraction

In contrast with TEM, X-ray diffraction (XRD) is a nondestructive characterization technique used to identify impurity elements and determine some structural properties such as the interlayer distance and strain (Arunkumar et al., 2018). XRD patterns of CNTs are quite similar to graphene; however, the carbon structure in graphene shows a random orientation (Rasel et al., 2015). The XRD peak intensity of CNTs is a morphology-dependent factor. In more detail, a (002) peak with (hkl) reflections is produced when X-rays hit a wall of the SWCNT. Meanwhile, when X-rays pass in the empty core of the CNT, an extra hexagonal (hko) reflection is produced (Rasel et al., 2015). It is worth mentioning that the (hko) results because of the asymmetric curvature on CNT while the (hkl) results from the regular arrangement of layers.

4.8 Carbon nanotube applications

4.8.1 Drug delivery, genetic engineering

CNTs have great potential in drug delivery due to their large surface area. The drug can be integrated on walls and can cross the cell membrane, delivering drugs in a safe way (Chen et al., 2008). They also can be used in gene manipulation and atom tissue engineering (Kam et al., 2005). The nucleotides of DNA are connected with CNTs, changing their electrostatic properties, which increases the CNT potential application in diagnostics, therapy, and DNA analysis. In addition, CNTs can be used in gene therapy or cancer treatment due to the tabular nature, where it was found that CNTs imposed by DNA release the DNA before being destroyed by the self-defense system. Some in vitro studies have shown the destruction of cancer cells by hyperthermia using CNTs as a result of the high thermal conductivity (Kam et al., 2005).

4.8.2 Artificial implants and biomedical applications

The body exhibits painful reactions against implants; however, CNTs with external implants can easily be attached with proteins to decrease this pain. CNTs can also be used as bone structures due to their strong tensile strength, and sometimes can be filled with calcium (Ding et al., 2001). In addition to the high sensitivity of CNTs in DNA detection, CNTs can also be used in pressure sensors, which are important for eye surgery and respiratory devices (Qiu et al., 2015). Moreover, polymer-based CNTs are used in artificial muscle devices because of their strength (Foroughi et al., 2016). CNTs also are used to detect cancer cells in their early stages, where they are hard to recognize by normal means such as X-ray and magnetic resonance imaging (MRI). Therefore, CNTs penetrating tissues enhances the diagnostic process in the early stages of cancer disease.

4.8.3 Field emission source in electronic devices

Electrons are emitted from CNT tips when an electric field is applied between the CNT surface and the anode. Because of that, CNTs can be used in multiple electronic devices such as X-ray sources (Rinzler et al., 1995). The advantages of CNTs as a field emission source are a long life time, the stability of the emitted field does not need a high vacuum to work, and a high current density, where it can achieve $4\,A\,cm^{-2}$.

4.8.4 Sensors

Biosensor efficiency can be enhanced by CNTs attached on top, where they can attract functional groups (Wong et al., 1998). Therefore, it is possible to fabricate different sensors with CNTs that are sensitive to gases so they can detect leakage. Where there is a change in the CNT electrical resistance when the surrounding environment changes, for example, SWCNTs are very sensitive (Collins et al., 2000) and MWCNTs are sensitive to NH_2, H_2O, CO, and CO_2. In 2002, high-sensitivity microwave resonant sensors were fabricated using SWCNTs and MWCNTs for NH_3 detection (Chopra et al., 2002).

4.9 Supercapacitor

Several materials have been investigated for energy storage devices; however, there are some struggles to manipulate the efficiency with geometry, where the good devise should interest with flexibility and lightweight that are struggling with the high efficiency. Supercapacitors are interesting tools in energy storage in comparison with others due to faster charge discharge rates, higher power density, a superlong life cycle, and pollution-free operation (Sheila and Bahareh, 2018). Textile supercapacitors are novel energy storage devices designed by coating the fibers with functional thin film layers such as carbon nanotubes (Islam et al., 2017). Jiang et al. in 2015 found that MnO_2 nanoflakes can be electrodeposited by wrapping MWCNTs on cotton textiles and treating with acid. This is an interesting conductive textile, exhibiting high flexibility and strong adhesion between the CNT and the textiles of interest. Moreover, conductive textile-based supercapacitors were found to show high specific and capacity retention of 94.7% that can be maintained at 2000 continuous charge-discharge cycles (Jiang et al., 2015).

4.10 Water treatment

CNTs show a high sorption ability that can be affected by the attached functional group. Using that, CNTs can be used in water treatment to adsorb different types of heavy metals (Gul et al., 2019). For example, CNTs with COO^- functional groups can be used to adsorb mercury, which is a toxic heavy metal (Pokhrel et al., 2017). In a similar way, the different morphology CNTs works as adsorptive membrane for the Ni ions (Tofighy and Mohammadi, 2015). In addition, CNTs can remove harmful organic groups, where they are considered a good absorbent for toxic organic dyes (Santhosh et al., 2016). For instance, methyl orange (MO) can be easily removed using MWCNTs supported with palladium nanocubes (Cano et al., 2017). In water treatment, CNTs are used as electrode materials, Table 1 shows CNTs and functional groups used in heavy metal detection. This table shows the direct detection of Zn, Cu, and Cd heavy metals even with simple CNTs; however, catalyst-free CNTs show some

Table 1 Types of CNTs and functional groups used in heavy metal detection.

Adsorbent	Absorbed heavy metal ion	Ref.
CNTs	Pb^{2+}	Mubarak et al. (2016)
MWCNTs	Ni^{2+}	Lu and Liu (2006)
CNT-COO	Hg^{2+}	Anitha et al. (2015)
CNT-OH	Cu^{2+}	Anitha et al. (2015)
SWCNT	Zn^{2+}	Long and Yang (2001)
CNT-COO	Cd^{2+}	Anitha et al. (2015)
CNTs	Cr^{6+}	Atieh (2011)
CNT fibers	Ag^{1+}	Jahandari et al. (2013)
CNTs	Pb^{2+}	Pérez et al. (2016)

limitations. Another potential application of CNTs in water treatment is oil removal, which is an important area due to water pollution from some industrial waste or ship leakage. Two main kinds of CNTs are used in oil detection: P-CNTs and C-CNTs. It is worth mentioning that P-CNTs show higher efficiency in oil detection (97%) compared to C-CNTs (87%) (Fard et al., 2016).

4.10.1 Toxicity of CNTs

In spite of the wide range of CNT applications, a toxic side effect is registered (Francis and Devasena, 2018). CNT toxicity is controlled and determined through factors such as morphology, size, shape, and purity (Luo et al., 2013). This is because CNTs penetrate cells, inducing a cytotoxic response where the contact between the CNT with a cell results in a strain on the cytoskeleton of the phagocyte. In accordance, some studies have shown that longer-length and larger-diameter CNTs have more of a toxicity effect than smaller ones (Lamberti et al., 2014). Moreover, CNT purity is an additional important factor, where metallic impurities lead to cell death (Meng et al., 2013).

5 Conclusion

Carbon nanotubes are of wide interest due to their electrical and physical properties. They are produced in different synthesis methods such as arc discharge and laser ablation according to the required physical properties such as diameter and length. The large surface area has made them good drug delivery systems; therefore, a purification process is required after synthesis to avoid toxicity.

Acknowledgment

The first author would like to thank Basma F. for the continuous support.

References

Anitha, K., Namsani, S., Singh, J., 2015. Removal of heavy metal ions using a functionalized single-walled carbon nanotube: a molecular dynamics study. J. Phys. Chem. A 119, 8349–8358.

Arepalli, S., Nikolaev, P., Gorelik, O., Hadjiev, V., Holmes, W., Files, B., Yowell, L., 2004. Protocol for the characterization of single-wall carbon nanotube material quality. Carbon 42, 1783–1791.

Arunkumar, T., Karthikeyan, R., Ram, R., Viswanathan, K., Anish, M., 2018. Synthesis and characterisation of multi-walled carbon nanotubes (MWCNTs). Int. J. Ambient Energy https://doi.org/10.1080/01430750.2018.1472657. 2162-8246.

Atieh, M., 2011. Removal of chromium (VI) from polluted water using carbon nanotubes supported with activated carbon. Procedia Environ. Sci. 4, 281–293.

Belin, T., Epron, F., 2004. Characterization methods of carbon nanotubes. Mater. Sci. Eng. B 119, 105–118.

Berber, S., Kwon, Y., Tomanek, D., 2000. Unusually high thermal conductivity of carbon nanotubes. Phys. Rev. Lett. 84, 4613–4616.

Bhaskar, A., Deshmukh, V., Prajapati, L., 2013. Carbon nanotube as a drug delivery system. Int. J. Pharm. Technol. 5, 2695–2711.

Cano, O., González, C., Paz, J.H., Madrid, P., Casillas, P., Hernández, A., Pérez, C., 2017. Catalytic activity of palladium nanocubes/multi-walled carbon nanotubes structures for methyl orange dye removal. Catal. Today 282, 168–173.

Chen, J., Chen, S., Zhao, X., Kuznetsova, L., Wong, S., Ojima, I., 2008. Functionalized single-walled carbon nanotubes as rationally designed vehicles fortumor-targeted drug delivery. J. Am. Chem. Soc. 130 (49), 16778–16785.

Chopra, S., Pham, A., Gaillard, J., Parker, A., Rao, A., 2002. Carbon-nano- tube-based resonant-circuit sensor for ammonia. Appl. Phys. Lett. 80, 4632–4634.

Collins, P., Bradley, K., Ishigami, M., Zettl, A., 2000. Extreme oxygen sensitivity of electronic properties of carbon nanotubes. Science 287 (5459), 1801–1804.

Datta, S., 1995. Electronic Transport in Mesoscopic Systems. Cambridge University Press, New York.

Ding, R., Lu, G., Yan, Z., Wilson, M., 2001. Recent advances in the preparation and utilization of carbon nanotubes for hydrogen storage. J. Nanosci. Nanotechnol. 1, 17–29.

Dresselhaus, M., Dresselhaus, G., Jorio, A., 2008. Unusual properties and structure of carbon nanotubes. Annu. Rev. Mater. Res. 34, 247–278.

Fard, A., Mckay, G., Manawi, Y., Malaibari, Z., Hussien, M., 2016. Outstanding adsorption performance of high aspect ratio and super-hydrophobic carbon nanotubes for oil removal. Chemosphere 164, 142–155.

Foroughi, J., Spinks, G., Aziz, S., Mirabedini, A., Jeiranikhameneh, A., Wallace, G., Kozlov, M., Baughman, R., 2016. Knitted carbon-nanotube-sheath/spandex core elastomeric yarns for artificial muscles and strain sensing. ACS Nano 10, 9129–9135.

Francis, A., Devasena, T., 2018. Toxicity of carbon nanotubes. Toxicol. Ind. Health 34 (3), 200–210.

Gul, R., Zainab, N., Asad, M., Salma, B., Anwar, A., Shabeer, A., Gulam, A., 2019. An overview of the recent progress in the synthesis and applications of carbon nanotubes. J. Carbon Res. 5, 3–33.

Hyung, K., Mei, W., Stephanie, K., Junmo, K., Jae-Do, N., Lijie, C., Jonghwan, S., 2017. Tensile properties of millimeterlong multi-walled carbon nanotubes. Sci. Rep. 7, 9512–9518.

Islam, M., Deng, Y., Tong, L., Roy, A., Faisal, S., Hassan, M., Gomes, V., 2017. In-situ direct grafting of graphene quantum dots onto carbon fibre by low temperature chemical synthesis for high performance flexible fabric supercapacitor. Mater. Today Commun. 10, 112–119.

Jahandari, S., Taher, M., Fazelirad, H., Sheikhshoai, I., 2013. Anodic stripping voltammetry of silver(I) using a carbon paste electrode modified with multi-walled carbon nanotubes. Microchim. Acta 180, 347–354.

Jiang, Y., Ling, X., Jiao, Z., Li, L., Ma, Q., Wu, M., Zhao, B., 2015. Flexible of multi-walled carbon nanotubes/manganese dioxide nanoflake textiles for high-performance electrochemical capacitors. Electrochim. Acta 153, 246–253.

Kam, N., O'Connell, M., Wisdom, J., Dai, H., 2005. Carbon nanotubes as multifunctional biological transporters and near infrared agents for selective cancer cell destruction. Proc. Natl. Acad. Sci. U. S. A. 102, 11600–11605.

Kittel, C., 1976. Introduction to Solid State Physics, fifth ed. John Wiley, New York.

Lamberti, M., Zappavigna, S., Sannolo, N., Stefania, P., Michele, C., 2014. Advantages and risks of nanotechnologies in cancer patients and occupationally exposed workers. Expert Opin. Drug Deliv. 11, 1087–1101.

Lee, G., Jurng, J., Hwang, J., 2004. Formation of Ni-catalyzed multi walled carbon nanotubes and nanofibers on a substrate using an ethylene inverse diffusion flame. Combust. Flame 139, 167–175.

Long, Q., Yang, R., 2001. Carbon nanotubes as superior sorbent for dioxin removal. J. Am. Chem. Soc. 123, 2058–2059.

Lu, C., Liu, C., 2006. Removal of nickel (II) from aqueous solution by carbon nanotubes. J. Chem. Technol. Biotechnol. 81, 1932–1940.

Luo, E., Song, G., Li, Y., Shi, P., Hu, J., Lin, Y., 2013. The toxicity and pharma- cokinetics of carbon nanotubes as an effective drug carrier. Curr. Drug Metab. 14, 879–890.

Meng, L., Jiang, A., Chen, R., Li, C.Z., Wang, L., Qu, Y., Wang, P., Zhao, Y., Chen, C., 2013. Inhibitory effects of multi-wall carbon nanotubes with high iron impurity on viability and neuronal differentiation in cultured PC12 cells. Toxicology 313, 49–58.

Mintmire, J.W., Robertson, D.H., White, C.T., 1994. Fullerenes: Recent Advances in the Chemistry and Physics of Fullerenes and Related Materials. The Electrochemical Society, Pennington, NJ.

Monthioux, M., Kuznetsov, V., 2006. Who should be given the credit for the discovery of carbon nano-tubes? Carbon 44, 1621–1623.

Mubarak, N., Sahu, J., Abdullah, E., Jayakumar, N., 2016. Rapid adsorption of toxic Pb (II) ions from aqueous solution using multi-wall carbon nanotubes synthesized by microwave chemical vapor deposition technique. J. Environ. Sci. 45, 143–155.

Pérez, C., Serrano, N., Díaz, C., Ariño, C., Esteban, M., 2016. Glutathione modified screen-printed carbon nanofiber electrode for the voltammetric determination of metal ions in natural samples. Talanta 155, 8–13.

Pokhrel, L., Ettore, N., Jacobs, Z., Zarr, A., Weir, M., Scheuerman, P., Kanel, S., Dubey, B., 2017. Novel carbon nanotube (CNT)-based ultrasensitive sensors for trace mercury(II) detection in water: a review. Sci. Total Environ. 574, 1379–1388.

Qiu, W., Xu, H., Takalkar, S., Gurung, A., Liu, B., Zheng, Y., Guo, Z., Baloda, M., Baryeh, K., Liu, G., 2015. Carbon nanotube-based lateral flow biosensor for sensitive and rapid detection of DNA sequence. Biosens. Bioelectron. 46, 367–372.

Rao, A., Chen, J., Richter, E., Schlecht, U., Eklund, P.C., Haddon, R.C., Venkateswaran, U., Kwon, Y.-K., Tománek, D., 2001. Effect of van der walls interactions on the Raman modes in single walled carbon nanotubes. Phys. Rev. Lett. 86, 3895–3898.

Rasel, D., Sharifah, B., Eaqub, A., Ramakrishna, S., Yongzhi, W., 2015. Carbon nanotubes characterization by X-ray powder diffraction. Curr. Nanosci. 11, 1–13.

Rinzler, A.G., Hafner, J.H., Nikolaev, P., Nordlander, P., Colbert, D.T., Smalley, R.E., Lou, L., Kim, S.G., Tomanek, D., 1995. Unraveling nanotubes: feld emission from an atomic wire. Science 269 (5230), 1550–1553.

Ruoff, R.S., 1992. SRI Report. CA MP 92–263.

Saito, R., Dresselhaus, G., Dresselhaus, M.S., 1998. Physical Properties of Carbon Nanotubes. Imperial College Press, London.

Santhosh, C., Velmurugan, V., Jacob, G., Jeong, S., Grace, A., Bhatnagar, A., 2016. Role of nanomaterials in water treatment applications. A review. Chem. Eng. J. 306, 1116–1137.

Sheila, S., Bahareh, M., 2018. Carbon nanotube and its applications in textile industry. J. Textil. Inst. 109 (12), 1653–1666.

Tersoff, J., 1992. Energies fullerenes. Phys. Rev. B 46, 15546–15549.

Tofighy, M.A., Mohammadi, T., 2015. Nickel ions removal from water by two different morphologies of induced CNTs in mullite pore channels as adsorptive membrane. Ceram. Int. 41, 5464–5472.

Wasel, W., Kuwana, K., Reilly, P., Saito, K., 2007. Experimental characterization of the role of hydrogen in CVD synthesis of MWCNTs. Carbon 45, 833–838.

Wong, S.S., Joselevich, E., Woolley, A.T., Cheung, C.L., Lieber, C.M., 1998. Covalently functionalized nanotubes as nanometre- sized probes in chemistry and biology. Nature 394 (6688), 52–57.

Further Reading

Baughman, R., 2016. Knitted carbon-nanotube-sheath/spandex-core elastomeric yarns for artificial muscles and strain sensing. ACS Nano 10, 9129–9135.

Li, W., Xie, S., Qian, L., Chang, B., Zou, B., Zhou, W.Y., Zhao, R., Wang, G., 1996. Large–scale synthesis of aligned carbon nanotubes. Science 274 (5293), 1701–1703.

Graphene-based nanocomposites: Synthesis, characterizations, and their agri-food applications

3

Khamis Youssef[a], Ayat F. Hashim[b], Sergio Ruffo Roberto[c], Said Fatouh Hamed[b], Kamel A. Abd-Elsalam[a]

[a]Plant Pathology Research Institute, Agricultural Research Center (ARC), Giza, Egypt, [b]Fats and Oils Department, National Research Centre, Egypt, [c]Agricultural Research Center, Department of Agronomy, Londrina State University, Londrina, Brazil

1 Concepts of graphene

Graphene, in its pristine form, is composed of a sole layer of carbon atoms arranged in an sp^2-bonded aromatic structure. It is naturally established as the graphite building block, where p-stacking of graphene sheets holds the lamellar graphite structure strongly in place, with an interlayer spacing of 3.34 Å (Angstrom) between the sheets (Hontoria-Lucas et al., 1995). Micromechanical exfoliation is the successive cleavage of graphite to graphene with very high quality using adhesive tape (Zhu et al., 2010), ideal for research purposes. This simple approach led to the Nobel Prize in Physics in 2010, as Geim and Novoselov produced pristine single-layer graphene sheets in their pivotal work on the electronic properties of graphene (Geim, 2009; Geim and Novoselov, 2007). Graphene can be exfoliated in organic solvents by graphite ultrasonication; however, this process was found to generate relatively low yields (Hernandez et al., 2008; Zhu et al., 2010). To create tall quality graphene on a mechanical scale, the foremost common approach is to straightforwardly synthesize graphene sheets by thermal decomposition of SiC or epitaxial growth of graphene on transition metals (Ni, Pd, Cu, Ru, Ir) via chemical vapor deposition (CVD) of alcohols or hydrocarbons (Bae et al., 2010; Diaye et al., 2008; Guermoune et al., 2011; Kim et al., 2009; Kwon et al., 2009; Li et al., 2009). CVD has been highlighted as the foremost promising, cheap, and adaptable methodology to create high-quality graphene (Bae et al., 2010; Li et al., 2009).

2 Properties of graphene

Graphene has several properties that make it attractive for environmental applications.

Carbon Nanomaterials for Agri-food and Environmental Applications. https://doi.org/10.1016/B978-0-12-819786-8.00003-7

2.1 Electronic properties

The electronic properties of graphene are probably the most studied aspect (Geim, 2009; Neto et al., 2009). Electrons were found to have high versatility in graphene, reaching 10,000–50,000 $cm^2V^{-1}s^{-1}$ at room temperature, with an intrinsic mobility limit of 4,200,000 $cm^2V^{-1}s^{-1}$ (Bolotin et al., 2008; Sreeprasad and Berry, 2013). Graphene can support current densities up to six orders of size higher than copper (Geim, 2009). These remarkable electronic properties of graphene were achieved under an ideal environment with mechanically exfoliated graphene under vacuum (Bolotin et al., 2008). Several factors hinder graphene electronic properties (the sheet size and flatness, and the substrate nature) (Sreeprasad and Berry, 2013; Yang et al., 2010). Nevertheless, the promising electronic properties of graphene have stimulated inquiries about the material as well as advancement for its utilize in novel electronic gadgets (Geim and Novoselov, 2007), photocatalytic materials (Yang et al., 2010), environmental sensors (Yang et al., 2010; Zhao et al., 2012), and energy production and storage (Sun et al., 2011).

2.2 Mechanical properties

Graphene is the strongest material measured in spite of being one atom in thickness with a Young's modulus of $E = 1.0$ TPa and an intrinsic strength of 130 GPa in its atomically perfect pristine form (Lee et al., 2008). These exceptional mechanical properties have excited interest in graphene used as a filler to strengthen the mechanical properties of softer materials (Potts et al., 2011). Graphene can present even superior improvement of the polymer mechanical properties, due to much better intelligence between the sheets and the polymer lattice coming about from the large surface zone of the planar graphene sheets (Rafiee et al., 2009).

2.3 Surface area

Graphene has a high specific surface area like all nanoscale materials. In reality, graphene speaks to the foremost extraordinary case of high-surface materials because each iota of a single-layer graphene sheet is uncovered from both sides to its environment (Sanchez et al., 2012). It has the highest specific surface area of all materials, with a theoretical value of 2630 m^2g^{-1}. Graphene is an ideal candidate for processes relating to adsorption or surface reactions due to its high surface area. Thus, graphene speaks to amazing bolster to stay chemical functionalities or nanomaterials thus, graphene-based nanocomposites have been an active area of research for novel materials (Compton and Nguyen, 2010). One of the most common approaches to graphene-based nanomaterials is an oxidized form of graphene (graphene oxide) due to its lower production costs. Graphene oxide (GO) shows a high density of oxygen functional groups (carboxyl, carbonyl, hydroxyl and epoxy) in the carbon lattice. Graphene oxide can be carried by chemical oxidation of graphite to graphite oxide and subsequent exfoliation by ultrasonication (Dreyer et al., 2010). Oxidation of the graphite structure increases the interlayer distance from 0.34 to 0.65 nm. The hydrophilic nature of oxidized graphite will also permit water to absorb into the lamellar structure, further increasing the distance of the interlayer to 1.15 nm (Lerf et al., 2006). The physicochemical properties of graphene are vital for the production of graphene-based composite materials. In combination with other functional materials, graphene-based composites were found to enhance the act of photocatalytic, biocidal, electroactive, and adsorbent materials (Compton and Nguyen, 2010; Kemp et al., 2013; Sun et al., 2011; Xiang et al., 2012). It looks that, the made strides execution of graphene-based composites frequently depends on synergetic intuitive between the properties of graphene, in its decreased shape, and of the materials joined to the graphene sheets. Careful optimization of the production method is essential for optimal performance of graphene-based nanocomposites.

3 Preparation of graphene-based nanocomposites

3.1 Graphene synthesis from foodstuffs

Chemical vapor deposition is used to synthesize thin films of various nanomaterials using thermochemical vapor-phase reactions in a vacuum furnace to achieve deposition of the preferred material on a substrate at great temperatures (~1000°C) (Muñoz and Gómez-Aleixandre, 2013). Graphene can also be synthesized on a metal-film substrate via CVD by flowing hydrogen and methane gases (carbon source) with a metal catalyst; the methane decomposes, leaving carbon atoms deposited on the substrate to form graphene layers. Ruan et al. (2011) used food materials as the carbon source to make monolayer graphene. They obtained a high-quality graphene film on a Cu foil at 1050°C under H_2/Ar flow. These films did not exhibit significant disorder D bands in their Raman spectra, signifying the presence of few defects. Also, the huge 2 D/G ratio advised that the graphene created was a monolayer film. Calcinations of glucose with dicyandiamide produced freestanding monolayers to oligolayered graphene, and layered graphic carbon nitride (g-C_3N_4) acted as a sacrificial template that underwent complete thermolysis at 750°C. However, graphene obtained from this method contains nitrogen atoms in the graphene lattice, mostly in the procedure of "pyridinic" nitrogen and a negligible amount of "graphitic" nitrogen (Li et al., 2012).

Qu et al. (2013) produced graphene from alfalfa plants by treating alfalfa shoots with nitric acid at 70°C for 300 min and obtained a black graphene precipitate. It is interesting to note that carbon nanotubes (CNTs) were produced in the initial stages from plant-cell walls, and were later converted to graphene by opening the nanotubes with the application by nitric acid. Kalita et al. (2012) utilized solid camphor as the precursor material in a microwave surface-wave plasma CVD process in which graphene layers were deposited on a Cu foil at relatively low temperatures< 600°C). Furthermore, synthesized graphene film could be easily removed to a clear plastic substrate by wet etching, and had a sheet resistance of 8.23 kΩ/square. Gupta et al. (2012) produced a graphene-coated composite to remove organics in contaminated water by warming a combination of river sand and sugar in a furnace at 750°C under an N_2 atmosphere.

3.2 Preparation of reduced graphene oxide

The reduction of graphene oxide can be reached using chemical reducing agents, thermal annealing, photoreduction, or microwave-assisted reduction (Chua and Pumera, 2014; Pei and Cheng, 2012). Even though GO reduction can remove a large fraction of its oxygen content, with the C:O ratio increasing from 2:1 to up to 246:1, complete reduction of GO is challenging (Pei and Cheng, 2012). More importantly, reduction of GO results in a changed chemical structure with carbon vacancies, residual oxygen content, and clustered pentagon and heptagon carbon structures (Bagri et al., 2010; Gómez-Navarro et al., 2010). Because of these defects, GO reduction only shows a fractional restoration of its mechanical and electronic properties in comparison with pristine graphene. Graphene oxide was synthesized from graphite powder by a modified Hummers method (Hummers and Offeman, 1985). Graphite powder (1 g), $NaNO_3$ (1 g), and H_2SO_4 (50 mL) were mixed together at 0°C. Then, $KMnO_4$ (an oxidizing agent) was added gradually into the reaction mixture with constant stirring. The mixture was heated to 35°C and stirred for 12 h, and then 90 mL of water was slowly added under vigorous stirring. Then, a 30% H_2O_2 solution was added to reduce the residual MnO_2. The mixture was then washed by acidified water (3%) and then with double distilled water three times followed by filtration and drying. Yellowish brown solid (RGO) sheets were then obtained. The yellowish brown solid powder (RGO) was dispersed in ultrapure H_2O. Graphite oxide was exfoliated by sonicating the graphite oxide in the water at room temperature for 1 h, generating uniform graphene oxide dispersions. Finally, GO nanosheets were filtered and dried at 35°C under a vacuum oven.

3.3 Preparation of nickel or cobalt oxide/graphene nanocomposites

Graphene oxide and a suitable amount of nickel oxide/graphene nanocomposite $Ni(NO_3)_2 \cdot 6H_2O$ salt were mixed together with double distilled water to have a nickel oxide loading of 10 wt.%. The pH of the solution was adjusted at 10 using 1 M NaOH solution and the mix was stirred for 4 h, and then 50 mL of 0.1 M sodium borohydride or hydrazine hydrate solution was added. The mixture was stirred for 3 h in the air, then the resulting material was filtered and washed several times with double distilled water and dried in an oven at 80°C. The dried mixture was then calcined at 400°C for 3 h in the air (Elzatahry et al., 2012). For the cobalt oxide/graphene nanocomposites, an appropriate amount of $CoCl_2 \cdot 6H_2O$ and graphene oxide were mixed with double dH_2O to achieve a 10 wt.% loading of cobalt oxide. The cobalt loading was fixed at 10 wt.%. The solution pH was adjusted at 8 and the mixture was stirred constantly for 4 h. An amount of 50 mL of 0.05 M sodium borohydride or hydrazine hydrate solution was added and the reaction was completed by stirring the mixture for 3 h in air. Then the product was filtered and washed with double distilled water at least three times and then dried in an oven at 80°C (Elzatahry et al., 2012).

3.4 Preparation of graphene oxide chitosan nanocomposite

A total of 0.10 g of chitosan was dissolved in 100 mL 1% acetic acid solution and stirred for 1 h. GO (0.15 g) was dispersed in 20 mL water and 0.15 g of 1-[3-(dimethylamino) propyl]-3-ethylcarbodiimide hydrochloride (EDC) and 0.1 g of *N*-hydroxysuccinimide (NHS) was added to the GO suspension and stirred for 3 h. Then, the GO suspension was drop-wise to the chitosan solution and stirred for 12 h. Finally, the product was collected by centrifugation at 15,000 rpm for 10 min and dried at 35°C under vacuum oven (solution 1). The mix was sonicated in ultrasonication and stirred for 10 min. Then, zinc acetate dehydrates (0.7 g) were dissolved in 30 mL methanol (solution 2). Solution 1 was drop-wise to solution 2, sonicated, and stirred for 15 min. Then, the mixture was moved to a 100 mL round-bottom flask. Potassium hydroxide solution was prepared by adding 0.3 g of KOH in 10 mL of methanol and added slowly to the above solution; the reaction continued to 8 h at 60°C. The mixture was centrifuged at 15,000 rpm for 10 min and washed three times by a methanol-water mixture to remove impurities (Chowdhuri et al., 2015).

4 Characterization of graphene-based nanocomposite

4.1 Fourier Transform infrared spectrophotometer

Fourier Transform infrared spectrophotometer (FTIR) is a tool used to observe specimens, both to identify the presence of target compounds and to measure the quantities of the compounds (quantification).

4.2 X-ray diffraction

X-ray diffraction (XRD) is a dominant and regular technique for identifying the crystal structure of crystalline materials (Murray et al., 2000).

4.3 Average particle size and zeta potential measurements

The dynamic light scattering (DLS) particle size and zeta potential analyzer is a user-friendly system for colloidal, nanoparticulate, and macromolecular characterization. It can define the distribution of particle size, the particle zeta (related to the electrical charge magnitude at the surface of the particle), and molecular weight of large polymeric substances dispersed in water.

4.4 Morphological studies of nanomaterials

There are a number of experimental techniques that can be used to characterize structural and surface properties of nanomaterials, either directly or indirectly.

4.5 Transmission electron microscopy

The advances made in transmission electron microscopy (TEM) have allowed direct imaging of atomic structures in solids and surfaces. Particles in the nanometer size are usually present in many different types of materials and utilizing TEM allows gathering information about particle size, shape, and any surface layers or absorbates (Brydson and Brown, 2008; Henry, 2005). More recently, changes in the nanoparticle structure as a result of interactions with gas, liquid, or solid-phase substrates can now be monitored by this technique (Howe et al., 2008).

4.6 Scanning electron microscopy

Scanning electron microscopy (SEM) is a powerful and common technique for imaging the surfaces of any material with a resolution down to about 1 nm (Wang, 2000; Wang et al., 2000). Both TEM and SEM techniques can supply topological, morphological, and compositional information about the specimen. TEM can provide crystallographic information as well. Also, TEM allows for diffraction patterns to be revealed that also have useful crystallographic information about the sample.

4.7 Atomic force microscopy

Atomic force microscopy (AFM) is able to directly supply detailed information on the scale of the atom, which is important for understanding the chemical bonding and electronic structure of atoms and molecules.

5 The importance of graphene-based nanocomposites

Graphene is the most promising candidate to be a major filling agent for composite applications due to its exceptional thermal, mechanical, and electronic properties (Compton and Nguyen, 2010; Geim and Novoselov, 2007; Potts et al., 2011). At very low loading, graphene-based nanocomposites display significant enhancements in their multifunctional applications, in comparison with conventional composites and their materials. This not only makes the substance lighter with simple processing, but also makes it stronger for various multifunctional aspects (Potts et al., 2011; Premkumar and Geckeler, 2012; Winey and Vaia, 2007). This assists in strengthening and increasing the interfacial bonds between the graphene layers and the matrix of the host. It is this bonding that fills the cumulative emergence properties of graphene in reinforced nanocomposites (Kuilla et al., 2010).

In the composites world, theoretical study plays a vital role in understanding their mechanisms, physical properties, molecular interactions, and potential applications. A number of simulation tools supply cumulative or specific results for composite analysis. In recent years, there have been some important articles on graphene-based composites with a polymer matrix. Jang and Zhamu (2008) reviewed the fabrication of composite materials from the processing of graphene nanoplatelets (GNPs). Mack et al. (2005) prepared nanocomposites of polyacrylonitrile (PAN) nanofibers strengthened by GNP, which they confirmed to have improved mechanical qualities. Hansma et al. (2007) optimized successfully the amount, adhesives, and high-strength nanostructure (graphene) combination required to yield a strong, lightweight, low-density, and damage-resistant composite material. Ramanathan et al. (2008) notified an unprecedented shift in the glass transition temperature (*Tg*) of a polymer nanocomposite with functionalized graphene sheets. Their comparative study concluded that monolayered functionalized graphene serves as the best nanofiller among all tested nanofillers. Das et al. (2009) used the nanoindentation method to the graphene-reinforced nanocomposite fabricated by using polyvinyl alcohol (PVA) and polymethyl methacrylate (PMMA). The results displayed important enhancements in crystallinity, elastic modulus, and hardness through the addition of only 0.06 wt.% of graphene. The authors recommended that the improvement was due to nearby mechanical contact between the polymer matrix and the graphene layers. This interaction sequentially supplies and fills a better load transfer within the host matrix and the nanofiller.

Yu et al. (2007) recognized that epoxy-based few-layer graphene nanocomposites display interesting properties for the electronics industry, perfect for the expansion of thermal interface-based materials. Luechinger et al. (2008) introduced metal nanoparticles in the host matrix along with graphene, this method was attracted due to its advantage by facilitating an improved interparticle contact (i.e., contact between the host matrix and the metal particle). Watcharotone et al. (2007) fabricated a transparent electrical conductor by employing a simple sol-gel, chemical reduction, spin-coating and thermal-curing route. The graphene oxide sheets were mixed in the silica solution to get metal-encapsulated graphene nanocomposites. Chen et al. (2008) fabricated graphene conducting paper that was electrically conducting, biocompatible, and mechanically strong. They dispersed graphene sheets in a solution uniformly using vacuum filtration, followed by moderate thermal annealing. Cheng et al. (2013) prepared carbon-coated tin (II) oxide (SnO)-graphene sheet composites in a green process via a single-pot hydrothermal route. The composite was fabricated as an anode material for a lithium-ion battery, and it displayed high storage capacity and enhanced cyclic performance. The great number of application possibilities suggests that future research and prospects for graphene-based nanocomposites are likely to grow vastly in many disciplines and products in the store.

6 Applications of graphene-based nanocomposites

Graphene has an excessive number of uses including engineering, electronics, medicine, energy, industrial and household design, etc. (Avouris and Dimitrakopoulos, 2012; Huang et al., 2011a). The majority of graphene applications deal with electronic/sensor-oriented applications, to simplify the wide applications of graphene and graphene-based nanocomposites into their respective disciplines (Fig. 1). Shen et al. (2012) extensively studied the biomedical applications of graphene, including gene delivery, drug delivery, cancer therapy, bioimaging and biosensing, GO-based antibacterial materials, and scaffolds for tissue/cell culture.

6.1 *Biological applications*

Graphene has revealed potential applications in medical/biological fields, especially related to toxicity. Hu et al. (2010) showed the antibacterial activity of two forms

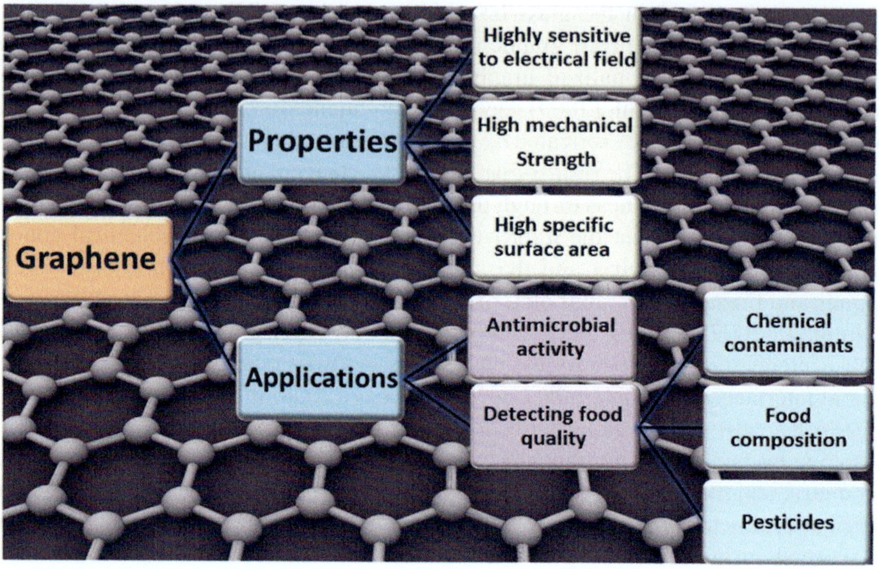

Fig. 1 Current progress on properties and applications of graphene-based nanocomposites in agrifood sectors.

of water-dispersible graphene with minimum cytotoxic effects against *Escherichia coli* on human participants. It was determined that GO paper can be effectively used for one day in various biological and environmental applications. Liao et al. (2011) demonstrated the cytotoxicity effect of graphene and graphene oxide materials under controlled physicochemical parameters. The results indicated that GO was more severely hemolytic than graphene and displayed high activity under an extremely small size. When chitosan was coated on graphene oxide, the hemolytic action disappeared totally, showing the composite biocompatibility for erythrocytes. They concluded that the biological or toxicological responses of the material were dependent on the particle size, state, and quality as well as the surface charge and the oxygen threshold.

Similarly, Liu et al. (2011) compared four diverse kinds of graphene substances—graphite (Gt), graphite oxide (GtO), GO, and reduced graphene oxide (rGO)—against *E. coli* to study the toxicity effects. The signals of the membrane and oxidative stress were used to measure the intensity of toxicity. The results showed that GO was the most severely toxic, followed in descending order by reduced graphene oxide, graphite, and graphite oxide. Santos et al. (2012) described the design, fabrication, and antimicrobial application of a graphene-poly-*N*-vinylcarbazole (PVK) nanocomposite, resulting in more than 80% microbial suppression and toxicity toward a broad group of bacteria. Carpio et al. (2012) investigated the toxicity effects of a PVK-GO nanocomposite on planktonic microbial cells, *E. coli*, *Bacillus subtilis, Cupriavidus metallidurans, Rhodococcus opacus*, biofilms, and mammalian fibroblast cells (NIH 3T3). The results indicated that PVK-GO presented a stronger antimicrobial effect than pristine GO. It was detected that the PVK-GO was significantly neutral toward the fibroblast cells, demonstrating a huge potential of the composite material in biomedical and industrial applications. Liu et al. (2013) produced a hydroxyapatite-GO nanocomposite as a biocompatible prosthetic. They detected that the (300) and (002) plane hydroxyapatite nanorods in the graphene matrix performed a crucial role in maintaining the composite's mechanical properties. Given its superior mechanical property, they suggested the nanocomposite's potential in composite and biomedical industries (Peng et al., 2012). It has been shown that GO possesses antibacterial properties against *E. coli* (Akhavan and Ghaderi, 2010). In addition, enhanced antibacterial properties have been observed on Ag-functionalized graphene materials (Xu et al., 2011). It was proved that water-dispersible graphene derivatives, GO, and rGO nanosheets could effectively inhibit the growth of *E. coli* while showing minimal cytotoxicity (Hu et al., 2010). After 2 h incubation with GO nanosheets of $20\,\mu g\,mL^{-1}$ at 37°C, the cell metabolic activity for *E. coli* decreased from ~70% to ~13%. In the same way, the metabolic activity of *E. coli* DH5α cells was reduced to ~24% on treatment with rGO nanosheets of $85\,\mu g\,mL^{-1}$ at 37°C for 2 h, and colony counting showed that <10% of *E. coli* survived. TEM studies revealed that rGO nanosheets destroyed the cell membrane of *E. coli*. However, the cell viability of A549 (a mammalian cell line) was reduced to ~47% and ~15% with rGO nanosheets of 20 and $85\,\mu g\,mL^{-1}$, respectively (Hu et al., 2010).

The antibacterial effect of rGO nanosheets was less than that of GO nanosheets while the cytotoxicity of rGO nanosheets was significantly greater than that of GO nanosheets. The mechanism of such irreversible GO-induced or rGO-induced cellular damage of *E. coli* might arise from the effects of oxidative stress or their cell membrane is damaged by direct contact with very sharp edges of graphene sheets (Akhavan and Ghaderi, 2010; Hu et al., 2010). Chook et al. (2012) synthesized Ag NPs on GO with a narrow size distribution using a microwave-irradiation method. Both AgNP and Ag-GO samples exhibited antibacterial activity that was stronger against gram-negative than against gram-positive bacteria. They reported that the antibacterial performance using Ag-GO with a lower concentration of Ag was better than using AgNPs. Thus, the synergistic effect between GO and AgNPs allows the use of a smaller amount of AgNPs without compromising the antibacterial properties. This alleviates concerns about excessive Ag usage and makes the Ag-GO composite a potential material for food-packaging and wound-dressing applications. The effectiveness of the biofunctionalized graphene as a reinforcing filler (4 wt.%) in a chitosan polymer matrix is verified by the dramatic enhancement of the mechanical properties (breaking stress = 330%, Young's modulus = 243%) and the electrical conductivity ($0.3\,Sm^{-1}$) without much loss in the elongation at break. In addition, this strong graphene-filled composite film could be used as an antimicrobial packaging material for food and biomedical devices (Yadav et al., 2013). When sheets of GO are fixed onto cotton fabrics, a strong antibacterial property and a great laundering durability were informed. These flexible, foldable, and reusable graphene oxide-based antibacterial cotton fabrics were done by direct adsorption, radiation-induced cross-linking, and chemical cross-linking. Antibacterial assessments of all these GO-containing fabrics displayed strong antibacterial activity and inactivation of 98% bacteria. These fabrics were found to kill more than 90% of bacteria and caused no irritation to the skin in experiments on rabbits (Zhao et al., 2013).

GO photothermal treatment using near-infrared laser described the surface activation of GO nanoflakes resulting in the killing of *Staphylococcus aureus* and *Pseudomonas aeruginosa* (Shahnawaz et al., 2015). TEM images showed that the cell wall and membrane of *Streptococcus mutans, Fusobacterium nucleatum,* and *Porphyromonas gingivalis* (He et al., 2015) lost their integrity and the intracellular contents leaked out when treated with graphene oxide (Fig. 2). The GO edge

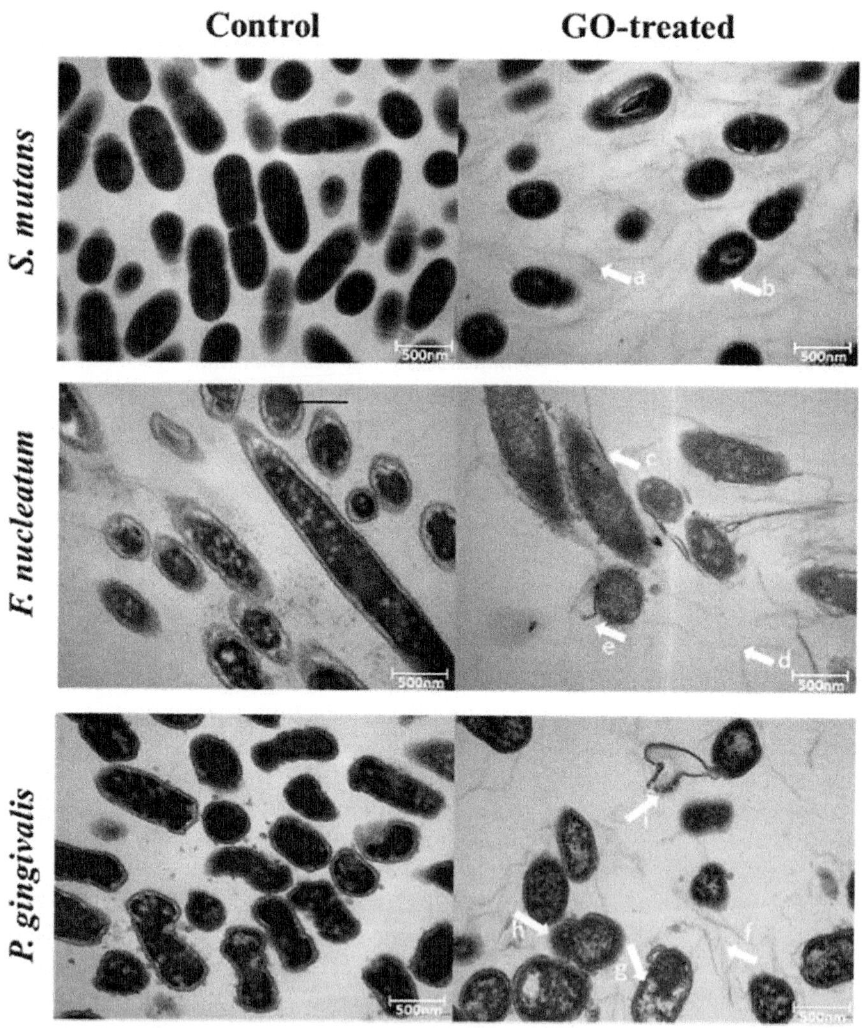

Fig. 2 TEM images of *S. mutans, F. nucleatum,* and *P. gingivalis* cells after incubation with GO nanosheet dispersion (right side) for 2 h, and after incubation with saline solution for 2 h as the control (left side). All treated cases had the same GO dose of 80 μg mL^{-1}. Scale bar = 500 nm. Reprinted from He, J., Zhu, X., Qi, Z., Wang, C., Mao, X., Zhu, C., He, Z., Li, M., Tang, Z., 2015. Killing dental pathogens using antibacterial graphene oxide. ACS Appl. Mater. Interfaces 7, 5605–5611, with permission from American Chemical Society.

nanowalls play an essential part in its antimicrobial mechanism (Singh, 2016). However, using the Langmuir-Blodgett technique to immobilize flat graphene oxide sheets on a PET substrate to detect its antibacterial activity showed that contact with the edges is not a vital factor of the mechanism (Mangadlao et al., 2015). The antibacterial activity of GO might also be because a superoxide radical anion was produced that leads to cell death as seen in *P. aeruginosa*, which was additionally confirmed through nuclear fragmentation (Gurunathan et al., 2012).

Many factors may also affect graphene nanomaterial's biological interactions with cells. It was seen that the antibacterial effect of graphene oxide sheets toward E. coli cells is lateral size dependent (Liu et al., 2012). Larger sheets of GO reveal stronger antibacterial activity than smaller ones, as they more easily cover cells resulting in the cell viability loss as detected in the colony counting test. In contrast, small sheets of GO adhere to the surfaces of bacteria. They cannot effectively isolate cells from the environment and show different concentration and time-dependent antibacterial properties. Therefore, it is important to tailor the lateral GO dimension sheets to adjust the application potential with low hazards for environmental health and safety (Liu et al., 2012). Also, bare GO sheets were intrinsically bactericidal, yielding a bacterial survival percentage of <1% at $200\,\mu g\,mL^{-1}$ (Hui et al., 2014) whereas when saline was supplemented with Luria-Bertani (LB) broth, it gradually deactivated its activity. An amount of 10% Luria-Bertani made graphene oxide totally inactive (Hui et al., 2014), concluding that noncovalent adsorption of LB components on GO basal planes may be a global deactivation action for GO's cytotoxicity.

A rapid decrease in environmental salinity and a biocompatible block copolymer (Pluronic F-127) were found to increase GO antibacterial activity. Hypo-osmotic stress prompted by water shock makes gram-negative pathogens more susceptible to Pluronic-GO that can populate around bacterial envelopes favoring the interactions between bacteria and GO. The GO mixture is both effective and safe in bacterial disinfection applications (Karahan et al., 2015). Cystamine-conjugated GO with low cytotoxicity but high antibacterial activity and strong reactive oxygen species effects were prepared, displaying minimum inhibitory concentration values of $1\,\mu g\,mL^{-1}$ against *E. coli* and *Salmonella typhimurium*, $4\,\mu g\,mL^{-1}$ against *B. subtilis,* and $6\,\mu g\,mL^{-1}$ *against Enterococcus faecalis* (Nanda et al., 2015).

The reduced graphene oxide is usually considered as one kind of chemically derived graphene that is done from the reduction of GO by thermal, chemical, or electrical treatments. Hence, some other names have been given to reduced graphene oxide, such as functionalized graphene chemically modified graphene, chemically converted graphene, or reduced graphene (Eda and Chhowalla, 2010). Different processes of reduction result in varied properties that in turn affect the final material performance or devices composed of rGO. RGO antibacterial activity results from the production of a superoxide radical anion, which induced loss of cell viability, oxidative stress, and DNA fragmentation in *P. aeruginosa* (Gurunathan et al., 2012). Some studies recommend that antimicrobial actions of reduced graphene oxide are contributed both by oxidation stress and membrane contact.

6.2 Graphene in detecting food quality

6.2.1 Detection of chemical contaminants

Surface-enhanced Raman spectroscopy (SERS) is an analytical technique used to investigate and identify biomolecules or chemicals by their improved (of the order of 104–1014) bands in the Raman spectra (Yaffe and Blanch, 2008). The intensity of Raman peaks of the molecules could be improved on the substrate (metal NPs) due to electromagnetic or chemical mechanisms (Gunnarsson et al., 1999). Graphene is a good substrate to investigate the chemical enhancement of SERS (Schedin et al., 2010). Liu et al. (2012) prepared a graphene nanomesh with controlled size and density of holes by first depositing a Cu film on graphene and then removing the copper film by annealing to leave plenty of hole rims around the copper nanoparticles. They used this graphene nanomesh to detect rhodamine B (RhB) by SERS. The intensity of the Raman bands on a single-layer graphene nanomesh (SGNM) (with different Cu films) was significantly greater than those on single-layer graphene (SLG). They attributed this result to the ability of edges in the SGNM to improve Raman bands by increasing the local charge transfer and to absorb RhB molecules.

Xie et al. (2012) developed a SERS sensor for detecting prohibited colorants (e.g., amaranth, erythrosine, lemon yellow, and sunset yellow) in foodstuffs using graphene/silver nanoparticles as substrates. The characteristic Raman peaks were significantly more enhanced on the graphene/Ag substrate than on the Ag substrate alone, due to the excellent adsorbent effect of graphene and AgNPs. The limits of detection (LODs) of this graphene-based SERS were 10^{-5} M for amaranth, lemon yellow, and sunset yellow and 10^{-7} M for erythrosine. The aqueous dispersion prepared of the homogeneous gum arabic (GA)-stabilized graphene [1.7–2.1 nm thick (i.e., fewer than four layers)] was stable for more than a month in a sealed bottle. GA could also reduce Ag^+ ions and immobilize them on graphene sheets for fabricating AgNP/GA-graphene hybrid (Fan et al., 2012). Concentrated electromagnetic fields are the reason for SERS when there are two or more coupled NPs or nanostructured surfaces with closely spaced features (Lee et al., 2006).

The SERS activity of the AgNP/GA graphene hybrid was investigated using aminothiophenol (4-ATP) as a Raman-probe molecule. There was no confirmation that the Raman peak of 4-ATP was discovered on the surface of the GA-graphene hybrid. However, the SERS spectrum of 4-ATP on the Ag/GA graphene hybrid exhibited four b2 modes at 1145, 1392, 1445, and 1572 cm^{-1} and one a1 mode at 1072 cm^{-1}. This indicated that the Ag/GA-graphene hybrid film was suitable for SERS detection of 4-ATP (Fan et al., 2012). Compared with the spectrum of the solid 4-ATP, there are distinct frequency shifts due to the changes in the band intensity. The deformation vibration of the carbon-sulfur (C-S) bond shift (1092–1074 cm^{-1}), and another frequency shift (1591–1574 cm^{-1}) were observed due to the formation of strong Ag-S bonds between the Ag and -SH groups of 4-ATP (Hunyadi and Murphy, 2006). Oxalic acid (OA) is naturally present in plants, fruits, vegetables, and nuts, and could form a less soluble complex with general cations (Ca^{2+}) (Massey et al., 1993).

When present in high levels in the diet, oxalic acid leads to irritation of the digestive system, especially of the stomach and kidneys (Ivandini et al., 2006).

Chen et al. (2013a,b) used a platinum NP (PtNP)-loaded graphene nanosheet (GNsh) (PtNP-GNsh)-modified electrode to detect OA in the range 0.1–50 mM. The PtNP-GNsh film exhibited high electrochemical activity with well-defined oxidation peaks of OA and decreased over potential compared to bare or graphene-modified electrodes. Hydrogen peroxide (H_2O_2) is a byproduct of a large number of oxidase enzymes, and also an essential mediator in food, pharmaceutical, clinical, industrial, and environmental analyses (Preston et al., 2001). A palladium NP/graphene nanosheet (PdNP)/GNsh film-coated glassy carbon electrode (GCE) showed enhanced electrochemical reduction of H_2O_2 from 0.1 to 1000 μM with a LOD of 0.05 μM. In addition, aPdNP/GNsh-based sensor allowed the selective detection of H_2O_2in the presence of ascorbic acid, glucose, and dopamine (Chen et al., 2013a,b). Synthetic aromatic azo dyes are considered harmful because they could cause allergies, migraines, eczema, anxiety, diarrhea, and even cancer, if consumed excessively (Rowe and Rowe, 1994; Stevenson et al., 1986). Sunset yellow (SY) and tartrazine (TT) are azo colorants that may be present in common foodstuffs (Alves et al., 2008).

Gan et al. (2012) prepared a graphene layer-wrapped phosphotungstic acid (PTA) hybrid film, which provides good selectivity and high sensitivity for the simultaneous determination of SY and TT by exhibiting well-defined oxidation peaks in differential pulse voltammetry with a peak potential separation of ~260 mV. The LODs were 0.5 $\mu g L^{-1}$ for SY and 30.0 $\mu g L^{-1}$ for TT. Moreover, the graphene-PTA/GCE allowed simultaneous determination of both dyes without the interference of glucose, sucrose and glycine, citric acid, Fe^{3+}, Fe^{2+}, and Ca^{2+}; or amaranth, ponceau R, allurared, and quinoline yellow. Using a mesoporous TiO_2-decorated graphene-based hybrid film, they were able to detect Sudan I and Orange II (toxic dyes, food colorants) in the range 3.3 nM–0.66 μM with LODs of 0.60 nM (Sudan I) and 2.85–28.54 nM (Orange II), respectively. This sensor was also tested for measuring Sudan I and Orange II in food extracts with reliable measurements (Gan et al., 2013a,b,c).

The high extraction efficiency of GNs has adsorbent in solid-phase microextraction (SPME) method and then was demonstrated for UV filters (benzophenone, octyl salicylate, homosalate,3-(4-methylbenzylidene) camphor and 2-hydroxy-4-methoxybenzophenone) (Zhang and Lee, 2012). Malachite green is a dye and an antimicrobial agent used in the food industry. It has emerged as a controversial agent in aquaculture and is considered highly toxic. Hou et al. (2013) proposed a graphene quantum dot (GQD)-AuNP-modified GCE for enhanced detection of malachite green with an LOD of $1.0 \times 10^{-7} mol L^{-1}$. Sheets of graphene suspended in a poly[*N*-(1-one-butyric acid)benzimidazole] (PBI-BA) polyelectrolyte solution (Hua et al., 2011) were coated on an Au electrode and dramatically improved the detection performance for H_2O_2 (from 2.5 μM to 5 mM with a sensitivity of $1056 mA mM^{-1} cm^{-2}$). A nanocomposite made of tin dioxide (SnO_2) and graphene (Lim et al., 2012) showed high catalytic activity and 10-fold enhanced reduction current of mercury(II), suggesting that the SnO_2/graphene nanocomposite could be employed as a mercury(II) sensor. Graphene sheets with lateral dimensions <100 nm are known as GQDs, which have very low toxicity and exhibit high stability, excellent solubility, stable photoluminescence, and good biocompatibility.

Upon adding GO, the absorbance intensity of the system increased due to the intrinsic peroxidase property of GO, which can catalyze H_2O_2-mediated oxidation of TMB and produced the oxidation product with the color change. GQDs have increased the absorbance intensity of the system with the concentration of H_2O_2 from 10 nM to 0.5 mM. Bisphenol A (BPA), 2, 2-bis (4-hydroxyphenyl) propane, is a monomer in the production of epoxy resins and polycarbonates in the polymer industry. BPA has been linked to adverse estrogenic effects on human health. BPA has been used in food-packaging materials that come in direct contact with food products. Considering the organic components of foodstuffs, nonpolymerized monomer residues could be soluble in food products. Macrolides (MLs) (e.g., lincomycin, azithromycin, tilmicosin, erythromycin, and roxithromycin) are synthetic antibiotics used in animal husbandry, aquaculture, and human medicine. The presence of even a trace level of macrolides in food products would lead to side effects (e.g., allergic reactions, the appearance of resistant bacteria, and even cross-resistance to other antibiotics).

6.2.2 Detection of food composition

High cholesterol content in the blood is one of the major controllable risk factors for coronary heart disease, heart attack, and stroke. Many foods contain cholesterol, as animal brains, egg yolks, meat, seafood, and milk-based products are known for their high cholesterol content. Biosensors have been developed to determine the cholesterol content in food. Cao et al. (2013a) proposed a biosensor using a TiO_2-graphene-Pt-Pd nanocomposite coated with AuNPs. This biosensor was appropriate for the sensitive recognition of cholesterol (0.017 μM) in foods (e.g., eggs, meat, margarine, and fish oil). Glucose may be an essential vitality supplement; its concentration within the blood reflects the vitality status of the body, and a change of hypothalamic glucose in people is related to diabetes (Tonon et al., 2013).

It is clinically important to do fast, accurate measurement of glucose concentration in biological or food samples. Highly crystalline manganese (II, III) oxide (Mn_3O_4) was deposited on a three-dimensional graphene foam (3DGF) to formulate a flexible, free-standing biosensor for nonenzymatic determination of glucose. Using Mn_3O_4/3DGF, Si et al. detected glucose in the range 0.1–8 mM with a high sensitivity of 360 μA mM cm^{-2} (Si et al., 2013). Generally, to retain the biological activity of glucose oxidase (GOD) on the electrode surface, it has to be immobilized on a biocompatible substance that keeps the native state of the enzyme, and that was possible using immobilized GOD on a graphene-AuNP film.

The immobilized GOD displayed a couple of stable, well-defined redox peaks with an electron transfer frequency constant of 3.25 s^{-1}. Glucose was detected in the range 0.02–2.26 mmol L^{-1} using a graphene-Au biosensor with high sensitivity (3.844 μA mmol L^{-1} cm^{-2}) (Cao et al., 2013b). β-lactoglobulin (BLG), which represents 10% of the proteins in cow's milk, is considered the most important milk allergen, especially for children (Chen et al., 2011). An amine-functionalized graphene surface was linked to anti-BLG antibodies using glutaraldehyde to detect BLG. In the presence of $[Fe(CN)_6]$3-/4- in an aqueous solution, the peak current of $[Fe(CN)_6]$3-/4- linearly decreased with increasing BLG concentration (from 1 pg mL^{-1} to 100 ng mL^{-1}) due to the formation of an antibody-antigen complex on the electrode surface (Eissa et al., 2012).

Bovine hemoglobin (BHb)-imprinted polydopamine-functionalized rGO (PDA@rGO) was done by Luo et al. (2013) by using immersing graphene oxide in a weak alkaline solution of dopamine-containing BHb. A thin, adherent polydopamine (PDA) film fixed with BHb was obtained on the surface of GNsh to produce the BHb imprinted PDA@rGO nanomaterial, which displayed much higher binding capacity toward BHb than control PDA@rGO because of nonspecific adsorption at the surface (Luo et al., 2013). GO can also distinguish single-stranded DNA (ss-DNA) from double-stranded DNA (ds-DNA) because of its significantly different adsorption affinities to ss-DNA and ds-DNA (He et al., 2010). Based on the specific binding ability of graphene, a colorimetric method for the detection of DNA damage was developed using hemin-GNsh (HGNsh), which keeps the capability of graphene to distinguish damaged DNA from intact DNA in the range 5–60 nM with an LOD of 1 nM (Wei et al., 2013).

An assay of salbutamol (SAL) is very important for animal tissues and biological fluids (Rajkumar et al., 2013). An ultrasensitive immunosensor was created on AuNPs, Prussian blue, polyaniline/poly (acrylic acid,) and an Au-hybrid graphene nanocomposite (AuGN). An Au-hybrid graphene nanocomposite film was used to restrain chitosan and the HRP-anti-SAL antibody (HRP-AAb). The resulting nanostructure (AuGN-HRP-AAb) was used as a label for the immunosensor. The high surface-area-to-volume ratio of graphene allows the immobilization of a high level of chitosan, AuNPs, and HRP-AAb, and its good electrical conductivity and biocompatibility enhance electron transfer between HRP, H_2O_2, and the electrode. This graphene-based immunosensor showed a linear plot for SAL in the range of 0.08–1000 ng mL^{-1} with an LOD of 0.04 ng mL^{-1} (Huang et al., 2011b). A squamous cell carcinoma antigen (SCC-Ag) was detected in the uterine cervical SCC and the serum level of SCC-Ag parallels the growth of the tumor size (Schedel et al., 2011). For diagnosis, it is clinically important to measure the serum level of SCC-Ag. Studies suggest that chemical substances stimulate estrogenic activity and cause adverse effects in humans and animals or environmental organisms (Tyler and Routledge, 1998). The core chemicals causing these effects are estrone (E1), estradiol (E2), estriol (E3), and synthetic estrogens. Detection of estrogens is vital to evaluate and forecast the estrogenic potency of foods or surface waters. Fe_3O_4-NPs with a composite made from $TiO_2$and GO (Fe_3O_4@TiO_2/GO) were used to extract estrogens E1, E2, and E3 from milk samples. The estrogens were then eluted from the beads with methanol and analyzed with HPLC (Tian et al., 2013a). The novel colorimetric technique for glucose detection with improved indicator strength was developed using nanocomposites containing $ZnFe_2O_4$-carbon nanotube-glucose oxidase (Wang et al., 2019). In the current method, glucose detection based on the coconjugation of GOD and $ZnFe_2O_4$ reduction dissemination of the generated hydrogen peroxide by transporting it nonstop to the adjacent $ZnFe_2O_4$, which in return increases the sensitivity. Temporarily, as CNTs are a first-rate electron transportation medium and have a wide large surface area, the electron transmission reaction was similarly enhanced, resulting in improved biosensor response. Therefore, a "turn-on" mechanism was approved for fast detection of glucose in human serum samples. The suggested mode of action of glucose detection is described in Fig. 3.

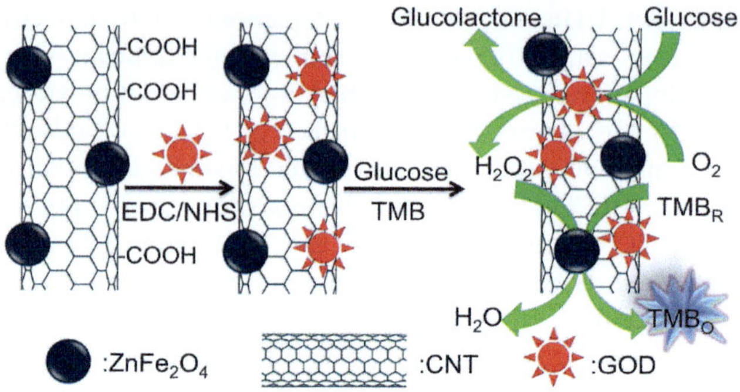

Fig. 3 Description of the $ZnFe_2O_4$-CNT-GOD colorimetric-based method applied for detection of glucose human serum samples.
Reprinted from Wang, C., Li, J., Tan, R., Wang, Q. and Zhang, Z., 2019. Colorimetric method for glucose detection with enhanced signal intensity using $ZnFe_2O_4$ carbon nanotube-glucose oxidase composite material. Analyst 144(5), 1831–1839, with permission from RSC Publishing, Royal Society of Chemistry.

6.2.3 Detection of pesticides

Pesticides (e.g., pymetrozine, thidiazuron, diuron, carbofuran, carbaryl, pirimicarb, diethofencarb, procymidone, folpet, vinclozolinandditalimfos) are commonly used in agriculture to thwart diseases (Guan et al., 2013; Li et al., 2013). Use of these chemicals has proved very effective, but several concomitant risks remain. The pesticide residues in foods adversely affect human health because of their high biological activity and inherent toxicity. The US Environmental Protection Agency (EPA) lists 12 of the 26 most widely used pesticides in the United States as carcinogens. Improper and excessive use of these pesticides has led to them being slowly accumulated in the food chain; these poisonous chemicals also contaminate natural water sources (Tankiewicz et al., 2010). According to the Council of the European Union, the total maximum miscible concentration of pesticides is no more than $0.5\,\mu g\,L^{-1}$ (EU Council, 1998). Wang et al. (2013) developed a graphene-based magnetic nanocomposite as the adsorbent for the extraction of seven triazole fungicides (triadimefon, paclobutrazol, hexaconazole, myclobutanil, diniconazole, propiconazole, and tebuconazole) in cucumber, cabbage, and tomato samples prior to gas chromatography-mass spectrometry (GC-MS) detection. The LODs for the analyses were $0.01–0.10\,ng\,g^{-1}$. However, monitoring lipophilic pesticides at trace levels is challenging and extensive sample extraction and clean-up procedures were required (Koesukwiwat et al., 2010).

Common methods [e.g., liquid-liquid extraction (LLE) and SPE] were required to preconcentrate trace analytes from environmental or food samples (Chen et al., 2010). However, liquid-liquid extraction needs large amounts of organic solvents and SPE is considered very tedious. Graphene was used as an adsorbent in LLE and SPE to preconcentrate and extract pesticide compounds in environmental samples due to its extremely large surface area and excellent adsorption capacity. Guan et al. (2013) proposed to use amine-modified graphene to clean up fatty acids and other interfering substances from acetonitrile extracts of oil plants. This amine-modified graphene was the best of the tested sorbents as a reversed-dispersive SPE (r-DSPE) clean-up sorbent and was used effectively to determine 31 pesticides in oil crops. Li et al. (2013) proposed another method based on magnetic particle-modified adsorbents in magnetic SPE (MSPE), which offered a simple way to extract analytes (fungicides) from contaminated samples without extensive centrifugation and filtration steps. In some cases, a magnetic graphene nanocomposite (G-Fe_3O_4) was used as the adsorbent to extract fungicides from water and grape-juice samples to be analyzed with GC.

Zhao et al. (2011a) extended the study on G-Fe_3O_4 magnetic NPs (G-Fe_3O_4 MNPs) for the MSPE of some triazine herbicides (atrazine, prometon, propazine, and prometryn) in water. After the extraction, the adsorbent could be conveniently separated from the aqueous samples by an external magnet and almost complete desorption of the analytes from the sorbent could be achieved using acetone as a desorption solvent. The sensitivity of the method is better than that of the other MSPE methods that do not use graphene. In addition, the G-Fe_3O_4 is a very efficient adsorbent and it could be reused at least 15 times without a significant decrease in extraction efficiency. GNsh-coated SPME fiber was also prepared by immobilizing GNsh on a stainless steel wire. The extraction efficiency of the GNsh-SPME method was better for five organochlorine pesticides (OCPs) [1,1-dichloro-2,2-bis-(p′-chlorophenyl) ethylene, 1,1-dichloro-2,2-bis-(p′-chlorophenyl) ethane,1,1,1-trichloro-2,2-bis(p′-chlorophenyl) ethane, dieldrin and endrin) (Ponnusamy and Jen, 2011), and carbamate pesticides (carbofuran, carbaryl, pirimicarb, and diethofencarb) (Zhao et al., 2011b), compared with the conventional water-bath microwave-assisted headspace-SPME method. For further information, there are recent review articles on sample preparation using metallic NPs (Tian et al., 2013b) and carbon nanomaterials (Zhang et al., 2013).

7 Conclusion and future perspectives

Graphene is the most promising candidate to be a major filling agent for composite applications due to its exceptional thermal, mechanical, and electronic properties. The great number of application possibilities suggests that forthcoming investigation and prospects for graphene-based nanocomposites are expected to vastly develop in many disciplines and products in the store. Graphene can be utilized to make improved food packaging, which could help lessen food waste and increase food safety. Adaptable, straightforward, and dampness-impermeable materials are basic for packaging applications in the food industry. Packing can extend the shelf life of products by integrating a layer of graphene that considerably decreases penetrability.

Acknowledgment

Sincere thanks to Conselho Nacional de Desenvolvimento Científico e Tecnológico (CNPq) Grant#315844/2018-3 to Dr. Youssef Khamis.

References

Akhavan, O., Ghaderi, E., 2010. Toxicity of graphene and graphene oxide nanowalls against bacteria. ACS Nano 4, 5731–5736.

Alves, S.P., Brum, D.M., Branco de Andrade, É.C., Pereira Netto, A.D., 2008. Determination of synthetic dyes in selected foodstuffs by high-performance liquid chromatography with UV-DAD detection. Food Chem. 107, 489–496.

Avouris, P., Dimitrakopoulos, C., 2012. Graphene: synthesis and applications. Mater. Today 15 (3), 86–97.

Bae, S., Kim, H., Lee, Y., Xu, X., Park, J.-S., Zheng, Y., Balakrishnan, J., Lei, T., Kim, H.R., Song II, Y., Kim, Y.-J., Kim, K.S., Ozyilmaz, B., Ahn, J.-H., Hong, B.H., Iijima, S., 2010. Roll-to-roll production of 30-inch graphene films for transparent electrodes. Nat. Nanotechnol. 5, 574–578.

Bagri, A., Mattevi, C., Acik, M., Chabal, Y.J., Chhowalla, M., Shenoy, V.B., 2010. Structural evolution during the reduction of chemically derived graphene oxide. Nat. Chem. 2, 581–587.

Bolotin, K.I., Sikes, K.J., Jiang, Z., Klima, M., Fudenberg, G., Hone, J., Kim, P., Stormer, H.L., 2008. Ultrahigh electron mobility in suspended graphene. Solid State Commun. 146, 351–355.

Brydson, R., Brown, A., 2008. An investigation of the surface structure of nanoparticulate systems using analytical electron microscopes corrected for spherical aberration. In: Turning Points in Solid-State, Material and Surface Science. RSC Publishing, London, pp. 778–791.

Cao, S., Zhang, L., Chai, Y., Yuan, R., 2013a. An integrated sensing system for detection of cholesterol based on TiO_2-graphene-Pt-Pd hybrid nanocomposites. Biosens. Bioelectron. 42, 532–538.

Cao, X., Ye, Y., Li, Y., Xu, X., Yu, J., Liu, S., 2013b. Self-assembled glucose oxidase/graphene/gold ternary nanocomposites for direct electrochemistry and electrocatalysis. J. Electroanal. Chem. 697, 10–14.

Carpio, M.I.E., Santos, C.M., Wei, X., Rodrigues, D.F., 2012. Toxicity of a polymer-graphene oxide composite against bacterial planktonic cells, biofilms, and mammalian cells. Nanoscale 4 (15), 4746–4756.

Chen, H., Muller, M.B., Gilmore, K.J., Wallace, G.G., Li, D., 2008. Mechanically strong, electrically conductive, and biocompatible graphene paper. Adv. Mater. 20 (18), 3557–3561.

Chen, J., Zou, J., Zeng, J., Song, X., Ji, J., Wang, Y., Ha, J., Chen, X., 2010. Preparation and evaluation of graphene-coated solid-phase microextraction fiber. Anal. Chim. Acta 678, 44–49.

Chen, H., Zhang, B., Cui, Y., Liu, B., Chen, G., Tang, D., 2011. One-step electrochemical immunoassay of biomarker-based on nanogold-functionalized grapheme sensing platform. Anal. Methods 3, 1615–1621.

Chen, X., Chen, X.-M., Cai, Z.-X., Huang, Z.-Y., Oyama, M., Jiang, Y.-Q., 2013a. Ultrafine palladium nanoparticles grown on graphene nanosheets for enhanced electrochemical sensing of hydrogen peroxide. Electrochim. Acta 97, 398–403.

Chen, H., Gao, Z., Cui, Y., Chen, G., Tang, D., 2013b. Nanogold-enhanced graphene nanosheets as multienzyme assembly for sensitive detection of low abundance proteins. Biosens. Bioelectron. 44, 108–114.

Cheng, J., Xin, H., Zheng, H., Wang, B., 2013. One-pot synthesis of carbon coated-SnO_2/graphene-sheet nanocomposite with highly reversible lithium storage capability. J. Power Sources 232, 152–158.

Chook, S.W., Chia, C.H., Zakaria, S., Ayob, M.K., Chee, K.L., Huang, N.M., Neoh, H.M., Lim, H.N., Jamal, R., Rahman, R.M.F.R.A., 2012. Antibacterial performance of Ag nanoparticles and AgGO nanocomposites prepared via rapid microwave-assisted synthesis method. Nanoscale Res. Lett. 7, 1–7.

Chowdhuri, A.R., Tripathy, S., Chandra, S., Roy, S., Sahu, S.K., 2015. A ZnO decorated chitosan-graphene oxide nanocomposite shows significantly enhanced antimicrobial activity with ROS generation. RSC Adv. 5, 49420–49428.

Chua, C.K., Pumera, M., 2014. Chemical reduction of graphene oxide: a synthetic chemistry viewpoint. Chem. Soc. Rev. 43, 291–312.

Compton, O.C., Nguyen, S.T., 2010. Graphene oxide, highly reduced graphene oxide, and graphene: versatile building blocks for carbon-based materials. Small 6, 711–723.

Das, B., Eswar, K.P., Ramamurty, U., Rao, C.N.R., 2009. Nano-indentation studies on polymer-matrix composites reinforced by a few-layer graphene. Nanotechnology 20 (12), 125705.

Diaye, N.'.A.T., Coraux, J., Plasa, T.N., Busse, C., Michely, T., 2008. Structure of epitaxial graphene on Ir(111). New J. Phys. 10, 043033.

Dreyer, D.R., Park, S., Bielawski, C.W., Ruoff, R.S., 2010. The chemistry of graphene oxide. Chem. Soc. Rev. 39, 228–240.

Eda, G., Chhowalla, M., 2010. Chemically derived graphene oxide: towards large-area thin-film electronics and optoelectronics. Adv. Mater. 22 (22), 2392–2415.

Eissa, S., Tlili, C., L'Hocine, L., Zourob, M., 2012. Electrochemical immunosensor for the milk allergen β-lactoglobulin based on electrografting of organic film on graphene modified screen-printed carbon electrodes. Biosens. Bioelectron. 38, 308–313.

Elzatahry, A.A., Abdullah, A.M., Salah El-Din, T.A., Al-Enizi, A.M., Maarouf, A.A., Galal, A., Hassan, H.K., El-Ads, E.H., Al-Theyab, S.S., Al-Ghamdi, A.A., 2012. Nanocomposite graphene-based material for fuel cell applications. Int. J. Electrochem. Sci. 7, 3115–3126.

EU Council, Directive 98/83/EC, 1998. On the quality of water intended for human consumption. Off. J. Eur. Commun. L. 330, 32–54.

Fan, J., Shi, Z., Ge, Y., Wang, J., Wang, Y., Yin, J., 2012. Gum arabic assisted exfoliation and fabrication of Ag-graphene-based hybrids. J. Mater. Chem. 22, 13764–13772.

Gan, T., Sun, J., Cao, S., Gao, F., Zhang, Y., Yang, Y., 2012. One-step electrochemical approach for the preparation of graphene wrapped-phosphotungstic acid hybrid and its application for simultaneous determination of sunset yellow and tartrazine. Electrochim. Acta 74, 151–157.

Gan, T., Sun, J., He, M., Wang, L., 2013a. Highly sensitive electrochemical sensor for Sudan I based on graphene decorated with mesoporous TiO_2. Ionics 1–7.

Gan, T., Sun, J., Lin, Z., Li, Y., 2013b. Highly sensitive determination of Orange II based on the dual amplified electrochemical signal of graphene and mesoporous TiO_2. Anal. Methods 5, 2964–2970.

Gan, T., Sun, J., Meng, W., Song, L., Zhang, Y., 2013c. Electrochemical sensor based on graphene and mesoporous TiO_2 for the simultaneous determination of trace colourants in food. Food Chem. 141, 3731–3737.

Geim, A.K., 2009. Graphene: status and prospects. Science 324, 1530–1535.

Geim, A.K., Novoselov, K.S., 2007. The rise of graphene. Nat. Mater. 6 (3), 183–191.

Gómez-Navarro, C., Meyer, J.C., Sundaram, R.S., Chuvilin, A., Kurasch, S., Burghard, M., Kern, K., Kaiser, U., 2010. Atomic structure of reduced graphene oxide. Nano Lett. 10, 1144–1148.

Guan, W.B., Li, Z.N., Zhang, H.Y., Hong, H.J., Rebeyev, N., Ye, Y., Ma, Y.Q., 2013. Amine modified graphene as reversed-dispersive solid phase extraction materials combined with liquid chromatography-tandem mass spectrometry for pesticide multi-residue analysis in oil crops. J. Chromatogr. A 1286, 1–8.

Guermoune, A., Chari, T., Popescu, F., Sabri, S.S., Guillemette, J., Skulason, H.S., Szkopek, T., Siaj, M., 2011. Chemical vapor deposition synthesis of graphene on copper with methanol, ethanol, and propanol precursors. Carbon 49 (13), 4204–4210.

Gunnarsson, L., Petronis, S., Kasemo, B., Xu, H., Bjerneld, J., Käll, M., 1999. Optimizing nanofabricated substrates for surface enhanced raman scattering. Nanostruct. Mater. 12, 783–788.

Gupta, S.S., Sreeprasad, T.S., Maliyekkal, S.M., Das, S.K., Pradeep, T., 2012. Graphene from sugar and its application in water purification. ACS Appl. Mater. Interfaces 4, 4156–4163.

Gurunathan, S., Han, J.W., Dayem, A.A., Eppakayala, V., Kim, J.H., 2012. Oxidative stress-mediated antibacterial activity of graphene oxide and reduced graphene oxide in Pseudomonas aeruginosa. Int. J Nanomed. 7, 5901–5914.

Hansma, P.K., Turner, P.J., Ruoff, R.S., 2007. Optimized adhesives for strong, lightweight, damage-resistant, nanocomposite materials: new insights from natural materials. Nanotechnology 8 (4), 044026.

He, S., Song, B., Li, D., Zhu, C., Qi, W., Wen, Y., Wang, L., Song, S., Fang, H., Fan, C., 2010. A graphene nanoprobe for rapid, sensitive, and multicolour fluorescent DNA analysis. Adv. Funct. Mater. 20, 453–459.

He, J., Zhu, X., Qi, Z., Wang, C., Mao, X., Zhu, C., He, Z., Li, M., Tang, Z., 2015. Killing dental pathogens using antibacterial graphene oxide. ACS Appl. Mater. Interfaces 7, 5605–5611.

Henry, C.R., 2005. Morphology of supported nanoparticles. Prog. Surf. Sci. 80 (3–4), 92–116.

Hernandez, Y., Nicolosi, V., Lotya, M., Blighe, F.M., Sun, Z., De, S., McGovern, I.T., Holland, B., Byrne, M., Gun'Ko, Y.K., Boland, J.J., Niraj, P., Duesberg, G., Krishnamurthy, S., Goodhue, R., Hutchison, J., Scardaci, V., Ferrari, A.C., Coleman, J.N., 2008. High-yield production of graphene by liquid-phase exfoliation of graphite. Nat. Nanotechnol. 3, 563–568.

Hontoria-Lucas, C., Lo′pez-Peinado, A.J., de Lopez-Gonzalez, J.D., Rojas-Cervantes, M.L., Martín-Aranda, R.M., 1995. Study of oxygen-containing groups in a series of graphite oxides: physical and chemical characterization. Carbon 33 (11), 1585–1592.

Hou, J., Bei, F., Wang, M., Ai, S., 2013. Electrochemical determination of malachite green at graphene quantum dots-gold nanoparticles multilayers-modified glassy carbon electrode. J. Appl. Electrochem. 43, 689–696.

Howe, J.M., Mori, H., Wang, Z.L., 2008. In situ high-resolution transmission electron microscopy in the study of nanomaterials and properties. MRS Bull. 33 (2), 115–121.

Hu, W., Peng, C., Luo, W., Lv, M., Li, X., Li, D., Huang, Q., Fan, C., 2010. Graphene-based antibacterial paper. ACS Nano 4 (7), 4317–4323.

Hua, M.-Y., Chen, H.-C., Tsai, R.-Y., Leu, Y.-L., Liu, Y.-C., Lai, J.-T., 2011. Synthesis and characterization of carboxylated polybenzimidazole and its use as a highly sensitive and selective enzyme-free H_2O_2 sensor. J. Mater. Chem. 21, 7254–7262.

Huang, X., Yin, Z., Wu, S., Qi, X., He, Q., Zhang, Q., Yan, Q., Boey, F., Zhang, H., 2011a. Graphene-based materials: synthesis, characterization, properties, and applications. Small 7 (14), 1876–1902.

Huang, J., Lin, Q., Zhang, X., He, X., Xing, X., Lian, W., Zuo, M., Zhang, Q., 2011b. Electrochemical immunosensor based on polyaniline/poly (acrylic acid) and Au-hybrid graphene nanocomposite for sensitivity-enhanced detection of salbutamol. Food Res. Int. 44, 92–97.

Hui, L., Piao, J.G., Auletta, J., Hu, K., Zhu, Y., Meyer, T., Liu, H., Yang, L., 2014. Availability of the basal planes of graphene oxide determines whether it is antibacterial. ACS Appl. Mater. Interfaces 6, 13183–13190.

Hummers, W.S., Offeman, R.E., 1985. Preparation of graphitic oxide. J. Am. Chem. Soc. 80 (6), 1339.
Hunyadi, S.E., Murphy, C.J., 2006. Bimetallic silver-gold nanowires: fabrication and use in surface-enhanced Raman scattering. J. Mater. Chem. 16, 3929–3935.
Ivandini, T.A., Rao, T.N., Fujishima, A., Einaga, Y., 2006. Electrochemical oxidation of oxalic acid at highly boron-doped diamond electrodes. Anal. Chem. 78, 3467–3471.
Jang, B.Z., Zhamu, A., 2008. Processing of nano graphene platelets (NGPs) and NGP nanocomposites: a review. J. Mater. Sci. 43 (15), 5092–5101.
Kalita, G., Sharma, S., Wakita, K., Umeno, M., Hayashi, Y., Tanemura, M., 2012. Synthesis of graphene by surface wave plasma chemical vapor deposition from camphor. Phys. Status Solidi A Appl. Mater. 209, 2510–2513.
Karahan, H.E., Wei, L., Goh, K., Wiraja, C., Liu, Z., Xu, C., Jiang, R., Wei, J., Chen, Y., 2015. Synergism of water shock and a biocompatible block copolymer potentiates the antibacterial activity of graphene oxide. Small 12 (7), 951–962.
Kemp, K.C., Seema, H., Saleh, M., Le, N.H., Mahesh, K., Chandra, V., Kim, K.S., 2013. Environmental applications using graphene composites: water remediation and gas adsorption. Nanoscale 5, 3149–3171.
Kim, K.S., Zhao, Y., Jang, H., Lee, S.Y., Kim, J.M., Kim, K.S., Ahn, J.-H., Kim, P., Choi, J.-Y., Hong, B.H., 2009. Large-scale pattern growth of graphene films for stretchable transparent electrodes. Nature 457, 706–710.
Koesukwiwat, U., Lehotay, S.J., Mastovska, K., Dorweiler, K.J., Leepipatpiboon, N., 2010. Extension of the QuEChERS method for pesticide residues in cereals to flaxseeds, peanuts, and doughs. J. Agric. Food. Chem. 58, 5950–5958.
Kuilla, T., Bhadra, S., Yao, D., Kim, N.H., Bose, S., Lee, J.H., 2010. Recent advances in graphene based polymer composites. Prog. Polym. Sci. 35 (11), 1350–1375.
Kwon, S.-Y., Ciobanu, C.V., Petrova, V., Shenoy, V.B., Bareno, J., Gambin, V., Petrov, I., Kodambaka, S., 2009. Growth of semiconducting graphene on palladium. Nano Lett. 9 (12), 3985–3990.
Lee, S.J., Morrill, A.R., Moskovits, M., 2006. Hot spots in silver nanowire bundles for surface-enhanced Raman spectroscopy. J. Am. Chem. Soc. 128, 2200–2201.
Lee, C., Wei, X., Kysar, J.W., Hone, J., 2008. Measurement of the elastic properties and intrinsic strength of monolayer graphene. Sci. 321, 385–388.
Lerf, A., Buchsteiner, A., Pieper, J., Schöttl, S., Dekany, I., Szabo, T., Boehm, H.P., 2006. Hydration behavior and dynamics of water molecules in graphite oxide. J. Phys. Chem. Solids 67, 1106–1110.
Li, X., Cai, W., An, J., Kim, S., Nah, J., Yang, D., Colombo, L., Ruoff, R.S., 2009. Large-area synthesis of high-quality and uniform graphene films on copper foils. Sci 324, 1312–1314.
Li, Y., Peng, H., Li, G., Chen, K., 2012. Synthesis and electrochemical performance of sandwich-like polyaniline/graphene composite nanosheets. Eur. Polymer J. 48 (8), 1406–1412.
Li, Z.J., Yang, B.C., Yun, G.Q., Zhang, S.R., Zhang, M., Zhao, M.X., 2013. Synthesis of Sn nanoparticle decorated graphene sheets for enhanced field emission properties. J. Alloys Comp. 550, 353–357.
Liao, K., Lin, Y., MacOsko, C.W., Haynes, C.L., 2011. Cytotoxicity of graphene oxide and graphene in human erythrocytes and skin fibroblasts. ACS Appl. Mater. Interface 3 (7), 2607–2615.
Lim, H.N., Nurzulaikha, R., Harrison, I., Lim, S.S., Tan, W.T., Yeo, M.C., Yarmo, M.A., Huang, N.M., 2012. Preparation and characterization of tin oxide, SnO_2 nanoparticles decorated graphene. Ceram. Int. 38, 4209–4216.

Liu, S., Zeng, T.H., Hofmann, M., Burcombe, E., Wei, J., Jiang, R., Kong, J., Chen, Y., 2011. Antibacterial activity of graphite, graphite oxide, graphene oxide, and reduced graphene oxide: membrane and oxidative stress. ACS Nano 5, 6971–6980.

Liu, S., Hu, M., Zeng, T.H., Wu, R., Jiang, R., Wei, J., Wang, L., Kong, J., Chen, Y., 2012. Lateral dimension-dependent antibacterial activity of graphene oxide sheets. Langmuir 28 (33), 12364–12372.

Liu, J.-J., Zhang, X.-L., Cong, Z.-X., Chen, Z.-T., Yang, H.-H., Chen, G.-N., 2013. Glutathione functionalized graphene quantum dots as selective fluorescent probes for phosphate-containing metabolites. Nanoscale 5, 1810–1815.

Luechinger, N.A., Booth, N., Heness, G., Bandyopadhyay, S., Grass, R.N., Stark, W.J., 2008. Surfactant-free, melt-processable metal-polymer hybrid materials: use of graphene as a dispersing agent. Adv. Mater. 20 (16), 3044–3049.

Luo, J., Jiang, S., Liu, X., 2013. Efficient one-pot synthesis of mussel-inspired molecularly imprinted polymer coated graphene for protein-specific recognition and fast separation. J. Phys. Chem. C 117, 18448–18456.

Mack, J.J., Viculis, L.M., Ali, A., Luoh, R., Yang, G., Hahn, H.T., Ko, F.K., Kaner, R.B., 2005. Graphite nanoplatelet reinforcement of electrospun polyacrylonitrile nanofibers. Adv. Mater. 17 (1), 77–80.

Mangadlao, J.D., Santos, C.M., Felipe, M.J., de Leon, A.C., Rodrigues, D.F., Advincula, R.C., 2015. On the antibacterial mechanism of graphene oxide (GO) Langmuir-Blodgett films. Chem. Commun. (Camb.) 51, 2886–2889.

Massey, L.K., Roman-Smith, H., Sutton, R.A.L., 1993. Effect of dietary oxalate and calcium on urinary oxalate and risk of formation of calcium oxalate kidney stones. J. Am. Diet. Assoc. 93, 901–906.

Muñoz, R., Gómez-Aleixandre, C., 2013. Review of CVD synthesis of graphene. Chem. Vap. Deposition 19, 297–322.

Murray, C.B., Kagan, C.R., Bawendi, M.G., 2000. Synthesis and characterization of monodisperse nanocrystals and close-packed nanocrystal assemblies. Ann. Rev. Mater. Sci. 30, 545–610.

Nanda, S.S., An, S.S., Yi, D.K., 2015. Oxidative stress and antibacterial properties of a graphene oxide-cystamine nanohybrid. Int. J Nanomed. 10, 549–556.

Neto, A.H.C., Peres, N.M.R., Novoselov, K.S., Geim, A.K., 2009. The electronic properties of grapheme. Rev. Mod. Phys. 81, 109–162.

Pei, S., Cheng, H.-M., 2012. The reduction of graphene oxide. Carbon 50, 3210–3228.

Peng, E., Choo, E.S., Chandrasekharan, P., Yang, C.T., Ding, J., Chuang, K.H., Xue, J.M., 2012. Synthesis of manganese ferrite/graphene oxide nanocomposites for biomedical applications. Small 8 (23), 3620–3630.

Ponnusamy, V.K., Jen, J.F., 2011. A novel graphene nanosheets coated stainless steel fiber for microwave assisted headspace solid phase microextraction of organochlorine pesticides in aqueous samples followed by gas chromatography with electron capture detection. J. Chromatogr. A 1218, 6861–6868.

Potts, J.R., Dreyer, D.R., Bielawski, C.W., Ruoff, R.S., 2011. Graphene-based polymer nanocomposites. Polymer 52, 5–25.

Premkumar, T., Geckeler, K.E., 2012. Graphene-DNA hybrid materials: assembly, applications, and prospects. Prog. Polym. Sci. 37 (4), 515–529.

Preston, T.J., Muller, W.J., Singh, G., 2001. Scavenging of extracellular H_2O_2 by catalase inhibits the proliferation of HER-2/Neu-transformed rat-1 fibroblasts through the induction of a stress response. J. Biol. Chem. 276, 9558–9564.

Qu, J., Luo, C., Zhang, Q., Cong, Q., Yuan, X., 2013. Easy synthesis of graphene sheets from alfalfa plants by treatment of nitric acid. Mater. Sci. Eng. B Adv. Funct. Solid State Mater. 178, 380–382.

Rafiee, M.A., Rafiee, J., Wang, Z., Song, H., Yu, Z., Koratkar, N., 2009. Enhanced mechanical properties of nanocomposites at low graphene content. ACS Nano 3, 3884–3890.

Rajkumar, M., Li, Y.-S., Chen, S.-M., 2013. Electrochemical detection of toxic ractopamine and salbutamol in pig meat and human urine samples by using poly taurine/zirconia nanoparticles modified electrodes. Colloids Surf. B Biointerfaces 110, 242–247.

Ramanathan, T., Abdala, A.A., Stankovich, S., Dikin, D.A., Herrera-Alonso, M., Piner, R.D., Adamson, D.H., Schniepp, H.C., Chen, X., Ruoff, R.S., Nguyen, S.T., Aksay, I.A., Prud'Homme, R.K., Brinson, L.C., 2008. Functionalized graphene sheets for polymer nanocomposites. Nat. Nanotechnol. 3 (6), 327–331.

Rowe, K.S., Rowe, K.J., 1994. Synthetic food colouring and behavior: a dose-response effect in a double-blind, placebo-controlled, repeated-measures study. J. Pediatr. 125, 691–698.

Ruan, G., Sun, Z., Peng, Z., Tour, J.M., 2011. Growth of graphene from food, insects, and waste. ACS Nano. 5, 7601–7607.

Sanchez, V.C., Jachak, A., Hurt, R.H., Kane, A.B., 2012. Biological interactions of graphene-family nanomaterials: an interdisciplinary review. Chem. Res. Toxicol. 25, 15–34.

Santos, C.M., Mangadlao, J., Ahmed, F., Leon, A., Advincula, R.C., Rodrigues, D.F., 2012. Graphene nanocomposite for biomedical applications: fabrication, antimicrobial and cytotoxic investigations. Nanotechnology 23 (39). ID 395101.

Schedel, F., Pries, R., Thode, B., Wollmann, B., Wulff, S., Jocham, D., Wollenberg, B., Kausch, I., 2011. mTOR inhibitors show promising in vitro activity in bladder cancer and head and neck squamous cell carcinoma. Oncol. Rep. 25, 763–768.

Schedin, F., Lidorikis, E., Lombardo, A., Kravets, V.G., Geim, A.K., Grigorenko, A.N., Novoselov, K.S., Ferrari, A.C., 2010. Surface-enhanced Raman spectroscopy of graphene. ACS Nano 4, 5617–5626.

Shahnawaz, K.M., Abdelhamid, H.N., Wu, H.F., 2015. Near infrared (NIR) laser mediated surface activation of graphene oxide nanoflakes for efficient antibacterial, antifungal and wound healing treatment. Colloids Surf. B Biointerfaces 127, 281–291.

Shen, H., Zhang, L., Liu, M., Zhang, Z., 2012. Biomedical applications of graphene. Theranostics 2 (3), 283–294.

Si, P., Dong, X.-C., Chen, P., Kim, D.-H., 2013. A hierarchically structured composite of Mn_3O_4/3D graphene foam for flexible nonenzymatic biosensors. J. Mater. Chem. B 1, 110–115.

Singh, Z., 2016. Applications and toxicity of graphene family nanomaterials and their composites. Nanotechnol. Sci. Appl. 9, 15–28.

Sreeprasad, T.S., Berry, V., 2013. How do the electrical properties of graphene change with its functionalization? Small 9, 341–350.

Stevenson, D.D., Simon, R.A., Lumry, W.R., Mathison, D.A., 1986. Adverse reactions to tartrazine. J. Allergy Clin. Immunol. 78, 182–191.

Sun, Y., Wu, Q., Shi, G., 2011. Graphene based new energy materials. Energy Environ. Sci. 4, 1113–1132.

Tankiewicz, M., Fenik, J., Biziuk, M., 2010. Determination of organophosphorus and organonitrogen pesticides in water samples. TrAC Trend Anal. Chem. 29, 1050–1063.

Tian, J.Y., Xu, J.Q., Zhu, F., Lu, T.B., Su, C.Y., Ouyang, G.F., 2013a. Application of nanomaterials in sample preparation. J. Chromatogr. A 1300, 2–16.

Tian, M.M., Feng, W., Ye, J.J., Jia, Q., 2013b. Preparation of Fe_3O_4@TiO_2/graphene oxide magnetic microspheres for microchip-based preconcentration of estrogens in milk and milk powder samples. Anal. Methods 5, 3984–3991.

Tonon, M.C., Lanfray, D., Castel, H., Vaudry, H., Morin, F., 2013. Hypothalamic glucosesensing: role of glia-to-neuron signaling. Horm. Metab. Res. 45, 955–959.

Tyler, C., Routledge, E., 1998. Oestrogenic effects in fish in English rivers with evidence of their causation. Pure Appl. Chem. 70, 1795–1804.

Wang, Z.L., 2000. Transmission electron microscopy of shape-controlled nanocrystals and their assemblies. J. Phys. Chem. B. 104, 1153–1175.

Wang, Z.L., Poncharal, P., de Heer, W.A., 2000. Proceedings of the carbon nanotubes by in situ TEM. Microscopy Microanalysis 6, 224–230.

Wang, Y.-W., Fu, Y.-Y., Wu, L.-J., Li, J., Yang, H.-H., Chen, G.-N., 2013. Targeted photothermal ablation of pathogenic bacterium, Staphylococcus aureus, with nanoscale reduced graphene oxide. J. Mater. Chem. B 1, 2496–2501.

Wang, C., Li, J., Tan, R., Wang, Q., Zhang, Z., 2019. Colorimetric method for glucose detection with enhanced signal intensity using $ZnFe_2O_4$ carbon nanotube-glucose oxidase composite material. Analyst 144 (5), 1831–1839.

Watcharotone, S., Dikin, D.A., Stankovich, S., Piner, R., Jung, I., Dommett, G.H., Evmenenko, G., Wu, S.E., Chen, S.F., Liu, C.P., Nguyen, S.T., Ruoff, R.S., 2007. Graphene-silica composite thin films as transparent conductors. Nano Lett. 7 (7), 1888–1892.

Wei, W., Zhang, D.M., Yin, L.H., Pu, Y.P., Liu, S.Q., 2013. Colorimetric detection of DNA damage by using hemin-graphene nanocomposites. Spectrochim. Acta A Mol. Biomol. Spectrosc. 106, 163–169.

Winey, K.I., Vaia, R.A., 2007. Polymer nanocomposites. MRS Bull. 32 (4), 314–319.

Xiang, Q., Yu, J., Jaroniec, M., 2012. Graphene-based semiconductor photocatalysts. Chem. Soc. Rev. 41, 782–796.

Xie, Y., Li, Y., Niu, L., Wang, H., Qian, H., Yao, W., 2012. A novel surface-enhanced Raman scattering sensor to detect prohibited colorants in food by graphene/silver nanocomposite. Talanta 100, 32–37.

Xu, W.-P., Zhang, L.-C., Li, J.-P., Lu, Y., Li, H.-H., Ma, Y.-N., Wang, W.-D., Yu, S.-H., 2011. Facile synthesis of silver@graphene oxide nanocomposites and their enhanced antibacterial properties. J. Mater. Chem. 21, 4593–4597.

Yadav, S.K., Jung, Y.C., Kim, J.H., Ko, Y.-I., Ryu, H.J., Yadav, M.K., Kim, Y.A., Cho, J.W., 2013. Mechanically robust, electrically conductive biocomposite films using antimicrobial chitosan-functionalized graphenes. Part. Part. Syst. Charact. 30, 721–727.

Yaffe, N.R., Blanch, E.W., 2008. Effects and anomalies that can occur in SERS spectra of biological molecules when using a wide range of aggregating agents for hydroxylamine-reduced and citrate-reduced silver colloids. Vib. Spectrosc. 48, 196–201.

Yang, W., Ratinac, K.R., Ringer, S.P., Thordarson, P., Gooding, J.J., Braet, F., 2010. Carbon nanomaterials in biosensors: should you use nanotubes or graphene? Angew. Chem. Int. Ed. Engl. 49, 2114–2138.

Yu, A., Ramesh, P., Itkis, M.E., Bekyarova, E., Haddon, R.C., 2007. Graphite nanoplatelet-epoxy composite thermal interface materials. J. Phys. Chem. C 111 (21), 7565–7569.

Zhang, H., Lee, H.K., 2012. Simultaneous determination of ultraviolet filters in aqueous samples by plunger-in-needle solid-phase microextraction with grapheme based sol-gel coating as sorbent coupled with gas chromatography-mass spectrometry. Anal. Chim. Acta 742, 67–73.

Zhang, B.-T., Zheng, X., Li, H.-F., Lin, J.-M., 2013. Application of carbon-based nanomaterials in sample preparation: a review. Anal. Chim. Acta 784, 1–17.

Zhao, G.Y., Song, S.J., Wang, C., Wu, Q.H., Wang, Z., 2011a. Solid-phase microextraction with a novel graphene-coated fiber coupled with high-performance liquid chromatography for the determination of some carbamates in water samples. Anal. Methods 3, 2929–2935.

Zhao, G.Y., Song, S.J., Wang, C., Wu, Q.H., Wang, Z., 2011b. Determination of triazine herbicides in environmental water samples by high-performance liquid chromatography using graphene-coated magnetic nanoparticles as adsorbent. Anal. Chim. Acta 708, 155–159.

Zhao, G., Wen, T., Chen, C., Wang, X., 2012. Synthesis of graphene-based nanomaterials and their application in energy-related and environmental-related areas. RSC Adv. 2, 9286–9303.

Zhao, J., Deng, B., Lv, M., Li, J., Zhang, Y., Jiang, H., Peng, C., Li, J., Shi, J., Huang, Q., Fan, C., 2013. Graphene oxide-based antibacterial cotton fabrics. Adv. Healthc. Mater. 2, 1259–1266.

Zhu, Y., Murali, S., Cai, W., Li, X., Suk, J.W., Potts, J.R., Ruoff, R.S., 2010. Graphene and graphene oxide: synthesis, properties, and applications. Adv. Mater. 22, 3906–3924.

Novel trends for synthesis of carbon nanostructures from agricultural wastes

4

Nady A. Fathy[a], *Altaf H. Basta*[b], *Vivian F. Lotfy*[b]
[a]Physical Chemistry Department, National Research Centre, Giza, Egypt, [b]Cellulose and Paper Department, National Research Centre, Giza, Egypt

1 Introduction

Carbon is a fascinating element of the periodic table because of its atomic structure ($1s^2$, $2s^2$, $2p^2$), which gives this element unique bonding possibilities both with other elements and with itself, thus resulting in a wide spectrum of compounds and allotropic forms (Delhaes, 1998; Marsh, 1997). For a long time, activated carbons (ACs) have been extensively used for water treatment and gas adsorption due to their well-developed porous structure with a large internal surface area and pore volume (Marsh, 1997). The new generation of carbon nanostructures such as nanotubes, fullerenes, nanocones, graphene, graphene nanoribbons, nanodiamonds, nanofibers, carbon gels, etc., with relatively large surface areas and novel electronic and chemical properties provides new horizons for achieving superior performance in environmental and industrial applications such as separation, purification, recovery, catalysis, and energy storage (Menendez-Diaz and Martin-Gullon, 2006).

Particularly, carbon nanotubes (CNTs) are a cylindrical carbon nanostructure discovered by Iijima (1991). Since then, CNTs have attracted remarkable attention in many applications such as electrodes in supercapacitors for energy storage, polymer reinforcements, gas storage materials, sensors, electronics, catalysis, and separation (De Volder et al., 2013; Yan et al., 2015). CNTs consist of one-dimensional (1D) graphene sheets rolled into a tubular form with a radius as small as a few nanometers and lengths up to several micrometers. The walls of these tubes are constructed of a hexagonal lattice of carbon atoms and capped by fullerene-like structures. This unique structure of CNTs provides them with superior electronic, optoelectronic, mechanical, and chemical properties, so they can be recognized as semiconducting or metallic materials. CNTs have some extraordinary properties such as being stronger than steel and harder than diamond with electrical conductivity higher than copper and thermal conductivity higher than diamond (Kumar and Ando, 2010).

There are two main types of nanotubes: the single-walled carbon nanotubes (SWCNTs) and the multiwalled carbon nanotubes (MWCNTs). There is also another structure called the double-walled carbon nanotube (DWCNT). MWCNTs consist of a single sheet of graphite rolled around itself (like a rolled-up newspaper) or multiple layers of graphite arranged in concentric cylinders like a Russian doll (De Volder et al., 2013).

Carbon Nanomaterials for Agri-food and Environmental Applications. https://doi.org/10.1016/B978-0-12-819786-8.00004-9

Currently, three major approaches are used in CNT synthesis: arc discharge (Iijima, 1991), laser ablation (Thess et al., 1996), and chemical vapor deposition (Ren et al., 1998; Yacamàn et al., 1993). Other technologies have been employed to prepare CNTs, including electron or ion beam irradiation (Chernozatonskii et al., 1995), flame synthesis (Richter et al., 1996), pyrolysis (Cho et al., 1996), electrolysis (Hsu et al., 1996), template methods (Kyotani et al., 1996), and solar approaches (Laplaze et al., 1998). However, although these methods can yield high-quality CNTs, each method requires high energy and expensive apparatuses. Specifically, chemical vapor deposition (CVD) is known as the most widely used route to manufacture CNTs on an industrial scale. The CVD process offers the benefit of significantly lower synthesis temperatures than other techniques to produce CNTs at 600–900°C and ambient pressure (Eatemadi et al., 2014; Yan et al., 2015). In general, the CVD technique has several advantages compared to the arc discharge and laser ablation processes, such as (i) it is a simple and economic technique for synthesizing CNTs at low temperature and ambient pressure, (ii) the crystallinity of SWCNTs can be grown by CVD at temperature >900°C, (iii) it is easy to control the morphology of CNTs, (iv) plenty of hydrocarbons in any state (solid, liquid, or gas) can be decomposed over various catalysts/substrates, and (v) it allows CNT growth in a variety of forms such as powder, thin or thick films, aligned or entangled, straight or coiled nanotubes, and bamboo nanotubes (Kumar and Ando, 2010). It also allows the manufacturing of CNT hybrids with another carbonaceous material such as carbon gels (Fathy et al., 2017a). Hence, this method continues to be the best process for CNT production.

2 History of carbon nanotubes

The history of CNT preparation did not just start in 1991 with Ijima, but actually began much earlier by other scientists. In 1952, Radushlevich and Lukyanovich noticed the formation of carbon atoms as filaments exhibiting a continuous inner cavity and forming tubes of 50 nm diameter during the thermal decomposition of carbon monoxide on an iron catalyst (Fe). Hillert and Lange (1958) reported that carbon nanotubes can take the morphologies of bamboo texture, branched nanotubes, and coiled nanotubes. After that, Oberlin et al. (1976) showed that hollow carbon tubes with nanometer-scale diameters could be produced using a vapor-growth technique. The formation of tubular carbon atoms as filaments via the catalytic decomposition of acetylene gas over iron metal catalysts supported on a porous substrate at 700°C was first reported by Yacamàn et al. (1993). Nearly 40 years later came Iijima (1991), who gets the credit for preparing multiwalled carbon nanotubes (MWCNTs) through an arc discharge of C_{60} carbon molecule graphite rods.

3 Chemical vapor deposition

CVD is carried out in a flow furnace at atmospheric pressure. The catalyst is placed in a quartz boat that is put into a quartz tube. Then, the hydrocarbon gas is pushed by

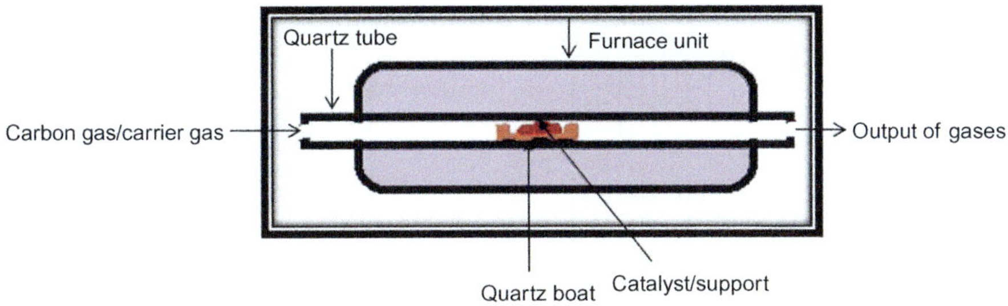

Fig. 1 Schematic representation of the experimental setup used for the catalytic decomposition of hydrocarbons.

a carrier gas to deposit on the heated surface of the catalyst (Fig. 1). In CVD, CNT synthesis mainly requires carbonaceous feedstocks, catalysts, substrates, and the appropriate temperature. Using this method, four structural forms of carbon can be obtained: amorphous carbon layers on the surface of the catalyst nanoparticles, filaments of amorphous carbon, graphite layers covering metal particles, and MWCNTs made from well-crystallized graphite layers usually covered with amorphous carbon on their outer layer (Journet and Bernier, 1998).

The growth mechanism of CNTs through CVD generally involves the dissociation of carbon precursors into carbon atoms, the dissolution and saturation of these atoms in the catalyst metal particles, and further precipitation into tubular carbon deposits. Accordingly, the type of carbon precursor, the furnace temperature, the type and particle size of the catalyst, and the type of support greatly affect the morphology and final properties of carbon nanotubes. For example, the formation of SWCNTs or MWCNTs is governed by the size of the catalyst particle. Finally, other parameters include the type of carbon precursor, the catalyst, the atmosphere in the reactor, and the temperature that affect the growth, yield, and diameter of carbon nanotubes in the CVD reactor. In addition, various CVD systems such as catalytic CVD-either by thermal- or plasma-enhanced (PE), oxygen-assisted CVD, water-assisted CVD, microwave plasma (MPECVD), radiofrequency CVD (RF-CVD), or hot filament (HFCVD)–have been established for CNT production (Eatemadi et al., 2014).

Among CVD processes, metallic catalysts such as Fe, Co, Ni, and their alloys have been extensively used as catalysts for CNT synthesis via the high-temperature catalytic decomposition of hydrocarbons under conventional heating. Normally, substrates such as Si, silica, graphite, activated carbon, or Al_2O_3 are necessary for the conventional preparation of CNTs using carbon source gases, such as CH_4, C_2H_4, and C_2H_2, to effectively form a growing CNT deposit on the substrates (Kumar and Ando, 2010; Yao et al., 2010).

It has been reported that the growth mechanism of CNTs that was produced via CVD of such hydrocarbons may be governed by the metal catalyst-substrate interaction (Kumar and Ando, 2010). When the interaction between both is strong, hydrocarbons decompose into carbon atoms that diffuse upward on the surface of the metal. Therefore, CNTs grow up with the catalyst particle rooted on its base. This mechanism is known as "base growth" of CNTs. If there is a weak interaction between catalyst

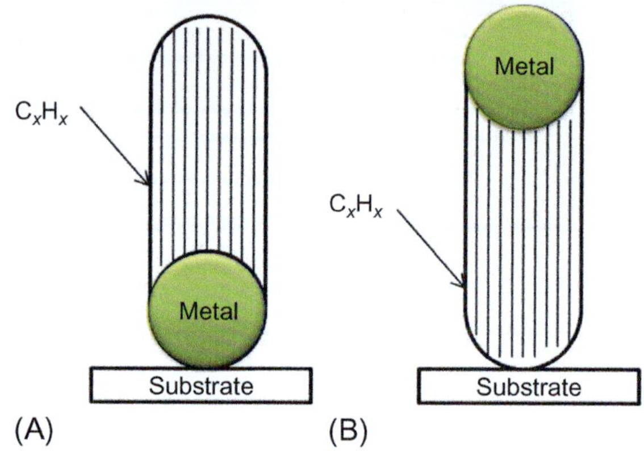

Fig. 2 Growth mechanisms for CNTs via decomposition of hydrocarbons (C_xH_x) by base-growth (A) and tip-growth models (B).

and substrate, the growth of CNTs follows a "tip-growth" mechanism, as shown in Fig. 2.

Therefore, it is of great significance to look for the low-cost carbon precursors and technologies for the synthesis of CNT materials. This chapter focuses on the synthesized CNTs from natural and agricultural feedstocks, where the type of carbon precursor considerably affects the structure properties of carbon deposits.

3.1 Synthesis of CNTs from natural carbon precursors

Based on growing environmental concern and increasing CNT demand, many researchers have attempted to develop green technology and increase the mass production of CNTs. For this purpose, ecofriendly botanical materials such as *camphor* (Andrews et al., 2006; Fathy et al., 2017a; Kumar and Ando, 2003, 2008), *oiled-petroleum* wastes such as turpentine oil (Chatterjee et al., 2003), *deoiled asphalt* (Liu et al., 2007), and *Eucalyptus oil* (Ghosh et al., 2007) have been tried. Moreover, natural oils such as coconut (Paul and Samdarshi, 2011), *neem* (Kumar et al., 2011), *palm oil* (Suriani et al., 2009), and the waste cooking of palm oil (Suriani et al., 2011) were studied as starting materials to prepare CNTs through the CVD process.

3.2 Synthesis of CNTs from agro-residues precursors

Agricultural wastes consisting of cellulose, hemicellulose, and lignin with high carbon content are used widely as carbon precursors. For a long time, several agricultural byproducts as activated carbon precursors have been investigated and continue to receive renewed attention (Alslaibi et al., 2013). As well, rice byproducts, sugarcane bagasse, cotton stalks, etc., are regarded as available and inexpensive alternative sources of natural wood in the manufacturing of engineered wood, filler for rubber composites, and paper sheet products (Basta et al., 2014a,b, 2016; El Nashar et al., 2004). However, developing sustainable technologies for the production of carbon nanomaterials (CNMs) through utilization of agroresidues has attracted considerable interest in recent years so as to optimize their cost production. CNTs, graphene, and carbon nanofibers are classified as CNMs. Practically, CNTs have gained more attention recently due to their extraordinary mechanical, thermal, and electrical properties, thus they have huge potential in environment and energy applications.

Goodell et al. (2008) reported that CNTs with outer diameters from 10 to 20 nm can be synthesized from *wood fiber* via a continuous oxidation process at 240°C followed by several cyclic oxidations at 400°C using a controlled amount of oxygen gas, whereas no CNTs were produced from pure cellulose and lignin. They proposed that the obtained morphology of tubular carbon atoms is closely related to the hexagonal arrangement of carbon atoms in the graphite sheets as a result of the spatial and molecular arrangement of the cell wall.

3.3 CVD-assisted thermal treatment of agroresidues as substrates

Sevilla and Fuertes (2010) synthesized graphitic carbon nanostructures with coil-like shapes via hydrothermal treatment of *cellulose* at 250°C followed by thermal treatment of impregnated hydrothermal char with a nickel salt at 900°C.

Alvesa et al. (2011) reported the synthesis of carbon nanomaterials from pyrolysis of both *sugarcane bagasse* and corn grains via catalytic conversion of bioethanol byproducts in horizontal stainless steel at temperatures ranging from 600 to 1000°C. They observed that pyrolysis of sugarcane bagasse produced MWCNTs with straight and longer tubes with lengths of 50 μm and diameters that varied from 20 to 50 nm. Pyrolysis of corn grains at 1000°C gave different forms of carbon nanomaterials that have long, entangled, rope-like structures with rugged walls and diameters in the range of 100–300 nm.

Zhu et al. (2012) used bamboo charcoal as a catalyst and ethanol vapor as a carbon source to prepare CNTs using CVD at temperatures between 1200°C and 1400°C. They observed that calcium silicate in the bamboo charcoal acted as a catalyst to accelerate the nucleation of MWCNTs through the tip mechanism.

Recently, Fathy (2017) synthesized CNT bundles through two subsequent routes: (i) hydrothermal treatment of pretreated rice straw impregnated with iron and/or nickel metals at 250°C for 2 h to produce hydrochar used as a substrate, and (ii) the CVD process of camphor as a carbonaceous gas. CNT growth with outer diameters of 22–66 nm was achieved when the hydrochar obtained from the rice straw was carbonized at 800°C; meanwhile, the carbon gas was evaporated from the cracked camphor at 250°C for 2 h. The results showed that the combination of Fe and Ni oxides as catalysts led to an enhancement in the graphitization process of CNTs onto the carbonized hydrochar, which was hydrothermally obtained from pretreated rice straw. The obtained yield of CNTs reached 44% with a specific surface area of ~30 m^2/g.

The role of natural and synthetic polymers on the performance of CNT growth was recently investigated (Fathy et al., 2017b). In this respect, rice straw (RS) and organic xerogels (OXG) were used as carbon supports for loading tri-metal oxides (Fe-Ni-Cu) and pyrolyzed at 750°C for 60 min; meanwhile, camphor as a carbon gas was cracked at 250°C. Two chain-length aldehydes (formaldehyde and butyraldehyde) were used in preparing OXG via a sol-gel reaction with resorcinol. The results revealed that xerogels as carbon supporting materials, prepared from formaldehyde and butyraldehyde, produced the higher yield of CNTs than that obtained over RS, as shown in Fig. 3. The yields of CNTs over carbonized xerogels varied from 85% to 320% while the yield over RS was 20%. Moreover, the produced CNTs from xerogels have a relatively higher specific surface area (>100 m^2/g) with irregular diameter ranging from 20 to 40 nm over that obtained from carbonized rice straw as a support (~8 m^2/g). The evaluation of the adsorption properties for the produced CNTs was carried out using iodine and methylene blue adsorbates. The synthesized formaldehyde-resorcinol xerogel loaded with tri-metallic catalysts (Fe-Ni-Cu) was found to be an efficient support because it provided CNTs with the best features, such as high yield of CNTs (320%), large surface area (117 m^2/g), reasonable morphology, and high adsorption capacity toward both iodine (117.5 mg/g) and methylene blue (88 mg/g) adsorbates (Fathy et al., 2017b).

3.4 Microwave plasma irradiation technology

Recently, microwave irradiation has been employed to prepare carbon nanomaterials (CNMs) such as carbon nanofibers (Menéndez et al., 2010), carbon nanowalls (Mori et al., 2011), and CNTs (Shi et al., 2014; Zeng et al., 2010) based on its ability to create a high-energy-density plasma.

Of late, Wang et al. (2015) reported synthesized graphenated-CNT (g-CNTs) materials from rice husk waste (RH) by microwave plasma irradiation (MPI). The obtained nanocarbons were found to be graphenated-carbon nanotubes (g-CNTs) composed of graphene standing on the sidewalls of CNTs, in which the graphene sheets mainly consisted of 2–6 layers and the CNTs had several tens of micrometers in length and 50–200 nm of diameter.

3.5 Direct pyrolysis in a column reactor

From an economic and environmental point of view, the pyrolysis of agricultural wastes in a tubular reactor is a viable process to prepare CNMs if compared to the CVD method in order to optimize sustainable development. Alves et al. (2012) developed a novel pyrolysis reactor for converting waste sugarcane bagasse into MWCNTs with lengths of 50 μm and diameters in the range of 20–50 nm. In this study, the carbon content of gases generated from the pyrolysis of *sugarcane bagasse* was catalytically converted into CNTs on the stainless steel meshes AISI 304. It has been found that the chemical components in the stainless steel mesh were Cr (18%–20%), Ni (8%–10.5%), Mn (≤2%), Si (≤1%), N (≤0.1%), C (≤0.08%), P (≤0.05%), S (≤0.03%), and Fe (balance). Likewise, Bernd et al. (2017) synthesized different carbon nanostructures of CNTs and carbon nanofibers (CNFs) during pyrolysis of *wood sawdust*. This process included a tubular reactor containing a reducing agent (commercial zinc), calcite (bed material), and a catalyst (ferrocene or Fe/Mo/MgO) arranged in the column reactor, and then heated until 750°C for 3 h without blowing any gases such as nitrogen, air, or external carbon gas. The obtained CNT and CNF samples were grown with different sizes and shapes.

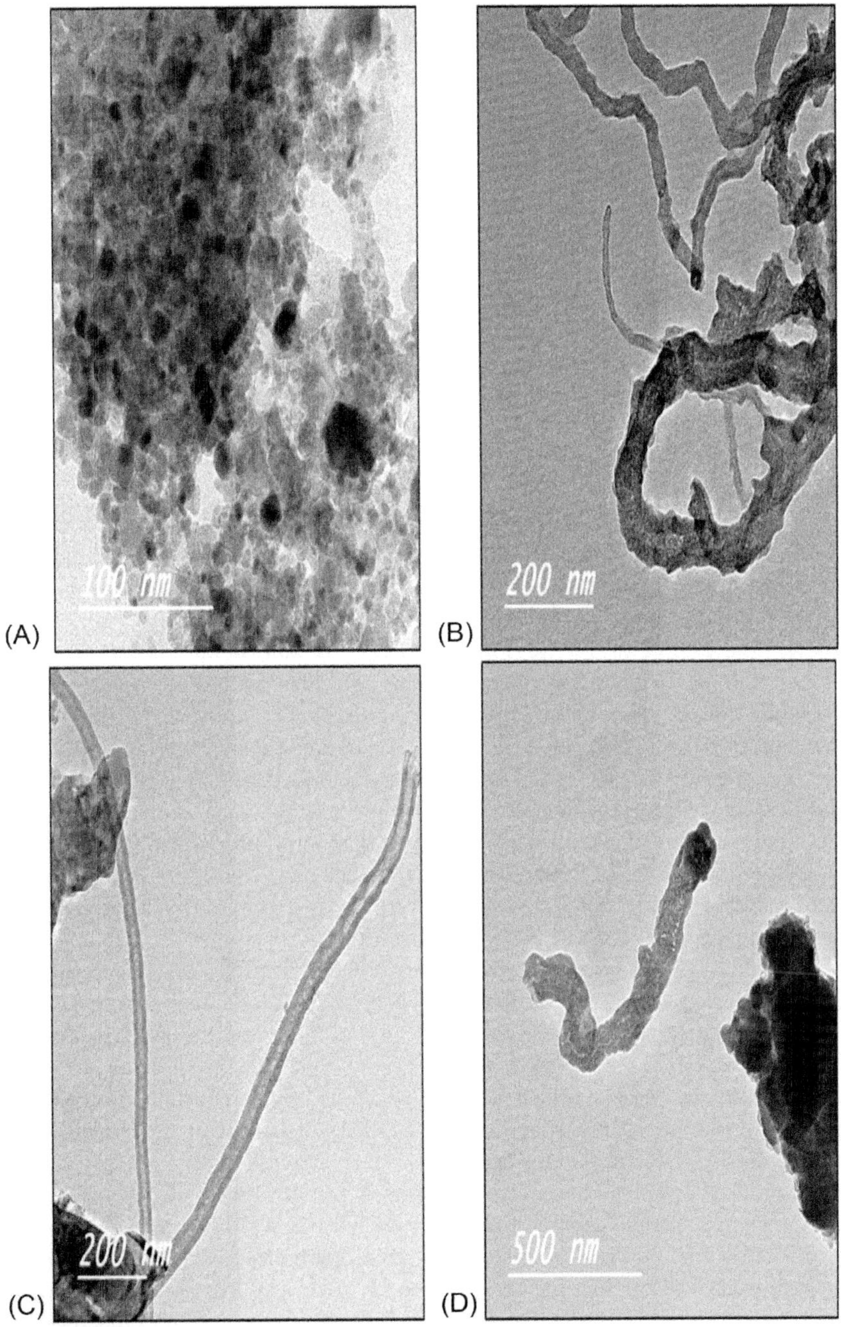

Fig. 3 TEM photographs of the as-synthesized materials (A) carbon xerogels-BX (B) XF-CNTs, (C) BX-CNTs, and (D) RS-CNTs (Fathy et al., 2017b).

Lotfy et al. (2018) studied a novel approach for synthesizing different shapes of CNTs from rice straw residues during catalytic hydrothermal of RS precursors followed by pyrolysis in a vertical bed reactor at 830 °C for 60 min without passing any external gases (e.g., N_2, H_2, or Ar). In this work, three types of RS pulps-alkaline pulp (RS-AP), sulfite pulp (RS-SP), and neutral pulp (RS-NP)-were synthesized and evaluated as carbon sources of CNTs in comparison with raw RS. Transmission electron microscopy (TEM) revealed that the RS precursor type had a significant effect on the shape and size of the produced CNTs. MWCNT bundles were produced from pyrolysis of hydrochars of RS, RS-SP, and RS-NP with diameters ranging from 15 to 40 nm, from 4 to 8 nm, and from 14.6 to 47.9 nm, respectively, as shown in Fig. 4. The RS-AP hydrochar precursor provided almost SWCNTs with diameters that varied from 3.8 to 4.5 nm. The obtained hydrochar from unpulped rice straw could produce CNTs with relatively higher surface areas (188 m^2/g) than others. On the other hand, CNT samples produced from the pyrolysis of RS-SP hydrochar provided the greatest methylene blue dye adsorption capacity (283.3 mg/g), which was related to its highest mesoporosity percent.

Additional TEM images on the synthesized CNTs (RS-NP-CNTs) from pyrolysis of RS-NP hydrochar are shown in Fig. 5. It has been found that different CNT shapes (bamboo and straight shapes) were grown over an Fe-Ni/Al_2O_3 catalyst during pyrolysis of RS-NP hydrochar.

The pulping of rice straw (RS) using base, acid, and neutral agents led to RS precursors with different chemical constituents. These constituents could be affected on decomposing the variety of hydrocarbons (e.g., CO, CO_2, CH_4, etc.) through its pyrolysis in the presence of an Fe-Ni/Al_2O_3 catalyst (Alvesa et al., 2011; Lotfy et al. 2018). Such hydrocarbons give rise to growing different CNT shapes via a dissolution-precipitation mechanism in which amorphous carbon is dissolved in the catalyst nanoparticles and then precipitated as graphitic sheets or as CNT bundles. However, the unpulped rice straw exhibited the best structure of CNTs and the highest yield (~80%) as compared to the other rice straw pulped by alkaline, sulfite, and neutral, respectively. The CNT yields from the pyrolysis of RS-AP, RS-SP, and RS-NP was 50%, 40%, and 65%, respectively. These results confirm the feasibility of agrohydrochar obtained from unpulped rice straw to prepare well-structured CNTs with the highest specific surface area (188 m^2/g) and the greatest methylene blue dye adsorption capacity (283.3 mg/g).

Table 1 summarizes the synthesized CNTs from natural and agroresidues with their corresponding diameters and specific surface areas. It has been concluded that the technology used and the preparation parameters (e.g., type of carbon precursor, type of catalyst, porous support, temperature, and time of growth) used in CNT synthesis has significantly impacted the quality and yield of CNTs as well as the morphology and porosity characteristics. Finally, it is feasible to prepare effective CNTs derived from agricultural wastes through some recent methods such as CVD-assisted pyrolysis of biomass waste, microwave-induced plasma, and pyrolysis in a column reactor.

A recent investigation on the synthesis of graphene oxide from sugarcane bagasse has been reported by Somanathan et al. (2015). The authors carried out a new method of graphene oxide (GO) synthesis via single-step reforming of sugarcane bagasse

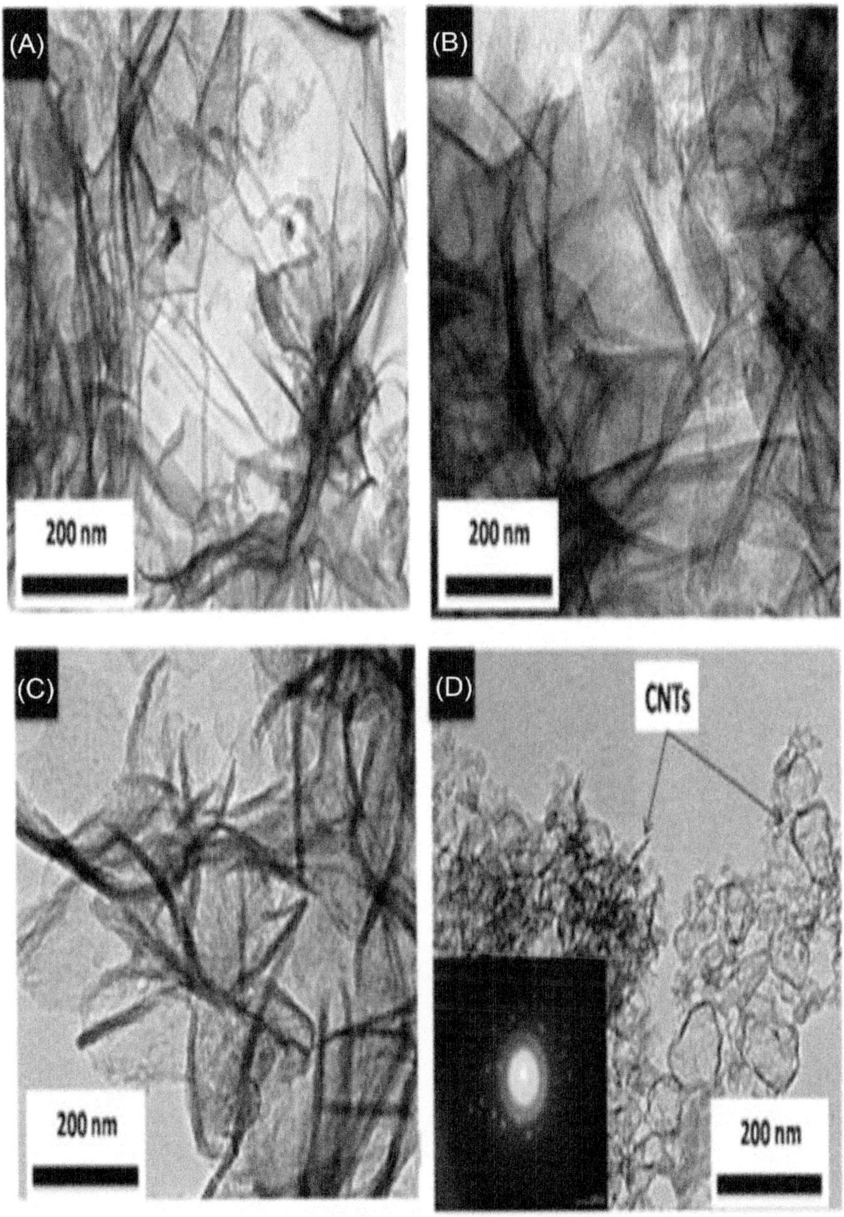

Fig. 4 TEM photographs of the prepared CNT bundles from pyrolysis of different hydrochars over Fe-Ni/Al_2O_3 in a column reactor: RS (A), RS-AP (B), RS-SP (C), and RS-NP (D) with insert image of selected area electron diffraction pattern (Lotfy et al., 2018).

Fig. 5 TEM images of the prepared CNT bundles from pyrolysis of RS-NP hydrochar over Fe-Ni/Al_2O_3 in a column reactor (Lotfy et al., 2018).

Table 1 Synthesized CNTs from various carbonaceous precursors by using different methods.

Type of CNTs	Method of preparation	Carbon source	Catalyst/temperature/growth time	Diameters (nm)	Specific surface area (m^2/g)	Ref.
MWCNTs	CVD	Camphor	Fe-Ni/rice straw hydrochar as substrate/800°C/120 min	22–66	30	Fathy (2017)
MWCNTs	CVD	Camphor	Fe-Ni-Cu/rice straw or organic xerogel as substrate/750°C/60 min	20–40	8.37–117	Fathy et al. (2017b)
g-CNTs	Microwave plasma irradiation (MPI)	Rice husk	Plasma irradiation at pressure of 300 mTorr, power of 900 W(2.45 GHz)/10–38 min	50–200	–[a]	Wang et al. (2015)
MWCNTs	Microwave-induced pyrolysis	Gumwood	SiC/300 W/500°C/30 min	50	–	Shi et al. (2014)
CNTs/CNFs	Pyrolysis in tubular reactor	Sawdust	Reducing agent (commercial zinc), calcite (bed material), and the catalyst (ferrocene or Fe/Mo/MgO) arranged in the column reactor, and then heated until 750°C for 3 h	–	–	Bernd et al. (2017)
Mixture of SWNTs and MWNTs	Catalytic hydrothermal-pyrolysis	Rice straw	Fe-Ni/Al_2O_3, 830°C/60 min	3.8–47	10.1–188	Lotfy et al. (2018)
MWCNTs (bamboo-like)	Catalytic hydrothermal-pyrolysis	Cotton stalks	Fe-Ni-Cu/Al_2O_3, 830°C/60 min	50–100	110	Azzam et al. (2019)

[a] (–) is not detected in the published paper.

agricultural waste by oxidation with ferrocene under a muffled atmosphere at 300°C for 10 min. Sun et al. (2013) synthesized porous graphene nanosheets with large surface areas via a simultaneous activation-graphitization route of a coconut shell as a kind of biomass waste with a graphitic catalyst precursor ($FeCl_3$) and an activating agent ($ZnCl_2$) at 900°C for 1 h. Therefore, agrowastes are not only valuable for sustainable energy resources such as bio fuels, but they also have the potential for use as carbon sources for the fabrication of valuable CNMs (Deng et al., 2016).

4 Agricultural wastes for synthesis of CNMs: Benefits and limitations

Recycling low-value agricultural wastes into high-value products is markedly feasible, which has inspired researchers in the last 20 years. Although a wide range of synthesis technologies has been developed to produce CNMs, green synthesis methods with either energy-saving reactions or low-value waste materials are earning more interest. The disposal of waste materials, ranging from field and household wastes to industrial residues, has detrimental effects on the environment and human health. Such wastes can be exploited as utility resources to prepare CNMs. Among these, agricultural wastes highly accumulate in huge amounts worldwide every year.

In 2011, the global production of CNMs was around 3500 tons, and that is projected to expand at a compound annual growth rate of 30.6%. Expanding markets for applications of CNMs such as nanotubes, graphene, and nanofibers would place ever-increasing demands on lowering their production costs (Alves et al., 2012; Zhuo et al., 2012).

A variety of synthesis methods have been used to produce CNMs (Deng et al., 2016; Kerdnawee et al., 2017; Ngu et al., 2016; Onishchenko et al., 2013; Sangeetha et al., 2017; Wang et al., 2018). Although CNMs with high quality can be produced through those conventional methods mentioned above, one crucial challenge is to achieve environmentally friendly production in a large scale with low cost. CNM manufacture starting from waste is likely to simultaneously reduce the amount of solid waste and lower the fabrication cost. More importantly, using agricultural waste as CNM feedstock is an encouraging step for recycling technology. However, the negative effects associated with CNMs cannot be overlooked due to their nm-range dimensions. The residues left during CNM synthesis can create environmental problems.

5 Conclusion and future perspectives

Carbon nanomaterials signify one of the most diverse classes of porous materials with outstanding properties, which makes them suitable for applications in widespread areas of research. There is an incisive need to look for new carbon nanostructures via sustainable and low-cost technologies to realize their role in energy and environmental applications. Most of the traditional synthesis strategies include energy-intensive circumstances and hence a high cost of production. In this regard, the green synthesis of carbon nanomaterials via cost-effective techniques has great importance.

Natural wastes with high carbon content, especially agricultural residues, are theorized as pronounced precursors for making carbon nanomaterials (CNMs). Carbon nanotubes (CNTs) as CNMs, principally, have attracted priority research in nanotechnology. CNTs are light, robust, and highly conductive, both electrically and thermally, while still being chemically stable. These properties may convert them into an important material for the future. The trouble is that the nanotube manufacturing process is not as sustainable and cost effective. This chapter highlights the previous and recent approaches for producing advanced carbon nanomaterials such as CNTs and graphene products. The agro-based CNMs, in addition to their unique morphological and structural properties, are also regarded as cost-effective and environmentally acceptable products.

In particular, CNTs have promising future applications based on their surface modification to act as precursors for further functionalization with various biomolecules, which improves biocompatibility and initiates CNT implementation in the field of nanomedicine and targeted drug delivery. Moreover, the high specific strength and excellent thermal and electrical properties of CNTs promote their reinforced metal matrix to be widely used in structural and functional applications such as automobiles, aerospace, sports, micro/electromechanical systems (MEMS), sensors, batteries, and energy storage. But unfortunately, these applications need further study to improve the dispersion of well-aligned and intact CNTs inside the matrix without chemical treatment.

References

Alslaibi, T.M., Abustan, I., Ahmad, M.A., Abu Foulc, A., 2013. A review: production of activated carbon from agricultural byproducts via conventional and microwave heating. J. Chem. Technol. Biotechnol. 88, 1183–1190.

Alves, J.O., Tenório, J.A.S., Zhuo, C., Levendis, Y.A., 2012. Characterization of nanomaterials produced from sugarcane bagasse. J. Mater. Res. Technol. 1, 31–34.

Alvesa, J.O., Zhuo, C., Levendis, Y.A., Tenório, J.A.S., 2011. Catalytic conversion of wastes from the bioethanol production into carbon nanomaterials. Appl. Catal. B Environ. 106, 433–444.

Andrews, R.J., Smith, C.F., Alexander, A.J., 2006. Mechanism of carbon nanotubes growth from camphor and camphor analogs by chemical vapor deposition. Carbon 44, 341–347.

Azzam, E.M.S., Fathy, N.A., El-Khouly, S.M., Sami, R.M., 2019. Enhancement the photocatalytic degradation of methylene blue dye using fabricated CNTs/TiO_2/AgNPs/Surfactant nanocomposites. J. Water Process Eng. 28, 311–321.

Basta, A.H., El-Saied, H., Lotfy, V.F., 2014a. Performance assessment of deashed and dewaxed rice straw on improving the quality of RS-based composites. RSC Adv. 4, 21794–21801.

Basta, A.H., El-Saied, H., Lotfy, V.F., 2014b. Performance of rice straw-based composites using environmentally friendly polyalcoholic polymers-based adhesive system. Pigment Resin Technol. 42, 24–33.

Basta, A.H., El-Saied, H., Baraka, A.M., Lotfy, V.F., 2016. Comparative evaluation of xerogel-based activated carbons synthesized from aliphatic aldehydes of different chain lengths. Soft Mater. 14, 297–308.

Bernd, M.G.S., Braganca, S.R., Heck, N., da Silva Filho, L.C.P., 2017. Synthesis of carbon nanostructures by the pyrolysis of wood sawdust in a tubular reactor. J. Mater. Res. Technol. 6, 171–177.

Chatterjee, A.K., Sharon, M., Banerjee, R., Neumann-Spallart, M., 2003. CVD synthesis of carbon nanotubes using a finely dispersed cobalt catalyst and their use in double layer electrochemical capacitors. Electrochim. Acta 48, 3439–3446.

Chernozatonskii, L.A., Kosakovskaja, Z.J., Fedorov, E.A., Panov, V.I., 1995. New carbon tubelite-ordered film structure of multi-layer nanotubes. Phys. Lett. A 197, 40–46.

Cho, W.S., Hamada, E., Kondo, Y., Takayanagi, K., 1996. Synthesis of carbon nanotubes from bulk polymer. Appl. Phys. Lett. 69, 278–279.

Delhaes, P., 1998. In: Rand, B., Appleyard, S.P., Yardim, M.F. (Eds.), Design and control of structure of advanced carbon materials for enhanced performance. NATO Science Series, Series E: Applied Sciences, vol. 347. Kluwer Academic Publishers, Netherlands, pp. 3–27.

Deng, J., You, Y., Sahajwalla, V., Joshi, R.K., 2016. Transforming waste into carbon-based nanomaterials. Carbon 96, 105–115.

De Volder, M.F.L., Tawfick, S.H., Baughman, R.H., Hart, A.J., 2013. Carbon nanotubes: present and future commercial applications. Science 339, 535–539.

Eatemadi, A., Daraee, H., Karimkhanloo, H., Kouhi, M., Zarghami, N., Akbarzadeh, A., Abasi, M., Hanifehpour, Y., Joo, S.W., 2014. Carbon nanotubes: properties, synthesis, purification, and medical applications. Nanoscale Res. Lett. 9, 1–13.

El Nashar, D.E., Abd El-Messieh, S.L., Basta, A.H., 2004. Newsprint paper waste as a fiber reinforcement in rubber composites. J. Appl. Polym. Sci. 9, 3410–3420.

Fathy, N.A., 2017. Carbon nanotubes synthesis using carbonization of pretreated rice straw through chemical vapor deposition of camphor. RSC Adv. 7, 28535–28541.

Fathy, N.A., Annamalai, K.P., Tao, Y., 2017a. Effects of phosphoric acid activation on the nanopore structures of carbon xerogel/carbon nanotubes hybrids and their capacitance storage. Adsorption 23, 355–360.

Fathy, N.A., Lotfy, V.F., Basta, A.H., 2017b. Comparative study on the performance of carbon nanotubes prepared from agro- and xerogels as carbon supports. J. Anal. Appl. Pyrol. 128, 114–120.

Ghosh, P., Afre, R.A., Soga, T., Jimbo, T., 2007. A simple method of producing single-walled carbon nanotubes from a natural precursor: Eucalyptus oil. Mater. Lett. 61, 3768–3770.

Goodell, B., Xie, X., Qian, Y., Daniel, G., Peterson, M., Jellison, J., 2008. Carbon nanotubes produced from natural cellulosic materials. J. Nanosci. Nanotechnol. 8, 2472–2474.

Hillert, M., Lange, N., 1958. The structure of graphite filaments. Z. Kristallogr. 111, 24–34.

Hsu, W.K., Terrones, M., Hare, J.P., Terrones, H., Kroto, H.W., Walton, D.R.M., 1996. Electrolytic formation of carbon nanotubes. Chem. Phys. Lett. 262, 161–166.

Iijima, S., 1991. Helical microtubules of graphitic carbon. Nature 354, 56–58.

Journet, C., Bernier, P., 1998. Production of carbon nanotubes. Appl. Phys. A Mater. Sci. Process. 67, 1–9.

Kerdnawee, K., Termvidchakorn, C., Yaisanga, P., Pakchamsai, J., Chookiat, C., Eiad-Ua, A., Wongwiriyapan, W., Chaiwat, W., Ratchahat, S., Faungnawakij, K., Suttiponparnit, K., 2017. Present advancement in production of carbon nanotubes and their derivatives from industrial waste with promising applications. KONA Powder Particle J. 34, 24–43.

Kumar, M., Ando, Y., 2003. Camphor-a botanical precursor producing garden of carbon nanotubes. Diam. Relat. Mater. 12, 998–1002.

Kumar, M., Ando, Y., 2008. Gigas growth of carbon nanotubes. Defence Sci. J. 58, 496–503.
Kumar, M., Ando, Y., 2010. Chemical vapor deposition of carbon nanotubes: A review on growth mechanism and mass production. J. Nanosci. Nanotechnol. 10, 3739–3758.
Kumar, R., Tiwari, R.S., Srivastava, O.N., 2011. Scalable synthesis of aligned carbon nanotubes bundles using green natural precursor: neem oil. Nanoscale Res. Lett. 6, 92. https://doi.org/10.1186/1556-276X-6-92.
Kyotani, T., Tsai, L.-F., Tomita, A., 1996. Preparation of ultrafine carbon tubes in nanochannels of an anodic aluminum oxide film. Chem. Mater. 8, 2109–2113.
Laplaze, D., Bernier, P., Maser, W.K., Flamant, G., Guillard, T., Loiseau, A., 1998. Carbon nanotubes: The solar approach. Carbon 36, 685–688.
Liu, X., Yang, Y., Liu, H., Ji, W., Zhang, C., Xu, B., 2007. Carbon nanotubes from catalytic pyrolysis of deoiled asphalt. Mater. Lett. 6, 3916–3919.
Lotfy, V.F., Fathy, N.A., Basta, A.H., 2018. Novel approach for synthesizing different shapes of carbon nanotubes from rice straw residue. J. Environ. Chem. Eng. 6 (5), 6263–6274.
Marsh, H., 1997. In: Marsh, H., Heintzz, E.A., Rodriguez-Reinoso, F. (Eds.), Introduction to Carbon Technologies. Universidad de Alicante, pp. 1–34.
Menéndez, J.A., Arenillas, A., Fidalgo, B., Fernández, Y., Zubizarreta, L., Calvo, E.G., Bermúdez, J.M., 2010. Microwave heating processes involving carbon materials. Fuel Process. Technol. 91, 1–8.
Menendez-Diaz, J.A., Martin-Gullon, I., 2006. Types of carbon adsorbents and their production. In: Bandosz, T.J. (Ed.), Activated Carbon Surfaces in Environmental Remediation. Elsevier Publisher, p. 37.
Mori, S., Ueno, T., Suzuki, M., 2011. Synthesis of carbon nanowalls by plasma-enhanced chemical vapor deposition in a CO/H_2 microwave discharge system. Diam. Relat. Mater. 20, 1129–1132.
Ngu, P.Z.Z., Chia, S.P.P., Fong, J.F.Y., Ng, S.M., 2016. Synthesis of carbon nanoparticles from waste rice husk used for the optical sensing of metal ions. New Carbon Mater. 31 (2), 135–143.
Oberlin, A., Endo, M., Koyama, T., 1976. Filamentous growth of carbon through benzene decomposition. J. Cryst. Growth 32, 335–349.
Onishchenko, D.V., Reva, V.P., Voronov, B.A., 2013. Farm crop waste as a promising resource for forming carbon nanotubes. Russ. Agric. Sci. 39 (5–6), 540–543.
Paul, S., Samdarshi, S.K., 2011. A green precursor for carbon nanotubes synthesis. New Carbon Mater. 26, 85–88.
Radushkevich, L.V., Lukyanovich, V.M., 1952. The structure of carbon forming in thermal decomposition of carbon monoxide on an iron catalyst. Russ. J. Phys. Chem. 26, 88–95.
Ren, Z.F., Huang, Z.P., Xu, J.W., Wang, J.H., Bush, P., Siegal, M.P., Provencio, P.N., 1998. Synthesis of large arrays of well-aligned carbon nanotubes on glass. Science 282, 1105–1107.
Richter, H., Hernadi, K., Caudano, R., Fonseca, A., Migeon, H.N., Nagy, J.B., Schneider, S., Vandooren, J., Van Tiggelen, P.J., 1996. Formation of nanotubes in low pressure hydrocarbon flames. Carbon 34, 427–429.
Sangeetha, J., Thangadurai, D., Hospet, R., Purushotham, P., Manowade, K.R., Mujeeb, M.A., Mundaragi, A.C., Jogaiah, S., David, M., Thimmappa, S.C., Prasad, R., 2017. Production of bionanomaterials from agricultural wastes. In: Nanotechnology. Springer, Singapore, pp. 33–58.
Sevilla, M., Fuertes, A.B., 2010. Graphitic carbon nanostructures from cellulose. Chem. Phys. Lett. 490, 63–68.
Shi, K., Yan, J., Lester, E., Wu, T., 2014. Catalyst-free synthesis of multiwalled carbon nanotubes via microwave-induced processing of biomass. Ind. Eng. Chem. Res. 53, 15012–15019.

Somanathan, T., Prasad, K., Ostrikov, K., Saravanan, A., Krishna, V.M., 2015. Graphene oxide synthesis from agro waste. Nanomaterials 5, 826–834.

Sun, L., Tian, C., Li, M., Meng, X., Wang, L., Wang, R., Yin, J., Fu, H., 2013. From coconut shell to porous graphene-like nanosheets for high power supercapacitors. J. Mater. Chem. A 1 (21), 6462–6470.

Suriani, A.B., Azira, A.A., Nik, S.F., Md Nor, R., Rusop, M., 2009. Synthesis of vertically aligned carbon nanotubes using natural palm oil as carbon precursor. Mater. Lett. 63, 2704–2706.

Suriani, A.B., Md Nor, R., Rusop, M., 2011. Vertically aligned carbon nanotubes synthesized from waste cooking palm oil. J. Ceram. Soc. Jpn 118, 963–968.

Thess, A., Roland, L., Nikolaev, P., Dai, H., Petit, P., Robert, J., Xu, C., Lee, Y.H., Gon, K.S., Rinzler, A.G., Colbert, D.T., Scuseria, G.E., Tománek, D., Fischer, J.E., Smalley, R.E., 1996. Crystalline ropes of metallic carbon nanotubes. Science 273, 483–487.

Wang, Z., Ogata, H., Morimoto, S., Ortiz-Medina, J., Fujishige, M., Takeuchi, K., Muramatsu, H., Hayashi, T., Terrones, M., Hashimoto, Y., Endo, M., 2015. Nanocarbons from rice husk by microwave plasma irradiation: from graphene and carbon nanotubes to graphenated carbon nanotubes hybrids. Carbon 94, 479–484.

Wang, Z., Shen, D., Wu, C., Gu, S., 2018. State-of-the-art on the production and application of carbon nanomaterials from biomass. Green Chem. 20 (22), 5031–5057.

Yacamàn, M.J., Yoshida, M.M., Rendon, L., Santiesteban, J.G., 1993. Catalytic growth of carbon microtubules with fullerene structure. Appl. Phys. Lett. 62, 202–204.

Yan, Y., Miao, J., Yang, Z., Xiao, F.-X., Yang, H.B., Liu, B., Yang, Y., 2015. Carbon nanotube catalysts: recent advances in synthesis, characterization and applications. Chem. Soc. Rev. 44, 3295–3346.

Yao, Y., Xu, F., Chen, M., Xu, Z., Zhu, Z., 2010. Adsorption behavior of methylene blue on carbon nanotubes. Bioresour. Technol. 101, 3040–3046.

Zeng, X., Fu, D., Sheng, H., Xie, S., Li, X., Hu, Q., Zou, J., 2010. Growth and morphology of carbon nanostructures by microwaveassisted pyrolysis of methane. Phys. E: Low-Dimensional Syst. Nanostruct. 42, 2103–2108.

Zhu, J., Jia, J., Kwong, F.L., Ng, D.H., Tjong, S.C., 2012. Synthesis of multiwalled carbon nanotubes from bamboo charcoal and the roles of minerals on their growth. Biomass Bioenergy 36, 12–19.

Zhuo, C., Alves, J.O., Tenório, J.A.S., Levendis, Y.A., 2012. Synthesis of carbon nanomaterials through up-cycling agricultural and municipal solid wastes. Ind. Eng. Chem. Res. 51, 2922–2930.

Part Two

Environmental applications

Improving diesel engine performance using carbon nanomaterials

5

Ahmed I. EL-Seesy[a,b], *Hamdy Hassan*[c], *Zhixia He*[a], *S. Ookawara*[d]
[a]Institute for Energy Research, Jiangsu University, Zhenjiang, China, [b]Department of Mechanical Engineering, Benha Faculty of Engineering, Benha University, Benha, Egypt, [c]Department of Energy Resources Engineering, Egypt-Japan University of Science and Technology (E-JUST), Alexandria, Egypt, [d]Department of Chemical Science and Engineering, School of Materials and Chemical Technology, Tokyo Institute of Technology, Tokyo, Japan

1 Introduction

The fast decline of fossil fuels has stimulated research and development on alternative renewable biofuels. Meanwhile, conventional fuels have caused an increase in global warming, which is recognized to be the main reason for rising global temperatures. Biodiesel oils, principally nonedible oils, are widely acknowledged to be potential feedstocks for energy that can substitute for fossil fuels because of their similar characteristics to diesel fuel. They are renewable, and they are readily available. An additional benefit of producing biodiesel from these feedstocks is that they are inappropriate for human consumption because of toxic compounds, so there will be no issues associated with food versus fuel. Beside this, arid lands and recycled wastewater can be utilized to cultivate these plants.

Jatropha curcas is considered a promising feedstock for biodiesel production that can be cultivated in many types of soils including arid land, which diminishes competition with edible oil plants (Gonzáles, 2016). Fruit production can begin the next year after establishment, but in a limited amount. Nevertheless, beginning at the fifth year, a tree can yield 4–5 kg of fruits if planted appropriately, and its productivity can keep going up to 40–50 years from the day of implantation (Soliman and He, 2015). Mechanical pressing followed by solvent extraction or solvent extraction only are commonly utilized to extract Jatropha oil (Abu-Arabi et al., 2000). Its oil properties are comparable to those of rapeseed oil as a fuel (Achten et al., 2008); they have >75% unsaturated fatty acid that impacts their physiochemical features (Achten et al., 2008). Its seeds hold about 30% oil that can yield biodiesel using the transesterification technique (Achten et al., 2008). Subsequently, it has substantial potential to be employed in biodiesel preparation (Achten et al., 2008; Gonzáles, 2016). Nevertheless, the primary issue that hinders straight utilization of Jatropha oil in different combustion systems is its high viscosity, which is significantly higher than that of diesel fuel. This fact is ascribed to its great molecular mass and chemical structure (Attia and Hassaneen, 2016).

Carbon Nanomaterials for Agri-food and Environmental Applications. https://doi.org/10.1016/B978-0-12-819786-8.00005-0

Thus, to overcome this problem, the transesterification process is utilized by most scientists, who have considered it to be the best technique for solving the viscosity issue of raw oil (Jiaqiang et al., 2017; Tamilselvan et al., 2017). It has many advantages as it is performed under normal conditions and yields better quality biodiesel. The biodiesel production via transesterification relies upon many parameters, including the type and amount of alcohol and catalyst, the reaction time, and the reaction temperature.

Although the transesterification process can produce biodiesel comparable with diesel fuel, the viscosity of the biodiesel product is still high compared to diesel fuel. Thus, higher alcohols can be employed to overcome this problem. For instance, the academic community has been significantly focused on using butanol as an oxygenated additive with diesel fuel in diesel engines because of its higher ability to mix with diesel fuel (Lee et al., 2011; Trindade and Santos, 2017; Yilmaz and Davis, 2016). In recent years, there have been a large number of studies that focused on the use of various butanol-diesel mixtures in compression ignition engines (Chen et al., 2014; Işık et al., 2017; Zhang and Balasubramanian, 2016; Zheng et al., 2016). The studies showed that the addition of butanol with diesel fuel led to increased brake specific consumption while the thermal efficiency and exhaust emissions depended on the comparative proportion of utilized butanol as well as the engine working situation. Moreover, Lee et al. (2011) and Chen et al. (2014) studied the combustion and exhaust emission characteristics of a diesel engine run with different mixtures of *n*-butanol-di-methyl ether and *n*-butanol-diesel fuel. They reported that the engine could be run efficiently for a varied range of engine loads up to 30% *n*-butanol blends by mass. The results revealed that engine emissions rose with the growing *n*-butanol ratio in fuel. Babu et al. (2017) and Trindade and Santos (2017) have done a comprehensive survey on the use of butanol and propanol in diesel engines. They reported that butanol can be employed in compression ignition engines without any adjustment. Also, it helps to reduce soot and NO_x emissions simultaneously without any deterioration in efficiency, irrespective of the category of engine utilized. They ascribed these findings to the oxygen content of butanol, which aids in reducing soot formation, while it's heating value drops the cylinder temperature and hence diminishes NO_x formation. The main physiochemical properties of diesel, methanol, ethanol, and higher alcohols are illustrated in Table 1. From the table, the physiochemical parameters probably emphasize that *n*-butanol has some fascinating properties, which illustrate that it has the possibility to overcome some drawbacks carried by other lower alcohols utilized as fuel or fuel additives. The drawbacks include ignition difficulties in low-temperature weather, higher fuel consumption, low lubricity, and corrosion. Those characteristics of *n*-butanol are as follows:

- Regarding the calorific value, the low calorific value of alcohol increases with enlarged carbon content. Higher alcohols start at four-carbon, a double number of carbon atoms as compared to lower alcohols, and contain 50% more energy density in volume. Therefore, an engine working with higher alcohols is projected to present a lower fuel consumption when compared to ethanol.
- Regarding the fuel handling, the volatility of alcohols diminishes with the rise of carbon content, which means that higher alcohols will have less propensity to vaporize. Also, it's flashpoint is higher in comparison to lower alcohols. These two features show that higher alcohols are possibly safer when thinking about transportation and utilization in hot conditions.

Table 1 Properties of some alcohols and diesel fuel (Trincade and Santos, 2017).

	Diesel	Methanol	Ethanol	*n*-Butanol	1-Heptanol	1-Octanol
Molecular formula	C_{12}-C_{25}	CH_3OH	C_2H_5OH	C_4H_9OH	$C_6H_{13}OH$	$C_8H_{18}O$
Molecular weight	198.4	32.04	46.06	74.11	116.19	130.21
Cetane number	40–55	3	8	25	23	39
Oxygen content (% weight)	–	50	34.8	21.6	15.7	15.7
Density (g/mL) at 20°C	0.82–0.86	0.796	0.79	0.808	0.822	0.827
Autoignition temperature (°C)	210	470	434	385	–	–
Flash point (°C) at closed cup	65–88	12	8	35	100	59
Lower heating value (MJ/kg)	42.5	19.9	26.8	33.1	34.65	52.94
Boiling point (°C)	180–370	64.5	78.4	117.7	175	195
Latent heating (kJ/kg) at 25°C	270	1109	904	582	575	486
Viscosity (mm^2/s) at 40°C	1.9–4.1	0.59	1.08	2.63	3.32	5.8

- Concerning ignition quality, engines operating with higher alcohols are expected to have fewer difficulties with a low-temperature start than the same engine working on lower alcohols (assuming the same air-to-fuel relation). This is accredited to the fact that the heat of vaporization of higher alcohols is less than half that of lower alcohols.
- Regarding the fuel mixture stability, higher alcohols are easier to mix with diesel fuel while lower carbon numbered alcohols such as ethanol and methanol are more polar and more solvable in water than nonpolar hydrocarbons. This means that it could be a challenge when it is added with diesel fuel, which that reduces the upper limit of mixing in diesel fuel without the use of external additives. Nevertheless, on the other hand, these alcohols are significantly less polar because of their lengthier nonpolar hydrocarbon structures. They are therefore easier to be mixed with nonpolar hydrocarbons and have a lower affinity for water.
- Regarding higher viscosity and lubricity, the viscosity of alcohols rises as the number of carbons in the molecule rises. Normal butanol may possibly keep some parts of the engine safe that have straight contact with fuel (such as fuel pumps, fuel rails, injectors) versus wear issues.

Carbon nanomaterials have drawn interest across many disciplines due to their interesting features. They could be environmentally friendly additives due to their free-metal nature. Moreover, their ability to diminish diesel emissions and enhance engine fuel consumption more than metal-based nanoparticles owes to their superior thermal and chemical properties. For example, the high heat of combustion (i.e., high exothermic heat release), high surface-to-volume ratio, and numerous oxygen functional groups (i.e., epoxides, hydroxides, and carboxylates) of carbon nanomaterials can enhance chemically active sites for complete combustion (Ooi et al., 2018). Ooi et al. (2016, 2017) investigated the combustion behavior of diesel and biodiesel fuel droplets containing graphene oxide nanoparticles (GO). They reported that there were significant enhancements on the droplet combustion behaviors of diesel and biodiesel fuel with the addition of GO. The findings illustrated that the ignition delay is diminished by adding these additives to diesel and biodiesel fuel. This is ascribed to boosted mass transfer and heightened heat transfer by the nanoparticles (Ooi et al., 2016). Consequently, the droplet temperature is reduced while the heat is transferred by the nanoparticles at the fuel vapor and air blending area to be burnt quickly. Another factor could be ascribed to the exothermic heat released from the decomposition of carbon nanomaterials reducing the ID of the fuel (Ooi et al., 2018).

A microexplosion is a substantial factor that can boost droplet breakup into smaller droplets through the combustion process, allowing minor droplets to combust more quickly (Ooi et al., 2016). Adding nanoparticles to diesel fuel droplets resulted in an improved microexplosion frequency and consequently sped up the combustion process. This can be described by the energetic decomposition of GO particles at high temperature (Ooi et al., 2018). Furthermore, the burn rate constant is increased by adding nanoparticles, owing to the increased burning rate among the air and fuel vapor inside the diffusion-controlled combustion (Ooi et al., 2016).

In recent decades, there have been many investigations that have focused on nanoparticle material as additives in diesel/biodiesel blends, owing to their substantial influences on the combustion characteristics of diesel/biodiesel blends

(Ali et al., 2013; Basha and Gopal, 2012; Khalife et al., 2017; Saxena et al., 2017). This is ascribed to the positive combustion characteristics (Attia et al., 2014; EL-Seesy et al., 2017).

The impact of MWCNT, graphene, silver, and cerium oxide nanoparticle additives on diesel engine performance were studied by Basha and Anand (2011), Tewari et al. (2013), Banapurmath et al. (2014), Mirzajanzadeh et al. (2015), and Karthikeyan and Prathima (2016). They stated that there was a significant reduction in exhaust emissions of CO, UHC, NO_x, and soot by about 50%, 50%, 25%, and 20%, respectively. Furthermore, BTE was heightened by 12%. Moreover, the same observations were achieved by Selvan et al. (2014), Ghafoori et al. (2015), Senthilkumar and Prabu (2015), and Heydari-Maleney et al. (2017), who examined the impact of the addition of cerium oxide nanoparticles (CERIA) and CNTs into diesel-biodiesel-ethanol fuel mixtures on diesel engine performance.

Recently, Hosseini et al. (2017), Ghanbari et al. (2017), and EL-Seesy et al. (2018a, b, c, d, e) studied the diesel engine performance run with diesel-biodiesel mixtures, CNTs, and silver nanoparticles. They reported that thermal efficiency was boosted by 4%. Furthermore, there was a significant reduction in engine emissions where CO, UHC, and soot were diminished by 65%, 45%, and 30%, respectively, while NO_x emissions were amplified by 27%. Table 2 summarizes the key results on the impact of nanoparticle additives on diesel/biodiesel mixtures. It is obvious from the presented outcomes that the engine performance and emission parameters are enhanced considerably by adding nanoparticles.

From this literature review, it is confirmed that Jatropha oil is commonly acknowledged as a promising source to produce biodiesel via the transesterification process (Achten et al., 2008). Furthermore, compared with lower alcohols, the higher alcohols have attractive features that make them a significant fuel source for diesel engines. There has been substantial attention on the utilization of different metal oxide-based nanoparticle additives such as zinc oxide and cerium oxide to enhance the burning rate of diesel-biodiesel-alcohol blends and to reduce diesel engine emissions (EL-Seesy et al., 2018a, b, c, d, e; Nanthagopal et al., 2017; Nour et al., 2018; Saxena et al., 2017). Despite these benefits, these additives and their compounds generated from burning byproducts have been stated to be harmful to the atmosphere (Horie and Fujita, 2011; Merwe and Pickrell, 2012). For example, some investigations have stated that these additives can lead to several categories of health issues such as breathing issues, lung-related problems, and skin allergies (Goswami et al., 2017; Jeng and Swanson, 2006; Pope et al., 2002). Nevertheless, carbon nanomaterials, namely, graphene nanoplatelets (GNPs), graphene oxide (GO), and multiwalled carbon nanotubes (MWCNTs), can be an environmentally cordial fuel additive for proficiently boosting the combustion process of diesel/biodiesel mixtures, attributed to their low toxicity, high energy density, and high thermal conductivity (EL-Seesy et al., 2018a, b, c, d, e; Ooi et al., 2016, 2017; Zhu et al., 2010). The beneficial characteristics of these additives have been emphasized in numerous applications such as in batteries, chemical sensors, heat transfer, and transparent conductors (Esfahani et al., 2016; Song et al., 2015; Zhu et al., 2010). The ultrathin and layered nanostructure of carbon nanomaterials and the existence of numerous oxygen functional groups in carbon nanomaterials

Table 2 Performance, combustion, and emission characteristics of diesel engines fueled by different fuel blends containing nanoparticles.

Engine	Base fuel	Operating condition	Nanofluid additives	Composition	Performance results	Combustion results	Emission results	Reference
Single cylinder, 661 cc	Jatropha biodiesel water emulsion	Constant speed with different loads	Carbon nanotubes (CNTs)	25, 50, and 100 ppm	Increased BTE and decreased bsfc	Decreased cylinder pressure (p), heat release rate (HRR), and ignition delay (ID)	Reduced NO_x, UHC, CO, and smoke opacity	Basha and Anand (2014)
Single cylinder, 550 cc	Karanja biodiesel	Constant speed with different loads	Multiwalled CNTs	20, 40, 60, 80, and 100 ppm	Increased BTE and decreased bsfc	–	Increased smoke opacity and CO; reduced NO_x and UHC	Senthilkumar and Prabu (2015)
Single cylinder, 661.5 cc	Diesel water emulsion	Constant speed with different loads	CNTs	50, 100, and 150 ppm	Increased BTE and decreased bsfc	Decreased p, HRR, and ID	Reduced NO_x, UHC, CO, and smoke opacity	Singh and Bharj (2015a, b)
Single cylinder, 552 cc	Diesel-biodiesel-ethanol blends	Constant speed with different loads	Cerium oxide nanoparticles and CNTs	25, 50, and 100 ppm	Increased BTE and decreased bsfc	Increased p; decreased HRR and ID	Reduced UHC and smoke opacity; no change in CO and NO_x	Selvan et al. (2014)
Single cylinder, 715 cc	Diesel and honge biodiesel	Constant speed with different loads	Multiwalled CNTs	25 and 50 ppm	Decreased BTE	–	Reduced NO_x, UHC, CO, and smoke opacity	Tewari et al. (2013)
Single cylinder, 661 cc	Diesel water emulsion	Constant speed with different loads	CNTs	25 and 50 ppm	Increased BTE and decreased bsfc	Decreased p, HRR, and ID	Reduced NO_x, UHC, CO, and smoke opacity	Basha and Anand (2011)

deliver great surface-area-to-volume ratios and chemically active sites for boosting complete combustion (Ooi et al., 2016, 2017). Ooi et al. (2017) showed that adding graphite oxide at a concentration of 0.1% in diesel/biodiesel fuel considerably boosted the linear burning rate and diminished the ignition temperature. The key factor for that is ascribed to the high exothermic heat release and the extra oxygen provided by the graphite oxide. Moreover, they reported that adding this additive could diminish engine emissions and enhance fuel consumption by boosting the burning rate of fuel.

Concerning the influence of carbon nanomaterials on engine parts, there are some studies that have examined the friction phenomena with the presence of carbon nanomaterials. For instance, the outstanding friction and wear reduction characteristics of these additives have been commonly recognized in the literature (Masood et al., 2017; Zhai et al., 2017). The interesting mechanical parameters, low surface energy (Kim et al., 2011), and high thermal and electrical conductivity (Geim and Novoselov, 2007) of these additives make them attractive for both nanoscale and macroscale tribological applications. Lee et al. (2009, 2010) examined the adhesion and friction forces on carbon nanomaterial films manufactured by chemical vapor deposition and converted onto an SiO_2/Si substrate. They reported that the adhesion and friction during microscale and nanoscale contacts decreased when diamond probes were used as counterparts under ambient situations.

Consequently, from this review, it is seen that the nanoparticle additives with diesel/biodiesel mixtures seem like an advanced approach to accomplish a substantial improvement in engine performance and exhaust emissions. However, to the author's best knowledge, there is no earlier study on the utilization of higher alcohols/diesel/biodiesel mixtures blended with GO, GNPs, and MWCNTs as fuel. Therefore, the purpose of the present study is to scrutinize the addition of these additives into diesel/biodiesel/alcohol blends on engine performance and exhaust emission parameters.

2 Experimental setups and methodology

The test rig, Model GUNT-CT100.22, utilized in the current study is shown in Table 3. The actual image of the test rig is shown in Fig. 1. More details about the test rig used can be found in the literature (EL-Seesy et al., 2016, 2017, 2018a, b, c, d, e).

Table 4 shows the exhaust gas analyzer, Model BACHARACH-ECA-450, employed to measure the engine emissions of CO, UHC, NO_x, and O_2. The steps by Holman (2001) were utilized to evaluate the uncertainty of the present measurements. The uncertainty in the measurements of bsfc, brake power, EGT, and engine speed was 1.7%, 1.4%, 0.5%, and 0.25% (±3 rpm), respectively.

2.1 *Jatropha biodiesel preparation*

The Jatropha methyl ester preparation was done in three steps: (i) oil extraction, (ii) esterification, and (iii) transesterification. Jatropha seeds were provided by Afforestation Management, Ministry of Agriculture, Cairo, Egypt. A summary of the parameters needed for both the esterification and transesterification processes is given in

Table 3 Specifications of diesel engine.

Engine parameter	Specification
Engine model	HATZ-1B30-2
Number of cylinders	1
Bore, mm	80
Stroke, mm	69
Crank length, mm	34.5
Rod length, mm	114.5
Displacement, cm^3	347
Compression ratio	21.5:1
Rated power, kW/rpm	5.4/3600
Idle speed, rpm	1000
Type of injection	Direct injection
Type of cooling	Air cooling
Starting up	Electrical
Oil capacity max-min, L	1.1–0.6

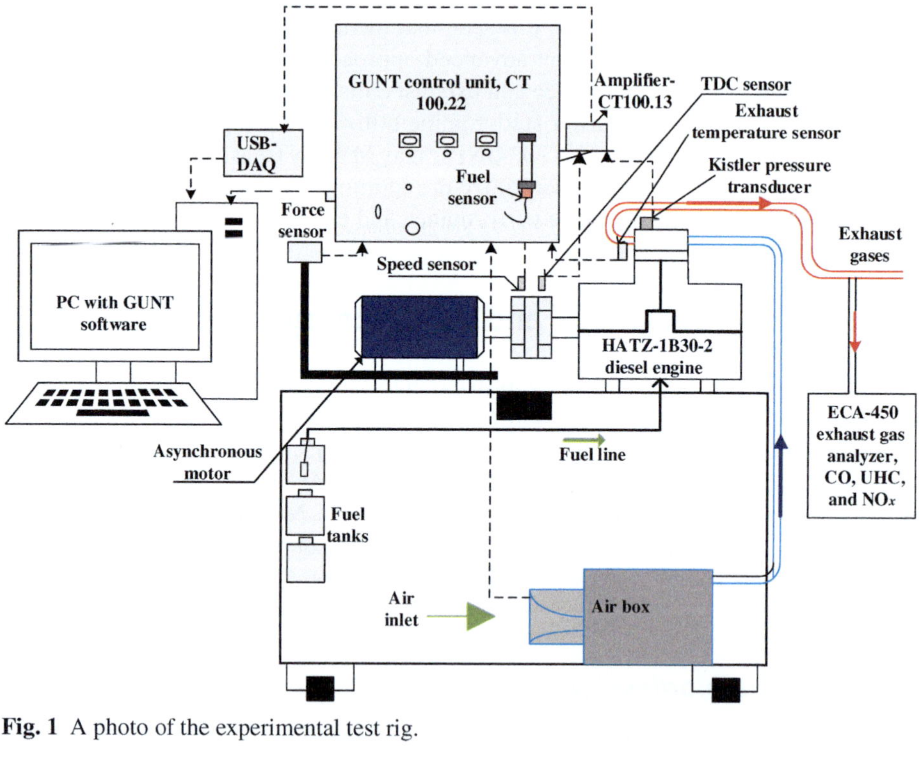

Fig. 1 A photo of the experimental test rig.

Table 4 Exhaust gas analyzer specifications.

Gas	Measuring range	Resolution	Accuracy
CO	0–4000 ppm	1 ppm	±5% of reading or ±10 ppm
CO_2	0–20% by vol.	0.1% by vol.	0.5% of reading
UHC	0–10% by vol.	0.01 by vol.	±0.3% of reading
O_2	0–20.9% by vol.	0.01% by vol.	±0.3%
NO_x	0–4000 ppm	1 ppm	±5% of reading or ±5 ppm
Stack temperature	−20°C to 1315°C	1°C	±2°C
Probe tip temperature	800°C max	–	–

Table 5 Summary of the conditions for both esterification and transesterification processes.

Process	Methanol to oil molar ratio	Catalyst type	Catalyst amount	Reaction condition	Stirring speed (rotor-stator mixer)
Esterification	6:1 molar	H_2SO_4	1% (vol/vol)	60°C, 90 min	8000 rpm
Transesterification	8:1 molar	KOH	1% (wt. of oil)	60°C, 60 min	8000 rpm

Table 5 (Tiwari et al., 2007). More details about Jatropha production and its characterization can be found in the literature (EL-Seesy and Hassan, 2018).

2.2 Carbon nanomaterial-diesel-biodiesel-alcohol blend preparation

Carbon nanomaterials, typically GO, GNPs, and MWCNTs, were mixed into diesel/biodiesel/alcohol blends according to Table 6. These blended fuels were dispersed using an ultrasonic technique (Hielscher ultrasonic UP400S, 400 W, 24 kHz) set at a frequency of 24 kHz for 30 min to enhance the dispersion of the fuel mixtures (EL-Seesy et al., 2018a, b, c, d, e; Khalife et al., 2017).

Table 6 Representative concentrations of carbon nanomaterials blending with diesel-biodiesel-alcohol mixtures.

	JB20	JME40B
MWCNTs	25, 50, 75, and 100 mg/L	50 mg/L
GNPs		50 mg/L
GO		50 mg/L

The sedimentation method is the simplest method to determine the stability of nanofluids under the impact of an outside force field through settling. There are a number of investigations (Li et al., 2009; Ma et al., 2013; Wei et al., 2002) that used this method to inspect the stability features of nanofluids. This technique was thus utilized to examine the stability of the resulting dispersion of nanofluids. All carbon nanomaterial-diesel-biodiesel blend samples were poured in a long glass tube under stationary situations to perceive mixture stability, and it was noticed that there was no blend separation for approximately 5 days. The agglomeration of nanoparticles in the diesel-biodiesel blends is probably due to the weak van der Waals forces among the nanoparticles that attract each other. This problem can be resolved by employing a surfactant (Saxena et al., 2017). However, the addition of a surfactant is not used in this examination because the surfactant can impact the burning parameters of the dosed diesel-biodiesel blends and the influences of adding numerous studied carbon nanomaterials cannot be quantified separately. Therefore, tested fuels were mixed without any surfactant for all experiments to identify the important contributions of individual carbon nanomaterial types to the combustion characteristics of diesel-biodiesel blends.

Table 7 shows that the thermal conductivity of GO, GNPs, and MWCNTs at room temperature is >2000 times greater than that of diesel-biodiesel blends. Naddaf and Heris (2018) examined the thermal conductivity and electrical conductivity of the diesel/GNP/MWCNT blends. The results showed that the thermal conductivity and electrical conductivity were enlarged considerably for all blends. Therefore, there is a significant enhancement in thermal conductivity with the addition of carbon nanomaterials to diesel-biodiesel blends. Thus, the ignition delay was shorted, owing to the enhancement in evaporation rate of fuel droplets. Similarly, the addition of carbon nanomaterials to diesel fuel will improve the heat transfer rate, owing to the high surface-area-to-volume ratio of carbon nanomaterials.

Table 7 Specifications of GO, GNPs, and MWCNTs.

Item	GO	GNPs	MWCNTs
Manufacturer	Nanotech Egypt for photo-electronics	Strem Chemicals, Inc., France	Arkema Company, France
Chemical name	Graphene oxide nanopowder, sheets, 90%	Graphene nanoplatelets nanopowder, the carbon content of >99.5 wt%	Multiwalled carbon nanotubes, nanopowder, 90%
Average particle size	~1.2 nm average thickness 7 μm wide	6–8 nm thickness, 5 μm wide	10–15 nm diameter, 1–10 μm length
Surface area (SSA)	~900	~300 m^2/g	~350 m^2/g
Purity	>90%	>99.5%	>90%
Appearance	Brown black	Black	Black
Density	~0.05 g/cm^3	0.03–0.1 g/cm^3	~0.05–0.17 g/cm^3
Thermal conductivity	~5000 W/m K	3000 W/m K	~3000 W/m K
Mechanical tensile modulus	~1000 GPa	~1000 GPa	~1000 GPa

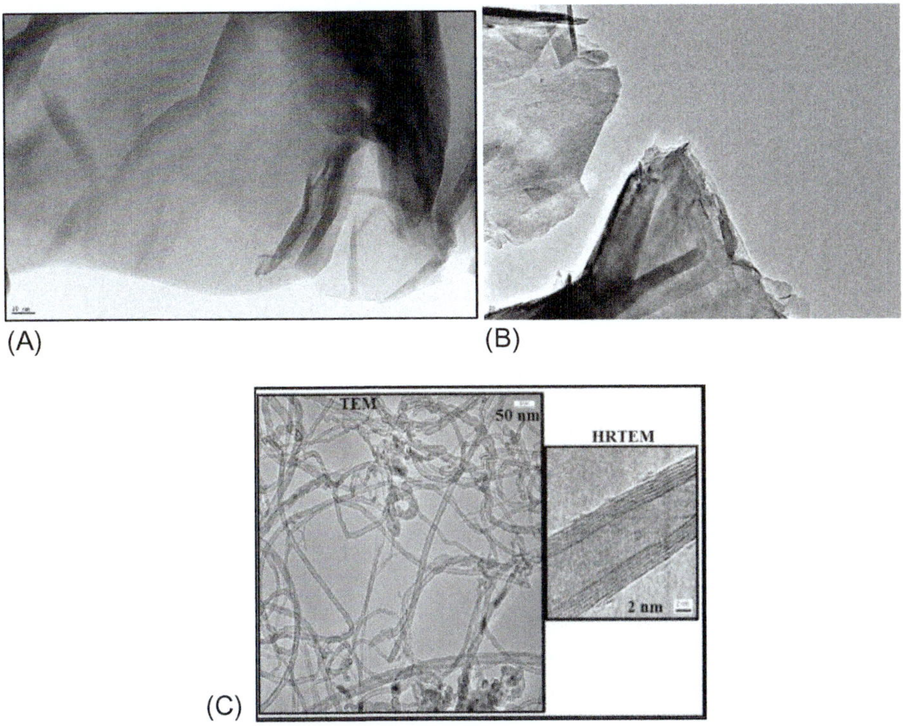

Fig. 2 TEM (A) GO, (B) GNPs, and (C) MWCNTs.

A transmission electron microscope (TEM), MODEL JEOL-JEM-2100F, was used to explore the GO, GNP, and MWCNT sample as given in Fig. 2A–C. Fig. 2A shows that the GO is consistently distributed, and the typical few-layered structure is well separated. From Fig. 2B, it is noticed that the existence of the graphene nanosheet is very obvious, and it also illustrates the lattice plane growth. Fig. 2C shows the image of MWCNTs, where the MWCNTs appear in bundles with smooth structures.

2.3 Heat release rate model

The gross heat release rate ($dQ_{gross}/d\theta$) is calculated using the measured cylinder pressure $p(\theta)$ and it is estimated employing the following equation (Heywood, 1988).

$$\frac{dQ_{gross}}{d\theta} = \frac{\gamma_{(T)}}{\gamma_{(T)} - 1} \times p \times \frac{dV}{d\theta} + \frac{1}{\gamma_{(T)} - 1} \times V \times \frac{dp}{d\theta} + \frac{dQ_{wall}}{d\theta} \tag{1}$$

To minimize cycle-to-cycle variation, the cylinder pressure was collected every 0.5-degree crank angle on an average of 50 cycles. The coefficient of cyclic variation of IMEP (COV_{IMEP}) ranged between 0.9% and 1.2% for various engine loads and speeds, which is within the drivability limit (i.e., COV_{IMEP} below 5%) (Heywood, 1988).

The specific heat ratio $\gamma_{(T)}$ is considered employing the following equation (EL-Seesy et al., 2017) where their values were similar to those informed by Heywood (1988):

$$\gamma_{(T)} = 1.35 - 6\times10^{-5}\times T + 10^{-8}\times T^2 \quad (2)$$

Eq. (3) is used to estimate the mean gas temperature as follows (Shehata, 2010).

$$T = \frac{T_r \times p \times V}{p_r \times V_r} \quad (3)$$

All thermodynamic properties (pressure p_r, temperature T_r, and volume V_r) were estimated at a given reference condition such as IVC (p_{IVC}, T_{IVC}, V_{IVC}). T_{IVC} and p_{IVC} are the temperatures and the pressure at IVC and they are equal 350 K and 1.013×10^5 Pa, respectively.

Eq. (3) is used to estimate the rate of heat transfer to the cylinder wall (Heywood, 1988).

$$\frac{dQ_{\text{wall}}}{d\theta} = h_c \times A_{(\theta)} \times (T - T_{\text{wall}}) \times \left(\frac{1}{6N}\right) \quad (4)$$

The wall temperature is presumed to be 450 K, which gives satisfactory findings (Shehata, 2010) and it is estimated using Eq. (5) as follows (Hohenberg, 1979).

$$h_c = C_1 \times V^{-0.06} \times p^{0.8} \times T^{0.4} \times (C_2 + V_m)^{0.8} \quad (5)$$

where p is the instantaneous pressure in bar, $C_1 = 130$, and $C_2 = 1.4$.

3 Results and discussions

The impact of adding carbon nanomaterials to diesel/biodiesel/alcohol blends was explored employing a CI engine. The diesel engine combustion characteristics, namely cylinder pressure and gross heat release rate, were investigated. The BTE and bsfc as well as engine emissions were also inspected.

3.1 *Effect of adding GNPs to Jatropha biodiesel-diesel blend*

The ASTM standards procedures are used to measure the characteristics of JAME, JB20 with and without GNPs, as illustrated in Table 8. It can be noticed that the kinematic viscosity is increased by about 25% by adding GNPs into the JB20 mixture. This could be ascribed to the fact that adding GNPs led to a slight rise in cohesion among the fuel molecules, leading to a small increase in resistance to their motion and therefore a rise in their viscosity. This trend was observed by several researchers in their studies (Basha and Anand, 2011; Singh and Bharj, 2015a, b; Tewari et al., 2013).

Table 8 Properties of JAME (B100) and JB20 without and with GNP samples.

Property	**Testing method**	**JAME (B100)**	**JB20**	**$JB20_{25GNPs}$**	**$JB20_{50GNPs}$**	**$JB20_{75GNPs}$**	**$JB20_{100GNPs}$**	**ASTM D6751-12**
Calorific value, kJ/kg	ASTM D-240	40,630	41,142	41,160	41,190	41,210	41,230	–
Kinematic viscosity at 40°C, mm^2/s	ASTM D-445	5.26	3.33	4.05	4.18	4.21	4.22	1.9–6
Specific gravity at 15.56°C	ASTM D-1298	0.883	0.8493	0.8501	0.8502	0.8503	0.8504	0.86–0.9
Cetane index	ASTM D-976	52	51.6	52.3	53.5	55.2	57.4	Min 47

The cetane number is heightened by 11% by adding GNPs into the JB20 mixture. This is credited to the higher thermal conductivity of the GNPs, which enhanced the vaporization rate of fuel droplets, resulting in a reduction in ignition delay and therefore improving the cetane number. This finding is comparable to those accomplished by the numbers of the literature (Hoseini et al., 2017; Khond and Kriplani, 2016; Saxena et al., 2017). The Jatropha biodiesel has about 6.5% by mass oxygen content.

3.2 Cylinder pressure

Fig. 3 illustrates the variation in cylinder pressure with the crank angle at 1500 rpm and 13.5 N m loads for the various tested fuels. Adding GNPs to the JB20 mixture advanced the combustion process, leading to a higher cylinder pressure. This is ascribed to the higher thermal conductivity of the GNPs, which is about 2000 times superior to that of JB20 mixtures. Additionally, the cetane number was boosted by about 11%, ascribed to the addition of GNPs. Accordingly, there is a significant enhancement in the vaporization rate of fuel droplets, resulting in a decrease in ignition delay. Adding GNPs also boosted the heat of reaction and improved peak pressures, as informed by Banapurmath et al. (2014), Ghafoori et al. (2015), and Mirzajanzadeh et al. (2015) in their investigations. At a 1500 rpm engine speed, the peak cylinder pressures were noted as 70.47, 71.26, 72.56, 72.9, and 71.56 bar for JB20, $JB20_{25GNPs}$, $JB20_{50GNPs}$, $JB20_{75GNPs}$, and $JB20_{100GNPs}$ fuels, respectively.

Fig. 3 illustrates the variation in the gross heat release rate ($dQ_g/d\theta$) with the crank angle for tested fuels at 13.5 N m loads and a speed of 1500 rpm. The findings illustrate that the heat release rate is negative through the ignition delay region. This is mostly ascribed to the cooling impact produced by fuel evaporation and heat losses from the engine cylinder walls (Basha and Anand, 2011). Moreover, at the engine speed of 1500 rpm, the gross heat release rate declines with adding GNPs. This could be

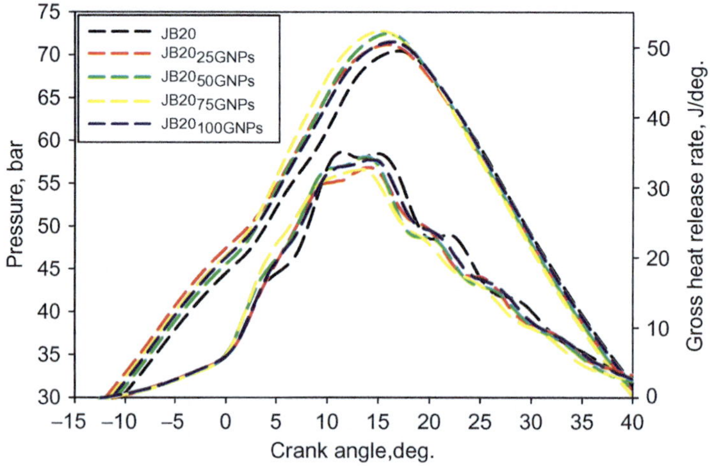

Fig. 3 The change of cylinder pressure and gross heat release rate against crank angle at 1500 rpm and 13.5 N m load for different tested fuels.

ascribed to the decrease in the mean cylinder temperature as mentioned in the previous part, which resulted in a small decrease in the peak heat release rate at low speed. The peak gross heat release rate and its location for tested fuels at all engine speeds and 13.5 N m loads are illustrated in Fig. 3. It can be observed that for both engine speeds of 2000 and 2500 rpm, the gross heat release rate is higher for the GNP-JB20 mixture compared to the pure JB20. This could be ascribed to the boosting impact of the large surface-area-to-volume ratio, higher thermal conductivity, and the enhanced ignition characteristics. Thus, there is an advancement in the combustion process resulting in a heightened peak value of the gross heat release rate compared to that of the JB20 mixture.

The peak pressure (p_{max}) and its crank angle location (θ_{max}) provide a real indication about the combustion process. Figs. 4A–C show the peak pressure, the position of maximum pressure, and the percentage of pressure increase, respectively, for the various test fuels and engine speeds at a constant load of 13.5 N m. The findings illustrate that the locations where the maximum cylinder pressure was accomplished for the tested fuels have the same inclinations with the engine speed but with different values. Additionally, for all the tested fuels, the position of p_{max} reduces with the rise in engine speed. This is ascribed to higher turbulence intensities, which result in an enhanced mixing and combustion process. Increasing the engine speed resulted in a reduction in the ignition delay period, which led to an advanced combustion process and earlier

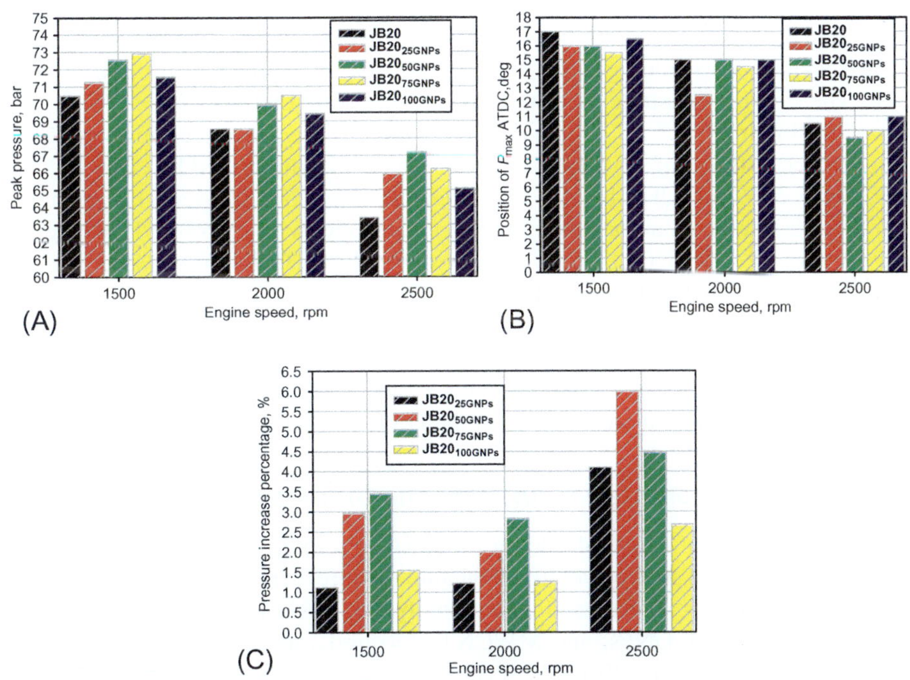

Fig. 4 The variation of pressure with engine speed at 13.5 N m load (A) peak pressure, (B) position of maximum pressure, and (C) pressure percentage increase.

occurrence of p_{max}. These findings are emphasized by those of Shehata and Abdel Razek (2011). From Fig. 4B, adding GNPs into the JB20 mixture led to an advanced position of p_{max}. These findings are ascribed to the boosted evaporation rate and high atomization rates leading to a shortened ignition delay. The maximum advancement of the position of peak pressure happened at a GNP dose level of 50–75 mg/L.

From Fig. 4A, at the engine speed of 2000 rpm, the peak cylinder pressure for the pure JB20 fuel was 68.56 bar, whereas it was 66.88, 69.94, 70.5, and 69.43 bar for the JB20-GNP mixture of 25, 50, 75, and 100 mg/L dose levels, respectively. The behavior was attained for the engine speed of 2500 rpm, where the maximum cylinder pressure for the JB20 fuel was 63.4 bar while it was 66, 67.2, 66.24, and 65.1 bar for the JB20-GNP mixtures of 25, 50, 75, and 100 mg/L dose levels, respectively. From Fig. 4C, it is noticed that the percentage rise in the cylinder pressure at the engine speed of 1500 rpm was 1.12%, 2.96%, 3.45%, and 1.55% for the JB20-GNP mixtures of 25, 50, 75, and 100 mg/L, respectively, compared to that of pure JB20 fuel. Moreover, at the engine speed of 2000 rpm, the percentage rise in the cylinder pressure was 1.22%, 2.01%, 2.83%, and 1.2690%, whereas at the speed of 2500 rpm it was 4.1%, 5.99%, 4.48%, and 2.68% for the JB20–GNP mixtures of 25, 50, 75, and 100 mg/L dose levels, respectively, compared to that of the JB20 mixture.

3.3 Summarization of engine performance and emission characteristics

Figs. 5 and 6 summarize the findings of the influence on the engine performance and exhaust emissions by adding GNP to the JB20 mixture compared to those of pure JB20. The findings illustrate that at an engine speed of 1500 rpm for all loads, the peak reduction in bsfc, NO_x, CO, and UHC emissions are accomplished at the GNP additive concentration of 25–50 mg/L, where the bsfc is diminished by 20%, NO_x by 23%, CO by 55%, and UHC by 50%. At the engine speed of 2000 rpm, a larger engine performance is attained at the GNP additive concentration of 25–50 mg/L, where the bsfc declines by 18%, NO_x by 55%, CO by 65%, and UHC by 50%. Moreover, the peak decrease in bsfc by 20%, NO_x by 40%, CO by 55%, and UHC by 65% is noticed for an engine speed of 2500 rpm and a 25–50 mg/L GNP dose level. Moreover, based on the earlier findings, it can be confirmed that the 50–75 mg/L dose level of GNPs gives the peak values for p_{max}, up/$d\theta_{max}$, and $dQ_g/d\theta_{max}$. Generally, the highest engine performance and emission are achieved at the GNP dose level of 50 mg/L.

3.4 Carbon nanomaterials in Jatropha biodiesel-n-butanol blend

In this section, the influence of adding *n*-butanol to biodiesel oil (60% by volume Jatropha biodiesel and 40% by volume *n*-butanol mixture, denoted by JME40B) on the engine performance is investigated. Additionally, this study purports also to explore the influences of GO, GNPs, and MWCNTs (50 mg/L) as additives on the performance of a CI engine run with a JME40B mixture.

Table 9 shows the characteristics of JME40B with and without carbon nanomaterials, which are measured regarding ASTM standards. Adding carbon nanomaterials into the

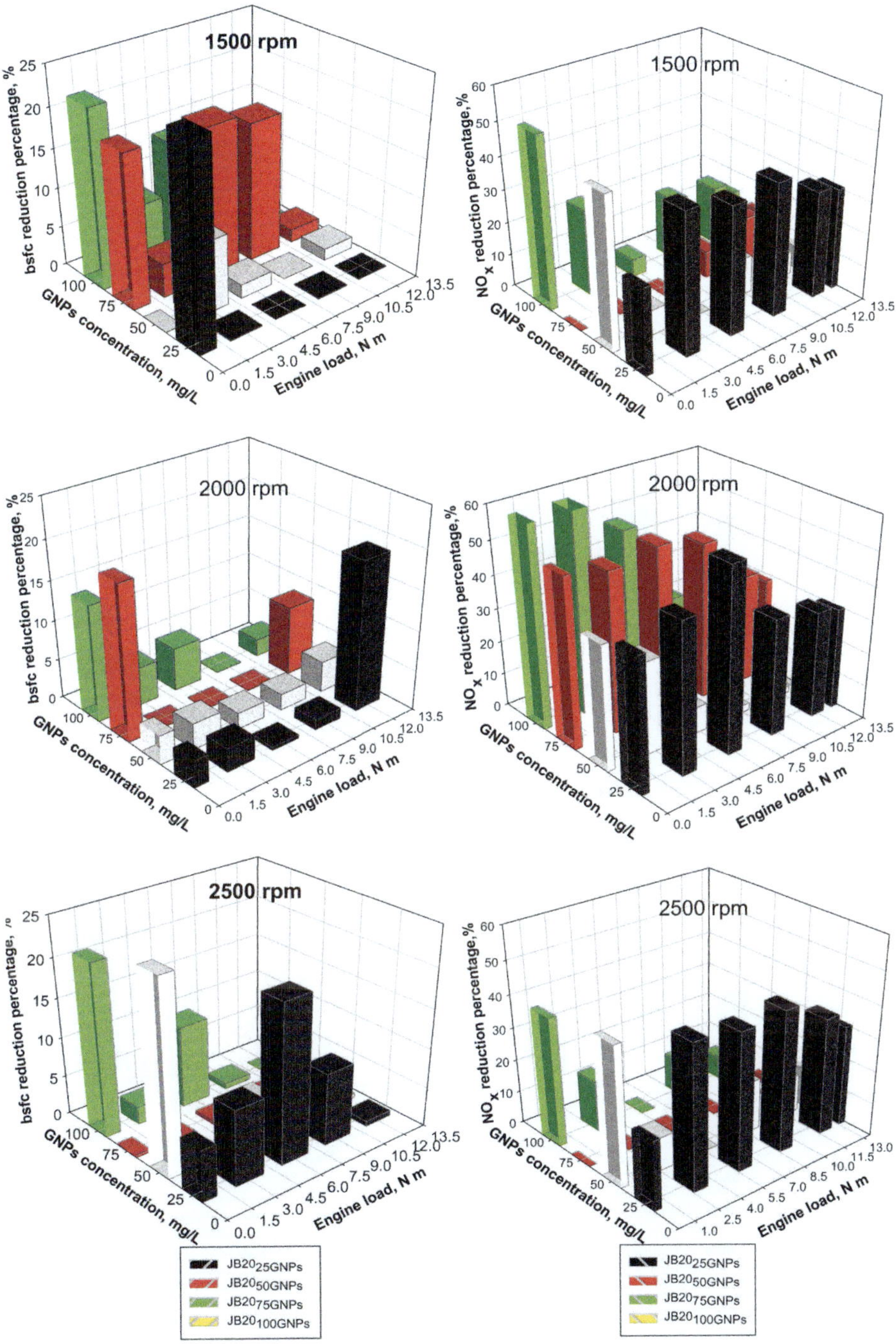

Fig. 5 The reduction percentage of bsfc and NO_x emissions for different engine loads and speeds versus GNP concentrations.

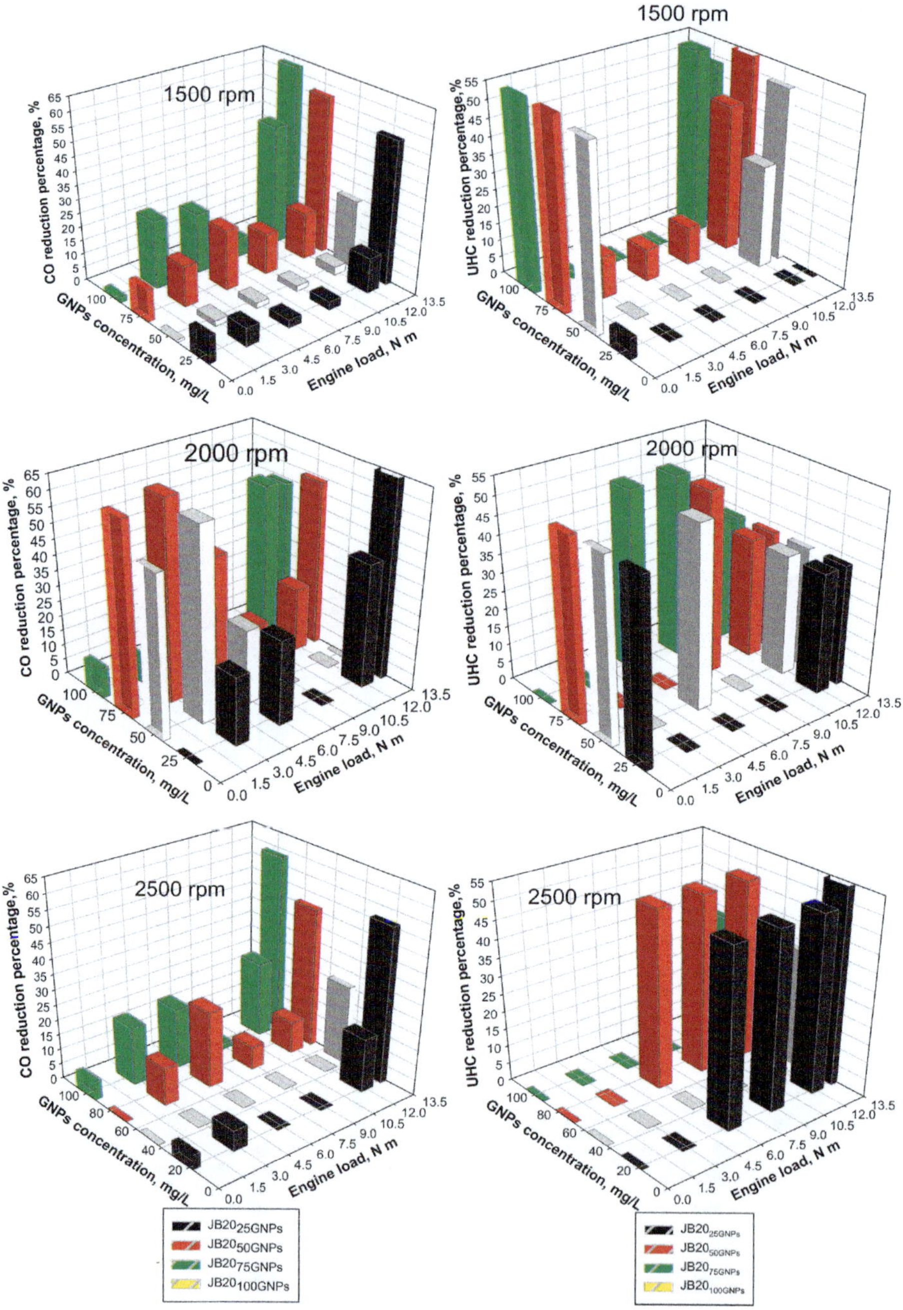

Fig. 6 The reduction percentage of the CO and UHC emissions for different engine loads and speeds versus GNP concentrations.

Table 9 Properties of D100, JME, JME40B without and with GO, GNP, and MWCNT samples.

Property	Testing method	D100	JME	*n*-Butanol	JME40B	JME40B 50GO	JME40B 50GNPs	JME40B 50MWCNTs
Calorific value, kJ/kg	ASTM D-240	45,448	40,630	33100[a]	37,532	37,555	37,562	37,565
Kinematic viscosity at 40°C, mm^2/s	ASTM D-445	3.34	5.26	2.22[a]	3.73	3.65	3.68	3.69
Specific gravity at 15.56°C	ASTM D-1298	0.8421	0.883	0.810[a]	0.8499	0.8510	0.8511	0.8511
Cetane index	ASTM D-976	49.5	52	~25[a]	43.53	48.10	47.95	47.98

[a] Data have been provided by the supplier.

JME40B mixture led to a reduction in the kinematic viscosity by 2.5%. Similar remarks are stated in the literature (Heydari-Maleney et al., 2017) on the addition of carbon nanomaterials in the alcohol-diesel-biodiesel mixture. Moreover, the cetane number is heightened by 10% by adding carbon nanomaterials into the JME40B mixtures. An improvement in the cetane number is credited to the higher thermal conductivity of the carbon nanomaterials, which enhances the evaporation rate of fuel droplets, leading to a reduced ignition delay and thus increasing the cetane number. The O_2 percentage in Jatropha biodiesel and *n*-butanol are 6.5% and 22%, respectively (Zheng et al., 2016).

Fig. 7A–C illustrates the peak pressure, the location of maximum pressure, and the percentage of pressure increase, respectively, for the various tested fuels and engine loads. Generally, the peak cylinder pressure improves with growing engine load, as given in Fig. 7A. This is ascribed to the rise in the amount of injected fuel as the engine load is enlarged, resulting in an enhancement in the cylinder pressure. The findings also show that the locations where the peak cylinder pressure is accomplished for the tested fuels have the same inclinations with the engine load, as given in Fig. 7B. Moreover, for all the tested fuels, the location of p_{max} is enlarged with the rise in engine speed. This could be ascribed to the rise in the amount of injected fuel with increasing engine load, which results in an increase of the time needed to burn the fuel blends, and thus the retarded location of peak pressure. For the JME40B mixture, the peak pressure is higher compared to pure diesel fuel and its location is retarded.

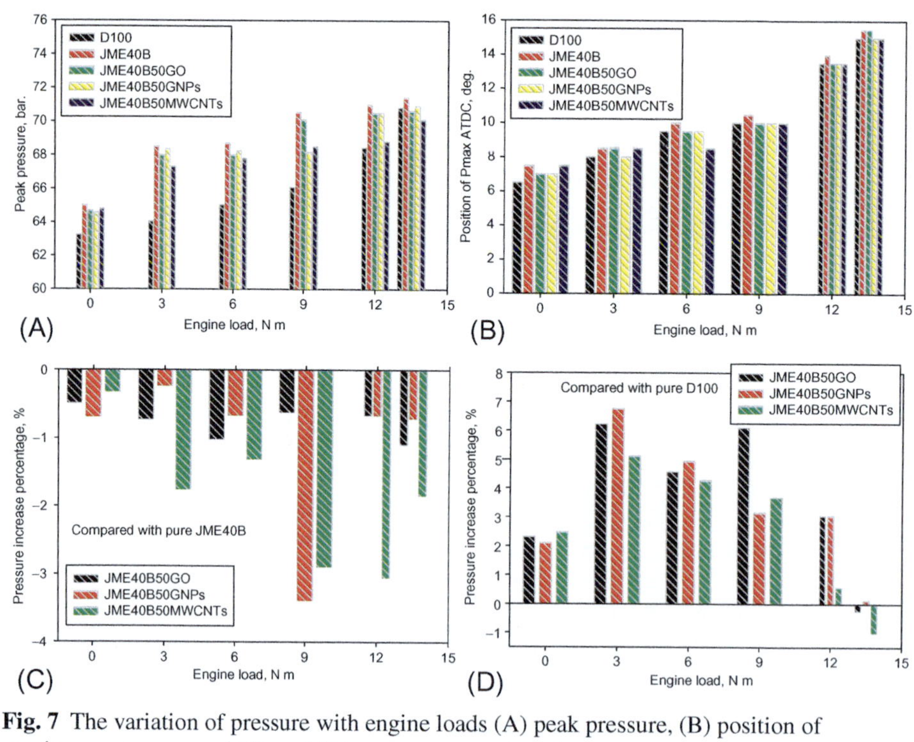

Fig. 7 The variation of pressure with engine loads (A) peak pressure, (B) position of maximum pressure, and (C) and (D) pressure percentage increase.

This could be attributed to the lower cetane number of the JME40B mixture, resulting in prolonged ignition delay. From Fig. 7A, B, adding carbon nanomaterials into JME40B mixtures drops the peak pressure and advances its location. These findings are ascribed to the enhanced evaporation rate and high atomization rates resulting in a reduction of the ignition delay.

From Fig. 7A, at the engine load of 9 N m, the highest cylinder pressure for pure D100 fuel is 66.07 bar while it is 70.54, 70.1, 68.15, and 68.5 bar for JME40B, JME40B50GO, JME40B50GNPs, and JME40B50MWCNTs, respectively. From Fig. 7C, it is realized that adding carbon nanomaterials to JME40B leads to a slight decrease in the peak pressure, which is diminished by 1%, 3%, and 3.2% for the JME40B50GO, JME40B50GNP, and JME40B50MWCNT mixtures, respectively, compared to that of the pure JME40B mixture. Moreover, from Fig. 7D, there is a significant enhancement in the peak pressure by 6%, 7%, and 5% for the JME40B50GO, JME40B50GNP, and JME40B50MWCNT mixtures, respectively, compared to that of pure D100 fuel.

Fig. 8 summarizes the findings of the influence of adding GO, GNPs, and MWCNTs to JME40B blends compared to those of a pure JME40B mixture on the engine performance and emissions. The findings illustrate that at an engine speed of 2000 rpm for all loads, the bsfc, NO_x, CO, and UHC emissions are diminished by 35%, NO_x by 45%, CO by 55%, and UHC by 50%. Moreover, compared to those of pure diesel, the bsfc is diminished by 25% whereas the NO_x, CO, and UHC emissions are considerably declined by 65%, 70%, and 50%, respectively.

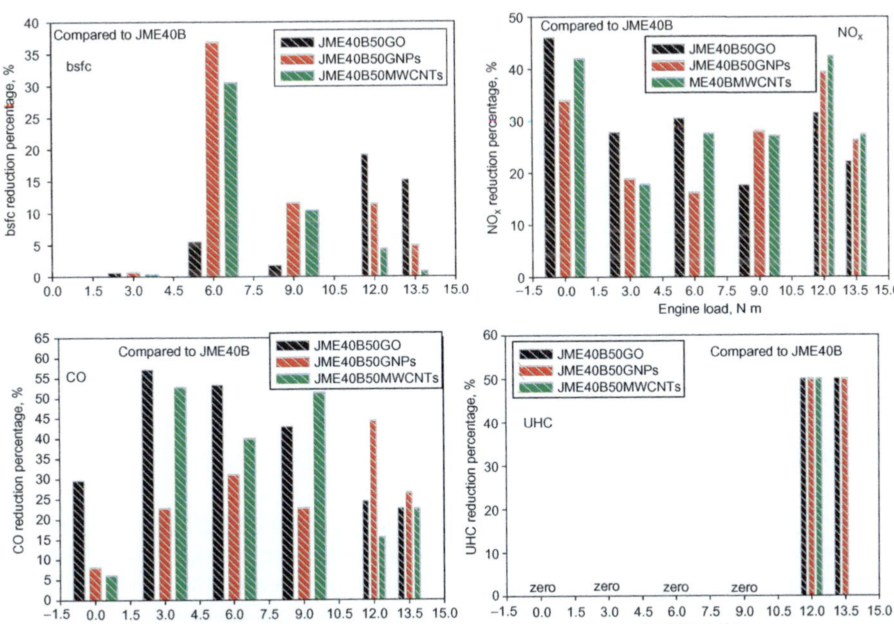

Fig. 8 Compared to JME40B, the reduction percentage of the bsfc, NO_x, CO, and UHC emissions for different engine loads.

4 Conclusion and future perspectives

The compression ignition engine test rig is utilized to scrutinize the influence of adding carbon nanomaterials to the Jatropha biodiesel-diesel blend and the Jatropha biodiesel-*n*-butanol blend. The MWCNT, GNP, and GO additives were mixed with these fuels to examine their effects on diesel engine performance, combustion, and emission characteristics at various engine loads and speeds. From these studies, it can be determined that:

- The p_{max}. and $dp/d\theta$ are boosted by adding GNPs to the JB20 mixture and their positions are advanced. Their peak values are accomplished at a GNP dose level of 50–75 mg/L.
- The ignition delay is decreased, which increases the gross heat release rate and its location is advanced.
- The BTE is amended by 25% and the bsfc is decreased by 20% by adding GNPs at the 25–50 mg/L dose level.
- The CO, NO_x, and UHC emissions are diminished substantially by 65%, 55%, and 65%, respectively, with adding the 25–50 mg/L concentration of GNPs.
- The recommended GNP dose level to accomplish the greatest substantial improvement in engine global performance is 50 mg/L (p_{max} is enhanced by 6%, bsfc is diminished by 20%, NO_x emission is declined by 40%, CO by 60%, and UHC by 50%).
- There is an increase in both the peak cylinder pressure and pressure rise rate by adding *n*-butanol to JME. Their positions are retarded due to prolonging the ignition delay.
- Adding carbon nanomaterials to the JME40B mixture results in a shortened ignition delay, which leads to a reduction in cylinder pressure and gross heat release rate as well as advancing their locations.
- The BTE is boosted by 25% and the bsfc is diminished by 35% by adding carbon nanomaterials to JME40B, compared to that of a pure JME40B mixture.
- The CO, NO_x, and UHC emissions are decreased pointedly by 55%, 45%, and 50%, respectively, by adding carbon nanomaterials to JME40B, compared to that of pure JME40B mixture.
- The CO, NO_x, and UHC emissions are diminished by 65%, 70%, and 50%, respectively, by adding carbon nanomaterials to JME40B, in comparison to pure D100 fuel.

Before applying these additives in a commercial way, the following investigations are needed to emphasize their impacts on long-term engine operation:

- Investigation of the long-term effect of carbon nanomaterial additives on engine components.
- Study of the instability issue for carbon nanomaterials in fuel mixtures for long operation.
- Study of the effect of using EGR with carbon nanomaterials in fuel mixtures to reduce NO_x emission.

Acknowledgments

The author would like to acknowledge the Mission Department of the Ministry of Higher Education (MoHE) of Egypt for providing a scholarship to conduct this study. Also, he would like to deeply acknowledge the Technology Management Department (TMD) at Egypt Japan University of Science and Technology (E-JUST) as well as the Japan International Cooperation Agency (JICA) for offering some of the facilities, tools, and equipment used to conduct this research work. The author would also like to thank Eng. Khalid El-Feky for his immense technical support in this study.

References

Abu-Arabi, M.K., Allawzi, M.A., Al-Zoubi, H.S., Tamimi, A., 2000. Extraction of jojoba oil by pressing and leaching. Chem. Eng. J. 76 (1), 61–65. https://doi.org/10.1016/S1385-8947(99)00119-9.

Achten, W.M.J., Verchot, L., Franken, Y.J., Mathijs, E., Singh, V.P., Aerts, R., Muys, B., 2008. Jatropha bio-diesel production and use. Biomass Bioenergy 32 (12), 1063–1084. https://doi.org/10.1016/j.biombioe.2008.03.003.

Ali, O.M., Mamat, R., Faizal, C.K.M., 2013. Review of the effects of additives on biodiesel properties, performance, and emission features. J. Renew. Sustain. Energy 5 (1). https://doi.org/10.1063/1.4792846.

Attia, A.M.A., Hassaneen, A.E., 2016. Influence of diesel fuel blended with biodiesel produced from waste cooking oil on diesel engine performance. Fuel 167, 316–328. https://doi.org/10.1016/j.fuel.2015.11.064.

Attia, A.M.A., EL-Seesy, A.I., El-Batsh, H.M., Shehata, M.S., 2014. Effects of alumina nanoparticles additives into jojoba methyl ester-diesel mixture on diesel engine performance. In: Proc. ASME 2014 Int. Mech. Eng. Congr. Expo. Montr. Quebec, Canada, pp. 1–10. https://doi.org/10.1115/IMECE2014-39988.

Babu, V.M., Murthy, K.M., Rao, A.P.G., 2017. Butanol and pentanol: the promising biofuels for ci engines—a review. Renew. Sust. Energ. Rev. 78, 1068–1088. https://doi.org/10.1016/j.rser.2017.05.038.

Banapurmath, N.R., Sankaran, R., Tumbal, A.V., Narasimhalu, T.N., Hunashyal, A.M., Ayachit, N.H., 2014. Experimental investigation on direct injection diesel engine fuelled with graphene, silver and multiwalled carbon nanotubes-biodiesel blended fuels. Int. J. Autom. Eng. Technol. 3 (4), 129–138.

Basha, J.S., Anand, R.B., 2011. An experimental investigation in a diesel engine using carbon nanotubes blended water-diesel emulsion fuel. Proc. Inst. Mech. Eng. Part A—J. Power Energy 225, 279–288. https://doi.org/10.1177/2041296710394247.

Basha, S.J., Anand, R.B., 2014. Performance, emission and combustion characteristics of a diesel engine using carbon nanotubes blended Jatropha methyl ester emulsions. Alex. Eng. J. 53 (2), 259–273. https://doi.org/10.1016/j.aej.2014.04.001.

Basha, S.A., Gopal, K.R., 2012. A review of the effects of catalyst and additive on biodiesel production, performance, combustion and emission characteristics. Renew. Sust. Energ. Rev. 16 (1), 711–717. https://doi.org/10.1016/j.rser.2011.08.036.

Chen, Z., Wu, Z., Liu, J., Lee, C., 2014. Combustion and emissions characteristics of high *n*-butanol/diesel ratio blend in a heavy-duty diesel engine and EGR impact. Energy Convers. Manag. 78, 787–795. https://doi.org/10.1016/j.enconman.2013.11.037.

EL-Seesy, A.I., Hassan, H., 2018. Investigation of the effect of adding graphene oxide, graphene nanoplatelet, and multiwalled carbon nanotube additives with *n*-butanol-Jatropha methyl ester on a diesel engine performance. Renew. Energy 132, 558–574. https://doi.org/10.1016/j.renene.2018.08.026.

EL-Seesy, A.I., Abdel-Rahman, A.K., Bady, M., Ookawara, S., 2016. The influence of multi-walled carbon nanotubes additives into non-edible biodiesel-diesel fuel blend on diesel engine performance and emissions. Energy Procedia 100, 166–172. https://doi.org/10.1016/j.egypro.2016.10.160.

EL-Seesy, A.I., Abdel-rahman, A.K., Bady, M., Ookawara, S., 2017. Performance, combustion, and emission characteristics of a diesel engine fueled by biodiesel-diesel mixtures with multi-walled carbon nanotubes additives. Energy Convers. Manag. 135, 373–393. https://doi.org/10.1016/j.enconman.2016.12.090.

EL-Seesy, A.I., Attia, A.M.A., El-Batsh, H.M., 2018a. The effect of aluminium oxide nanoparticles addition with jojoba methyl ester-diesel fuel blend on a diesel engine performance, combustion and emission characteristics. Fuel 224, 147–166. https://doi.org/10.1016/j.fuel.2018.03.076.

EL-Seesy, A.I., Hassan, H., Ookawara, S., 2018b. Effects of graphene nanoplatelet addition to Jatropha biodiesel–diesel mixture on the performance and emission characteristics of a diesel engine. Energy 147, 1129–1152. https://doi.org/10.1016/j.energy.2018.01.108.

EL-Seesy, A.I., Hassan, H., Ookawara, S., 2018c. Performance, combustion, and emission characteristics of a diesel engine fueled with Jatropha methyl ester and graphene oxide additives. Energy Convers. Manag. 166, 674–686. https://doi.org/10.1016/j.enconman.2018.04.049.

EL-Seesy, A.I., Hassan, H., Ookawara, S., Attia, A.M.A., Abd-Elbar, A.R., 2018d. The effect of nanoparticles addition with biodiesel–diesel fuel blend on a diesel engine performance. In: The 5th IEEE Int. Conf. Renew. Energy Gener. Appl, pp. 98–101.

EL-Seesy, A.I., Hassan, H., Dawood, A., Attia, A.M.A., Kosaka, H., Ookawara, S., 2018e. Investigation of the impact of adding titanium dioxide to jojoba biodiesel-diesel-*n*-hexane mixture on the performance and emission characteristics of a diesel engine. In: ASME 2018 Intern. Combust. Engine Div. Fall Tech. Conf. San Diego, California, USA. vol. 1, pp. 1–9. https://doi.org/10.1115/ICEF2018-9647.

Esfahani, M.R., Languri, E.M., Nunna, M.R., 2016. Effect of particle size and viscosity on thermal conductivity enhancement of graphene oxide nanofluid. Int. Commun. Heat Mass Transf. 76, 308–315. https://doi.org/10.1016/j.icheatmasstransfer.2016.06.006.

Geim, A.K., Novoselov, K.S., 2007. The rise of graphene. Nat. Mater. 6, 183–191. https://doi.org/10.1038/nmat1849.

Ghafoori, M., Ghobadian, B., Najafi, G., Layeghi, M., Rashidi, A., Mamat, R., 2015. Effect of nano-particles on the performance and emission of a diesel engine using a biodiesel-diesel blend. Int. J. Automot. Mech. Eng. 12, 3097–3108. https://doi.org/10.15282/ijame.12.2015.23.0258.

Ghanbari, M., Najafi, G., Ghobadian, B., Yusaf, T., Carlucci, A.P., Kiani, M.K.D., 2017. Performance and emission characteristics of a CI engine using nanoparticles additives in biodiesel-diesel blends and modelling with GP approach. Fuel 202, 699–716. https://doi.org/10.1016/j.fuel.2017.04.117.

Gonzáles, N.F.C., 2016. International experiences with the cultivation of *Jatropha curcas* for biodiesel production. Energy 112, 1245–1258. https://doi.org/10.1016/j.energy.2016.06.073.

Goswami, L., Kim, K., Deep, A., Das, P., Bhattacharya, S.S., Kumar, S., Adelodun, A.A., 2017. Engineered nano particles: nature, behavior, and effect on the environment. J. Environ. Manag. 196, 297–315. https://doi.org/10.1016/j.jenvman.2017.01.011.

Heydari-Maleney, K., Taghizadeh-alisaraei, A., Ghobadian, B., Abbaszadeh-mayvan, A., 2017. Analyzing and evaluation of carbon nanotubes additives to diesohol-B2 fuels on performance and emission of diesel engines. Fuel 196, 110–123. https://doi.org/10.1016/j.fuel.2017.01.091.

Heywood, J.B., 1988. Internal combustion engine fundamentals. McGrawHill Series in Mechanical Engineering, USA.

Hohenberg, G.F., 1979. Advance Approaches for Heat Transfer Calculation. Soc. Autom. Eng. SAE Tech. Pap. No. 790825, pp. 2788–2798.

Holman, J.P., 2001. Experimental methods for engineers, eighth ed. McGraw-Hill Companies, Inc., New York, NY.

Horie, M., Fujita, K., 2011. Toxicity of metal oxides nanoparticles. Adv. Mol. Toxicol. first ed. https://doi.org/10.1016/B978-0-444-53864-2.00004-9.

Hoseini, S.S., Najafi, G., Ghobadian, B., Mamat, R., Sidik, N.A.C., Azmi, W.H., 2017. The effect of combustion management on diesel engine emissions fueled with biodiesel-diesel blends. Renew. Sust. Energ. Rev. 73, 307–331. https://doi.org/10.1016/j.rser.2017.01.088.

Hosseini, S.H., Taghizadeh-Alisaraei, A., Ghobadian, B., Abbaszadeh-Mayvan, A., 2017. Performance and emission characteristics of a ci engine fuelled with carbon nanotubes and diesel-biodiesel blends. Renew. Energy 111, 201–213. https://doi.org/10.1016/j.renene.2017.04.013.

Işık, M.Z., Bayındır, H., İscan, B., Aydın, H., 2017. The effect of *n*-butanol additive on low load combustion, performance and emissions of biodiesel-diesel blend in a heavy duty diesel power generator. J. Energy Inst. 90 (2), 174–184. https://doi.org/10.1016/j.joei.2016.02.006.

Jeng, H.A., Swanson, J., 2006. Toxicity of metal oxide nanoparticles in mammalian cells. J. Environ. Sci. Health A Tox. Hazard. Subst. Environ. Eng. 41 (12), 2699–2711. https://doi.org/10.1080/10934520600966177.

Jiaqiang, E., Pham, M., Zhao, D., Deng, Y., Le, D., Zuo, W., Zhu, H., Liu, T., Peng, Q., Zhang, Z., 2017. Effect of different technologies on combustion and emissions of the diesel engine fueled with biodiesel: a review. Renew. Sust. Energ. Rev. 80, 620–647. https://doi.org/10.1016/j.rser.2017.05.250.

Karthikeyan, S., Prathima, A., 2016. Effects of novel microemulsion fuel additive Ce–Ru–O catalysts on diesel engine emissions. Energy Sources, Part A Recover. Util. Environ. Eff. 38 (22), 3397–3403. https://doi.org/10.1080/15567036.2016.1154900.

Khalife, E., Tabatabaei, M., Demirbas, A., Aghbashlo, M., 2017. Impacts of additives on performance and emission characteristics of diesel engines during steady-state operation. Prog. Energy Combust. Sci. 59, 32–78. https://doi.org/10.1016/j.pecs.2016.10.001.

Khond, V.W., Kriplani, V.M., 2016. Effect of nanofluid additives on performances and emissions of emulsified diesel and biodiesel-fueled stationary CI engine: a comprehensive review. Renew. Sust. Energ. Rev. 59, 1338–1348. https://doi.org/10.1016/j.rser.2016.01.051.

Kim, K.S., Lee, H.J., Lee, C., Lee, S.K., Jang, H., Ahn, J.H., Kim, J.H., Lee, H.J., 2011. Chemical vapor deposition-grown graphene: the thinnest solid lubricant. ACS Nano 5 (6), 5107–5114. https://doi.org/10.1021/nn2011865.

Lee, H., Lee, N., Seo, Y., Eom, J., Lee, S.W., 2009. Comparison of frictional forces on graphene and graphite. Nanotechnology 20 (32), 320–332. https://doi.org/10.1088/0957-4484/20/32/325701.

Lee, C., Li, Q., Kalb, W., Liu, X., Berger, H., Carpick, R.W., Hone, J., 2010. Frictional characteristics of atomically thin sheets. Science 80, 76–80. https://doi.org/10.1126/science.1184167.

Lee, S., Oh, S., Choi, Y., Kang, K., 2011. Performance and emission characteristics of a CI engine operated with *n*-butane blended DME fuel. Appl. Therm. Eng. 31, 1929–1935. https://doi.org/10.1016/j.applthermaleng.2011.02.039.

Li, Y., Zhou, J., Tung, S., Schneider, E., Xi, S., 2009. A review on development of nanofluid preparation and characterization. Powder Technol. 196, 89–101. https://doi.org/10.1016/j.powtec.2009.07.025.

Ma, Y., Zhu, M., Zhang, D., 2013. The effect of a homogeneous combustion catalyst on exhaust emissions from a single cylinder diesel engine. Appl. Energy 102, 556–562. https://doi.org/10.1016/j.apenergy.2012.08.028.

Masood, M.T., Papadopoulou, E.L., Heredia-Guerrero, J.A., Bayer, I.S., Athanassiou, A., Ceseracciu, L., 2017. Graphene and polytetrafluoroethylene synergistically improve the tribological properties and adhesion of nylon 66 coatings. Carbon 123, 26–33. https://doi.org/10.1016/j.carbon.2017.07.026.

Merwe, D.V., Pickrell, J.A., 2012. Toxicity of nanomaterials. Vet. Toxicol. 80, 383–390. https://doi.org/10.1016/B978-0-12-385926-6.00024-7.

Mirzajanzadeh, M., Tabatabaei, M., Ardjmand, M., Rashidi, A., Ghobadian, B., Barkhi, M., Pazouki, M., 2015. A novel soluble nano-catalysts in diesel–biodiesel fuel blends to improve diesel engines performance and reduce exhaust emissions. Fuel 139, 374–382. https://doi.org/10.1016/j.fuel.2014.09.008.

Naddaf, A., Heris, S.Z., 2018. Experimental study on thermal conductivity and electrical conductivity of diesel oil-based nanofluids of graphene nanoplatelets and carbon nanotubes. Int. Commun. Heat Mass Transf. 95, 116–122. https://doi.org/10.1016/j.icheatmasstransfer.2018.05.004.

Nanthagopal, K., Ashok, B., Tamilarasu, A., Johny, A., Mohan, A., 2017. Influence on the effect of zinc oxide and titanium dioxide nanoparticles as an additive with calophyllum inophyllum methyl ester in a ci engine. Energy Convers. Manag. 146, 8–19. https://doi.org/10.1016/j.enconman.2017.05.021.

Nour, M., EL-Seesy, A.I., Abdel-Rahman, A.K., Bady, M., 2018. Influence of adding aluminum oxide nanoparticles to diesterol blends on the combustion and exhaust emission characteristics of a diesel engine. Exp. Thermal Fluid Sci. 98, 634–644. https://doi.org/10.1016/j.expthermflusci.2018.07.009.

Ooi, J.B., Ismail, H.M., Swamy, V., Wang, X., Swain, A.K., Rajanren, J.R., 2016. Graphite oxide nanoparticle as a diesel fuel additive for cleaner emissions and lower fuel consumption. Enegy Fuel 30 (2), 1341–1353. https://doi.org/10.1021/acs.energyfuels.5b02162.

Ooi, J.B., Rajanren, J.R., Ismail, H.M., Swamy, V., Wang, X., 2017. Improving combustion characteristics of diesel and biodiesel droplets by graphite oxide addition for diesel engine applications. Int. J. Energy Res. 70, 1–10. https://doi.org/10.1002/er.3787.

Ooi, J.B., Ismail, H.M., Tan, B.T., Wang, X., 2018. Effects of graphite oxide and single-walled carbon nanotubes as diesel additives on the performance, combustion, and emission characteristics of a light-duty diesel engine. Energy 161, 70–80. https://doi.org/10.1016/j.energy.2018.07.062.

Pope III, C.A., Burnett, R.T., Thun, M.J., Calle, E.E., Krewski, D., Thurston, G.D., 2002. Lung cancer, cardiopulmonary mortality, and long-term exposure to fine particulate air pollution. J. Am. Med. 287 (1), 132–141. https://doi.org/10.1001/jama.287.9.1132.

Saxena, V., Kumar, N., Saxena, V.K., 2017. A comprehensive review on combustion and stability aspects of metal nanoparticles and its additive effect on diesel and biodiesel fuelled C.I. engine. Renew. Sust. Energ. Rev. 70, 563–588. https://doi.org/10.1016/j.rser.2016.11.067.

Selvan, V.A.M., Anand, R.B., Udayakumar, M., 2014. Effect of cerium oxide nanoparticles and carbon nanotubes as fuel-borne additives in diesterol blends on the performance, combustion and emission characteristics of a variable compression ratio engine. Fuel 130, 160–167. https://doi.org/10.1016/j.fuel.2014.04.034.

Senthilkumar, R., Prabu, M., 2015. Performance and emission characteristics of single cylinder CI diesel engine using multiwalled nanotubes and karanja oil methyl ester. J. Chem. Pharm. Sci. 6, 109–114.

Shehata, M.S., 2010. Cylinder pressure, performance parameters, heat release, specific heats ratio and duration of combustion for spark ignition engine. Energy 35 (12), 4710–4725. https://doi.org/10.1016/j.energy.2010.09.027.

Shehata, M.S., Abdel Razek, S.M., 2011. Experimental investigation of diesel engine performance and emission characteristics using jojoba/diesel blend and sunflower oil. Fuel 90 (2), 886–897. https://doi.org/10.1016/j.fuel.2010.09.011.

Singh, N., Bharj, R., 2015a. Experimental investigation on the role of indigenous carbon nanotube emulsified fuel in a four-stroke diesel engine. Proc. Inst. Mech. Eng. Part C J. Mech. Eng. Sci. 77, 1–14. https://doi.org/10.1177/0954406215586021.

Singh, N., Bharj, R.S., 2015b. Effect of CNT-emulsified fuel on performance emission and combustion characteristics of four stroke diesel engine. Int. J. Curr. Eng. Technol. 55 (11), 2277–4106.

Soliman, W.M., He, X., 2015. The potentials of Jatropha plantations in Egypt: a review. Mod. Econ. 6, 190–200. https://doi.org/10.4236/me.2015.62016.

Song, Y., Yu, J., Yu, L., Alam, F.E., Dai, W., Li, C., Jiang, N., 2015. Enhancing the thermal, electrical, and mechanical properties of silicone rubber by addition of graphene nanoplatelets. Mater. Design 88, 950–957. https://doi.org/10.1016/j.matdes.2015.09.064.

Tamilselvan, P., Nallusamy, N., Rajkumar, S., 2017. A comprehensive review on performance, combustion and emission characteristics of biodiesel fuelled diesel engines. Renew. Sust. Energ. Rev. 79, 1134–1159. https://doi.org/10.1016/j.rser.2017.05.176.

Tewari, P., Doijode, E., Banapurmath, N.R., Yaliwal, V.S., 2013. Experimental investigations on a diesel engine fuelled with multiwalled carbon nanotubes blended biodiesel fuels. Int. J. Emerg. Technol. Adv. Eng. 3 (3), 72–76.

Tiwari, A.K., Kumar, A., Raheman, H., 2007. Biodiesel production from Jatropha oil (*Jatropha curcas*) with high free fatty acids: an optimized process. Biomass Bioenergy 31, 569–575. https://doi.org/10.1016/j.biombioe.2007.03.003.

Trindade, W.R.d.S., Santos, R.G.d., 2017. Review on the characteristics of butanol, its production and use as fuel in internal combustion engines. Renew. Sust. Energ. Rev. 69, 642–651. https://doi.org/10.1016/j.rser.2016.11.213.

Wei, D., Dave, R., Pfeffer, R., 2002. Mixing and characterization of nanosized powders: an assessment of different techniques. J. Nanopart. Res. 4, 21–41. https://doi.org/10.1023/A:1020184524538.

Yilmaz, N., Davis, S.M., 2016. Polycyclic aromatic hydrocarbon (PAH) formation in a diesel engine fueled with diesel, biodiesel and biodiesel/*n*-butanol blends. Fuel 181, 729–740. https://doi.org/10.1016/j.fuel.2016.05.059.

Zhai, W., Srikanth, N., Kong, L.B., Zhou, K., 2017. Carbon nanomaterials in tribology. Carbon 119, 150–171. https://doi.org/10.1016/j.carbon.2017.04.027.

Zhang, Z.H., Balasubramanian, R., 2016. Investigation of particulate emission characteristics of a diesel engine fueled with higher alcohols/biodiesel blends. Appl. Energy 163, 71–80. https://doi.org/10.1016/j.apenergy.2015.10.173.

Zheng, Z., Wang, X.F., Zhong, X., Hu, B., Liu, H., Yao, M., 2016. Experimental study on the combustion and emissions fueling biodiesel/*n*-butanol, biodiesel/ethanol and biodiesel/2,5-dimethylfuran on a diesel engine. Energy 115, 539–549. https://doi.org/10.1016/j.energy.2016.09.054.

Zhu, Y., Murali, S., Cai, W., Li, X., Suk, J.W., Potts, J.R., Ruoff, R.S., 2010. Graphene and graphene oxide: synthesis, properties, and applications. Adv. Mater. 22 (35), 3906–3924. https://doi.org/10.1002/adma.201001068.

Further reading

Gnanasikamani, B., Cheralathan, M., 2015. Effect of CNT as additive with biodiesel on the performance and emission characteristics of a di diesel engine. Int. J. ChemTech Res. 7 (3), 1230–1236. https://doi.org/10.1016/j.fuel.2013.03.042.

Prabu, A., Anand, R.B., 2015. Performance, combustion and emission characteristics of a D.I. diesel engine fuelled with nanoparticle blended Jatropha biodiesel. Period. Polytech. Mech. Eng. 59 (2), 88–93. https://doi.org/10.3311/PPme.7766.

Saraee, S.H., Taghavifar, H., Jafarmadar, S., 2017. Experimental and numerical consideration of the effect of CeO_2 nanoparticles on diesel engine performance and exhaust emission with the aid of artificial neural network. Appl. Therm. Eng. 113, 663–672. https://doi.org/10.1016/j.applthermaleng.2016.11.044.

Biosorbents for heavy metal removal from dilute aqueous solution

6

George Z. Kyzas[a], *Kostas A. Matis*[b]
[a]Department of Chemistry, International Hellenic University, Kavala, Greece, [b]Division of Chemical Technology and Industrial Chemistry, Department of Chemistry, Aristotle University of Thessaloniki, Thessaloniki, Greece

1 Introduction: The metal ions

Many industrial wastewater streams with large flows often contain toxic metals that have to be removed prior to water reuse (recycling), indirect discharge into the sewage system, or direct discharge into surface waters (Peleka and Matis, 2011). For instance, cadmium is a heavy metal with a great hazard potential to human beings (and other organisms in the environment). It may end up in our food chain through agricultural soil as noted with the case of rice, a staple food crop for much of China (Liu et al., 2009); soil samples were collected from a paddy field, tested, and revealed to contain 0.12 mg/kg of cadmium. Arsenic in drinking water also poses a serious threat to the health of people in many countries. A common example of such a country is Bangladesh (Leupin et al., 2005), where millions of people were consuming water from wells with arsenic concentrations exceeding guidelines. On the other hand, new economic activities are in fact surprisingly dependent on traditional raw materials (Humphreys, 2000); that is, a personal computer typically contains around 30 mineral ingredients and electrical power, required for our appliances, is transmitted by copper wires and cables. So, metal recycling is advisable and contributes to sustainable development. A good example for metal recycling today is waste electrical and electronic equipment (Zhang and Xu, 2016). Elsewhere, cadmium was finally recovered from the eluate by electrolysis using a rotating cathode cell (Butter et al., 1998); a characteristic conceptual flow diagram was presented (see Fig. 1A). Cd^{2+}, for instance, is released into the environment by electroplating, plastic manufacturing, smelting, mining, paint pigments, cadmium nickel batteries, stabilizers, and alloys.

Sorption (a general term describing the attachment or bonding of charged species from a solution to a coexisting solid surface) is a well-known and promising separation technique for metal ion removal (Kyzas et al., 2016). Research has been currently focused on the development of highly selective sorbents usually possessing a large surface area with different surface functional groups, and also fast reaction kinetics suitable for the removal of heavy metal ions. The latter may lead to a better use of the capacity of these materials, resulting in smaller units and low residual concentration of toxic metals in the treated water streams, to comply with individual standards for

Carbon Nanomaterials for Agri-food and Environmental Applications. https://doi.org/10.1016/B978-0-12-819786-8.00006-2

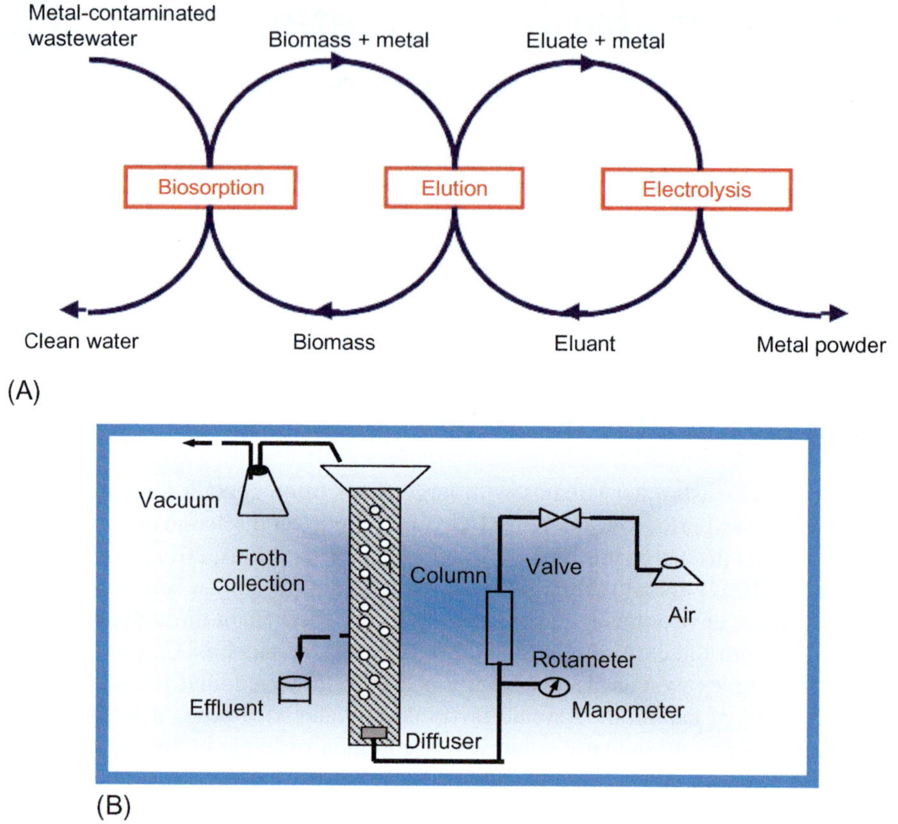

Fig. 1 (A) Diagram showing recycle routes of a total process; the second cycle represents the applied flotation separation. (B) A typical dispersed-air flotation column arrangement in bench scale.
(A) Reprinted with permission, copyright Elsevier Butter, T.J., Evison, L.M., Hancock, I.C., Holland, F.S., Matis, K.A., Philipson, A., Sheikh, A.I., Zouboulis, A.I., 1998. The removal and recovery of cadmium from dilute aqueous solutions by biosorption and electrolysis at laboratory scale. Water Res. 32, 400–406.

water reuse or discharge; the various activated carbons is here probably the best example and perhaps, recently graphene oxide (Kyzas et al., 2013). For instance, a magnetic cyclodextrin chitosan/graphene oxide material was tested (Fan et al., 2013), among others (see the chapter on carbon nanotubes). In order to improve the performance of grapheme oxide-based adsorbents in the acidic condition of actual water pollution, usually the surfaces were modified by some functional materials.

As an alternative to conventional wastewater treatment facilities, a biotechnological method of environmental control, biosorption, has been recently proposed and will be analyzed in the following chapters. Microorganisms have a high surface area-to-volume ratio because of their small size. Therefore, they can provide a large contact interface that would interact with metals from the surrounding environment.

Free mycelium was not considered suitable for commercial application in industrial wastewater treatment due to its physical limitations such as small particle size, low density and rigidity, poor mechanical strength, solid/liquid separation problems, and organic leaching during the adsorption process (Park et al., 1999).

1.1 Biosorption

Microorganisms and microbial products can be highly efficient bioaccumulators of soluble and particulate forms of metals, especially from dilute external concentrations. Also, microbe-related technologies may provide an alternative or adjunct to conventional techniques of metal removal/recovery (Gadd, 1988). Microorganisms with known potential for metal biosorption include various bacteria, cyanobacteria, yeasts, fungi, and algae. Generally, metal recovery or removal from solution may involve the following pathways: (i) the binding of metal cations to cell surfaces or within the cell wall, where microprecipitation may enhance uptake; (ii) translocation of the metal into the cell, possibly by active (metabolic energy-dependent) transport–the active uptake or concentration of metal by living microbial cells is often termed bioaccumulation; (iii) the formation of metal-containing precipitates by reaction with extracellular polymers or microbially produced anions, such as sulfide or phosphate (bioprecipitation); and (iv) the volatilization of the metal by biotransformation (Hughes and Poole, 1989). Maintaining (and feeding) a viable biomass during the metal removal process can be rather difficult due to toxicity; further, a higher difficulty in biomass separation could be encountered. So, biosorption was termed the process that makes use of dead biomass, usually from fermentation wastes or byproducts (Loukidou et al., 2011). It was found that nonliving *Actinomycetes* biomass showed greater binding capacities for cadmium ion than the living one (Kefala et al., 1999). In another application of *Bacillus* laterosporus and *Bacillus* licheniformis strains (isolated from polluted by metals soil) to remove cadmium or chromates (that is, oxyanion), similar results were published concerning the above comparison; nonliving cells showed higher sorption capacities (Zouboulis et al., 2004). Although in contrast, a live species of algal was tested to remove two metal cations (Dixit and Singh, 2013). There are several parameters that can influence the hydrophobicity of microorganisms. However, surface phenomena, including surface tension, electrokinetic, and contact angle, have been rather seldom examined for the effective removal of toxic metals from aqueous solution (Zouboulis and Matis, 2009). Parallel laboratory measurements for these major physicochemical parameters influencing the studied system were found to correlate quite satisfactorily with the effectiveness of the process. Also, floatability and hydrophobicity are interrelated.

Hydrophobicity is considered to be a result of complex interactions (chemical, bonding, specific properties of solid/liquid and solid/liquid/air interfaces, solid crystal structure, reactions of solid surface in water, etc.). Negative gradually decreasing values of zeta-potential were observed over the whole of the studied pH range. The point of zero charge was determined for Streptomyces and fungal biomass to be around pH 3 while for yeast it occurred at a lower pH value. The presence of metals as well as a flotation collector (or surfactant) were found to change the zeta-potentials toward less negative values. Surface charge neutralization and reversal toward positive values were also observed when the cationic polyelectrolyte was added. The main mechanisms of adsorption (and desorption) or binding of metal ions on microorganisms are often based on the hydrophobicity of cell walls, that is, on the properties and physicochemical characteristics of cell outer membranes. The latter depend largely on the presence of various constituents such as polysaccharides, proteins, and lipids, usually forming a layer of biopolymer on the cell wall. These properties can be altered by appropriately varying the conditions (Zouboulis and Matis, 2010).

Biosorption isotherm and kinetic models were reviewed to highlight some crucial parameters that possibly affect selectivity (Karapantsios et al., 2005; Kyzas et al., 2014). In the latter, it was argued that selective biosorption could be achieved using in synthesis an innovative technique termed molecular imprinting. Various inhibitions, including the multimechanistic role of the biosorption technology, were identified and have played (according to the authors) a contributory role to its noncommercialization (Gupta et al., 2015). Nevertheless, several commercial biosorbents were proposed in the 1990s for the removal of heavy metal ions from industrial or mining wastewaters, among them BIOFIX, AMTBIOCLAIM, and AlgaSORB (Michalak et al., 2013); attention was also paid to patents in the field and pilot-scale systems. It was said that the application of biosorptive processes could reduce capital costs by 20%, operating costs by 36%, and total treatment costs by 28%, as compared with convenient ion exchange systems (Volesky, 2001). It has been argued that immobilizing biomass, that is, in a polymer matrix, may improve biomass performance and capacity (Zouboulis et al., 2003). Models for fixed-bed columns for biosorption considered the process to be similar to typical adsorption (Trujillo et al., 1991); in the latter, work conducted at the US Bureau of Mines with sphagnum peat moss biomass was reported. *Sargassum fluitans* seaweed biomass in a flow-through bed was found capable of effectively removing toxic heavy metals (Naja and Volesky, 2006). It should be stressed that biosorptive treatments need not necessarily replace existing methodologies, but may act as polishing systems to processes that are not completely efficient, complementing them in multidisciplinary optimized processes (Kefala et al., 1999). The difficulty of the subsequent (to biosorption) solid/liquid separation stage was also stressed (Brierley et al., 1989). The plausible reason that hindered the application of this process on the industrial scale was related to the low stability and low mechanical resistance of the biomass (Michalak et al., 2013). If the biosorbents are used as a free cell suspension of dead biomass (which was the case extensively studied in our lab), then flotation for downstream separation may be the solution; the combined process was termed biosorptive flotation. In this way, advantages of this innovation are: (i) the otherwise needed immobilization is avoided, and (ii) the opportunity of biomass nanostructures (in the broad sense) is exploited.

2 Flotation-separation-regeneration

Flotation involves the preliminary abstraction or scavenging of metal ions using proper "sorbents," which exist at a fine particle size range, followed by a subsequent flotation stage for the separation of metal-loaded sorbents from the treated solution. As a result, purified water is produced as underflow as well as foam concentrate; the recovery of removed species is possible from the latter, leading to an overall clean technology. If required, the sorbents may be recycled after the application of a suitable metal eluant; this could be considered to contribute (to a certain extent) to the disposal problem of solid wastes (Zouboulis and Matis, 1997). In this, a photograph of the continuous-flow, laboratory-scale, dispersed-air flotation apparatus was also presented. The typical arrangement, which can be seen in Fig. 1B, usually includes the following parts: a conditioning/feed tank with a mechanical mixer, a peristaltic pump, liquid rotameters, the dispersed-air flotation cell (or column) with a weir (on the top), a foam collection tank, an air compressor connected through a needle valve, a washing trap and air rotameter, a porous diaphragm at the bottom of the cell, a mercury U-tube manometer and effluent tank, a pH meter, and a chemical dosing pump. Flotation constitutes a gravity separation process that certainly originated from mineral processing but has found other wide applications, including industrial wastewater treatment (Peleka et al., 2018).

Bubble columns are quite familiar as bioreactors, but are rather less familiar in their use for harvesting or separating biological materials. For instance, the recovery of proteins and other microorganisms from fermentation media is an important application in downstream processing. In many fermentation processes, gas bubbles are either formed by intensive agitation (air) or produced as a result of a microbial metabolism; if the bacteria have a hydrophobic surface, they will attach to the bubbles (Kyzas and Matis, 2014). A comparison of the floatability of various types of biomass was conducted over the whole pH range. Four types of biomass were evaluated: *Streptomyces rimosus*, *Penicillium chrysogenum*, *Saccharomyces carlsbergensis*, and grape stalks. Improved biomass flotation recoveries were observed, some reaching almost 100%. Hence, biomass separation by flotation is possible and optimum conditions have been found for the different systems examined. Two typical flotation techniques for bubble generation exist, that is, dispersed air and dissolved air (Matis and Lazaridis, 2002). The first generates bubbles by introducing air directly into the flotation tank or cell. In large agitated cells used typically in mineral processing, air is usually introduced through the bottom of the agitator shaft and small bubbles are obtained mainly by the shearing effect of impellers. In the rest, however, and in flotation columns, an air sparger or diffuser is often used. Dissolved-air flotation is based on the varying solubility of air in water, according to the pressure in the saturation vessel; the arrangement typically follows the idea of a recycle chemical reactor configuration. When a comparison between these techniques was attempted, dispersed-air flotation produced higher removals during biosorption with fungi (*Penicillium chrysogenum*) for all the examined metals, compared with dissolved-air flotation. The reason for this was due to the fact that in the dispersed-air technique cell, the hydrodynamics of the complex solid/liquid/air system were more intensive. This was presumably due to the existence of air bubbles having larger diameters as well as to their greater number; therefore, the resultant highly turbulent mixing conditions favored biosorption (Zouboulis et al., 1999). The analytical techniques in biosorption research were highlighted (Fomina and Gadd, 2014). These include atomic absorption spectrophotometry, ion selective electrodes, spectrophotometry, scanning or transmission electron microscopy coupled with energy dispersive X-ray spectroscopy, Fourier-transform infrared spectroscopy, X-ray absorption spectroscopy, X-ray diffraction, titration, electron spin resonance spectroscopy, nuclear magnetic resonance, X-ray photoelectron spectroscopy, thermogravimetric analysis, and differential scanning calorimetry. A process was developed by using the synergy between sulfate-reducing bacteria, *Bacillus cereus* and Camellia oleifera cake, assisted by flotation (Wu et al., 2017). Foam separation, an efficient downstream processing unit operation, was elsewhere tested as a posttreatment technique for phenol removal after biosorption (Saravani et al., 2017). Copper adsorption was also carried out using a material known as air-filled emulsion, with bovine serum albumen (Nazari et al., 2015). As found appropriate through extensive laboratory research, the polluted aqueous system and the pretreated biomass were contacted countercurrently, as presented in Fig. 2A, where for simplification the biosorption

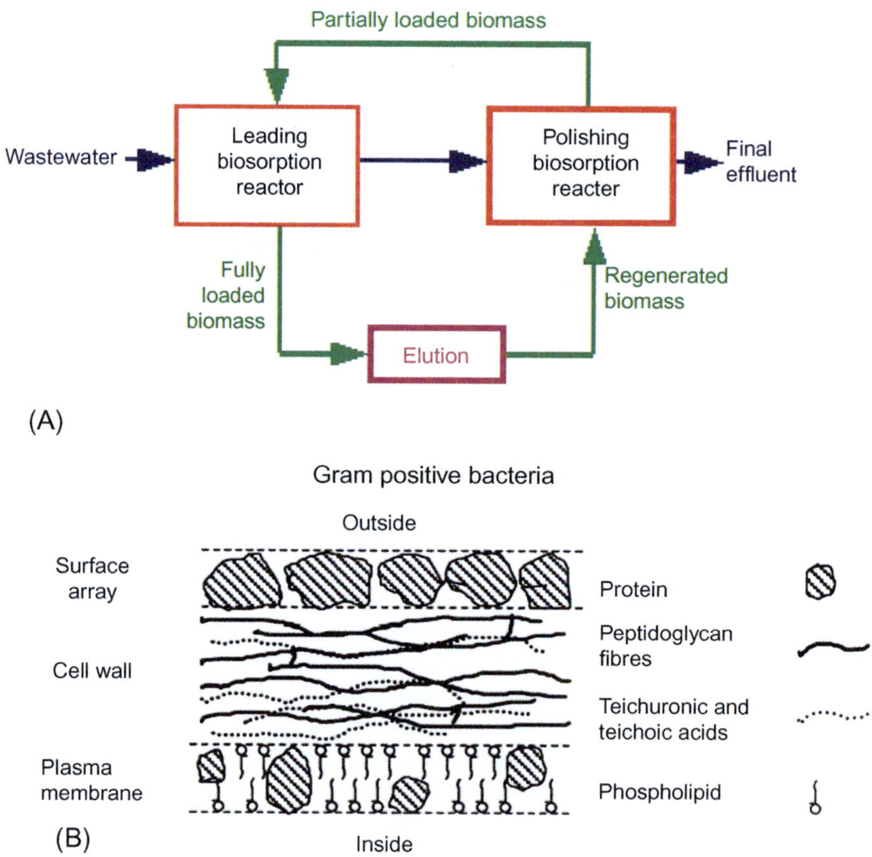

Fig. 2 (A) Flowsheet of the proposed two-stage countercurrent biosorption/regeneration process; separation by flotation was included in each biosorption stage. (B) Schematic outline of the cell wall structures of gram$^+$ bacteria.
(A) Modified, copyright of the original Soc. Chem. Ind., Zouboulis, A.I., Lazaridis, N.K., Matis, K.A., 2002. Removal of toxic metal ions from aqueous systems by biosorptive flotation. J. Chem. Technol. Biotechnol. 77, 958–964. (B) Reprinted with permission; copyright Elsevier, Volesky, B., 2007. Biosorption and me. Water Res. 41, 4017–4029.

and flotation are shown in one square; another (perhaps more complicated) ikon was published elsewhere (Zamboulis et al., 2011).

Two feed solutions containing different metal concentrations were prepared and treated in series to simulate the process diagram: a high concentration (strong) solution, representing the first stage of the pilot-scale cycle, and a low concentration (weak) solution, representing the second stage, simulating the partially treated metal-containing aqueous stream. The use of a multiple stage, countercurrent configuration is among the possible options available to improve the performance of metal removal when applying biosorption. An advantage of the proposed biosorptive process is that its basic unit processes are conventional and widely applied. The application of combined technology may be novel, but the underlying methodology is familiar. The contact and mixing time for biosorption was kept constant at 15 min, a time found to be sufficient to establish system equilibrium. Flotation removals, following laboratory-scale experiments, were found to be on the order of 100%, 85%, and 70% for copper, zinc, and nickel, respectively. The biomass was more efficient in separating copper than the other ions. The aqueous mixture of metals treated contained the following metals (in mmol/dm^3): Zn 0.765, Cu 0.157, Ni 0.034, Ca 2.49, and Na 4.35, in deionized water. The aqueous metal speciation usually has great impact on the applied process (by different techniques) for metal ion removal from dilute aqueous solution. The bioavailability and toxicity of dissolved metals are closely linked to the chemical speciation of metals in solution (Matis and Zouboulis, 2001). Hence, pH is an important parameter in biosorption because meanwhile the ionic charge of the various functional groups is affected such as carboxyl, hydroxyl, sulphydryl, amino, phosphate, etc., which exist on the biomass surface. It is worth pointing out that the main mechanisms by which biological materials accumulate metals are: (i) precipitation, (ii) complexation, (iii) adsorption, (iv) ion exchange, and (v) active transport across the cell membrane. In the pilot-scale experiments that followed, biomass sorption capacities were determined as 25 for copper, 81 for zinc, and 7 mol/dm^3 for nickel, respectively. The order of biomass affinity regarding the studied metals was Cu > Zn > Ni. A short retention time and high effectiveness were suggested. Grape stalks, a byproduct of the wine industry, were used as a biosorbent material, among others. The biomass was ground by a homogenizer to small sizes and fractionated; the smallest particle dimension (<0.001 dm) was selected. The elution of metals from the floated biomass was carried out using an aqueous mixture of sodium sulfate (1 mol/dm^3) and sodium citrate (0.1 mol/dm^3) (Zouboulis et al., 2002). Two consecutive steps were applied for the modification of the used biomass: during the first one, the contact of biomass with the flocculant Zetag-64 was conducted; in the second step, the biomass was washed as a filter cake, under mild vacuum, by the elution solution. The pilot flotation column had a volume of 10 dm^3 and the inside diameter was 0.10 dm; a diffuser (1016 m porosity) was used for bubble generation. The air flowrate for flotation was set at 0.6 dm^3/min and the flotation time was 10 min. Similar work with fungi (*Penicillium chrysogenum*) was also published (Matis et al., 2003b) as well as with *Saccharomyces cerevisiae* and *Saccharomyces carlsbergensis* biomasses from two brewing industries (Zouboulis et al., 2001b). The biomass elution with EDTA was earlier investigated (Matis et al., 1994a). With increasing EDTA concentration in the desorption stage, a deterioration of cadmium removal was detected. Further, filter aids are used by industry to improve the permeability of the filter bed and achieve better filtration characteristics. Celite (calcined diatomaceous earth) was tested and used with biomass but was found to not really contribute to metal biosorption; however, its separation by flotation was not a problem (Matis et al., 1996). In the latter, also an inhibition in the next cycles by the flotation surfactant applied was reported, that is, hexadecyl trimethyl-ammonium bromide (HDTMA-Br). The problem was solved by the application of a primary amine flotation collector, such as dodecylamine (DA) with a concentration of 10^{-4} M in ethanolic solution. The use of treated olive tree pruning as the biosorbent of lead ions was examined by successive biosorption–desorption cycles in a fixed-bed column (Ronda et al., 2015); HCl was chosen as the optimal eluting agent. *Eichhornia crassipes* leaves also showed that they could be used for (at least) seven cycles without any apparent physical change, eluted with dilute NaHCO3 solution; the biosorbent was found effective for the dye Acid Red 27 (Ramírez-Rodríguez et al., 2018).

3 Actinomycetes

Gram-positive and gram-negative strains may display differential affinity of metal biosorption in their cell components. The so-called "gram reaction" refers to the response of bacteria (that is, prokaryotes) to a rapid staining test involving a complex of the dye crystal violet with iodine; bacteria retaining the color after the process are called gram-positive (Bailey and Ollis, 1986). The prokaryotic cells (in contrast to eukaryotic) are known to contain a membrane-enclosed nucleus and usually exist as small, relatively simple unicellular organism. The metal binding properties of gram-positive bacteria are largely due to specific anionic polymers (for example, peptidoglycan, teichoic, or teichuronic acids) in the cell wall structure. Due to this high fixed anion content, they exhibit high sorption capacities, important for the industrial application of these bacteria as biosorbents (Hancock, 1988). Another important advantage of these bacteria, including the Actinomycetes, is the ease with which the properties of their thick anionic wall can be controlled, that is, a matrix that is particularly well suited for metal accumulation (see also Fig. 2B).

They also have a filamentous morphology suitable for effective floc formation. The great significance to life is certainly their use in antibiotic production. Hence, Actinomycetes form a large group of bacteria, naturally occurring in soil, that surpasses all other microorganisms in their capacity to produce biologically active substances (Boularbah et al., 1992). Microbial biomass in its natural form consists of small particles of low density, low mechanical strength, and low rigidity; biomass is generally available in the form of individual cells or cell aggregates. In this way, two Actinomycetes, *Streptomyces clavuligerus* and *Streptomyces griseus*, were examined; the typical concentration was 1.14 kg (dry wt)/m^3. *S. clavuligerus* is a branched filamentous actinomyces that is used for the production of clavulanic acid (penicillin potentiator). The bacteria were harvested and heat killed at the industrial site (SmithKline Beecham). The initial biomass contained soluble material of the order of 70%, mainly composed of culture-medium residues. It was washed exhaustively

Table 1 Cadmium cation maximum removal capacity, q_{max} (according to the Langmuir model*) for various sorbents; a comparison with values from the literature.

Sorbent	q_{max} (mg/g)	Reference
Bacillus laterosporus	160	Zouboulis et al. (2004)
Aeromonas caviae	155	Loukidou et al. (2005)
Actinomyces	45	Kefala et al. (1999)
Rhizopus arrhizus	27	Holan et al. (1995)
Zooglea ramigera	26	Norberg and Persson (1984)
Sargassum fluitans	101	Fourest and Volesky (1996)
Streptomyces lunalinharesii	25	Veneu et al. (2012)
Alginate/Fucus vesiculosus/PEI	98	Demey et al. (2018)
Pectin gel	34	Mata et al. (2010)
Lignin	25	Guo et al. (2008)
Graphene oxide/SiO_2/Fe_3O_4	167	Wang et al. (2013b)
Carbon nanotobes	135	Anitha et al. (2015)
Hydroxyapatite	49	Mandjiny et al. (1995)
Red mud	43	Yang et al. (2018a)
Alumina nanoparticles	2	Koju et al. (2018)

with distilled water and was dispersed by a hand homogenizer (with 76 m clearance) in the appropriate medium. Formaldehyde was added at 2% and the biomass sludge was stored at 4°C to prevent contamination. *S. griseus*, a fast-growing strain, came from the culture collection of the Microbiology Department, University of Newcastle-upon-Tyne, the United Kingdom (UK). (Matis et al., 1994b). The Cd(II) removal capacity according to the Langmuir model was compared for various sorbents from the literature (see Table 1), and in quite similar experimental conditions, although a direct comparison is often rather difficult.

The Langmuir equation refers to monolayer sorption onto surfaces containing a finite number of accessible sites while the empirical Freundlich equation (mentioned below) accounts macroscopically for sorption on heterogenous surfaces. Two novel biosorbents were also produced in the laboratory toward the removal of toxic cadmium cations. Also, the subsequent metal-laden biomass floatability was studied to identify the limiting parameters for such a metal treatment process (Kefala et al., 1999). The specifically isolated strains of Actinomycetes were isolated from toxic metal-contaminated soils located in an old mining activity area of North England. It was observed that due to the specific external structure of used Actinomycetes strains and to their (naturally occurring) hydrophobic properties, promoting their aggregation/flocculation, larger entities were formed that were more readily captured by air bubbles and more readily floated than the small single cells. Therefore, the examined biomass was found to float relatively easy (>80%), even without the presence of a surfactant. The second process downstream of flotation separation, that is, of cadmium-loaded biomass, was earlier investigated in depth (Matis and Zouboulis, 1994). The fundamental aspects of a combined biosorption/bioflotation system were published, and applied to cadmium removal from aqueous solutions using a Streptomyces lunalinharesii strain (Veneu et al., 2012). The removal of heavy metals from aqueous solution through Streptomyces rimosus, produced from the pharmaceutical industry as solid waste, was reviewed (Sahmoune, 2018).

From a theoretical speciation study (with the Mineql+ computer program), it was shown that in the used aqueous mixture of metals (mentioned in the previous chapter), copper is the first metal that precipitates out as hydroxide at a pH of about 5.9 (with a phase change to tenorite over this pH), followed by zinc at pH 7.3 (zincite) and then by nickel at pH 7.8; calcium does not precipitate at the range investigated.

A comparison was tried between biosorption with Streptomyces rimosus (an actinomyces bacterium resulting from tetracycline production, supplied by Pfizer Pharmaceuticals, UK) and other metal separation methods (as centrifugation, filtration, etc.). It was found that in terms of removal efficiency and applicability, biosorption was favored under acidic conditions, as shown in Fig. 3; hence, its advantage. It can be seen that biosorption moves the pH front toward the left while in conventional processes, the metal cation removal is mainly due to their precipitation as hydroxides, due to pH alteration toward alkaline values.

So, the baseline (on the right) actually represents what is termed as precipitate flotation of the first kind, or in the broad sense ion flotation. Fig. 4 presents multiple cycles; in between cycles, elution of the flotation concentrate (that is, separated metal-loaded biomass) was involved. No difference in biosorption of metals was obtained. The results proved to be very promising. In the following work, a dodecylamine collector was applied in the first, third, and fifth cycles; in other words, every other biosorption/flotation cycle that was found to be good enough to give flotation recoveries near 100%. It was also found, as expected, that by increasing the biomass addition in the solution to be treated, the quantity of biosorbed metals also increased. This, however, reduces the loading on the biomass surface, thereby making elution less effective because a greater volume of eluant would be used to treat the increased amount of biomass.

Although the sorptive processes are not known for being very selective (particularly for species with similar properties), there were indications of selective separation of copper from the mixture with the other metals. A possible explanation for this behavior can be found by considering a thermodynamic equilibrium diagram of this aqueous

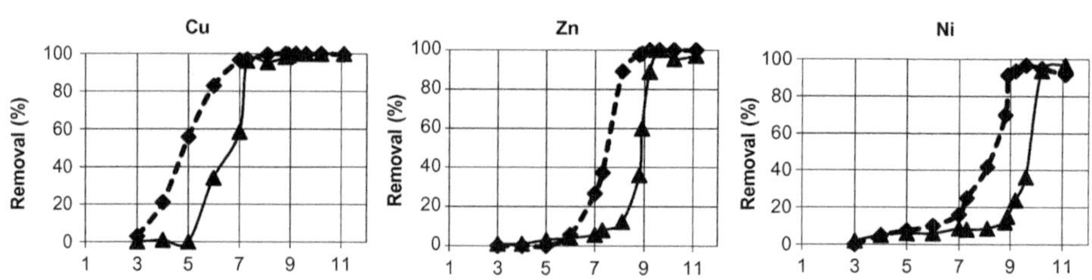

Fig. 3 Effect of pH on the removal of heavy metals by biosorption or precipitation (the latter shown with the triangle sign): a comparison of the two processes where in both cases flotation separation by dodecylamine followed downstream.
Modified from the original; Copyright Soc. Chem. Ind., Zouboulis, A.I., Rousou, E.G., Matis, K.A., Hancock, I.C., 1999. Removal of toxic metals from aqueous mixtures: part 1 biosorption. J. Chem. Technol. Biotechnol. 74, 429–436.

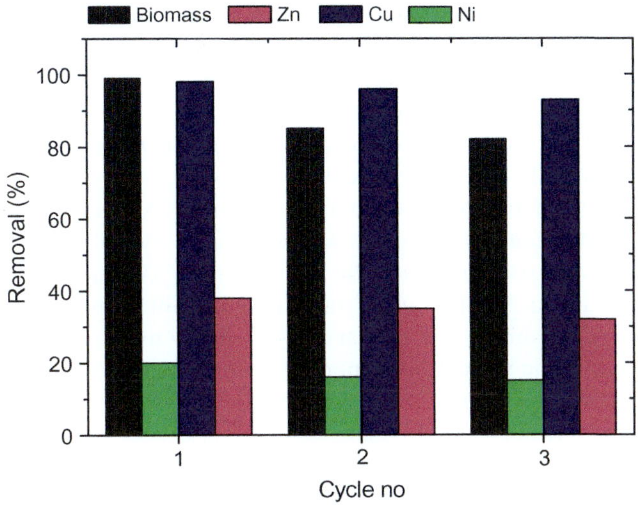

Fig. 4 Multiple-cycle operation with *Streptomyces rimosus*, at pH 7 (with flotation surfactant addition in the first cycle).
Reprinted with permission; copyright IWA Publishing, Zouboulis, A.I., Matis, K.A., Rousou, E.G., Kyriakidis, D.A., 2001a. Biosorptive flotation for metal ions recovery. Water Sci. Technol. 43, 123–129.

system. At the pH range around 6–8.5, the polymeric ionic species $Cu_2(OH)_2^{2+}$ becomes predominant, noting that the formation of these complexes depends upon the initial metal concentration (Baes Jr. and Mesmer, 1976). The hydrolyzed species undergo several hydrolytic reactions, and ultimately result in the creation of precipitates.

4 Fungi

Fungi can accumulate metal and radionuclide species by physicochemical and biological mechanisms, including extracellular binding by metabolites and biopolymers, binding to specific polypeptides, and metabolism-dependent accumulation (Tobin et al., 1994). However, to date the most promising approach to metal removal by fungi is biosorption. Fungal biomass can be cheaply and easily grown or procured in rather substantial quantities as a byproduct from established industrial fermentation processes. Zinc biosorption characteristics of locally isolated *Aspergillus flavus* were examined (Aftab et al., 2013). The fungal cell wall is thought to have two main components: interwoven skeletal framework microfibrils, usually of chitin, embedded in an amorphous layer of proteins and various polysaccharides (Volesky, 1987). Sorption capacity variations between different biosorbent types could be related to their acidity; pH neutralization during the sorption reaction considerably enhanced zinc chelation. It has also been reported that prior treatment of mycellial wastes with NaOH increased their capacity for metal sorption (Fourest and Roux, 1992). *Filamentous* fungi are used in fermentation industries to produce varied metabolites such as enzymes, flavorings, and antibiotics. These industrial byproducts are able to chelate various heavy metals. The main chemical groups in biomass that are able to partake in biosorption are electronegative groups such as hydroxyl or sulphydryl groups, anionic groups such as carboxyl or phosphate groups, and nitrogen-containing groups such as amino groups (Zouboulis et al., 1997). The cells of *Candida utilis* and *Candida tropicalis* were immobilized by calcium alginate gel and used for the investigation of biosorption parameters for the removal of zinc ions from aqueous solution (Ahmad et al., 2013). Waste mycelium of *Penicillium chrysogenum* was entrapped by polyvinyl alcohol, then coated by chitosan and so a new adsorbent was prepared (Xiao et al., 2014). Chitosan, produced by partial deacetylation of chitin, is a hydrophilic natural biopolymer with abundant amino and hydroxyl groups that can sorb heavy metal ions.

Experiments with the same aforementioned metals mixture were also carried out by applying as biosorbent *Penicillium chrysogenum*, an industrial waste of filamentous fungal biomass (trade name Mycan), unmodified except for washing. See as an example Fig. 5, where the influence of a cationic polyelectrolyte (from Ciba) presence on the biosorptive flotation of loaded fungi was examined, noting that effective flocculation is a prerequisite mainly for dissolved-air flotation. The choice for a cationic surfactant was based on previous zeta-potential measurements. The results of biomass recovery by flotation were found to be substantially increased while the removal of toxic metals was either unaffected or slightly affected. The residual turbidity (after flotation) was also remarkably improved (from 126 decreased to 0.4 NTU) when polyelectrolyte concentrations were increased to 10 mg/dm^3.

Although the literature contains many studies of the ability of biosorbents to bind metal cations, the binding of anions (as the pentavalent arsenic oxyanions) by biosorbents has not been much reported. The pretreatment of biomass with common

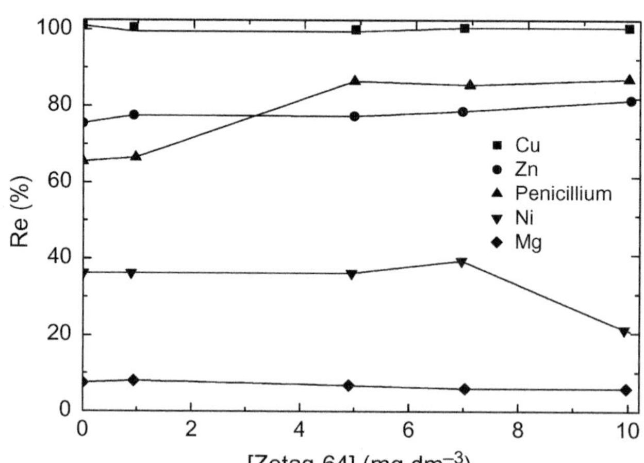

Fig. 5 Influence of flocculant concentration on biosorption and flotation of *Penicillium chrysogenum* fungi (by dodecylamine at pH 6) and on the removal of toxic metals. Reprinted with permission; copyright Elsevier, Matis, K.A., Zouboulis, A.I., Lazaridis, N.K., Hancock, I.C., 2003a. Sorptive flotation for metal ions recovery. Int. J. Miner. Process. 70, 99–108.

surfactants (such as hexadecyl-trimethylammonium bromide or dodecylamine) and a cationic polyelectrolyte was found to improve biosorption efficiency (Loukidou et al., 2003). A speciation diagram was presented to predict the As(V) species with the solution pH. Potentiometric titration was conducted in order to characterize the biomass, being Mycan; electrokinetic studies and relative FTIR spectral analyses followed to assist the investigation of the biosorption results. In Fig. 6A, the effect of solution pH on the combined biosorptive flotation process is presented. In a rather wide pH range, sufficient separation of biomass was achieved (>80%) while the removal of arsenate anions was further improved, up to 75%. An addition of a small concentration of ethanol (0.25% v/v, as a frother) further improved the floatability to >90%. The need for biomass modification was apparent. Electrophoretic mobility measurements were performed and the respective zeta-potential values were calculated in order to examine the effect of specific pretreatment; as demonstrated (Fig. 6B), the surface charge of treated biomass was substantially altered when the biomass was examined alone or in the presence of a surfactant/flocculant.

Biosorption by *Saccharomyces* yeast followed by separation of zinc-loaded biomass using a hybrid flotation-microfiltration unit was investigated (Peleka and Matis, 2009); and hence, the production of a clean water stream. It was demonstrated that it was possible to reduce substantially the fouling process by combining membrane microfiltration with flotation, with the membrane module submerged inside the flotation cell. As far as the zinc ions are concerned, their removal was almost always total for the tested experimental conditions. Laboratory experimental attempts of biomass regeneration and reuse for metal removal purposes were also carried out in five subsequent treatment cycles, using 0.1 M sodium hydroxide. The zinc desorption was almost 100% while during the subsequent treatment cycles, the brewery yeast could remove Zn(II) with about the same high effectiveness. Apparently, no energy was added for fouling control of the membranes further to that required for flotation, leading to low operational costs of the process compared to conventional membrane filtration. In today's world of economical and environmental constraints for chemical and related industries, process intensification may be a path for the future of chemical and process technology (Charpentier, 2007).

5 Industrial wastes

Sludge-based adsorbents are widely used for the removal of various pollutants from water and wastewater systems and the available data are much diversified (Devi and Saroha, 2017). The production of municipal sewage sludge in the European Union was estimated to be 11.5 million tons in 2010. Presently, the main methods for sewage sludge disposal include incineration, landspreading, and landfilling; nevertheless, waste microbial biomass may well be an alternative, a more economic metal sorbent in comparison with more conventional ones. Extracellular biopolymers that usually present in activated sludge processes mainly include polysaccharides, but proteins and nucleic acids can also be found and may play an important role in metal-microorganism interactions (Solari et al., 1996). Metal uptake by living sewage sludge seems to be

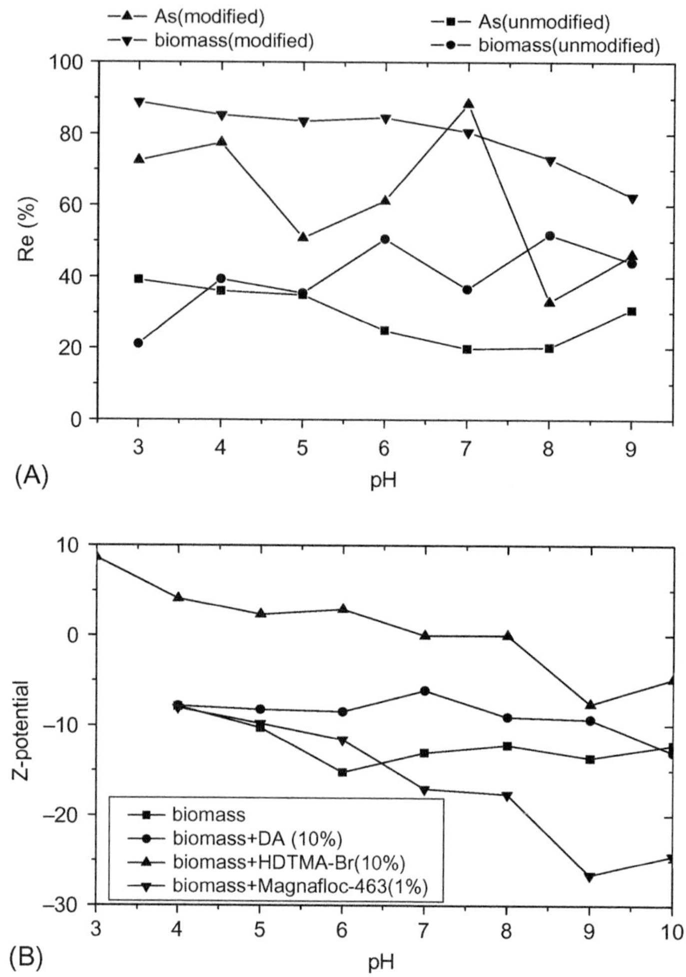

Fig. 6 (A) As(V) biosorption removal and flotation of arsenic-loaded fungal biomass, with or without previous modification: influence of pH. (B) Zeta-potential measurements; biomass unmodified and modified by the two surfactants or a polyelectrolyte.
(A) Reprinted with permission; copyright GVC-VDI, Loukidou, M.X., Matis, K.A., Zouboulis, A.I., 2001. Removal of arsenic from contaminated dilute aqueous solutions using biosorptive flotation. In: 3rd Europ. Congress Chemical Engineering. DECHEMA, Nuremberg, Germany. (B) Reprinted with permission; copyright Elsevier, Loukidou, M.X., Matis, K.A., Zouboulis, A.I., Liakopoulou-Kyriakidou, M., 2003. Removal of As(V) from wastewaters by chemically modified fungal biomass. Water Res. 37, 4544–4552.

more dependent on physical activity (such as adsorption and ion exchange) and on chemical interactions (such as chemisorption, complexation, and surface precipitation) between the bacterial surface and the metal ions rather than on metabolic activity, confirming the process to be mainly a passive one (Nelson et al., 1981). Therefore, sewage sludge could be equally applicable as a nonliving system, avoiding many problems (including, among others, the negative influence of pH when it is outside a certain range, usually 6.5–7.5) associated with a living system. An alternative but effective flotation technique, mainly in a smaller scale, is electroflotation (perhaps it should be correctly called electrolytic flotation), owing its name to the bubble-generation method it uses, that is, electrolysis of the aqueous medium. When an effluent is brought between two electrodes, one of which is the positive anode and the other is the negative cathode, and electricity is supplied to the electrodes, an electric field is built up between them through the use of the suspension conductivity. Even without any other addition of chemical reagents, a preliminary coagulation occurs within the particulate matter of the effluent, which results in the grouping of the negative and positive particles together. The electroflotation technique was reviewed indepth, including its application in sludge thickening (Matis and Peleka, 2010).

Sufficient biomass separation (around 90%) was observed in batch experiments, approximately from the pH value of 7 (see Fig. 7). This result was found by applying a small flotation time (10 min) and at a low current density (65 A/m^2). The two curves of cadmium biosorption on biomass and of floated (Cd-loaded) biomass almost coincide with the studied pH variation. Doubling the current density, the same curves were obtained but effectively shifted to the left by about 2 pH units. Also, in the same figure, the substantially lower removal of Cd ions, but without the presence of biomass (simply by the influence of pH), is indicated.

To separate the metal-laden sludge, dissolved-air flotation was also used; the reported results collectorless (with 30% recycle ratio) were of the order of 80% regarding the recovery of biomass. The nonliving, anaerobically digested activated sludge was obtained from the Central Sewage Treatment Plant of the city of Thessaloniki. Ethanol was used as a convenient frothing agent during flotation at 0.05% *v*/v concentration,

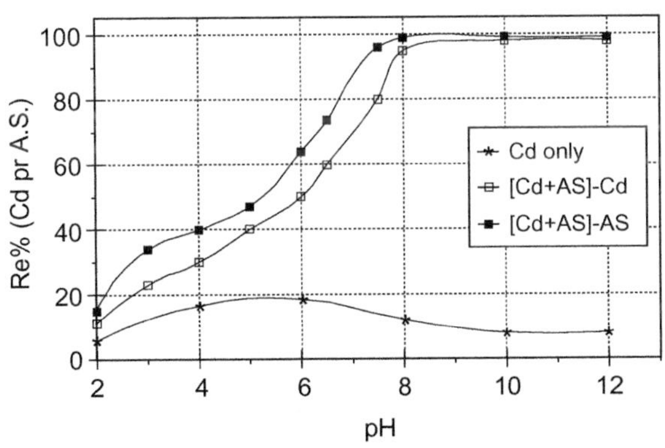

Fig. 7 Effect of solution pH on cadmium biosorption onto activated sludge (AS), and separation by electroflotation; with aluminum alloy electrodes, current density 65 A/m^2 and 600 s retention time.
Reprinted with permission; copyright Inderscience Enterprises, Zouboulis, A.I., Matis, K.A., 2012. Cadmium ion removal by electroflotation onto sewage sludge biomass. Int. J. Environ. Waste Manag. 9, 245–256.

considering also that ethanol-treated yeast cells were elsewhere examined as capable metal biosorbents (Göksungur et al., 2005).

In fact, most of the biosorbents are industrial wastes or byproducts, such as for instance yeasts available from fermentation industries. Yeast has been used by humans for >6000 years for a variety of applications, and it is presently one of the most important commercial microorganisms. Nevertheless, its significance and industrial utilization could be further increased; nonliving yeast cells may be used in many ways and forms for metal removal. A bioelectrospraying technique to immobilize *Saccharomyces cerevisiae* onto the surface of poly(−caprolactone)/chitosan/rectorite ternary composite-based nanofibrous mats, in order to be recycled and reused (Xin et al., 2017). The effectiveness of sugar-beet pectin xerogels was reported for the removal of heavy metals, with multiple batch sorption–desorption cycles (Mata et al., 2010). The spent compost of *Hypsizygus marmoreus* (an edible mushroom) was used as a biosorbent to remove Pb(II) ions from aqueous solution, after modification by NaOH (Yang et al., 2018b). The ability of spent black tea leaves to remove lead from aqueous solution using the biosorptive-flotation process was examined (Mohammed et al., 2016).

Therefore, activated carbons produced from various agricultural wastes, for example potato peels (Kyzas et al., 2016), could also be included (extensively published), particularly for developing countries to find a cheap and available feedstock for the preparation of suitable sorbent (activated carbon) for use in industry, drinking water purification, and wastewater treatment.

6 Carbon nanotubes

Graphene oxide, particularly as magnetic particles, has recently been used as an adsorbent for wastewater treatment in applications such as heavy metal separation and also organics such as antibiotics, dyes, etc. (Kyzas et al., 2013). The published results generally are of importance for the environmental application of graphene oxide nanocomposites, based on their advantages (that is, a two-dimensional layer structure, a large surface area, pore volume, and the presence of surface functional groups in these materials), for enrichment or removal processes in the large volumes of aqueous solutions and effluents. The magnetic separation of the loaded adsorbent could also contribute to convenient solid–liquid separation, required downstream.

Nanoparticle use as adsorbents has attracted widespread interest (Kyzas and Matis, 2015). Nanomaterials possess a series of unique physical and chemical properties, and one of the most important is that most of the atoms that have high chemical activity and adsorption capacity are on the surface of the nanomaterials. Among their key properties, which make these materials particularly attractive for use as separation media for water treatment, are the following: (i) larger surface areas than bulk particles; (ii) ability for functionalization using different chemical groups to enhance their affinity to a given compound; and (iii) use as high selectivity/capacity recyclable ligands for toxic elements and organic and inorganic solutes in aqueous media. The mechanism and kinetics of adsorption on various adsorbents depend apparently on their chemical nature and the various physicochemical experimental conditions such as the solution pH, the initial metal concentration, the adsorbent dosage, and the temperature of the system. The present information regarding the biological methods, which are employed to fabricate greener, safer, and environmentally sustainable nanosynthesis routes, was summarized (Saratale et al., 2018).

In a recent review, the development of various functionalized carbon nanotubes and grapheme, which are used to remove heavy metals from contaminated water, was assessed (Xu et al., 2018). Similarly, the adsorption of noxious heavy metal ions from wastewater effluents was reviewed using various conventional adsorbents and nanostructures (fullerenes, carbon nanotubes, graphenes) (Burakov et al., 2018). The efficiency of the developed materials for adsorption of heavy metal was discussed, along with a comparison of their maximum adsorption capacity in tabular form (see also Fig. 8). Two graphene nanosheets and an oxide were tested and found to exhibit

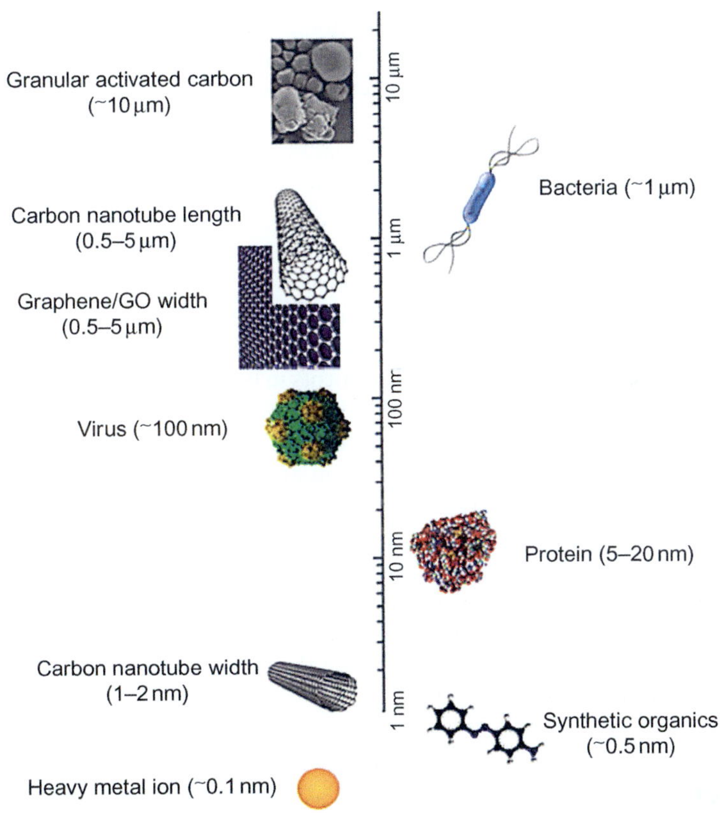

Fig. 8 Scale showing relative sizes of major classes of carbon nanomaterials and common water contaminants.
Reprinted with permission; copyright Elsevier, Smith, S.C., Rodrigues, D.F., 2015. Carbon-based nanomaterials for removal of chemical and biological contaminants from water: a review of mechanisms and applications. Carbon 91, 122–143.

comparable or better adsorption capacities than an SWCNT, an MWCNT, or a coal-based granular activated carbon (but for phenanthrene and biphenyl removal in the presence of natural organic matter) (Apul et al., 2013).

Nanocomposite beads were synthesized for cobalt ion removal from amide group functionalized MWCNTs imprinted in the network of sodium alginate containing hydroxyapatite (Karkeh-abadi et al., 2016). The synthesis of nanocomposite film based on functionalized MWCNTs incorporated with polyvinyl alcohol in the presence of citric acid as a cross-linking agent was reported (Youssef et al., 2019). Iron-oxide filled (that is, in order to become magnetic) MWCNTs were also amino-functionalized for the covalent immobilization of an enzyme (Ranjan et al., 2019). A phosphorylated MWCNT-cyclodextrin/silver-doped titania (that is, a nanosponge biopolymer) was synthesized and tried (Taka et al., 2018). Polymeric nanocomposites were shown to enhance the adsorption efficiency because of the electron-rich functional groups, which are present on the polymer backbone. Two-dimensional sheets of sodium alginate bound to functional carbon nanotubes were synthesized (Allaboun et al., 2016). Carbon nanotubes were taken as the Trojan horse in the nanotechnology race. It has been often argued that several separation techniques have limitations, such as high operational costs; however, possibly the synthesis of certain adsorbents proposed in the literature may seem quite extraordinary. Accurate cost assessments are generally difficult to obtain for any treatment process due to commercial confidentiality.

In a chapter, the potential of microorganisms as biosorbents used in the bioremediation of toxic heavy metal pollution of water systems was elaborated, exploring the possibility of enhancing their performance through immobilization on adequate support matrices, namely carbon nanotubes (Fosso-Kankeu et al., 2014). However, a major concern when using carbon nanotubes in water could be their relative toxicity to humans. Carbon nanotubes were said to present a number of features that make them attractive for immobilization of microbial sorbents to be used in biosorption processes. So, the latter idea was investigated in this scientific area, that is, the combination of a biosorbent immobilized on carbon nanotube. Examples of this include *Mucor circinelloides* (fungi) applied to azo dyes (Azin et al., 2017), and *Bacillus mojavensis* (thermo-tolerant bacterium) (Özdemir et al., 2017) and *Pseudomonas aeruginosa* (gram-negative bacteria) (Tuzen et al., 2008), both for uranium. The latter, being a source in nuclear energy applications, is an environmental contaminant with certain long-time toxic effects. Also, experiments were performed on ion flotation combined with MWCNTs as adsorbents to investigate the continuous removal of nickel ions from aqueous solution (Dehghani et al., 2017); the concentration 2 g/L of MWCNTs increased the removal efficiency to 98%.

7 Other

Extracellular polymeric substances produced by microorganisms during their growth were found to be effective for the removal of heavy metals (Wang et al., 2013a). Marine algae have been used as suitable biosorbents, often modified, for the same purpose (Demey et al., 2018; Lou et al., 2015; Senthilkumar et al., 2017). The quoted prices for two commercially available algae, Chlorella and Spirulina, were said to be the highest among those for a variety of biomass products (Wilde and Benemann, 1993). Fabrication costs and process optimization were discussed for biogenic nanoparticles (using microbes/microorganisms) in the treatment of toxic and emerging pollutants and the survival of conventional wastewater treatment technologies (Ali et al., 2019). Finding the better possibility for the valorization of exhausted biosorbents (loaded with metal ions) may contribute to the best way of using these low-cost materials, in accordance with the principles of the circular economy (Bădescu et al., 2018). Biosorption has been also investigated in a well-stirred batch reactor by using the biomass of *Aeromonas caviae*, a gram-negative bacteria isolated from water wells near Thessaloniki; Fig. 9A shows a photograph of it. Potentiometric titration, zeta-potential measurements, and FTIR studies for biomass characterization provided useful information. As the pH value was increased (but kept far from the precipitation value of metal hydroxide), the carboxyl and phosphate groups gave an overall negative charge to the biomass, which was able to remove positively charged heavy metal ions and biosorb them onto the cell surface (see Fig. 9B). The biosorption of cadmium ions to A. caviae was of an endothermic nature. The Freundlich and Langmuir adsorption models were employed for the mathematical description of biosorption equilibrium data for varying temperatures and biomass concentrations. The results demonstrated that both models were suitable for describing the biosorption equilibrium of Cd(II) by the biomass in the examined concentration range.

A kinetic investigation was also conducted (Loukidou et al., 2005), analyzing various kinetic rate expression mechanisms; several diffusion (external and intraparticle) kinetic models were examined. Predictions based on the so-called (and typical) pseudo second-order rate expression were found in satisfactory accordance with experimental data; see also (Ho et al., 2000). Evidence was provided that the sorption of metal ions on A. caviae is a complex process, concluding that possibly biosorption, in this case (at least), was more correctly described by more than one model–as is often the case with the sorption of metal ions (Smith, 1996). The advantage of the present biosorbent was its ability to also remove metal oxyanions, that is, Cr(VI) (at pH 2.5 and without any modification), as studied in depth (Loukidou et al., 2004). The removal of hexavalent chromium oxyanions at low pH values may be attributed to their attraction/affinity toward positively charged surface groups of biomass or reduction of Cr(VI) to Cr(III), followed by its bonding to the negatively charged biosorbent. The Cr(VI) removal capacity found for *Aeromonas caviae* was compared with values of various sorbents (from the literature), for example in order just to get an idea and in quite similar experimental conditions, as presented in Table 2. Either of the comparison tables presented here for two specific toxic metals (a cation and an oxyanion), although only from selected literature papers, may reveal the promise of this treatment process, due to the biosorbent efficiency for metal removal. A relative extended comparative chapter of biosorption versus adsorption has been also published (Loukidou et al., 2011).

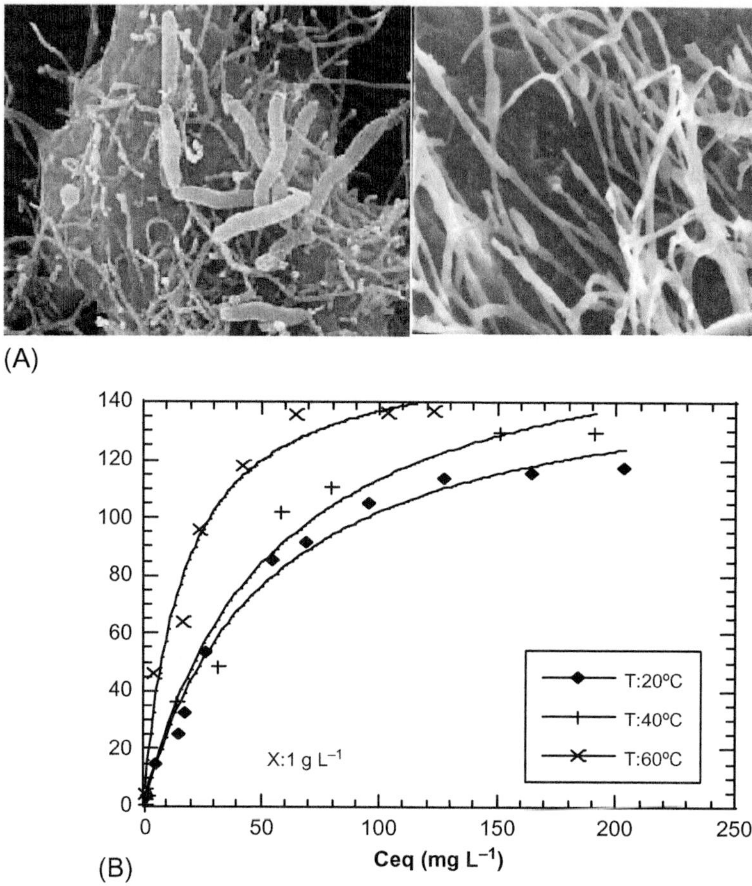

Fig. 9 (A) Left: Electronic microscope picture of *A. caviae* (Gram− bacteria); the peptidoglycan layer is much thinner, the outer membrane consists of lipopolysaccharide, no teichoic acid in wall. Thanks to Dr. Maria Loukidou (from her Ph.D. thesis at AUTh, 2003). On the right, for comparison, an *Actinomyces* (Gram$^+$). (B) Application of equilibrium Langmuir model to cadmium biosorption (at pH 7) onto *Aeromonas caviae* at different temperatures.
Reprinted with permission; copyright Taylor and Francis, Loukidou, M.X., Karapantsios, Th.D., Zouboulis, A.I., Matis, K.A., 2005. Cadmium(II) biosorption by Aeromonas caviae: kinetic modeling. Sep. Sci. Technol. 40, 1293–1311.

8 Conclusions

Over the past decades, with the rapid development of the fermentation industry for antibiotics, enzymes, beverages, and food additives, >80 million tons of waste mycelia (inactive biomass) are generated every year, and that's only in China. The use of biomass (nonliving microorganisms) to remove and possibly recover toxic or precious metals from industrial wastewaters has gained credibility during recent years because

Table 2 Chromium oxyanion removal capacity q_{max} (according to Langmuir) for various sorbents; an indicative comparison with values from the literature.

Sorbent	q_{max} (mg/g)	Reference
Waste sludge	27	Li et al. (2004)
Sewage sludge	16	Rozada et al. (2008)
Biopolymer	15	Donner et al. (2019)
Banana peel	132	Memon et al. (2009)
Jackfruit peel	64	Saranya et al. (2018)
Streptomyces rimosus	29	Ammar (2009)
Aeromonas caviae	124	Loukidou et al. (2004)
Chlorella vulgaris	24	Veglio and Beolcini (1997)
Rhizopus arrhizus	62	Prakasham et al. (1999)
Pinus sylvestris	202	Ucun et al. (2002)
Graphene oxide	44	Yang et al. (2014)
Carbon nanotube/Al_2O_3	265	Sankararamakrishnan et al. (2014)
Calcined hydrotalcite	120	Lazaridis et al. (2004)
Activated carbon	147	Aksu et al. (2002)
Activated sludge	19	Aksu et al. (2002)

of the good performance and low cost of these sorbent materials. The natural affinity of biological materials for metallic elements could contribute to economically purifying metal-loaded wastewater, a fact that has already been proven in several cases and by many researchers.

A logical approach to the treatment of metal-contaminated waters is to combine metal removal for clean up purposes with metal recovery, making the metal then available for reuse in industry. Flotation has proved promising for the solid/liquid separation of the metal-laden biosorbent downstream. Biosorption is an area of endeavor that occupies an interface between biology, chemistry, and engineering. Progress depends on the successful interaction between these disciplines (which often has not been appropriate).

References

Aftab, K., Akhtar, K., Jabbar, A., Bukhari, I.H., Noreen, R., 2013. Physico-chemical study for zinc removal and recovery onto native/chemically modified *Aspergillus flavus* NA9 from industrial effluent. Water Res. 47, 4238–4246.

Ahmad, M.F., Haydar, S., Quraishi, T.A., 2013. Enhancement of biosorption of zinc ions from aqueous solution by immobilized *Candida utilis* and *Candida tropicalis* cells. Int. Biodeterior. Biodegrad. 83, 119–128.

Aksu, Z., Gönen, F., Demircan, Z., 2002. Biosorption of chromium(VI) ions by Mowital B30H resin immobilized activated sludge in a packed bed: comparison with granular activated carbon. Process Biochem. 38, 175–186.

Ali, I., Peng, C., Khan, Z.M., Naz, I., Sultan, M., Ali, M., Abbasi, I.A., Islam, T., Ye, T., 2019. Overview based fabricated biogenic nanoparticles for water and wastewater treatment. J. Environ. Manag. 230, 128–150.

Allaboun, H., Fares, M.M., Al-Rub, F.A.A., 2016. Removal of uranium and associated contaminants from aqueous solutions using functional carbon nanotubes-sodium alginate conjugates. Minerals 6, 9–21.

Ammar, S., 2009. Biosorption of Cr^{6+} from aqueous solution by a Streptomyces rimosus biomass. New Biotechnol. 25S. Article ID S274.

Anitha, K., Namsani, S., Singh, J.K., 2015. Removal of heavy metal ions using a functionalized single-walled carbon nanotube: a molecular dynamics study. J. Phys. Chem. A 119, 8349–8358.

Apul, O.G., Wang, Q., Zhou, Y., Karanfil, T., 2013. Adsorption of aromatic organic contaminants by graphene nanosheets: comparison with carbon nanotubes and activated carbon. Water Res. 47, 1648–1654.

Azin, E., Moghimi, H., Taheri, R.A., 2017. Development of carbon nanotube-mycosorbent for effective Congo red removal: optimization, isotherm and kinetic studies. Desalin. Water Treat. 94, 222–230.

Bădescu, I.S., Bulgariu, D., Ahmad, I., Bulgariu, L., 2018. Valorisation possibilities of exhausted biosorbents loaded with metal ions—a review. J. Environ. Manag. 224, 288–297.

Baes Jr., C.F., Mesmer, R.E., 1976. The Hydrolysis of Cations. Wiley, New York, USA.

Bailey, J.E., Ollis, D.F., 1986. Biochemical Engineering Fundamentals. McGraw-Hill, New York, USA.

Boularbah, A., Morel, J.L., Bitton, G., Guckert, A., 1992. Cadmium biosorption and toxicity to six cadmium-resistant gram-positive bacteria isolated from contaminated soil. Environ. Toxicol. Water Qual. 7, 237–246.

Brierley, C.L., Brierley, J.A., Davidson, M.S., 1989. Applied microbial processes for metals recovery and removal from wastewater. In: Beveridge, T.J., Doyle, R. (Eds.), Metal Ions and Bacteria. Wiley, Chichester, U.K, pp. 359–382.

Burakov, A.E., Galunin, E.V., Burakova, I.V., Kucherova, A.E., Agarwal, S., Tkachev, A.G., Gupta, V.K., 2018. Adsorption of heavy metals on conventional and nanostructured materials for wastewater treatment purposes: A review. Ecotoxicol. Environ. Saf. 148, 702–712.

Butter, T.J., Evison, L.M., Hancock, I.C., Holland, F.S., Matis, K.A., Philipson, A., Sheikh, A.I., Zouboulis, A.I., 1998. The removal and recovery of cadmium from dilute aqueous solutions by biosorption and electrolysis at laboratory scale. Water Res. 32, 400–406.

Charpentier, J.-C., 2007. In the frame of globalization and sustainability, process intensification, a path to the future of chemical and process engineering (molecules into money). Chem. Eng. J. 134, 84–92.

Dehghani, M.H., Karimi, B., Rafieepour, A., 2017. Continuous removal of toxic nickel from industrial wastewater by flotation-adsorption with multi-walled carbon nanotubes: kinetic and adsorption isotherms. Desalin. Water Treat. 71, 107–115.

Demey, H., Vincent, T., Guibal, E., 2018. A novel algal-based sorbent for heavy metal removal. Chem. Eng. J. 332, 582–595.

Devi, P., Saroha, A.K., 2017. Utilization of sludge based adsorbents for the removal of various pollutants – A review. Sci. Total Environ. 578, 16–33.

Dixit, S., Singh, D.P., 2013. Phycoremediation of lead and cadmium by employing *Nostoc muscorum* as biosorbent and optimization of its biosorption potential. Int. J. Phytoremed. 15, 801–813.

Donner, M.W., Arshad, M., Ullah, A., Tariq Siddique, T., 2019. Unravelled keratin-derived biopolymers as novel biosorbents for the simultaneous removal of multiple trace metals from industrial wastewater. Sci. Total Environ. 647, 1539–1546.

Fan, L., Luo, C., Sun, M., Qiu, H., Li, X., 2013. Synthesis of magnetic β-cyclodextrin-chitosan/graphene oxide as nanoadsorbent and its application in dye adsorption and removal. Colloids Surf. B 103, 601–607.

Fomina, M., Gadd, G.M., 2014. Biosorption: current perspectives on concept, definition and application. Bioresour. Technol. 160, 3–14.
Fosso-Kankeu, E., Mulaba-Bafubiandi, A.F., Mishra, A.K., 2014. Prospects for immobilization of microbial sorbents on carbon nanotubes for biosorption: Bioremediation of heavy metals polluted water. In: Mishra, A.K. (Ed.), Application of Nanotechnology in Water Research. Scrivener Publishing, Wiley, USA, pp. 37–61.
Fourest, E., Volesky, B., 1996. Contribution of sulfonate groups and alginate to heavy metal biosorption by the dry biomass of *Sargassum fluitans*. Environ. Sci. Technol. 30, 277–282.
Fourest, E., Roux, J.-C., 1992. Heavy metal biosorption by fungal mycelial by-products: mechanisms and influence of pH. Appl. Microbiol. Biotechnol. 37, 399–403.
Gadd, G.M., 1988. Accumulation of metals by microorganisms and algae. In: Rehm, H.J. (Ed.), Biotechnology. vol. 6b. VCH Publishers, Weinheim, Germany, pp. 401–433.
Göksungur, Y., Űren, S., Gűvenç, U., 2005. Biosorption of cadmium and lead ions by ethanol treated waste baker's yeast biomass. Bioresour. Technol. 96, 103–109.
Guo, X.Y., Zhang, S.Z., Shan, X.Q., 2008. Adsorption of metal ions on lignin. J. Hazard. Mater. 151, 134–142.
Gupta, V.K., Nayak, A., Agarwal, S., 2015. Bioadsorbents for remediation of heavy metals: Current status and their future prospects. Environ. Eng. Res. 20, 1–18.
Hancock, I.C., 1988. The use of Gram-positive bacteria for the removal of metals from aqueous solution. In: Thompson, R. (Ed.), Trace Metal Removal from Aqueous Solution. Royal Soc. Chem., London, UK, pp. 25–43. spec. publ. 61.
Ho, Y.S., Ng, J.C.Y., McKay, G., 2000. Kinetics of pollutant sorption by biosorbents: Review. Sep. Purif. Rev. 29, 189–232.
Holan, Z.R., Volesky, B., Prasetyo, I., 1995. Biosorption of cadmium by biomass of marine algae. Biotechnol. Bioeng. 41, 819–825.
Hughes, M.N., Poole, R.K., 1989. Metals and Microorganisms. Chapman and Hall, London, U.K., pp. 328–395.
Humphreys, D., 2000. New horizons for an old industry. In: *MassMin2000 Conf.* Austral. IMM, Brisbane, Australia.
Karapantsios, T.D., Loukidou, M.X., Matis, K.A., 2005. Sorption kinetics. In: Lehr, J. (Ed.), Water Encyclopedia. vol. 4. Wiley, Hoboken, N.J., USA, pp. 564–569.
Karkeh-abadi, F., Saber-Samandari, S., Saber-Samandari, S., 2016. The impact of functionalized CNT in the network of sodium alginate-based nanocomposite beads on the removal of Co(II) ions from aqueous solutions. J. Hazard. Mater. 312, 224–233.
Kefala, M.I., Zouboulis, A.I., Matis, K.A., 1999. Biosorption of cadmium ions by *Actinomycetes* and separation by flotation. Environ. Pollut. 104, 283–293.
Koju, N.K., Song, X., Wang, Q., Hu, Z.H., Colombo, C., 2018. Cadmium removal from simulated groundwater using alumina nanoparticles: behaviors and mechanisms. Environ. Pollut. 240, 255–266.
Kyzas, G.Z., Matis, K.A., 2015. Nanoadsorbents for pollutants removal: a review. J. Mol. Liq. 203, 159–168.
Kyzas, G.Z., Matis, K.A., 2014. Flotation of biological materials. Processes 2, 293–310.
Kyzas, G.Z., Deliyanni, E.A., Matis, K.A., 2016. Activated carbons produced by pyrolysis of waste potato peels: cobalt ions removal by adsorption. Colloids Surf. A 490, 74–83.
Kyzas, G.Z., Fu, J., Matis, K.A., 2014. New biosorbent materials: selectivity and bioengineering insights. Processes 2, 419–440.
Kyzas, G.Z., Deliyanni, E.A., Matis, K.A., 2013. Graphene oxide and its application as adsorbent to wastewater treatment. J. Chem. Technol. Biotechnol. 89, 196–205.

Lazaridis, N.K., Pandi, T.A., Matis, K.A., 2004. Chromium(VI) removal from aqueous solutions by Mg-Al-CO_3 hydrotalcite: sorption-desorption kinetic and equilibrium studies. Ind. Eng. Chem. Res. 43, 2209–2215.

Leupin, O.X., Hug, S.J., Badruzzaman, A.B.M., 2005. Arsenic removal from Bangladesh tube well water with filter columns containing zero-valent iron filings and sand. Environ. Sci. Technol. 39, 8032–8037.

Li, Y., Liu, C., Chiou, C., 2004. Adsorption of Cr(III) from wastewater by wine processing waste sludge. J. Colloid Interface Sci. 273, 95–101.

Liu, C., Li, F., Luo, C., Liu, X., Wang, S., Liu, T., Li, X., 2009. Foliar application of two silica sols reduced cadmium accumulation in rice grains. J. Hazard. Mater. 161, 1466–1472.

Lou, Z., Wang, J., Jin, X., Wan, L., Wang, Y., Chen, H., Shan, W., Xiong, Y., 2015. Brown algae based new sorption material for fractional recovery of molybdenum and rhenium from wastewater. Chem. Eng. J. 273, 231–239.

Loukidou, M.X., Peleka, E.N., Karapantsios, T.D., Matis, K.A., 2011. Biosorption of metal ions. Trends Chem. Eng. 13, 53–64.

Loukidou, M.X., Karapantsios, T.D., Zouboulis, A.I., Matis, K.A., 2005. Cadmium(II) biosorption by *Aeromonas caviae*: kinetic modeling. Sep. Sci. Technol. 40, 1293–1311.

Loukidou, M.X., Karapantsios, T.D., Zouboulis, A.I., Matis, K.A., 2004. Diffusion kinetic study of chromium(VI) biosorption by *Aeromonas caviae*. Ind. Eng. Chem. Res. 43, 1748–1755.

Loukidou, M.X., Matis, K.A., Zouboulis, A.I., Liakopoulou-Kyriakidou, M., 2003. Removal of As(V) from wastewaters by chemically modified fungal biomass. Water Res. 37, 4544–4552.

Mandjiny, S., Zouboulis, A.I., Matis, K.A., 1995. Removal of cadmium from dilute solutions by hydroxyapatite. Part I. Sorption studies. Sep. Sci. Technol. 30, 2963–2978.

Mata, Y.N., Blázquez, M.L., Ballester, A., González, F., Muñoz, J.A., 2010. Studies on sorption, desorption, regeneration and reuse of sugar-beet pectin gels for heavy metal removal. J. Hazard. Mater. 178, 243–248.

Matis, K.A., Peleka, E.N., 2010. Alternative flotation techniques for wastewater treatment: focus on electroflotation. Sep. Sci. Technol. 45, 2465–2474.

Matis, K.A., Lazaridis, N.K., 2002. Flotation techniques in water technology for metals recovery: dispersed-air vs. dissolved-air flotation. J. Min. Metall. A 38, 1–27.

Matis, K.A., Zouboulis, A.I., 2001. Flotation techniques in water technology for metals recovery: the impact of speciation. Sep. Sci. Technol. 36, 3777–3800.

Matis, K.A., Zouboulis, A.I., 1994. Flotation of cadmium-loaded biomass. Biotechnol. Bioeng. 44, 354–360.

Matis, K.A., Zouboulis, A.I., Lazaridis, N.K., 2003b. Heavy metals removal by biosorption and flotation. Water Air Soil Pollut. 3, 143–151.

Matis, K.A., Zouboulis, A.I., Grigoriadou, A.A., Lazaridis, N.K., Ekateriniadou, L.V., 1996. Metal biosorption—flotation. Application to cadmium removal. Appl. Microbiol. Biotechnol. 45, 569–573.

Matis, K.A., Zouboulis, A.I., Hancock, I.C., 1994a. Waste microbial biomass for cadmium ion removal: application of flotation for downstream separation. Bioresour. Technol. 49, 253–259.

Matis, K.A., Zouboulis, A.I., Hancock, I.C., 1994b. Biosorptive flotation in metal ions recovery. Sep. Sci. Technol. 29, 1055–1071.

Memon, J.R., Memon, S.Q., Bhanger, M.I., El-Turki, A., Hallam, K.R., Allen, G.C., 2009. Banana peel: a green and economical sorbent for the selective removal of Cr(VI) from industrial wastewater. Colloids Surf. B 70, 232–237.

Michalak, I., Chojnacka, K., Witek-Krowiak, A., 2013. State of the art for the biosorption process—a review. Appl. Biochem. Biotechnol. 170, 1389–1416.

Mohammed, A.A., Abed, F.I., Al-Musawi, T.J., 2016. Biosorption of Pb(II) from aqueous solution by spent black tea leaves and separation by flotation. Desalin. Water Treat. 57, 2028–2039.

Naja, G., Volesky, B., 2006. Multi-metal biosorption in a fixed-bed flow-through column. Colloids Surf. A 281, 194–201.

Nazari, A.M., Cox, P.W., Waters, K.E., 2015. Biosorptive flotation of copper ions from dilute solution using BSA-coated bubbles. Miner. Eng. 75, 140–145.

Nelson, P.O., Chung, A.K., Hudson, M.C., 1981. Factors affecting the fate of heavy metals in the activated sludge process. J. Water Pollut. Control Fed. 53, 1323–1333.

Norberg, A.B., Persson, H., 1984. Accumulation of heavy metal ions by Zooglea ramigera. Biotechnol. Bioeng. 26, 239–246.

Özdemir, S., Kadir Oduncu, M., Kilinc, E., Soylak, M., 2017. Tolerance and bioaccumulation of U(VI) by *Bacillus mojavensis* and its solid phase preconcentration by *Bacillus mojavensis* immobilized multiwalled carbon nanotube. J. Environ. Manag. 187, 490–496.

Park, J.K., Jin, Y.B., Chang, H.N., 1999. Reusable biosorbents in capsules from *Zoogloea ramigera* cells for cadmium removal. Biotechnol. Bioeng. 63, 116–121.

Peleka, E.N., Gallios, G.P., Matis, K.A., 2018. A perspective on flotation: a review. J. Chem. Technol. Biotechnol. 93, 615–623.

Peleka, E.N., Matis, K.A., 2011. Water separation processes and sustainability. Ind. Eng. Chem. Res. 50, 421–430.

Peleka, E.N., Matis, K.A., 2009. Bioremoval of metal ion and water treatment in a hybrid unit. Sep. Sci. Technol. 44, 3597–3614.

Prakasham, S., Merre, J.S., Sheela, R., Saswathi, N., Ramakrishna, S., 1999. Biosorption of chromium VI by free and immobilized *Rhizopus arrhizus*. Environ. Pollut. 104, 421–427.

Ramírez-Rodríguez, A.E., Reyes-Ledezma, J.L., Chávez-Camarillo, G.M., Cristiani-Urbina, E., Morales-Barrera, L., 2018. Cyclic biosorption and desorption of Acid Red 27 onto *Eicchornia crassipes* leaves. Rev. Mexic. Ing. Quím. 17, 1121–1134.

Ranjan, B., Pillai, S., Permaul, K., Singh, S., 2019. Simultaneous removal of heavy metals and cyanate in a wastewater sample using immobilized cyanate hydratase on magnetic-multiwall carbon nanotubes. J. Hazard. Mater. 363, 73–80.

Ronda, A., Calero, M., Blázquez, G., Pérez, A., Martín-Lara, M.A., 2015. Optimization of the use of a biosorbent to remove heavy metals: Regeneration and reuse of exhausted biosorbent. J. Taiwan Chem. Eng. 51, 109–118.

Rozada, F., Otero, M., Morán, A., García, A.I., 2008. Adsorption of heavy metals onto sewage sludge-derived materials. Bioresour. Technol. 99, 6332–6338.

Sahmoune, M.N., 2018. Performance of *Streptomyces rim*osus biomass in biosorption of heavy metals from aqueous solutions. Microchem. J. 141, 87–95.

Sankararamakrishnan, N., Jaiswal, M., Verma, N., 2014. Composite nanofloral clusters of carbon nanotubes and activated alumina: an efficient sorbent for heavy metal removal. Chem. Eng. J. 235, 1–9.

Saranya, N., Ajmani, A., Sivasubramanian, V., Selvaraju, N., 2018. Hexavalent chromium removal from simulated and real effluents using *Artocarpus heterophyllus* peel biosorbent—batch and continuous studies. J. Mol. Liq. 265, 779–790.

Saratale, R.G., Karuppusamy, I., Saratalec, G.D., Pugazhendhid, A., Kumare, G., Parka, Y., Ghodakef, G.S., Bharagavag, R.S., Banuh, J.R., Shinc, H.S., 2018. A comprehensive review on green nanomaterials using biological systems: recent perception and their future applications. Colloids Surf. B 170, 20–35.

Saravani, N., Arulmozhi, M., Anbuthangam, A., Manimozhi, M., 2017. Abstraction of phenol onto *Pseudomonas putida* and cetyl trimethyl ammonium bromide. Cell. Mol. Biol. 63, 33–41.

Senthilkumar, R., Reddy Prasad, D.M., Govindarajan, L., Saravanakumar, K., Naveen Prasad, B.S., 2017. Green alga-mediated treatment process for removal of zinc from synthetic solution and industrial effluent. Environ. Technol. 40, 1262–1270.

Smith, E.H., 1996. Uptake of heavy metals in batch systems by a recycled iron-bearing material. Water Res. 30, 2424–2434.

Solari, P., Zouboulis, A.I., Matis, K.A., Stalidis, G.A., 1996. Removal of toxic metals by biosorption onto nonliving sewage sludge. Sep. Sci. Technol. 31, 1075–1092.

Taka, A.L., Fosso-Kankeu, E., Pillay, K., Mbianda, X.Y., 2018. Removal of cobalt and lead ions from wastewater samples using an insoluble nanosponge biopolymer composite: adsorption isotherm, kinetic, thermodynamic, and regeneration studies. Environ. Sci. Pollut. Res. 25, 21752–21767.

Tobin, J.M., White, C., Gadd, G.M., 1994. Metal accumulation by fungi: applications in environmental biotechnolcgy. J. Ind. Microbiol. 13, 126–130.

Trujillo, E.M., Jeffers, T.H., Ferguson, C., Stevenson, H.Q., 1991. Mathematically modeling the removal of heavy metals from wastewater using immobilized biomass. Environ. Sci. Technol. 25, 1559–1565.

Tuzen, M., Saygi, K.O., Usta, C., Soylak, M., 2008. *Pseudomonas aeruginosa* immobilized multiwalled carbon nanotubes as biosorbent for heavy metal ions. Bioresour. Technol. 99, 1563–1570.

Ucun, H., Bayhan, Y.K., Kaya, Y., Cakici, A., Algur, O.F., 2002. Biosorption of chromium(VI) from aqueous solution by cone biomass of *Pinus sylvestris*. Bioresour. Technol. 85, 155–158.

Veglio, F., Beolcini, F., 1997. Removal of metals by biosorption: a review. Hydrometallurgy 44, 301–316.

Veneu, D.M., Pino, G.A.H., Torem, M.L., Saint'Pierre, T.D., 2012. Biosorptive removal of cadmium from aqueous solutions using a *Streptomyces lunalinharesii* strain. Miner. Eng. 29, 112–120.

Volesky, B., 2001. Detoxification of metal-bearing effluents: biosorption for the next century. Hydrometallurgy 59, 203–216.

Volesky, B., 1987. Biosorbents for metal recovery. Trends Biotechnol. 5, 96–101.

Wang, L., Yang, J., Chen, Z., Liu, X., Ma, F., 2013a. Biosorption of Pb(II) and Zn(II) by extracellular polymeric substance (EPS) of *Rhizobium radiobacter*: equilibrium, kinetics and reuse studies. Arch. Environ. Prot. 39, 129–140.

Wang, Y., Liang, S., Chen, B., Guo, F., Yu, S., Tang, Y., 2013b. Synergistic removal of Pb(II), Cd(II) and humic acid by Fe3O4@mesoporous silica-graphene oxide composites. PLoS ONE 8. Aricle ID e65634.

Wilde, E.W., Benemann, J.R., 1993. Bioremoval of heavy metals by the use of microalgae. Biotechnol. Adv. 11, 781–812.

Wu, M., Liang, J., Tang, J., Li, G., Shan, S., Guo, Z., Deng, L., 2017. Decontamination of multiple heavy metals-containing effluents through microbial biotechnology. J. Hazard. Mater. 337, 189–197.

Xiao, G., Lan, K., Su, H., Tan, T., 2014. Preparation of a modified chitosan-mycellium adsorbent with polyvinyl alcohol. Sep. Sci. Technol. 49, 1279–1288.

Xin, S., Zeng, Z., Zhou, X., Luo, W., Shi, X., Wang, Q., Deng, H., Du, Y., 2017. Recyclable *Saccharomyces cerevisiae* loaded nanofibrous mats with sandwich structure constructing via bio-electrospraying for heavy metal removal. J. Hazard. Mater. 324, 365–372.

Xu, J., Zhen Cao, Z., Zhang, Y., Yuan, Z., Lou, Z., Xinhua Xu, X., Wang, X., 2018. A review of functionalized carbon nanotubes and graphene for heavy metal adsorption from water: preparation, application, and mechanism. Chemosphere 195, 351–364.

Yang, S., Li, L., Pei, Z., Li, C., Lv, J., Xie, J., Wen, B., Zhang, S., 2014. Adsorption kinetics, isotherms and thermodynamics of Cr(III) on graphene oxide. Colloids Surf. A 457, 100–106.
Yang, T., Sheng, L., Wang, Y., Wyckoff, K.N., He, C., He, Q., 2018a. Characteristics of cadmium sorption by heat-activated red mud in aqueous solution. Sci. Rep. 8, 1–13.
Yang, Y., Lin, E., Tao, X., Hu, K., 2018b. High efficiency removal of Pb(II) by modified spent compost of *Hypsizygus marmoreus* in a fixed-bed column. Desalin. Water Treat. 102, 220–228.
Youssef, A.M., El-Naggar, M.E., Malhat, F.M., El Sharkawi, H.M., 2019. Efficient removal of pesticides and heavy metals from wastewater and the antimicrobial activity of *f*-MWCNTs/PVA nanocomposite film. J. Clean. Prod. 206, 315–325.
Zamboulis, D., Peleka, E.N., Lazaridis, N.K., Matis, K.A., 2011. Metal ion separation and recovery from environmental sources using various flotation and sorption techniques. J. Chem. Technol. Biotechnol. 86, 335–344.
Zhang, L., Xu, Z., 2016. A review of current progress of recycling technologies for metals from waste electrical and electronic equipment. J. Clean. Prod. 127, 19–36.
Zouboulis, A.I., Matis, K.A., 2010. Hydrophobicity in biosorptive flotation for metal ion removal. Int. J. Environ. Technol. Manag. 12, 192–201.
Zouboulis, A.I., Matis, K.A., 2009. Biosorptive flotation for metal ions removal: the influence of surface tension. Desalination 253, 1–13.
Zouboulis, A.I., Matis, K.A., 1997. Removal of metal ions from dilute solutions by sorptive flotation. Crit. Rev. Environ. Sci. Technol. 27, 195–235.
Zouboulis, A.I., Loukidou, M.X., Matis, K.A., 2004. Biosorption of toxic metals from aqueous solutions by bacteria strains isolated from metal-polluted soils. Process Biochem. 39, 909–916.
Zouboulis, A.I., Matis, K.A., Loukidou, M.X., Šebesta, F., 2003. Metal biosorption by PAN-immobilized fungal biomass in simulated wastewaters. Colloids Surf. A 212, 185–195.
Zouboulis, A.I., Lazaridis, N.K., Matis, K.A., 2002. Removal of toxic metal ions from aqueous systems by biosorptive flotation. J. Chem. Technol. Biotechnol. 77, 958–964.
Zouboulis, A.I., Matis, K.A., Lazaridis, N.K., 2001b. Removal of metal ions from simulated wastewater by *Saccharomyces* yeast biomass; combining biosorption and flotation processes. Sep. Sci. Technol. 36, 349–365.
Zouboulis, A.I., Rousou, E.G., Matis, K.A., Hancock, I.C., 1999. Removal of toxic metals from aqueous mixtures: part 1 biosorption. J. Chem. Technol. Biotechnol. 74, 429–436.
Zouboulis, A.I., Matis, K.A., Hancock, I.C., 1997. Biosorption of metals from dilute aqueous solutions. Sep. Purif. Method 26, 255–295.

Further reading

Cayllahua, J.E.B., Torem, M.L., 2011. Biosorptive flotation of nickel and aluminum ions from aqueous solution. Desalination 279, 195–200.
Loukidou, M.X., Matis, K.A., Zouboulis, A.I., 2001. Removal of arsenic from contaminated dilute aqueous solutions using biosorptive flotation. In: *3rd Europ. Congress Chemical Engineering*. DECHEMA, Nuremberg, Germany.
Matis, K.A., Zouboulis, A.I., Lazaridis, N.K., Hancock, I.C., 2003a. Sorptive flotation for metal ions recovery. Int. J. Miner. Process. 70, 99–108.
Smith, S.C., Rodrigues, D.F., 2015. Carbon-based nanomaterials for removal of chemical and biological contaminants from water: a review of mechanisms and applications. Carbon 91, 122–143.

Volesky, B., 2007. Biosorption and me. Water Res. 41, 4017–4029.
Zouboulis, A.I., Matis, K.A., 2012. Cadmium ion removal by electroflotation onto sewage sludge biomass. Int. J. Environ. Waste Manag. 9, 245–256.
Zouboulis, A.I., Matis, K.A., Rousou, E.G., Kyriakidis, D.A., 2001a. Biosorptive flotation for metal ions recovery. Water Sci. Technol. 43, 123–129.

Carbon nanomaterial applications in air pollution remediation

7

Esraa Gabal[a], Somenath Chatterjee[b], Farah K. Ahmed[c], Kamel A. Abd-Elsalam[d]
[a]Agriculture Science and Resource Management in the Tropics and Subtropics, Faculty of Agriculture, Bonn University, Bonn, Germany, [b]Centre for Materials Science and Nanotechnology, Sikkim Manipal Institute of Technology, Sikkim Manipal University, Gangtok, India, [c]Biotechnology English Program, Faculty of Agriculture, Ain Shams University, Shoubra El-Kheima, Cairo, Egypt, [d]Plant Pathology Research Institute, Agricultural Research Center (ARC), Giza, Egypt

1 Introduction

As one of the most growing scientific areas, nanotechnology is often referred to as the "Next Industrial Revolution," holding within its aspects both research and technology means (Roco, 2005). Hence, nanotechnology is believed to be a promising solution for minimizing environmental hazards (Watlington, 2005). Additionally, nanoparticles (NPs) can modify their physicochemical properties, and hence, factors such as the chemical origin, size, and shape of NPs determine their application and efficiency. NPs encompass two classes: the organic NPs, which include carbon NPs (fullerenes), and inorganic NPs, representing magnetic noble metals and semiconductor nanomaterials (Monica and Cremonini, 2009). According to Okhovat et al. (2015), Large surface to volume ratio and size dependent optical properties enable nanomaterials to demonstrate their excellent applications. Carbon-based nanomaterials, having allotrope e.g., graphene, C60 fullerenes, and nanotubes, are very popular due its wide applications starting from electronics, drug delivery, sensors, energy storage, and many more. NPs are extensively used for the elimination of biological impurities (such as bacteria) and chemical contaminants, including organic pollutants. Carbon-based nanomaterials such as carbon nanotubes (CNTs), C60 fullerenes, graphene, and graphene oxides have been the focus of applied research for the last 20 years. They are envisaged to be promising tools for the potential implementation of wastewater treatment, detection system, and other remediation processes. Hence, they are believed to be effective at facing the challenges of filtration-related problems, water scarcity, and polluted resources (Kah and Hofmann, 2015) through being implemented in the development of highly selective, sensitive, and efficient adsorbent gears that can assist in the removal of a wide range of contaminants (biological, organic, and inorganic); they can also be applied to drug delivery systems and nanosensors (Bergmann and Machado et al., 2015). Recently, industrial development and trends towards agricultural practices boost the possibility of massive environmental pollution in the society. Topic related to resist the degradation of atmosphere is timely. In this regard, the carbon-based nanomaterials show its competencies to removal of airborne gas pollutants. The well

Carbon Nanomaterials for Agri-food and Environmental Applications. https://doi.org/10.1016/B978-0-12-819786-8.00007-4

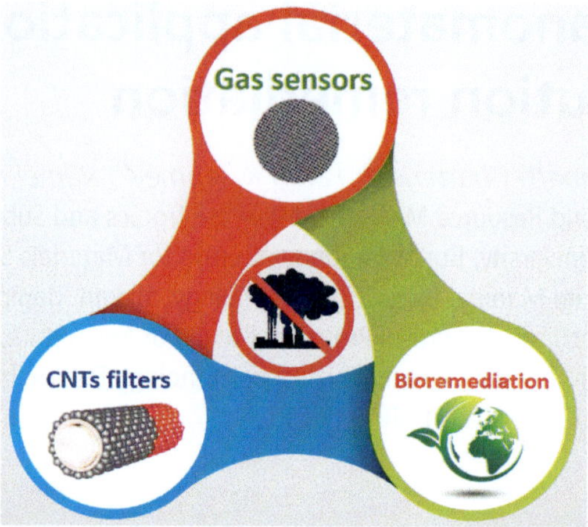

Fig. 1 The three main applications of carbon nanomaterials in air pollution include remediation of air pollution, CNT filters, and CNM-based gas sensors.

known characteristics of carbon nanomaterials, i.e., large specific surface area, mesoporous nature ensure them for the application of air pollution monitoring system. In this chapter, we spotlight the applications of various types of CNMs regarding the remediation of air pollution, CNT filters, CNM-based gas sensors, their mechanisms, and potential implications in the future (Fig. 1).

2 Nanobioremediation

An emerging technique that involves the application of nanoparticles and nanomaterials in the removal of environmental pollutants such as heavy metals and both organic and inorganic contaminants from polluted sites is widely called nanobioremediation. This technique guarantees efficient remediation of the environmental spots with economic and ecofriendly means. Hence, it can be considered the most preferable and sustainable tool for environmental clean-up rather than physical and chemical remediation and incineration (Singh and Walker, 2006). Traditional bioremediation includes the implementation of plants, microbes, and enzymes. Several reports may be found to rule out atmospheric contaminants using nanomaterials to act as adsorbents (Bergmann and Machado, 2015; Gupta and Saleh, 2013). Although the aforementioned biological factors showed efficiency, advancing their implementation with nanoparticles exhibited enhanced pesticide degradation. For instance, pesticides such as atrazine, molinate, and chlorpyrifos may be degraded by nano-sized zerovalent iron (Ghormade et al., 2011) However, several complex organic compounds, for example, long-chain hydrocarbons and organochlorines, could be degraded through a combined implementation of both nanoparticles and biocontrol agents such as nano-encapsulated enzymes. This can result in smaller-sized compounds that can be easily degenerated by the joint activity of plants and microbes. This could overcome the limitations of the microbial and phytochemical inability to degrade complex organic molecules (Yadav et al., 2017). It is envisaged that nanotechnology can positively affect the functionality of plants in bioremediation. This is due to the stimulating influence of some nanoparticles such as CeO_2-, TiO_2-, and ZnO-NPs on plant growth through enhancing both shoot and root development (López-Moreno et al., 2010a,b). The findings proved the promoting effect of nano-TiO_2 on enhancing several enzyme activities, along with its ability to accelerate the conversion of inorganic nitrogen into an organic form and stimulating nitrate adsorption (Hong et al., 2005; Yang et al., 2006a,b).

Carbon nanomaterials have gained more research attention with an increase in their upscaling due to their diverse functions, mainly in environmental remediation such as that of soil, water, and air with various forms. Depending on their construction mechanism, functionalization status, and the cleaning/purification methods employed, the properties of CNMs dramatically vary. Therefore, their environmental function/effect varies between their different forms and its determination is challenging (Nowack and Bucheli, 2007). In the scientific community, the challenge has been taken to synthesize carbon nanomaterials and their activities as a detector of gas pollutants at the ppb level (Terrones et al., 2010). High concern has been recently given to the remediation of air and water pollutants that, according to the US Environmental Protection Agency (2009), contain particulate matter (PM); nitrogen-based oxides (NOx); ground-level ozone; sulfur-based oxides (SOx); volatile organic compounds (VOCs), mainly benzene, toluene, ethylbenzene, and xylenes (BTEX); carbon monoxide; chlorofluorocarbons (CFCs); arsenic; lead; mercury; microbial pathogens; or endocrine disruptors believed to be easily dispersed over several distances (Llobet, 2012; US Environmental Protection Agency, 2009).

3 Carbon nanomaterials (CNMs): forms and mode of action

Organic carbon nanomaterials include three main classes with different applications depending on their various functional mechanisms, represented in fullerene (Kroto et al., 1985), carbon nanotubes (CNTs) (Iijima, 1991; Iijima and Ichihashi, 1993), and the graphene family (Novoselov et al., 2004). Both CNTs and the graphene family were found to possess a high adsorption capacity that enables them to remove various contaminants, particularly in water treatment, while fullerene showed a lower intensity in the adsorptive activity compared to the other two CNM classes (Bergmann and Machado et al., 2015; Tonel et al., 2015) (see Fig. 2). For example, CNTs are able to isolate methane from carbon dioxide in an exhaust system due to their different pore sizes. They then trap these greenhouse gases, which are common outcomes as air pollutants in coal mining and power generation regions. Studies demonstrated the efficiency of graphene oxide and its reduced form in the successful treatment of various pollutants encompassing antibiotics (Chowdhury and Balasubramanian, 2014), heavy metals (Huang et al., 2011), and synthetic dyes (Li et al., 2013). Various properties

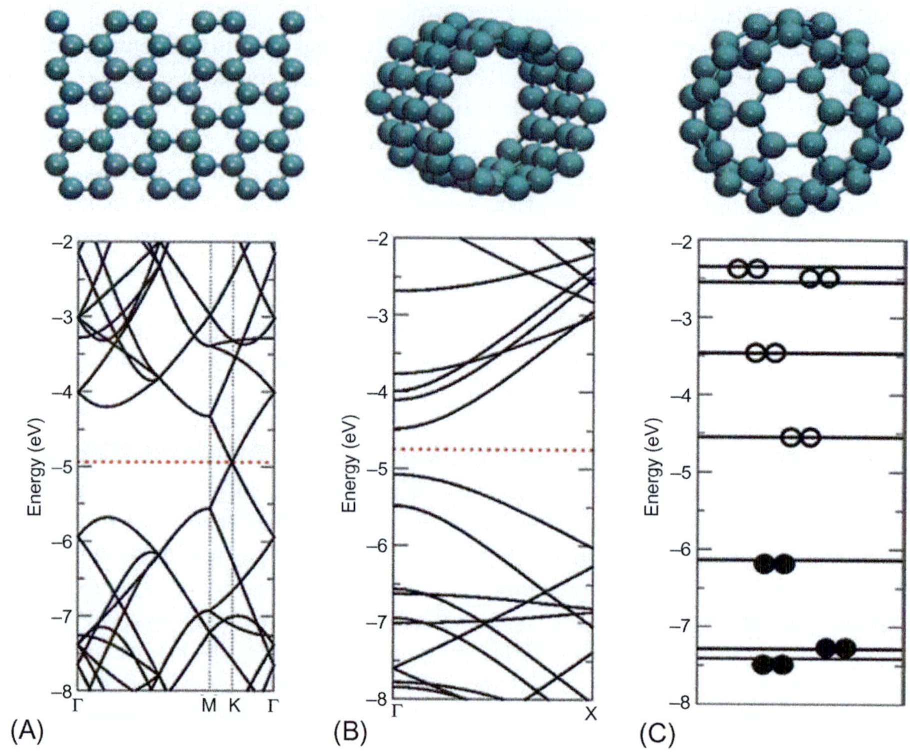

Fig. 2 Various forms of carbon nanomaterials (CNMs) with their respective electronic levels. The horizontal dashed red lines correspond to the Fermi energy. (A) graphene, (B) carbon-nanotube (CNTs), and (C) fullerene.
Adapted from Tonel, M.Z., de Menezes, V.M., Zanella, I., Fagan, S.B., 2015. Molecules with biological interest adsorbed on carbon nanostructures. In: Carbon Nanomaterials as Adsorbents for Environmental and Biological Applications. Springer, Cham, pp. 107–122, with permission from Springer.

support CNM application in physical/chemical treatment, along with the development of efficient, selective, and sensitive adsorbent devices for environmental sanitization. According to several studies, CNMs are characterized by chemical stability, low density, structure diversification, accountability for being produced at a large scale. The ideal textile properties regarding their average pore diameter, total pore volume, and the high surface-to-volume ratio was enhanced their application in the studies concerning water treatment and various remediation processes (Machado et al., 2011, 2012; Prola et al., 2013; Wang et al., 2011). Additionally, along with the high adsorptive sites existing on the surface of CNM structures, their surface is known to be amphoteric where it can be deprotonated (negatively charged) or protonated (positively charged). This features the possibility of adding more functional groups with oxygen to the CNM surface, which enhances their chemical adsorption activity (Gupta and Saleh, 2013; Ren et al., 2013).

3.1 Carbon nanotubes (CNTs)

This form of carbon nanomaterial possesses a wide spectrum of properties, supporting its ideal functionality in adsorption practices. Studies have clarified that carbon nanotubes can undergo interactions with organic molecules via noncovalent forces such as hydrogen bonding, π–π stacking, electrostatic forces, van der Waals forces, and hydrophobic interactions (Rengaraj et al., 2007). CNTs have a short intraparticle distance, tunable surface chemistry, assessable adsorption sites, and high specificity. Additionally, CNT construction tolerates the combination of other surface functional groups (OH, COOH, C=O), which may boost the selectivity, fidelity, and efficacy of the extreme remediation ability of the resulting system (Gupta and Saleh, 2013) compared to the bulk form of carbon, that is, activated carbons. CNTs shaped with cylindrical pores and air-pollutant molecules interact with their carbon atoms on the surrounding walls. The dimension and geometry of pores depend on the interaction between molecules and the solid surface during synthesis. Moreover, the adsorptive surface of the CNTs is hydrophobic in nature (Ren et al., 2013). Because of the existence of oxygen holding functional groups with a high surface area, CNTs are able to chemically adsorb the heavy metal ions (arsenic, phosphate, mercury, etc.) and organic compounds (Polycyclic aromatic hydrocarbons (PAHs), Dichlorodiphenyltrichloroethane (DDT), dioxin, (Mueller and Nowack, 2010; Pan et al., 2010; Rickerby and Morrison, 2007; Stafiej and Pyrzynska, 2007). In addition, CNTs have high chemical stability, low toxicity, and strong antimicrobial activity, which may support them executing the disinfection process to remove diamines, phenols, formaldehyde, hydrogen peroxide, silver ions, halogens, glutaraldehyde, and acridines (Amin et al., 2014; Mcdonnell and Russell, 1999). The high permeability, photocatalytic activity, specificity and selectivity, hydrophilicity, and the stable mechanical and chemical nature of CNTs assist their functionality in membrane penetration and sweeping away chemicals such as chlorinated compounds, polyethylene, 1,2-dichlorobenzene, (in) organic solutes, and halogenated organic solvents (Amin et al., 2014; Mcdonnell and Russell, 1999). Carbon nanotubes are widely known for their application in water remediation due to their highly significant adsorption capacity for many aqueous pollutants (organic and inorganic) (see Fig. 3) (Gupta and Saleh, 2013; Machado et al., 2011). Two types of CNTs were developed representing multiwalled carbon nanotubes (MWCNTs) and single-walled carbon nanotubes (SWCNTs), which differ in their morphology and porous nature. Structure-wise, CNTs are rolling up a single-layer of graphitic carbon into a hollow cylindrical form with a diameter in the nanoscale. The SWCNTs contain a cylindrical single layer of graphitic carbon, whereas MWCNTs hold a number of concentric cylindrical single layers of graphitic carbon (Zhao and Stoddart, 2009). Both types are managed in their structure by the influence of Van der Waals forces, which control CNT adherence to one another, forming aggregations (Upadhyayula et al., 2009). Moreover, the adsorptive activity of individual SW/MWCNTs was found to be different from that of their aggregated forms as a consequence of the difference in adsorption sites (Babaa et al., 2004; Machado et al., 2012; Ren et al., 2011).

In the case of SWCNT bundles, adsorption can occur at four sites: outside the bundles, the interstitial channels between the tubes in bundles, in the pores of the

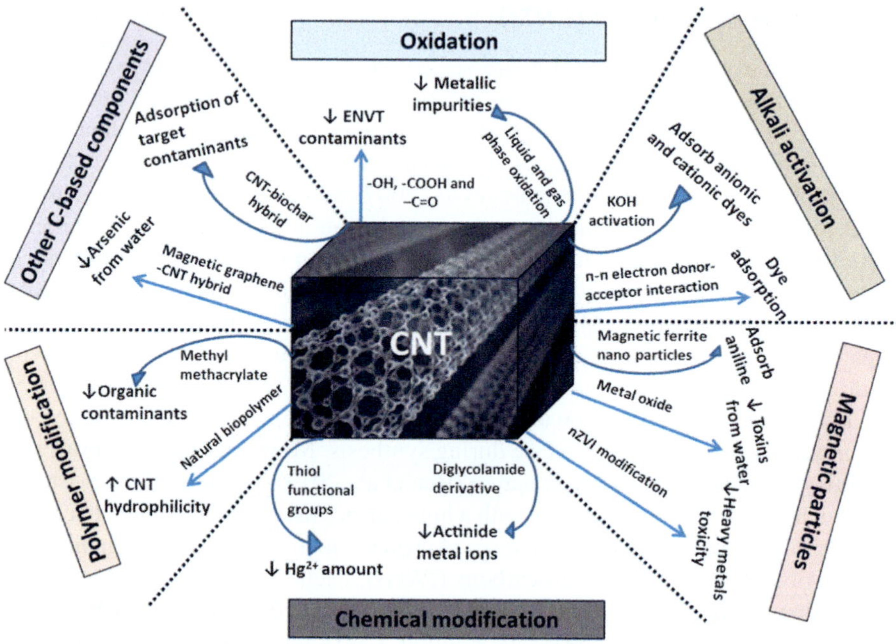

Fig. 3 Schematic diagram showing various modifications of CNTs for contaminant removal from water and wastewater (C: carbon; CNT: carbon nanotube; ENVT: environmental; Hg: mercury; KOH: potassium hydroxide).
Adapted from Sarkar, B., Mandal, S., Tsang, Y.F., Kumar, P., Kim, K.H., Ok, Y.S., 2018. Designer carbon nanotubes for contaminant removal in water and wastewater: A critical review. Sci. Total Environ. 612, 561–581 with permssion from Elsevier.

nanotube, and the grooves between adjacent tubes (Ren et al., 2011; Serp et al., 2003). On the other hand, the adsorptive sites of MWCNT aggregated forms are represented in the pores aggregated in their inner surface and at the outer one because MWCNT aggregations do not take the bundle shape (Machado et al., 2011; Yang et al., 2001). Hence, SWCNT bundles are highly effective for the adsorption of small molecules such as gases through their interstitial channels (Agnihotri et al., 2005) while MWCNTs are effective for the adsorption of large biomolecules such as bacteria and viruses (Upadhyayula et al., 2009). The increase in aromatic ring number within MWCNTs vigorously advances the sorption affinity of phenolic compounds on MWCNTs due to the enhanced π–π interactions, which rely on the number, shape, and size of aromatic molecules (Chen et al., 2007). In general, CNTs are considered attractive components for constructing gas sensors involved in environmental monitoring that are small in size and wearable due to their easy application in miniaturized sensors and on flexible substrates, along with their elasticity when being operated at room temperature (Leghrib et al., 2010a). They are considered the most promising of all nanomaterials developed (Giles, 2006). To enhance the photocatalytic degradation mechanism, a combination of carbon-based nanostructures, that is, CNTs or graphene nanosheets, with TiO_2 resulting in the formation of the composite may be applied, because electrons can be easily transferred through the CNTs and restrict the electron-hole recombination (Low et al., 2017).

3.2 Fullerenes

Among CNM forms, fullerenes (commonly known as the chemical compound C60; buckminsterfullerene) have gained wide attention within the research and commercial sectors due to their thermal and chemical characteristics that enable fullerenes to be efficient electron acceptors possessing various environmental applications (Pan et al., 2008). Fullerene has diversified its structure because the ball-shaped C60 molecule has a number of carbon atoms ranging between 20, 60, 70, 82, 100, and up to 960, arranged in 12 pentagons and 20 hexagons (Inagaki and Kang, 2014). Interestingly, the surface area of fullerenes is highly affected by the surrounding area because studies proved the incidence of modification in the fullerene surface area in response to its exposure to both chemical and physical treatments. For instance, the functional groups of hydroxyl (-OH) and methyl (-CH_3) existing on the surface of fullerenes were found to be eliminated, along with the release of carbon monoxide, upon heat-treating fullerenes with carbon dioxide (Silva et al., 1999). Similar to CNTs, fullerenes molecule has a unique physical adsorptive activity for organic and organometallic compounds (Ballesteros et al., 2000), polycyclic aromatic hydrocarbons (Yang et al., 2006a,b), naphthalene, and 1,2-dichlorobenzene (Cheng et al., 2005), strongly believed to be controlled by the dispersive interaction forces (Yang et al., 2006a,b). The interaction robustly occurs between the fullerene surface and organic molecules through π–π stacking, supported by the coverage of the inner and outer surface of fullerenes by homogenous π electrons (Pan et al., 2008).

3.3 Graphene and its derivatives

Being a potential alternative to water pollution adsorbents with unique physical-chemical characteristics, the graphene (monolayer of graphite in hexagonal form, sp^2-hybridized two-dimensional allotrope of carbon) family, including graphene, graphene oxide (GO), and reduced graphene oxide (rGO), has been intensively investigated (Bergmann and Machado et al., 2015; Novoselov et al., 2004; Terrones et al., 2010). Synthesized graphene is known for its large surface area due to the existence of large amounts of micro- and mesopores (Seresht et al., 2013). The layers of pristine graphene generally stay in an agglomeration state, which results in the reduction of surface area. The introduction of metals or other compounds into pristine graphene (called functionalization) will increase the effective surface area and consequently that is better for pollutant removal (Gupta Chatterjee et al., 2015). The surface of graphene oxide was scrutinized, showing that it possessed a high number of reactive chemical groups, including carboxyl (–COOH) and carbonyl groups (–C=O), located at the sheet edges while the basal plane of the sheet is the site of epoxy (C–O–C) and hydroxyl (–OH) (Stankovich et al., 2006). The incidence of the aforementioned functional groups in GO sheets enhances GO negative charge density, and therefore

GO represents a vigorous adsorbent for heavy metals of positive charges (Mishra and Ramaprabhu, 2011; Ren et al., 2013), synthetic dyes (Yang et al., 2011), and cationic dyes (Ramesha et al., 2011), along with its adsorptive activity for organic compounds such as antibiotics (Gao et al., 2012) due to its aromatic nature (Bergmann and Machado et al., 2015). Similarly, the reduced form of graphene oxide (rGO) has a wide range of micro- and mesopores on its surface that supports its implementation as an adsorbent with environmental applications (Pan et al., 2013; Ramesha et al., 2011; Tiwari et al., 2013). Studies illustrated the adsorptive effectivity of rGO for effluents containing radionuclides (Pan et al., 2013), oil and organic solvents (Bi et al., 2012), metallic ions (Wang et al., 2012a,b), and synthetic dyes (Bai et al., 2013). Unlike graphene oxide, rGO lacks the high negative charge intensity at its surface, which advances its applicability for the adsorption of anionic dyes (Ramesha et al., 2011). The adsorptive activity of rGO for dyes is processed through two mechanisms, either the π–π stacking that occurs between sp2 regions of rGO and the aromatic part of the reacted dye, or the electrostatic interactions between the charged dye and the oxygen-containing functional groups existing at the rGO surface (Bai et al., 2013).

4 Applications

4.1 Carbon nanomaterials and removal of air pollutants

The steps involved in sensing and controlling air pollutants, even with a lesser quantity in the environment, using nanomaterials (NMs) are adsorption by NMs, degradation by nanocatalysts, and isolation/filtration by nanofillers (Mohamed, 2017). Among the environmental applications of CNMs, air remediation involving the eradication of air pollutants was proven through the development of CNT-supported air filters. The detection of air pollution can occur when contact happens between the CNTs and polluted gas that causes a change in CNT conductance. Possessing a high adsorption capacity and a large specific area, CNTs are ideal for the development of air pollution filters (Ong et al., 2010), and their membranes were reported in the filtration of carbon dioxide from power plant emissions (Lin et al., 2008). Relying on their adsorptive capability, CNTs represent ideal sorbents for improving air remediation, particularly the removal of hazardous air pollutants (HAP), due to their high-energy binding sites (Gangupomu et al., 2014). However, implementing CNTs in pollutant removal requires a pretreatment of the CNTs with chemical, physical, or chemical-physical means, which permits the passage of pollutants within CNTs and the removal of impurities by providing them with an access (window) to the CNT interior structure (Gangupomu et al., 2014). One common method of the physical pretreatment applied to CNTs is heat treatment (HT) at the existence of an oxidizing gas such as oxygen, carbon dioxide, or ozone. Despite the loss of some CNTs during that method, studies proved its functionality in the removal of amorphous carbon and graphitic materials (Karthikeyan et al., 2009; Yudasaka et al., 2003). Interestingly, this method, despite its efficiency, is determined by three main factors, including the oxidizing gas that helps in the release of some functional groups during the treatment, the treatment temperature, and the time of the CNT exposure to the HT. These three factors, as reviewed by Gangupomu et al. (2014), determine the number and the size of the windows formed within the CNTs, which in turn control the functionality of CNTs. The chemical pretreatment can be performed using acids such as nitric, sulfuric, and hydrochloric or nonacids such as ammonium hydroxide/hydrogen peroxide (NH_4OH/H_2O_2) and potassium permanganate/hydrogen peroxide (K_2MnO_4/H_2O_2), and this method depends on the type and strength of the chemical applied, the treatment duration, and the temperature (Gangupomu et al., 2014).

Due to the highly significant adsorptive properties, CNTs were reported to be a highly efficient loading carrier for anticancer agents such as epirubicin hydrochloride (EPI) through the interactions between functional carboxyl groups at the surface of MWCNTs and the EPI through π-π bonding (Chen et al., 2011). Following the same mechanism, a study clarified the incidence of interaction between two benzene rings in dioxin with the CNT surface, which yielded a high sorption energy compared to that from the active carbon application, with a higher amount of adsorbed dioxin by 1035 times (Long and Yang, 2001; Matthews et al., 2000). Hence, the reports indicated the effectiveness of π-π bonding mechanisms in the function of CNTs, particularly with the removal of aromatic compounds and other pollutants (Long and Yang, 2001; Matthews et al., 2000). Concerning clean energy, CNTs, in particular SWCNTs, have shown a potential base for developing fuel cell-powered vehicles due to their impressive performance in storing large amounts of hydrogen. As a result of a practical study, up to 4.2 wt.% of hydrogen could be stored by the SWCNTs under room temperature and a pressure of 10 mega Pascale (Liu et al., 1999) where the adsorption occurred at both the exterior walls and the interstitial channels of the SWCNTs (Kim et al., 2005). Moreover, the reversible performance concerning the release of stored hydrogen from SWCNTs happened through the reduction of the experimenting pressure by 1 atm, which resulted in the release of 78% of the adsorbed gas. However, the reversible operation is debatable (Prakash and Yeom, 2014). From the above discussion, the role of nanomaterials as excellent adsorbents, catalysts, and sensors is distinct. A glimpse of treatment techniques and conditions to detect and isolate the emission of greenhouse gases as airborne pollutants using carbon-based nanomaterials is shown in Table 1. The removal mechanisms of the pollutants are also specified in the table with the references (Mohamed, 2017).

4.2 Carbon nanotube filters

Development of nanostructured membranes to capture several gas pollutants is a vital scope of research today. In air purification (e.g., particle removal) filtration of this inlet airflow for gasoline turbines processed at industrial plants, the nanofiber-coated filter media has recently been put to use (Muralikrishnan et al., 2014). Compared with microfibers, the most significant advantage of nanofibers is the high specific surface area (Misslitz et al., 2013). Likewise, high-efficiency particulate airflow (HEPA) filter systems and activated carbon (AC) filter systems are trusted to cope with inside PM and ozone pollution, respectively. Presently, the most popular nanofibers found in air filtration are polymer nanofibers (Misslitz et al., 2013). A high microporus and adequate

Table 1 Treatment of greenhouse gases and other air pollutants by nanoadsorptive structures (Mohamed, 2017, Open Access Journal)

Nanoadsorptive materials	Types of nanoparticles	Target pollutant gases	Removal mechanism	References
Carbon nanotubes (CNTs)	(SWCNTs and MWCNTs)	NO*x* (mixture of NO and NO_2)	NO and O_2 pass through CNTs and NO is oxidized to NO_2 and then adsorbed onto the surface of nitrate species.	Zhang et al. (2012)
	(CNTs-APTS), Modified CNTs using 3aminopropyltrietho xysilane (APTS)	CO_2	Surface of CNTs with abundant amine groups that provide numerous chemical sites for CO_2 adsorption, which makes CNTs adsorb more CO_2 gases at a low temperature range (20–100°C).	Su et al. (2009)
	SWNTs/NaClO	Isopropyl vapor	Physical adsorption by van Der Waals forces and chemical adsorption onto adsorbent surface functional groups	Hsu and Lu (2007)
	CNTs deposited on quartz filters	VOCs	Carried out by π-π interactions.	Amade et al. (2014)
	Si-doped and boron-doped SWCNTs	CO and CH_3OH gases	Physisorption or chemisorption, the electronic properties of SWCNTs significantly improve gas adsorption	Azama et al. (2017)
Fullerene	Fullerene B_{40}	CO_2	High adsorption capacity for CO_2 by strong chemisorptions	Dong et al. (2015)
	Fullerene-like boron nitride nanocage	N_2O	Adsorption and decomposition of N_2O	Esrafili (2017)
Graphene	Graphene oxide (GO)/ nanocomposites	CO_2, NH_3, SO_2, H_2S, and N_2	Functional groups on GO are responsible for adsorption of gases and synergetic effect between the metal centers, GO surface promotes gas adsorption	Seredych and Bandosz (2012), Wu et al. (2013)

nitrogen-containing functional groups as effective adsorption sites are the characteristics of electrospun polyacrylonitrile (PAN)-based activated carbon nanofiber (ACNF), which is capable to adsorb a reasonable quantity of formaldehyde even at a low concentration (Lee et al., 2010). Different designated forms of CNT-based atmospheric filters were suggested for PM purification, such as CNT coating (Soares and Trindade, 2019), free-standing CNT film, aligned CNT sheet (Nasibulin et al., 2011), hierarchical CNT framework, and agglomerated CNT fluidized bed (Wang et al., 2014). Included in this, the hierarchical CNT construction shows better efficiency because of the macroporous setup and substantial structural stability of this porous substrate with a straightforward synthesis process. Reports have handled micron-sized fibrous material filters (Li et al., 2014), glass fiber filters (Park and Hwang, 2014), and ceramic filters (Parham et al., 2013) as substrates for CNT development to acquire hierarchical CNT-based air filters. CNT sheets, attracted from millimeter-sized CNT arrays, have been integrated between classic microfiber fabrics to produce aerosol filters. The outcomes show that these novel CNT filtration systems have filtration features that are much like electrospun fabrics, positioning them as a feasible option for potential filtration applications (Yildiz and Bradford, 2013). For mass-scale production, a method to develop the CNT/quartz fiber (QF) filter with gradient nanostructures using an aerosol technique is reported elsewhere (Li et al., 2014). Due to the introduction of CNTs into the QF filter, the filtration efficiency of the air filter was increased (Li et al., 2014). The manufacturing of high-quality air filters considers the advantages of each structural part, which includes microfibers with high structural stability and CNTs with a high definite surface area, as reported in Fig. 4. A combination of CNT/QF layers acts as the high dust holding capacity by the QF layer and the high filtration efficiency of the CNT-rich layer (Li et al., 2014). Similarly, a hybrid architecture formed by an arrangement of graphene with a variety of metal oxides is studied by (Gupta Chatterjee et al., 2015). Some researchers supported outcomes that showed improved gas sensing ability, particularly the sensitivity and selectivity of the graphene-based hybrid structure at room temperature.

4.3 CNM-supported gas sensors

Reports revealed the efficiency of carbon-based nanomaterials for some relevant chemical species such as CO, NO_x, and VOCs in the environmental state. A study by Leghrib et al. (2010a, 2010b) and Leghrib and Llobet (2011) employed carbon nanotube sensors supported by various metal nanoparticles in detecting air benzene traces. As reported, various metal NPs differently react with benzene and this performance is a consequence of the change in the charge conductivity between the metal NP and the CNTs when adsorption of benzene particles occurs on the surface of the metal nanoparticle. With the implementation of interdigitated electrodes, the change in charge conductivity can be measured (Leghrib and Llobet, 2011; Llobet, 2012; Zanolli et al., 2011). Moreover, carbon nanotubes are involved in the development of sensors depending on the hybrid nanomaterials such as MWCNTs decorated with tin oxide (SnO_2), constructed depending on the ratio between the amount of plasma-treated nanotubes and the tin oxide precursors employed. Such sensors were implemented in

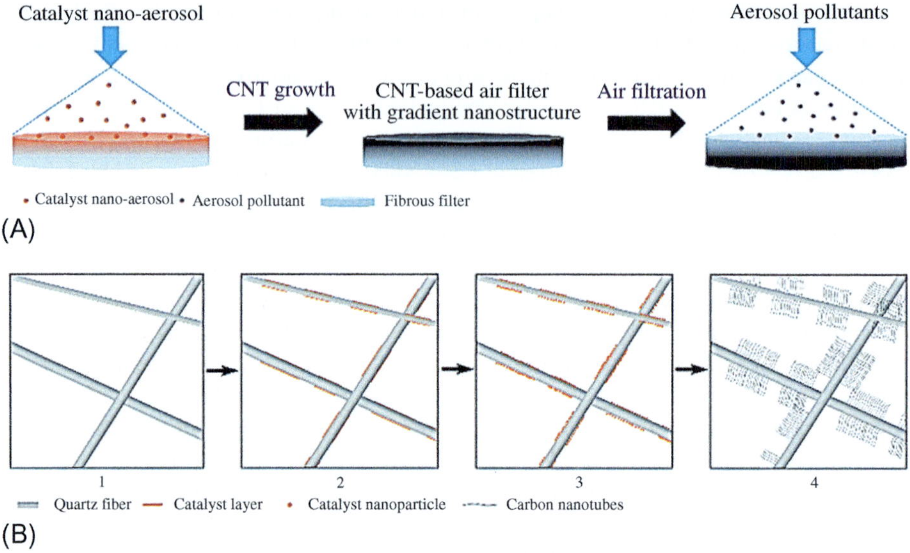

Fig. 4 The synthetic process used to prepare the CNT/quartz fiber filter. (A) Schematic illustrating the catalyst loading on a quartz fiber filter using an aerosol technique and the growth of CNTs with its constant decrease along the thickness of the filter to form a CNT-based air filter with a gradient structure. The CNT-rich side will be placed facing downstream for air filtration. (B) Schematic further illustrating the CNT growth process on the QF filter. `1' Stands for the pristine QF filter, `2' stands for the catalyst layer formed on the QFs after catalyst nanoaerosol filtration by the QF filter, `3' stands for the catalyst nanoparticles formed on the QFs after annealing, and 4 stands for CNT growth on the QF after the CVD process. Reprinted from Li, P., Wang, C., Li, Z., Zong, Y., Zhang, Y., Yang, X., Li, S., Wei, F., 2014. Hierarchical carbon-nanotube/quartz-fiber films with gradient nanostructures for high efficiency and long service life air filters. RSC Adv. 4, 54115–54121, with permission from Royal Society of Chemistry.

the detection of nitrogen oxide in the ppb range and carbon monoxide in the low ppm range and showed significant performance along with their application in reducing moisture cross-sensitivity (Leghrib et al., 2010b; Llobet, 2012). Different carbon materials, such as carbon nanotubes (CNT), charcoal, graphene, graphene oxide (GO), etc., have been shown to be useful as chemical and biosensors due to the ease in tailoring their sensitivity by simple chemical treatments (Gupta Chatterjee et al., 2015).

The detection of hazardous gaseous species was reported using nitrogen or boron-substituted carbon nanotubes that have pyridine type sites on their surface that support the N-doped CNT functionality (Llobet, 2012). By initiating an electric current/voltage through the flow of the charged carrier induced in any material, the target particles become adsorbed on the surface of the CNT, which drives the modulation of the electric current charges. This mechanism is the main step for operating CNT-based sensors that are able to detect temperature, air pressure, chemical gases (such as CO and NH_3), molecular pressure, strain, etc. (Kaushik and Majumder, 2015). NO_2-CNT-based gas sensors showed function in fire detection and exhaust gas monitoring applications. Furthermore, reports showed the extended applications of CNT-based sensors as well in the detection of organic vapors such as methyl ethyl ketone, toluene, dimethyl methyl phosphonate, hexane, acetone, chloroform, acetonitrile, and methanol (Attar and Ranveer, 2015). CNTs have been widely implemented as the basis for the development of various types of gas sensors due to their extremely high adsorptive capacity and the excellent sensitivity to atoms and molecules (Boyd et al., 2014; Zaporotskova et al., 2016; Zhang and Zhang, 2009). According to the study, about four types of CNT-based gas sensors have been developed: sorption, ionization, capacitance, and resonance frequency shift gas sensors (Zaporotskova et al., 2016). Concerning the sorption sensors, they rely on the aforementioned mechanism related to the adsorption of gas molecules on the surface of the nanotube causing a change in its conductivity through losing/acquiring an electron. This type of sensor is the largest among all CNT-based gas sensors, with various forms through modifying the CNTs with metals, polymers, or metal oxides (Li et al., 2003; Zhang and Zhang, 2009; Zhao et al., 2002). In an attempt to reveal the potential functionality of CNT sensors, both Li et al. (2003) and Kong et al. (2000) examined the gas detection of NO_2 and nitrotoluene, but used different designed sensors. For the sensor designed within the first study, the gas detection under room temperature was limited, representing as low as 44 ppb for NO_2 and 262 ppb for nitrotoluene, and the sensor requires up to 10 hours to recover due to the high energy bond existing between the CNT and the adsorbed gas particles. The recovery time could be reduced by up to 10 minutes upon its exposure to UV (Chen et al., 2001). However, in the study by Kong et al. (2000), which applied a field-effect transistor with one semiconducting monolayer CNT, the gas detection was significant, recording 100–1000 ppm for NO_2. Moreover, the device response for the gas occurred within a few seconds. The study of Sun et al. (2007) tested the detection of chlorine (Cl_2) through applying an electrochemical gas sensor supported with modified multilayered CNT films. The red color was detcted when the cathode (sensor surface) is exposed to chlorine gas, and as a consequence, a galvanic effect is generated and later measured. During that process, the nanotubes functioned as microelectrodes (Sun et al., 2007; Zaporotskova et al., 2016).

Reports illustrated how efficient the modification of CNT can be by the usage of functional groups through enhancing their selectivity and response, which can be of importance in the development of gas sensors with detectable ability to specific gases (Zaporotskova et al., 2016). For instance, the modification of CNT using a carboxyl functional group (-COOH) within the gas sensor could help in the detection of carbonyl (CO) with a detection rate of 1 ppm while the gas sensor supported with pure CNT lacked the sensitivity to CO and hence, no detection occurred (Fu et al., 2008). As illustrated, the addition of a carboxyl group to CNT increased its functionality by creating more reactive sites at the sidewalls and the termination of CNT considered the interaction spots for CNT with the adsorbed molecules (Zaporotskova et al., 2016). However, gas detection can be impossible under the case of their low adsorptive energy and hence, the ionization gas sensors can be implemented to overcome this limitation. This relies on the ion collision between the sensor and the gas molecules, and thereafter, the parameters of gas ionization can be evaluated (Zaporotskova et al., 2016). Moreover, when gases such as NH_3, CO, N_2, He, O_2, and Ar exist at a low concentration that might limit their detection with the sorption sensors, resonance frequency shift sensors can be applied. This kind of sensor depends on the change in the electrical properties of CNT when the interaction occurs between the nanotube and the gas particles (Chopra et al., 2002, 2003). As being reviewed by Prakash and Yeom (2014), a study was performed on MWCNTs with an N_2-BET surface area and a pore volume of 462 m^2/g and 0.42 cm^3/gm, respectively, to detect their adsorption for three gases including NO, SO_2, and CO_2. The study confirmed the high adsorptive capacity of MWCNTS for NO, which is shown to be higher compared to that of the best-known NO adsorbent, FeOOH-impregnated activated carbon fiber, while their adsorption for SO_2 and CO_2 was moderate and low, respectively. As indicated by the authors, the high adsorption of NO is a consequence of the incidence of NO, NO_2, and N_2O_4 species on the carbon nanotubes (Long and Yang, 2001). Furthermore, MWCNTs were assumed to be a potential candidate to be involved in the development of capacitive- and resistance-based sensors that will help in the detection of carbonyl, carbon dioxide, water vapor, and ammonia gases because the MWCNT-based sensors proved to be sensitive to water vapor and ammonia (Varghese et al., 2001). Other studies with both experimental and computational simulation techniques assumed the potential adsorptive applicability of CNTs with gases such as methane, oxygen, and nitrogen. According to the authors, the pretreatment of CNTs with heat, in air, and at 350°C helped in opening the ends of the nanotubes, which doubled the adsorptive activity of CNTs for oxygen and nitrogen (Fujiwara et al., 2001; Prakash and Yeom, 2014). Carbon nanomaterials used for CO_2 arresting and also nanocatalysts that are responsible for the catalytical conversion of CO_2 and H_2O to fuels are reported elsewhere (Raj et al., 2017). In the modern era, carbon nanotube-based sensors can be appropriate for sensing various gases such as NH_3, NO_2, or O_3 (Azama et al., 2017; Ye et al., 2015). During the monitoring of these gases, the electrical resistance of the nanotubes is significantly changed. The change in resistance values with and without gases provides the existence and quantification of respective gases in the environment.

5 Future prospective

Recent challenges for researchers are to manufacture commercial prototypes using CNMs to detect and remove airborne pollutants:

1. The intrinsic CNMs are often inert toward environmental contaminants.
2. For better efficiency, the CNMs must be functionalized with other metal oxides (reactive materials) having appropriate functional groups or charges. Hence, hybrid structures are able to accumulate multiple functions into one template in order to increase the desired performance (Guerra et al., 2018).
3. Identify suitable CNMs to detect and remove air pollutants based on their selectivity and sensitivity measurements.
4. A cost-effective process is required to evolve to synthesize the CNMs and fabricate the device to detect airborne pollutants.

6 Conclusion

In this chapter, the authors focused on the different types of CNMs and their potential practical applications toward monitoring airborne pollutants. The different types of CNMs, for example, CNTs, fullerenes, and graphene and its derivatives, are clearly enlightened here. The proper mechanisms for monitoring air pollutants using CNM-based materials are reviewed in detail. The growth of CNT-based gas sensors results in high sensitivity with rapid sensor response toward pollutant gases. The development of nanostructured membranes to capture several gas pollutants requires attention. The government should frame policies to control air pollution by proper monitoring in industries and other sources of air pollution.

References

Agnihotri, S., Mota, J.P.B., Rostam-Abadi, M., Rood, M.J., 2005. Structural characterization of single-walled carbon nanotube bundles by experiment and molecular simulation. Langmuir 21, 896–904.

Amade, R., Hussain, S., Ocana, I.R., Bertran, E., 2014. Growth and functionalization of carbon nanotubes on quartz filter for environmental applications. J. Environ. Eng. Ecol. Sci. 3 (2), 1–7.

Amin, M.T., Alazba, A.A., Manzoor, U., 2014. A review of removal of pollutants from water/wastewater using different types of nanomaterials. Adv. Mater. Sci. Eng. 2014, 1–24.

Attar, S., Ranveer, A., 2015. Carbon nanotubes and its environmental applications. Environ. Sci. Comput. Sci. Eng. Technol. 4, 304–311.

Azama, M.A., Aliasa, F.M., Tacka, L.W., Amalina, R.N., Mohamad, R.S., Taibb, F.M., 2017. Electronic properties and gas adsorption behaviour of pristine, silicon and boron-doped (8, 0) single-walled carbon nanotube: A first principles study. J. Mol. Graph. Model. 75, 85–93.

Babaa, M.R., Dupont-Pavlovsky, N., McRae, E., Masenelli-Varlot, K., 2004. Physical adsorption of carbon tetrachloride on as-produced and on mechanically opened single walled carbon nanotubes. Carbon 42, 1549–1554.

Bai, S., Shen, X., Zhu, G., Yuan, A., Zhang, J., Ji, Z., Qiu, D., 2013. The influence of wrinkling in reduced graphene oxide on their adsorption and catalytic properties. Carbon 60, 157–168.

Ballesteros, E., Gallego, M., Valcárcel, M., 2000. Analytical potential of fullerene as adsorbent for organic and organometallic compounds from aqueous solutions. J. Chromatogr. A 869 (1–2), 101–110.

Bergmann, C.P., Machado, F.M. (Eds.), 2015. Carbon nanomaterials as adsorbents for environmental and biological applications. Springer, Berlin, Germany, pp. 1–122.

Bi, H., Xie, X., Yin, K., Zhou, Y., Wan, S., He, L., Xu, F., Banhart, F., Sun, L., Ruoff, R.S., 2012. Spongy graphene as a highly efficient and recyclable sorbent for oils and organic solvents. Adv. Funct. Mater. 22 (21), 4421–4425.

Boyd, A., Dube, I., Fedorov, G., Paranjape, M., Barbara, P., 2014. Gas sensing mechanism of carbon nanotubes: From single tubes to high-density networks. Carbon 69, 417–423.

Chen, R.J., Franklin, N.R., Kong, J., Cao, J., Tombler, T.W., Zhang, Y., Dai, H., 2001. Molecular photodesorption from single-walled carbon nanotubes. Appl. Phys. Lett. 79 (14), 2258–2260.

Chen, W., Duan, L., Zhu, D., 2007. Adsorption of polar and nonpolar organic chemicals to carbon nanotubes. Environ. Sci. Technol. 41 (24), 8295–8300.

Chen, Z., Pierre, D., He, H., Tan, S., Pham-Huy, C., Hong, H., Huang, J., 2011. Adsorption behavior of epirubicin hydrochloride on carboxylated carbon nanotubes. Int. J. Pharm. 405 (1–2), 153–161.

Cheng, X., Kan, A.T., Tomson, M.B., 2005. Uptake and sequestration of naphthalene and 1,2-dichlorobenzene by C 60. J. Nanopart. Res. 7 (4–5), 555–567.

Chopra, S., Pham, A., Gaillard, J., Parker, A., Rao, A.M., 2002. Carbon-nanotube-based resonant-circuit sensor for ammonia. Appl. Phys. Lett. 80 (24), 4632–4634.

Chopra, S., McGuire, K., Gothard, N., Rao, A.M., Pham, A., 2003. Selective gas detection using a carbon nanotube sensor. Appl. Phys. Lett. 83 (11), 2280–2282.

Chowdhury, S., Balasubramanian, R., 2014. Recent advances in the use of graphene-family nanoadsorbents for removal of toxic pollutants from wastewater. Adv. Colloid Interface Sci. 204, 35–56.

Dong, H., Lin, B., Gilmore, K., Hou, T., Lee, S.-T., Li, Y., 2015. B40 fullerene: An efficient material for CO_2 capture, storage and separation. Curr. Appl. Phys. 15, 1084–1089.

Esrafili, M.D., 2017. N_2O reduction over a fullerene-like boron nitride nanocage: A DFT study. Phys. Lett. A 381 (25–26), 2085–2091.

Fu, D., Lim, H., Shi, Y., Dong, X., Mhaisalkar, S.G., Chen, Y., Moochhala, S., Li, L.J., 2008. Differentiation of gas molecules using flexible and all-carbon nanotube devices. J. Phys. Chem. C 112 (3), 650–653.

Fujiwara, A., Ishii, K., Suematsu, H., Kataura, H., Maniwa, Y., Suzuki, S., Achiba, Y., 2001. Gas adsorption in the inside and outside of single-walled carbon nanotubes. Chem. Phys. Lett. 336 (3–4), 205–211.

Gangupomu, R.H., Sattler, M.L., Ramirez, D., 2014. Carbon nanotubes for air pollutant control via adsorption: a review. Rev. Nanosci. Nanotechnol. 3, 149–160.

Gao, Y., Li, Y., Zhang, L., Huang, H., Hu, J., Shah, S.M., Su, X., 2012. Adsorption and removal of tetracycline antibiotics from aqueous solution by graphene oxide. J. Colloid Interface Sci. 368 (1), 540–546.

Ghormade, V., Deshpande, M.V., Paknikar, K.M., 2011. Perspectives for nano-biotechnology enabled protection and nutrition of plants. Biotechnol. Adv. 29 (6), 792–803.

Giles, J., 2006. Top five in physics. Nature. https://doi.org/10.1038/441265a.

Guerra, F., Attia, M., Whitehead, D., Alexis, F., 2018. Nanotechnology for environmental remediation: materials and applications. Molecules 23, 1760.

Gupta Chatterjee, S., Chatterjee, S., Ray, A.K., Chakraborty, A.K., 2015. Graphene–metal oxide nanohybrids for toxic gas sensor: a review. Sensors Actuat. B: Chem. 221, 1170–1181.

Gupta, V.K., Saleh, T.A., 2013. Sorption of pollutants by porous carbon, carbon nanotubes and fullerene- An overview. Environ. Sci. Pollut. Res. 20 (5), 2828–2843.

Hong, F., Yang, F., Liu, C., Gao, Q., Wan, Z., Gu, F., Wu, C., Ma, Z., Zhou, J., Yang, P., 2005. Influences of nano-TiO2 on the chloroplast aging of spinach under light. Biol. Trace Elem. Res. 104 (3), 249–260.

Hsu, S., Lu, C., 2007. Modification of single-walled carbon nanotubes for enhancing isopropyl alcohol vapor adsorption from water streams. Sep. Sci. Technol. 42, 2751–2766.

Huang, Z.H., Zheng, X., Lv, W., Wang, M., Yang, Q.H., Kang, F., 2011. Adsorption of lead(II) ions from aqueous solution on low-temperature exfoliated graphene nanosheets. Langmuir 27 (12), 7558–7562.

Iijima, S., 1991. Helical microtubules of graphitic carbon. Nature 354 (6348), 56.

Iijima, S., Ichihashi, T., 1993. Single-shell carbon nanotubes of 1-nm diameter. Nature 363 (6430), 603.

Inagaki, M., Kang, F., 2014. Materials Science and Engineering of Carbon: Fundamentals, second ed. Elsevier, Oxford.
Kah, M., Hofmann, T., 2015. The Challenge: Carbon nanomaterials in the environment: New threats or wonder materials? Environ. Toxicol. Chem. 34, 954.
Karthikeyan, S., Mahalingam, P., Karthik, M., 2009. Large scale synthesis of carbon nanotubes. J. Chem. 358 (6383), 220.
Kaushik, B.K., Majumder, M.K., 2015. Carbon nanotube based VLSI interconnects: Analysis and design. Springer India, New Delhi, pp.1–14.
Kim, D.K., Park, S.H., Kim, B.C., Chin, B.D., Jo, S.M., Kim, D.Y., 2005. Electrospun polyacrylonitrile-based carbon nanofibers and their hydrogen storages. Macromol. Res. 13 (6), 521–528.
Kong, J., Franklin, N.R., Zhou, C., Chapline, M.G., Peng, S., Cho, K., Dai, H., 2000. Nanotube molecular wires as chemical sensors. Science 287 (5453), 622–625.
Kroto, H., Heath, J., O'Brien, S., Curl, R., Smalley, R., 1985. C60: Buckminsterfullerene 62–63.
Lee, K.J., Shiratori, N., Lee, G.H., Miyawaki, J., Mochida, I., Yoon, S.-H., Jang, J., 2010. Activated carbon nanofiber produced from electrospun polyacrylonitrile nanofiber as a highly efficient formaldehyde adsorbent. Carbon 48 (15), 4248–4255.
Leghrib, R., Llobet, E., 2011. Quantitative trace analysis of benzene using an array of plasma-treated metal-decorated carbon nanotubes and fuzzy adaptive resonant theory techniques. Anal. Chim. Acta 708 (1-2), 19–27.
Leghrib, R., Felten, A., Demoisson, F., Reniers, F., Pireaux, J.J., Llobet, E., 2010a. Room-temperature, selective detection of benzene at trace levels using plasma-treated metal-decorated multiwalled carbon nanotubes. Carbon 48 (12), 3477–3484.
Leghrib, R., Pavelko, R., Felten, A., Vasiliev, A., Cané, C., Gràcia, I., Pireaux, J.J., Llobet, E., 2010b. Gas sensors based on multiwall carbon nanotubes decorated with tin oxide nano-clusters. Sens. Actuators B 145 (1), 411–416.
Li, J., Lu, Y., Ye, Q., Cinke, M., Han, J., Meyyappan, M., 2003. Carbon nanotube sensors for gas and organic vapor detection. Nano Lett. 929–933.
Li, Y., Du, Q., Liu, T., Peng, X., Wang, J., Sun, J., Wang, Y., Wu, S., Wang, Z., Xia, Y., Xia, L., 2013. Comparative study of methylene blue dye adsorption onto activated carbon, graphene oxide, and carbon nanotubes. Chem. Eng. Res. Des. 91 (2), 361–368.
Li, P., Wang, C., Li, Z., Zong, Y., Zhang, Y., Yang, X., Li, S., Wei, F., 2014. Hierarchical carbon-nanotube/quartz-fiber films with gradient nanostructures for high efficiency and long service life air filters. RSC Adv. 4, 54115–54121.
Lin, Y.C., Li, J.Y., Yen, W.T., 2008. Low temperature ITO thin film deposition on PES substrate using pulse magnetron sputtering. Appl. Surf. Sci. 254 (11), 3262–3268.
Liu, C., Fan, Y.Y., Liu, M., Cong, H.T., Cheng, H.M., Dresselhaus, M.S., 1999. Hydrogen storage in single-walled carbon nanotubes at room temperature. Science 86 (5442), 1127–1129.
Llobet, E., 2012. Carbon Nanomaterials for Environmental Monitoring Sensors. 14th Int. Meet. Chem. Sensors 384–386.
Long, R., Yang, R., 2001. Carbon nanotubes as a superior sorbent for nitrogen oxides. Ind. Eng. Chem. Res. 40 (20), 4288–4291.
López-Moreno, M.L., De La Rosa, G., Hernández-Viezcas, J.A., Castillo-Michel, H., Botez, C.E., Peralta-Videa, J.R., Gardea-Torresdey, J.L., 2010a. Evidence of the differential biotransformation and genotoxicity of ZnO and CeO_2 nanoparticles on soybean (*Glycine max*) plants. Environ. Sci. Technol. 44 (19), 7315–7320.
López-Moreno, M.L., De La Rosa, G., Hernández-Viezcas, J.A., Peralta-Videa, J.R., Gardea-Torresdey, J.L., 2010b. X-ray absorption spectroscopy (XAS) corroboration of the uptake and storage of CeO_2 nanoparticles and assessment of their differential toxicity in four edible plant species. J. Agric. Food Chem. 58 (6), 3689–3693.

Low, J., Cheng, B., Yu, J., 2017. Surface modification and enhanced photocatalytic CO_2 reduction performance of TiO2: a review. Appl. Surf. Sci. 392 (15), 658–686.
Machado, F.M., Bergmann, C.P., Fernandes, T.H.M., Lima, E.C., Royer, B., Calvete, T., Fagan, S.B., 2011. Adsorption of Reactive Red M-2BE dye from water solutions by multi-walled carbon nanotubes and activated carbon. J. Hazard. Mater. 192 (3), 1122–1131.
Machado, F.M., Bergmann, C.P., Lima, E.C., Royer, B., De Souza, F.E., Jauris, I.M., Calvete, T., Fagan, S.B., 2012. Adsorption of reactive blue 4 dye from water solutions by carbon nanotubes: experiment theory. Phys. Chem. Chem. Phys. 14 (31), 11139–11153.
Matthews, E., Amann, C., Bringezu, S., 2000. The weight of nations: material outflows from industrial economies, 2000. World Resources Institute, Washington, DC.
Mcdonnell, G., Russell, A.D., 1999. Antiseptics and disinfectants: Activity, action, and resistance. Clin. Microbiol. Rev. 12 (1), 147–179.
Mishra, A.K., Ramaprabhu, S., 2011. Removal of metals from aqueous solution and sea water by functionalized graphite nanoplatelets based electrodes. J. Hazard. Mater. 185 (1), 322–328.
Misslitz, H., Kreger, K., Schmidt, H.-W., 2013. Networks: Supramolecular Nanofiber Webs in Nonwoven Scaffolds as Potential Filter Media (Small 12/2013). Small 9, 2025.
Mohamed, E.F., 2017. Nanotechnology: future of environmental air pollution control. Environ. Manag. Sustain. Development 6, 429–454.
Monica, R.C., Cremonini, R., 2009. Nanoparticles and higher plants. Caryologia. 62 (2), 161–165.
Mueller, N.C., Nowack, B., 2010. Nanoparticles for remediation: Solving big problems with little particles. Elements 6 (6), 395–400.
Muralikrishnan, R., Swarnalakshmi, M., Nakkeeran, E., 2014. Nanoparticle-membrane filtration of vehicular exhaust to reduce air pollution—A review. Int. Res. J. Environment Sci. 3 (4), 82–86.
Nasibulin, A.G., Kaskela, A., Mustonen, K., Anisimov, A.S., Ruiz, V., Kivisto, S., Rackauskas, S., Timmermans, M.Y., Pudas, M., Aitchison, B., Kauppinen, M., 2011. Multifunctional free-standing single-walled carbon nanotube films. ACS Nano 5 (4), 3214–3221.
Novoselov, K.S., Geim, A.K., Morozov, S.V., Jiang, D., Zhang, Y., Dubonos, S.V., Grigorieva, I.V., Firsov, A.A., 2004. Electric field in atomically thin carbon films. Science 306 (5696), 666–669.
Nowack, B., Bucheli, T.D., 2007. Occurrence, behavior and effects of nanoparticles in the environment. Environ. Pollut. 150 (1), 5–22.
Okhovat, N., Hashemi, M., Golpayegani, A.A., 2015. Photocatalytic decomposition of metronidazolein aqueous solutions using titanium dioxide nanoparticles. J. Mater. Environ. Sci. 6, 792–799.
Ong, Y.T., Ahmad, A.L., Zein, S., Huat Tan, S., 2010. A review on carbon nanotubes in an environmental protection and green engineering perspective, Brazil. J. Chem. Eng. 27 (2), 227–242.
Pan, B., Lin, D., Mashayekhi, H., Xing, B., 2008. Adsorption and hysteresis of bisphenol A and 17α-ethinyl estradiol on carbon nanomaterials. Environ. Sci. Technol. 42 (15), 5480–5485.
Pan, G., Li, L., Zhao, D., Chen, H., 2010. Immobilization of non-point phosphorus using stabilized magnetite nanoparticles with enhanced transportability and reactivity in soils. Environ. Pollut. 158 (1), 35–40.
Pan, N., Deng, J., Guan, D., Jin, Y., Xia, C., 2013. Adsorption characteristics of Th(IV) ions on reduced graphene oxide from aqueous solutions. Appl. Surf. Sci. 130, 166–175.
Parham, H., Bates, S., Xia, Y., Zhu, Y., 2013. A highly efficient and versatile carbon nanotube/ ceramic composite filter. Carbon 54, 215–223.

Park, K.-T., Hwang, J., 2014. Filtration and inactivation of aerosolized bacteriophage MS2 by a CNT air filter fabricated using electro-aerodynamic deposition. Carbon 75, 401–410.

Prakash, S., Yeom, J., 2014. Energy and environmental applications of carbon nanotubes. Nanofluidics Microfluid. 47, 241–269.

Prola, L.D.T., Machado, F.M., Bergmann, C.P., de Souza, F.E., Gally, C.R., Lima, E.C., Adebayo, M.A., Dias, S.L.P., Calvete, T., 2013. Adsorption of Direct Blue 53 dye from aqueous solutions by multi-walled carbon nanotubes and activated carbon. J. Environ. Manage. 130, 166–175.

Raj, B., Van de Voorde, M., Mahajan, Y., Ibram, G., 2017. Nanomaterials for the conversion of carbon dioxide into renewable fuels and value-added products. In: Nanotechnology for Energy Sustainability. Wiley-VCH, Weinheim, Germany, pp. 1107–1138.

Ramesha, G.K., Vijaya Kumara, A., Muralidhara, H.B., Sampath, S., 2011. Graphene and graphene oxide as effective adsorbents toward anionic and cationic dyes. J. Colloid Interface Sci. 361 (1), 270–277.

Ren, X., Chen, C., Nagatsu, M., Wang, X., 2011. Carbon nanotubes as adsorbents in environmental pollution management: a review. Chem. Eng. J. 170, 395–410.

Ren, X., Li, J., Tan, X., Wang, X., 2013. Comparative study of graphene oxide, activated carbon and carbon nanotubes as adsorbents for copper decontamination. Dalton Trans. 42 (15), 5266–5274.

Rengaraj, S., Yeon, J.W., Kim, Y., Kim, W.H., 2007. Application of Mg-mesoporous alumina prepared by using magnesium stearate as a template for the removal of Nickel: Kinetics, isotherm, and error analysis. Ind. Eng. Chem. Res. 46 (9), 2834–2842.

Rickerby, D., Morrison, M., 2007. Report from the Workshop on Nanotechnologies for Environmental Remediation JRC Ispra, 16.

Roco, M.C., 2005. The emergence and policy implications of converging new technologies integrated from the nanoscale. J. Nanopart. Res. 7 (2–3), 129–143.

Seredych, M., Bandosz, T.J., 2012. Manganese oxide and graphite oxide/MnO composites as reactive adsorbents of ammonia at ambient conditions. Micropor. Mesopor. Mat. 150, 55–63.

Seresht, J.R., Jahanshahi, M., Rashidi, A., Ghoreyshi, A.A., 2013. Synthesize and characterization of graphene nanosheets with high surface area and nano-porous structure. Appl. Surf. Sci. 276, 672–681.

Serp, P., Corrias, M., Kalck, P., 2003. Carbon nanotubes and nanofibers in catalysis. Appl. Catal. Gen. 253 (2), 337–358.

Silva, S.A.M., Perez, J., Torresi, R.M., Luengo, C.A., Ticianelli, E.A., 1999. Surface and electrochemical investigations of a fullerene soot. Electrochim. Acta 44 (20), 3565–3574.

Singh, B.K., Walker, A., 2006. Microbial degradation of organophosphorus compounds. FEMS Microbiol. Rev. 30 (3), 428–471.

Soares, F., Trindade, D.-d.-S., 2019. Trimethyl chitosan/siloxane-hybrid coated Fe_3O_4 nanoparticles for the uptake of sulfamethoxazole from water. Molecules 24, 1958.

Stafiej, A., Pyrzynska, K., 2007. Adsorption of heavy metal ions with carbon nanotubes. Sep. Purif. Technol. 58 (1), 49–52.

Stankovich, S., Dikin, D.A., Dommett, G.H.B., Kohlhaas, K.M., Zimney, E.J., Stach, E.A., Piner, R.D., Nguyen, S.B.T., Ruoff, R.S., 2006. Graphene-based composite materials. Nature 442 (7100), 282.

Su, F., Lu, C., Cnen, W., Bai, H., Hwang, J.F., 2009. Capture of CO_2 from flue gas via multiwalled carbon nanotubes. Sci. Total Environ. 407 (8), 3017–3023.

Sun, G., Liu, S., Hua, K., Lv, X., Huang, L., Wang, Y., 2007. Electrochemical chlorine sensor with multi-walled carbon nanotubes as electrocatalysts. Electrochem. Commun. 9 (9), 2436–2440.

Terrones, M., Botello-Méndez, A.R., Campos-Delgado, J., López-Urías, F., Vega-Cantú, Y.I., Rodríguez-Macías, F.J., Elías, A.L., Muñoz-Sandoval, E., Cano-Márquez, A.G., Charlier, J.C., Terrones, H., 2010. Graphene and graphite nanoribbons: Morphology, properties, synthesis, defects and applications. Nano Today. 5 (4), 351–372.

Tiwari, J.N., Mahesh, K., Le, N.H., Kemp, K.C., Timilsina, R., Tiwari, R.N., Kim, K.S., 2013. Reduced graphene oxide-based hydrogels for the efficient capture of dye pollutants from aqueous solutions. Carbon 56, 173–182.

Tonel, M.Z., de Menezes, V.M., Zanella, I., Fagan, S.B., 2015. Molecules with biological interest adsorbed on carbon nanostructures. In: Carbon Nanomaterials as Adsorbents for Environmental and Biological Applications. Springer, Cham, pp.107–122.

Upadhyayula, V.K.K., Deng, S., Mitchell, M.C., Smith, G.B., 2009. Application of carbon nanotube technology for removal of contaminants in drinking water: A review. Sci. Total Environ. 408 (1), 1–13.

US Environmental Protection Agency, 2009. Long-Chain Perfluorinated Chemicals (PFCs) Action Plan. Environmental Protection Agency, U.S.

Varghese, O.K., Kichambre, P.D., Gong, D., Ong, K.G., Dickey, E.C., Grimes, C.A., 2001. Gas sensing characteristics of multi-wall carbon nanotubes. Sensors Actuat. B Chem. 81 (1), 32–41.

Wang, Y., Li, Z., Wang, J., Li, J., Lin, Y., 2011. Graphene and graphene oxide: Biofunctionalization and applications in biotechnology. Trends Biotechnol. 29 (5), 205–212.

Wang, Y., Gao, S., Zang, X., Li, J., Ma, J., 2012a. Graphene-based solid-phase extraction combined with flame atomic absorption spectrometry for a sensitive determination of trace amounts of lead in environmental water and vegetable samples. Anal. Chim. Acta 716, 112–118.

Wang, Y.K., Gao, S.T., Ma, J.J., Li, J.C., 2012b. Application of graphene as a sorbent for simultaneous preconcentration and determination of trace amounts of cobalt and nickel in environmental water and vegetable samples. J. Chinese Chem. Soc. 59 (11), 1468–1477.

Wang, C., Li, P., Zong, Y., Zhang, Y., Li, S., Wei, F., 2014. A high efficiency particulate air filter based on agglomerated carbon nanotube fluidized bed. Carbon 79, 424–431.

Watlington, K., 2005. Emerging nanotechnologies for site remediation and wastewater treatment. Report, US Environ. Prot. Agency, Off. Solid Waste Emerg. Response, Off. Superfund Remediat. Technol. Innov. Technol. Innov. F. Serv. Div. Washington, DC LK.

Wu, S., He, Q., Tan, C., Wang, Y., Zhang, H., 2013. Graphene-based electrochemical sensors. Small 9, 1160–1172.

Yadav, K.K., Singh, J.K., Gupta, N., Kumar, V., 2017. A review of nanobioremediation technologies for environmental cleanup: A novel biological approach. J. Mater. Environ. Sci. 8 (2), 740–757.

Yang, Q.H., Hou, P.X., Bai, S., Wang, M.Z., Cheng, H.M., 2001. Adsorption and capillarity of nitrogen in aggregated multi-walled carbon nanotubes. Chem. Phys. Lett. 345 (1–2), 18–24.

Yang, F., Hong, F., You, W., Liu, C., Gao, F., Wu, C., Yang, P., 2006a. Influences of nano-anatase TiO2 on the nitrogen metabolism of growing spinach. Biol. Trace Elem. Res. 110 (2), 179–190.

Yang, K., Zhu, L., Xing, B., 2006b. Adsorption of polycyclic aromatic hydrocarbons by carbon nanomaterials. Environ. Sci. Technol. 40 (6), 1855–1861.

Yang, S.T., Chen, S., Chang, Y., Cao, A., Liu, Y., Wang, H., 2011. Removal of methylene blue from aqueous solution by graphene oxide. J. Colloid Interface Sci. 359 (1), 24–29.

Ye, Y., Guo, Y., Yue, Y., Zhang, Y., 2015. Facile colorimetric detection of nitrite based on anti-aggregation of gold nanoparticles. Anal. Methods 7, 4090–4096.

Yildiz, O., Bradford, P.D., 2013. Aligned carbon nanotube sheet high efficiency particulate air filters. Carbon 64, 295–304.
Yudasaka, M., Ichihashi, T., Kasuya, D., Kataura, H., Iijima, S., 2003. Structure changes of single-wall carbon nanotubes and single-wall carbon nanohorns caused by heat treatment. Carbon 41 (6), 1273–1280.
Zanolli, Z., Leghrib, R., Felten, A., Pireaux, J.J., Llobet, E., Charlier, J.C., 2011. Gas sensing with au-decorated carbon nanotubes. ACS Nano 5 (6), 4592–4599.
Zaporotskova, I.V., Boroznina, N.P., Parkhomenko, Y.N., Kozhitov, L.V., 2016. Carbon nanotubes: Sensor properties. A review. Mod. Electron. Mater. 2, 95–105.
Zhang, W.-D., Zhang, W.-H., 2009. Carbon nanotubes as active components for gas sensors. J. Sensors. 16, Article ID 160698. https://doi.org/10.1155/2009/160698.
Zhang, X.X., Bing, Y., Dai, Z.Q., Luo, C.C., 2012. The gas response of hydroxyl modified SWCNTs and carboxyl modified SWCNTs to H_2S and SO_2. Przeglad Elektrotechniczny 88, 311–314.
Zhao, Y.-L., Stoddart, J.F., 2009. Noncovalent functionalization of single-walled carbon nanotubes. Acc. Chem. Res. 42 (8), 1161–1171.
Zhao, J., Buldum, A., Han, J., Lu, J.P., 2002. Gas molecule adsorption in carbon nanotubes and nanotube bundles. Nanotechnology 13 (2), 195.

Further reading

Sarkar, B., Mandal, S., Tsang, Y.F., Kumar, P., Kim, K.H., Ok, Y.S., 2018. Designer carbon nanotubes for contaminant removal in water and wastewater: A critical review. Sci. Total Environ. 612, 561–581.
Young, S.J., Lin, Z.D., Hsiao, C.H., Huang, C.S., 2012. Ethanol gas sensors composed of carbon nanotubes with adsorbed gold nanoparticles. Int. J. Electrochem. Sci. 7, 11634–11640.

Carbon-based sponges for oil spill recovery

8

Elpida Piperopoulos[a], *Luigi Calabrese*[a], *Emanuela Mastronardo*[b,c], *Candida Milone*[a], *Edoardo Proverbio*[a]
[a]Department of Engineering, University of Messina, Messina, Italy, [b]Thermochemical Processes Unit, IMDEA Energy Institute, Madrid, Spain, [c]Materials Science and Engineering, Northwestern University, Evanston, IL, United States

1 Introduction

Over the past few decades, many oil spills have occurred during oil extraction, transportation, and storage. Many of them occurred in seawater as a consequence of the intensive oil transportation carried out via the sea route. For example, in a huge oil spill in the Gulf of Mexico in 2010, about 500,000 tons of oil were spilled, surpassing the estimated 454,000 tons of oil spilled into the Bay of Campeche by the Mexican Ixtoc I in 1979. That was an explorative oil well being drilled by the semisubmersible Sedco 135, and was previously considered the world's largest oil spill. This environmentally catastrophic accident impacted 82 marine species, including protected species and barrier reefs (Campagna et al., 2011). More recently, in January 2018, around 136,000 tons of oil spilled into the waters of the East China Sea. The Panamanian-flagged, Iranian-owned tanker Sanchi, with a full natural-gas condensate cargo, was sailing from Iran to South Korea when it collided with the Hong Kong-flagged cargo ship CF Crystal 160 nautical miles (300 km) off Shanghai. But these disasters do not promise to stop. The last one occurred in February 2019, when around 60 tons of oil spilled from the Solomon Trader that ran aground on a coral reef in the Solomon Islands. Therefore, preparation in an emergency is a strategic element. Local and national planning tools are more effective when they are well known and easily feasible. In particular, when the emergency concerns hydrocarbons, knowledge of the substance characteristics, mutations, and polluting capacities is a fundamental element to support the decisions to be taken.

The current intervention methods, used to combat marine oil pollution, are:

- Containment and recovery of the product with the use of containment creams, skimmers, and pump.
- Application of absorbent products.
- Application of dispersing action products.

The choice of the most appropriate methodology for the specific case, or a combination of them, is fundamental in determining the success of the intervention, maximizing the amount of recovered product, and minimizing the environmental, economic and social consequences of pollution. In fact, an effective cleanup operation

Carbon Nanomaterials for Agri-food and Environmental Applications. https://doi.org/10.1016/B978-0-12-819786-8.00008-6

greatly reduces the environmental loss and highly improves the economic resources of the damaged area (Edwards and White, 2013). However, it is necessary to take into account that it will never be possible to recover 100 wt% of the spilled product; the statistics related to the analysis of past events at the international level show, in fact, that a part of the spilled substance is collected while the remainder evaporates, disperses, sinks, or reaches the coasts. Among the above-mentioned techniques, the use of absorbent materials is considered to be one of the most effective and reliable approaches thanks to their easy, flexible, and cost-effective manufacturing technology; high selectivity; environmental friendliness; and worthy recyclability. In order to offer high oil spill recovery efficiency, the absorbent products, when immersed in water, must absorb hydrocarbons in a preferential manner, functioning as a selective sponge recovered together with the oily product. For this reason, the promising candidates for oil spill recovery must possess both hydrophobicity and oleophilicity in order to selectively absorb organic solvents, repelling water completely. Many natural absorbents have been widely studied for this purpose such as sawdust (Annunciado et al., 2005), wool fiber (Wang et al., 2012), and zeolite (Sakthivel et al., 2013). But these materials present some drawbacks such as poor absorption capacities and recyclability; moreover, they absorb both oil and water. In recent years, numerous polymers and resins (Atta et al., 2006; Shan et al., 2003a,b; Zhou and Chuai, 2010) have also been investigated for oil-recovery applications. But among the most promising sorbent materials, carbon-based materials have been considered the most suitable for superhydrophobic and superoleophilic behavior, being environmentally acceptable, easily achievable in different and more handling forms (e.g., sponges), and reusable for many cycles.

2 Oil spill recovery: Advancement in carbon-based materials and systems

An ideal absorbent material must have multiple qualities, such as high porosity, a large surface area, chemical inertness, high selectivity toward oil, and outstanding recyclability. In recent years, researchers have employed microporous polymers (Atta et al., 2007; Farag and El-Saeed, 2008; Li et al., 2011) as absorbent materials, but they require a long time for degradation, which causes environmental and ecological issues. Furthermore, these materials have low mechanical properties and the manufacturing process is relatively long and not easily scalable. Carbon-based superhydrophobic materials can be produced by various methods. These materials can be easily synthesized through chemical deposition using suitable metal catalysts and carbon sources (Donato et al., 2009; Messina et al., 2008). Furthermore, the high surface area, low density, excellent flexibility, good chemical stability, great hydrophobicity and oleophilicity, environmental compatibility, and the large pore volume of these materials make them suitable for oil-recovery applications. Recently, research has been focused on carbon aerogels (Lu et al., 2019; Wu et al., 2013b), forests of carbon nanotubes (CNTs) (Dong et al., 2012), graphene foams or sponges (Fragouli and Athanassiou, 2017), carbonaceous coatings (Jiang et al., 2016), activated carbon

(Maulion et al., 2015), and porous carbon nanoparticles (Dai et al., 2013) that have been widely used as absorbents. Although these nanomaterial-based absorbents show promising results, the use of nanomaterials exclusively for large-scale purposes will be expensive, complex, and time-consuming. Alternatively, the process can be simplified by incorporating nanomaterials onto a polymer matrix to form a smart selective sponge sorbent able to combine the sorption performances of the carbonaceous materials and the cost-effectivity and feasibility of polymer technology.

2.1 Carbon fillers for oil spill recovery sponges

Sponges offer several critical advantages over current standard industrial technologies. Sponges are reusable by simply wringing the reclaimed oil into a holding tank. In this way a lower amount of absorbent is requested and the waste resulting from the clean-up process decreases. Existing sorbent technologies absorb oil in a single use, and then saturated materials must be disposed. A sponge is environmentally friendly, being harmless to sea life, animals, or the larger environment, a key advantage compared to conventional chemical dispersants or burning techniques that are usually applied. The oil and other pollutants that are recovered by the sponge can be recovered or stored for future use. To increase foam selectivity toward oil, an effective approach is to use carbon materials as fillers, exploiting their hydrophobic and oleophilic properties. Zhao et al. (2011) prepared a sponge-like absorbent material, placing aligned CNT arrays into exfoliated vermiculite (EV) layers using different CNT content. It was observed that the oil absorption capacities of the as-obtained sponge-like hybrids were significantly enhanced compared with original vermiculite particles due to the large number of pores produced by the intercalated growth of aligned CNT. The highest absorption capacity of the EV/CNT hybrids was $26.7\,g\,g^{-1}$ for diesel oil, using a percentage of CNT fillers of 91.0 wt%, which also exhibited good reutilizing performance. Ong et al. (2018) proposed an innovative nanoscale surface modification approach able to enhance the oil/water selectivity of porous poly-(dimethylsiloxane) (PDMS). The surface modification method was simple and low cost by using sugar as a sacrificial template for temporary adhering of CNTs before the addition of the PDMS prepolymer to capture the CNT filler, according to the preparation scheme reported in Fig. 1.

This composite material demonstrated a great improvement in absorption capacity and a faster absorption rate compared to conventional absorbents composed of pure PDMS. The enhanced absorption rate was ascribed to the incorporation of CNT, which roughened the surface of the polymer at the nanoscale and lowered the surface energy of porous PDMS. At the same time, the addition of a CNT filler increased the absorbent hydrophobicity and oleophilicity. This property made the new absorbent effective in absorbing only oil but rejecting water at the same time. Novel CNT-reinforced polyurethane sponges with superhydrophobic, superoleophilic, and high mechanical properties were synthesized by Wang et al. (2015). The proposed synthesis approach consisted of the oxidative self-polymerization of dopamine and the reaction of polydopamine (PDA) with octadecylamine (ODA), characterized by hydrophic and hydrophobic properties, respectively. The mechanical strength of these sponges was enhanced

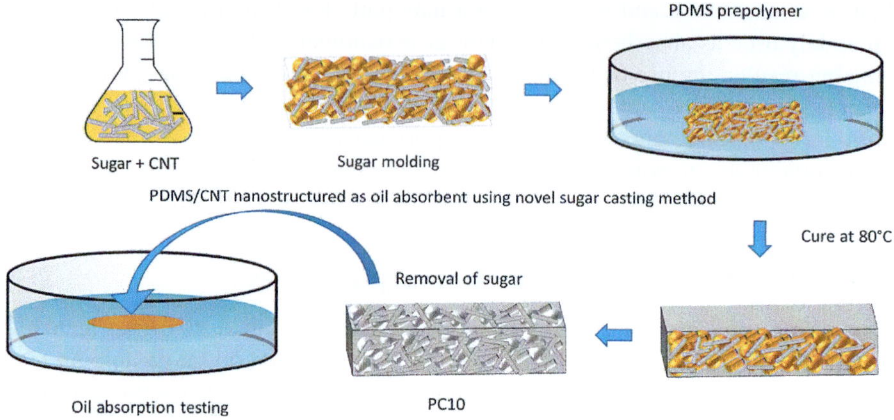

Fig. 1 Schematic illustration of absorbent material preparation.
From Ong, C.C., Sundera Murthe, S., Mohamed, N.M., Perumal, V., Mohamed Saheed, M.S., 2018. Nanoscaled surface modification of poly(dimethylsiloxane) using carbon nanotubes for enhanced oil and organic solvent absorption. ACS Omega 3, 15,907–15,915.

thanks to the CNT filler anchored on the sponge structure, acting in this way as structural reinforcement. The sponge is able to rapidly and selectively absorb several oils up to about 35 times its own weight. Furthermore, the absorbed oils could be stored by a simple squeezing process of the sponge. The recovered sponge can be reused for further separation of the oil–water mixture up to 150 times without evidence of absorption capacity reduction. Nguyen et al. (2012) realized superhydrophobic and superoleophilic graphene-based sponges as reliable absorbent materials for a wide range of pollutants (such as oils and organic solvents) offering effective selectivity and recyclability while showing an absorption capacity up to 165 times their own weight. The foam was produced via a simple and inexpensive dip-coating method on commercially available melamine sponges. Wu et al. (2013a) realized a composite graphene/flexible polymer matrix foam by self-assembly of graphene sheets on a three-dimensional (3D) polyurethanic skeleton, presenting excellent mechanical properties including compression, bend, and twist. A superhydrophobic foam by the grafting polymerization method was prepared by Liu et al. (2013a). The graphene functionalized with an amine group was grafted with a covalent bond with a nitrile group of the polyurethane in order to generate superhydrophobic behavior in the foam. The as-proposed composite foam exhibited a high water contact angle (WCA) (159.1 ± 2.3 degrees) compared to unfilled foam (121.4 ± 3.2 degrees), due to the foam surface roughness and its low surface energy. Because the modified foam was extremely hydrophobic and preferentially absorbed oil rather than water, the sorption capacity of the modified foam for toluene, gasoline, and diesel was 41, 27, and $26\,g\,g^{-1}$, respectively. Piperopoulos et al. (2018a) synthesized a composite foam containing CNTs by a simple foaming solution constituted by a silicone PDMS matrix and CNT filler (5.6 wt%) in the presence of an Sn-based catalyst. Pristine and functionalized CNTs were used. Each foam was tested in four commonly used oils (kerosene, pump oil, naphtha, and crude oil). Among all, the foam filled with pristine CNT showed the highest sorption capacity (800 wt% in virgin naphtha) and excellent reusability (up to 10 times). The main result related to all the above-mentioned foams is the reusability, which decreases the economic and environmental impact of the carbon-filled foams, allowing good selectivity and a high absorption capacity.

2.2 Carbon sponges for oil spill recovery

Considering the intrinsic properties of CNT, characterized by high hydrophobic and oleophilic behavior, the realization of pure CNT macro- or microporous structures represents an effective approach to obtain functional absorbent materials with high oil selectivity.

In such a context, CNT (Gui et al., 2010) and magnetic CNT sponged (Gui et al., 2013) (Turco et al., 2015) were prepared by chemical vapor deposition (CVD) using ferrocene and dichlorobenzene. Nanotubes were self-assembled into a 3D interconnected framework. The sponges were very light, highly porous (pore sizes ranged from several nanometers to a few micrometers), hydrophobic in pristine form, and could be elastically and cyclically deformed without mechanical cracks. The sponges could float on water and were able to absorb large amounts of oil films on the seawater surface. The CNT and magnetic CNT sponges possessed hydrophobic properties with WCAs of 156 degrees and 140 degrees, respectively. The densities of the CNT sponge and the magnetic CNT sponge were measured to be 5–10 $mg\,cm^{-3}$ and 15 $mg\,cm^{-3}$, respectively. The absorption capacities of the magnetic sponge and CNT sponge were very high, ranging from 49 to 56 to 80 to 180 times their own weight for different oils and organic solvents, respectively. There are several benefits of magnetic sorbents in the absorption of oil and organic solvents from water. After saturated absorption, the magnetic CNT-based sorbent can be easily collected by using a magnetic apparatus. In some applications, where not prevented damage to the sorbent occurs, all the parts generated during its damaging can be collected in a controllable way more safely by a magnetic field. In addition, these sponges could be used >1000 times without a significant change in their structure, WCA, or absorption capacities. Liu et al. (2013b) realized a CNTs foam by a low-temperature chemical fusion (LTCF) technique by using CNTs as a skeleton. It used dextrose as a carbon source and citric acid as a carboxyl group donor to allow the reaction with the hydroxyl group present in dextrose. The obtained composite had a tridimensional interconnected porous structure with a high surface area (>350 $m^2\,g^{-1}$) and tunable meso- and macroporosity favored by the addition of ammonium carbonate into the preparation mixture followed by a direct thermal decomposition. This CNT foam, furthermore, exhibited an acceptable mechanical strength that facilitated its handling, transport, and disposal while the nanoscopic morphology of the CNT significantly reduced the problem of diffusion and contributed to an improvement of the effective surface area for subsequent applications. Hashim et al. (2012) synthesized tridimensional macroscale nanotube elastic solids applying a doping with boron approach directly during chemical vapor deposition. That allowed obtaining atomic-scale "elbow" links and nanotube covalent interconnections in the material bulk. Exploiting this material's properties, strongly oleophilic sponge-like solids can be obtained, demonstrating a unique reusable sorbent capability able to suitably and repeatedly remove oil from contaminated seawater. This class of material showed an absorption capacity of 22–180 $g\,g^{-1}$ for different pollutants. Furthermore, a 3D CNT network for water treatment was synthesized by Camilli et al. (2014). The CNT sponge was obtained by a sulfur-addition method during an ambient-pressure CVD process, with ferrocene as the catalyst. The obtained 3D sponge material effectively and selectively absorbed from a water solution organic solvents such as dichlorobenzene, about 3.5 times higher than a pure CNT filler. This approach is effective to enhance CNT interaction to lead to the formation of the synthesized porous networks with reliable flexibility and elasticity (Fig. 2).

Furthermore, the superhydrophobic and superoleophilic hybrid composite foam based on graphene and CNTs for selective spills of oils or organic solvents was synthesized by Dong et al. (2012). In their research activity, a monolith 3D hybrid material was realized by two steps of chemical vapor deposition. Its absorption capacity ranged from 80 to 130 $g\,g^{-1}$. Bi et al. (2012) reported a graphene absorbent sponge with a very effective absorption rate of 0.57 $g\,g^{-1}\,s^{-1}$ of dodecane and an absorption capacity 20–86 times its own weight. The graphene sponge, characterized by microporous needle-like structures, was prepared by a chemical reduction in suspension of graphene oxide (GO) followed by a molding step. In particular, the work indicated that it was difficult to completely remove all the functional groups. This small number of hydrophilic groups increased the interaction between polar solvents and graphene surfaces and thus enhanced the absorption capacity. Zhao et al. (2012) proposed an ultralow weight foam (Fig. 3C), 3D, nitrogen-doped graphene framework produced from graphite oxide and pyrrole using the hydrothermal method followed by freeze drying and annealing. The graphene aerogel presented a porous interconnected, 3D structure, characterized by an ultralight density ($2.1 \pm 0.3\,mg\,cm^{-3}$), which can be surprisingly comparable to the air density ($1.2\,mg\,cm^{-3}$) at room temperature. The pyrrole also prevented restacking of reduced GO sheets, resulting in the formation of a

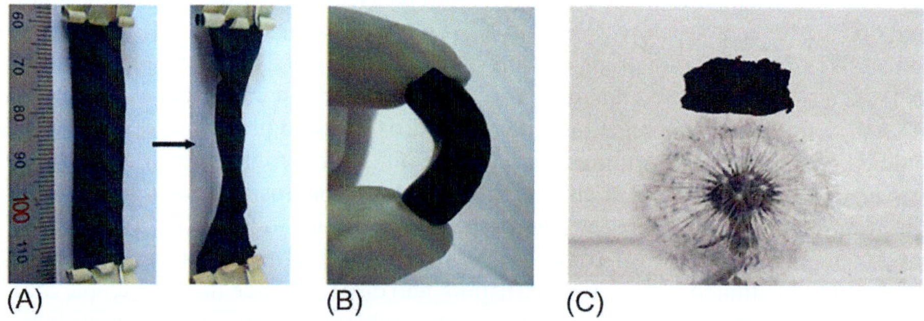

Fig. 2 Twisting (A) and bending (B) of a CNT sponge at a large angle. Ultralight nitrogen-doped graphene framework (C).
From (A and B) Gui, B.X., Wei, J., Wang, K., Cao, A., Zhu, H., Jia, Y., Shu, Q., Wu, D., 2010. Carbon nanotube sponges. Adv. Mater. 22(5), 617–621. (C) Zhao, Y., Hu, C., Hu, Y., Cheng, H., Shi, G., Qu, L., 2012. A versatile, ultralight, nitrogen-doped graphene framework. Angew. Chem. Int. Ed. 51, 11,371–11,375.

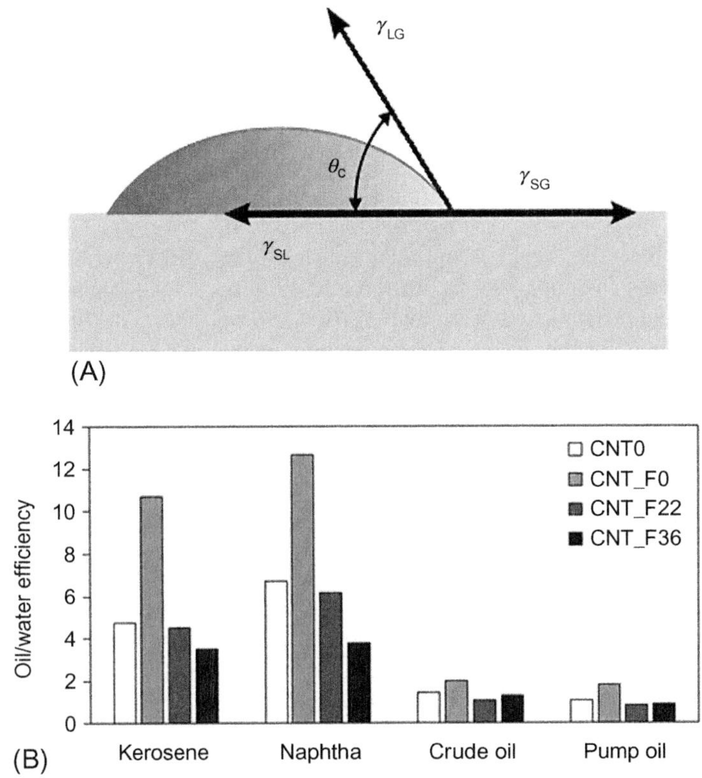

Fig. 3 A liquid drop on a surface (A). Selectivity oil/water of all composite silicone foams (B). From (B) Piperopoulos, E., Calabrese, L., Mastronardo, E., Proverbio, E., Milone, C., 2018a. Synthesis of reusable silicone foam containing carbon nanotubes for oil spill remediation. J. Appl. Polym. Sci. 135, 46,067–46,078.

highly porous volume graphene network. The graphene network showed outstanding mechanical and thermal stability as well as a high absorption rate of 41.7 $g\,g^{-1}\,s^{-1}$. Furthermore, the sorbent material showed a high absorption capacity (about 200–600 times its own weight in different pollutants). Three different types of GO foams were obtained by He et al. (2013) using three kinds of freezing approaches (e.g., unidirectional freezing drying (UDF), nondirectional freezing drying, and air freezing drying), and the equivalent reduced graphene oxide (RGO) foams were fabricated by the thermal reduction of GO foams. The RGO foams evidenced a hydrophobic behavior showing, furthermore, an extremely high absorbtion capability for organic liquids. The absorption capacity of the RGO foams made by UDF was higher than 100 $g\,g^{-1}$ for all the tested oils (gasoline, diesel oil, pump oil, lubricating oil, and olive oil) and had the highest value of about 122 $g\,g^{-1}$ for olive oil. The oil absorption capacity of the GO foams was lower than the RGO foams. In particular, the olive oil absorption capacity was still higher than 70 $g\,g^{-1}$, which is higher than that of most oil absorbents. In light of the rapid research progress in the field of carbon-absorbent materials for oil recovery, it can be asserted that future development will be oriented to the industrialization and marketing of these types of materials for environmental applications.

3 Performance evaluation and characterization of oil spill recovery materials

As reported before, the ideal sorbent material must have specific characteristics such as high absorption capacity, fast absorption rate, high porosity, large surface area, and great mechanical resistance in order to be reused several times. All this information is deduced from suitable characterization protocols carried out on the absorbent material/absorbed oil couple, before and after the sorption phase. Below, some of the most important characterization techniques to define and understand the performances of the material and its interaction with the polluting oil under examination are reported.

3.1 Absorption testing

To calculate the absorption capacity of the carbon-based sponges, generally, polluting oils (such as kerosene, pump oil, naphtha, dichlorobenzene, crude oil) are selected.

The sorbent sponges are put onto the tested oil at room temperature and under slow stirring. After sorption, the foam is held to allow the residual liquid to drip away and then its weight is recorded. The mass sorption capacity is normally calculated as:

$$Q_t = \frac{m_t - m_0}{m_0} \tag{1}$$

where Q_t (%) is the sorption capacity of the foam at t sorption time, m_t (g) is the weight of the sample after t sorption time, and m_0 (g) is the initial weight of the sample. When the sorption capacity does not change with time, its value is identified as the saturation sorption capacity (Piperopoulos et al., 2018a). The same procedure can be carried out with water in order to assess sponge hydrophobicity and consequently its selectivity toward oil pollutants. Further information about sponge hydrophobicity and oleophilicity is provided by the water/oil contact angle measurement. WCA θ_c is used to measure the wettability of a surface, which depends on the solid-vapor, solid–liquid, and liquid–vapor surface tensions, and can be expressed by Young's equation:

$$\cos\theta_c = \frac{\gamma_{SV} - \gamma_{SL}}{\gamma_{LV}} \tag{2}$$

where γ_{SV}, γ_{SL}, and γ_{LV} are the interfacial tensions between solid and vapor, solid and liquid, and liquid and vapor, respectively, as shown in Fig. 3A.

Depending on the value of the contact angle, surface properties are categorized as hydrophilic, hydrophobic, or superhydrophobic. If the WCA (θ_c) is <90 degrees, the surface is described as hydrophilic; if θ_c is between 90 and 150 degrees hydrophobic and if θ_c is above 150 degrees, the surface is described as superhydrophobic (Gupta and Tai, 2016). Piperopoulos et al. realized a hydrophobic CNT-filled PDMS with a

WCA of 115.9 degrees. The addition of pristine CNT enhanced the hydrophobic behavior of the PDMS surface, which had a WCA of 100.6 degrees. Moreover, the kerosene contact angle (KCA) of this composite material was 0 degrees, compared with 30 degrees for unfilled PDMS foam, reaching an absorption capacity of $6830\,mg\,g^{-1}$ (Piperopoulos et al., 2018a). This increased selectivity toward polluting oil is represented in Fig. 3B. The functionalized CNT, on the contrary, increased foam hydrophilicity due to the polar functional groups of CNT, decreasing oil selectivity.

Ge et al. (2013) produced a magnetically CNT-filled PTFE composite material that could selectively absorb oil spreading on the water's surface due to its superhydrophobicity and superoleophilicity (CA contact angle in Fig. 4: 158 degrees), and could be easily collected from water under a magnetic field.

Foam morphology is a focal feature that seriously influences the absorption capacity of carbon sponges. In fact, the porosity of the sponge affects the foam density. In relation to its density, the foam can float, remaining in contact with the oil on the surface of the sea water, or it can partially sink, not using the sunk part of its volume for oil recovery. Furthermore, the size of the pores and the roughness of the foam also influence the absorption rate, which should not be underestimated due to a sudden intervention in the event of marine pollution. For carbon filled sponges, they usually present macropores as reported in Fig. 5A, where a CNT-filled PDMS foam structure is shown (Piperopoulos et al., 2018b), and CNTs are incorporated in the siloxane matrix (Fig. 5B).

In carbon sponges, conversely, the pore size is lower, belonging to the micropore scale. In fact, in these materials, the nanotube entanglement defines the tridimensional

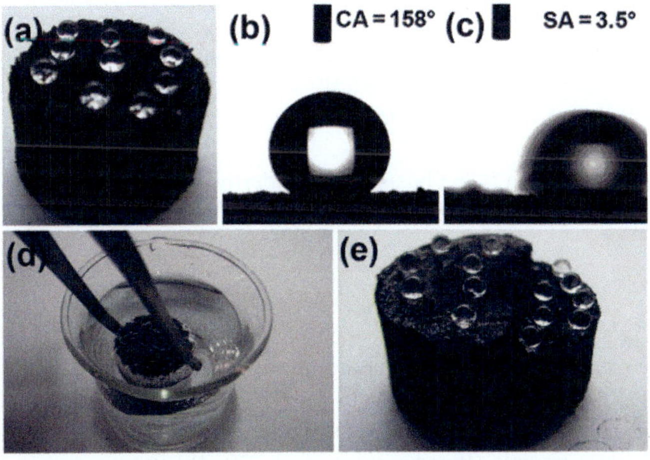

Fig. 4 Optical images (A), WCA profile (B) sliding angle (C) the mirror-like phenomenon on the bulk material immersed in water (D) and spherical water droplets on the material surface (E).
From Ge, B., Zhang, Z., Zhu, X., Ren, G., Men, X., Zhou, X., 2013. A magnetically superhydrophobic bulk material for oil removal. Colloids Surf. A Physicochem. Eng. Asp. 429, 129–133.

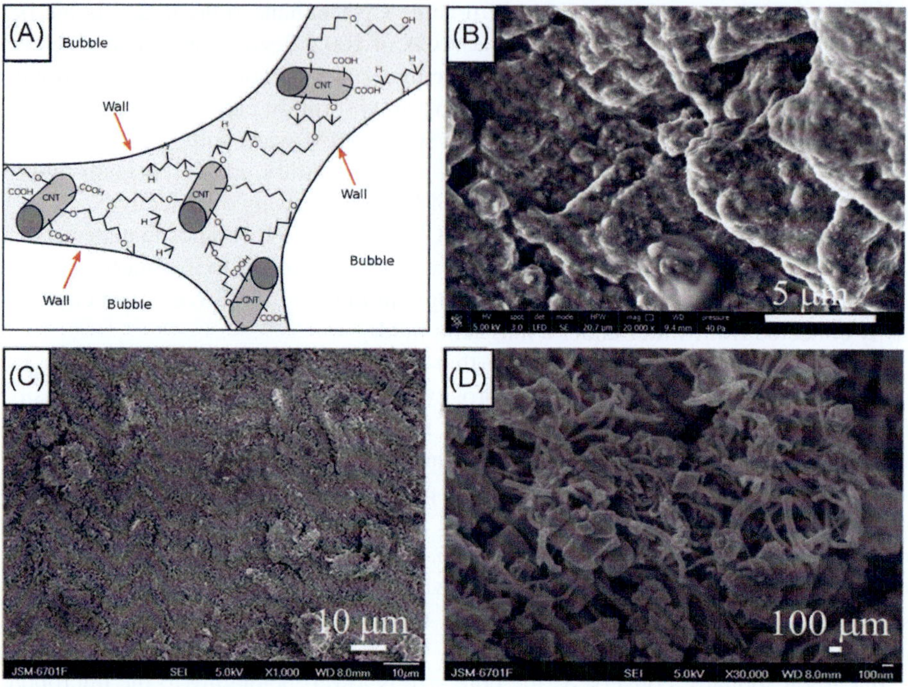

Fig. 5 Scheme of the CNT silicone composite sponge structure (A) and SEM image of pristine CNT sponge wall (B) SEM images of the sorbent foam at low (C) and high (D) magnification.
From (A and B) Piperopoulos, E., Calabrese, L., Mastronardo, E., Proverbio, E., Milone, C., 2018a. Synthesis of reusable silicone foam containing carbon nanotubes for oil spill remediation. J. Appl. Polym. Sci. 135, 46,067–46,078. (C and D) Ge, B., Zhang, Z., Zhu, X., Ren, G., Men, X., Zhou, X., 2013. A magnetically superhydrophobic bulk material for oil removal. Colloids Surf. A Physicochem. Eng. Asp. 429, 129–133.

texture of the foam, which is the basis of the superhydrophobicity of these materials, as reported in Fig. 5C and D by Ge et al. (2013).

3.2 *Kinetics*

Sorption rate is a significant feature to be taken into account to assess the sorption process and sorption capabilities of the composite foams. A deep study and a right comprehension of sorption kinetic parameters could explain the competing mechanisms occurring on liquid/solid phases during sorption phenomena. In this concern, usually four kinetic models (Piperopoulos et al., 2018b) are applied in oil retention applications:

- Pseudo first-order model.
- Pseudo second-order model.
- Elovich model.
- Intraparticle diffusion model.

For each model, the correlation coefficient, R^2, between experimental and fitting data is normally obtained in order to evaluate the effectiveness of the applied approach.

3.2.1 Pseudo first-order kinetic model

The pseudo first-order kinetic model (Yaneva and Koumanova, 2006; Yaneva et al., 2012). developed by Lagergren in 1898 is widely used to describe the sorption phenomena at the solid/liquid interface based on the sorption capacity, according to the equation:

$$\frac{dq_t}{dt} = K_1 \left(q_e - q_t\right) \tag{3}$$

where q_e (mg g^{-1}) and q_t (mg g^{-1}) are the sorption capacities (expressed as amount of sorbate on sorbent unit mass) at equilibrium conditions and at time t, respectively. K_1 (s^{-1}) is the rate constant of the pseudo first-order sorption reaction. Integrating Eq. (3) with boundary conditions $q_t=0$ at $t=0$ and $q_t=q_t$ at $t=t$, a linear relation can be obtained as:

$$\ln\left(q_e - q_t\right) = \ln\left(q_e\right) - K_1 t \tag{4}$$

Using Eq. (4), K_1 can be calculated as the slope of $\ln(q_e - q_t)$ versus time t. The initial sorption rates $h_{0,1}$ (g g^{-1} min) can be calculated from Eq. (5) for $q_t \rightarrow 0$, as:

$$h_{0,1} = K_1 q_e \tag{5}$$

3.2.2 Pseudo second-order kinetic model

The pseudo second-order kinetic model (Yaneva et al., 2012) is defined agreeing with the following equation:

$$\frac{dq_t}{dt} = K_2 \left(q_e - q_t\right)^2 \tag{6}$$

By the integration of Eq. (6) at boundary conditions ($q_t=0$ at $t=0$ and $q_t=q_t$ at $t=t$) the pseudo second-order kinetics may be expressed in a linear form as:

$$\frac{t}{q_t} = \frac{1}{K_2 q_e^2} + \frac{t}{q_e} \tag{7}$$

where K_2 (g mg^{-1} min^{-1}) is its rate constant. The slope and intercept of linear t/q_t versus time plot can be used to determine the second-order rate constant, K_2, and maximum equilibrium sorption, q_e, respectively. Furthermore, the initial sorption rates $h_{0,2}$ (mg g^{-1} s^{-1}) and half sorption time (s) can be calculated as:

$$h_{0,2} = K_2 q_e^2. \tag{8}$$

$$t_{1/2} = \frac{1}{K_2 \cdot q_e} \tag{9}$$

considering that from Eq. (6), dq_t/dt approaches kq_e^2 when $q_t \rightarrow 0$.

3.2.3 Elovich model

Elovich's equation (Yaneva et al., 2012), based on the sorption capacity, is defined by the following expression:

$$\frac{dq_t}{dt} = \alpha e^{-\beta q_t} \tag{10}$$

where q_t (mg g^{-1}) is the sorption capacity at time t. The coefficient α (mg g^{-1} s^{-1}) can be considered the initial sorption rate, considering that dq_t/dt approaches α when $q_t \rightarrow \beta$ (g mg^{-1}), is the absorption constant during an experiment and it is related to the extent of surface coverage and the activation energy for chemisorption. Integrating, considering that $q_t = q_t$ at $t = t$ and $q_t = 0$ at $t = 0$ and assuming also that α, β, and $t >> 1$, Eq. (11) is obtained.

$$q_t = \frac{\ln(\alpha\beta)}{\beta} + \frac{\ln(t)}{\beta} \tag{11}$$

Therefore, plotting q_t versus $\ln(t)$, a straight line should be obtained. According to the Elovich kinetic model, the constants α and β can be calculated from the slope and the intercept of the interpolating line.

3.2.4 Intraparticle diffusion model

The intraparticle diffusion model (Yaneva and Koumanova, 2006; Yaneva et al., 2012) is required to obtain further information concerning the diffusion mechanisms affecting sorption kinetics. Weber's intraparticle diffusion model is broadly used to assess the main aspects of diffusion mechanisms. This model is defined as:

$$q_t = K_{id} t^{½} + C \tag{12}$$

where k_{id}, the intraparticle diffusion rate constant (mg g^{-1} s$^{-1/2}$), is the slope of the linear plot of q_t versus $t^{1/2}$. C is the intercept and it is related to the thickness of the boundary layer. In particular, a higher value of C indicates a higher boundary layer effect and consequently a higher contribution of the surface sorption in the rate-controlling step. Consequently, if the trend of q_t versus $t^{1/2}$ gives a single straight line and its intercept is near zero, then the sorption mechanism can be considered affected only by the interparticle diffusion mechanism. But, if the data show a multilinear trend, this indicates that several sorption steps occur, influencing the kinetic sorption process. It is assumed that the external resistance to mass transfer surrounding the particles is significant only in the early stages of sorption: this is represented by the first sharper portion. The second linear portion is the gradual sorption stage with controlling intraparticle diffusion.

Among the kinetics mechanisms reported above, the pseudo second-order appears the most suitable as reported in the literature (Piperopoulos et al., 2018b). Gui et al. (2011) described by a second-order kinetic model the sorption process of a CNT sponge, which had sorption capacities >100 g g^{-1} for a lot of organics and oils (mineral oil, diesel oil, octane, and ethyl acetate) with viscosities of 3–200 cP, as reported in Fig. 6.

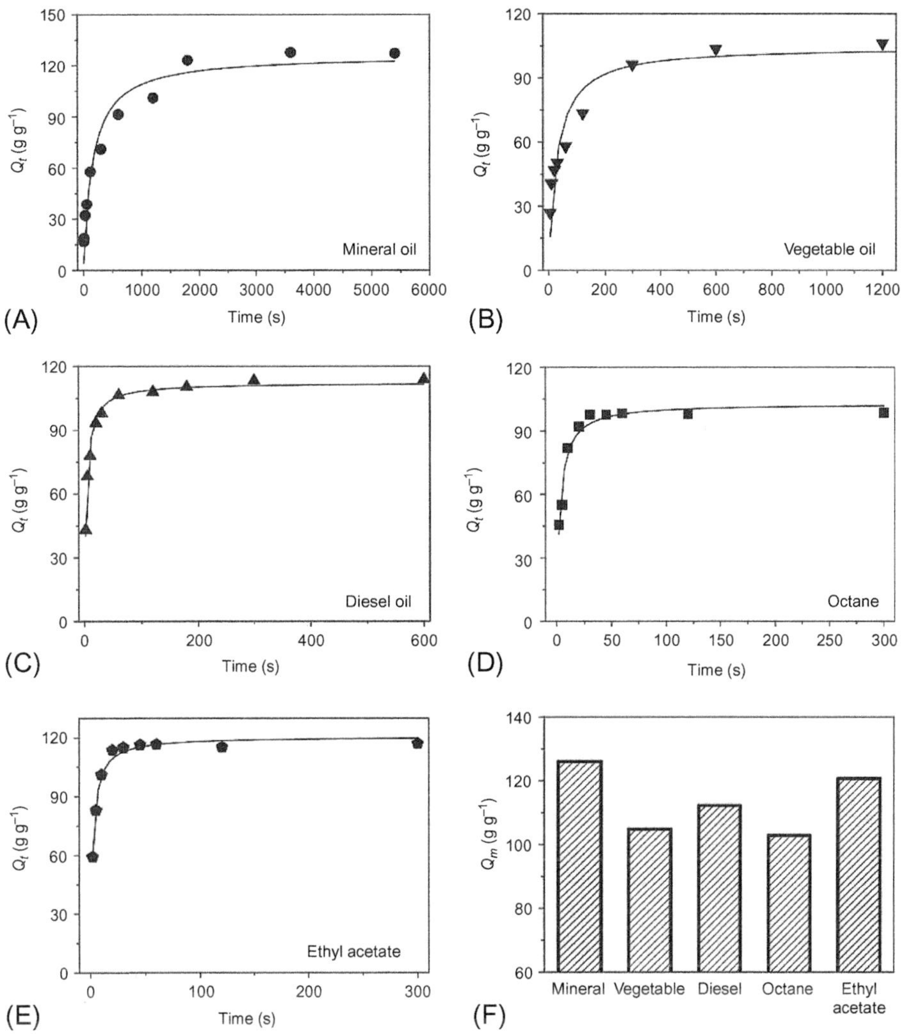

Fig. 6 Sorption kinetics of different pollutants absorbed by the CNT sponge: mineral oil (A), vegetable oil (B), diesel oil (C), octane (D), and ethyl acetate (E), saturated sorption capacities of different pollutants (F).
From Gui, X., Li, H., Wang, K., Wei, J., Jia, Y., Li, Z., Fan, L., Cao, A., Zhu, H., Wu, D., 2011. Recyclable carbon nanotube sponges for oil absorption. Acta Mater. 59, 4798–4804.

Also, Piperopoulos et al. (2018b) confirmed that the pseudo second-order model showed the highest degree of correlation ($R^2 \sim 1$) with the experimental data, assuming that chemisorption played a relevant role in the oil sorption phenomena, after having analyzed all the above-mentioned kinetics methods for pristine CNT-filled foam. Based on this consideration, for carbon-based sponges the authors affirmed that the accuracy of the kinetic models was classified in the order of pseudo second-order > Elovich > pseudo first-order kinetic model. In particular, the pseudo second-order model showed a correlation index of 0.999, despite the Elovich and pseudo first-order kinetic models that showed a correlation index in the range 0.925–0.980. These results suggested that the chemisorption phenomena could be the rate-controlling factors on oil sorption. Furthermore, according to the pseudo second-order kinetics affinity, it was possible to assess that the occupation rate of the binding sites suitable to interact with sorbed oils was proportional to the square number of unoccupied active sites on sorbent surfaces.

3.3 Reusability

The reusability of the absorbent sponges and the pollutant recovery are the key factors for its practical and cost-effective use for oil clean-up applications. To evaluate the reusability of the sorbent sponges, a sponge sample after the sorption tests is usually mechanically regenerated and then characterized by simple sorption and squeezing steps. In other cases, especially for CNT sponges, absorbed oil is burned off in air (Gui et al., 2011). The reusability experiments are always carried out by immersing the foam in oil up to complete saturation, then leaving it to drip off surface oil. This is followed by weighing it to determine the mass sorption capacity and then squeezing it mechanically by compressible cycles to desorb the absorbed oil or burning it. The restored sponge is subsequently reused for further sorption tests. Gui et al. (2011) burned off in air their CNT sponge and tested the samples for 10 cycles of sorption/burning. They characterized also the recyclability of the CNT sponge in diesel oil by the squeezing method and they compared the two different approaches (Fig. 7A).

These results showed that 33 $mg\,cm^{-3}$ (~4 wt%) of diesel pollutant oil still remain in the CNT sponges after the squeezing step. Furthermore, at increasing sorption/squeezing cycles, the saturation sorption capacity remains about 20 $g\,g^{-1}$, indicating that the absorption efficiency of the CNT sponges does not noticeably change after a high number of cycles. Instead, due to the direct burning method, the saturation sorption capacity of CNT sponges increases. The limit of this approach is that the waste must be sent to a power station to allow its efficient conversion to heat energy. Afterward, the sponges can be reused for environmental cleaning. The recyclability of CNT sponges by burning for other pollutants is summarized in Fig. 7B. These results clearly confirm the good reusability of CNT sponges. The squeezing method is a simpler and more convenient approach and it can be carried out onsite after oil absorption. The saturated sorption capacity remains almost the same also at high squeezing cycles, indicating that this procedure is suitable and cost-effective for on-site spill recovery.

Furthermore, the results shown in Fig. 7C indicate that >96 wt% of the absorbed pollutants can be removed (mainly converted to heat energy) from the CNT sponges. The main result from these considerations is that the carbon sponges allow high reusability efficiency, despite conventional and commercial oil recovery products. In such a context, these new functional materials are able to be used for hundreds of cycles without structural collapse, owing to their mechanical robustness and elasticity.

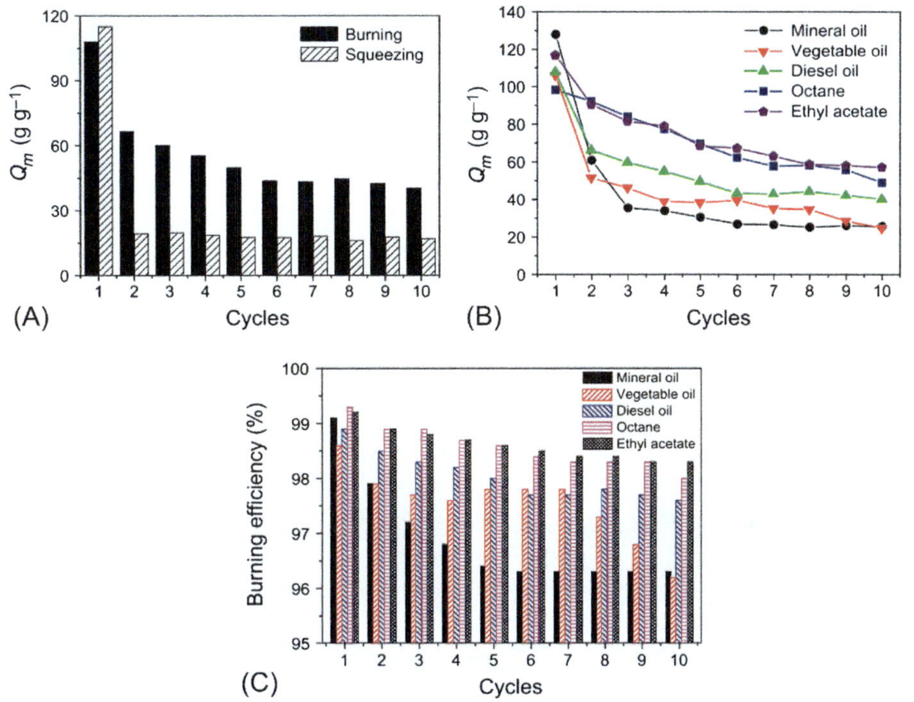

Fig. 7 Reusability of the CNT sponges: saturated sorption capacities of diesel oil after burning and squeezing regeneration methods (A); saturated sorption capacities of investigated pollutants after 10 cycles of testing (B); burning efficiency of different pollutants (C).
From Gui, X., Li, H., Wang, K., Wei, J., Jia, Y., Li, Z., Fan, L., Cao, A., Zhu, H., Wu, D., 2011. Recyclable carbon nanotube sponges for oil absorption. Acta Mater. 59, 4798–4804.

In fact, mechanical performance is a relevant point of sponge capabilities (Gui et al., 2010). A low mechanical strength may degrade the materials, inducing premature fractures with a subsequent decrease in sorption capacity. Therefore, it is required to define mechanical aging tests, such as dynamic mechanical analyses (DMA), in order to assess performance stability and durability (Gui et al., 2013). Hashim et al. (2012) conducted the mechanical investigation of their covalently bonded 3D CNT sponges by means of dynamic mechanical tests at 60% strain. They found that the complex modulus was $E = 1.26$ MPa and the tanδ (the ratio of the loss modulus to storage modulus) $= 0.058$. The high tanδ value is indicative of the material's high energy sorption and rubber-like damping capacity, due to the casually orientated entangled CNT network. As a reference, the evolution of compressive stress versus strain at increasing cycles is reported (Fig. 8A), confirming the high stability of the carbon sponge and indicating this class of materials, thanks to their mechanical damping properties, as a good candidate for oil recovery applications.

Mechanical behavior and stability are strictly related to the sponge morphology and its microstructure, as underlined by Piperopoulos et al. (2018a). Based on SEM

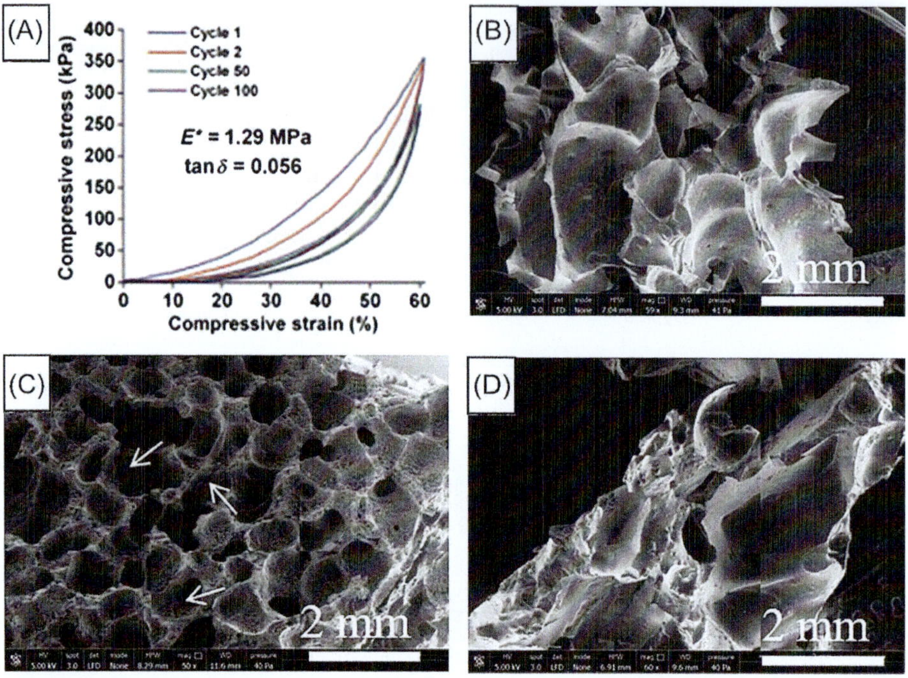

Fig. 8 Stress–strain curves up to 60% strain at 0.5 Hz and room temperature for cycles 1 (*blue*), 2 (*red*), 50 (*green*), and 100 (*purple*) (A). SEM images of squeezed foams after absorption test in kerosene: PDMS foam (B), pure CNT foam (C), and functionalized CNT foam (D).

From (A) Hashim, D.P., Narayanan, N.T., Romo-Herrera, J.M., Cullen, D.A., Hahm, M.G., Lezzi, P., Suttle, J.R., Kelkhoff, D., Muñoz-Sandoval, E., Ganguli, S., Roy, A.K., Smith, D.J., Vajtai, R., Sumpter, B.G., Meunier, V., Terrones, H., Terrones, M., Ajayan, P.M., 2012. Covalently bonded three-dimensional carbon nanotube solids via boron induced nanojunctions. Sci. Rep. 2, 363. (B–D) Piperopoulos, E., Calabrese, L., Mastronardo, E., Proverbio, E., Milone, C., 2018a. Synthesis of reusable silicone foam containing carbon nanotubes for oil spill remediation. J. Appl. Polym. Sci. 135, 46,067–46,078.

observations, they highlighted that pure silicone (Fig. 8B) and functionalized CNT-PDMS (Fig. 8D) sponges suffer local damage in cell walls that significantly modify foam density and sorption performances. On the contrary, pristine CNT-filled sponges showed few cracks (*white arrows* in Fig. 8C), but the foam remained structurally stable with still-effective sorption capabilities and reusability.

It is worth noting that the reuse of carbon sponges is a key point for efficient management of oil recovery costs. The recovery of the oil and the reusability of the materials increase the environmental friendliness and could decrease the cost of the oil recovery methodology. A good and stable reuse of the sorbent material induces a large amount of recovered pollutants for each sorption cycle, giving practical and economical advantages in oil clean-up applications and adding one more step toward sustainability.

4 Recent developments

Recently, Song et al. (2018) proposed blow spinning as a new promising technology to produce large-scale porous foams constituted by carbon microfibers with relevant oil-absorption capacity. Also, for large-scale production, the material showed effective performances compatible with lab-scale tests in terms of absorption and reusability. Moreover, the carbon microfiber preparation process is very environmentally friendly and cost-effective. After all, it is of considerable importance in future developments to pay particular attention to this aspect to improve the sustainability and diminish the costs of the product, which may be marketed. With the purpose of maximizing the material performances and simplifying the manufacturing procedure, in the last year, new research approaches in this field are moving toward the synthesis of innovative sponges of other carbon forms, confirming the relevance of this topic. Shi et al. (2019) realized an antiflaming and durable melamine superhydrophobic sponge with a "lotus leaf-mimic" surface that exhibited an absorption capacity of 69–153 times its own weight for a wide range of oils and organic solvents. Again, Wang and Deng (2019) presented an easy and low-cost method for constructing magnetic, durable, and superhydrophobic melamine carbon sponges. The so-obtained carbon sponges featured rapid magnetic response, suitable superhydrophobicity and superoleophilicity, high chemical resistance, environmental stability, mechanical durability, and high oil/water separation efficiency and oil absorbency. Li et al. (2019) also studied melamine employment in this field, fabricating a superhydrophobic melamine-formaldehyde sponge with excellent sorption capacity (up to 78–172 times its own weight) and outstanding reusability. Xie et al. (2019) concentrated on an easy and green manufacture of a superhydrophobic melamine sponge for constant oil/water separation from hard environments. In particular, they obtained the sorbent material by a green, bio-inspired, glue-assisted, sol–gel strategy. These sponges showed effective compressibility, wear resistance, and harsh environment resistance. Among the new investigated sorbents, cellulose-based absorbents were investigated thanks to their environmental friendliness, renewability, biodegradability, suitable mechanical performance, low density, high porosity, high absorption capacity, and cost-effectiveness. Jiang et al. (2019) recently realized a zwitterionic bacterial cellulose aerogel for water remediation with superior performance in recovering its original shape. But the main achievable goal in this research would be the use of waste material for oil absorption in environmental remediation. Waste materials represent environmental problems themselves and their reuse is preferable to reduce disposal costs.

In such a context, the authors started a research activity focused on used char waste for the synthesis of oil recovery foams. The char sponge presented an absorption capacity in heavy oils of 130 wt%. But the main result is the reuse of waste char, decreasing the economic and environmental impacts and respecting the circular economy.

5 Conclusions and future trends

Carbon-based sorbents have shown great and promising capabilities in oil-spill recovery applications, owing to their superior properties. Furthermore, they are biodegradable and chemically inert, which make them superior to polymer-based sorbents. Carbon (CNT, graphene, graphite, or hybrids)-filled polymer sponges have also been widely used to produce superhydrophobic and superoleophilic materials. In fact, the presence of carbon fillers is able to alter the polymer's intrinsic nature to absorb only oil while repelling water completely while also improving their thermal and mechanical stability. A relevant aspect that needs to be taken into consideration is the flexibility and scalability of carbon sponge production techniques in order to make their use valuable in the industrial field. The recent results represent an incentive to deeply investigate this context. The promising properties of carbon-based sponges are a valid prerequisite for developing new onsite research activities. This approach will make it possible to validate the manufacturing technique on a real scale. It also allows assessing the application context, defining possible issues not only related to the absorption or reusability performances of the functional material, but also to the layout, disposal, and logistics problems linked to the oil recovery process chain.

References

Annunciado, T.R., Sydenstricker, T.H.D., Amico, S.C., 2005. Experimental investigation of various vegetable fibers as sorbent materials for oil spills. Mar. Pollut. Bull. 50, 1340–1346.

Atta, A.M., El-Ghazawy, R.A.M., Farag, R.K., Abdel-Azim, A.A.A., 2006. Crosslinked reactive macromonomers based on polyisobutylene and octadecyl acrylate copolymers as crude oil sorbers. React. Funct. Polym. 66, 931–943.

Atta, A.M., El-Hamouly, S.H., Alsabagh, A.M., Gabr, M.M., 2007. Crosslinked poly(octadecene-alt-maleic anhydride) copolymers as crude oil sorbers. J. Appl. Polym. Sci. 105, 2113–2120.

Bi, H., Xie, X., Yin, K., Zhou, Y., Wan, S., He, L., Xu, F., Banhart, F., Sun, L., Ruoff, R.S., 2012. Spongy graphene as a highly efficient and recyclable sorbent for oils and organic solvents. Adv. Funct. Mater. 22, 4421–4425.

Camilli, L., Pisani, C., Gautron, E., Scarselli, M., Castrucci, P., D'Orazio, F., Passacantando, M., Moscone, D., De Crescenzi, M., 2014. A three-dimensional carbon nanotube network for water treatment. Nanotechnology 25, 065701.

Campagna, C., Short, F.T., Polidoro, B.A., McManus, R., Collette, B.B., Pilcher, N.J., Sadovy de Mitcheson, Y., Stuart, S.N., Carpenter, K.E., 2011. Gulf of Mexico Oil Blowout Increases Risks to Globally Threatened Species. BioScience 61, 393–397.

Dai, W., Kim, S.J., Seong, W.K., Kim, S.H., Lee, K.R., Kim, H.Y., Moon, M.W., 2013. Porous carbon nanoparticle networks with tunable absorbability. Sci. Rep. 3, 2524.

Donato, M.G., Faggio, G., Galvagno, S., Lanza, M., Messina, G., Milone, C., Piperopoulos, E., Pistone, A., Santangelo, S., 2009. Influence of gas-mixture composition on yield, purity and morphology of carbon nanotubes grown by catalytic isobutane-decomposition. Diamond Relat. Mater. 18, 360–363.

Dong, X., Chen, J., Ma, Y., Wang, J., Chan-Park, M.B., Liu, X., Wang, L., Huang, W., Chen, P., 2012. Superhydrophobic and superoleophilic hybrid foam of graphene and carbon nanotube for selective removal of oils or organic solvents from the surface of water. Chem. Commun. 48, 10,660–10,662.

Edwards, R., White, I., 2013. The sea empress oil spill: environmental impact and recovery. Int. Oil Spill Conf. Proc. 1999, 97–102.

Farag, R.K., El-Saeed, S.M., 2008. Synthesis and characterization of oil sorbers based on docosanyl acrylate and methacrylates copolymers. J. Appl. Polym. Sci. 109, 3704–3713.
Fragouli, D., Athanassiou, A., 2017. Oil spill recovery: graphene heaters absorb faster. Nat. Nanotechnol. 12 (5), 406–407.
Ge, B., Zhang, Z., Zhu, X., Ren, G., Men, X., Zhou, X., 2013. A magnetically superhydrophobic bulk material for oil removal. Colloids Surf. A Physicochem. Eng. Asp. 429, 129–133.
Gui, B.X., Wei, J., Wang, K., Cao, A., Zhu, H., Jia, Y., Shu, Q., Wu, D., 2010. Carbon Nanotube Sponges. Adv. Mater. 22 (5), 617–621.
Gui, X., Li, H., Wang, K., Wei, J., Jia, Y., Li, Z., Fan, L., Cao, A., Zhu, H., Wu, D., 2011. Recyclable carbon nanotube sponges for oil absorption. Acta Mater. 59, 4798–4804.
Gui, X., Zeng, Z., Lin, Z., Gan, Q., Xiang, R., Zhu, Y., Cao, A., Tang, Z., 2013. Magnetic and highly recyclable macroporous carbon nanotubes for spilled oil sorption and separation. ACS Appl. Mater. Interfaces 5 (12), 5845–5850.
Gupta, S., Tai, N.H., 2016. Carbon materials as oil sorbents: a review on the synthesis and performance. J. Mater. Chem. A 4, 1550–1565.
Hashim, D.P., Narayanan, N.T., Romo-Herrera, J.M., Cullen, D.A., Hahm, M.G., Lezzi, P., Suttle, J.R., Kelkhoff, D., Muñoz-Sandoval, E., Ganguli, S., Roy, A.K., Smith, D.J., Vajtai, R., Sumpter, B.G., Meunier, V., Terrones, H., Terrones, M., Ajayan, P.M., 2012. Covalently bonded three-dimensional carbon nanotube solids via boron induced nanojunctions. Sci. Rep. 2, 363.
He, Y., Liu, Y., Wu, T., Ma, J., Wang, X., Gong, Q., Kong, W., Xing, F., Liu, Y., Gao, J., 2013. An environmentally friendly method for the fabrication of reduced graphene oxide foam with a super oil absorption capacity. J. Hazard. Mater. 260, 796–805.
Jiang, J., Zhu, J., Zhang, Q., Zhan, X., Chen, F., 2019. A shape recovery zwitterionic bacterial cellulose aerogel with superior performances for water remediation. Langmuir. https://doi.org/10.1021/acs.langmuir.8b04180.
Jiang, Z.R., Ge, J., Zhou, Y.X., Wang, Z.U., Chen, D., Yu, S.H., Jiang, H.L., 2016. Coating sponge with a hydrophobic porous coordination polymer containing a low-energy CF3-decorated surface for continuous pumping recovery of an oil spill from water. NPG Asia Mater. 8, e253.
Li, A., Sun, H.X., Tan, D.Z., Fan, W.J., Wen, S.H., Qing, X.J., Li, G.X., Li, S.Y., Deng, W.Q., 2011. Superhydrophobic conjugated microporous polymers for separation and adsorption. Energ. Environ. Sci. 4, 2062–2065.
Li, M., Bian, C., Yang, G., Qiang, X., 2019. Facile fabrication of water-based and non-fluorinated superhydrophobic sponge for efficient separation of immiscible oil/water mixture and water-in-oil emulsion. Chem. Eng. J. 368, 350–358.
Liu, H.D., Liu, Z.Y., Yang, M.B., He, Q., 2013a. Surperhydrophobic polyurethane foam modified by graphene oxide. J. Appl. Polym. Sci. 130, 3530–3536.
Liu, Y., Ba, H., Nguyen, D.L., Ersen, O., Romero, T., Zafeiratos, S., Begin, D., Janowska, I., Pham-Huu, C., 2013b. Synthesis of porous carbon nanotubes foam composites with a high accessible surface area and tunable porosity. J. Mater. Chem. A 1, 9508–9516.
Lu, Y., Niu, Z., Yuan, W., 2019. Multifunctional magnetic superhydrophobic carbonaceous aerogel with micro/nano-scale hierarchical structures for environmental remediation and energy storage. Appl. Surf. Sci. 480, 851–860.
Maulion, R.V., Abacan, S.A., Allorde, G.G., Cherrielyne, M., Umali, S., 2015. Oil spill adsorption capacity of activated carbon tablets from corncobs in simulated oil–water mixture. Asia Pac. J. Multidiscip. Res. 3, 146–151.
Messina, G., Modafferi, V., Santangelo, S., Tripodi, P., Donato, M.G., Lanza, M., Galvagno, S., Milone, C., Piperopoulos, E., Pistone, A., 2008. Large-scale production of high-quality multi-walled carbon nanotubes: role of precursor gas and of Fe-catalyst support. Diamond Relat. Mater. 17, 1482–1488.

Nguyen, D.D., Tai, N.H., Lee, S.B., Kuo, W.S., 2012. Superhydrophobic and superoleophilic properties of graphene-based sponges fabricated using a facile dip coating method. Energ. Environ. Sci. 5, 7908–7912.

Ong, C.C., Sundera Murthe, S., Mohamed, N.M., Perumal, V., Mohamed Saheed, M.S., 2018. Nanoscaled surface modification of poly(dimethylsiloxane) using carbon nanotubes for enhanced oil and organic solvent absorption. ACS Omega 3, 15,907–15,915.

Piperopoulos, E., Calabrese, L., Mastronardo, E., Proverbio, E., Milone, C., 2018a. Synthesis of reusable silicone foam containing carbon nanotubes for oil spill remediation. J. Appl. Polym. Sci. 135, 46,067–46,078.

Piperopoulos, E., Calabrese, L., Mastronardo, E., Rahim, S.H.A., Proverbio, E., Milone, C., 2018b. Assessment of sorption kinetics of carbon nanotube-based composite foams for oil recovery application. J. Appl. Polym. Sci. 136, 47374.

Sakthivel, T., Reid, D.L., Goldstein, I., Hench, L., Seal, S., 2013. Hydrophobic high surface area zeolites derived from fly ash for oil spill remediation. Environ. Sci. Technol. 47, 5843–5850.

Shan, G.R., Xu, P.Y., Weng, Z.X., Huang, Z.M., 2003a. Synthesis and properties of oil absorption resins filled with polybutadiene. J. Appl. Polym. Sci. 89, 3309–3314.

Shan, G.R., Xu, P.Y., Weng, Z.X., Huang, Z.M., 2003b. Oil-absorption function of physical crosslinking in the high-oil-absorption resins. J. Appl. Polym. Sci. 90, 3945–3950.

Shi, J., Tian, Y., Li, W., Zhao, Y., Wu, Y., Jiang, Z., 2019. Plant polyphenol-inspired nanoengineering topological and chemical structures of commercial sponge surface for oils/organic solvents clean-up and recovery. Chemosphere 218, 559–568.

Song, J., Wang, H., Li, Z., Long, Y., Liu, Z., Wang, H., Li, X., Fang, M., Li, B., Wu, H., 2018. Large-scale blow spinning of carbon microfiber sponge as efficient and recyclable oil sorbent. Chem. Eng. J. 343, 638–644.

Turco, A., Malitesta, C., Barillaro, G., Greco, A., Maffezzoli, A., Mazzotta, E., 2015. A magnetic and highly reusable macroporous superhydrophobic/superoleophilic PDMS/MWNT nanocomposite for oil sorption from water. J. Mater. Chem. A 3, 17,685–17,696.

Wang, H., Wang, E., Liu, Z., Gao, D., Yuan, R., Sun, L., Zhu, Y., 2015. A novel carbon nanotubes reinforced superhydrophobic and superoleophilic polyurethane sponge for selective oil–water separation through a chemical fabrication. J. Mater. Chem. A 3, 266–273.

Wang, J., Zheng, Y., Wang, A., 2012. Superhydrophobic kapok fiber oil-absorbent: preparation and high oil absorbency. Chem. Eng. J. 213, 1–7.

Wang, N., Deng, Z., 2019. Synthesis of magnetic, durable and superhydrophobic carbon sponges for oil/water separation. Mater. Res. Bull. 115, 19–26.

Wu, C., Huang, X., Wu, X., Qian, R., Jiang, P., 2013a. Mechanically flexible and multifunctional polymer-based graphene foams for elastic conductors and oil–water separators. Adv. Mater. 25, 5658–5662.

Wu, Z.Y., Li, C., Liang, H.W., Chen, J.F., Yu, S.H., 2013b. Ultralight, flexible, and fire-resistant carbon nanofiber aerogels from bacterial cellulose. Angew. Chem. Int. Ed. 52, 2925–2929.

Xie, A., Chen, Y., Cui, J., Lang, J., Li, C., Yan, Y., Dai, J., 2019. Facile and green fabrication of superhydrophobic sponge for continuous oil/water separation from harsh environments. Colloids Surf. A Physicochem. Eng. Asp. 563, 120–129.

Yaneva, Z., Koumanova, B., 2006. Comparative modelling of mono- and dinitrophenols sorption on yellow bentonite from aqueous solutions. J. Colloid Interface Sci. 293, 303–311.

Yaneva, Z., Koumanova, B., Georgieva, N., 2012. Study of the mechanism of nitrophenols sorption on expanded perlite—equilibrium and kinetics modelling. Maced. J. Chem. Chem. Eng. 31, 101–114.

Zhao, M.Q., Huang, J.Q., Zhang, Q., Luo, W.L., Wei, F., 2011. Improvement of oil adsorption performance by a sponge-like natural vermiculite-carbon nanotube hybrid. Appl. Clay Sci. 53, 1–7.

Zhao, Y., Hu, C., Hu, Y., Cheng, H., Shi, G., Qu, L., 2012. A versatile, ultralight, nitrogen-doped graphene framework. Angew. Chem. Int. Ed. 51 (11), 371–11,375.

Zhou, X.M., Chuai, C.Z., 2010. Synthesis and characterization of a novel high-oil-absorbing resin. J. Appl. Polym. Sci. 115, 3321–3325.

Graphene and activated graphene as adsorbents for removal of heavy metals from water resources

9

Katerina Atkovska[a], *Perica Paunovik*[a], *Aleksandar Dimitrov*[a], *Kiril Lisichkov*[a], *Mousa Alghuthaymi*[b], *Anita Grozdanov*[a]

[a]Faculty of Technology and Metallurgy, University Ss. Cyril and Methodius, Skopje, Macedonia, [b]Department of Biology, Science and Humanities College, Shaqra University, Alquwayiyah, Saudi Arabia

1 Introduction

One of the greatest problems that the world is facing today is that of environmental pollution, which is increasing with every passing year and causing grave and irreparable damage to the Earth. Factors such as industrial development, population growth, and urbanization have intensified air, water, and soil pollution and caused the deterioration of ecosystems by the accumulation of many pollutants, especially heavy metals. The impacts of heavy metal pollution on living organisms are very serious. Heavy metals are bioaccumulative and toxic when present in water at concentrations higher than the permitted limits, thus leading to health hazards and environmental degradation. Therefore, the elimination of heavy metals from wastewater before it is released into the environment is of great importance. Different methods such as adsorption, ion exchange, reverse osmosis, chemical precipitation, solvent extraction, electrolysis, and membrane filtration are commonly used for heavy metal removal from water and wastewater (Bakalár et al., 2009; Fu and Wang, 2011; Rahul, 2013; Thakur and Parmar, 2013; Vinodh et al., 2011; Wanees et al., 2012). Among those, adsorption is ranked as one of the most preferred methods because it is a highly efficient and low-cost technique. Activated carbon (Bernard and Jimoh, 2013; Kouakou et al., 2013), clay minerals (Atkovska et al., 2016; Ghormi et al., 2013; Memedi et al., 2017), zeolites (Jakupi et al., 2016; Shaheen et al., 2012), and biosorbents (Aikpokpodion et al., 2010; Atkovska et al., 2017; George, 2012) have been studied as popularly used adsorbents for years. Besides the above-mentioned conventional materials, in recent years, there has been great interest in creating and upgrading new sorption materials for effective removal of heavy metals from water systems. The development of nanoscience and nanotechnology has enabled nanomaterials, especially carbon-based materials such as graphene and carbon nanotubes, to find wide application in wastewater treatments (Wang et al., 2012).

Graphene is a two-dimensional carbon nanomaterial with a single layer of sp^2-hybridized carbon atoms arranged in six-membered rings. It has excellent electrical, thermal, and mechanical properties (Huang et al., 2012; Wang et al., 2013).

Carbon Nanomaterials for Agri-food and Environmental Applications. https://doi.org/10.1016/B978-0-12-819786-8.00009-8

Having a high specific surface area of about 2300 m^2/g, graphene has an exceptional adsorption capacity and can be exploited for heavy metal ion removal from aqueous solutions (Rossaria et al., 2015). Over the last few years, researchers have indicated that graphene and its modified forms are effective sorbents for heavy metal removal due to their unique structure and surface characteristics.

Few-layered graphene oxide (GO) was synthesized and its adsorptive characteristics have been tested for removal of Cd^{2+} and Co^{2+} (Zhao et al., 2011a), Pb^{2+} (Zhao et al., 2011b), and U(VI) (Zhao et al., 2012) from aqueous solutions. Maximal adsorption capacities of GOs at different working conditions in this article were obtained and are 106.3, 68.2, 842, and 97.5 mg/g for Cd^{2+}, Co^{2+}, Pb^{2+}, and U(VI), respectively.

Other researchers have investigated the adsorption capacity of graphene for Zn(II) removal and it was found to be 126 mg/g (Hou et al., 2013). The adsorptive properties of GO toward divalent metal ions (copper, zinc, cadmium, and lead) were investigated and the competitive adsorption experiments showed the affinity of GO in the order of Pb(II) > Cu(II) > Cd(III) > Zn(II) (Sitko et al., 2013). Graphene can be modified or inserted into other materials such as polymers and metals, thus obtaining nanocomposites with improved adsorption properties that are more efficient in the removal of heavy metal ions (Sangit et al., 2013; Xiaohong et al., 2015; Yongji et al., 2014). The aim of this article is to investigate and compare the adsorption properties of graphene, G-ASP2, synthesized in our laboratory of nanomaterials and its acid activated form, AG-ASP2, for removal of Ni(II), Fe(II), and Pb(II) ions from aqueous solutions.

2 Materials and methods

2.1 Synthesis of graphene G-ASP2

Graphene is produced by high-temperature electrolysis in a molten salt using a nonstationary current regime. It should be underlined that during electrolysis, the cation is reduced at the electrode intercalates at the graphite surface and generates high mechanical stress that causes the exfoliation of the cathode. This phenomenon enables the electrochemical synthesis of graphene to be performed. The electrochemical route also offers a possibility for accurate control of various parameters such as voltage, current density, temperature, electrolyte, and the morphology of the starting material.

2.2 Activation of G-ASP2

Ten g of graphene are activated with a total of 1 L of concentrated H_2SO_4 (96%) and HNO_3 (62%). The volume ratio of the acids is $H_2SO_4/HNO_3 = 3/1$, which means that 750 mL of H_2SO_4 and 250 mL of HNO_3 are used. A mixture of 10 g graphene and 1 L acids is placed in a 2 L glass and put on a magnetic stirrer at a temperature of 70°C for 3 h in a digester. After activation, the graphene is separated by centrifugation for 15 min at 4000 rpm. It is then rinsed with distilled water and finally dried in a drying oven at 70°C at the beginning, and later at 50°C. The dried graphene is collected in a plastic container and stored in a desiccator.

2.3 Reagents and chemicals

Standard solutions of $Ni(NO_3)_2$ and $Pb(NO_3)_2$ with a concentration of 1000 mg/L were used to prepare Ni(II) and Pb(II) solutions with the desired initial concentrations. $FeCl_2$ salt (98% purity) was used for the preparation of Fe(II) solutions. H_2SO_4 (96%) and HNO_3 (62%) were applied for graphene activation. The pH of the solutions was adjusted using 0.1M HCl or 0.1M HNO_3 and 0.1M NaOH. All the reagents and chemicals were analytical grade.

2.4 Characterization methods

The morphologies of G-ASP2 and AG-ASP2 were observed by scanning electron microscopy (SEM) (TESCAN VEGA3LMU model) and transmission electron microscopy (TEM) (FEI Technai G2 Spirit TWIN). Raman spectroscopic analysis of the examined graphene was performed using a Jobin Yvon HR640 instrument. Thermogravimetric analysis (TGA/DTA) was carried out under N_2 flow using a Perkin Elmer Dyamond PE-7 with a heating rate of 20°C/min. The specific surface area was determined using the BET method (ASAP 20101, Micromeritics).

2.5 Adsorption experiment

The batch adsorption technique was applied to conduct the adsorption experiment. A laboratory glass reactor that contains a mixture of 2 L of the metal solution and 2.5 g/L G-ASP2 or AG-ASP2 as the sorbent was placed on the magnetic stirrer at 400 rpm at room temperature for 90 min, sufficient time to reach equilibrium. The initial concentrations of Ni(II) and Pb(II) solutions were 0.5 mg/L and the Fe(II) solution had an initial concentration of 1.0 mg/L. The samples were taken at a particular time (5, 10, 20, 30, 60, and 90 min), then centrifuged. The remaining concentrations of Ni(II), Pb(II), and Fe(II) ions were determined using an atomic absorption spectrophotometer (AAS Perkin Elmer model AA700Exp) and a UV/VIS spectrophotometer (Prove 600). The removal of Ni(II) ions was carried out at pH 9, based on data from the literature showing that graphene has better efficiency in the elimination of this metal from solutions at higher pH (Elif et al., 2016; Yueming et al., 2011). The experiments for the removal of the Pb(II) and Fe(II) ions were performed at pH 5 (Xiaohong et al., 2015).

3 Results and discussion

3.1 Characterization of graphene

SEM and TEM images show the morphology of the G-ASP2, Fig. 1A and B, and AG-ASP2, Fig. 1C. The figures clearly show that the graphene sheets were obtained, that is, exfoliated, from the graphite electrode. Fig. 1D presents the morphology of the activated graphene after the performed adsorption, and it can be seen that the graphene sheets are covered, that is, attached as a result of the adsorption of metal ions.

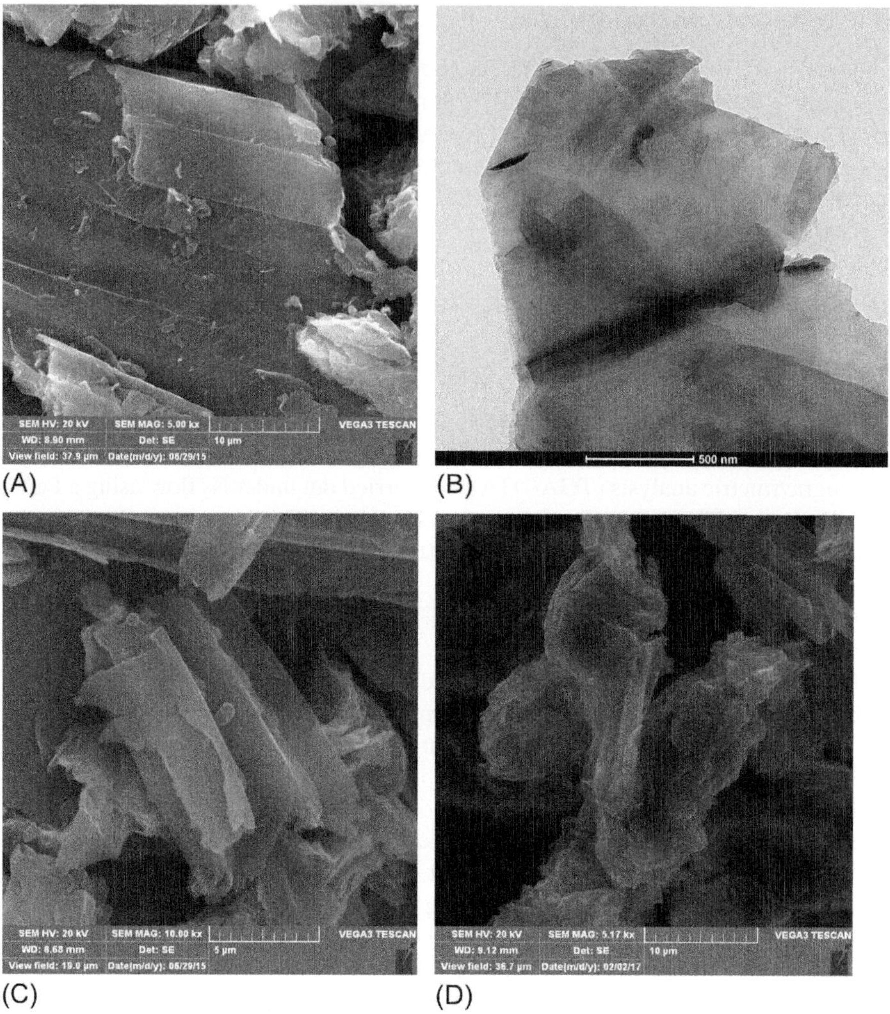

Fig. 1 (A) SEM of graphene (G-ASP2), (B) TEM of graphene (G-ASP2), (C) SEM of activated graphene (AG-ASP2), and (D) SEM of AG-ASP2 after adsorption.

The specific surface area determined using the BET method is 363 and 2657 m^2/g for graphene and activated graphene, respectively.

The Raman spectrum of the G-ASP2 is shown in Fig. 2A. The spectrum shows a peak at ~1340 cm^{-1} at the D-band and ~1580 cm^{-1} at the G-band. The D-band is related to the vibration of sp^3 defective or out-of-order graphitic carbon atoms. The G-band is associated with the vibration of sp^2 atoms in the crystal carbon nanostructure. At ~2680 cm^{-1}, the 2D-band appears. The greater the intensity of this peak, the lower the number of graphene layers. The thermogram of the G-ASP2 is presented in Fig. 2B. It shows that at a temperature of 200°C, the water is released and the graphene is thermally stable to ~680°C, when the thermal decomposition of the carbon structure begins.

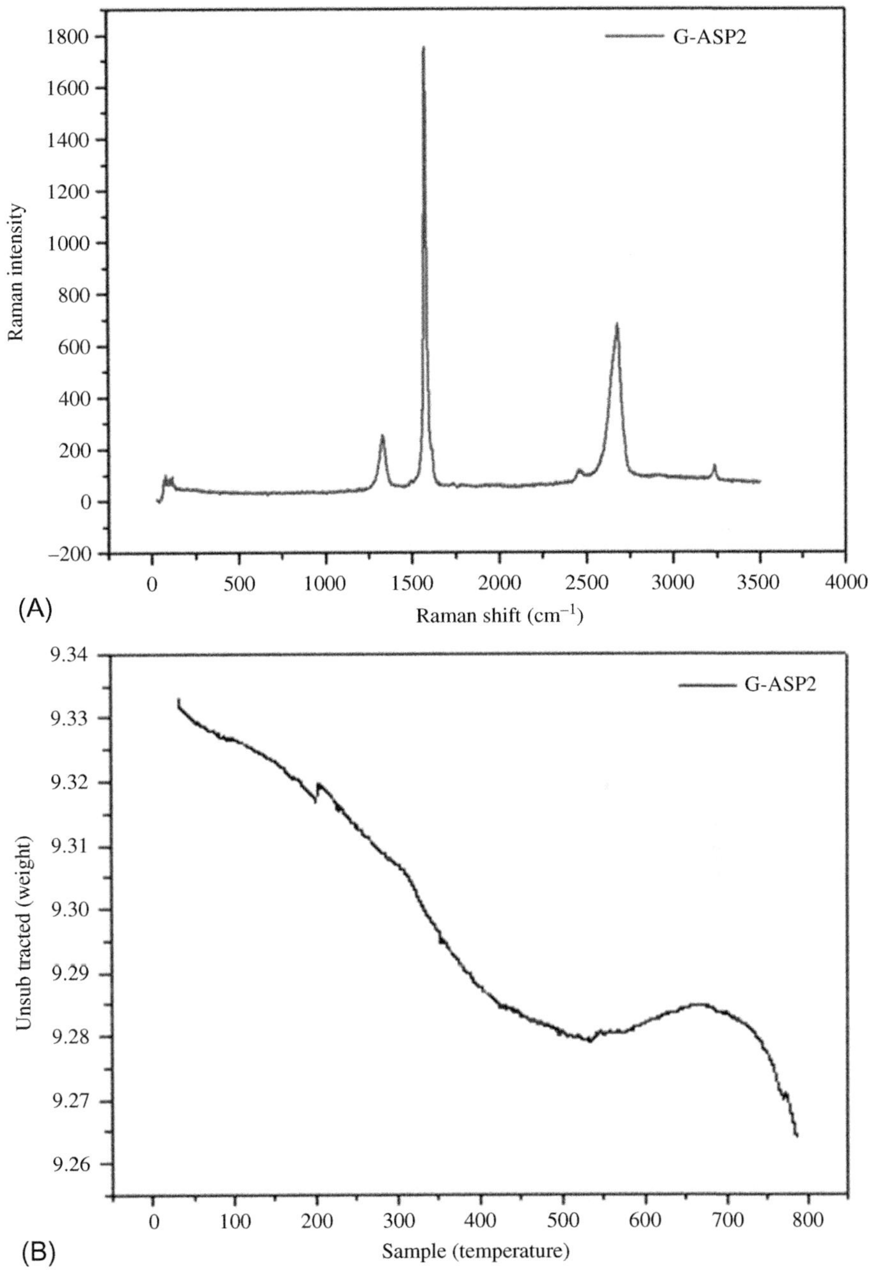

Fig. 2 (A) Raman spectrum of G-ASP2 and (B) TGA of G-ASP2.

3.2 Adsorption studies

The dependence of the concentration of Ni(II), Pb(II), and Fe(II) ions from the time of adsorption applying graphene and activated graphene as sorbents is represented in Fig. 3A–C, respectively.

The dependence of the adsorbed amount of Ni(II), Pb(II), and Fe(II) ions from the time of adsorption for G-ASP2 and AG-ASP2 is given in Fig. 4A–C, respectively. The figures show that the adsorption process, in all considered systems,

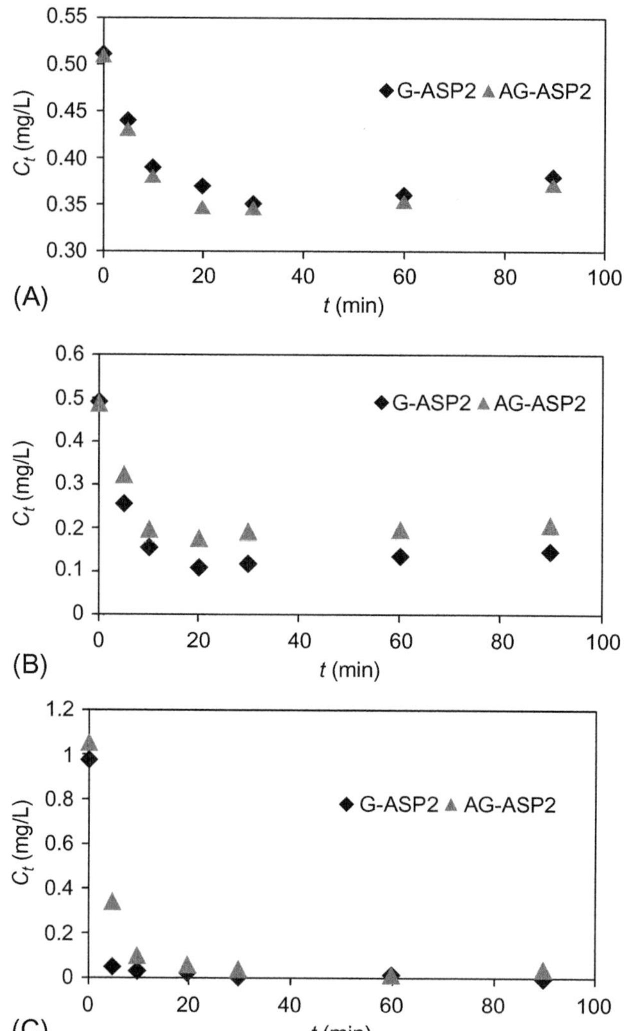

Fig. 3 (A) Dependence of Ni(II) concentration from time of adsorption, (B) dependence of Pb(II) concentration from time of adsorption, and (C) dependence of Fe(II) concentration from time of adsorption.

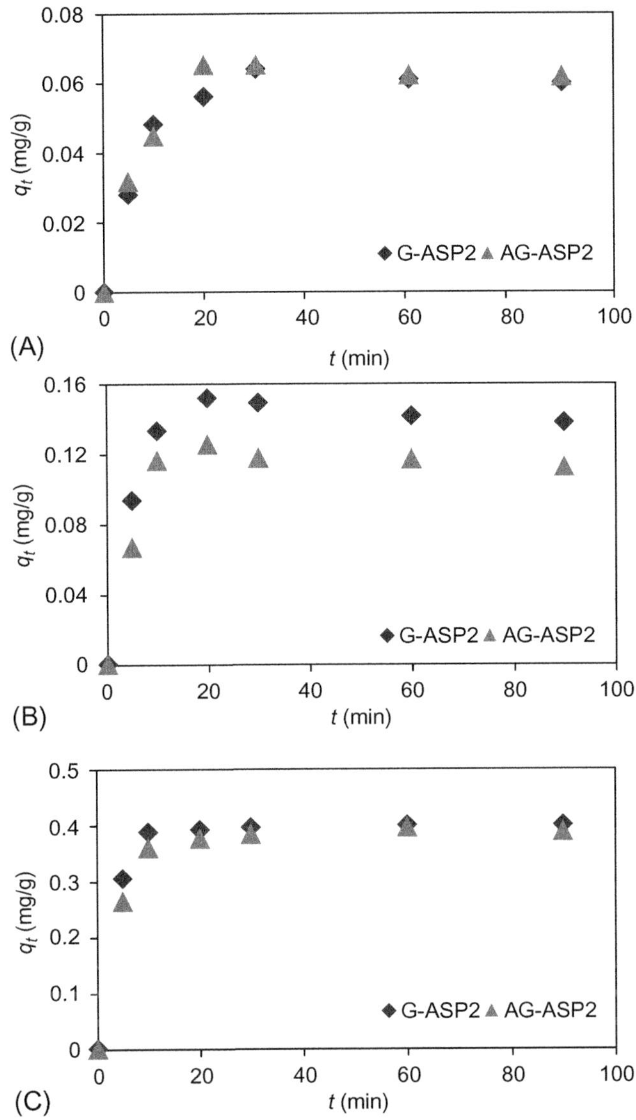

Fig. 4 (A) Dependence of adsorbed amount of Ni(II) from time of adsorption, (B) dependence of adsorbed amount of Pb(II) from the time of adsorption, and (C) dependence of adsorbed amount of Fe(II) from time of adsorption.

can be divided into two phases, fast and slow. The fast phase is at the beginning of the process, from 20 to 30 min of adsorption, when the highest efficiency of the removal of Ni(II) and Pb(II) ions is achieved using G-ASP2 and AG-ASP2. The specified period of 20–30 min for the greatest adsorption efficiency is the same as the period reported in the literature data for Hg(II) ion adsorption on graphene

(Rahimi et al., 2014) while it is slightly longer than the maximum efficiency period of 15 min for Cd(II) and Co(II) ion removal (Minghan et al., 2014). It can be noticed from Figs. 3C and 4C that for Fe(II) ions, the rapid phase happens even faster and in 10 min of adsorption, for both sorbents, maximum removal of Fe(II) is obtained. Therefore, it can be concluded that 30 min is sufficient time to reach adsorption equilibrium for all examined systems in this article. After that, the slow phase of the process follows and until the end of the treatment (90 min), there is no significant change in concentration or the adsorbed amount of Ni(II), Pb(II), and Fe(II) ions.

This way of conducting the adsorption process with time arises as a result of the fact that at the beginning of the treatment, the concentration of metal ions in the solutions is high and the number of free active centers on the surface of the sorbents is also large. It has a favorable effect on the speed of the process, which is mostly controlled by the diffusion of the ions. Reducing the concentration of metal ions leads to a decrease in the driving force of the absorption process, the speed of which depends on the diffusion of ions through the pores of the sorbents. Also, by saturation of the active centers on the surface of the sorbents with time, the probability of binding of the metal ions is reduced and the overall speed of adsorption decreases too. A comparison of the efficiency of graphene and activated graphene in the removal of Ni(II), Pb(II), and Fe(II) ions at the same operating conditions is shown in Fig. 5.

The maximum percentage of removal of Ni(II) ions with G-ASP2 and AG-ASP2 is 30.0% and 33.3%, respectively. Graphene shows better efficiency of removal than its activated form for the other two metals. The percentages of elimination of Pb(II) ions are 77.8% and 64.4% for G-ASP2 and AG-ASP2, respectively. Removal of Fe(II) ions by graphene is complete, that is, it is 100%. Activated graphene shows slightly less efficiency and removes iron ions at 99.1%.

It is well known that adsorption strongly depends on the pore structure and surface area as well as the surface functionality of adsorbents. In spite of the fact that activated graphene has a much higher specific surface area, due to the removal of functional groups on its surface by acid activation, AG-ASP2 loses affinity to some heavy metal ions.

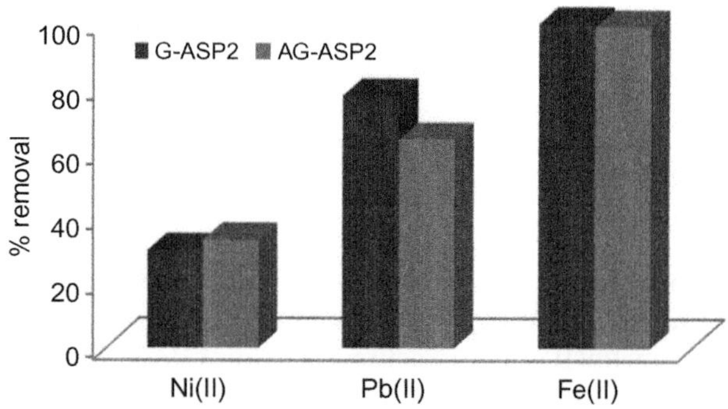

Fig. 5 The efficiency of the removal of metal ions of G-ASP2 and AG-ASP2.

Based on these results, it can be concluded that the acid activation of G-ASP2 did not meet the expectations for improving the absorption capacity of graphene and thus achieving better efficiency for the elimination of heavy metal ions. Considering that the AG-ASP2 has improved efficiency over G-ASP2 for nickel ions of only 3.3%, this acid activation is practically and economically not justified.

3.3 Kinetic studies

Fitting the experimental data into different kinetic models enables studying the adsorption rate, modeling the process and predicting information about adsorbent/adsorbate interaction (physisorption or chemisorption) (Amanda et al., 2013). In this study, pseudo first-order and pseudo second-order kinetic models were applied.

The linear forms of the pseudo first-order (Eq. 1) and the pseudo second-order model (Eq. 2) can be expressed as:

$$\ln(q_e - q_t) = \ln q_e - k_1 t \tag{1}$$

$$\frac{t}{q_t} = \frac{1}{k_2 q_e^2} + \frac{1}{q_e} t \tag{2}$$

where k_1 (min^{-1}) and k_2 (g/mg min) are rate constants for pseudo first-order and pseudo second-order models, respectively, q_t (mg/g) is the adsorbed amount of metal ions at time t, q_e (mg/g) is the adsorbed amount of metal ions at equilibrium, and $k_2 q_e^2$ or h (mg/g min) is the initial adsorption rate.

k_1 and q_e for Ni(II), Pb(II), and Fe(II) ions for G-ASP2 and AG-ASP2 were calculated using the slope and intercept of the plots of $\ln(q_e - q_t)$ versus t, Fig. 6A and B. Kinetic parameters for the pseudo first-order model and coefficients of correlation for examined metal ion sorption on the graphene and activated graphene are listed in Table 1. The pseudo second-order adsorption parameters in Eq. (2) were determined by plotting t/q_t versus t, Fig. 7A and B, and are listed in Table 1.

The correlation coefficients for the pseudo first-order kinetic model for all studied systems were low. Also, the calculated q_e values differ a lot from the experimentally obtained equilibrium adsorptions. This suggests that the pseudo first-order model is not suitable to describe the kinetic sorption of Ni(II), Pb(II), and Fe(II) ions on graphene and activated graphene.

The straight lines in the plots (Fig. 6A and B), confirmed by high values of coefficients of correlation (Table 1), show the good agreement of experimental data with the pseudo second-order kinetic models for all analyzed systems. In most systems, the correlation coefficients were higher than 0.99. Moreover, the calculated q_e values agreed very well with the experimental data. As such, in comparison to pseudo first-order kinetic models, the pseudo second-order model is considered more appropriate to represent the kinetic data in the investigated systems and the chemisorption is the rate-limiting step. Similar phenomena are also observed for

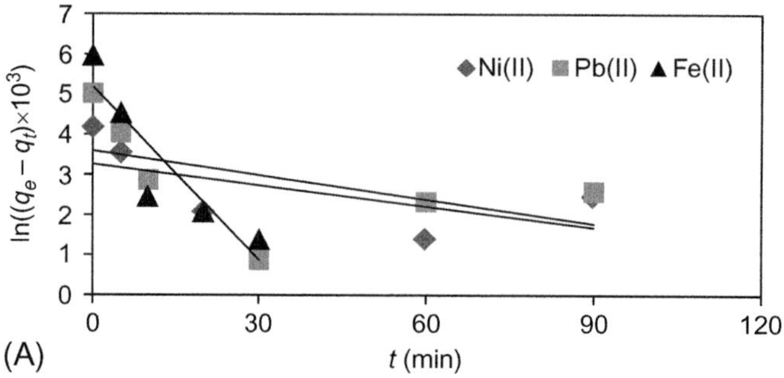

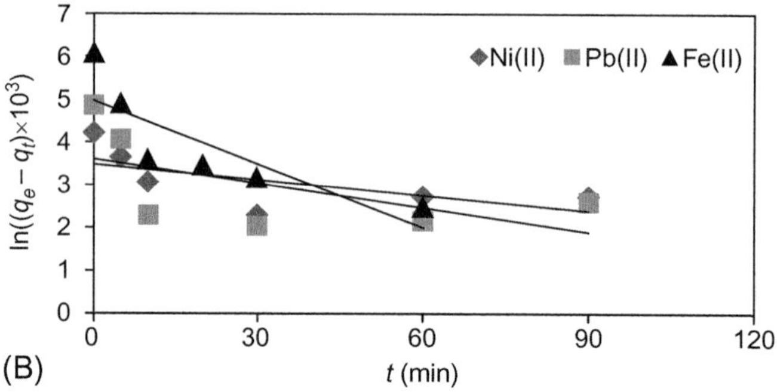

Fig. 6 (A) Pseudo first-order model for Ni(II), Pb(II), and Fe(II) adsorption on G-ASP2 and (B) Pseudo first-order model for Ni(II), Pb(II), and Fe(II) adsorption on AG-ASP2.

the adsorption of nickel ions using a graphene nanocomposite (Sangit et al., 2013) and the adsorption of Cu(II), Cd(II), and Ni(II) ions by GO (Ping et al., 2015).

3.4 Competitive adsorption on G-ASP2 and AG-ASP2 in single and multimetal systems

The working conditions for multicomponent system adsorption experiments are the same as mentioned above in order to make the comparison of the percentage of removal of metal ions in the single and multimetal systems using G-ASP2 and AG-ASP2 as sorbents (Fig. 8A and B).

The total adsorption was generally higher for the single-metal systems than for the multimetal systems. This clearly demonstrates that in a combination of more metal ions in the solutions, an antagonistic effect appears. The reason for this is the competition of the Ni(II), Pb(II), and Fe(II) ions for the absorption centers on the surface of these sorbents. Fig. 8B shows the increased efficiency of removal of lead ions in the three-component system than in the mono-component one. This means that synergism arises for Pb(II) ions using activated graphene as the sorbent. The inhibitory effect of Pb(II) ions on the sorption of Ni(II) and Fe(II) ions can be explained by the greater ionic radius and the greater electronegativity of lead than nickel and iron. As a result of that, by the mechanism of chemisorption, there is a more efficient binding of Pb(II) ions to the active centers of AG-ASP2, which leads to an increase in the percentage of removal.

Table 1 Adsorption kinetic model parameters and coefficients of correlation.

			Pseudo first-order			Pseudo second-order			
Adsorbent	**Metal ion**	q_e, **exp (mg/g)**	q_e, **cal. (mg/g)**	k_1 **(min^{-1})**	R^2	q_e, **cal. (mg/g)**	k_2 **(g/mg min)**	h **(mg/g min)**	R^2
G-ASP2	Ni(II)	0.064	0.027	0.017	0.391	0.055	63.29	0.191	0.982
	Pb(II)	0.151	0.037	0.020	0.247	0.140	21.01	0.412	0.997
	Fe(II)	0.400	0.176	0.143	0.825	0.397	2.81	0.443	0.999
AG-ASP2	Ni(II)	0.065	0.033	0.012	0.398	0.051	37.73	0.098	0.991
	Pb(II)	0.125	0.036	0.018	0.327	0.115	24.17	0.320	0.996
	Fe(II)	0.396	0.142	0.048	0.693	0.411	1.15	0.194	0.999

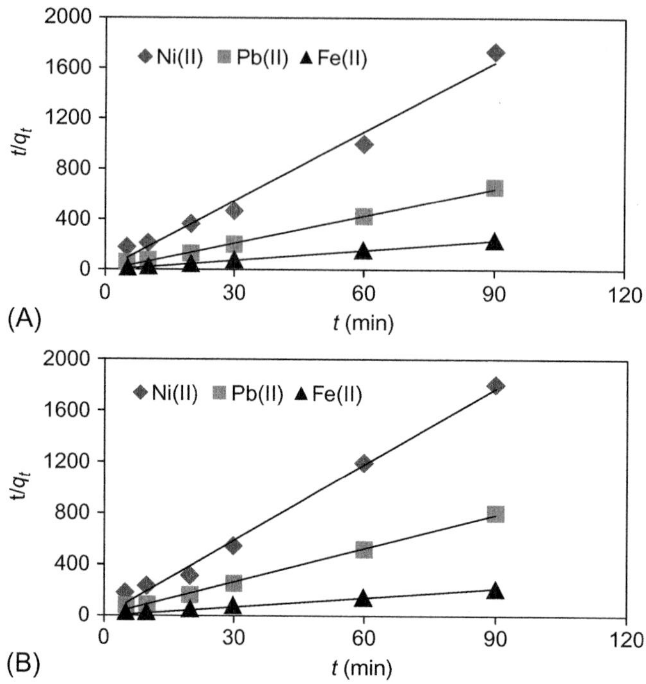

Fig. 7 (A) Pseudo second-order model for Ni(II), Pb(II), and Fe(II) adsorption on G-ASP-2 and (B) Pseudo second-order model for Ni(II), Pb(II), and Fe(II) adsorption on AG-ASP2.

4 Conclusion

In this article, the adsorption of Ni(II), Pb(II), and Fe(II) ions from aqueous solutions using graphene and activated graphene was investigated. The comparison of the efficiency of graphene and its activated form in the removal of metal ions suggested that the applied acid activation of graphene is not justified. G-ASP2 proved to be more efficient for Pb(II) and Fe(II) ions, with the exception of nickel ions when AG-ASP2 showed only a 3.3% higher percentage of removal. A total of 30 min of adsorption was sufficient time to reach equilibrium for all analyzed systems. The sorption kinetics of Ni(II), Pb(II), and Fe(II) ions on both sorbents were well described by the pseudo second-order model. The efficiency of removal was generally higher for single-metal systems than for the multimetal systems, which indicates that the antagonistic effect appears. Only for Pb(II) ions when using activated graphene as the sorbent was there an increased removal effect in multimetal solutions, that is, synergism. The results of this study indicated that graphene synthesized in our laboratory has excellent heavy metal removal capabilities, especially for Fe(II) ions with an efficiency of 100%. The use of active systems and graphene nanomaterials can deliver great advantages as a heavy metal adsorbent from water resources.

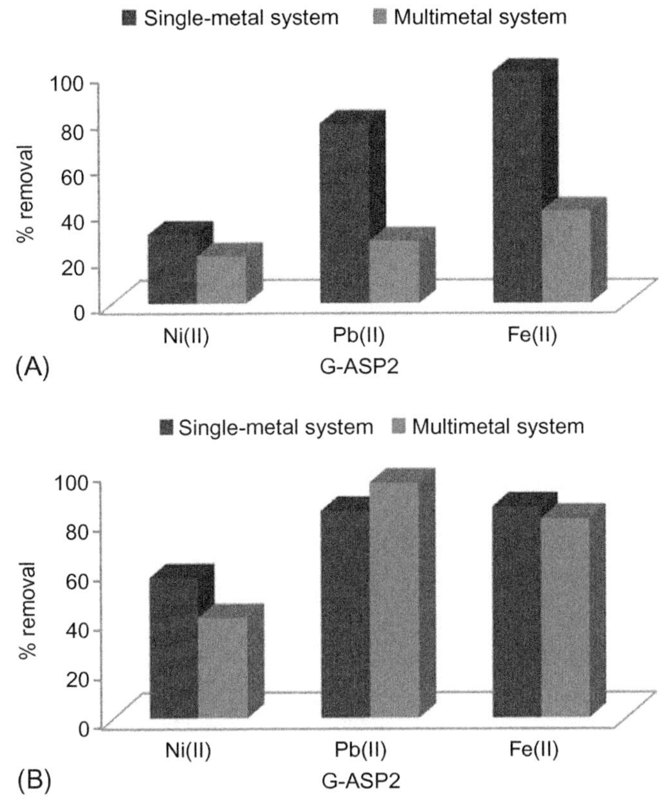

Fig. 8 (A and B) Comparison of the efficiency of G-ASP2 and AG-ASP2 of the removal of metal ions in single and multimetal systems.

References

Aikpokpodion, P.E., Ipinmoroti, R.R., Omotoso, S.M., 2010. Biosorption of nickel(II) from aqueous solution using waste tea (*Camella cinecis*) materials. Am.–Euroasian J. Toxicol. Sci. 2 (2), 72–82.

Amanda, E.M., Milene, S.P., Alexandre, O.J., Marco, A.U.M., Rafael, I.V.S., Margarida, J.S., Gustavo, R.C., 2013. The reactive surface of castor leaf (*Ricinus communis* L.) powder as a green adsorbent for the removal of heavy metals from natural river water. Appl. Surf. Sci. 276, 24–30.

Atkovska, K., Bliznakovska, B., Ruseska, G., Bogoevski, S., Boskovski, B., Grozdanov, A., 2016. Adsorption of Fe(II) and Zn(II) ions from landfill leachate by natural bentonite. J. Chem. Technol. Metall. 51 (2), 215–222.

Atkovska, K., Memedi, H., Lisichkov, K., Kuvendziev, S., Marinkovski, M., Ruseska, G., Grozdanov, A., 2017. Biosorption of nickel ions from aqueous solutions by natural and modified peanut husks: equilibrium and kinetics. Water Environ. J. https://doi.org/10.1111/wej.12325.

Bakalár, T., Búgel, M., Gajdošová, L., 2009. Heavy metal removal using reverse osmosis. Acta Montan. Slovaca 14 (3), 250–253.

Bernard, E., Jimoh, A., 2013. Adsorption of Pb, Fe, Cu and Zn from industrial electroplating wastewater by orange peel activated carbon. Int. J. Eng. Appl. Sci. 4, 95–103.

Elif, C.S., Jiabin, W., Daniel, J.L.C., Lidija, S., 2016. Enhanced removal of nickel (II) ions from aqueous solutions by SDS–functionalized graphene oxide. Sep. Sci. Technol. 51 (8).

Fu, F., Wang, Q., 2011. Removal of heavy metal ions from wastewaters: a review. J. Environ. Manage. 92 (3), 407–418.

George, Z.K., 2012. Commercial coffee wastes as materials for adsorption of heavy metals from aqueous solutions. Materials 5, 1826–1840.

Ghormi, F., Lahsini, A., Laajeb, A., Addaou, A., 2013. The removal of heavy metal ions (copper, zinc, nickel and cobalt) by natural bentonite. Larhyss J. 12, 37–54.

Hou, W., Xingzhong, Y., Yan, W., Huajun, H., Guangming, Z., Yan, L., Xueli, W., Ningbo, L., Yu, Q., 2013. Adsorption characteristics and behaviors of graphene oxide for Zn(II) removal from aqueous solution. Appl. Surf. Sci. 279, 432–440.

Huang, X., Qi, X., Boey, H., Zhang, H., 2012. Graphene-based composites. Chem. Soc. Rev. 41, 666–686.

Jakupi, S., Atkovska, K., Lisichkov, K., Golomeova, M., Marinkovski, M., Kuvendziev, S., 2016. Removal of Ni(II) ions from aqueous solutions by nanoporous material. Quality of Life 7 (1–2), 29–35.

Kouakou, U., Ello, A.S., Yapo, J.A., Trokourey, A., 2013. Adsorption of iron and zinc on commercial activated carbon. J. Environ. Chem. Ecotoxicol. 5 (6), 168–171.

Memedi, H., Atkovska, A., Lisichkov, K., Marinkovski, M., Kuvendziev, S., Bozinovski, Z., Reka, A., 2017. Separation of Cr(VI) from aqueous solutions by natural bentonite: equilibrium study. Quality of Life 8 (1–2), 41–47.

Minghan, X., Jing, C., Nantao, H., Da, H., Yuxi, W., Xiaolu, H., Hao, W., Zhi, Y., Yafei, Z., 2014. Facile synthesis of soluble functional graphene by reduction of graphene oxide via acetylacetone and its adsorption of heavy metal ions. Nanotechnology 25, 395602.

Ping, T., jian, S., Yongyou, H., Zheng, F., Qi, B., Yuancai, C., Jianhua, C., 2015. Adsorption of Cu^{2+}, Cd^{2+} and Ni^{2+} from aqueous single metal solutions on graphene oxide membranes. J. Hazard. Mater. 297, 251–260.

Rahimi, R., Rouholah, Z-D., Koohi, A., Zargari, S. (2014). Synthesis of graphene oxide–porphyrin nanocomposite and its application in removal of toxic metals, In Proceedings of the 18th International Electronic Conference on Synthetic Organic Chemistry, 1–30 November 2014; Sciforum Electronic Conference Series Vol. 18(2017), a031.

Rahul, K.J., 2013. Application of electro–dialysis (ED) to remove divalent metals ions from wastewater. Int. J. Chem. Sci. Appl. 4 (1), 68–72.

Rossaria, C., Nan, Z., Min-Quan, Y., Francesco, M., Yi-Jun, X., Mario, P., 2015. Commercialization of graphene–based technologies: a critical insight. Chem. Commun. 51, 7090–7095.

Sangit, V., Dhanashree, S., Sagar, W., Bhanvase, B.A., Deosarkar, M.P., 2013. Removal of nickel from waste water using graphene nanocomposite. Int. J. Chem. Phys. Sci. 2, 132–139.

Shaheen, S.M., Derbalah, a.S., Moghanm, F.S., 2012. Removal of heavy metals from aqueous solutions by zeolite in competitive sorption system. Int. J. Environ. Sci. Dev. 3 (4), 362–367.

Sitko, R., Turek, E., Zawisza, B., Malicka, E., Talik, E., Heimann, J., Gagor, A., Feist, B., Wrzalik, R., 2013. Adsorption of divalent metal ions from aqueous solutions using graphene oxide. Dalton Trans. 42 (16), 5682–5689.

Thakur, L.S., Parmar, M., 2013. Adsorption of heavy metal (Cu^{2+}, Ni^{2+} and Zn^{2+}) from synthetic waste water by tea waste adsorbent. Int. J. Chem. Phys. Sci. 2 (6), 6–19.

Vinodh, R., Padmavathi, R., Sangeetha, D., 2011. Separation of heavy metals from water samples using anion exchange polymers by adsorption process. Desalination 267 (2–3), 267–276.

Wanees, S.A., Ahmed, A.M.M., Adam, M.S., Mohamed, M.A., 2012. Adsorption studies on the removal of hexavalent chromium–contaminated wastewater using activated carbon and bentonite. Chem. J 2 (3), 95–105.

Wang, H., yuan, X., Wu, Y., Huang, H., Peng, X., Zeng, G., Zhong, H., Liang, J., Ren, M., 2013. Graphene–based materials: fabrication, characterization and application for the decontamination of wastewater and wastegas and the hydrogen storage/generation. Adv. Colloid Interface Sci. 195–196 (19–40).

Wang, X., Guo, Y., Yang, L., Han, M., Zhao, J., Cheng, X., 2012. Nanomaterials as sorbents to remove heavy metal ions in wastewater treatment. J. Environ. Anal. Toxicol. 7 (2), 154.

Xiaohong, L., Zheng, W., Qinq, L., Jiaxian, M., Mingzi, Z., 2015. Preparation, characterization and application of mesoporous silica–grafted graphene oxide for highly selective lead adsorption. Chem. Eng. J. 273, 630–637.

Yongji, Z., Hui, J.C., Wen, H.Z., Youyi, S., Qing, L., Yu, G., Riya, J., 2014. Highly efficient adsorption of copper ions by a PVP–reduced graphene oxide based on a new adsorption mechanism. Nano-Micro Letters 6 (1), 80–87.

Yueming, R., Ni, Y., Qing, W., Zhuangjun, F., Tong, W., Milin, Z., Jun, M., 2011. Graphene/δ–MnO_2 composite as adsorbent for the removal of nickel ions from wastewater. Chem. Eng. J. 175 (1–7).

Zhao, G., Li, J., Ren, X., Chen, C., Wang, X., 2011a. Few–layered graphene oxide nanosheets as superior sorbents for heavy metal ion pollution management. Environ. Sci. Technol. 45, 10454–10462.

Zhao, G., Ren, X., Gao, X., Tan, X., Li, J., Chen, C., Huang, Y., Wang, X., 2011b. Removal of Pb(II) ions from aqueous solutions on few–layered graphene oxide nanosheets. Dalton Trans. 40, 10945–10952.

Zhao, G., Wen, T., Yang, X., Yang, S., Liao, J., Hu, J., Shao, D., Wang, X., 2012. Preconcentration of U(VI) ions on few–layered graphene oxide nanosheets from aqueous solutions. Dalton Trans. 41, 6182–6188.

Graphene quantum dot-based nanostructures for water treatment

10

Gcina Mamba[a], Lerato Moss[a], Gumani Gangashe[a], Sourbh Thakur[b,c], Velluchamy Muthuraj[d], Sethumathavan Vadivel[e], Gcina D. Vilakati[f], Thabo T.I. Nkambule[a]
[a]Nanotechnology and Water Sustainability Research Unit, College of Science, Engineering and Technology, University of South Africa, Roodepoort, South Africa, [b]Department of Chemistry, PSG College of Technology, Coimbatore, India, [c]Institute of Materials Science of Kaunas University of Technology, Kaunas, Lithuania, [d]School of Chemistry, Shoolini University, Solan, India, [e]Department of Chemistry, VHNSN College, Virudhunagar, India, [f]Department of Chemistry, University of Swaziland, Kwaluseni, Swaziland

1 Introduction

1.1 The global water demand

The demand for a clean, adequate, and reliable water supply has continued to increase over the years, owing to rapid population growth and industrialization. However, the global water supply, especially in developing countries, is severely strained as a result of adverse weather conditions (droughts and high temperatures) and pollution of freshwater sources. According to the United Nations World Water Development Report (2018), there has been a 1.0% increase in water demand per year, which could be related to industrial developments, population growth, and changing consumption patterns. This is expected to continue over the next two decades, with the demand for industrial and domestic water consumption growing faster than the agricultural needs (UN-Water, 2018). In Africa alone, it has been projected that up to 26 countries could be experiencing either water stress or scarcity by 2025, which could translate to nearly 50% of the African population (Hameeteman, 2013). Globally, it is predicted that for at least 1 month in a year, nearly 3.6 billion people experience water scarcity and the number could reach 5.7 billion by 2050 (UN-Water, 2018). Agricultural activities continue to draw the largest portion of the water supply and this is key to the provision of an adequate food supply. Therefore, an adequate supply of good quality water is key to ensuring food security.

1.2 Water pollution: Sources and impact

With a rapidly increasing global population, industrial activities, and development, pollution challenges are inevitable. Unfortunately, pollution contributes to water

Carbon Nanomaterials for Agri-food and Environmental Applications. https://doi.org/10.1016/B978-0-12-819786-8.00010-4

shortages that are already experienced as a result of climate change. Various anthropogenic activities such as municipalities, hospitals, agricultural activities, mining, pharmaceuticals, textile industries, etc., contribute to the direct or indirect release of polluted wastewater into the environment. Such wastewater contains a wide range of potentially toxic organic compounds, inorganics, and microbial species. For example, a wide range of pesticides, herbicides, and fertilizers emanates from agricultural activities and may be toxic to the aquatic environment or potentially carcinogenic; also, fertilizers result in the eutrophication of freshwater sources (Lapworth et al., 2012). The textile industry is known for its colored wastewater, which is an obvious indicator of pollution and contains toxic and potentially carcinogenic and mutagenic dyestuffs (Crini, 2006). Mining wastewater is rich in heavy metals that are known to be toxic to aquatic life and mankind while hospitals and pharmaceuticals could release wastewater loaded with drugs such as antibiotics, resulting in the development of drug-resistant bacteria (Islam et al., 2015; Le-Minh et al., 2010). Antibiotics belong to a class of pollutants known as pollutants of everyday concerns, which have captured global interest because their fate in the environment is not yet fully understood. Despite the diversity and complexity of the wastewater matrices, the treatment regimes are standard and have stayed the same over the years. Consequently, some of the emerging pollutants are able to sneak past the treatment train and end up in our tap water, where they could cause various negative health effects.

Besides compromising the quality of drinking water, the presence of various pollutants in irrigation water could lead to the uptake and accumulation of these pollutants in crops that are eventually harvested as food. Moreover, both organic and inorganic pollutants may undergo transformation or modifications when they are exposed to various environmental conditions, leading to potentially toxic complexes or metabolites. Therefore, priority has been given to the development of complementary treatment protocols to ensure efficient water treatment. The emergence of various carbon nanomaterials has allowed tailoring different kinds of nanocomposites for application in the photocatalytic removal of pollutants, adsorption, and the membrane filtration process. In this review, we provide a detailed discussion of graphene quantum dot (GQD)-based nanostructures for water treatment with a major emphasis on photocatalytic applications. Moreover, a discussion of GQD-based nanostructures beyond photocatalysis (adsorption and membrane filtration) is given.

2 Graphene quantum dots: Structure, preparation, and properties

Often regarded as the "new kid on the block," GQDs have emerged as a new member of the nanocarbon family. Generally, GQDs can be viewed as smaller fragments of graphene. By definition, a GQD is a 0D, sp^2 bonded carbon atom arranged in a flat, honey-comb structure like graphene, with lateral dimensions of <10 nm. Like graphene, GQDs could be a single layer or a few layered structures, typically <10 layers (Bianco et al., 2013; Feng et al., 2019; Tian et al., 2018). Graphene quantum dots display properties that are derived from both graphene and quantum dots. There is often confusion between carbon dots (CDs) and GQDs, which arises from the fact that they are both carbon nanomaterials with lateral dimensions <10 nm. However, there is a clear distinction between these carbon nanomaterials with CDs being spherical and largely amorphous in nature. On the other hand, GQDs are crystalline in nature and have a sheet-like structure (Kalluri et al., 2018; Valail et al., 2017). However, both CDs and GQDs are rich in oxygen-containing functional groups, which makes them attractive materials for various modifications. Graphene quantum dots possess numerous unique physical, chemical, and electronic properties that have resulted in their extensive investigation. These include chemical and thermal stability, solubility, a large specific surface area, low cytotoxicity, reactive functional groups that allow easy modification, and high photoluminescence properties. Furthermore, unlike its "parent" graphene that has a zero band gap, a GQD has a nonzero band gap that can be modified by controlling the size of the dots, among other strategies (Alidad et al., 2018; Tian et al., 2018; Valail et al., 2017; Wang et al., 2019a; Zeng et al., 2018). Modification of GQDs with heteroatoms has been shown to improve their electronic properties, photoluminescence, and catalytic behavior (Feng et al., 2019; Wang et al., 2019a).

Graphene quantum dots are generally obtained via two approaches: a bottom-up and a top-down approach. In the top-down route, GQDs are obtained from the fragmentation of larger graphene sheets while in the bottom-up route, GQDs are synthesized from the fusion of molecular precursors. Typical top-down synthesis routes include liquid exfoliation, electrochemical exfoliation, electron beam lithography, hydrothermal/solvothermal cutting, microwave-assisted cutting, and ultrasonic shearing. Bottom-up routes, on the other hand, include intermolecular coupling, precursor pyrolysis, and cage opening of fullerenes (Abbas et al., 2018; Shen et al., 2012; Valail et al., 2017). For details on the synthesis and properties of GQDs, readers may consult a number of rich reviews that have been published in recent years (Abbas et al., 2018; Tian et al., 2018; Yan et al., 2019; Zeng et al., 2018).

Owing to their remarkable properties, GQDs have found extensive application in various fields, including photocatalysis (Sajjadi et al., 2019), sensors (Chu et al., 2019), bioimaging (Wang et al., 2019b), anticorrosion materials (Pourhashem et al., 2018), tissue engineering (Mallakpour and Khadem, 2018), solar cells (Subramanian et al., 2017), adsorption of pollutants (Khojasteh et al., 2017), lithium-ion batteries (Lijuan et al., 2016), organic synthesis (Mahyari et al., 2016), and membrane filtration (Xu et al., 2019). This chapter provides an overview of the application of GQD-derived nanostructures in water treatment, which is crucial toward the provision of clean and adequate water to ensure food security, among other important sectors. A detailed discussion of the role of GQDs in influencing the pollutant-removal capability of nanostructures will be discussed in detail.

3 Graphene quantum dots in water treatment

Graphene quantum dots have found wide exploitation toward the fabrication of nanostructures for removal of various pollutants in water. Several reports have emerged on the application of doped GQDs and GQD-derived nanocomposites in the photocatalytic degradation of pollutants (Deng et al., 2017), the photoelectrocatalytic degradation of pollutants (Wang et al., 2017), the adsorption of pollutants (Agarwal et al., 2016), and membrane filtration (Bi et al., 2019). In all the applications, incorporation of GQDs led to a significant improvement in the pollutant-removal capability of the nanocomposites. Pollutants such as dyes, emerging pollutants, and heavy metals have been removed from contaminated water using various GQD-derived nanocomposites. Photocatalytic water treatment presents the potential to simultaneously remove both organic and inorganic pollutants as well as microbial contaminants from water. However, issues such as the fast recombination of charge carriers, a small surface area, poor visible light utilization, and aggregation of nanophotocatalysts in water are major bottlenecks of the process. Coupling various semiconductors with carbon nanomaterials including graphene, graphene oxide, reduced graphene oxide, carbon quantum dots, and graphene quantum dots presents an attractive strategy to improve the overall activity of various semiconductors. Graphene quantum dots are among the most attractive carbon nanomaterials in photocatalysis, owing to their visible light absorption, photostability, good adsorption properties, and charge separation capability (Xie et al., 2018).

4 Graphene quantum dot heterostructures for degradation of organic pollutants

Photocatalytic dye degradation is one of the most investigated applications of GQD-derived nanostructures. Dye pollution presents a massive challenge to the environment where it imparts color to freshwater and could be toxic to aquatic life. Furthermore, some dyes have been found to be carcinogenic, mutagenic and cause kidney problems, skin irritation and various allergies (Zhang et al., 2012). In a recent study, hydrothermally synthesized N-doped GQDs (NGQDs) were coupled with $BiVO_4$ and probed for visible light-assisted degradation of methylene blue (MB). It was observed that in the presence of appropriate amounts of NGQDs (5 wt%), the photocatalytic removal efficiency reached about 90% in 200 min compared to around 60% observed over pure $BiVO_4$ (Fig. 1A).

This could be explained in terms of the multiple roles of NGQDs in enhancing visible light absorption, charge separation efficiency, and transfer. This was evident from the lower photoluminescence (PL) signal, the higher photocurrent (PC) response, and a smaller semicircle in the electrochemical impedance measurements (EIS), indicative of a smaller charge transfer resistance. Moreover, any increase in NGQD loading beyond 5 wt% was accompanied by a decrease in activity, which could be due to the excess NGQDs blocking the photocatalytic $BiVO_4$ (Wu et al., 2019). Similarly, a significant enhancement in RhB degradation was reported over NGQDs coupled with bismuth oxyhalides (BiOX, X = Br, Cl) obtained via a solvothermal route. Both BiOX nanocomposites (NGQDs/BiOCl and NGQDs/BiOBr) displayed the highest RhB removal when 7 wt% NGQDs were incorporated, reaching nearly 100% in 75 min and 60 min, respectively. This was largely attributed to the improvement in visible light

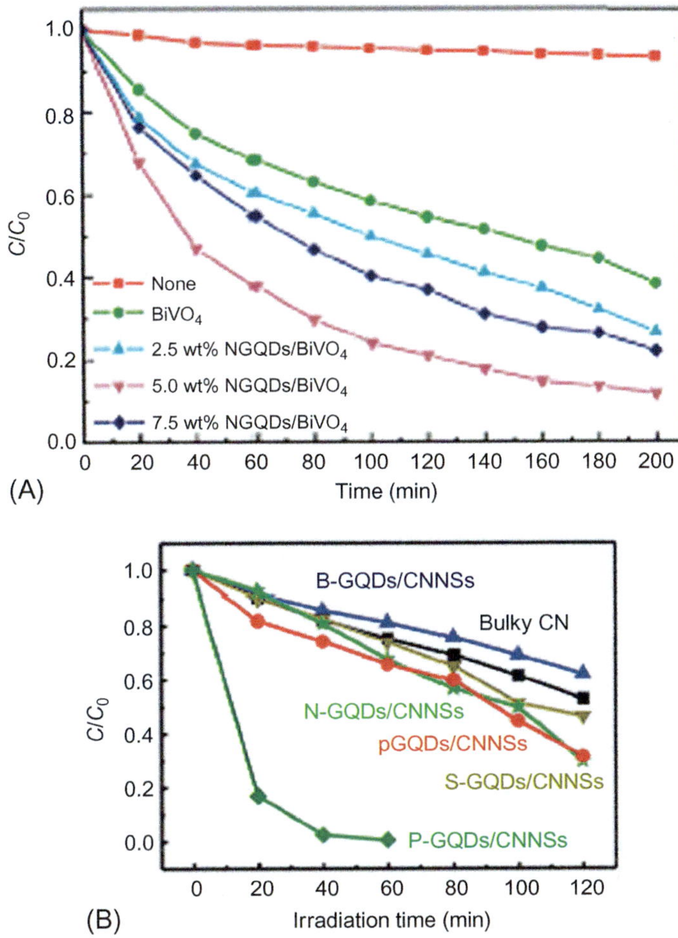

Fig. 1 (A) Effect of NGQDs loading on MB removal over NGQD/$BiVO_4$ and (B) RhB removal over nonmetal doped GQDs/g-C_3N_4 (CNNSs) nanostructures.
(A) Reproduced with permission from Wu, C., Chen, R., Ma, C., Cheng, R., Gao, X., Wang, T., Liu, Y., Huo, P., Yan, Y., 2019. Construction of upconversion nitrogen doped graphene quantum dots modified BiVO4 photocatalyst with enhanced visible-light photocatalytic activity. Ceram. Int. 45, 2088–2096. Copyright 2018, Elsevier. (B) Reproduced with permission from Qian, J., Yan, J., Shen, C., Xi, F., Xiaoping Dong, X., Liu, J., 2018. Graphene quantum dots-assisted exfoliation of graphitic carbon nitride to prepare metal-free zero-dimensional/two-dimensional composite photocatalysts. J. Mater. Sci. 53, 12103–12114. Copyright 2018, Springer.

absorption and charge separation due to the incorporation of NGQDs (Zhang et al., 2019). Shafaee and coworkers demonstrated the effect of the photocatalyst preparation method on its activity by using ultrasound and hydrothermal synthesis to couple GQDs with TiO_2 nanoparticles for rhodamine B (RhB) removal. Generally, it was observed that both nanocomposites had superior activity toward RhB compared to pure TiO_2. Moreover, the binary nanostructures prepared via the hydrothermal route showed slightly higher activity compared to the ultrasound-derived materials. This could be attributed to the presence of numerous defects induced by ultrasound, which could act as recombination centers for the charge carriers. Notably, the binary nanostructures degraded nearly 100% RhB in <30 min with >80% TOC removal achieved in 135 min. Apart from altering the TiO_2 band gap, which improved visible light absorption and positively influenced charge separation, the presence of GQDs also enhanced RhB adsorption via the π–π interactions (Shafaee et al., 2018).

Qian et al. utilized an ultrasonic exfoliation synthesis route to tailor nonmetal doped (P, N, S, B) GQDs/g-C_3N_4 nanostructures and evaluated their visible light-assisted decomposition of RhB. Interestingly, there was not an apparent enhancement in the activity of g-C_3N_4 when coupled with pure GQDs, S-GQDs, and N-GQDs, with the activity of B-GQDs even lower than that of bulk g-C_3N_4 (Fig. 1B). This was credited to poor visible light absorption and weaker charge separation. However, coupling g-C_3N_4 with P-GQDs showed a significant improvement in RhB degradation, reaching 100% in just 40 min, with a degradation rate that was 17 times that of bulk g-C_3N_4 (Qian et al., 2018b). Similarly, enhanced photocatalytic activity was reported for RhB over PGQDs/g-C_3N_4 under visible light exposure. Improved visible light utilization and formation of an intimate p-n heterojunction, which ensured efficient charge separation, were the main reasons for the higher activity of the composite compared to pristine g-C_3N_4 (Qian et al., 2018a).

Elsewhere, enhanced photocatalytic degradation of methyl orange (MO) (Luo et al., 2018) and MB and RhB (Tang et al., 2018) has been reported. This improvement in activity could be attributed to the incorporation of GQDs, which positively influenced the optical properties and charge separation efficiency. Meanwhile, reports on the photocatalytic capability of nonmetal doped and pristine GQDs have emerged, suggesting the potential application of the metal-free photocatalyst toward dye degradation (Huang et al., 2018; Kumar et al., 2017). More than 80% basic fuchsin (BF) removal was observed in 120 min over S-GQDs, which was much higher than the 18% observed over GQDs. This was linked to better visible light utilization and well-matched energy levels between S-GQDs and BF (Huang et al., 2018). In another work, pristine GQDs could only attain 45% MB removal in 100 min of sunlight exposure. Despite the low removal efficiency, the photocatalyst showed good stability over three cycles with negligible loss of activity (Kumar et al., 2017). The remarkable stability and utilization of sunlight could compensate for the somewhat lower activity.

An indepth analysis of the photodegradation results coupled with optical response analysis, identification of the predominant reactive species, and the band structure of the photocatalyst, provides key information leading to understanding the charge transfer mechanism. Obviously, the charge transfer pathway gets more complicated as the photocatalyst becomes complex. Accordingly, the degradation mechanism is more complicated in ternary semiconductor nanocomposites compared to binary nanostructures. For example, in NGQDs/$BiVO_4$, upon visible light excitation, the electrons occupied the conduction band of $BiVO_4$, leaving holes in the valence band. Thereafter, the electrons transferred to NGQDs where they were trapped by O_2 to form the superoxide radicals while the holes reacted with H_2O to form hydroxyl radicals. Both radical species were responsible for the degradation of the dye molecules (Wu et al., 2019).

A similar mechanism was proposed for RhB degradation over BiOBr/NGQDs and BiOCl/NGQDs, with the addition of a dye sensitization route. According to the dye sensitization mechanism, the RhB molecules could absorb visible light, leading to the formation of an excited state that could inject electrons into the conduction band of BiOX. Notably, this mechanism was proposed as the sole reason for the activity of BiOCl, which is not sensitive to visible light (Zhang et al., 2019). In another scenario, the band structure of GQDs/ZnO consisted of the conduction band of GQDs and ZnO positioned at −3.09 eV and −4.202 eV, versus the vacuum energy level, respectively, while the valence bands were located at −5.62 eV and −7.39 eV versus the vacuum level, respectively (Fig. 2). Due to these matching bands, subsequent to photoexcitation, electrons transferred from the conduction band of the GQDs to the conduction band of ZnO while the holes transferred in the opposite direction. The electrons on ZnO could be involved in reduction reactions to form H_2O_2, which could be decomposed by sunlight to form the hydroxyl radicals. Accordingly, the holes could directly oxidize the dye molecules to form degradation byproducts (Kumar et al., 2018). Similar charge transfer pathways were proposed in Ti^{3+}-TiO_2/GQDs (Tang et al., 2018) and S-GQDs/TiO_2 (Luo et al., 2018). A clear understanding of the charge transfer pathway in a photocatalyst is crucial in designing materials with high charge separation efficiency, desirable optical properties, and matching band potentials. This could ensure higher photodegradation performances.

Apart from dye pollution, emerging pollutants (EPs) or micropollutants (MPs) is another important category of organic pollutants which has received tremendous attention due to their lack of regulation and unknown fate in the environment. This category of pollutants comprises new or existing substances and their metabolites,

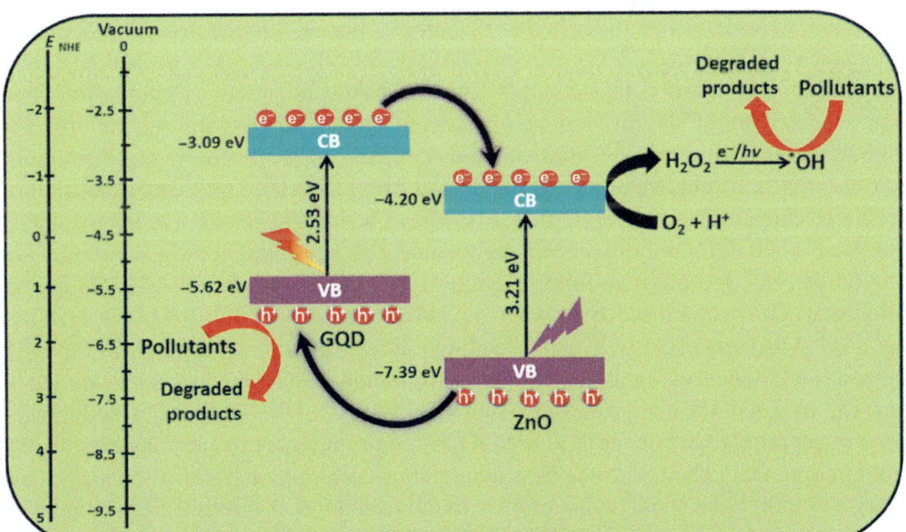

Fig. 2 Proposed charge transfer mechanism in GQD/ZnO leading to dye degradation. Reproduced with permission from Kumar, S., Dhiman, A., Sudhagar, P., Krishnan, V., 2018. ZnO-graphene quantum dots heterojunctions for natural sunlight-driven photocatalytic environmental remediation. Appl. Surf. Sci. 447, 802–815. Copyright 2018, Elsevier.

which have been detected in the environment and are poorly or not regulated at all. It is worth noting that most EPs are useful substances and may be nontoxic, but once introduced into the environment, they may undergo transformations into potentially toxic compounds. Examples of EPs include antibiotics, fungicides, herbicides, pesticides, personal care products, detergents, hormones, antiviral drugs, fire retardants, and drugs of abuse. Besides some of these compounds being linked to carcinogenic and mutagenic behavior, the presence of antibiotics in the environment has led to the emergence of drug-resistant bacteria (superbugs), which pose a serious threat to health care (Fagan et al., 2016; Kümmerer, 2009). Photocatalytic degradation of EPs in water using GQD-derived nanostructures has been widely explored in recent years.

For example, Nie et al., demonstrated the stimulated solar light-mediated simultaneous degradation of three EPs (4-nitrophenol (4-NP), diethyl phthalate (DEP), and Ciprofloxacin (CIP)) and the evolution of hydrogen over GQDs/Mn-N-TiO_2/g-C_3N_4 with Pt as a cocatalyst. This was an interesting study in that it demonstrated the potential to solve two major problems at the same time, that of pollution mitigation and fuel generation. It was observed that the optimized photocatalyst consisting of 5 wt% GQDs and 40 wt% g-C_3N_4 (S/GQDs/TCN-0.4) showed the highest pollutant removal rates and H_2 evolution. Interestingly, the H_2 evolution rate in the pollutant solution was higher than in pure water, which was indicative of a synergy between the two processes. Moreover, DFT calculations and GC–MS analysis revealed the involvement of electrons in the degradation of 4-NP but not CIP and DEP. Hence, lower H_2 evolution was recorded in the 4-NP solution (Nie et al., 2018). Meanwhile, Deng and coworkers probed the influence of the light source on the degradation of tetracycline (TC) using Ag/NGQDs/g-C_3N_4 ternary nanostructures. Under optimized conditions, the ternary nanocomposites demonstrated higher tetracycline (TC) removal under both NIR and full-spectrum light irradiation, which was five times higher than that of pure g-C_3N_4. Tetracycline removal efficiencies of 92.8% and 32.3% were recorded under full-spectrum light and NIR light illumination, respectively. This could be ascribed to the cooperative effect of the three components, with g-C_3N_4 providing a platform to anchor NGQDs and Ag, thereby preventing aggregation while all three materials contributed to the efficient utilization of the full-spectrum light. Moreover, NGQDs and AgNPs ensured efficient separation of the charge carriers (Deng et al., 2017). Elsewhere, a sunlight-responsive photocatalyst comprised of NGQDs hybridized with nanocubic TiO_2 was obtained via a combination of hydrothermal and physical mixing synthesis. Complete bisphenol A (BPA) degradation was achieved in just 30 min when the photocatalyst containing 0.5 wt% NGQDs was used. Most importantly, NGQD loadings above 0.5 wt% resulted in lower BPA degradation kinetics, owing to the excessive formation of oxygen vacancies that led to a recombination of the charge carriers (Lim and Leong, 2019). Similarly, BPA degradation was reported over BiOCl/$BiVO_4$/NGQDs ternary nanocomposites under a visible light environment. The ternary nanostructure showed higher activity compared to the binary and individual semiconductors due to the formation of multiple heterojunctions that improved charge separation (Zhu et al., 2017).

In terms of the photocatalytic degradation mechanism, in GQDs/Mn-N-TiO_2-g-C_3N_4, radical scavenging experiments revealed the holes and hydroxyl radicals as the active species with the hydroxyl radicals being the most dominant. With such information coupled with the band structure of the photocatalyst and optical response, it was proposed that under solar light irradiation, both g-C_3N_4 and Mn-N-TiO_2 were excited, resulting in electrons being promoted to their conduction bands and holes left in their valence bands. In addition, GQDs could also absorb light of a longer wavelength (800 nm) and emit light of a shorter wavelength (500 nm), which was absorbed by both g-C_3N_4 and Mn-N-TiO_2 generating more electron/hole pairs. Thereafter, the electrons on g-C_3N_4 could transfer to the conduction band of Mn-N-TiO_2, with the holes transferring in the opposite direction. The electrons could then be captured by the surface-deposited Pt nanoparticles where the reduction of 4-NP and H_2 evolution took place, as depicted in Fig. 3A. Accordingly, the holes in g-C_3N_4 could oxidize hydroxide ions to form hydroxyl radicals and also, the holes in Mn-N-TiO_2 were positive enough to oxidize H_2O to form more hydroxyl radicals. Moreover, the holes could directly attack the organic pollutants to form degradation byproducts (Fig. 3A) (Nie et al., 2018). A similar charge transfer route was proposed for TC degradation over Ag/NGQDs/g-C_3N_4 (Deng et al., 2017). Meanwhile, an internal Z-scheme mechanism was put forward to explain the charge transfer route in BiOCl/$BiVO_4$/NGQDs. Typically, multiple heterojunctions were formed between the three semiconductors and upon visible light irradiation, only the conduction bands of $BiVO_4$ and NGQDs were populated with electrons (Fig. 3B). However, it was possible for the electrons from the valence band of BiOCl to transfer and recombine with the valence band holes in $BiVO_4$ while the electrons on $BiVO_4$ could transfer to the valence band of the NGQDs. This Z-scheme route preserved the highly reductive electrons in the conduction band of NGQDs and highly oxidative holes on BiOCl, allowing them to undergo redox reactions to form the hydroxyl and superoxide radicals. These radical species resulted in the degradation of the emerging pollutant (Zhu et al., 2017).

The complexity of wastewater requires materials that are capable of remediating a wide range of pollutants. Moreover, the development of multifunctional materials could offset the sometimes costly synthesis/precursors. Numerous materials have shown the potential to remove both dye molecules and emerging pollutants (Bu et al., 2018; Chen et al., 2018; Liu et al., 2017; Xia and Li, 2018; Zhu and Li, 2017). For example, photocatalytic degradation of RhB, TC, and BPA was observed over N,S-GQDs coupled with $(BiO)_2CO_3$ (NS-GQDs/$(BiO)_2CO_3$) and the removal rates were significantly higher compared to those of pristine $(BiO)_2CO_3$. More importantly, the photocatalyst showed activity under light and in the dark environment. The unusual activity in the dark was attributed to the biomimetic effect of N,S-GQDs which could decompose H_2O_2 to form hydroxyl radicals in a similar way as the enzyme peroxidase (Xia and Li, 2018). Meanwhile, the versatility of black TiO_2/N,S-GQDs was demonstrated through sunlight-assisted degradation of several dyes, including RhB, MO, Fluorescein (Flu), MB, bromophenol blue (Bpb), methyl red (MR), and phenolphthalein (Pph), and EPs such as phenol (Phe), aniline (Ani), nitrobenzene (Nbe), dimethyl *o*-phthalate (Dop), acrylonitrile (AH), formaldehyde (Fd), and chlorobenzene (Chh). Moreover, the nanocomposite was also investigated for sewage treatment. Besides demonstrating higher degradation rates ($k > 0.68\,min^{-1}$) for all the pollutants, the photocatalyst showed remarkable stability, maintaining 95% of its initial efficiency after 30 cycles. In terms of the sewage treatment, nearly 85% TOC removal was recorded in just 30 min, signifying a large potential in the application of the photocatalysts in wastewater treatment (Bu et al., 2018).

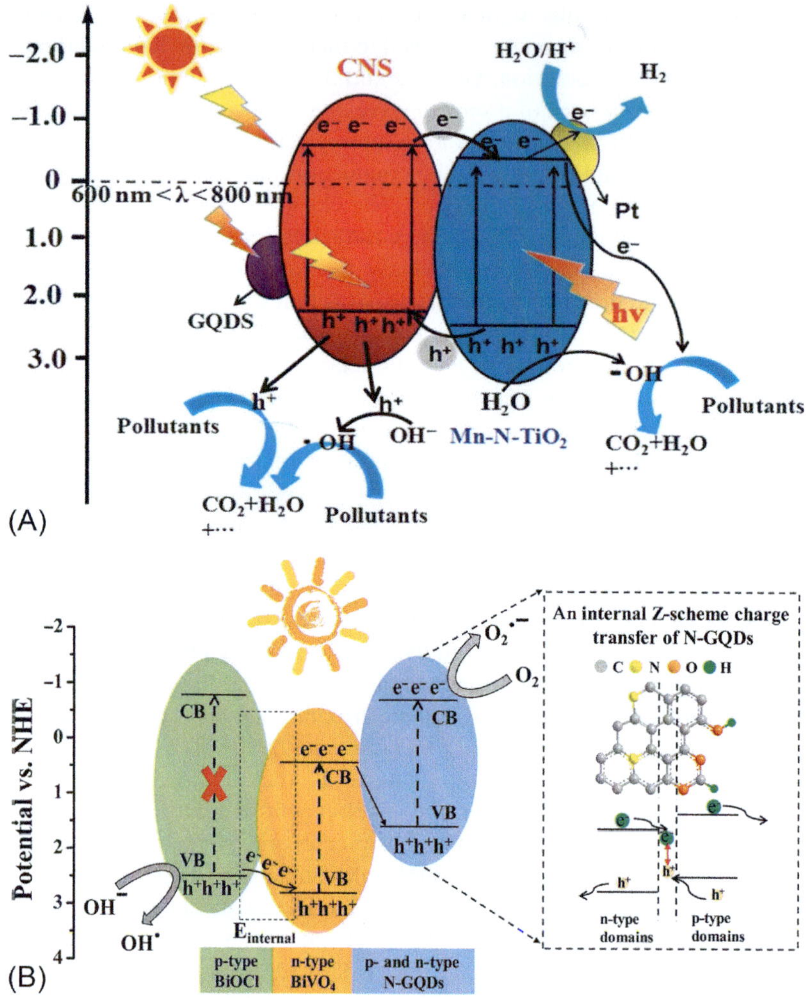

Fig. 3 Charge transfer mechanism in (A) GQDs/Mn-N-TiO_2-g-C_3N_4 and (B) BiOCl/$BiVO_4$/NGQDs.

(A) Reproduced with permission from Nie, Y., Yu, F., Wang, L., Xing, Q., Liu, X., Pei, Y., Zou, J., Dai, W., Li, Y., Sui, S.L., 2018. Photocatalytic degradation of organic pollutants coupled with simultaneous photocatalytic H2 evolution over graphene quantum dots/Mn-N-TiO2/g-C3N4 composite catalysts: performance and mechanism. Appl. Catal. B 227, 312–321. Copyright 2018, Elsevier. (B) Reproduced with permission from Zhu, M., Liu, Q., Chen, W., Yin, Y., Ge, L., Li, H., Wang, K., 2017. Boosting the visible-light photoactivity of BiOCl/BiVO4/N-GQD ternary heterojunctions based on internal Z-scheme charge transfer of N-GQDs: simultaneous band gap narrowing and carrier lifetime prolonging. ACS Appl. Mater. Interfaces 9, 38832–38841, Copyright 2017, ACS.

Despite the impressive results obtained for the degradation of both dyes (colored substances) and EPs, the degradation experiments were carried out separately, whereas in real water samples, these pollutants coexist and interact extensively. Therefore, it would paint a better picture to study the degradation process in mixed pollutant solutions, in order to evaluate the effect of one pollutant on the other. However, the future still looks promising in terms of tailoring and using GQD-derived nanostructures for photocatalytic removal of organic pollutants in water. A summary of other reports on the exploitation of GQD-based nanocomposites for photocatalytic degradation of organic pollutants is presented in Table 1.

Table 1 A summary of GQDs derived nanocomposites toward degradation of organic pollutants.

Photocatalyst	Application	GQD loading	Efficiency	Reference
GQDs/ZnO	MB/solar light	0.4 wt%	78%/180 min	Ebrahimi et al. (2017)
ZnO/S,N-GQDs	RhB/Vis	0.2 wt%	82.4%/90 min	Cai et al. (2017a,b)
g-C_3N_4/N,S-GQDs	RhB/Vis	3 wt%	96%/90 min	Cai et al. (2017a,b)
GQDs/TiO_2	RhB/Vis	No data	100%/120 min	Bian et al. (2017)
GQDs/TiO_2	RhB/Vis	6.5 wt%	90%/240 min	Zhang et al. (2017)
NGQDs/TiO_2	MB/UV	No data	85%/70 min	Safardoust-Hojaghan and Salavati-Niasari (2017)
GQDs/red P	RhB/Vis	2 wt%	90%/60 min	Chan et al. (2017)
SnO_2/GQDs	MB/UV–vis	2.92 wt%	83.19%/150 min	Quan et al. (2016)
GQDs-PVP-CdS	MO/Vis	No data	92.3%/180 min	Fan et al. (2016)
GQDs/TiO_2	MO/UV–vis	2.5 wt%	94.64%/20 min	Qu et al. (2016)
GQDs/ZnS	RhB/Vis	No data	90%/40 min	Ham et al. (2016)
GQDs/$AgVO_3$	Ibuprofen (IBP)/Vis	3 wt%	90%/120 min	Lei et al. (2016)
GQDs/Bi_2MoO_6	CIP/BPA/ MB/TC/RhB/ simulated solar	No data	90%/90 min/RhB 69%/30 min/BPA 80%/30 min/CIP 65%/30 min/TC 90%/30 min/ Phenol 75%/30 min/MB	Hao et al. (2016)

5 Graphene quantum dot-derived nanostructures for water disinfection and heavy metal reduction

Disinfection is one of the most important steps in water treatment, ensuring that the water is safe from microbial contamination. Without the disinfection step, pathogenic microbes such as bacteria, fungi, algae, and viruses could find their way into the drinking water distribution system and cause major waterborne diseases and outbreaks. Moreover, the heavy usage and release of antibiotics into the environment have led to the development of drug-resistant bacteria (superbugs), which pose a huge threat to health care. Therefore, water needs to be sufficiently disinfected to combat the spread of these superbugs (Ganguly et al., 2018; Ván and Kristina, 2015). Photocatalytic water disinfection offers a promising alternative/complementary method to chlorine disinfection, which often leads to the formation of potentially toxic disinfection byproducts. As a complementary process, photocatalytic disinfection can destroy microbes as well as the disinfection byproducts formed during chlorination (Wang et al., 2015).

Recently, Teymourinia and coworkers used a solvothermal synthesis route to tailor a TiO_2/Sb_2S_3/GQDs ternary nanostructure and investigated its antibacterial properties toward *Escherichia coli* (E. coli) and Staphylococcus aureus (S. aureus) under a visible light environment. It was observed that coupling GQDs with TiO_2 (GQDs/TiO_2) and Sb_2S_3 (GQDs/Sb_2S_3) significantly lowered the minimum inhibition concentration (MIC) for both *E. coli* and *S. aureus*. This was due to improved visible light utilization and charge separation efficiency. Furthermore, the TiO_2/Sb_2S_3/GQDs ternary nanostructure showed the lowest MIC values for both bacterial strains compared to pure TiO_2, GQDs/TiO_2, and GQDs/Sb_2S_3. Minimum inhibition concentration values of 0.03 and 0.1 were recorded for *E. coli* and *S. aureus*, respectively. This could be linked to the cosensitization effect of GQDs and Sb_2S_3 on TiO_2 as well as improved charge separation and formation of the reactive species responsible for bacterial growth inhibition. Fig. 4A shows the decrease in bacterial growth with increasing irradiation time, until nearly no growth could be observed after 24 h (Teymourinia et al., 2019). Meanwhile, Kholikov et al., evaluated the antimicrobial performance of GQDs combined with methylene blue (MB) under visible light illumination. Coupling GQDs and MB was found to increase the formation of singlet oxygen, with maximum production observed when the GQDs:MB ratio was 1:1. Under dark conditions, the presence of GQDs did not have any effect on the cell viability, suggesting the potential applicability of MB-GQDs in photodynamic therapy. Moreover, it was noted that upon illumination with 660 nm light, both *Micrococcus luteus* (*M. luteus*) and *E. coli* were nearly completely disinfected after just 5 min. Notably, *M. luteus* (Fig. 4B) was disinfected much faster than *E. coli* (Fig. 4C). This was attributed to the fact that the gram-negative *E. coli* had a more complex and thicker cell wall that provides extra protection compared to the gram-positive *M. luteus*. The disinfection process resulted from the synergy between GQDs that could disrupt the bacterial cell envelope, thereby opening tunnels through which MB entered the cell and generated singlet oxygen that altered protein synthesis and DNA replication (Kholikov et al., 2018). The differences in bacterial inactivation rates between gram-positive and gram-negative bacteria seem to be somewhat complicated and involve more than just the differences in cell envelope

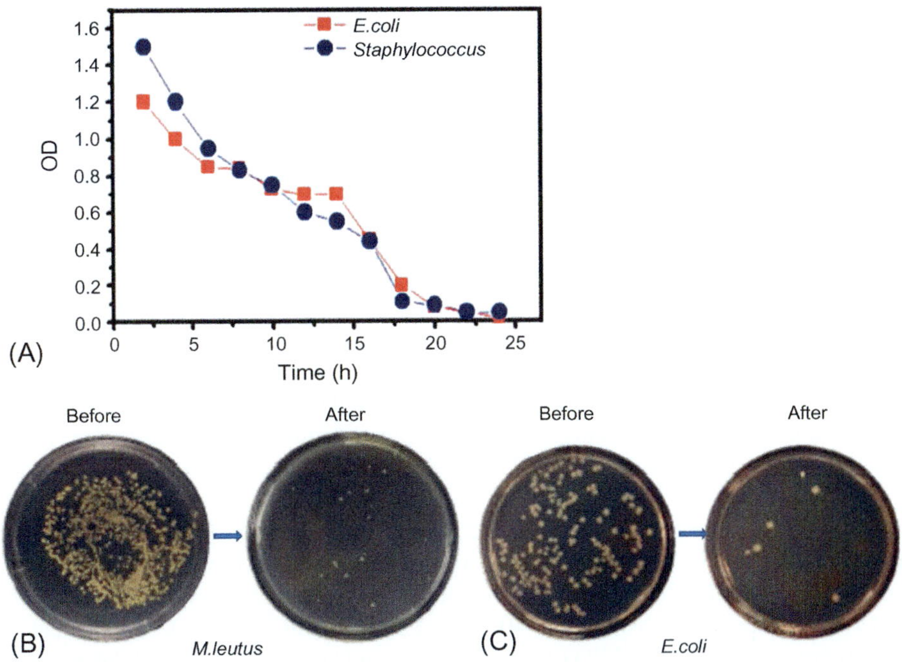

Fig. 4 Bacterial growth inhibition in the presence of (A) TiO_2/Sb_2S_3/GQDs over 24h under visible light, (B) GQDs/MB for 2min, and (C) GQDs/MB for 5min under 660nm light. (A) Reproduced with permission from Teymourinia, H., Salavati-Niasaria, M., Amiri, O., Fatemeh Yazdian, F., 2019. Application of green synthesized TiO2/Sb2S3/GQDs nanocomposite as high efficient antibacterial agent against *E. coli* and *Staphylococcus aureus*. Mater. Sci. Eng. C 99, 296–303. Copyright 2019, Elsevier. (B) and (C) Adapted from Kholikov, K., Ilhom, S., Sajjad, M., Smith, M.E., Monroe, J.D., San, O., Er, A.O., 2018. Improved singlet oxygen generation and antimicrobial activity of sulphur-doped graphene quantum dots coupled with methylene blue for photodynamic therapy alications. Photodiagnosis Photodyn. Ther. 24, 7–14. Copyright 2018, Elsevier.

structure. For example, the surface charge of the photocatalyst relative to the bacterial cell could play an important part in the bacteria-photocatalyst interactions. In case of a negatively charged photocatalyst surface, the repulsive interaction with gram-negative bacteria could result in lower inactivation kinetics compared to gram-positive bacteria.

Heavy metal pollution presents another serious challenge in the quest to provide clean water for drinking and irrigation purposes. Heavy metals such as Cr(VI), Hg(II), and Pb(II), among others, have been reported in the environment in various concentrations; they mainly originate from industrial activities. Some of these heavy metals have been labeled as embryotoxic, mutagenic, carcinogenic, teratogenic, toxic to the reproductive system; they also have been linked to cardiac diseases, liver damage, and many other health problems (Maia et al., 2019; World Health Organization, 2003). Moreover, heavy metals can be introduced to crops via irrigation with polluted water and eventually reach humans upon consumption of such crops. Catalytic removal of heavy metals presents a promising solution due to the nonselective nature of the process and the potential to transform toxic heavy metal ions to nontoxic metal species. For example, photocatalytic reduction of Cr(VI) was probed over Bi_2S_3/GQDs/TiO_2 nanowires in the presence of MO under visible light illumination. Interestingly, higher Cr(VI) reduction was observed in the presence of MO (97%) compared to without MO (92%). Meanwhile, MO removal was nearly 52% in the presence of Cr(VI) compared to 43% without Cr(VI). This could be ascribed to the efficient trapping of the photo-generated electrons by Cr(VI), which reduced it to Cr(III), leaving the holes free to undergo oxidation reactions leading to the degradation of MO. This significantly improved the removal of MO in the presence of Cr(VI), which acted as an electron scavenger. The presence of GQDs not only ensured improved visible light response but also enhanced charge separation efficiency in the nanocomposite (Geng et al., 2018).

In another work, the photocatalytic (PC), electrocatalytic (EC), and photoelectrocatalytic (PEC) capability of the ternary Fe_2O_3-GQDs/NF-TiO_2 nanostructure was probed for Cr(VI) reduction and EDTA degradation in aqueous solutions under visible light irradiation. Significantly higher PEC activity was observed over the ternary nanostructure compared to the binary nanostructures. Moreover, the PEC process showed superior performance compared to the EC and PC processes. In terms of the apparent rate constant, the PEC activity was 7.67 times and 4.6 times higher than that of PC and EC, respectively. Simultaneous removal of Cr(VI) and EDTA was observed with the Cr(IV) removal increasing from 40% to 91% in 80 min of the PEC experiment. A synergy between Cr(VI) and EDTA was observed and also between Cr(VI)/phenol and Cr(VI)/MB, which signified the suitability of Fe_eO_3-GQDs/NF-TiO_2 toward the PEC removal of organic heavy metal wastewater. The GQDs were found to function mainly as electron accelerators between Fe_2O_3 and NF-TiO_2 (Wang et al., 2017). The synergy between organics and heavy metals during removal provides an attractive prospect toward the treatment of industrial wastewater containing such pollutant species.

6 Graphene quantum dot-derived nanocomposites in adsorption and membrane filtration

Adsorption is one of the most important tools in water treatment for the removal of a wide range of organic–inorganic and microbial pollutants. Industrially, activated carbon has been long used as an adsorbent during water treatment, owing to its higher specific surface area and adsorption properties. Furthermore, various adsorbents based on graphene and graphene family members have also been explored, displaying remarkable adsorption properties. Among these graphene materials, GQD-based adsorbents have recently emerged as potential adsorbents, owing to their large specific surface area, biocompatibility, and dense functional groups (Agarwal et al., 2016). However, GQDs are highly soluble in water, which means they need to be modified before being explored toward the adsorption of pollutants in the aquatic environment. Yao and coworkers obtained GQDs confined in two-dimensional (2D) hydrophobic space by the simultaneous intercalation of citrate and dodecyl sulfate (SDS) within the layered structure of layered double hydroxides (LDH). The (GQDs + SDS)-LDH adsorbent was evaluated for the removal of 2,4,6-trichlorophenol (2,4,6-TCP). It is worth noting that 80% 2,4,5-TCP removal was attained using the nanocomposite, which was 65% and 40% higher than the removal over GQDs-LDHs and SDS-LDHs, respectively. The adsorption process followed pseudo second-order kinetics and the adsorption data correlated well with the Langmuir adsorption model with an equilibrium adsorption capacity of 119.00 mg/g. Furthermore, it was noted that the adsorption process was as a result of the synergy between π–π interactions, hydrophobic interactions, and hydrogen bonding between the pollutant and adsorbent, as illustrated in Fig. 5. Simply, the intercalation process resolves the issues relating to the aggregation and solubility of GQDs in water (Yao et al., 2017).

Meanwhile, GQDs hybridized with an eggshell (GQDs/eggshell) were tailored via a hydrothermal synthesis route and employed as a solid phase extraction material for polycyclic aromatic hydrocarbons (PAHs) in a synthetic aqueous medium. It was observed that pH had no significant influence on the extraction of PAHs, due to the fact that the PAHs were nonionizable organics. Furthermore, maximum extraction was obtained when 150 mg of the nanosorbent was utilized. The GQDs/eggshell nanosorbent was further investigated for the extraction of PAHs from real water samples spiked with known amounts of analytes to validate the effectiveness of the extraction procedure. The procedure showed good extraction efficiencies ranging from 92.4% to 112.8%. This demonstrated the potential applicability of this adsorbent in the solid-phase extraction of PAHs. The combined adsorption properties of GQDs and eggshells ensured good overall extraction efficiencies (Razmi and Abdollahi, 2016). Elsewhere, GQDs have been employed as adsorbents for MB and RhB (Hu et al., 2018) and oxamyl (Agarwal et al., 2016) but the separation and recovery of the GQD from the aqueous is a challenge when used in its pristine form due to its solubility in water. Centrifugation was generally employed to separate the sorbent from the treated solution. However, the unmodified GQD does not present an attractive option as an adsorbent due to the difficulty to recover and recycle pristine GQDs compared to GQD-based nanocomposites. Tailoring water-insoluble GQD-derived nanostructures remains the best option for utilizing GQDs as adsorbents in water treatment applications.

Recently, studies on GQD-modified membranes have emerged, demonstrating the vital role of these nanocarbons in influencing the morphology and membrane performance (Bi et al., 2019; Xu et al., 2019). For example, interfacial polymerization was employed to fabricate polyethyleneimine (PEI) modified with GQDs in the presence of trimesoyl chloride (TMC) anchored on polyacrylonitrile (PAN) ultrafiltration supports, as illustrated in Fig. 6A. GQDs with an average size of 2.19 nm and 1–3 layers thick were covalently bonded to PEI chains. Under optimized conditions (0.05 wt% GQDs), the mixed-matrix membrane displayed good hydrophilicity, a neutral surface, high water flux, and good antifouling properties. More importantly, the membrane showed good salt ($MgCl_2$) rejection ($96.8 \pm 0.4\%$), which signifies the desalination capability of the membrane. Moreover, the membrane could be exploited in wastewater treatment (Xu et al., 2019). Elsewhere, in situ polymerization of TMC and GQDs within ultrafiltration (UF) membrane pores and subsequent thermal treatment were employed to engineer nanofiltration membranes. The addition of the GQDs altered the membrane porosity, chemical stability, and water flux, among other parameters. It was observed that the NF membrane had improved water flux and rejection properties toward orange II (OGII), MB, Alcian Blue (AB), and Congo red (CR) while Na_2SO_4

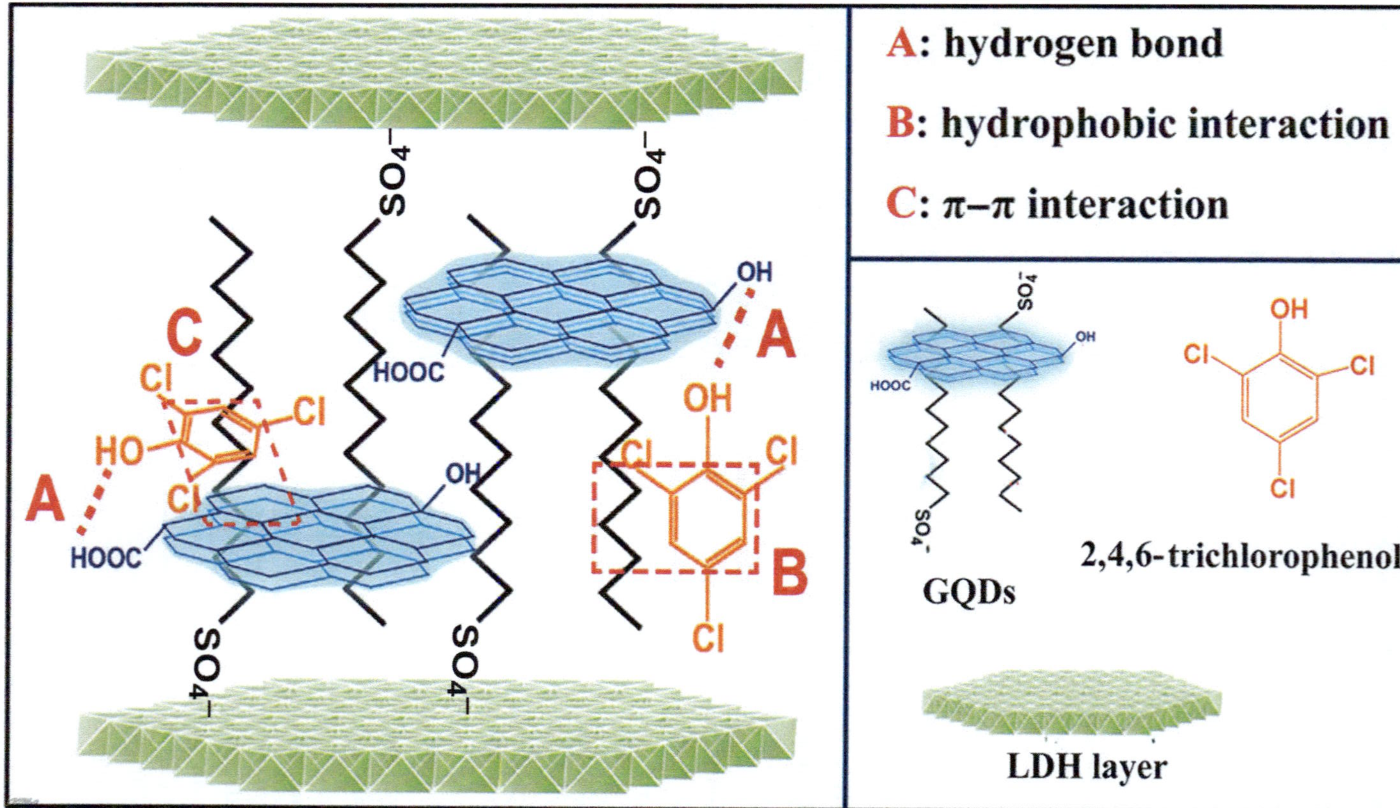

Fig. 5 Illustration of the possible interaction between 2,4,6-TCP and (GQDs + SDS)-LDH during adsorption.
Reproduced with permission from Yao, Q., Wang, S., Shi, W., Lu, C., Liu, G., 2017. Graphene quantum dots in two-dimensional confined and hydrophobic space for enhanced adsorption of nonionic organic adsorbates. Ind. Eng. Chem. Res. 56, 583–590. Copyright 2016, ACS.

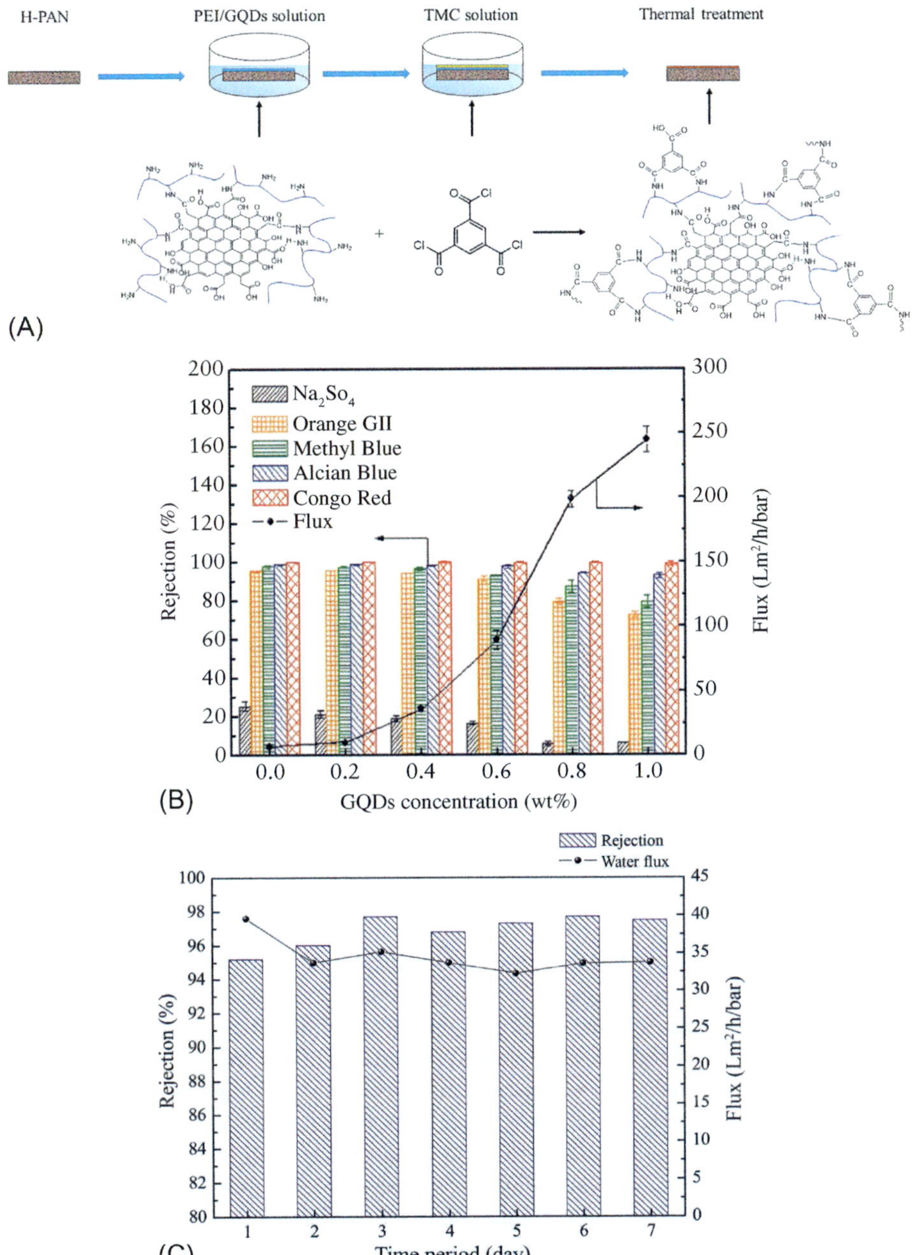

Fig. 6 (A) Schematic illustration of the preparation of the GQD-incorporated membranes, (B) water flux measurements and dye and salt rejection studies of the PES/TMC/GQDs NF membrane as a function of the GQD loading and (C) stability test of the PES/TMC/GQDs NF by monitoring the water flux and AB rejection over 1 week.
(A) Reproduced with permission from Xu, S., Li, F., Su, B., Hu, M.Z., Gao, X., Gao, C., 2019. Novel graphene quantum dots (GQDs)-incorporated thin film composite (TFC) membranes for forward osmosis (FO) desalination. Desalination 451, 219–230. Copyright 2018, Elsevier. (B) and (C) Reproduced with permission from Bi, R., Zhang, R., Shen, J., Liua, Y., He, M., Xinda You, X., Su Y., Jiang, Z., 2019. Graphene quantum dots engineered nanofiltration membrane for ultrafast molecular separation. J. Membr. Sci., 572, 504–511. Copyright 2018, Elsevier.

rejection was poor (Fig. 6B). The rejection for CR and AB was over 90%, which could be attributed to physical sieving through the membrane nanochannels and possibly repulsive interactions between the membrane surface and dye molecules. It was noted that an increase in the GQDs from 0.6 to 0.1 wt% was accompanied by a decrease in the rejection of MB and OGII from 92.9% to 79.4% and 91.2% to 72.6%, respectively. A combination of size exclusion and electrostatic repulsions was responsible for the higher dye rejection capability of the NF membranes. Moreover, the NF membranes showed remarkable stability over 1 week of AB filtration, maintaining high rejection and relatively consistent water flux (Fig. 6C) (Bi et al., 2019). The recent studies have clearly demonstrated the positive effects of modifying membranes with GQDs for water desalination and treatment. However, careful control of the GQD amount is necessary to ensure that they are well dispersed with a sufficient modification of the membrane pore structure as well as hydrophilicity and surface charge properties. These properties alter the filtration capability of the membranes, fouling properties, and the membrane pollutant interactions.

7 Conclusion

Water pollution by organic, inorganic, and microbial species presents a threat to the water supply as well as food safety and security. Graphene quantum dot-derived nanostructures have emerged as potential solutions toward water pollution mitigation. Such materials have been successfully prepared and evaluated for the catalytic removal of organic pollutants such as dyes and emerging pollutants, the adsorption of pollutants, filtration, and disinfection. Incorporation of GQDs in various nanocomposites resulted in the modification of the composite properties and improved the removal efficiencies of different pollutants. Careful control and optimization of the amounts of GQDs incorporated are necessary to ensure a positive influence in the pollutant-removal efficiencies of the different nanocomposites. Despite the promising developments around GQD-based nanostructures, more work still remains in order to ensure the design and utilization of such materials in large-scale applications. Nonetheless, GQD-derived nanostructures have potential as pollution remediation tools, owing to the nontoxicity, biodegradable, and abundant functional groups. However, there is a need to develop optimized synthesis conditions that will yield GQDs that are uniform in terms of size and surface functionalities and also develop synthesis routes that will ensure proper distribution of the GQDs within the nanocomposite matrix. This could contribute to consistency in terms of the reported performances of various GQD-derived nanostructures in water pollution abatement.

Acknowledgments

Funding from the College of Science, Engineering, and Technology (CSET) of the University of South Africa is highly appreciated.

References

Abbas, A., Tuti, L., Phan, A.N., 2018. Biomass-waste derived graphene quantum dots and their applications. Carbon 140, 77–99.

Agarwal, S., Sadeghi, N., Tyagi, I., Gupta, V.K., Fakhri, A., 2016. Adsorption of toxic carbamate pesticide oxamyl from liquid phase by newly synthesized and characterized graphene quantum dots nanomaterials. J. Colloid Interface Sci. 478, 430–438.

Alidad, F., Navik, R., Gai, Y., Zhao, Y., 2018. Production of pristine graphene quantum dots from graphite by a shear-mixer in supercritical CO_2. Chem. Phys. Lett. 710, 64–69.

Bi, R., Zhang, R., Shen, J., Liua, Y., He, M., Xinda You, X., Su, Y., Jiang, Z., 2019. Graphene quantum dots engineered nanofiltration membrane for ultrafast molecular separation. J. Membr. Sci. 572, 504–511.

Bian, S., Zhou, C., Li, P., Liu, J., Dong, X., Xi, F., 2017. Graphene quantum dots decorated titania nanosheets heterojunction: efficient charge separation and enhanced visible-light photocatalytic performance. ChemCatChem 9, 3349–3357.

Bianco, A., Cheng, H., Enoki, T., Gogotsi, Y., Hurt, R.H., Koratkar, N., Kyotani, T., Monthioux, M., Park, C.R., Tasco, J.M.D., Zang, J., 2013. All in the graphene family—a recommended nomenclature for two-dimensional carbon materials. Carbon 65, 1–6.

Bu, X., Yang, S., Bu, Y., He, P., Yang, Y., Wang, G., Li, H., Wang, P., Wang, X., Ding, G., Yang, J., Xie, X., 2018. Highly active black TiO_2/N-doped graphene quantum dots nanocomposites for sunlight driven photocatalytic sewage treatment. Chem. Select 3, 201–206.

Cai, A., Wang, Q., Chang, Y., Wang, X., 2017a. Graphitic carbon nitride decorated with S, N co-doped graphene quantum dots for enhanced visible-light-driven photocatalysis. J. Alloys Compd. 692, 183–189.

Cai, A., Wang, X., Qi, Y., Ma, Z., 2017b. Hierarchical ZnO/S, N:GQD composites: biotemplated synthesis and enhanced visible-light-driven photocatalytic activity. Appl. Surf. Sci. 391, 484–490.

Chan, D.K.L., Yu, J.C., Li, Y., Hu, Z., 2017. A metal-free composite photocatalyst of graphene quantum dots deposited on red phosphorus. J. Environ. Sci. 60, 91–97.

Chen, Z., Zhao, Z., Liu, J., Zhu, X., Wang, L., Xu, Y., Xu, H., Li, H., 2018. Graphene quantum dots modified Ag_3PO_4 for facile synthesis and the enhanced photocatalytic performance. J. Chinese Adv. Mater. Soc. 6, 255–269.

Chu, X., Dai, P., Liang, S., Bhattacharya, A., Dong, Y., Epifani, M., 2019. The acetone sensing properties of $ZnFe_2O_4$-graphene quantum dots (GQDs) nanocomposites at room temperature. Phys. E 106, 326–333.

Crini, G., 2006. Non-conventional low-cost adsorbents for dye removal: a review. Bioresour. Technol. 97, 1061–1085.

Deng, Y., Tang, L., Feng, C., Zeng, G., Wang, J., Lu, Y., Yani Liu, Y., Yu, J., Chen, S., Zhou, Y., 2017. Construction of plasmonic Ag and nitrogen-doped graphene quantum dots codecorated ultrathin graphitic carbon nitride nanosheet composites with enhanced photocatalytic activity: full-spectrum response ability and mechanism insight. ACS Appl. Mater. Interfaces 9, 42816–42828.

Ebrahimi, M., Samadi, M., Yousefzadeh, S., Soltani, M., Rahimi, A., Chou, T., Chen, L., Chen, K., Moshfegh, A.Z., 2017. Improved solar-driven photocatalytic activity of hybrid graphene quantum dots/ZnO nanowires: a direct Z-scheme mechanism. ACS Sustain. Chem. Eng. 5, 367–375.

Fagan, R., McCormack, D.E., Dionysiou, D.D., Pillai, S.C., 2016. A review of solar and visible light active TiO_2 photocatalysis for treating bacteria, cyanotoxins and contaminants of emerging concern. Mater. Sci. Semicond. Process. 42, 2–14.

Fan, T., Li, Y., Shen, J., Ye, M., 2016. Novel GQD-PVP-CdS composite with enhanced visible-light-driven photocatalytic properties. Appl. Surf. Sci. 367, 518–527.

Feng, J., Dong, H., Pang, B., Chen, Y., Yu, L., Dong, L., 2019. Tuning the electronic and optical properties of graphene quantum dots by selective boronization. J. Mater. Chem. C 7, 237.

Ganguly, P., Byrne, C., Breen, A., Pillai, S.C., 2018. Antimicrobial activity of photocatalysts: fundamentals, mechanisms, kinetics and recent advances. Appl Catal B 225, 51–75.

Geng, H., Du, P., Zhang, Z., Yao, L., Cao, K., Li, S., Sheng, P., 2018. Architecting Bi_2S_3/graphene quantum dots/TiO_2 photoelectrodes for aqueous Cr(VI)/methyl orange removal. Mater. Lett. 214, 146–149.

Ham, S., Kim, Y., Park, M.J., Hong, B.H., Jang, D., 2016. Graphene quantum dots-decorated ZnS nanobelts with highly efficient photocatalytic performances. RSC Adv. 6, 24115–24120.

Hameeteman, E., 2013. Future Water (In) Security: Facts, Figures, and Predictions. Global Water Institute.

Hao, Y., Dong, X., Wang, X., Zhai, S., Ma, H., Zhang, X., 2016. Controllable electrostatic self-assembly of sub-3 nm graphene quantum dots incorporated into mesoporous Bi_2MoO_6 frameworks: efficient physical and chemical simultaneous co-catalysis for photocatalytic oxidation. J. Mater. Chem. A 4, 8298–8307.

Hu, H., Quan, H., Zhong, B., Li, Z., Huang, Y., Wang, X., Zhang, M., Chen, D., 2018. A reduced graphene oxide quantum dot-based adsorbent for efficiently binding with organic pollutants. ACS Appl. Nano Mater. 1, 6502–6513.

Huang, B., He, J., Bian, S., Zhou, C., Li, Z., Xi, F., Liu, J., Dong, X., 2018. S-doped graphene quantum dots as nanophotocatalyst for visible light degradation. Chin. Chem. Lett. 29, 1698–1701.

Islam, M.S., Ahmed, M.K., Mohammad Raknuzzaman, M., Habibullah-Al-Mamun, M., Islam, M.K., 2015. Heavy metal pollution in surface water and sediment: a preliminary assessment of an urban river in a developing country. Ecol. Indic. 48, 282–291.

Kalluri, A., Debnath, D., Dharmadhikari, B., 2018. Graphene quantum dots: synthesis and applications in methods in enzymology. Methods Enzymol. 609, 335–354.

Khojastch, H., Salavati-Niasari, M., Safajou, H., Safardoust-Hojaghan, H., 2017. Facile reduction of graphene using urea in solid phase and surface modification by N-doped graphene quantum dots for adsorption of organic dyes. Diamond Relat. Mater. 79, 133–144.

Kholikov, K., Ilhom, S., Sajjad, M., Smith, M.E., Monroe, J.D., San, O., Er, A.O., 2018. Improved singlet oxygen generation and antimicrobial activity of sulphur-doped graphene quantum dots coupled with methylene blue for photodynamic therapy alications. Photodiagnosis Photodyn. Ther. 24, 7–14.

Kumar, S., Ojha, A.K., Ahmed, B., Kumar, A., Das, J., Materny, A., 2017. Tunable (violet to green) emission by high-yield graphene quantum dots and exploiting its unique properties towards sun-light-driven photocatalysis and supercapacitor electrode materials. Mater. Today Comm. 11, 76–86.

Kumar, S., Dhiman, A., Sudhagar, P., Krishnan, V., 2018. ZnO-graphene quantum dots heterojunctions for natural sunlight-driven photocatalytic environmental remediation. Appl. Surf. Sci. 447, 802–815.

Kümmerer, K., 2009. The presence of pharmaceuticals in the environment due to human use—present knowledge and future challenges. J. Environ. Manage. 90 (8), 2354–2366.

Lapworth, D.J., Baran, N., Stuart, M.E., Ward, R.S., 2012. Emerging organic contaminants in groundwater: a review of sources, fate and occurrence. Environ. Pollut. 163, 287–303.

Lei, Z., Wang, J., Wang, L., Yang, X., Xu, G., Tang, L., 2016. Efficient photocatalytic degradation of ibuprofen in aqueous solution using novel visible-light responsive graphene quantum dot/$AgVO_3$ nanoribbons. J. Hazard. Mater. 312, 298–306.

Le-Minh, N., Khan, S.J., Drewes, J.E., Stuetz, R.M., 2010. Fate of antibiotics during municipal water recycling treatment processes. Water Res. 44, 4295–4323.

Lijuan, K., Yongqiang, Y., Ruiyi, L., Zaijun, L., 2016. Phenylalanine-functionalized graphene quantum dot-silicon nanoparticle composite as an anode material for lithium ion batteries with largely enhanced electrochemical performance. Electrochim. Acta 198, 144–155.

Lim, P.F., Leong, K.H., 2019. Amalgamation of N-graphene quantum dots with nanocubic like TiO_2: an insight study of sunlight sensitive photocatalysis. Environ. Sci. Pollut. Res. 26, 3455.

Liu, J., Xu, H., Xu, Y., Song, Y., Lian, J., Zhao, Y., Wang, L., Huang, L., Ji, H., Li, H., 2017. Graphene quantum dots modified mesoporous graphite carbon nitride with significant enhancement of photocatalytic activity. Appl Catal B 207, 429–437.

Luo, Y., Li, M., Hu, G., Tang, T., Wen, J., Li, X., Wang, L., 2018. Enhanced photocatalytic activity of sulfur-doped graphene quantum dots decorated with TiO_2 nanocomposites. Mater. Res. Bull. 97, 428–435.

Mahyari, M., Bide, Y., Nasrollah, J., 2016. Iron (III) porphyrin suorted on S and N co-doped graphene quantum dot as an efficient photocatalyst for aerobic oxidation of alcohols under visible light irradiation. Appl. Catal. A 517, 100–109.

Maia, L.F.O., Hott, R.C., Ladeira, P.C.C., Batista, B.L., Andrade, T.G., Santos, M.S., Faria, M.C.S., Oliveira, L.C.A., Monteiro, D.S., Pereira, M.C., Rodrigues, J.L., 2019. Simple synthesis and characterization of L-Cystine functionalized ∂-FeOOH for highly efficient Hg (II) removal from contamined water and mining waste. Chemosphere 215, 422–431.

Mallakpour, S., Khadem, E., 2018. Construction of crosslinked chitosan/nitrogen-doped graphene quantum dot nanocomposite for hydroxyapatite biomimetic mineralization. Int. J. Biol. Macromol. 120, 1451–1460.

Nie, Y., Yu, F., Wang, L., Xing, Q., Liu, X., Pei, Y., Zou, J., Dai, W., Li, Y., Sui, S.L., 2018. Photocatalytic degradation of organic pollutants coupled with simultaneous photocatalytic H_2 evolution over graphene quantum dots/Mn-N-TiO_2/g-C_3N_4 composite catalysts: performance and mechanism. Appl Catal B 227, 312–321.

Pourhashem, S., Ghasemy, E., Rashidi, A., 2018. Corrosion protection properties of novel epoxy nanocomposite coatings containing silane functionalized graphene quantum dots. J. Alloys Compd. 731, 1112–1118.

Qian, J., Shen, C., Yan, J., Xi, F., Xiaoping Dong, X., Jiyang Liu, J., 2018a. Tailoring the electronic properties of graphene quantum dots by P doping and their enhanced performance in metal-free composite photocatalyst. J. Phys. Chem. C 122, 349–358.

Qian, J., Yan, J., Shen, C., Xi, F., Xiaoping Dong, X., Liu, J., 2018b. Graphene quantum dots-assisted exfoliation of graphitic carbon nitride to prepare metal-free zero-dimensional/two-dimensional composite photocatalysts. J. Mater. Sci. 53, 12103–12114.

Qu, A., Xie, H., Xu, X., Zhang, Y., Wen, S., Cui, Y., 2016. High quantum yield graphene quantum dots decorated TiO_2 nanotubes for enhancing photocatalytic activity. Appl. Surf. Sci. 375, 230–241.

Quan, B., Liu, W., Liu, Y., Zheng, Y., Yang, G., Ji, G., 2016. Quasi-noble-metal graphene quantum dots deposited stannic oxide with oxygen vacancies: synthesis and enhanced photocatalytic properties. J. Colloid Interface Sci. 481, 13–19.

Razmi, H., Abdollahi, V., 2016. Graphene quantum dots–eggshell nanocomposite to extract polycyclic aromatic hydrocarbons in water. Environ. Chem. Lett. 14, 521–526.

Safardoust-Hojaghan, H., Salavati-Niasari, M., 2017. Degradation of methylene blue as a pollutant with N-doped graphene quantum dot/titanium dioxide nanocomposite. J. Clean. Prod. 148, 31–36.

Sajjadi, S., Khataee, A., Reza Darvishi Cheshmeh Soltani, R.D.C., Hasanzadeh, A., 2019. N, S co-doped graphene quantum dot–decorated Fe_3O_4 nanostructures: Preparation, characterization and catalytic activity. J. Phys. Chem. Solid 127, 140–150.

Shafaee, M., Goharshadi, E.K., Mashreghi, M., Sadeghinia, M., 2018. TiO_2 nanoparticles and TiO_2@graphene quantum dots nancomposites as effective visible/solar light photocatalysts. J. Photochem. Photobiol. A 357, 90–102.

Shen, J., Zhu, Y., Yang, X., Li, C., 2012. Graphene quantum dots: emergent nanolights for bioimaging, sensors, catalysis and photovoltaic devices. Chem. Commun. 48, 3686–3699.

Subramanian, A., Pan, Z., Rong, G., Li, H., Zhou, L., Li, W., Qiu, Y., Xu, Y., Hou, Y., Zhaozhao Zheng, Z., Zhang, Y., 2017. Graphene quantum dot antennas for high efficiency for energy transfer based dye-sensitized solar cells. J. Power Sources 343, 39–46.

Tang, J., Liu, Y., Hu, Y., Lv, G., Yang, C., Yang, G., 2018. Carbothermal reduction induced Ti^{3+} self-doped TiO_2/GQD nanohybrids for high-performance visible light photocatalysis. Chem. A Eur. J. 24, 4390–4398.

Teymourinia, H., Salavati-Niasaria, M., Amiri, O., Fatemeh Yazdian, F., 2019. Application of green synthesized TiO_2/Sb_2S_3/GQDs nanocomposite as high efficient antibacterial agent against *E. coli* and *Staphylococcus aureus*. Mater. Sci. Eng. C 99, 296–303.

Tian, P., Tang, L., Teng, K.S., Lau, S.P., 2018. Graphene quantum dots from chemistry to applications. Mater. Today Chem. 10, 221–258.

Valail, M.O., Pillai, V.K., Alwaraan, S., 2017. Spotlighting graphene quantum dots and beyond: synthesis, properties and sensing applications. Appl. Mater. Today 9, 350–371.

Ván, R., Kristina, C., 2015. ZnO/graphite composites and its antibacterial activity at different conditions. J. Photochem. Photobiol. B 151, 256–263.

Wang, W., Huang, G., Yu, J.C., Wong, P.K., 2015. Advances in photocatalytic disinfection of bacteria: development of photocatalysts and mechanisms. J. Environ. Sci. 34, 232–247.

Wang, Q., Zhu, N., Liu, E., Zhang, C., Crittenden, J.C., Zhang, Y., Cong, Y., 2017. Fabrication of visible-light active Fe_2O_3-GQDs/NF-TiO_2 composite film with highly enhanced photoelectrocatalytic performance. Appl Catal B 205, 347–356.

Wang, R., Fan, H., Jiang, W., Ni, G., Qu, S., 2019a. Amino-functionalized graphene quantum dots prepared using high-softening point asphalt and their alication in Fe^{3+} detection. Appl. Surf. Sci. 467–468, 446–455.

Wang, G., He, P., Xu, A., Guo, Q., Li, J., Wang, Z., Liu, Z., Chen, D., Yang, S., Ding, G., 2019b. Promising fast energy transfer system between graphene quantum dots and the application in fluorescent bioimaging. Langmuir 35, 760–766.

World Health Organization, 2003. Chromium in drinking-Water: Background Document for Development of WHO Guidelines for Drinking-Water Quality, World Health Organization.

Wu, C., Chen, R., Ma, C., Cheng, R., Gao, X., Wang, T., Liu, Y., Huo, P., Yan, Y., 2019. Construction of upconversion nitrogen doped graphene quantum dots modified $BiVO_4$ photocatalyst with enhanced visible-light photocatalytic activity. Ceram. Int. 45, 2088–2096.

WWAP (United Nations World Water Assessment Programme)/UN-Water. 2018. The United Nations World Water Development Report 2018: Nature based solutions for water, Paris, UNESCO.

Xia, J., Li, H., 2018. S, N codoped graphene quantum dots embedded in $(BiO)_2CO_3$: incorporating enzymatic-like catalysis in photocatalysis. ACS Sustain. Chem. Eng. 6, 10229–10240.

Xie, Y., Yu, S., Zhong, Y., Zhang, Q., Zhou, Y., 2018. SnO_2/graphene quantum dots composited photocatalyst for efficient nitric oxide oxidation under visible light. Appl. Surf. Sci. 448, 655–661.

Xu, S., Li, F., Su, B., Hu, M.Z., Gao, X., Gao, C., 2019. Novel graphene quantum dots (GQDs)-incorporated thin film composite (TFC) membranes for forward osmosis (FO) desalination. Desalination 451, 219–230.

Yan, Y., Gong, J., Chen, J., Zeng, Z., Huang, W., Pu, K., Liu, J., Chen, P., 2019. Recent advances on graphene quantum dots: from chemistry and physics to applications. Adv. Mater. 31, e1808283.

Yao, Q., Wang, S., Shi, W., Lu, C., Liu, G., 2017. Graphene quantum dots in two-dimensional confined and hydrophobic space for enhanced adsorption of nonionic organic adsorbates. Ind. Eng. Chem. Res. 56, 583–590.

Zeng, Z., Chen, S., Tan, T.T.Y., Xiao, F., 2018. Graphene quantum dots (GQDs) and its derivatives for multifarious photocatalysis and photoelectrocatalysis. Catal. Today 315, 171–183.

Zhang, W., Liu, W., Zhang, J., Zhao, H., Zhang, Y., Quan, X., Jin, Y., 2012. Characterisation of acute toxicity, genotoxicity and oxidative stress posed by textile effluent on zebrafish. J. Environ. Sci. 24, 2019–2027.

Zhang, Y., Qi, F., Li, Y., Zhou, X., Sun, H., Zhang, W., Liu, D., Song, X., 2017. Graphene oxide quantum dot-sensitized porous titanium dioxide microsphere: visible-light-driven photocatalyst based on energy band engineering. J. Colloid Interface Sci. 498, 105–111.

Zhang, W., Fu, J., Wang, Y., Zhang, X., Li, J., 2019. A facile solvothermal method synthesis of nitrogen-doped graphene quantum dots/BiOX (X = Br, Cl) hybrid material for enhanced visible-light photoactivity. Optik 176, 448–456.

Zhu, W., Li, X., 2017. Graphene quantum dots/$LaCoO_3$/attapulgite heterojunction photocatalysts with improved photocatalytic activity. Appl. Phys. A Mater. Sci. Process. 123, 1–10.

Zhu, M., Liu, Q., Chen, W., Yin, Y., Ge, L., Li, H., Wang, K., 2017. Boosting the visible-light photoactivity of BiOCl/$BiVO_4$/N-GQD ternary heterojunctions based on internal Z-scheme charge transfer of N-GQDs: simultaneous band gap narrowing and carrier lifetime prolonging. ACS Appl. Mater. Interfaces 9, 38832–38841.

Carbon-based nanosensors: An efficient tool for use in the food industry and agricultural and environmental sectors

11

Anam Munawar, Sadia Zafar Bajwa, Waheed S. Khan
Nanobiotechnology Group, Industrial Biotechnology Division, National Institute for Biotechnology and Genetic Engineering (NIBGE), Faisalabad, Pakistan

1 Introduction

The well-being of the food industry is of prime importance in everyday life. Nourishment of agricultural land is based on the expanded use of sustenance, which infers the significance of nourishment (Parisi et al., 2015; Smolkova et al., 2015). The chain incorporates "from the homestead to the table" creation, appropriation, preparation, and promotion of horticultural sustenance items to buyers. Globalization of nourishment creation alongside customer concerns identified with sustenance quality and well-being have brought about interconnected and worldwide frameworks for the generation and dispersion of nourishment, trailed by a critical increment in sustenance benchmarks (Morris, 2011). This methodology needs a change from previously reported investigation methods. Quality confirmation is required at each progression of nourishment generation to guarantee safe sustenance (Wang et al., 2013).

Subsequently, strategies that can be used to ensure the quality and nourishment of agricultural require some advanced unique features for example, low recognition limits, high affectability and specificity, scaling down of instrumentation for compact utilize, POC devices (Restuccia et al., 2010). Different advanced techniques can be used to quantitatively analyze the nourishment level in agricultural products, comprising investigation of the fine instrument and cell culture. The primary impediments within these techniques will require a full examination time from a few hours to days, due to the various sample treatment steps (Khot et al., 2012). Still, sustenance security is undermined these days due to unrecognizable contaminants by using typical expository techniques, that is, pathogens (plant or environmental), formaldehyde, and over-the-top pesticide buildups (Jain et al., 2018). The deficiency of ordinary strategies to understand new difficulties in nourishment well-being prompts the advancement of new, scaled-down, and quick systematic procedures with low recognition limits. Nanomaterials having advanced features and coordinated with scientific devices can be a source for new gadgets or improvement in old ones (Szakal et al., 2014).

Furthermore, nanotechnology is a new emerging era of research. Nanomaterials with unique features (shape, size, electrical, mechanical, chemical and optical etc.

Carbon Nanomaterials for Agri-food and Environmental Applications. https://doi.org/10.1016/B978-0-12-819786-8.00011-6

properties) ensure to fulfill the significant requirements in nourishment besides horticulture segments. Nanotechnology provides new ways of sustenance creation, collecting and packaging, plant trimming for creature sustaining (Durán and Marcato, 2013). In the nourishment area, this advancement can upgrade sustenance helpfulness and value. Lamentably, a critical figure of nanotechnological instruments for sustenance in addition to the agribusiness area is on the investigation level (Chaudhry et al., 2010). Nanotechnology use is in sustenance packaging and nanosensors for acknowledgment of nourishment contaminants. Area of nanotechnology, nanominerals and nanosensors helps in the increasing the nourishment level of agriculture products (Das et al., 2011).

Creative improvements for the atomic treatment of maladies, quick illness identification, effective management of infections and harvest pathogens, upgrading the intensity of plants to ingest supplements, and bringing down the utilization measurements of pesticides and herbicides demonstrate that nanotechnology can possibly change the horticultural and nourishment industry (Ruiz-Garcia et al., 2009). Fig. 1 summarizes the list of various potential applications in different sectors based on carbon nanomaterials.

To comprehend and discourse the nanotechnology significance in the agriculture food segment, this chapter speaks to a survey of sorts and utilization of commonly used nanosensors in agribusiness in addition to the nourishment industry (Das et al.,

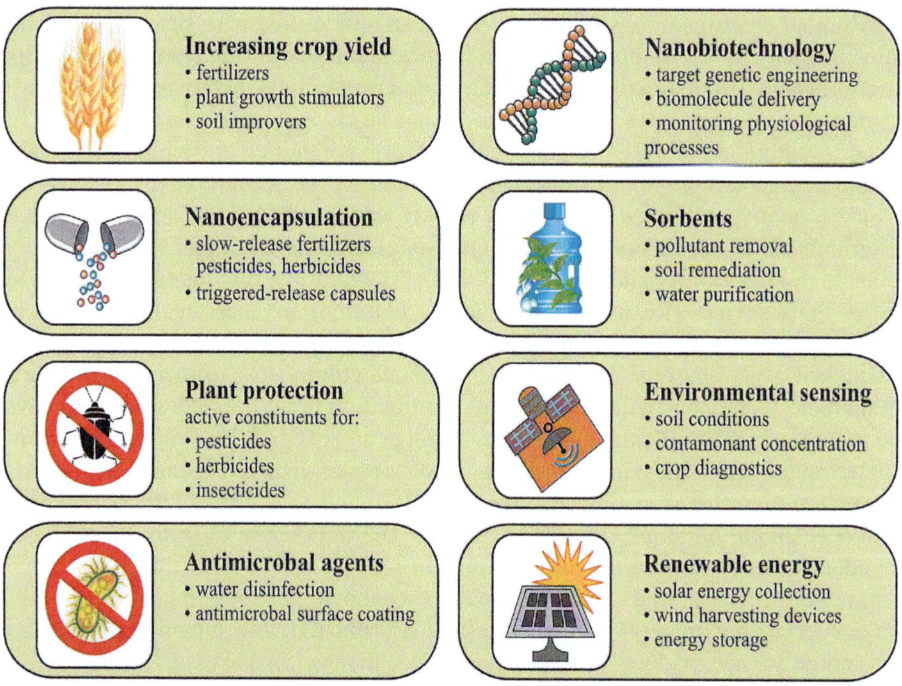

Fig. 1 Carbon-based nanomaterial applications in the agricultural and environmental sector. Reprint from reference Zaytseva, O., Neumann, G., 2016. Carbon nanomaterials: production, impact on plant development, agricultural and environmental applications. Chem. Biol. Technol. Agric., 3(1), 17 with permission from Springer.

2014). Indeed, nanotechnology provides a platform to achieve various goals in agriculture field (nourishment level, removal of toxins, diminishing contamination). Yet there are present numerous unsolved problems with respect to nanomaterials lethality and conceivable negative effect on the environment (Das et al., 2011). In addition to the potential focal points of nanotechnology applications in the agri-nourishment production network, concerns in regard to nanotoxicology and well-being will also be discussed.

2 Carbon-based nanomaterials

Some substance components can be arranged at various subatomic level from a similar kind of molecules—a novel element acknowledged as "allotropy." The distinctive physical and chemical properties of allotropic materials are dictated through the basic structure of iotas and the type of concoction limits present inside atoms (Dai, 2006). Within this specific situation, carbon is a standout among the supreme intriguing components, having a capacity to frame an extensive variety of structures, much of the time within a general sense of diverse properties. Established models of carbon allotropes contain a "hard" precious stone and a "delicate" graphite utilized in science and innovation as well as in an extensive variety of items, incorporating buyer merchandise in different zones of human movement (Pyun, 2011). In the most recent years of the twenty first century different field experimentation performed on allotropes of carbon was recorded. Unusual novel materials such as carbon nanotubes (CNTs), graphene, and fullerenes have ignited extraordinary enthusiasm in science and technology. Also, such materials have displayed an extensive variety of remarkable and innovative highlights as capable resources for various areas of applications (Gottschalk et al., 2013).

In view of these properties, they were more than once named "ponder materials" in logical writing. Novel carbon-based nanoparticles exist in different forms and in combination with other metals or metal oxides with unique properties. Subsequently, their accessibility does not rely upon common stores (e.g., precious stones), and hypothetically creation can be accomplished with boundless amounts in such a way as crude substances for amalgamation are accessible (Llobet, 2013). As indicated by most recent gauges, a consistent increment of generation volume is normal amid the following decade. Notwithstanding, regardless of the way in which carbon based nanomaterials have played significant role in mechanical advancement, there are presently different environmental issues related with the use of nanomaterials including health issues, extraction of nanomaterials from environment after use (Gottschalk et al., 2009).

Notwithstanding the expanding significance of down to earth applications (Yang and Xing, 2010), this part will center around two noteworthy angles related to the treatment of carbon-based nanomaterials in view of:

1. In natural and horticulture area carbon-based nanomaterials (morphology with improved features and use) have exceptional spotlight. On the other hand noteworthy changes were recorded to increase the nourishment level.
2. Improvement in the area of horticulture produces prospective effect on life forms, for example, a major segment of evolved ways of life is horticultural biological systems. Where the expanded contribution of carbon nanomaterials result as a purposeful expenditure in agrarian and ecological applications (Magnuson et al., 2011).

3 Classification of carbon nanomaterials

Carbon is among only a handful of substance components (counting silicone) having the capacity to become a polymer at the nuclear stage, in this manner framing long chains of carbon (Georgakilas et al., 2015). In an external electron layer, the presence of four electrons, carbon molecules with four valences, can be connected through covalent bonds onefold, twofold, or threefold, likewise with different components. Such structures can be named as "allotropic changes" or "allotropes" of a specific substance. For the above-mentioned reasons, carbon can be present in the scope of various subatomic structures. It exhibits distinctive material properties because of various structures. Such structures can be called "allotropic changes" or "allotropes" of a specific synthetic component. As of not long ago, just two normal allotropes of carbon were acknowledged, jewel and graphite. Then, different novel allotropic structures were depicted containing carbon nanomaterials (Cano-Marquez et al., 2009). When all is said and done, the characterization of nanomaterials will be as materials containing particles somewhere in the range of 1–100 nm. Fig. 2 shows carbon atoms with four valences having four electrons in the outer electron layer, and can be connected through one, two, or three covalent bonds.

All nanomaterials prepared from carbon atoms/molecules are named as carbon-based nanomaterials. The characterization of nanomaterials having carbon is usually executed by geometrical arrangement. Carbon nanostructures incorporate particles that can present in the form of tube-molded, horn-formed, round, or ellipsoidal shapes (Kim and Zachariah, 2006). Nanoparticles as tubes are named carbon nanotubes, horn-molded particles are nanohorns, and circles or ellipsoids are fullerenes. Meanwhile, carbon nanomaterials have various specialized applications including

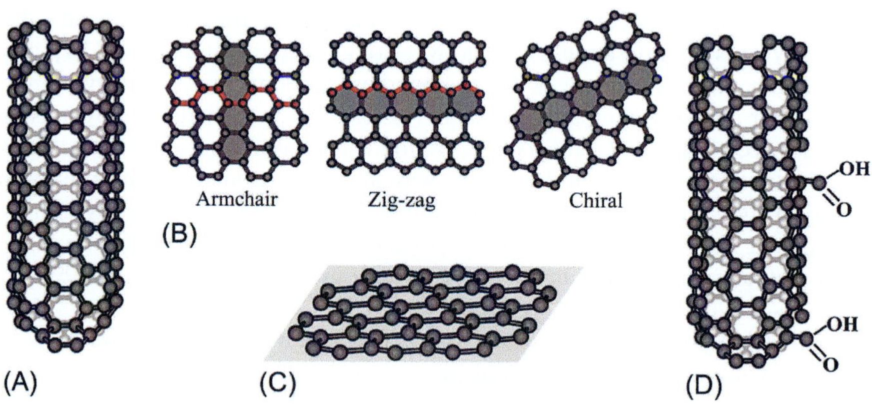

Fig. 2 Assembly of a single-walled carbon nanotube (A); categories of single-walled carbon nanotubes (B); assembly of a graphene sheet (C); configuration of an oxidized single-walled nanotube (D).
Reprinted from reference Zaytseva, O., Neumann, G., 2016. Carbon nanomaterials: production, impact on plant development, agricultural and environmental applications. Chem. Biol. Technol. Agric., 3(1), 17 with permission from Springer.

smaller-scale and nanohardware, gas stockpiling, plastic generation (conductive), gas biosensors, shows, paints (antifouling property), materials, batteries with enhanced toughness, etc. (Staii et al., 2005).

3.1 Fullerenes

Keeping aside the synthesized nanomaterials, normally fullerenes fabricated from carbon have been recognized. Velasco-Santos et al. provide detailed description on coal-oil blend having almost same properties as of carbon nanotubes. On behalf of the SWCNT blend utilizing the CVD strategy, Su and Mracek et al. utilized volcanic magma as a stimulus and materials comprising metal oxide particles. The procedure of CVD strategy can give a proof to conceivable arrangement of nanotubes under different conditions by ascending temperature to higher degree (De Volder et al., 2013). Alongside carbon nanotubes, this is additional proof for an event of topographical substance such as fullerenes. Fullerenes were distinguished by studying the characteristic of other carbon based nanomaterials in low concentrations (2% w/w). Strikingly, the round structure of fullerenes is by all accounts not just confined to carbon nanomaterials. As of late, structures having resemblance with fullerene had been portrayed in sheet of linked hexagonal rings (Chinese grains) which was applied in the area of agricultural field to increase the quality and nourishment level.

3.2 Carbon nanotubes (CNTs)

Between remaining nanomaterials based on carbon, CNTs are a type of carbon allotrope having extraordinary characteristics that are reasonable for specialized presentations. Carbon nanotubes are described as tube-shaped arrangements measuring just a few nanometers, comprising graphene sheets. Carbon nanotubes can change the length, width, and chirality (arrangement of graphite sheet) in the end quantity of sheets (Schnorr and Swager, 2010). As per their assembly, CNTs might be arranged in two primary gatherings: single-walled nanotubes (SWCNTs) and multiwalled nanotubes (MWCNTs). A few specialists also disconnect twofold-walled carbon nanotubes (DWCNTs) with different types of CNTs. For the most part, SWCNTs with breadths of about 1–3 nm have lengths of a couple of microns. MWCNTs can have distances across of 5–40 nm and lengths of about 10 μm. Still, as of late, the amalgamation of CNTs with a length of >550 mm has been accounted (Choi et al., 2001). Different arrangement of CNTs in nanocomposites impart excellent characteristics like high strength, density, and hardness. For example, CNTs display impressively higher angle proportions (length to distance across proportions) than different materials, and bigger viewpoint proportions for SWCNTs when contrasted with MWCNTs due to their little measurement. Also, CNTs demonstrate high warm and electrical conductivity contrasted with additional conductive constituents. SWCNTs possess an extraordinary electrical property that is dependent on chirality or hexagonal structure.

Appropriately, SWCNTs can be characterized by three subclasses: (I) rocker (electrical conductivity > copper), crisscross (semiconductive properties), and (iii) chiral (semiconductive characteristics) (Qadir et al., 2016). The complexity in the structure

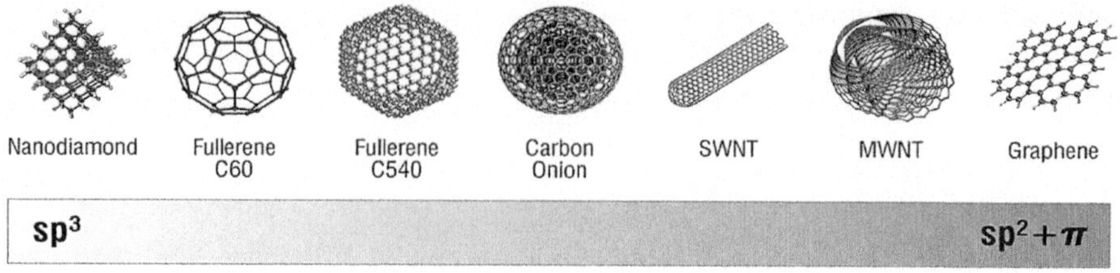

Fig. 3 Hybridization conditions of carbon-based nanomaterials. Numerous concoction and electronic properties of carbonaceous nanomaterials are dictated by the predominant hybridization condition of the carbon-carbon bonds.
Reprinted from reference Mauter, M.S., Elimelech, M., 2008. Environmental applications of carbon-based nanomaterials. Environ. Sci. Technol., 42(16), 5843–5859 with permission from American Chemical Society.

of MWCNTs (comprising of different carbon sheets) exhibits adjustable chirality that leads to display exceptional mechanical features and remarkable electrical qualities. Fig. 3 represents the hybridization conditions accounting for the numerous concoction of organic compounds besides the significant alterations among bulk configurations of carbon.

3.3 Graphene oxide

It is an allotropic (two-dimensional) type of carbon, and it is shaped due to a single sheet of carbon molecules. In graphene oxide, sp^2-hybridization is displayed by carbon molecules, associated by σ- and π-bonds in a two-dimensional hexagonal precious stone cross-section with a separation of 0.142 nm between neighboring particles of carbon hexagons. Graphene oxide additionally speaks to an auxiliary component of some other carbon allotropes, for example, graphite, fullerenes, and carbon nanotubes (Iijima and Ichihashi, 1993). Hypothetical examinations on graphene oxide started quite a while before the genuine material examples were achieved. The Canadian hypothetical physicist P.R. Wallace originally investigated the hypothesis of graphene oxide in 1947. In spite of the long history of hypothetical examination, the way that the genuine material has been achieved as of late infers that far-reaching thoughts on the properties of graphene oxide are as yet progressing. Graphene oxide has numerous interesting physical properties, for example, a very high mechanical unbending nature and a high warm steadiness. Additionally, the electric properties of this carbon allotrope are generally not quite the same as the properties of three-dimensional materials (Chen et al., 2016).

3.4 Graphene quantum dots

Graphene quantum dots (GQDs) have received increased critical enthusiasm of late because of their potential applications in the biomedical field, attributable to their tunable photoluminescence properties, high photostability, amazing physicochemical properties, small size, and great biocompatibility (Yang et al., 2014).

3.5 Carbon quantum dots

Carbon quantum dots (CQDs) are another class of nanoparticles that have fluorescence, and are carbon nanoparticles with a molecule size under 10nm. These fluorescent CQDs have tremendous applications in illness identification, bioimaging, and fabrication of biosensing. These are promising materials for nanobiotechnology because they have excitation wavelength subordinate photoluminescence (PL), small size, brilliant biocompatibility, photograph-incited electron exchange, substance latency, and low toxicity. These materials have magnificent fluorescent properties, for example, tight and tunable discharge spectra, expansive excitation spectra, and high photostability against photograph dying and squinting than other fluorescent semiconductor quantum spots (Liu et al., 2016).

4 Nanosensors and their types

Nanosensors have useful applications in various fields, for example, farming and nourishment generation (Zhang et al., 2015). They possess noteworthy changes in selectivity, specificity, and affectability contrasted with conventional concoction and organic strategies (Mehrzad-Samarin et al., 2017). Nanosensors can be employed for the detection of microorganisms, contaminations, and sustenance freshness (Rendler et al., 2017). The nanosensors utilized in nourishment investigations consolidate learning of science and nanotechnology and may be termed nanobiosensors. Present nanosensor procedures can be grouped into:

Nanostructured materials, for example, nanoparticle-based sensors (porous silicon), nanoelectromechanical systems (NEMS), nanowire nanosensors, nanoprobes, and cantilevers.

On the basis of their applications in different fields, nanosensors can be grouped into:

- Nanoparticle-based nanosensors.
- Electrochemical nanosensors.
- Optical nanosensors.

5 Nanosensor applications

Various nanosensors have been created for different applications in the horticultural and nourishment industries, either to rapidly distinguish dangers on account of suspected sustenance harming, or incorporated into bundling as nanotracers to demonstrate the historical backdrop of the sustenance item and whether it is of worthy quality at some random time (Kosaka et al., 2014). For example, the utilization of nanosensors as a tracer to identify the development of microorganisms and nanosensors connected in on-line process control for observing capacity conditions. Different kind of signals will be produced by nanosensors to observe the change in nourishment level. As another model, researchers are making gold nanoparticles and coating them with particles that can tie to substances such as pesticides.

Agriculturists could shower these nanoparticles on their fields to distinguish a synthetic such as a pesticide. Besides, nanosensors utilizing Raman Spectroscopy is a perfect combination to investigate the nourishment crime. Nourishment crime scene investigation will be an examination of the sustenance starting point, defilement, and pollution (Wei et al., 2017). Nanosensor applications in this add to the specificity of the strategy and permit utilization of different analytes that can be examined, extending from full-scale sustenance, lipids, proteins, and starches to minor parts, colors, shades, and additives (Wang et al., 2011).

It very well may be that nanosensors are valuable for detecting and detailing constant data with respect to the item from creation to the shopper. Nanosensors are a long way from being just a detached, data-gathering gadget. They can get data from prompt and remote settings and can dissect, record, and report information. They can be intended to deal with this at basic control focuses on the production network—from the point sustenance is delivered or bundled through the time it is devoured (Yola et al., 2014). The most recent advancements have brought about nanosensors that are very close to commercialization: nanosensors and nanoscale coatings to supplant thicker, more inefficient polymer coatings for averting erosion; nanosensors to locate oceanic poisons; nanoscale biopolymers for enhanced cleaning and reusing of substantial metals; and nanosensors ready to give quality affirmation by following microorganisms, poisons, and contaminants through the nourishment handling chain by utilizing information for programmed control capacities and documentation, among others.

Gold nanoparticles functionalized with cyanuric corrosive gatherings specifically tie to melamine, an adulterant used to falsely blow up the deliberate protein substance of pet sustenance and newborn child equations (Chen et al., 2017). A promising photograph-initiated pointer ink for in-bundle oxygen recognition dependent on nanosized TiO_2 or SnO_2 particles and a redox-color (methylene blue) has been created. Nanosensors dependent on nanoparticles have additionally been produced to recognize the nearness of dampness inside a nourishment bundling (Vo-Dinh et al., 2015).

6 Graphene-featured nanosensors for food-related applications

Nanotechnology manages nanomaterials that have somewhere around one measurement running from 1 to 100 nm. Tried nanomaterials in the sustenance industry incorporate inorganic (graphene and graphene oxide NPs), natural (for the most part common item NPs), and joined (for example, earth). Among every single metal NP, graphene NP is the one most economically made and connected because of its antimicrobial movement; it is broadly considered a sensor/identifier. Graphene oxide NPs are very much concentrated as purifying operators such as a nourishment added substance (white shading color) and a flavor enhancer. Common item NPs are ordinarily planned as a conveyance framework and furthermore included as fixings or enhancements in the nourishment industry (Chen et al., 2014). Fig. 4 shows the variety of concentrations of carbon nanomaterials that cause plant toxicity.

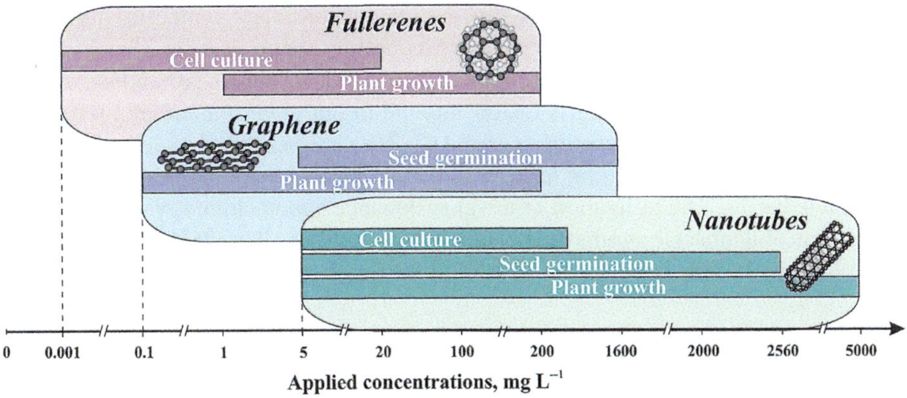

Fig. 4 Concentrations range of carbon-based nanomaterials applied in toxicity-related studies. Reprinted from reference Zaytseva, O., Neumann, G., 2016. Carbon nanomaterials: production, impact on plant development, agricultural and environmental applications. Chem. Biol. Technol. Agric., 3(1), 17 with permission from Springer Nature.

6.1 Pathogen detection

Different applications of nanosensor (having unique properties) includes the pesticides discovery, pathogens location, and poisons identification. An ongoing research found that ZnO quantum specks (QD) can be utilized to identify different pesticides, including aldrin, tetradifon, glyphosate, and atrazine. Quantum dots can detect pesticides residues even at 107 M^{-1} (Sahoo et al., 2018). Moreover, ZnO QD could likewise photocatalyze pesticides amid cooperation. "Dynamic bundling" and "shrewd pressing" presently have turned into prominent patterns to use nanomaterial-based sensors for observing the nature of sustenance (Chen et al., 2015).

6.2 Food packaging

Sustenance contact materials are proposed to legitimately contact nourishment items amid assembling, transportation, and capacity. Nanotechnology as a novel arrangement presently has been broadly considered and created for nourishment bundling in the sustenance industry. Nanomaterials intended for nourishment bundling have numerous points of interest when contrasted with traditional bundling materials. Among numerous novel nanomaterials, graphene is a standout among the most broadly utilized and contemplated for nourishment bundling due to its mechanical, warm, and boundary properties as well as minimal effort. For example, a 1 wt% bentonite dirt/poly (vinyl liquor) stacked nanocomposite layer altogether improved permeance with water permeance of 6500 gpu and a selectivity estimation of 46 (Chhipa, 2017).

6.3 Food safety

Nourishment nanotechnology has invaded numerous parts of client items, for example, sustenance bundling, added substances, and sustenance protection. The acknowledgement of this novel innovation has propelled the sustenance preparing and capacity in guaranteeing nourishment well-being. Numerous ordinary synthetic substances included as sustenance-added substances or bundling materials have likewise been found incompletely existing at the nanometer scale. For instance, TiO_2 NPs currently used as nourishment substance, its residues found in our environment are up to 40%. In spite of the fact that nanomaterials like TiO_2 NPs are perceived as less dangerous to our environment but still long term introduction to such nanomaterials may cause unfavorable harms. The utilization of novel sustenance nanotechnology, together with the nearness of nanoscale synthetic substances, has additionally pulled in open consideration in regard to the potential dangers (Chhipa and Joshi, 2016).

6.4 Food processing

Nanomaterials are all around, structured as shading or flavor-added substances, additives, or bearers for nourishment supplement (e.g., nanoencapsulation and nanoemulsion), including creature feed items. The interesting properties of designed nanomaterials offer incredible points of interest for example as nourishing agent or as a sensor to detect the harmful substances. Notwithstanding the nanomaterials used as nourishment items involve in serving people and industry. NMs as nourishing agents provide financial benefits and safe agricultural products all through the world. The utilization of dicopper oxide (Cu_2O or copper(I) oxide) is banned by the European Food Safety Authority (EFSA) Panel in additives and products or substances utilized in animal Feed (FEEDAP Panel). These dicopper oxides produce harmful effects for every single creature species (Handford et al., 2015).

7 Carbon quantum dot-based nanosensors and their exploitation in the agriculture sector

European Commission considers nanotechnology as one of its six "Key Enabling Technologies" that can fundamentally bring advancements in agricultural sector. Indeed, nanotechnology application to the agribusiness and nourishment divisions is at the incipient stage. Nanotechnology provides improvements in different areas like sensing of harmful substances, environmental change observation and increased nourishing agents (Dasgupta et al., 2015). Nanotechnology holds the possibility to change the worldwide sustenance framework, and there are significantly more applications in farming and nourishment that can be relied upon in the years to come. Novel horticultural and nourishment well-being frameworks, illness treatment conveyance strategies, instruments for atomic and cell science, sensors for identifying pathogens and pesticides, wise bundling materials, ecological insurance, and instruction of the general population and future workforce are some models of the imperative effect that nanotechnology could have in agrarian areas (Fraceto et al., 2016).

Cheap sensors, distributed computing, and canny programming together can upset the entire agri-sustenance segment. On account of PCs, worldwide satellite situating frameworks, and remote detecting gadgets, exact cultivating moves closer to reality. Observing and estimating ecological factors, settling on fitting and time choices, and performing focused on activity as per gathered information with the end goal to expand yield (i.e., trim yields) with ideal utilization of assets (i.e., composts, pesticides, herbicides, etc.) are the fundamental attributes of the exact cultivating vision (King et al., 2018). It is vital to feature different miniaturized scale nanobioframeworks, created as activities and used to acknowledge savvy agri-nourishment frameworks. Innovation improvement, especially in the Internet of Things (IoT) (Bogataj and Drobne, 2017), which is a developing reality where an ever-increasing number of gadgets are associated with clients and different gadgets by means of the Internet, fundamentally adds to the acknowledgement of savvy farming by expanding the quality, amount, supportability, and cost adequacy of horticultural generation. The imaginative utilization of nanotechnology in the IoT creates another worldview, in particular, the Internet of Nano Things (IoNT). Nanosensors, as a result of their small measurements, can gather data from various distinctive focuses. Nanosensors produced using nonorganic materials, for example, carbon nanotubes, have the capacity to detect and flag, going about as remote nano antennas (Dubey and Mailapalli, 2016).

Outer gadgets would then be able to coordinate the information to consequently produce an amazingly full detailed report and react to conceivably pulverizing changes in their condition. There are various astute nanosystems utilized in keen horticulture for yield checking, immobilization of supplements and their discharge in the soil, investigation of pesticides, dampness, and soil pH (Prasad et al., 2017). Systems of associated nanosensors for checking soil or plant conditions have the capacity to alarm consequently as indicated by conditions identified by sensors and accordingly impact more productive utilization of the water, manure, herbicide, pesticide, bug spray, and so forth (Afsharinejad et al., 2016). Fig. 5 shows the spread of a C/Cu suspension on a substrate made from glass used for the detection of gases.

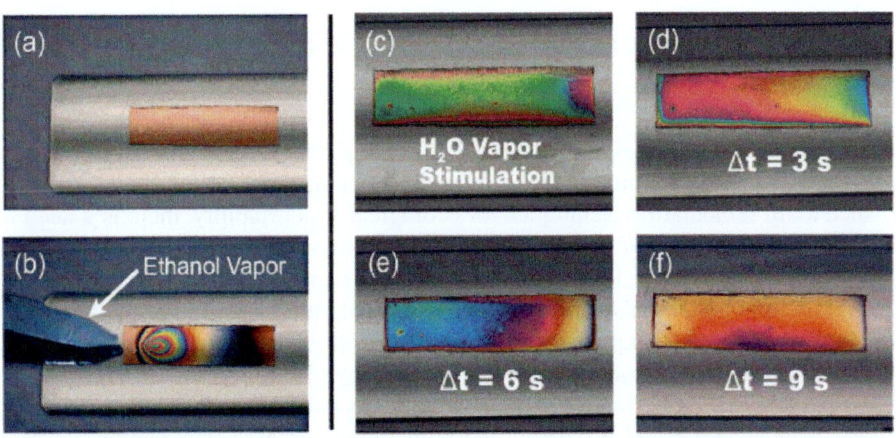

Fig. 5 Moisture sensor based on carbon-coated copper nanoparticles decorated on a polymer matrix (A); iridescent coloration produced due to ethanol vapor (B); polymer swell due to water vapor exposure (C); As moisture dissipates (D–F), the sensor reverts back to its native state and appearance.

Reprinted from reference Luechinger, N.A., Loher, S., Athanassiou, E.K., Grass, R.N., Stark, W.J., 2007. Highly sensitive optical detection of humidity on polymer/metal nanoparticle hybrid films. Langmuir, 23(6), 3473–3477 with permission from American Chemical Society.

Along these lines, continuous and thorough checking of the yield development, prompt precise and on-time choices, diminished expenses and waste, help to enhance the level of nourishment to maintain horticulture. As nourishment wellbeing is a fundamental concern these days, it is basic to empower foodstuff nutrients traceability and checking by using an entire natural way of life process. Nanosensors can be connected along the whole nourishment production network, with the end goal to recognize diverse focuses in rapid, delicate, and savvy behavior (Ranjan et al., 2016). This is essential during the time spent guaranteeing nourishment quality, well-being, freshness, genuineness, and traceability. For example, including nanosensors in the plan of keen or insightful bundling, deterioration or unsafe contaminants amid circulation or capacity can be distinguished by empowered exchange of data with respect to item conditions. Along these lines, sustenance makers, transportation, and retail organizations wind up associated as at no other time (Dasgupta et al., 2017).

The reaction created because of changes identified with inward or outside ecological factors are recorded through particular sensors, put away in the database, and by means of the Internet, we have daily accessible knowledge on the soil and yield well-being, sustenance pollution, quality, etc. Along these lines different properties like, fast reaction and responses towards specific analyte (at different parameters), are empowered. These kind of devices produces prompt reaction, easy to use and safe sustenance, have a great impact to human wellbeing (Liu et al., 2015). Still, to accomplish the maximum capacity IoNT brings to the table in farming and nourishment, numerous worries, considering information security, protection and nanomaterial toxicity and effect on the Earth and humans, must be considered (Wang et al., 2005).

8 Development of carbon nanosensors for environmental application

Ecological contamination is a major problem at the global level as numerous poisons debase urban and horticultural zones (Rameshaiah et al., 2015). With the end goal to enhance toxin remediation techniques for ecological supportability, there is a need to expand the effectiveness of customary strategies or to present imaginative methodologies (Servin et al., 2015). In this specific situation, nanotechnology, and particularly carbon-based nanomaterials, can significantly contribute in light of the fact that they have tremendous retention potential because of their high surface region.

A water sterilization strategy has been reported by (Sharma et al., 2015) in which there is a combination between functionalized CNTs and microwaves. First, aliphatic liquor 1-octadecanol ($C_{18}H_{38}O$) was used to functionalize CNTs. Application of CNTs in water filtration invention as well as water sterilization has been reported in the surveys of Stanisavljevic et al. (2015) and Al-Hakami et al. (2013). Fig. 6 indicates the different applications of carbon-based nanomaterials having exceptional properties in environmental organizations.

Furthermore, not just expulsion of poisons from various natural compartments, but also checking tainting levels has significance (Upadhyayula et al., 2009). Carbon-based

Physical properties	Environmental applications						
	Pollution prevention	Sorbents	Composite filters	Energy storage	Antimicrobial agents	Aligned CNT membranes	Sensors
Size		X	X		X	X	X
Shape					X	X	
Surface area	X	X	X	X			
Molecular specificity		X		X			X
Hydrophobicity		X	X		X	X	
Electrical conductivity				X			X
Optical activity		X					
Thermal conductivity	X			X			

Bulk properties	X
Individual properties	X

Fig. 6 Environmental application of carbonaceous nanomaterials with unique properties. Unique properties are possessed by carbonaceous nanomaterials in both individual and bulk configurations. Further characterization of carbon-based nanomaterial attributes, synthesis techniques, and purification procedures will expand their range of environmental applications. Reprinted from reference Mauter, M.S., Elimelech, M., 2008. Environmental applications of carbon-based nanomaterials. Environ. Sci. Technol., 42(16), 5843–5859 with permission from American Chemical Society.

nanomaterials can be applied to improve novel, exceptionally effective biochemical sensors for differentiation of low centralizations of concoction mixes in various conditions. Likewise, application potential is shown by these sensors in horticulture (Lam et al., 2004).

9 CNT/carbon dot/graphene-based nanosensors for heavy metal detection

Carbon-based nanomaterials with novel substance, physical, and mechanical properties are utilized to grow very delicate sensors and indicative gadgets for various horticultural and ecological applications (Guo et al., 2015). Nanosensors display various working standards, yet a typical method of activity is a change of physicochemical properties into signs (Cash et al., 2015). The high affectability of these gadgets is dictated by the nanoscale size of the detecting components, for example, carbon nanotubes. Along these lines, a couple of particles are every now and again adequate to impact the electrical properties (concoction to electrical transduction) of the nanoparticles. In addition, the considerable surface zone of carbon-based nanomaterials gives extensive spaces to communications with the detected atoms (Miao et al., 2014).

The advancement of novel advances dependent on nanomaterials that can be effectively utilized for simple, fast, and exceptionally touchy compound examination is archived in various investigations (Cayuela et al., 2015). A graphene-based sensor for the location of cadmium defilements in water, which works adequately at fixation levels of 0.25 μg L^{-1}, has been portrayed by (Sekhon, 2014). An exceedingly touchy sensor for the assurance of nickel, in natural examples as well as in sustenance, depends on changed nanotubes and has an identification point of confinement of 4.9 ng L^{-1}. CNT-based gas sensors for alkali discovery can be utilized to follow gas estimations. Sensors to recognize pesticides, herbicides, and their metabolites in natural examples have been produced dependent on changed MWCNTs or on GO. Besides, nanosensors have been utilized for observing soil dampness. Nanosensor-based checking frameworks for product well-being have likewise been depicted (Zhu et al., 2017). In this unique situation, CNT sensors were effectively utilized for in vivo checking of the ROS arrangement in plant tissues as pressure markers. An electronic gadget for ongoing detection of dangerous gases dependent on CNTs, which could be set onto creepy crawlies or plants, can be connected in remote detecting frameworks for plant illnesses. Covering a field site with various sensors can give a full image of the spatial dissemination of illness seriousness at a field scale and permits the identification of problem areas, requiring exceptional treatment, as a huge commitment to diminishing the contribution of agrochemicals (Sumner et al., 2006).

Nonetheless, at present, the lion's share of nanosensors are still in the item advancement stage and vast scale creation has not yet been set up because of various confinements. Above all else, in transit from the lab to the market, long haul testing is required for alignment and approval of the created gadgets (Adio et al., 2017). As of now, just a predetermined amount of analyte produces best results under specific conditions like temperature, humidity, and interfering agents. Difficulties related to up-scaling creation systems and the high expenses of the materials and hardware required for the generation of nanosensors have been guaranteed by different creators. Real difficulties still remain regarding issues of security, wel-lbeing perspectives, and hazard appraisal, requiring a whole life cycle investigation of an item. Nonetheless, in spite of the recorded issues, a few creators are expecting to increase the commercialization of nanosensor innovations. Because, by using this, different problems can be solved regarding our agricultural sector (sustainability of food nutrients).

10 Conclusion

Currently, nanotechnology research emphasizes extraordinary improvements in academic and modern levels and beyond this, its advancements in ICTs has established a definite influence in the agri-nourishment region. In ecological and rural divisions, carbon-based nanomaterials are delivered on an extensive scale for a vast range of purposes to be used in different fields. Although nanomaterials with different striking features have immense uses and applications in different areas yet these NMs can produce different hazardous effect on our local environment and living creators. Nanotechnology, by its applicable tools such as nanosensors and nanomaterials, has brought many advantages in rural and nutrition industries such as significant increases in plant development, precise cultivating appliances, improved soil nature, prolonged productivity, detection of nutrition quality among different generations. These are only a portion of their numerous advantages.

Nanosensors having unique properties ensure detection with remarkable specificity and selectivity for a variety of analytes such as toxic substances, pathogens, DNA, biomolecules, etc. Nanoparticles have added specific selectivity in diagnostic tools by offering better electrochemical properties in sensors and by providing a large surface area to aptamer immobilization. Although the nanomaterial industry has eased the burden of industry, some difficulties also exist in the advanced techniques of field-material nanosensors. Pretreatment systems, specific activity, and high costs are some parameters that still need to be fixed. Additional knowledge and research improvements can provide a solution to these issues. Moreover, advanced and innovative tools can help in the preparation of novel nanomaterials with modest and easily applicable sensing frameworks.

Increased learning of damage produced by nano-empowered items and improved vision can acknowledge the maximum capability of nanotechnology in the agribusiness and nutrition industries. This can also be done by proficiently considering all possible adverse effects of nanomaterials on the Earth and humans.

Future study will concern the development of novel and more reliable materials, methods, or systems with more easy and accurate devices on the nanoscale to the acknowledgement of IoNT. It further includes estimation of the effect of these tools to humans and the overall Earth environment. Green nanotechnology and green Internet of Things (IoT) will rapidly move towards a radically new world. Because it contains ecofriendly and safe nanoitems having a number of applications with nearly no threat for humans and the earth. Data will be recorded to analyze the effect of these different complicated nanomaterials (conformational changes). Aside from these impediments, some essential properties of carbon-based nanomaterials are important for plants growth.

- **i.** Mostly, plants uptake carbon nanomaterials that are regulated by translocation.
- **ii.** The physiological reactions are activated by little events of assimilated nanomaterials.
- **iii.** By decreasing the level of carbon based nanomaterials (<100 mg L^{-1}), it was observed that it is producing hindering impacts on plant development.
- **iv.** Enlistment of oxidative worry through development of ROS is by all accounts a noteworthy basic component of phytotoxicity instigated via carbon nanomaterials while gainful impacts are most likely dependent on hormesis. In other words, ameliorated seed germination lead to stir up the level of water take-up.
- **v.** Modification in concoction and confirmation of carbon nanomaterial have a great impact on poisonous quality potential.
- **vi.** In numerous investigations of CNM phytotoxicity, the tried measurements fundamentally surpass the normal ecological focuses.

Acknowledgment

The authors acknowledge the Higher Education Commission (HEC) of Pakistan for financial support under Research Grants No. 6114 and 6115. Dr. Waheed S. Khan is also grateful to the parent department *National Institute for Biotechnology and Genetic Engineering (NIBGE)* and Faisalabad Pakistan for providing a conducive environment for such writing.

References

Adio, S.O., Omar, M.H., Asif, M., Saleh, T.A., 2017. Arsenic and selenium removal from water using biosynthesized nanoscale zero-valent iron: a factorial design analysis. Process Saf. Environ. Prot. 107, 518–527.

Afsharinejad, A., Davy, A., Jennings, B., Brennan, C., 2016. Performance analysis of plant monitoring nanosensor networks at THz frequencies. IEEE Internet Things J. 3 (1), 59–69.

Al-Hakami, S.M., Khalil, A.B., Laoui, T., Atieh, M.A., 2013. Fast disinfection of *Escherichia coli* bacteria using carbon nanotubes interaction with microwave radiation. Bioinorg. Chem. Appl, 13.

Bogataj, D., Drobne, D., 2017. Control of perishable goods in cold logistic chains by bionanosensors. In: Applying Nanotechnology for Environmental Sustainability, IGI Global, pp. 376–402.

Cano-Marquez, A.G., Rodriguez-Macias, F.J., Campos-Delgado, J., Espinosa-González, C.G., Tristán-López, F., Ramírez-González, D., Cullen, D.A., Smith, D.J., Terrones, M., Vega-Cantú, Y.I., 2009. Ex-MWNTs: graphene sheets and ribbons produced by lithium intercalation and exfoliation of carbon nanotubes. Nano Lett. 9 (4), 1527–1533.

Cash, K.J., Li, C., Xia, J., Wang, L.V., Clark, H.A., 2015. Optical drug monitoring: photoacoustic imaging of nanosensors to monitor therapeutic lithium in vivo. ACS Nano 9 (2), 1692–1698.

Cayuela, A., Soriano, M.L., Valcárcel, M., 2015. Reusable sensor based on functionalized carbon dots for the detection of silver nanoparticles in cosmetics via inner filter effect. Anal. Chim. Acta 872, 70–76.

Chaudhry, Q., Castle, L., Watkins, D.R., 2010. Nanotechnologies in Food. Royal Society of Chemistry, London.

Chen, H., Seiber, J.N., Hotze, M., 2014. ACS select on nanotechnology in food and agriculture: a perspective on implications and applications. J. Agric. Food Chem. 62 (6), 1209–1212.

Chen, G., Ji, S., Li, H., Kang, X., Chang, S., Wang, Y., Yu, G., Lu, J., Claverie, J., Sang, Y., Liu, H., 2015. High-energy faceted SnO_2-coated TiO_2 nanobelt heterostructure for near-ambient temperature-responsive ethanol sensor. ACS Appl. Mater. Interfaces 7 (44), 24950–24956.

Chen, Y.M., Yu, L., Lou, X.W., 2016. Hierarchical tubular structures composed of Co_3O_4 hollow nanoparticles and carbon nanotubes for lithium storage. Angew. Chem. Int. Ed. 55 (20), 5990–5993.

Chen, J., Liu, J., Li, J., Xu, L., Qiao, Y., 2017. One-pot synthesis of nitrogen and sulfur co-doped carbon dots and its application for sensor and multicolor cellular imaging. J. Colloid Interface Sci. 485, 167–174.

Chhipa, H., 2017. Nanofertilizers and nanopesticides for agriculture. Environ. Chem. Lett. 15 (1), 15–22.

Chhipa, H., Joshi, P., 2016. Nanofertilisers, nanopesticides and nanosensors in agriculture. In: Nanoscience in Food and Agriculture. vol. 1, Springer, pp. 247–282.

Choi, W.B., Jin, Y.W., Kim, H.Y., Lee, S.J., Yun, M.J., Kang, J.H., Choi, Y.S., Park, N.S., Lee, N.S., Kim, J.M., 2001. Electrophoresis deposition of carbon nanotubes for triode-type field emission display. Appl. Phys. Lett. 78 (11), 1547–1549.

Dai, L.e., 2006. Carbon nanotechnology: Recent developments in chemistry, physics, materials science and device applications. In: Elsevier.

Das, M., Ansari, K.M., Tripathi, A., Dwivedi, P.D., 2011. Need for safety of nanoparticles used in food industry. J. Biomed. Nanotechnol. 7 (1), 13–14.

Das, R., Ali, M.E., Hamid, S.B.A., Ramakrishna, S., Chowdhury, Z.Z., 2014. Carbon nanotube membranes for water purification: a bright future in water desalination. Desalination 336, 97–109.
Dasgupta, N., Ranjan, S., Mundekkad, D., Ramalingam, C., Shanker, R., Kumar, A., 2015. Nanotechnology in agro-food: From field to plate. Food Res. Int. 69, 381–400.
Dasgupta, N., Ranjan, S., Ramalingam, C., 2017. Applications of nanotechnology in agriculture and water quality management. Environ. Chem. Lett. 15 (4), 591–605.
De Volder, M.F., Tawfick, S.H., Baughman, R.H., Hart, A.J., 2013. Carbon nanotubes: present and future commercial applications. Science 339 (6119), 535–539.
Dubey, A., Mailapalli, D.R., 2016. Nanofertilisers, nanopesticides, nanosensors of pest and nanotoxicity in agriculture. In: Sustainable Agriculture Reviews, Springer, pp. 307–330.
Durán, N., Marcato, P.D., 2013. Nanobiotechnology perspectives. Role of nanotechnology in the food industry: a review. Int. J. Food Sci. Technol. 48, 1127–1134.
Fraceto, L.F., Grillo, R., de Medeiros, G.A., Scognamiglio, V., Rea, G., Bartolucci, C., 2016. Nanotechnology in agriculture: which innovation potential does it have? Front. Environ. Sci. 4, 20.
Georgakilas, V., Perman, J.A., Tucek, J., Zboril, R., 2015. Broad family of carbon nanoallotropes: classification, chemistry, and applications of fullerenes, carbon dots, nanotubes, graphene, nanodiamonds, and combined superstructures. Chem. Rev. 115 (11), 4744–4822.
Gottschalk, F., Sonderer, T., Scholz, R.W., Nowack, B., 2009. Modeled environmental concentrations of engineered nanomaterials (TiO_2, ZnO, Ag, CNT, fullerenes) for different regions. Environ. Sci. Technol. 43 (24), 9216–9222.
Gottschalk, F., Sun, T., Nowack, B., 2013. Environmental concentrations of engineered nanomaterials: review of modeling and analytical studies. Environ. Pollut. 181, 287–300.
Guo, Y., Zhang, L., Zhang, S., Yang, Y., Chen, X., Zhang, M., 2015. Fluorescent carbon nanoparticles for the fluorescent detection of metal ions. Biosens. Bioelectron. 63, 61–71.
Handford, C.E., Dean, M., Spence, M., Henchion, M., Elliott, C.T., Campbell, K., 2015. Awareness and attitudes towards the emerging use of nanotechnology in the agri-food sector. Food Control 57, 24–34.
Iijima, S., Ichihashi, T., 1993. Single-shell carbon nanotubes of 1-nm diameter. Nature 363 (6430), 603.
Jain, A., Ranjan, S., Dasgupta, N., Ramalingam, C., 2018. Nanomaterials in food and agriculture: an overview on their safety concerns and regulatory issues. Crit. Rev. Food Sci. Nutr. 58, 297–317.
Khot, L.R., Sankaran, S., Maja, J.M., Ehsani, R., Schuster, E.W., 2012. Applications of nanomaterials in agricultural production and crop protection: a review. Crop Prot. 35, 64–70.
Kim, S.H., Zachariah, M.R., 2006. In-flight kinetic measurements of the aerosol growth of carbon nanotubes by electrical mobility classification. J. Phys. Chem. B 110 (10), 4555–4562.
King, T., Osmond-McLeod, M.J., Duffy, L.L., 2018. Nanotechnology in the food sector and potential applications for the poultry industry. Trends Food Sci. Technol. 72, 62–73.
Kosaka, P.M., Pini, V., Ruz, J.J., Da Silva, R.A., González, M.U., Ramos, D., Calleja, M., Tamayo, J., 2014. Detection of cancer biomarkers in serum using a hybrid mechanical and optoplasmonic nanosensor. Nat. Nanotechnol. 9 (12), 1047.
Lam, C.W., James, J.T., McCluskey, R., Hunter, R.L., 2004. Pulmonary toxicity of single-wall carbon nanotubes in mice 7 and 90 days after intratracheal instillation. Toxicol. Sci. 77 (1), 126–134.
Liu, J., Chen, Y., Wang, W., Feng, J., Liang, M., Ma, S., Chen, X., 2015. "Switch-on" fluorescent sensing of ascorbic acid in food samples based on carbon quantum dots–MnO_2 probe. J. Agric. Food Chem. 64 (1), 371–380.
Liu, Z.Q., Cheng, H., Li, N., Ma, T.Y., Su, Y.Z., 2016. $ZnCo_2O_4$ quantum dots anchored on nitrogen-doped carbon nanotubes as reversible oxygen reduction/evolution electrocatalysts. Adv. Mater. 28 (19), 3777–3784.

Llobet, E., 2013. Gas sensors using carbon nanomaterials: a review. Sensors Actuators B Chem. 179, 32–45.
Magnuson, B.A., Jonaitis, T.S., Card, J.W., 2011. A brief review of the occurrence, use, and safety of food-related nanomaterials. J. Food Sci. 76 (6), 126–R133.
Mehrzad-Samarin, M., Faridbod, F., Dezfuli, A.S., Ganjali, M.R., 2017. A novel metronidazole fluorescent nanosensor based on graphene quantum dots embedded silica molecularly imprinted polymer. Biosens. Bioelectron. 92, 618–623.
Miao, R., Mu, L., Zhang, H., She, G., Zhou, B., Xu, H., Wang, P., Shi, W., 2014. Silicon nanowire-based fluorescent nanosensor for complexed Cu^{2+} and its bioapplications. Nano Lett. 14 (6), 3124–3129.
Morris, V., 2011. Emerging roles of engineered nanomaterials in the food industry. Trends Biotechnol. 29, 509–516.
Parisi, C., Vigani, M., Rodríguez-Cerezo, E., 2015. Agricultural nanotechnologies: what are the current possibilities? Nano Today 10, 124–127.
Prasad, R., Bhattacharyya, A., Nguyen, Q.D., 2017. Nanotechnology in sustainable agriculture: recent developments, challenges, and perspectives. Front. Microbiol. 8, 1014.
Pyun, J., 2011. Graphene oxide as catalyst: application of carbon materials beyond nanotechnology. Angew. Chem. Int. Ed. 50 (1), 46–48.
Qadir, D., Mukhtar, H., Keong, L.K., 2016. Synthesis and characterization of polyethersulfone/carbon molecular sieve based mixed matrix membranes for water treatment applications. Proc. Eng. 148, 588–593.
Rameshaiah, G.N., Pallavi, J., Shabnam, S., 2015. Nano fertilizers and nano sensors—an attempt for developing smart agriculture. Int. J. Eng. Res. Gen. Sci. 3 (1), 314–320.
Ranjan, S., Dasgupta, N., Lichtfouse, E. (Eds.), 2016. Nanoscience in Food and Agriculture. Springer International Publishing, Cham.
Rendler, T., Neburkova, J., Zemek, O., Kotek, J., Zappe, A., Chu, Z., Cigler, P., Wrachtrup, J., 2017. Optical imaging of localized chemical events using programmable diamond quantum nanosensors. Nat. Commun. 8, 14701.
Restuccia, D., Spizzirri, U.G., Parisi, O.I., Cirillo, G., Curcio, M., Iemma, F., Puoci, F., Vinci, G., Picci, N., 2010. New EU regulation aspects and global market of active and intelligent packaging for food industry applications. Food Control 21 (11), 1425–1435.
Ruiz-Garcia, L., Lunadei, L., Barreiro, P., Robla, I., 2009. A review of wireless sensor technologies and applications in agriculture and food industry: state of the art and current trends. Sensors 9 (6), 4728–4750.
Sahoo, D., Mandal, A., Mitra, T., Chakraborty, K., Bardhan, M., Dasgupta, A.K., 2018. Nanosensing of pesticides by zinc oxide quantum dot: An optical and electrochemical approach for the detection of pesticides in water. J. Agric. Food Chem. 66 (2), 414–423.
Schnorr, J.M., Swager, T.M., 2010. Emerging applications of carbon nanotubes. Chem. Mater. 23 (3), 646–657.
Sekhon, B.S., 2014. Nanotechnology in agri-food production: an overview. Nanotechnol. Sci. Appl. 7, 31.
Servin, A., Elmer, W., Mukherjee, A., De la Torre-Roche, R., Hamdi, H., White, J.C., Bindraban, P. and Dimkpa, C., 2015. A review of the use of engineered nanomaterials to suppress plant disease and enhance crop yield. J. Nanopart. Res., 17(2), 92.
Sharma, T.K., Ramanathan, R., Rakwal, R., Agrawal, G.K., Bansal, V., 2015. Moving forward in plant food safety and security through NanoBioSensors: adopt or adapt biomedical technologies? Proteomics 15 (10), 1680–1692.
Smolkova, B., El Yamani, N., Collins, A.R., Gutleb, A.C., Dusinska, M., 2015. Nanoparticles in food. Epigenetic changes induced by nanomaterials and possible impact on health. Food Chem. Toxicol. 77, 64–73.

Staii, C., Johnson, A.T., Chen, M., Gelperin, A., 2005. DNA-decorated carbon nanotubes for chemical sensing. Nano Lett. 5 (9), 1774–1778.

Stanisavljevic, M., Krizkova, S., Vaculovicova, M., Kizek, R., Adam, V., 2015. Quantum dots-fluorescence resonance energy transfer-based nanosensors and their application. Biosens. Bioelectron. 74, 562–574.

Sumner, J.P., Westerberg, N.M., Stoddard, A.K., Fierke, C.A., Kopelman, R., 2006. Cu^{+}- and Cu^{2+}-sensitive PEBBLE fluorescent nanosensors using DsRed as the recognition element. Sensors Actuators B Chem. 113 (2), 760–767.

Szakal, C., Roberts, S.M., Westerhoff, P., Bartholomaeus, A., Buck, N., Illuminato, I., Canady, R., Rogers, M., 2014. Measurement of nanomaterials in foods: integrative consideration of challenges and future prospects. ACS Nano 8, 3128–3135.

Upadhyayula, V.K., Deng, S., Mitchell, M.C., Smith, G.B., 2009. Application of carbon nanotube technology for removal of contaminants in drinking water: a review. Sci. Total Environ. 408 (1), 1–13.

Vo-Dinh, T., Liu, Y., Fales, A.M., Ngo, H., Wang, H.N., Register, J.K., Yuan, H., Norton, S.J., Griffin, G.D., 2015. SERS nanosensors and nanoreporters: golden opportunities in biomedical applications. Wiley Interdiscip. Rev.: Nanomed. Nanobiotechnol. 7 (1), 17–33.

Wang, W.U., Chen, C., Lin, K.H., Fang, Y., Lieber, C.M., 2005. Label-free detection of small-molecule–protein interactions by using nanowire nanosensors. Proc. Natl. Acad. Sci. 102 (9), 3208–3212.

Wang, X., Shen, Y., Xie, A., Li, S., Cai, Y., Wang, Y., Shu, H., 2011. Assembly of dandelion-like Au/PANI nanocomposites and their application as SERS nanosensors. Biosens. Bioelectron. 26 (6), 3063–3067.

Wang, H., Du, L.J., Song, Z.M., Chen, X.X., 2013. Progress in the characterization and safety evaluation of engineered inorganic nanomaterials in food. Nanomedicine 8, 2007–2025.

Wei, Z., Li, X., Zhong, N., Tan, X., Zhang, X., Liu, H., Meng, H., Liang, R., 2017. Analogue electromagnetically induced transparency based on low-loss metamaterial and its application in nanosensor and slow-light device. Plasmonics 12 (3), 641–647.

Yang, K., Xing, B., 2010. Adsorption of organic compounds by carbon nanomaterials in aqueous phase: Polanyi theory and its application. Chem. Rev. 110 (10), 5989–6008.

Yang, F., Wang, X., Zhang, D., Yang, J., Luo, D., Xu, Z., Wei, J., Wang, J.Q., Xu, Z., Peng, F., Li, X., 2014. Chirality-specific growth of single-walled carbon nanotubes on solid alloy catalysts. Nature 510 (7506), 522.

Yola, M.L., Eren, T., Atar, N., 2014. Molecular imprinted nanosensor based on surface plasmon resonance: application to the sensitive determination of amoxicillin. Sensors Actuators B Chem. 195, 28–35.

Zhang, Y., Zheng, B., Zhu, C., Zhang, X., Tan, C., Li, H., Chen, B., Yang, J., Chen, J., Huang, Y., Wang, L., 2015. Single-layer transition metal dichalcogenide nanosheet-based nanosensors for rapid, sensitive, and multiplexed detection of DNA. Adv. Mater. 27 (5), 935–939.

Zhu, L., Peng, X., Li, H., Zhang, Y., Yao, S., 2017. On-off-on fluorescent silicon nanoparticles for recognition of chromium (VI) and hydrogen sulfide based on the inner filter effect. Sensors Actuators B Chem. 238, 196–203.

Further reading

Luechinger, N.A., Loher, S., Athanassiou, E.K., Grass, R.N., Stark, W.J., 2007. Highly sensitive optical detection of humidity on polymer/metal nanoparticle hybrid films. Langmuir 23 (6), 3473–3477.

Mauter, M.S., Elimelech, M., 2008. Environmental applications of carbon-based nanomaterials. Environ. Sci. Technol. 42 (16), 5843–5859.

Pech, D., Brunet, M., Durou, H., Huang, P., Mochalin, V., Gogotsi, Y., Taberna, P.L., Simon, P., 2010. Ultrahigh-power micrometre-sized supercapacitors based on onion-like carbon. Nat. Nanotechnol. 5 (9), 651.

Zaytseva, O., Neumann, G., 2016. Carbon nanomaterials: production, impact on plant development, agricultural and environmental applications. Chem. Biol. Technol. Agric. 3 (1), 17.

Toxic effects of engineered carbon nanoparticles on environment

12

Sayali S. Patil[a], *Utkarsha M. Lekhak*[b]
[a]Department of Environmental Science, Savitribai Phule Pune University, Pune, India,
[b]Department of Biochemistry, The Institute of Science, Mumbai, Mumbai, India

1 Introduction

Carbon (C), the sixth element of the periodic table, is a true element of life. It is most commonly obtained from coal deposits as a natural source (Kharisov and Kharissova, 2019). Carbon is one of the most abundant elements in the human body. It is a normal constituent of all body cells and thus exhibits very low toxicity in the human body. Hence, it has been explored tremendously for drug delivery applications. However, the inhalation of carbon soot or carbon coal dust can be harmful, leading to lung tissue irritation and pneumoconiosis. The overall quality of human life has been greatly improved by nanotechnology developments in the last few decades. More and more research is focused on natural and engineered nanomaterials that are widely applied in many areas (Mandal et al., 2006; Mansor et al., 2019; Patil et al., 2016; Shedbalkar et al., 2014). Nanoparticles usually have at least one dimension in the range of 1–100 nm. Carbon nanoparticles (CNPs) (molar mass: 12.01 g/mol; melting and boiling point above 3500°C) appear as a black powder and are spherical. Various types and sources of CNPs in the environment are depicted in Fig. 1.

1.1 CNP structure

The CNP structure is much less explored. Structure, surface morphology, and surface properties such as the surface charge of CNPs are largely dependent on the manufacturing technique used to create them. For example, CNPs created through activation present a porous surface, creating a large surface area, facilitates binding, chemical reactions or utilization of them in absorption functions.

Graphene, carbon nanotubes (CNTs), and fluorescent carbon quantum dots (CQDs) have gained much more importance in the scientific and engineering community due to their extraordinary physicochemical, thermal, optical, and mechanical properties. Knowing the CNP structure is useful in determining its various properties and ultimately to determine and decide its applications in various fields.

1.2 CNP synthesis

Considering wide spectrum of applications for CNPS, their economical synthesis remains an attractive challenge. CNPs are synthesized naturally and can also be

Carbon Nanomaterials for Agri-food and Environmental Applications. https://doi.org/10.1016/B978-0-12-819786-8.00012-8

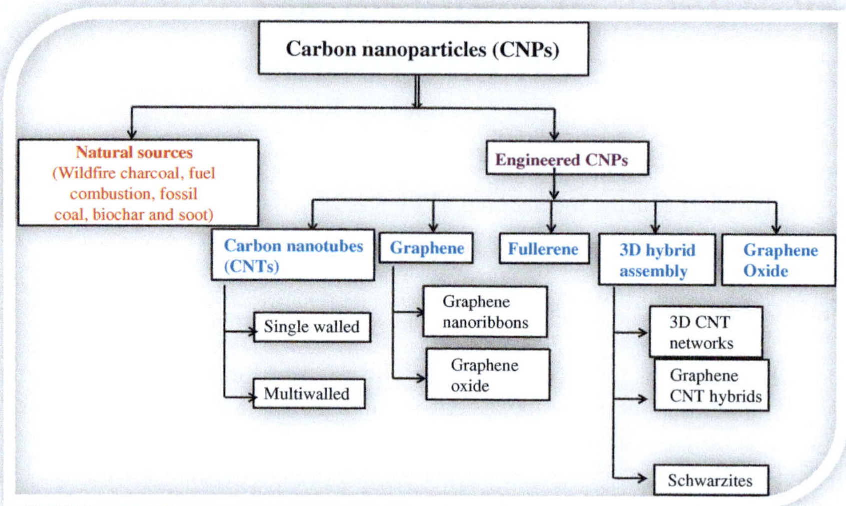

Fig. 1 Sources and various types of CNPs.

engineered. Natural synthesis includes fossil fuel combustion, forest fires, combustion of oils, automobile exhaust, etc. However, there is no any control over size, shape, or characteristics and thus on CNP properties. On the other hand, engineered CNPs can be synthesized by conventional methods of synthesis such as chemical vapor deposition, arc-discharge, laser ablation, etc., are used limitedly to solid state strategies that can tolerate relatively high energies and high temperatures, because of highly stable graphite structures of CNPs (Kang et al., 2003; Sun and Li, 2004). However, currently used CNP synthesis strategies face major challenges such as large-scale production of CNPs with low cost and high quality; control over the structure, monodispersity, and properties; and control over the mechanism of production. Moreover, such methods provide low yields and they require extreme conditions, meaning they have limitations in terms of large-scale and economical production.

1.3 Characterization of CNPs

CNP characterization is done by various methods. Spectroscopic and microscopic techniques are quite often used to characterize CNPs. Primary characterization is carried out by UV–Vis spectroscopy. Surface morphology of CNPs is detected by scanning electron microscopy (SEM), atomic force microscopy (AFM), transmission electron microscopy (TEM), and high-resolution transmission electron microscopy (HRTEM). Crystal structure, phase composition, and mean size are determined by an x-ray diffractometer. Elemental analysis, electronic state, and chemical characterization are done by energy dispersive spectrum (EDS), energy dispersive x-ray (EDX), x-ray photoelectron spectral technique (XPS), and Raman spectroscopy. Besides these techniques, thermogravimetric analysis is used to detect CNP thermal stability.

1.4 Properties of CNPs

Various types of CNPs have attracted much more interest from material scientists due to their exceptional properties. CNPs have distinctive physical and chemical properties, making them widely applied in many fields. Other physical properties of CNPs, Thus research on CNPs has gained a lot of importance over the last two decades (Iijima, 1991; Keller et al., 2002; Kroto et al., 1985). As CNPs are composed of pure carbon, they are environmentally friendly and exhibit high stability, good conductivity, and low toxicity to humans and animals. The CNP surface is highly available for electrolyte binding because of its mechanical properties and conductance nature. CNPs have linear geometry and pose exceptional electrical and anisotropic thermal conductivity. These properties make the carbon-based nanomaterials suitable candidates in advanced computing electronics.

1.5 Applications of CNPs

CNPs have been studied extensively for their applications in various fields (Fig. 2), due to their fascinating physical and chemical properties and their large surface area-to-volume ratio. CNPs can be surface functionalized in many ways, including organic molecules or polymers being chemically bound to the particle surface. In another instance, fictionalization of graphene oxide can fundamentally change its properties, making chemically modified graphenes potentially more useful for many applications. CNPs are well known as adsorbents (El-Sayed and Bandosz, 2005; Kim et al., 2008a, b; Sun and Li, 2004), composites (Kato and Ishibashi, 2008), sensors (Prasad et al., 2008), catalyst supports (Coloma et al., 1998; Oliveira et al., 2005; Rodriguez-Reinoso, 1998; Valente et al., 2001; Yoon et al., 2007), electronic materials (Kato and Ishibashi, 2008), drug delivery (Kim et al., 2008a, b; Ma et al., 2004), medical imaging (Sun and Li, 2004), optical imaging (Luo et al., 2013), cell delivery (Yan et al., 2006), etc. The wide applicability of CNPs in various biological fields may be because of their ability to be

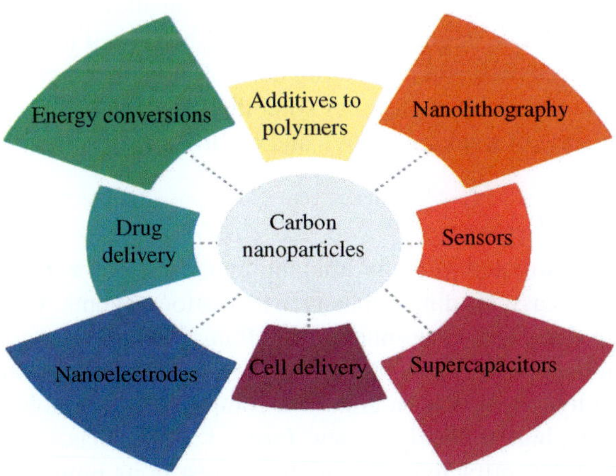

Fig. 2 Several applications of CNPs in different sectors.

functionalized, which allows them to undergo high-capacity binding to various biomolecules. CNPs, by coating with different polymers, can be loaded with multiple drugs at the same time and thus can be used for multitherapy treatments, meaning their use is not limited to one type of treatment. Further future work is also focused on CNP use for gene therapy and trace gene delivery, which will help in better understanding certain diseases. A few specific applications of CNPs are described below:

1.5.1 Cancer treatment

CNPs have proven to be a versatile platform both in vivo and in vitro for cancer targeting, treatment, and diagnosis because of their biocompatibility, nontoxicity, and fluorescent properties (Bao et al., 2018; Pardo et al., 2018; Schriber et al., 2009). CNP-based nanodrug delivery systems can deliver medicine to melanoma cells. Based on the same principles, CNPs can be used to deliver drugs to all types of cancer cells, and can present a promising alternative to current drug delivery systems based on inorganic nanoparticles. One of the major advantages of CNPs in cancer treatment is that they can be loaded with one or more drugs at the same time because of their surface properties, thus the drugs exhibiting various actions on carcinoma cells can be simultaneously loaded on CNPs to create an effective cancer therapeutic system (Bao et al., 2018). They can be conjugated and doped with other metal ions such as platinum to enhance their efficiency in cancer treatment (Bao et al., 2018).

1.5.2 Drug delivery

Drug and gene delivery applications of various inorganic and organic nanomaterials have gained a lot of interest in recent years. CNP applications as drug delivery vehicles are very common as they allow a slow, consistent, and prolonged release of the drug (Kushwaha et al., 2013; Tasis et al., 2006). The unique π-conjugated structure of six-atom rings increases the loading capability of a variety of drugs to be targeted and fluorescent probes. Furthermore, CNP surface modification allows their conjugation with targeting ligands to achieve targeted drug delivery. CNP-based drug delivery systems have several advantages over other nanodrug delivery systems, including easy surface modification, low cytotoxicity, high water solubility, excellent cell permeability, high photostability, and ease of tracking in vivo and in vitro.

1.5.3 Bioimaging

CNPs have been investigated in many imaging applications for the last two to three decades due to their easy binding capacity with various agents. CNPs can be used in fluorescence imaging (FL), two-photon FL, Raman imaging, magnetic resonance imaging (MRI), tomography (CT), photoacoustic imaging (PAI), computed positron emission tomography/single photon emission computed tomography (PET/SPECT), and multimodal imaging. Recently, a new form of carbon-based nanomaterial, the carbon quantum dot, has attracted tremendous interest in its bioimaging applications (Yan et al., 2016).

1.5.4 Energy storage and high-capacity lithium sulfur batteries

Hydrogen energy-storage devices such as batteries, capacitors, and fuel cells present a practical alternative to energy storage. They have gained more importance these days because of the depletion of natural energy reservoirs. Carbon-based materials are considered desirable materials for electrodes due to their high surface area, good electrical and thermal conductivity, and excellent energy-storage capacity.

Lithium sulfur batteries are ecofriendly and low-cost, next-generation energy storage devices (Chen et al., 2018). Instead of using conventional materials for electrodes in lithium sulfur batteries, the use of a sulfur-nanocarbon electrode has attracted the attention of the scientific fraternity. That's because that electrode enhances cycling stability and the lithium storage capacity of sulfur atoms while maintaining high electron mobility through the CNP matrix (Qi et al., 2019).

2 Sources of CNPs in the environment

CNPs are introduced into the environment during their production, use, and disposal. As CNP production and use is rising rapidly, studies need to be focused on the environmental occurrence of CNPs and the risks posed by them on the biotic community and all the abiotic factors of the ecosystem. As shown in Fig. 1, environmental CNPs are released from two different sources: natural and artificial, that is, engineered or man-made. Naturally, environmental CNPs are produced by wildfire charcoal, fuel combustion, fossil coal, biochar, and soot. They are elaborated upon below:

2.1 Natural sources of CNPs (nCNPs)

2.1.1 Wildfire charcoal

Charcoal is a carbon and ash residue produced after removing water during the slow pyrolysis of animal or vegetative substances. The formation of charcoal (charcoal burning) occurs in the absence of oxygen (Forbes et al., 2006). The properties (both physical and chemical) of wildfire charcoal are strongly dependent on the temperature of the fire heating duration and the type and size of the feedstock. Wildfire charcoal is less stable and prone to physical disintegration. About 15% of the total natural CNPs are derived from wildfire charcoal (Jaffé et al., 2013; Knicker et al., 2008; Randerson et al., 2012; Reisser et al., 2016; Uhelski and Miesel, 2017).

2.1.2 Fuel combustion

Fuel combustion, most commonly fossil fuel combustion, occurs in the presence of oxygen to generate heat, which is used in the operation of many industrial types of equipment such as boilers, furnaces, engines, etc. Besides heat, fossil fuel combustion generates CO_2 and H_2O as major byproducts. However, some of the minor byproducts such as carbon monoxide, sulfur dioxide, nitrogen oxide, lead, and particulate matter are also generated during combustion.

2.1.3 Fossil coal

Fossil coal combustion remains an important factor for energy generation in many countries. The amount of CNPs formed from fossil coal combustion is negligible; however, CNPs are released into the environment during the mining and transport of fossil coal (IEA, 2012, IEA, 2017; Sigmund et al., 2018).

2.1.4 Biochar

Biochar is obtained as result of the pyrolysis of biomass. Its occurrence is extensive in Terra Preta. The mechanism of biochar formation is poorly known. Biochar is very heterogeneous in nature and is dependent on the pyrolysis duration, pyrolysis temperature, steam, and feedstock. Thus, the amount of biochar generated is more (35%) in the case of slow pyrolysis compared to fast pyrolysis (12%) and gasification (10%). Biochar is widely used in agriculture as a fertilizer in combination with inorganic NPK fertilizer and compost. Biochar application rates for soil are yearly, half-yearly, or one-off. Polyaromatic hydrocarbons (PAHs), dioxins, heavy metals, and nanoparticles are the major biochar contaminants, among which CNPs present in biochar pose a great environmental risk to the soil (Alvarez-Puebla et al., 2005; Pulleman et al., 2000; Zackrisson et al., 1996).

2.1.5 Soot

Soot is produced naturally from smoldering candles, natural gas combustion, food caramels, etc. (Liu et al., 2007; Palashuddin et al., 2012; Tian et al., 2009). Carbon soot and soot-based CNP synthesis has attracted many researchers because of the simple preparation of luminescent CNPs (Gardner et al., 1996; Jang et al., 2004; Kim et al., 2008a,b; Liu et al., 2007; Okpalugo et al., 2005; Ray et al., 2009; Tian et al., 2009). CNPs derived from castor oil soot were modified and activated for their use in electrocatalytic applications (Bowling et al., 1989).

2.2 Engineered CNPs (eCNPs)

eCNPs that have attracted most researchers include carbon nanotubes (CNTs), fullerenes, graphene, graphene oxide, etc. (Bhatt and Tripathi, 2011; Nowack and Bucheli, 2007; Petersen et al., 2011). When compared to nCNPs (>99%; Tg per year range), a much smaller amount of eCNPs (< 1% Gg per year range) is released into the environment. For various applications, eCNPs can be activated or can be incorporated into other matrices. As eCNPs are used under very controlled conditions, a small proportion of eCNPs is expected to be released into the natural environment. Global eCNP production was reported as approximately 20 Gg in 2016, with annual production predicted to increase 20% per year until 2023 (De Volder et al., 2013; Sigmund et al., 2018). This section elaborates various types of eCNPs.

2.2.1 Carbon nanotubes (CNTs)

Carbon nanotubes are one-dimensional cylindrical nanostructures consisting of sp2-bonded carbon atoms, the properties of which are influenced by their size effects. CNT research has flourished since Iijima's original report (Iijima, 1991). CNTs have been recognized as fascinating materials with nanometer dimensions promising exciting new areas of carbon chemistry and physics. Early assessments of CNT manufacturing and handling practices revealed that airborne CNT concentrations could be as high as 53 micrograms/m^3 and the potential dermal exposures could be anywhere from 0.5–14.5 micrograms/cm^2 (Maynard et al., 2004). CNTs are tube-shaped carbon materials and are mainly of two types: single-walled carbon nanotubes (SWCNTs) and multiwalled carbon nanotubes (MWCNTs). SWCNTs are one-dimensional cylindrical materials consisting of graphene sheets rolled up to form hollow tubes (White and Mintmire, 2005). They have a diameter of <1 nm. Very few studies in the literature about the risks associated with SWCNT exposure are available. On the other hand, MWCNTs consist of a series of single-walled tubes nested within one another in many rolled layers (Holloway et al., 2008; White and Mintmire, 2005). As MWCNTs are among the most important CNPs, their ever-increasing production and use needs special attention focused on the environmental hazards and associated risks (Maynard et al., 2004). One of the studies using murine models has demonstrated that the environmental exposure of MWCNTs to CD1 mice led to potential damage to several organs by induction of macrophage recruitment, activation, and amyloid deposition (Albini et al., 2015).

2.2.2 Fullerene

Fullerenes are carbon allotropes in the form of hollow spheres or ellipsoids (Kroto et al., 1985). Fullerenes are generated by carbon combustion or high energy state carbon, which can be achieved by lighting, the massive collision of a meteoroid, or a massive fire (Becker et al., 2001). Since the start of the 1990s, fullerene research has blossomed in many different directions. However, very few reports mentioning the toxicity and environmental fate of fullerenes are available.

2.2.3 Graphene

Graphene is the thinnest two-dimensional material comprised of a one-atom-thick planar sheet of sp2-bonded carbon atoms arranged like a honeycomb lattice. It is an indefinitely large aromatic molecule in which a single layer of carbon is arranged in a hexagonal lattice. It has zero band gap and is a basic structural material for many other CNPs such as graphite, CNTS, fullerenes, and charcoal. Graphene exists in many forms such as graphene nanoribbons, graphene sheets, graphene nanoplates, bilayer graphene, etc. (Ho et al., 2015).

2.2.4 Graphene oxide

Graphene oxide is a flake of a monolayer or a few layers. In order to get the honeycomb hexagonal lattice (graphene sheet), graphene oxide needs to be reduced. It is produced by exfoliation of graphite oxide in water by sonication. It is easily dispersible in water and other organic solvents. Graphene oxide is a good electrical insulator.

2.2.5 3D hybrid assembly

Three-dimensional (3D) hybrid assemblies include 3D CNT networks, graphene CNT hybrids, and Schwarzites. Three-dimensional CNT hybrids can be fabricated with other metals and metal oxides such as CuO (Du et al., 2011; Zheng et al., 2008) and graphene (Yi et al., 2015; You et al., 2013). Schwarzites are the negative curvature carbon allotropes in which the ratio of different structural rings (pentagons, hexagons, etc.) are of fundamental importance to determine the material response to tensile and/or compression deformations (Sajadi et al., 2017).

3 Environmental transformation of CNPs: Release and fate

As discussed in earlier sections, nCNPs and eCNPs are released in the environment from different sources in varied proportions. Furthermore, these particles, even with few molecular similarities such as high electron acceptors, donor capacity, and aromaticity, exhibit differences in certain characteristics such as their structure, surface characteristics, and porosity. These molecular similarities and structural differences in CNPs may decide their fate and transformation after their release into the environment. Currently, very little research is focused on the environmental release and fate of CNPs, and thus the available published data on the release and transformation of nCNPs is scarce. Similarly, very few methods such as microscopy, laser fluorescence, and quantification measuring concentrations of characteristic metal catalyst impurities have been developed and put into use for the detection and quantification of CNPs. However, none of these methods can detect eCNPS and nCNPs separately (Petersen et al., 2011; Sigmund et al., 2018). Thus, there is an urgent need to develop methods for identification and quantification of eCNPS and nCNPs in the environment. Due to variations in the physical and chemical characteristics of bulk C materials and CNPs, CNPs behave differently in the environment compared to their bulk forms. Therefore, this is true in the behavior of eCNPs and nCNPs due to differences in their sources. After their release into the environment, CNPs can undergo several changes and transform into different forms, altering their physical and chemical properties by photochemical reactions, various environmental factors, chemical changes, and biological interactions. These transformations can be categorized as physical, chemical, and biological (Fig. 3). These transformations are explained in detail below.

3.1 Chemical transformation

Although there are very few reports available on the chemical transformation of eCNPs and nCNPs, the available literature suggests that chemical transformation of CNPs leads to changes in the surface characteristics such as surface charge. Because of this, the surface groups get altered, leading to the change in the surface functionality

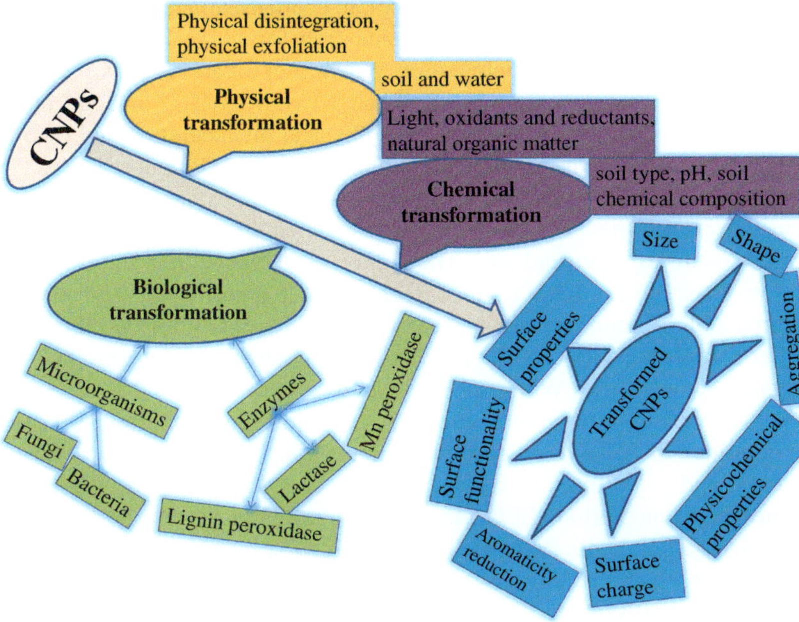

Fig. 3 Environmental transformation pathways of CNPs.

(Lowry et al., 2012; Sigmund et al., 2018). The main two reactions mediating the chemical transformation of CNPs are redox reactions and photolysis. Chemical transformation of CNPs in the environment is found to be affected by environmental factors such as the light presence of oxidants and reductants and natural organic matter (Chowdhury et al., 2015; Mitrano et al., 2015; Pulleman et al., 2000; Zhao et al., 2014). Moreover, the chemical transformation of carbonaceous matter in the environment has been found to be influenced by the presence of different inorganic and naturally occurring organic matter. External factors such as soil type, pH, and the chemical composition of soil are said to have an effect on the chemical transformation of CNPs, although the mechanism of which remains poorly understood. Current knowledge in the chemical transformation of eCNPs and nCNPs is scarce; therefore future research should be focused on the chemical transformation of CNPs in environments such as soil, aquatic systems, air, and sediment. The studies also need to be attentive to the environmental fate and transport of chemically transformed CNPs as well as their effects on the environment. Further, studies on the chemical transformation of CNPs should also focus on each type of eCNPs and nCNPs derived from different sources. Detailed studies on the mechanism of chemical transformation of CNPs need to be conducted.

3.2 Physical transformation

Physical transformation is naturally occurring and refers to the changes in the physical properties of CNPs such as particle size and porosity as well as to interactions with other particles as a result of abrasion, coagulation, condensation, changes in pH, ionic strength, and intrinsic composition. The physical transformation and disintegration of CNPs is mediated by physical disintegration, physical exfoliation, and CNP exposure to various environmental factors such as soil and water. nCNPs originated from fossil coal, biochar, or wildfires are transformed by physical disintegration and exfoliation is mediated chemically by the alkaline conditions of the soil (Braadbaart et al., 2009; Sigmund et al., 2018). Moreover, higher disintegration rates are observed in sandy soils than other soil types. Furthermore, CNP disintegration and breakdown is followed by exposure to water and soil. The degree of disintegration and formation of nCNPs is inversely proportional to the carbonization and aromaticity. Further, CNPs produced from high-lignin feedstocks (e.g., wood) were more prone to disintegration into smaller particles than those from high cellulose feedstocks (e.g., grass) (Liu et al. 2018; Sigmund et al., 2018). When released into the environment, CNPs interact with themselves and/or atmospheric moisture, sulfates, nitrates, sea salt, mineral dust, fly ash, and colloidal, dissolved, and particulate materials. This can lead to coagulation, condensation, and the filling of exposed cavities, thus stabilizing them, which may prevent further disintegration (Li et al., 2016; Sigmund et al., 2018; Spokas et al., 2014; Tiwari and Marr, 2010; Wang et al., 2017). nCNPs coagulation into initial spherical and later subsequent chain-like fractal clusters occurs in the environment when the number and polydispersity of CNPs is greater (Kerminen et al., 2004; Ni et al., 2014; Tiwari and Marr, 2010; Wang et al., 2017). Condensation also occurs when there is a sudden release of CNPs into the environment because of volcanic eruptions, nearby roadways, and industrial exhaust. nCNPs originating from soot are a result of the nucleation of compounds such as water, acetylene, and PAHs from the gas phase. Coagulation and condensation may lead to the growth of nCNPs to more than the nanoscale because of CNP aggregation and coating (Kulmala et al., 2004; Tiwari and Marr, 2010). nCNPs growth during and after their formation is counteracted by oxidation in the presence of O_2 and reactive oxygen species. The particle size, surface charge, aggregation, and shape of CNPs during their formation and transport in the environment is greatly affected by interactions with other constituents of the gas phase, changes in temperature, physical stress, and humidity (D'Anna, 2009; Lighty et al., 2000; Liu et al., 2018; Wang et al., 2017; Zhang et al., 2008a,b, 2014).

3.3 Biological transformation

Changes in the physicochemical properties of CNPs after their interaction with microorganisms, individual cells, and enzymes are referred to as biological transformations (Lowry et al., 2012). Biological transformation can occur both intracellularly and extracellularly in microorganisms. There are many reports by several researchers on the biotransformation of many highly aromatic compounds such as lignin by bacteria, fungi, actinomycetes, and enzymes obtained from microorganisms (Ruiz-Dueñas and Martínez, 2009) with the help of extra as well as intracellular ligninolytic/biotransformtion enzymes such as lignin peroxidase, laccase, Mn peroxidase, etc. (Czimczik and Masiello, 2007; Han et al., 2016). These enzymes are expected to play a major role in CNP biological transformation, even though not recorded so far for the same. All these enzymes belong to the class oxidoreductases and are thus expected to bring about CNP oxidation and reduction. They are also expected to carry out changes in CNP surface properties. So far, research in this field is limited to carbonaceous materials (Hilscher et al., 2009; Saquing et al., 2016) who demonstrated the reduction in aromaticity for some carbonaceous materials (Hilscher et al., 2009) and quinone groups in the carbonized phase of biochar can act both as electron donors and as acceptors for microbiota (Saquing et al., 2016). The enzymatic transformations of CNPs are very poorly investigated under laboratory conditions and thus they are not representative of enzymatic transformations in natural environments. A very limited study was conducted using horseradish peroxidase and Mn peroxidase to understand the key enzymatic transformation processes (Allen et al., 2009; Flores-Cervantes et al., 2014; Kotchey et al., 2011; Zhang et al., 2014). However, there is need of a detailed investigation on the biological transformation of nCNPs as well as eCNPs, as the knowledge about this is very limited. This is despite the fact that the role of fungal hyphae in the physical breakdown of large fossil coal and biochar particles and of fungal extracellular enzymes in their chemical decomposition is widely recognized (Spokas, 2010). The research in CNP biological transformation is very challenging and many factors need to be considered, including the physicochemical properties of CNPs, their aggregation and interactions, the microbial diversity and composition, the microbial activity, the effect of various physicochemical parameters and seasonal variations on microbial activity, the nutrient availability, and the nutrient accessibility. The studies in CNP biotransformation should also focus on the microbe CNP interactions and their cytotoxicity as well as finding newer and more efficient microorganisms or a microbial consortium for CNP biotransformation, standardization of optimum conditions for CNP biotransformation, characterization of the enzyme systems responsible for CNP biotransformation, and purification and characterization of the enzymes responsible for CNP biotransformation. Further investigation is also required for the transfer and possible cellular uptake of CNPs by microbial cells as well as the factors that drive the associated gene expression of CNP-degrading enzymes. A better understanding with the help of such systematically conducted studies will help to explore the facts about the factors affecting biological transformation processes as well as the fate and possible risks associated with CNPs.

4 Environmental perspective and eco-nanotoxicology: Hazards and risks

Several types of eNPs such as carbon nanoparticles (CNPs) have wide applications in the industrial, medical, agricultural, and environmental sectors with a consumer market that is increasing day after day. This may pose a greater risk to the environment and human health if handled haphazardly or in an irrational manner. Econanotoxicology provides insight into NP production pathways, their lifecycle, and potential impacts, hazards, and risks when released into the environment (Fig. 4); therefore proving an emerging discipline of nanotechnology. This section will highlight the ecotoxicological aspects of CNPs when exposed to different environmental resources and their subsequent ecological impacts.

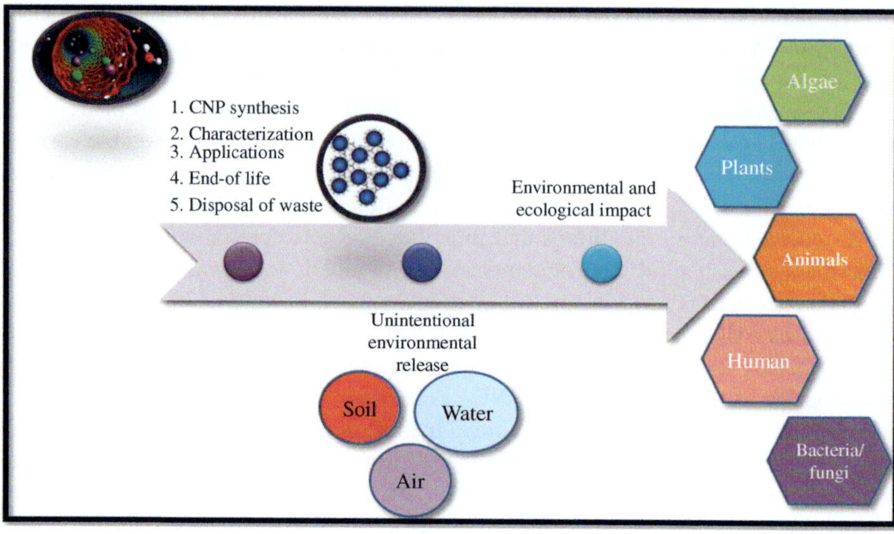

Fig. 4 CNP production life cycle, transport, and fate.

4.1 Impacts on aqua/marine ecosystem

Carbon nanoparticles can enter into the environment and ecosystem at any stage, including production, handling, application, and disposal. This can have a negative impact on the different biotic strata of the ecosystem by changing CNP behavior (QDs, fullerenes), which can influence their fate and impact the environment (Derfus et al., 2004; Mottier et al., 2017). CNTs are regarded as one of the least-biodegradable cNPs that are totally insoluble in pristine water form and lipophilic by nature, which tends to accumulate along the food chain of aquatic organisms (Wu et al., 2006). A study by Oberdörster et al. (2006) reported the presence of SWCNTs in the digestive tract of the suspension feeding worm, *Caenorbabditis elegans*. These SWCNTs could end up in the food chain if consumed by benthivores. SWCNTs are generally bioavailable to aquatic organisms. This was confirmed by the presence of unsolubalized SWCNTs in the fecal material collected from the digestive tract of the exposed fish; the fish mistook the unsolubalized SWCNTs floating on the water surface for food and ingested them.

The first nanotoxicity-based study on fish was conducted using fullerenes and CNTs on several fish species, namely the largemouth bass, the fathead minnow, the Japanese medaka, the rainbow trout, and the zebrafish. The C_{60} and SWCNTs (0–1 ppm dose) resulted in lipid peroxidation localized in the brain, increased aggressive behavior, gill irritation, and some changes in brain pathology (Federici et al., 2007; Oberdörster et al., 2005; Smith et al., 2007). In a similar case, the *Nitzchia Palea*, a benthic diatom observed in an aquatic environment, has exhibited drastic effects on trophic-level microorganisms when exposed to traces of MWCNTs and DWCNT particles (Verneuil et al., 2015). The results indicated an increased growth inhibition rate when CNTs were used as dispersants. SWCNTs and graphene oxides (GO) also exhibited toxicological effects on *Chlorella vulgarius*, green algae with a generation of reactive oxygen species from 0.01 mg/L, and oxidative damage to *Euglena gracilis*, a protozoan, when exposed to GO. The *Xenopus laevis*-based amphibian models, which are well characterized, are efficient in assessing and addressing the genotoxicological and ecotoxicological impacts of CNPs. This model showed nonrepairable DNA damage and oxidative stress when benthic diatoms were exposed to CNPs over 12 days, which indicated the modifications and disturbances undergone by organisms. Similarly, bacteria respond to nanoparticle-induced toxicity such as ROS production and dissolution of toxic ions, which leads to drastic variations in species diversity, their composition, the microbial community growth, and performance. Weisner et al. (2006) considered the effect of the aggregate size of fullerene nanoparticles (C_{60} and SW/MWCNTs) on ROS production and toxicity toward *Vibrio fischeri*, a bioluminescent marine bacterium. Changes in bacterial luminescence indicated bacterial toxicity, which is directly proportional to the respiration rate for *V. fischeri* and related to nanoparticle aggregate size. This resulted in direct oxidative damage of membrane proteins and lipids, indicating the toxic effects of fullerene nanoparticles on *V. fischeri*. In terms of algal nanotoxicity, CNP toxicity to algae has been found to be dependent on the CNP specific surface area and size (Navarro et al., 2008). Adsorption of larger CNP aggregates to cell walls has been observed to alter the cellular acquisition of essential nutrients, either through clogging of cell walls or nutrient adsorption (Fernandes et al., 2007). Similarly, aggregates of carbon black bound to sperm cells reduced the fertilization success of *Fucus serratus*, a marine seaweed. Generally, it is observed that CNP accumulation on the surface of photosynthetic organisms such as algae inhibits photosynthetic activity due to reduced light availability.

4.2 Impacts on terrestrial ecosystem

Plants, animals, bacteria, and fungi form an integral component of the terrestrial ecosystem at the macro and micro levels, both of which perform integral environmental functions. The dose–response relation (concentration, frequency at which eNPs are released into the environment) of eNPs provides insight into the toxicity of eNPs toward various components of the terrestrial ecosystem. To date, virtually no study exists on the effects of eNPs on soil (Navarro et al., 2008). Bacteria are ubiquitous members of ecosystems with particular importance in nutrient cycling. However, little evidence from ecotoxicity studies conducted on bacteria, fungi, and plants has shown negative effects in terms of threatening free-living, nitrogen-fixing bacteria and symbiotic relations involving mycorrhiza, rhizhobia in legumes, lichens, etc. This would result in reduced plant nutrient availability and oxidative stress while hindering significant ecological processes such as biomass production, organic matter breakdown, groundwater purification, and soil properties (Hilderbrandt et al., 2007). Dunphy Guzman et al. (2006) and Hardman (2006) suggested that some CNPs such as QDs may be taken up by bacteria due to their long residence time in the cell; they can then get transferred through the food web. This can lead to significant CNP bioaccumulation because of their strong partition into membranes.

Plants are particularly relevant in econanotoxicology as they are exposed to CNPs in atmospheric and terrestrial environments. Airborne CNPs get attached to leaves and other aerial plant parts, whereas roots will interact with waterborne, soil material-associated CNPs. Köstner (2001) reported that plant communities with higher leaf area indexes have a higher interceptional potential for airborne CNPs, thus increasing their entrance into trophic webs. An altered chemical composition and surface reactivity are indicators of the toxic effects of CNPs on organisms. The photo-induced electron transfer capacity of some cNPs such as fullerenes (Imahori et al., 2003) suggests that the photosynthetic or respiratory process will be impacted as a consequence of CNPs penetrating the cell wall and membrane, then reaching the cytosol. Another phytonanotoxicity study was carried out by Serag et al. (2011) to study the interaction between *Catharanthus roseus* cells and fluorescent MWCNTs (<500 nm in length) to characterize cellular uptake and localization. In another instance, the effects of aqueous suspensions of graphene oxide (500–2000 mg/L) were recorded on seedlings of cabbage, tomato, red spinach, and lettuce by monitoring their physiological parameters. Except in lettuce, plant growth was found to be impeded due to nanoparticle exposure at the earliest stages of plant development. Moreover, earthworms are considered as model detrivores in soil ecological studies (Maurer-Jones et al., 2013). One of the handful of the nanotoxicological studies conducted using ^{14}C-labeled CNTs reported DNA damage in earthworms as a result of higher bioaccumulation levels of CNTs in earthworms (Petersen et al., 2016).

4.3 Impacts on human health and well-being

Carbon nanotubes are becoming widely applied in the biomedical, pharmaceutical, electronic, and industrial sectors as well as for drug and gene delivery. For example, CNTs have been proposed as a possible new orthopedic/dental implant surface material because of their unique mechanical, electrical, and cytocompatibility properties (Price et al., 2004). Considering the increasing scope of eCNPs and their prevalent route of exposure, nanomaterials may pose a substantial risk to public health. Producers via occupational exposure and consumers of CNPs both are at equal risk, where chronic exposure to particulate matter of a nanosize is leading to several diseases associated with pulmonary toxicology. For example, SWCNT manufacturers could very easily get exposed through dermal or oral ingestion or inhalation where very low respirable CNPs are being prepared and processed in a liquid medium (Colvin, 2003). Zhang et al. (2008a, b) reported skin penetration as a major route of QD exposure in biological systems. Moreover, MWCNTs induce inflammatory and apoptosis responses in human T-cells (Bottini et al., 2006), and are responsible for human skin fibroblasts (Ding et al., 2005). Silva et al. (2005) reported ultrafine carbon particle penetration in the lungs and their ability to cross the blood–brain barrier to impact the central nervous system. However, the distinct properties of individual forms of CNPs can have variable effects on organisms based on their length, functionalization, coating and agglomeration properties, exposure dose, life cycle, etc. Once CNTs are taken up by humans or other species, they may cause oxidative stress, inflammation, cell damage, cell dysfunction, and even granulomas, fibrosis, and wall thickening over the long term (Helland et al., 2007). The negative effects of various forms of CNPs on human health are listed in Table 1.

Table 1 Toxicity effects of a few types of CNTs.

Mode of toxicity	Type of CNTs	Effects	Reference
Pulmonary toxicity	Fullerene soot	Change in pulmonary function	Grubek-Jaworska et al. (2006)
	MWCNTs	Multifocal granulomas; mild fibrosis in alveolar septa, inflammatory reactions of terminal/respiratory bronchioles	Muller et al. (2005)
	SWCNT soot	Transient inflammation, cell injury effects, multifocal granulomas	Grubek-Jaworska et al. (2006)
	Ultrafine carbon black	Granulomas in centrilobular locations	Muller et al. (2005)
	Ground and unground MWCNTs (5.9 to 0.7 μm)	Higher degree of pulmonary inflammation, fibrotic reaction, increased tumor necrosis	Muller et al. (2005)
	HipCO-produced SWCNTs	Increased number of alveolar type II cells	Shedova et al. (2005)
Dermal Cytotoxicity	MWCNTs	Induced cell cycle arrest and increased apoptosis and necrosis	Helland et al. (2007)
	HipCO-produced MWCNTs	Increased oxidative stress, reduced vitamin E level, reduced cell viability	Shedova et al. (2005)
Hispo-pathologic analysis	MWCNTs with varying lengths (825 nm)	Increased inflammation	Sato et al. (2005)
Pulmonary cytotoxicity	SWCNTs suspended in plumonic F108 (a nonionic surfactant)	Oxidative stress, disturbed intracellular equilibrium	Ding et al. (2005)

5 Way forward: Challenges and recommendations

The environmental use of various types of eNPs is increasing day by day. Their unintentional release and exposure to algae, plants, animals, fungi, bacteria, and human beings is becoming a reality. There is a greater need to assess the fate, transport, and effects of eNPs in aquatic systems (lakes, rivers, oceans, etc.) and soil (Kühnel et al., 2018) by developing a more complex mesocosm system. Lack of information on these key aspects is preventing better understanding and assessment of the toxicity and nanoecotoxicity of CNPs to key ecosystem organisms. Still today, many challenging questions remain answered (Ray et al., 2009), including:

(a) The relevant CNP concentrations in different aquatic, terrestrial, and aerial environments.
(b) The physical and chemical characterization of CNPs and exposure routes.
(c) The mechanisms allowing CNPs to pass through cellular membranes and cell walls.
(d) The specific properties that are related to CNP toxic effects.
(e) The mechanism underlying CNP trophic transfer.
(f) In situ CNP characterization in complex environments.

Methodologies based on "classical contaminants" had been the conventional way to assess the ecotoxicological effects of CNPs. These test methods played a vital role in assessing the toxicity thresholds of newly evolving eCNPs such as graphene, fullerenes, etc., in the environment. In the future, dedicated attention should be given to "nonclassical dose metrices" to unfold some toxicological processes and overcome the limitations of mass-based concentration methods. Mathematical models predicated no effect concentrations (PNEC) could come to a rescue to gather CNPs toxicology data for single species test within whole biota.

The significant aspects to consider to fully assess the environmental impacts of CNPs are transformations and aging (Romeo-Franco et al., 2017). Normalized tests, a classical ecotoxicological study based on a single species, could provide better understanding of toxicity mechanisms. However, critical parameters such as biotransformation, bioaccumulation, biomagnification, and the effects of abiotic factors must be taken into consideration. The potential impact of CNPs could be higher than expected, and hence it's high time that the classical toxicology approach needs a shift toward more appropriate ecoevaluation of the CNP impact. This can be achieved with the use of integrated biological approaches targeting biomarkers and biodegradation to monitor CNP-induced impacts, even at lower levels of organizations. The major five recommendations to fill current gaps in knowledge and address existing challenges are listed below.

Key recommendations:

- Functionalization in addition to CNT mobility influences toxicity; hence, a key parameter to consider.
- Integrating nanotoxicology with life cycle studies should be considered a prerequisite for safe nanotechnology applications.
- The pluridisciplinary approach between ecotoxicologists, toxicologists, biologists, chemists, biophysicists, and analytical researchers should be used to address key challenges.
- Designing frameworks that allow the extrapolation of in vitro results to natural systems based on previous knowledge.

- The standardization of CNP toxicity testing methods as a more feasible approach.
- Developing an integrative framework for assessing the potential risks of eCNPs.

6 Concluding remarks

In summary, this chapter critically discusses different types of eNPs, particularly CNPs, as well as their properties, structure, applications, fate, behavior, and toxicity when released into the environment. While several efforts have been put forth by the scientific community worldwide to address the ecological complications of CNPs, gaps in the current knowledge still exist. It is quintessential to (i) address the complexities of eCNPs and their transformations; (ii) integrate qualitative and quantitative data; (iii) facilitate the use of modeling tools to fill the data gaps and provide both qualitative and quantitative information; and (iv) minimize the reliance on expert judgment. To sustainably adopt nanotechnology and have zero impact on the environment, it is crucial for the expertise of a wide variety of disciplines to work together and contribute to the better use of nano-bio interactions to propel this emerging field in the right direction.

References

Albini, A., Pagani, A., Pulze, L., Bruno, A., Principi, E., Congiu, T., Gini, E., Grimaldi, A., Bassani, B., De Flora, S., de Eguileor, M., Noonan, D.M., 2015. Environmental impact of multi-wall carbon nanotubes in a novel model of exposure: systemic distribution, macrophage accumulation, and amyloid deposition. Int. J. Nanomedicine 10, 6133–6145.

Allen, B.L., Kotchey, G.P., Chen, Y., Yanamala, N.V.K., Klein-Seetharaman, J., Kagan, V.E., Star, A., 2009. Mechanistic investigations of horseradish peroxidase-catalyzed degradation of single-walled carbon nanotubes. J. Am. Chem. Soc. 131, 17194–17205.

Alvarez-Puebla, R.A., Goulet, P.J.G., Garrido, J.J., 2005. Characterization of the porous structure of different humic fractions. Colloids Surf. A Physicochem. Eng. Asp. 256, 129–135.

Bao, Y.W., Hua, X.W., Chen, X., Wu, F., 2018. Platinum-doped carbon nanoparticles inhibit cancer cell migration under mild laser irradiation: multi-organelle-targeted photothermal therapy. Biomaterials 183, 30–42.

Becker, L., Poreda, R.J., Hunt, A.G., Bunch, T.E., Rampino, M., 2001. Impact event at the Permian-Triassic boundary: evidence from extraterrestrial noble gases in fullerenes. Science 291, 1530–1533.

Bhatt, I., Tripathi, B.N., 2011. Interaction of engineered nanoparticles with various components of the environment and possible strategies for their risk assessment. Chemosphere 82, 308–317.

Bottini, M., Bucker, S., Nika, K., Bottini, N., Bellucci, S., Margini, A., et al., 2006. Multi-walled carbon nanotubes induce T lymphocyte apoptosis. Toxicol. Lett. 160 (2), 121–126.

Bowling, R., Packard, R.T., McCreery, R.L., 1989. Mechanism of electrochemical activation of carbon electrodes: role of graphite lattice defects. Langmuir 5, 683–688.

Braadbaart, F., Poole, I., van Brussel, A.A., 2009. Preservation potential of carcoal in alkaline environments: an experimental approach and implications for the archaeological record. J. Archaeol. Sci. 36, 1672–1679.

Chen, K., Sun, Z., Fang, R., Shi, Y., Cheng, H., Li, F., 2018. Metal–Organic Frameworks (MOFs)-Derived Nitrogen-Doped Porous Carbon Anchored on Graphene with Multifunctional Effects for Lithium–Sulfur Batteries. Adv. Funct. Mater. 28, 1707592. https://doi.org/10.1002/adfm.201707592.

Chowdhury, I., Hou, W.C., Goodwin, D., Henderson, M., Zepp, R.G., Bouchard, D., 2015. Sunlight affects aggregation and deposition of graphene oxide in the aquatic environment. Water Res. 78, 37–46.

Coloma, F., Narciso-Romero, J., Sepulveda-Escribano, A., Rodriguez-Reinoso, F., 1998. Gas phase hydrogenation of crotonaldehyde over platinum supported on oxidized carbon black. Carbon 36, 1011–1019.

Colvin, V.L., 2003. The potential environmental impact of engineered nanomaterials. Nat. Biotechnol. 21, 1166–1170.

Czimczik, C.I., Masiello, C.A., 2007. Controls on black carbon storage in soils. Global Biogeochem. Cycles 21, 1–8.

D'Anna, A., 2009. Combustion-Formed Nanoparticles. Proc. Combust. Inst. 32, 593–613.

Derfus, A.M., Chan, W.C.W., Bhatia, S.N., 2004. Probing the cytotoxicity of semiconductor quantum dots. Nano Lett. 4 (1), 11–18.

De Volder, M.F.L., Tawfick, S.H., Baughman, R.H., Hart, A.J., 2013. Carbon nanotubes: Present and future commercial applications. Science 339, 535–539.

Ding, L.H., Stilwell, J., Zhang, T.T., Elboudwarej, O., Jiang, H.J., Selegue, J.P., et al., 2005. Molecular characterization of the cytotoxic mechanism of multiwall carbon nanotubes and nano-onions on human skin fibroblast. Nano Lett. 5 (12), 2448–2464.

Du, F., Dingshan, Y., Liming, D., Ganguli, S., Varshney, V., Roy, A.K., 2011. Preparation of tunable 3D pillared carbon nanotube–graphene networks for high-performance capacitance. Chem. Mater. 23, 4810–4816.

Dunphy Guzman, K.A., Taylor, M.R., Banfield, J.F., 2006. Environmental risks of nanotechnology: National nanotechnology intiative funding, 2000-2004. Environ. Sci. Technol. 40, 1401–1407.

El-Sayed, Y., Bandosz, T.J., 2005. Role of surface oxygen groups in incorporation of nitrogen to activated carbons via ethylmethylamine adsorption. Langmuir 21, 1282–1289.

Federici, G., Shaw, B.J., Handy, R.D., 2007. Toxicity of titanium dioxide nanopaticles to Rainbow Trout. Aquat Toxicol 84, 415–430.

Fernandes, T., Nielsen, H., Burridge, T., Stone, V., 2007. Toxicity of nanoparticles to embryos of the marine macroalgae *Fucus serratus*. In: 2nd International conference on the Environmental Effects of Nanoparticles and Nanomaterials, London, England.

Flores-Cervantes, D.X., Maes, H.M., Schäffer, A., Hollender, J., Kohler, H.P.E., 2014. Slow Biotransformation of Carbon Nanotubes by Horseradish Peroxidase. Environ. Sci. Technol. 48, 4826–4834.

Forbes, M.S., Raison, R.J., Skjemstad, J.O., 2006. Formation, transformation and transport of black carbon (charcoal) in terrestrial and aquatic ecosystems. Sci. Total Environ. 370, 190–206.

Gardner, S.D., He, G., Pittman Jr., C.U., 1996. A spectroscopic examination of carbon fiber cross sections using XPS and ISS. Carbon 34, 1221–1228.

Grubek-Jaworska, H., Nejman, P., Czuminska, K., Przybylowski, T., Huczko, A., Lange, H., et al., 2006. Preliminary results on the pathogenic effects of intratracheal exposure to one-dimensional nanocarbons. Carbon 44 (6), 1057–1063.

Han, L., Sun, K., Jin, J., Xing, B., 2016. Some concepts of soil organic carbon characteristics and mineral interaction from a review of literature. Soil Biol. Biochem. 94, 107–121.

Hardman, R., 2006. A toxicologic review of quantum dots: toxicity depends on physicochemical and environmental factors. Environ. Health Perspect. 114, 165–172.

Helland, A., Wick, P., Koehler, A., Schmid, K., Som, C., 2007. Reviewing the environmental and human health knowledge base of carbon nanotubes. Environ. Health Perspect. 115 (8), 1125–1131.

Hilderbrandt, U., Regvar, M., Bothe, H., 2007. Arbuscular mycorrhiza and heavy metal tolerance. Phytochemistry 68, 139–146.

Hilscher, A., Heister, K., Siewert, C., Knicker, H., 2009. Mineralisation and structural changes during the initial phase of microbial degradation of pyrogenic plant residues in soil. Org. Geochem. 40, 332–342.

Ho, K.I., Mohamed, B., Su, C.Y., Moreddu, R., Marianathan, E., Sebastian, R., Montes, L., Lai, C.S., 2015. A self-aligned high-mobility graphene transistor: decoupling the channel with fluorographene to reduce scattering. Adv. Mater. 27, 6519–6525.

Holloway, A.F., Wildgoose, G.G., Compton, R.G., Shao, L., Green, M.L.H., 2008. The influence of edgeplane defects and oxygen-containing surface groups on the voltammetry of acid-treated, annealed and "super-annealed" multiwalled carbon nanotubes. J. Solid State Electrochem. 12, 1337. https://doi.org/10.1007/s10008-008-0542-2.

IEA International Energy Agency, 2017. World Energy Outlook 2017.

IEA Clean Coal Centre, 2012. Losses in the Coal Supply Chain.

Iijima, S., 1991. Synthesis of Carbon Nanotubes. *Nature* (London) 354, 56–58.

Imahori, H., Mori, Y., Matano, Y., 2003. Nanostructured artificial photosynthesis. J. Photochem. Photobiol. C-Photochem. Rev. 4, 51–83.

Jaffé, R., Ding, Y., Niggemann, J., Vähätalo, A.V., Stubbins, A., Spencer, R.G.M., Campbell, J., Dittmar, T., 2013. 2013. Global charcoal mobilization from soils via dissolution and riverine transport to the oceans. Science 340 (6130), 345–347.

Jang, J.W., Lee, C.E., Lyu, S.C., Lee, T.J., Lee, C.J., 2004. Structural study of nitrogen-doping effects in bamboo-shaped multiwalled carbon nanotubes. Appl. Phys. Lett. 84, 2877–2879.

Kang, Z.H., Wang, E.B., Gao, L., Lian, S.Y., Jiang, M., Hu, C.W., Xu, L., 2003. One-Step Water-Assisted Synthesis of High-Quality Carbon Nanotubes Directly from Graphite. J. Am. Chem. Soc. 125, 13652–13653.

Kato, M., Ishibashi, M., 2008. Carbon nanoparticle composite actuators, 4th World Congress on Biomimetics, Artificial Muscles and Nano-Bio, 2008. J. Phys. Conf. Ser. 2008, 012003.

Keller, N., Maksimova, N.I., Roddatis, V.V., Schur, M., Mestl, G., Butenko, Y.V., Kuznetsov, V.L., Schlögl, R., 2002. The catalytic use of onion-like carbon materials for styrene synthesis by oxidative dehydrogenation of ethylbenzene. Angew. Chem. Int. Ed. 41, 1885–1888.

Kerminen, V.M., Lehtinen, K.E.J., Anttila, T., Kulmala, M., 2004. Dynamics of atmospheric nucleation mode particles: A timescale analysis. Tellus B 56, 135–146.

Kharisov, B.I., Kharissova, O.V., 2019. General data on carbon allotropes. In: Allotropes, C. (Ed.), Metal-Complex Chemistry, Properties and Applications. Springer, Cham, pp. 1–8.

Kim, D.Y., Yang, C.M., Noguchi, H., Yamamoto, M., Ohba, T., Kanoh, H., Kaneko, K.K., 2008a. Enhancement of H_2 and CH_4 adsorptivities of single wall carbon nanotubes produced by mixed acid treatment. Carbon 46, 611–617.

Kim, S., Shibata, E., Sergiienko, R., Nakamura, T., 2008b. Purification and separation of carbon nanocapsules as a magnetic carrier for drug delivery systems. Carbon 46, 1523–1529.

Knicker, H., Hilscher, A., González-Vila, F.J., Almendros, G., 2008. A new conceptual model for the structural properties of char produced during vegetation fires. Org. Geochem. 39, 935–939.

Köstner, B., 2001. Evaporation and transpiration from forests in Central Europe relevance of patch-level studies for spatial scaling. Meteorol. Atmos. Phys. 76, 69–82.

Kotchey, G.P., Allen, B.L., Vedala, H., Yanamala, N., Kapralov, A.A., Tyurina, Y.Y., Klein-Seetharaman, J., Kagan, V.E., Star, A., 2011. The enzymatic oxidation of graphene oxide. ACS Nano 5, 2098–2108.

Kroto, H.W., et al., 1985. C60: buckminsterfullerene. Nature 318 (14), 162–163. https://doi.org/10.1038/318162a0.

Kühnel, D., Krug, H.F., Kukalj, A.J., 2018. Environmental impacts of engineered nanomaterials-imbalances in the safety assessment of selected nanomaterials. Materials 11, 1444.

Kulmala, M., Vehkamaki, H., Petaja, T., Maso, M.D., Lauri, A., Kermin-en, V.M., Birmili, W., McMurry, P.H., 2004. Formation and growth rates of ultrafi ne atmospheric particles: A review of observations. J. Aerosol. Sci. 35, 143–176.

Kushwaha, S.K.S., Ghoshal, S., Rai, A.K., Singh, S., 2013. Carbon nanotubes as a novel drug delivery system for anticancer therapy: a review. Braz. J. Pharm. Sci. 49, 629–644.

Li, W., Sun, J., Xu, L., Shi, Z., Riemer, N., Sun, Y., Fu, P., Zhang, J., Lin, Y., Wang, X., et al., 2016. A conceptual framework for mixing structures in individual aerosol particles. J. Geophys. Res. Atmos. 121, 13784–13798.

Lighty, J.A.S., Veranth, J.M., Sarofim, A.F., 2000. Combustion aerosols: Factors governing their size and composition and implications to human health. J. Air Waste Manage. Assoc. 50, 1565–1618.

Liu, G., Zheng, H., Jiang, Z., Zhao, J., Wang, Z., Pan, B., Xing, B., 2018. Formation and physicochemical characteristics of nano biochar: Insight into chemical and colloidal stability. Environ. Sci. Technol. 52, 10369–10379.

Liu, H., Ye, T., Mao, C., 2007. Fluorescent Carbon Nanoparticles Derived from Candle Soot. Angew. Chem. Int. Ed. 46, 6473–6475.

Lowry, G.V., Gregory, K.B., Apte, S.C., Lead, J.R., 2012. Transformations of nanomaterials in the environment. Environ. Sci. Technol. 46, 6893–6899.

Luo, P.G., Sahu, S., Yang, S.T., Sonkar, S.K., Wang, J., Wang, H., LeCroy, G.E., Cao, L., Sun, Y.P., 2013. Carbon "quantum" dots for optical bioimaging. J. Mater. Chem. B 1, 2116–2127.

Ma, Y.H., Manolache, S., Denes, F.S., Thamm, D.H., Kurzman, I.D., Vail, D.M., 2004. Plasma synthesis of carbon magnetic nanoparticles and immobilization of doxorubicin for targeted drug delivery. J. Biomater. Sci. Polym. Ed. 15, 1033–1049.

Mandal, D., Bolander, M.E., Mukhopadhyay, D., Sarkar, G., Mukherjee, P., 2006. The use of microorganisms for the formation of metal nanoparticles and their application. Appl. Microbiol. Biotechnol. 69, 485–492.

Mansor, M.R., Fadzullah, S.H.S.M., Masripan, N.A.B., Omar, G., Akop, M.Z., 2019. Comparison between functionalized graphene and carbon nanotubes: Effect of morphology and surface group on mechanical, electrical, and thermal properties of nanocomposites. In: Functionalized Graphene Nanocomposites and Their Derivatives Synthesis, Processing and Applications, pp. 177–204.

Maurer-Jones, M.A., Gunsolus, I.L., Murphy, C.J., Haynes, C.L., 2013. Toxicity of engineered nanoparticles in the environment. Anal. Chem. 85, 3036–3049.

Maynard, A.D., Baron, P.A., Foley, M., Shvedova, A.A., Kisin, E.R., Castranova, V., 2004. Exposure to carbon nanotube material: aerosol release during the handling of unrefined single-walled carbon nanotube material. J. Toxicol. Environ. Health A 67, 87–107.

Mitrano, D.M., Motellier, S., Clavaguera, S., Nowack, B., 2015. Review of nanomaterial aging and transformations through the life cycle of nano-enhanced products. Environ. Int. 77, 132–147.

Mottier, A., Moucher, F., Pinelli, E., Gauthier, L., Flahaut, E., 2017. Environmental impact of engineered carbon nanoparticles: from releases to effects on the aquatic biota. Curr. Opin. Biotechnol. 46 (1-6).

Muller, J., Huaux, F., Moreau, N., Mission, P., Heilier, J.-F., Delos, M., et al., 2005. Respiratory toxicity of multi-walled nanotubes. Toxicol. Appl. Pharmacol. 207 (3), 221–231.

Navarro, E., Baun, a., Behra, R., Hartmann, N., Filser, J., Mio, A.J., Quigg, A., Santschi, P., Sigg, L., 2008. Environmetal behavior and ecotoxicity of engineered nanoparticles to algae, plants and fungi. Ecotoxicology 17, 372–386.

Ni, M., Huang, J., Lu, S., Li, X., Yan, J., Cen, K., 2014. A review on black carbon emissions, worldwide and in China. Chemosphere 107, 83–93.

Nowack, B., Bucheli, T.D., 2007. Occurrence, behavior and effects of nanoparticles in the environment. Environ. Pollut. 150, 5–22.

Oberdörster, G., Oberdörster, E., Oberdörster, J., 2005. Nanotoxicology: an emerging discipline evolving from studies of ultrafine particles. Environ. Health Perspect. 113, 823–839.

Oberdörster, G., Zhu, S., Blickley, T.M., McClellan-Green, P., Haasch, M.L., 2006. Ecotoxicology of carbon-based engineered nanoparticles: effects of fullerene (C_{60}) on aquatic organisms. Carbon 44, 1112–1120.

Okpalugo, T.I.T., Papakostatinou, P., Murphy, H., McLaughlin, J.A., Brown, N.M.D., 2005. High resolution XPS characterization of chemical functionalised MWCNTs and SWCNTs. Carbon 43, 153–161.

Oliveira, P., Ramos, A.M., Fonseca, I., do Rego, A.B., Vital, J., 2005. Oxidation of limonene over carbon anchored transition metal Schiff base complexes: Effect of the linking agent. Catal. Today 102, 67–77.

Palashuddin, M.D., Jaiswal, A., Paul, A., Ghosh, S.S., Chattopadhyay, A., 2012. Presence of Amorphous Carbon Nanoparticles in Food Caramels. Sci. Rep. 2, 383.

Pardo, J., Peng, Z., Leblanc, R.M., 2018. Cancer Targeting and Drug Delivery Using Carbon-Based Quantum Dots and Nanotubes. Molecules 23, 378. https://doi.org/10.3390/molecules23020378.

Patil, S.S., Shedbalkar, U.U., Truskewycz, A., Chopade, B.A., Ball, A.S., 2016. Nanoparticles for environmental clean-up: A review of potential risks and emerging solutions. Environ. Technol. Innov. 5, 10–21.

Petersen, E.J., Flores-Cervantes, D.X., Bucheli, T.D., Elliott, L.C.C., Fagan, J.A., Gogos, A., Hanna, S., Kägi, R., Mansfield, E., Bustos, A.R.M., et al., 2016. Quantification of carbon nanotubes in environmental matrices: current capabilities, case studies, and future prospects. Environ. Sci. Technol. 50, 4587–4605.

Petersen, E.J., Zhang, L., Mattison, N.T., O'Carroll, D.M., Whelton, A.J., Uddin, N., Nguyen, T., Huang, Q., Henry, T.B., Holbrook, R.D., et al., 2011. Potential release pathways, environmental fate, and ecological risks of carbon nanotubes. Environ. Sci. Technol. 45, 9837–9856.

Prasad, K.S., Muthuraman, G., Zen, J.M., 2008. The role of oxygen functionalities and edge plane sites on screen-printed carbon electrodes for simultaneous determination of dopamine, uric acid and ascorbic acid. Electrochem. Commun. 10, 559–563.

Price, R.L., Haberstroh, K.M., Webster, T.J., 2004. Improved osteoblast viability in the presence of smaller nanometer dimensioned carbon fibres. Nanotechnology 15, 892–900.

Pulleman, M.M., Bouma, J., van Essen, E.A., Meijles, E.W., 2000. Soil organic matter content as a function of different land use history. Soil Sci. Soc. Am. J. 64, 689–693.

Qi, S., Sun, J., Ma, J., Sun, Y., Goossens, K., Li, H., Jia, P., Fan, X., Bielawski, C.W., Geng, J., 2019. Covalent bonding of sulfur nanoparticles to unzipped multiwalled carbon nanotubes for high-performance lithium–sulfur batteries. Nanotechnology 30, 024001 (11pp).

Randerson, J.T., Chen, Y., van der Werf, G.R., Rogers, B.M., Morton, D.C., 2012. Global burned area and biomass burning emissions from small fires. J. Geophys. Res. Biogeosci. 117, 1–23.

Reisser, M., Purves, R.S., Schmidt, M.W.I., Abiven, S., 2016. Pyrogenic carbon in soils: A literature-based inventory and a global estimation of its content in soil organic carbon and stocks. Front. Earth Sci. 4, 1–14.

Ray, P.C., Yu, H., Fu, P.P., 2009. Toxicity and environmental risks of nanomaterials: Toxicity and environmental risks of nanomaterilas: Challenges Future Needs. J. Environ. Sci. Health. C Environ. Carcinog. Ecotoxicol. Rev. 27 (1), 1–35.

Rodriguez-Reinoso, F., 1998. The role of carbon materials in heterogeneous catalysis. Carbon 36, 159–175.

Romeo-Franco, M., Godwin, H.A., Bilai, M., Cohen, Y., 2017. Needs and challenges for assessing the environmental impacts of engineered nanomaterials (ENMs). Beilsetein J. Nanotechnol. 8, 989–1014.

Ruiz-Dueñas, F.J., Martínez, Á.T., 2009. Microbial degradation of lignin: How a bulky recalcitrant polymer is efficiently recycled in nature and how we can take advantage of this. Microb. Biotechnol. 2, 164–177.

Sajadi, S.M., Samora, O.P., Steven, S., Woellner, C.F., Rodrigues, V., Vajtai, R., Lou, J., Galvão, D.S., Tiwary, C.S., Ajayan, P.M., 2017. Multiscale Geometric Design Principles Applied to 3D Printed Schwarzites. Adv. Mater. 30. https://doi.org/10.1002/adma.201704820.

Saquing, J.M., Yu, Y.H., Chiu, P.C., 2016. Wood-derived black carbon (biochar) as a microbial electron donor and acceptor. Environ. Sci. Technol. Lett. 3, 62–66.

Sato, Y., Yokoyama, A., Shibata, K., Akimoto, Y., et al., 2005. Influence on length on cytotoxicity of multi-walled carbon nanotubes against human acute monocytic leukemia cell line THP-1 in vitro and subcutaneous tissue of rats in vivo. Mol. Biosyst. 1 (2), 176–182.

Schriber, H.A., Prechl, J., Jiang, H., Zozulya, A., Fabry, Z., Denes, F.S., ... Sandor, M., 2009. In: Kawai, Y., Ikegami, H., Sato, N., Matsuda, A., Uchino, K., Kuzuya, M., ... Mizuno, A. (Eds.), Industrial Plasma Technology: Applications from Environmental to Energy Technologies. Wiley-VCH Verlag GmbH & Co. KGaA, Weinheim, pp. 141.

Serag, M.F., Kaji, N., Gaillard, C., Okamoto, Y., Terasaka, K., et al., 2011. Trafficking and subcellular localization of multi walled carbon nanotubes in plant cells. ACS Nano 5, 493–499.

Shedova, A.A., Kistin, E.R., Mercer, R., Murray, A.R., et al., 2005. Unusual inflammatory and fibrogenic pulmonary responses to single-walled carbon nanotubes in mice. Am. J. Physiol. Lung C 289 (5), L698–L708.

Shedbalkar, U.U., Singh, R., Wadhwani, S.A., Gaidhani, S.V., Chopade, B.A., 2014. Microbial synthesis of gold nanoparticles: Current status and future prospects. Adv. Colloids Interface Sci. 409 (40-48).

Sigmund, G., Jiang, C., Hofmann, T., Chen, W., 2018. Environmental transformation of natural and engineered carbon nanoparticles and implications for the fate of organic contaminants. Environ. Sci. Nano 5, 2500–2518.

Silva, V.H., Carson, N., Elder, A., Oberdörster, G., 2005. The rate ear vein model for investigating in vivo thrombogenicity of ultrafine particles (UFP). Toxicol. Sci. 85, 983.

Smith, C.J., Shaw, B.J., Handy, R.D., 2007. Toxicity of single walled carbon nanotubes to Rainbow Trout. Aquat. Toxicol. 82, 94–109.

Spokas, K.A., Novak, J.M., Masiello, C.A., Johnson, M.G., Colosky, E.C., Ippolito, J.A., Trigo, C., 2014. Physical disintegration of biochar: An overlooked process. Environ. Sci. Technol. Lett. 1, 326–332.

Spokas, K.A., 2010. Review of the stability of biochar in soils: predictability of O:C molar ratios. Carbon Manage. 1, 289–303.

Sun, X.M., Li, Y.D., 2004. Ga_2O_3 and GaN Semiconductor Hollow Spheres. Angew. Chem. Int. Ed. 43, 3827–3831.

Tasis, D., Tagmatarchis, N., Bianco, A., Prato, M., 2006. Chemistry of carbon nanotubes. Chem. Rev. 106, 1105–1136.

Tian, L., Ghosh, D., Chen, W., Pradhan, S., Chang, X.J., Chen, S.W., 2009. Nanosized carbon particles from natural gas soot. Chem. Mater. 21, 2803–2809.

Tiwari, A.J., Marr, J., 2010. The role of atmospheric transformations in determining environmental impacts of carbonaceous nanoparticles. J. Environ. Qual. 39, 1883–1895.

Uhelski, D., Miesel, J.R., 2017. Physical Location in the Tree during Forest Fire Influences Element Concentrations of Bark-Derived Pyrogenic Carbon from Charred Jack Pines (*Pinus Banksiana* Lamb.). Org. Geochem. 110, 87–91.

Valente, A., do Rego, A.M.B., Reis, M.J., Silva, I.F., Ramos, A.M., Vital, J., 2001. Oxidation of pinane using transition metal acetylacetonate complexes immobilised on modified activated carbon. Appl. Catal. A 207, 221–228.

Verneuil, L., Silverstre, J., Randrianjatovo, I., et al., 2015. Double walled carbon nanotubes promote the overproduction of extracellular protein-like polymers in *Nitzschia palea*: an adhesive response for an adaptive issue. Carbon 88, 113–125.

Wang, Y., Liu, F., He, C., Bi, L., Cheng, T., Wang, Z., Zhang, H., Zhang, X., Shi, Z., Li, W., 2017. Fractal dimensions and mixing structures of soot particles during atmospheric processing. Environ. Sci. Technol. Lett. 4, 487–493.

Weisner, M.R., Lowry, G.V., Alvarez, P., Dionysiou, D., Biswas, P., 2006. Assessing the risks of manufactured nanomaterials. Environ. Sci. Technol. 40, 4336–4345.

White, C.T., Mintmire, J.W., 2005. Fundamental properties of single-wall carbon nanotubes. J. Phys. Chem. B 109, 52–65.

Wu, Y., Hudson, J., Lu, Q., Moore, J., Mount, A., Rao, A., et al., 2006. Coating single walled-carbon nanotubes with phospholipids. J. Phys. Chem. B 110 (6), 2475–2478.

Yan, A.H., Lau, B.W., Weissman, B.S., Kulaots, I., Yang, N.Y.C., Kane, A.B., Hurt, R.H., 2006. Biocompatible, Hydrophilic, Supramolecular Carbon Nanoparticles for Cell Delivery. Adv. Mater. 18, 2373. https://doi.org/10.1002/adma.200600838.

Yan, Q.L., Gozin, M., Zhao, F.Q., Cohen, A., Pang, S.P., 2016. Highly energetic compositions based on functionalized carbon nanomaterials. Nanoscale 8 (9), 4799–4851.

Yi, H., Wang, H., Jing, Y., Peng, T., Wang, Y., Guo, J., He, Q., Guo, Z., Wang, X., 2015. Advanced asymmetric supercapacitors based on CNT@Ni$(OH)_2$core–shell composites and 3D graphene networks. J. Mater. Chem. A 3, 19545–19555.

Yoon, H., Ko, S., Jang, J., 2007. Nitrogen-doped magnetic carbon nanoparticles as catalyst supports for efficient recovery and recycling. Chem. Commun. (14), 1468–1470.

You, B., Wang, L., Yao, L., Yang, J., 2013. Three dimensional N-doped graphene–CNT networks for supercapacitor. Chem. Commun. 49, 5016–5018.

Zackrisson, O., Nilsson, M.C., Wardle, D.A., 1996. Key ecological function of charcoal from wildfires in the Boreal forest. Oikos 77, 10–19.

Zhang, C., Chen, W., Alvarez, P.J.J., 2014. Manganese peroxidase degrades pristine but not surface-oxidized (carboxylated) single-walled carbon nanotubes. Environ. Sci. Technol. 48, 7918–7923.

Zhang, R., Khalizov, A.F., Pagels, J., Zhang, D., Xue, H., McMurry, P.H., 2008a. Variability in morphology, hygroscopicity and optical properties of soot aerosols during atmospheric processing. Proc. Natl. Acad. Sci. U. S. A. 105, 10291–10296.

Zhang, L.W., Yu, W.W., Colvin, V.L., Monterio-Riviere, 2008b. Biological interactions of quantum dots nanoparticles in skin and in human epidermal karatinocytes. Toxicol. Appl. Pharmacol. 228, 200–211.

Zhao, J., Wang, Z., White, J.C., Xing, B., 2014. Graphene in the aquatic environment: adsorption, dispersion, toxicity and transformation. Environ. Sci. Technol. 48, 9995–10009.

Zheng, S.F., Hu, J., Zhong, L.S., Song, W.G., Wan, L.J., Guo, Y.J., 2008. Introducing Dual Functional CNT Networks into CuO Nanomicrospheres toward Superior Electrode Materials for Lithium-Ion Batteries. Chem. Mater. 20, 3617–3622.

Further reading

Lin, J., Chen, X., Huang, P., 2016. Graphene-based nanomaterials for bioimaging. Adv. Drug Delivery Rev. 105, 242–254.

Sharon, M., Soga, R., Afre, D., Sathiyamoorthy, K., Bhardwaj, S., Sharon, M., Jaybhaye, S., 2007. Hydrogen storage by carbon materials synthesized from oil seeds and fibrous plant materials. Int. J. Hydrogen Energy 32, 4238–4249.

Yue, Z.R., Economy, J., 2005. Nanoparticle and Nanoporous Carbon Adsorbents for Removal of Trace Organic Contaminants from Water. J. Nanopart. Res. 7 (2005), 477.

Carbon nanostructures: detection, controlling plant diseases and mycotoxins

13

Tahsin Shoala
Department of Environmental Biotechnology, College of Biotechnology, Misr University for Science and Technology, 6th of October City, Egypt

1 Introduction

The astonishing and versatile carbon with the atomic number (6), carbon is the 4th and most abundant element in the Universe by mass after (hydrogen, helium, and oxygen). The distinctive diversity of organic compounds and their uncommon polymer-forming capability at the temperatures normally found on Earth make this element the chemical origin of all known life. Carbon nanotubes (CNTs) are estimated as one of the most widely held types of nanomaterials, with distinctive morphological and superficial properties. They have been well studied in various applications in bionanotechnology as well as plant technology (Kwak et al., 2017; Min et al., 2017), gene delivery, tissue engineering (Harrison and Atala, 2007), and many biomedical applications (Oliveira et al., 2015). Recently, CNTs have been discovered as active nanomaterials against different microorganisms, therefore increasing interest in nanotechnology application for plant protection (Al-Jumaili et al., 2017). Additionally, numerous studies have shown the critical impact of CNT surface functional groups on their antibacterial activities (Chen et al., 2006; Meng et al., 2009).

Conversely to the previous studies on CNT antibacterial activities, studies associated with the antifungal activities of CNTs are underemphasized. Although more than 80% of phytopathogens occur by plant pathogenic fungi, further studies are needed to study the antifungal activities of CNTs according to their surface structure, chemistry, and length. In many circumstances, employing functionalized covalent and noncovalent chemical groups on the surface of nanotubes can increase their biological activities (Gu et al., 2008). On the other hand, before applying CNTs in plant protection, nanotoxicological studies should investigate the impact of CNTs on biological systems. The similar structure of CNTs to asbestos fibers has increased their potential environmental and health hazards (Donaldson et al., 2010). Herein, I have presented various applications of CNTs for detecting and managing phytopathogens and mycotoxins.

2 Types of carbon-based nanomaterials

Carbon has been known for a long time and has polymerization ability at the atomic level, which forms long carbon chains. Also, carbon is a nonmetallic and tetravalent structure

Carbon Nanomaterials for Agri-food and Environmental Applications. https://doi.org/10.1016/B978-0-12-819786-8.00013-X

with four free active electrons ready to form covalent chemical bonds. Furthermore, carbon termed as allotropes which is the abilities of carbon atoms to connect together and with other carbon atoms in different ways, such as diamond and graphite. Also, carbon in the nanostructure forms classified according to its geometrical shape, which varies from tube, horn shaped, spherical, and ellipsoidal. Nanohorns, spheres, and ellipsoids belong to the fullerene group, but the tube-shaped ones belong to CNTs.

2.1 Carbon nanotubes

CNTs, well characterized and synthesized as long thin cylinders, were revealed in 1991 by Sumio Iijima. These are macromolecules with distinctive and amazing physical properties related to their shape and size. These fascinating arrangements have attracted more researchers to try to understand and discover more characteristics of CNTs. CNTs have unique physical, chemical, electrical, thermal, and mechanical properties, making them among the most important and useful nanoparticles in many fields. As a result of the distinctive CNT properties, they are capable of infiltrating both the cell wall and the membrane. Also, they could provide a suitable delivery system for certain chemicals to the cells. CNTs could be structured as single-walled CNTs (SWCNTs) or multiwalled CNTs (MWCNTs) (Iijima, 1991).

2.1.1 Single-walled carbon nanotube structure

Normally, SWCNTs can be shaped in three diverse forms: armchair, chiral, and zigzag. The shapes rely on the way that the graphene is folded into a cylinder. For instance, imagine folding a sheet of paper from the corner or the edge, as it could give two different shapes. Different forms and structural designs have a straight effect on the electrical properties of CNTs (Fig. 1).

2.1.2 Multiwalled carbon nanotube structures

MWCNTs exist in two different structural models: a CNT surrounding another CNT, and a single graphene sheet rolled up around itself several times. MWCNTs have properties similar to single-walled nanotubes, but have a more stretchy strength than SWCNTs. Uniquely, MWCNTs protect the inner materials from interacting chemically with the external materials (Fig. 2).

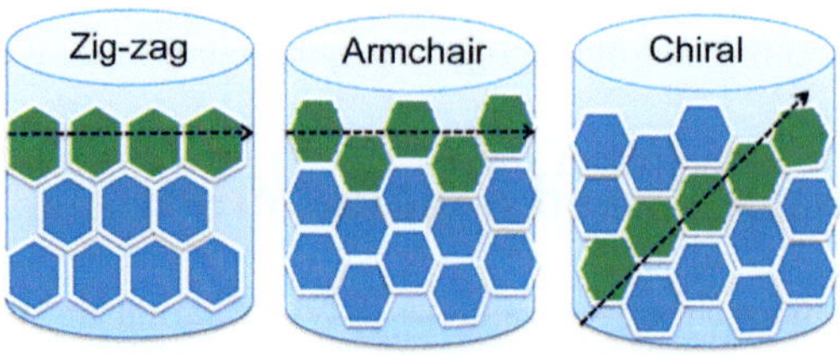

Fig. 1 Different types of single-walled CNTstructures (SWCNTs).

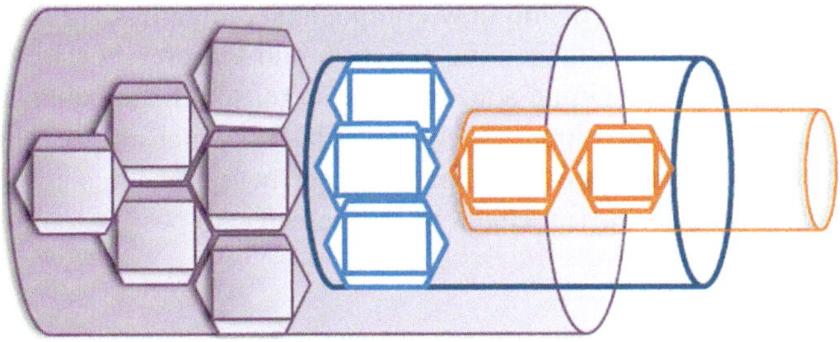

Fig. 2 MWCNT structures (MWCNTs).

2.2 *Fullerenes*

Fullerene atoms are organized in closed shells and are composed of 20 hexagonal and 12 pentagonal rings as the base of an icosahedral symmetry closed-cage structure. Also, every carbon atom connects to three other carbon atoms. Fullerenes are the molecular form of carbon. They contain carbon atoms on the spherical surface, which is also called allotrophic; this is the modified form of carbon. Fullerene C60 is the most known and superlative explored fullerenes. Fullerene C60 is a spherical molecule that is highly symmetric and made up of 60 carbon atoms with a diameter of 0.7 nm. It is positioned at the vertices of 20 hexagons and 12 pentagons (Yadav and Kumar, 2008).

2.3 *Graphene*

Graphene is the allotropic two-dimensional (2D) carbon form shaped by single layers of carbon atoms. Graphene is precisely nonmetal, but it is regularly denoted as a pseudometal according to its physical properties being similar to a semiconductor metal. Separately, every carbon atom is covalently connected (sp^2 hybridized) to three other carbon atoms in a hexagonal array, leaving one free electron per each carbon atom. Also, carbon atoms in the hexagonal crystal 2D shapes are connected by σ and π bonds and show sp^2 hybridization with 0.142 nm distance carbon atoms. Graphene has many distinctive and unique properties, making it as a perfect model material for use in electronic applications when compared to conventional materials. Graphene denotes a structural element of other carbon allotropes, for instance, graphite, CNTs, and fullerenes. Furthermore, graphene has different distinctive properties such as electronic, thermal, and mechanical strength and elasticity. The oxidation form of graphene nanosheets, graphene oxide (GO), could be the perfect choice as a carbon filler because it encloses epoxide, hydroxyl, and carboxylic acid groups (Huang et al., 2012).

3 Chemical features and properties of carbon-based nanoparticles

The chemical function and properties of carbon-based nanomaterials can show the ability of carbon nanomaterials in connection with or hosting other molecules that impact the physical properties of the new composites. New physical features of new nanocomposites may greatly enhance the function and features of the new structure. Ultrasonic treatment of CNTs in an acid mixture increased the attaching of sidewalls of CNTs with carboxylic functional groups (–COOH), which impact the functional mechanisms of the new structure. CNT oxidization increased the solubility in aqueous solutions with preserving mechanical and electrical possessions (Hernandez-Fernandez et al., 2010).

3.1 Features of carbon-based nanomaterials

Manipulation of carbon-based nanomaterials includes controlling the formation, construction, and assembly of different carbon-based nanomaterials, which include the size, length, and geometric properties.

3.2 Physical, electrical, and thermal properties of carbon-based nanomaterials

Carbon-based nanomaterials have variable distinctive physical, electrical, and thermal properties, according to changeable chemical and physical construction and properties. The main physical properties that impact positively or negatively on carbon-based nanomaterials include diameter, length, and physical shape (Saito et al., 1998). The diameter of carbon-based nanomaterials varies according to the synthesis methodology, which controls the physical properties and application of CNT nanostructures (Andreas, 2002). Different carbon-based nanomaterials encourage higher tension energies, electron orbital rehybridization, and modification of the σ and α bond, which boosts the mechanical, electronic, thermal, strength, rigidity, optical, and elastic features of SWCNTs and MWCNTs. For instance, CNTs show remarkably higher length-to-diameter ratios than other materials. Furthermore, graphene contains many distinctive physical possessions such as extreme thermal constancy and mechanical inflexibility.

3.3 Carbon-based nanomaterial interaction with attached other chemicals

Physical and chemical interaction features direct carbon-based nanomaterial connections with other materials. The interaction between carbon-based nanomaterials and other chemicals relies on the proposed physical and chemical models plus suggested theories. Molecule modeling at the nanoscale may provide more explanation regarding physical and chemical interactions. The proposed movements of electrons and excitations of carbonaceous nanomaterials have been defined previously, bearing in mind both van der Waals attractive forces and Pauli repulsion intensifying from electron orbitals overlapping at short parting distances (Hunter, 2001). The sorption activities and location of adsorbed material on carbon nanomaterials will be controlled by physical and chemical interactions such as hydrophobicity and capillarity. Physical–chemical interactions of carbon-based nanomaterials show the varying changes of many physical and chemical behaviors such as great sorption capacity, decreased sensitivity to pH value, speedy balanced rates, insignificant hysteresis in the detached nanoparticle, and stability (Hilding et al., 2001). However, the adsorption capability of the attached molecules to normal carbonaceous is restricted by surface-active site density, deliberate kinetics, and the adsorption imbalance in heterogeneous systems. Adsorption of large dimension materials decreases transference through low absorbency materials and obfuscates the subsurface remediation. Furthermore, the volume ratio and surface area of carbon-based nanoparticles increased compared to original form which positively impact on the physical and chemical properties of carbon-based nanocomposites.

4 Antimicrobial activity against plant pathogens

In general, different nanomaterials have great impact and perspective in phytopathogen management. Also, nanomaterials introduce distinctive strategies in plant disease management without harming the environment, to some extent (Prasad et al., 2014, 2017a,b). Carbon-based nanoparticles, predominantly CNTs, fullerene, and graphene, are the best potential examples presenting effective antimicrobial nanoactivity against different plant pathogens (Jung et al., 2011; Schmitt et al., 2015). Exposure of CNTs and aliphatic alcohol 1-octadecanol ($C_{18}H_{38}O$) composites to microwaves enhanced the antimicrobial activities of composites against different microorganisms because of the enhancement of the long carbon chain adsorption to microwaves by carbon nanomaterials (Al-Hakami et al., 2013).

SWCNTs have been applied successfully as nanofilters in eradicating bacteria and pathogens (Brady-Estevez et al., 2008). Also, sprayed pristine SWCNTs in a biocompatible surfactant solution showed robust nanoactivities against both gram-positive and gram-negative bacteria (Liu et al., 2009). Furthermore, biomaterials based on CNTs dispersed in poly (lactic-*co*-glycolic acid) have been applied successfully to control phytopathogens (Aslan et al., 2010). In addition, CNTs disabled the copper resistance of the phytopathogenic bacterium *Ralstonia solanacearum* (Wang et al., 2013a,b) and the fungi *Fusarium graminearum* and *F. oxysporum* (Wang et al., 2014c). Furthermore, carbon-based nanomaterials showed extraordinary deactivation of the phytopathogenic fungi *Fusarium graminearum* and *F. oxysporum* (Wang et al., 2013a,b). The application of CNTs increased successfully against phytopathogens because of their great antimicrobial activities. Additionally, CNT and aliphatic alcohol 1-octadecanol ($C_{18}H_{38}O$) composites revealed antifungal activities against the plant pathogenic fungi *Fusarium graminearum* and *Fusarium poae*, compared to fullerenes and GO nanomaterials (Zaytseva and Neumann, 2016). Previously, six carbon-based nanomaterials as well as SWCNTs, MWCNTs, GO, reduced graphene oxide (rGO), fullerenes (C60), and activated carbon (AC) were tested against the phytopathogenic fungi *Fusarium graminearum* and *Fusarium poae*. The tested carbon-based nanomaterials showed that SWCNTs had the highest antifungal activities compared to no significant antifungal activities for MWCNTs, GO, rGO, C60, and AC (Wang et al., 2014a,b,c).

SWCNT activities against microorganisms are higher than MWCNTs (Oyelami and Semple, 2015). Moreover, MWCNTs have perceptible robust antimicrobial activities against a broad range of microorganisms, wherever the coated MWCNTs with diverse metallic nanomaterials such as silver and zinc nanoparticles showed antifungal activities against the fungal growth of plant pathogenic fungi *Aspergillus fumigatus* and *A. ochraceus* (Fosso-Kankeu et al., 2016). Encapsulated fungicides in MWCNTs increased fungicide activities against plant pathogenic fungi *Alternaria alternata* compared to the noncapsulated fungicides with MWCNTs (Sarlak et al., 2014). The synthesized polyethylene films with CNT nanocomposite bases were prepared by a molding solution from boiling xylene. A polyethene-CNT nanocomposite revealed robust antifungal activities in packaged materials (Asgari et al., 2014). Most of the suggested theories regarding the mode of action for CNTs proposed that CNT antimicrobial activities rely on either the physical effects on the microorganism such as cell wall damage and cytoplasm separation or chemical effects such as oxidative stress and reactive oxygen species (ROS) (Wang et al., 2017).

The mode of action of GO nanomicrobial activities against plant pathogenic fungi and bacteria has been revealed as a result of microbial membrane damage and disruption of membrane strength, affecting electron transport and increasing the ROS. Also, the nanobacterial activities of GO nanoparticles could be reliant on both the size and grander sheets of GO. Large GO nanoparticles isolate bacteria from the external environment by enfolding bacterial cells and enhancing nanoactivities compared to small GO sheets (Chen et al., 2014; Hui et al., 2014; Liu et al., 2012; Mangadlao et al., 2015).

Plant treatment with CNTs affects many pathways, which may impact positively or negatively on plant growth (such as seed germination, water uptake, nutrient uptake, leaf area, and root length) and response to different stimuli (biotic and abiotic stresses). Mitogen-activated protein kinase (MAPK) genes control plant growth via increasing leaf area and root length (Wurzinger et al., 2011). CNTs may trigger MAPKs genes that, in turn, induce oxidative inducible signaling (Oxi1 serine–threonine) MAPKs. Consequently, the induction of oxi1 increases plant resistance to different stimuli. However, upregulation of the oxi1 gene induces plant cell death, but downregulation leads to cell susceptibilities in response to different stimuli (Shoala et al., 2018).

5 Plant pathogen diagnosis

Phytopathogens can be identified in many ways that vary according to cost, time consumption, and sensitivity. Serological and molecular methods are the best and most-sensitive methods for diagnosing different plant pathogens. Many different direct or indirect diagnostic techniques can be applied to screen, monitor, track, and identify phytopathogens such as enzyme-linked immunosorbent assay (ELISA), immunofluorescence, fluorescence in situ hybridization, flow cytometry, polymerase chain reaction (PCR), gas chromatography, thermography, fluorescence imaging, and hyperspectral techniques (Fang and Ramasamy, 2015). Also, a lateral flow kit (LFK) was successfully applied for detecting many variable plant pathogens such as *Ralstonia solanacearum* (*brown rot*), *Erwinia amylovora* (*fire blight*), *Phytophthora infestans* (*late blight*), *Pepino mosaic virus*, *Tomato mosaic virus*, *Potato virus Y*, and *Potato virus X* (Danks and Barker, 2000). Nowadays, we need specifically advanced, automated, sensitive, reliable, accurate, portable, and sensitive techniques to speed the diagnosis of phytopathogens. Currently, most research studies focus on developing phytopathogen diagnostic biosensors. Nanotechnology participates in developing onsite diagnostic techniques to speed up the detection and identification techniques. Nanoparticle-based sensors are effectively developed and applied to detect and identify phytopathogens. Many nanoparticles such as carbon and metal nanoparticles are successfully applied for nanodiagnosing and nanosensing phytopathogens (Fig. 4).

5.1 Nanodiagnosis and nanosensors through carbon nanoparticles

Nowadays, the quick detection of different pathogens plays a vital role in the early discovery of many phytopathogens. Onsite, real-time and sensor-based diagnostic techniques are increasingly attributable to active changes in plant pathogens and climate changes. The biosensor is a diagnostic device for different plant pathogens by using sensitive constituents and a transducer along with a sensor that transforms the recognition into an electrical signal (Mody, 2011). Even though sensors have been developed for sensing antibiotics and complementary DNA sequences as well as diagnosing different pathogens, the use of nanomaterials in designing biosensors has boosted their sensitivity. Diagnostic techniques have been developed to detect DNA or protein-functionalized gold NPs to be used as probes for a specific target, such techniques are established on the basis of direct charge transmission to conduct metric biosensor (Pal et al., 2008), CNTs (Serag et al., 2013), and silver and gold NPs (Sadowski, 2010). Many carbon nanomaterial-based enzymatic biosensors such as graphene, CNTs, nanofibers, and composites have been technologically developed.

Carbon-based nanosensors were designed and divided into different types of carbon materials, metal-based nanomaterials, and screen-printed electrodes that commonly apply the electrochemical mode of measurement and/or a microfluidic-constructed structure to attain modest and dense analytical devices. Carbon-based nanosensors can be applied in monitoring the environment and agriculture, diagnosing pathogens, and detecting toxins as well as for many applications in food. Carbon nanoparticles have unique features that successfully impact CNT application in plant growth and diagnostic techniques for plant pathogens. The distinctive electrochemical properties of carbon nanoparticles facilitate using CNTs as constituents in biosensors. Additionally, carbon nanoparticles have many other unique features such as greater length-to-diameter ratio, facilitation of fast electron transferal kinetics and the ability to bond with any chemical species (Balasubramanian and Burghard, 2006). This enables them to be applied in diagnostic techniques.

Carbon nanomaterials have been applied successfully for sensing metabolites, such as aromatic compounds, and diagnosing phytopathogens. Phenolic compounds are considered to be among the volatile metabolites and play a vital role in suppressing plant pathogens. Phenolic compounds can be easily detected using amperometric and potentiometric devices because of their oxidizing abilities. Conversely, electrochemical sensors harbor a foremost problem of electrode oxidization as a result of phenol exposure that forms dimeric or polymeric oxidation products (Balasubramanian and Burghard, 2006). The use of electrodes designed from carbon nanomaterials decreased the electrode oxidization effect by accumulating biomolecules such as nucleic acids (Wang et al., 2003a,b). Accordingly, the carbon nanomaterials either uniquely or in combination with enzymes could be applied in detecting phenolic compounds. Sensors have been developed for detecting indole-3-acetic acid (IAA), one of the important growth-regulator hormones, by coating DHP (dihydropyran)-stabilized MWCNT distribution onto a carbon electrode (Wu et al., 2003).

SWCNT radiometric sensors have been designed and applied for detecting H_2O_2 and NO in plant tissue. Also, composites of SWCNTs and metal/metal oxide nanoparticles were designed to detect and monitor agricultural byproducts such as ammonia and nitrogen oxide as well as agricultural pollutants (Giraldo et al., 2015). On the other side, CNTs were applied in a screen-printed immunosensor assay to diagnose the plant pathogenic fungus *Botrytis cinerea* with a sensitivity reaching $0.02\,\mu g\,mL^{-1}$. Additionally, this system was reformed by using micro magnetic beads attached to the carbon-based screen-printed electrodes, which enhanced the detection sensitivity to $0.008\,\mu g\,mL^{-1}$ for the plant pathogenic fungus *Botrytis cinerea* (Fernández-Baldo et al., 2009, 2010). Nanosensors of electrochemically functionalized SWCNT-based and metal/metal oxide nanoparticles or nanotubes were designed for detecting gases such as ammonia, nitrogen oxide, sulfur dioxide, hydrogen sulfide, and volatiles. Furthermore, this system could be successfully applied for monitoring agricultural pollutants and their influences on living organisms or health (Sekhon, 2014). More fascinating is a screen-printed gold electrode (SPGE) biosensor designed for detecting the cucumber mosaic virus (CMV) (Zulkifli et al., 2016) (Fig. 3).

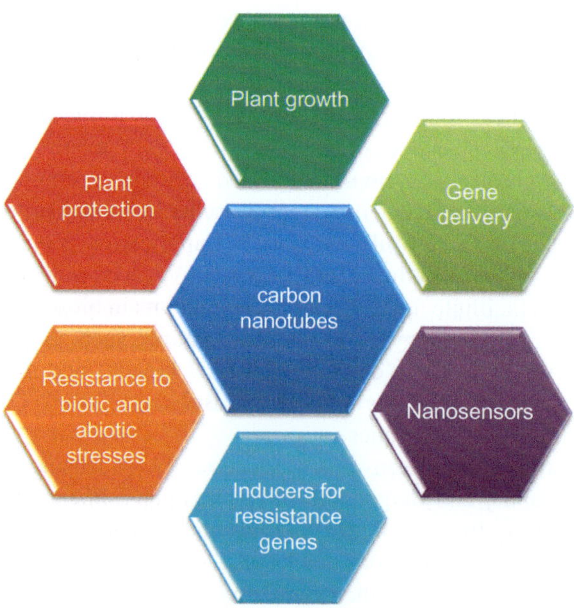

Fig. 3 Applications of nanocarbon in diagnosing phytopathogens and plant protection.

5.2 Graphene-DNA-based sensors

Graphene has distinctive features such as a large surface area and a unique sp^2 (sp^2/sp^3)-bonded network. Its other extraordinary characteristics have directed its speedy development for use in sensors, making it an exceptionally gifted material for biomolecule anchoring and detection (Suvarnaphaet and Pechprasarn, 2017). Detection and identification of different microorganisms and genetically modified organisms have been applied widely via specific DNA sequences. Accordingly, new inventive graphene-DNA nanosensors have been designed for diagnosing different pathogens, even though the graphene-nucleobase binding energy is slightly dissimilar through different approaches. One of those approaches may assure that ssDNA attached superficially on the graphene sheet and attaching crossover through number of interactive forces by using the nucleobases as the anchors. So that ssDNA binds to graphene stronger than dsDNA, and because of that, the nucleobases are captured and protected by the phosphate-deoxyribose backbones. Consequently, GO has been effectively applied as a platform to distinguish between DNA sequences. Alternatively, a graphene-DNA-based biosensor permits quick, modest, and inexpensive testing of genetic materials and phytopathogens (Eun and Wong, 2000). Diverse reports revealed the practical potentiality of diagnosing viral, bacterial, and fungal pathogens using nanosensors. Even though the DNA-based nanosensors presented quick solutions for diagnosing phytopathogens, the PCR technique may be performed before the probing process as a result of the trivial amount of nucleic acid in the bacterial cells (Fang and Ramasamy, 2015).

6 Mycotoxin detection using nanoparticles

Mycotoxins are toxic compounds formed by molds in response to stress conditions, such as abiotic stresses. However, the manifestation of grains with molds does not inevitably mean that mycotoxins are present. Additionally, the absence of molds on stored food does not assure that the grain is free of mycotoxins (Edwards, 2004; Simpson et al., 2001). So, the risk assessments of mycotoxin are complicated and require more attention and consideration. Mycotoxins can be determined using many developed reliable analytical methods. The most frequently applicable approaches in practice are high-pressure liquid chromatography (HPLC) and the mass detection and enzyme-linked immunoassay (ELISA) (Anfossi et al., 2016; Chauhan et al., 2016).

Nanoparticles have been successfully applied to detect and monitor mycotoxins in different samples. In practice, two different nanoparticle models have been distinguished in detecting mycotoxins. Nanoparticles can be applied at the receptor level to detect toxic molecules that require suitable specificity and reproducibility. Also, nanoparticles can be applied as enhancers for signal transduction to the detector (Rai et al., 2015; Rhouati et al., 2017). Nanoparticles such as CNTs, polymers, superparamagnetic NPs, quantum dots, and metal NPs have small molecules and high surface area-to-volume ratios, which enhance the nanoparticle binding process with a high concentration of mycotoxins (Gontero et al., 2017). Additionally, the use of nanoparticles facilitates different modifications with the specific ligands or superficial beautification by functional groups such as CH_3, $-OH$, $-COOH$, $-NH_2$, or $-CONH_2$.

6.1 Reduction of mycotoxins

Activated charcoal has been previously applied as an eliminator for mycotoxins. Accordingly, carbon nanoparticles have been applied as a gifted inheritor to AC. Also, fullerenes, graphene, GO, nanodiamonds, fibers, and nanotubes have unlimited potential to develop innovative adsorbents of mycotoxins. Nanocarbons have unique chemical structures and amphoteric structure properties while their surface could be protonated or deprotonated, which affects the linking ability to polar or nonpolar compounds. In general, CNT absorption similarity associates poorly with hydrophobicity but decreases in the order of nanopolar aliphatic < nonpolar aromatic < nitroaromatic functional groups (Chen et al., 2007). Fullerene adsorption activities are higher than AC.

Initial research studies the effect of fullerenol $C_{60}(OH)_{24}$ nanoparticles on *Aspergillus flavus* to monitor the mycelial growth, aflatoxin production, and oxidative stress modulation and showed that mycelial biomass weight slightly reduced but aflatoxin concentration in media significantly decreased aflatoxin concentration in the media. The application of 10 ng/mL of fullerenol decreased the assembly compared to untreated fungus, which resulted from the suppression of oxidative stress in the culture (Kovac et al., 2017).

Nanodiamonds have unique chemical structures that permit superficial functionalization as well as carboxylation, hydroxylation, and hydrogenation. Additionally, nanodiamonds can offer a binding attraction to different types of fungal mycotoxins. Nanodiamond aggregates (~40 nm) incline to adsorb Aflatoxin B1 (AFL B1) and Ochratoxin A (OTA) through electrostatic interactions that rely on different types of functional groups on the nanodiamond surface (Puzyr et al., 2007). The absorption ability of nanodiamonds reached 10 µg per mg for AFL B1 and 15 µg per mg for OTA (Gibson et al., 2011).

Magnetic graphene (MGO), which could be synthesized using iron oxide nanoparticles and GO, can be applied successfully to absorb fungal aflatoxin. Several oxygen functional groups on the surface of MGO facilitate interface with Fusarium mycotoxins (DON, T1, and T2). The application was performed by dissolving a palm kernel cake in water (6.2 pH for 5.2 h at 40.6°C), and consequently, MGO with constrained mycotoxins was successfully detached by using a magnet. The fungal mycotoxins T2 and DON were effectively reduced to 40% and 70%, respectively (Pirouz et al., 2017).

6.2 Mycotoxin purifications

MWCNTs have unique electronic, mechanical, and chemical properties that impact positively in enriching the ability of the MWCNT adsorbent for different toxic materials such as heavy metals (Kosa et al., 2012), pesticide remains (Qin et al., 2015), and type A trichothecenes (Dong et al., 2015). Magnetic CNTs have been investigated as operative sorbents for diverse types of mycotoxins because of their distinctive features as well as their easy separation, high surface area, superior hydrophobic property, and ability to be functionalized with covalent and noncovalent modification modes. MWCNTs have been applied effectively as a new solid-phase extraction (SPE) absorbent type to pretreated samples with mycotoxins as a result of their strong absorption capability, distinctive structure, and simplicity of being functionally reformed as carboxylic MWCNT (MWCNT–COOH) and hydroxyl MWCNT (MWCNT–OH) (Moazzen et al., 2013). Magnetic MWCNT combinations are mixtures between magnetic nanoparticles (MNPs) and MWCNTs. These combinations can be basically synthesized and incorporate the distinctive physical and chemical features of MWCNTs with the paramagnetic property of MNPs, allowing them to be valuable adsorption materials in the magnetic solid-phase extraction (M-SPE) procedure. The magnetic MWCNT compounds have been applied in combination with chromatographic techniques to define different types of environmental pollutants such as heavy metals, drug residues, pesticides, and bisphenol A (Jiao et al., 2012; Tarigh and Shemirani, 2013; Xu et al., 2013). Nanocomposites prepared from MNPs coated with a layer of octadecyl group-modified silica containing MWCNTs were optimized for the extraction of different types of zearalenone (ZEA) from maize samples (Fig. 4), prior to their determination by liquid chromatography-mass spectrometry. The current nanocomposites can be reused at least six times without losing their isolation efficiency (Moreno et al., 2016).

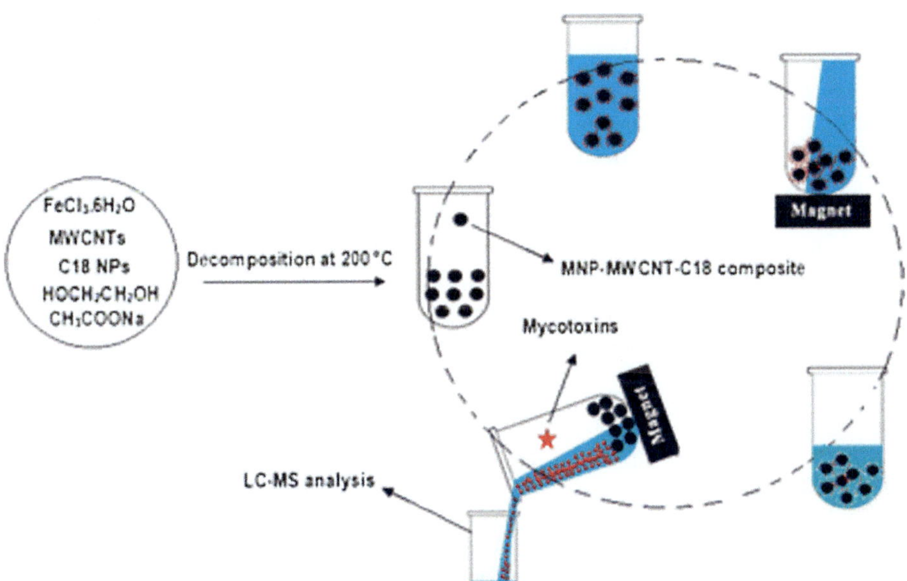

Fig. 4 Extraction and detection procedures of mycotoxin using magnetic MWCNTs and liquid chromatography-mass spectrometry (LC–MS).
Reprinted from Moreno, V., Zougagh, M., Ríos, Á., 2016. Hybrid nanoparticles based on magnetic multiwalled carbon nanotube-nano$C_{18}SiO_2$ composites for solid phase extraction of mycotoxins prior to their determination by LC-MS. Microchim. Acta 183(2), 871–880, with permission from Springer.

MNPs, MWCNTs, and nano $C_{18}SiO_2$ composites have been successfully applied to purify the fusarium mycotoxin ZEA and its products. MWCNTs have been successfully applied as sorbent nanomaterials for type A trichothecenes and ZEA ZEN mycotoxins in cereals (Dong et al., 2016; Jiang et al., 2017). However, the sample pretreatment process is inconvenient and time consuming. MWCNTs have been applied as adsorbents on the basis of one-step SPE clean up followed by UPLC-Ms/MS analysis to speed up the detection process of 21 mycotoxins in corn and wheat, including trichothecenes, ZEA ZEN and its derivatives, four aflatoxines, and two ochratoxins (Jiang et al., 2018). Mycotoxins were selectively extracted on magnetic MWCNT nano C18SiO composites and sensitively determined by liquid chromatography-mass spectrometry (LC–MS). The method meets the requirements established by the European Commission Directive 856/2005/EC (Moreno et al., 2016).

7 Conclusion

Carbon nanomaterials have unique features that allow them to be applied successfully in variable fields such as agriculture, biology, medicine, and chemical engineering. Carbon-based nanomaterials have many applications in agriculture, especially plant pathology and plant protection. Carbon-based nanomaterials can be applied successfully in plant pathology to offer nanodiagnostic techniques, control different phytopathogens, and decrease fungal aflatoxins. Magnetic CNTs have been investigated as operative sorbents for diverse types of mycotoxins because of their distinctive features as well as their easy separation, high surface area, superior hydrophobic property, and ability to be functionalized with covalent and noncovalent modification modes. Magnetic MWCNT combinations can be synthesized and incorporate the distinctive physical and chemical features of MWCNTs with the paramagnetic property of MNPs, allowing them to be valuable adsorption materials in the M-SPE procedure. Further studies on carbon nanomaterials are needed to discover new roles and modes of action of different nanoparticles. Nanotoxicity research studies are required of carbon nanoparticles to discover the possibilities of harming the environment as well as animals and humans.

References

Al-Hakami, S.M., Khalil, A.B., Laoui, T., Atieh, M.A., 2013. Fast disinfection of *Escherichia coli* bacteria using carbon nanotubes interaction with microwave radiation. Bioinorg. Chem. Appl. 45, 8943.

Al-Jumaili, A., Alancherry, S., Bazaka, K., Jacob, M.V., 2017. Review on the antimicrobial properties of carbon nanostructures. Materials 10, 1066.

Andreas, H., 2002. Functionalization of single-walled carbon nanotubes. Angew. Chem. Int. Ed. 41, 1853–1859.

Anfossi, L., Giovannoli, C., Baggiani, C., 2016. Mycotoxin detection. Curr. Opin. Biotechnol. 37, 120–126.

Asgari, P., Moradi, O., Tajeddin, B., 2014. The effect of nanocomposite packaging carbon nanotube base on organoleptic and fungal growth of mazafati brand dates. Int. Nano. Lett. 4, 1–5.

Aslan, S., Loebick, C.Z., Kang, S., Elimelech, M., Pfefferle, L.D., Van Tassel, P.R., 2010. Antimicrobial biomaterials based on carbon nanotubes dispersed in poly(lactic-co-glycolic acid). Nanoscale 2, 1789–1794.

Balasubramanian, K., Burghard, M., 2006. Biosensors based on carbon nanotubes. Anal. Bioanal. Chem. 385 (3), 452–468.

Brady-Estevez, A.S., Kang, S., Elimelech, M., 2008. A single-walled-carbon-nanotube filter for removal of viral and bacterial pathogens. Small 4, 481–484.

Chauhan, R., Singh, J., Sachdev, T., Basu, T., Malhotra, B.D., 2016. Recent advances in mycotoxins detection. Biosens. Bioelectron. 81, 532–545.

Chen, X., Tam, U.C., Czlapinski, J.L., Lee, G.S., Rabuka, D., Zettl, A., Bertozzi, C.R., 2006. Interfacing carbon nanotubes with living cells. J. Am. Chem. Soc. 128, 6292–6293.

Chen, W., Duan, L., Zhu, D.Q., 2007. Adsorption of polar and nonpolar organic chemicals to carbon nanotubes. Environ. Sci. Technol. 41, 8295–8300.

Chen, J., Peng, H., Wang, X., Shao, F., Yuan, Z., Han, H., 2014. Graphene oxide exhibits broad spectrum antimicrobial activity against bacterial phytopathogens and fungal conidia by intertwining and membrane perturbation. Nanoscale 6, 1879–1889.

Danks, C., Barker, I., 2000. On-site detection of plant pathogens using lateral-flow devices. EPPO Bull. 30 (3–4), 421–426.

Donaldson, K., Murphy, F.A., Duffin, R., Poland, C.A., 2010. Asbestos, carbon nanotubes and the pleural mesothelium: a review of the hypothesis regarding the role of long fibre retention in the parietal pleura, inflammation and mesothelioma. Part. Fibre Toxicol. 7, 5.

Dong, M., Si, W., Jiang, K., Nie, D., Wu, Y., Zhao, Z., De Saeger, S., Han, Z., 2015. Multi-walled carbon nanotubes as solid-phase extraction sorbents for simultaneous determination of type A trichothecenes in maize, wheat and rice by ultra-high performance liquid chromatography-tandem mass spectrometry. J. Chromatogr. A 1423, 177–182.

Dong, M., Si, W., Wang, W., Bai, B., Nie, D., Song, W., Zhao, Z., Guo, Y., Han, Z., 2016. Determination of type A trichothecenes in coix seed by magnetic solid-phase extraction based on magnetic multi-walled carbon nanotubes coupled with ultra-high performance liquid chromatography-tandem mass spectrometry. Anal. Bioanal. Chem. 408, 1–9.

Edwards, S.G., 2004. Influence of agricultural practices on fusarium infection of cereals and subsequent contamination of grain by trichothecene mycotoxins. Toxicol. Lett. 153, 29–35.

Eun, A.J.C., Wong, S.M., 2000. Molecular beacons: a new approach to plant virus detection. Phytopathology 90 (3), 269–275.

Fang, Y., Ramasamy, R.P., 2015. Current and prospective methods for plant disease detection. Biosensors 5 (3), 537–561.

Fernández-Baldo, M.A., Messina, G.A., Sanz, M.I., Raba, J., 2009. Screen-printed immunosensor modified with carbon nanotubes in a continuous-flow system for the *Botrytis cinerea* determination in apple tissues. Talanta 79 (3), 681–686.

Fernández-Baldo, M.A., Messina, G.A., Sanz, M.I., Raba, J., 2010. Microfluidic immunosensor with micromagnetic beads coupled to carbon-based screen-printed electrodes (SPCEs) for determination of *Botrytis cinerea* in tissue of fruits. J. Agric. Food Chem. 58 (21), 1201–11206.

Fosso-Kankeu, E., De Klerk, C.M., Botha, T.A., Waanders, F., Phoku, J., Pandey, S., 2016. The antifungal activities of multi-walled carbon nanotubes decorated with silver, copper and zinc oxide particles. In: International conference on advances in science, engineering, technology and natural resources (ICASETNR-16), Parys, South Africa, 24–25 November, pp 55–59.

Gibson, N.M., Luo, T.J.M., Brenner, D.W., Shenderova, O., 2011. Immobilization of mycotoxins on modified nanodiamond substrates. Biointerphases 6, 210–217.

Giraldo, J.P., Landry, M.P., Kwak, S., Jain, R.M., Wong, M.H., Iverson, N.M., Ben-Naim, M., Strano, M.S., 2015. A ratiometric sensor using single chirality near-infrared fluorescent carbon nanotubes: application to in vivo monitoring. Small 32, 3973–3984.

Gontero, D., Lessard-Viger, M., Brouard, D., Bracamonte, A.G., Boudreau, D., Veglia, A.V., 2017. Smart multifunctional nanoparticles design as sensors and drug delivery systems based on supramolecular chemistry. Microchem. J. 130, 316–328.

Gu, L., Luo, P.G., Wang, H., Meziani, M.J., Lin, Y., Veca, L.M., Cao, L., Lu, F., Wang, X., Quinn, R.A., et al., 2008. Single-walled carbon nanotube as a unique scaffold for the multivalent display of sugars. Biomacromolecules 9, 2408–2418.

Harrison, B.S., Atala, A., 2007. Carbon nanotube applications for tissue engineering. Biomaterials 28, 344–353.

Hernandez-Fernandez, P., Montiel, M., Ocón, P., de la Fuente, J.L.G., Garcia-Rodriguez, S., Rojas, S., Fierro, J.L., 2010. Functionalization of multi-walled carbon nanotubes and application as supports for electrocatalysts in proton exchange membrane fuel cell. Appl Catal B 99, 343–352.

Hilding, J., Grulke, E.A., Sinnot, S.B., Qian, D., Andrews, R., Jagtoyen, M., 2001. Sorption of butane on carbon multiwall nanotubes at room temperature. Langmuir 17, 7540–7544.

Huang, X., Xiaoying, Q., Boey, F., Zhang, H., 2012. Graphene-based composites. Chem. Soc. Rev. 41, 666–686.

Hui, L., Piao, J., Auletta, J., Hu, K., Zhu, Y., Meyer, T., Liu, H., Yang, L., 2014. Availability of the basal planes of graphene oxide determines whether it is antibacterial. ACS Appl. Mater. Interfaces 6, 13183–13190.

Hunter, R.J., 2001. Foundation of colloid science, second ed. Oxford University Press, Oxford; New York.

Iijima, S., 1991. Synthesis of carbon nanotubes. Nature 354, 56–58.

Jiang, K., Huang, P., Luan, L., Fan, K., Guo, W., Zhao, Z., Wu, Y., Han, Z., 2017. Iron (II, III) oxide/multi-walled carbon nanotube composite as solid-phase extraction sorbent followed by ultra-high-performance liquid chromatography-tandem mass spectrometry for simultaneous determination of zearalenone and type A trichothecenes in *Salviae miltiorrhizae* Radix et Rhizoma (Danshen). J. Chromatogr. A 1482, 1–10.

Jiang, D., Wei, D., Wang, L., Ma, S., Du, Y., Wan, G.M., 2018. Multiwalled carbon nanotube for one-step cleanup of 21 mycotoxins in corn and wheat prior to ultraperformance liquid chromatography-tandem mass spectrometry analysis. Toxins (Basel). 10, 10(10).

Jiao, Y., Ding, L., Fu, S., Zhu, S., Li, H., Wang, L., 2012. Determination of bisphenol A, bisphenol F and their diglycidyl ethers in environmental water by solid phase extraction using magnetic multiwalled carbon nanotubes followed by GC-MS/MS. Anal. Methods 4 (1), 291–298.

Jung, J.H., Hwang, G.B., Lee, J.E., Bae, G.N., 2011. Preparation of airborne Ag/CNT hybrid nanoparticles using an aerosol process and their application to antimicrobial air filtration. Langmuir 27, 10256–10264.

Kosa, S.A., Al-Zhrani, G., Salam, M.A., 2012. Removal of heavy metals from aqueous solutions by multi-walled carbon nanotubes modified with 8-hydroxyquinoline. Chem. Eng. J. 181, 159–168.

Kovac, T., Sarkanj, B., Klapec, T., Borisev, I., Kovac, M., Nevistic, A., Strelec, I., 2017. Fullerol $C_{60}(OH)_{24}$ nanoparticles and mycotoxigenic fungi: a preliminary investigation into modulation of mycotoxin production. Environ. Sci. Pollut. Res. 24, 16673–16681.

Kwak, S.Y., Wong, M.H., Tts, L., Bisker, G., Lee, M.A., Kaplan, A., Dong, J., Liu, A.T., Koman, V.B., Sinclair, R., 2017. Nanosensor technology applied to living plant systems. Annu. Rev. Anal. Chem. 10, 113–140.

Liu, S., Wei, L., Hao, L., Fang, N., Chang, M.W., Xu, R., Yang, Y., Chen, Y., 2009. Sharper and faster "nano darts" kill more bacteria: A study of antibacterial activity of individually dispersed pristine single-walled carbon nanotube. ACS Nano 3, 3891–3902.

Liu, S., Hu, M., Zeng, T.H., Wu, R., Jiang, R., Wei, J., Wang, L., Kong, J., Chen, Y., 2012. Lateral dimension-dependent antibacterial activity of graphene oxide sheets. Langmuir 28, 12364–12372.

Mangadlao, J.D., Santos, C.M., Felipe, M.J.L., De Leon, A.C.C., Rodrigues, D.F., Advincula, R.C., 2015. On the antibacterial mechanism of graphene oxide (GO) Langmuir-Blodgett films. Chem. Commun. 51, 2886–2889.

Meng, L., Fu, C., Lu, Q., 2009. Advanced technology for functionalization of carbon nanotubes. Prog. Nat. Sci. 19, 801–810.

Min, H.W., Giraldo, J.P., Kwak, S.Y., Koman, V.B., Sinclair, R., Lew, T.T.S., Bisker, G., Liu, P., Strano, M.S., 2017. Nitroaromatic detection and infrared communication from wild-type plants using plant nanobionics. Nat. Mater. 16, 264–272.

Moazzen, M., Ahmadkhaniha, R., Gorji, M.E.H., Yunesian, M., Rastkari, N., 2013. Magnetic solid-phase extraction based on magnetic multi-walled carbon nanotubes for the determination of polycyclic aromatic hydrocarbons in grilled meat samples. Talanta 115, 957–965.

Mody, H.R., 2011. Cancer nanotechnology: recent trends and developments. Int. J. Med. Updt. 6 (1), 3.

Moreno, V., Zougagh, M., Ríos, Á., 2016. Hybrid nanoparticles based on magnetic multiwalled carbon nanotube-nano$C_{18}SiO_2$ composites for solid phase extraction of mycotoxins prior to their determination by LC-MS. Microchim. Acta 183 (2), 871–880.

Oliveira, S.F., Bisker, G., Bakh, N.A., Gibbs, S.L., Landry, M.P., Strano, M.S., 2015. Protein functionalized carbon nanomaterials for biomedical applications. Carbon 95, 767–779.

Oyelami, A.O., Semple, K.T., 2015. Impact of carbon nanomaterials on microbial activity in soil. Soil Biol. Biochem. 86, 172–180.

Pal, S., Ying, W., Alocilja, E.C., Downes, F.P., 2008. Sensitivity and specificity performance of a direct-charge transfer biosensor for detecting *Bacillus cereus* in selected food matrices. Biosyst. Eng. 99 (4), 461–468.

Pirouz, A.A., Selamat, J., Iqbal, S.Z., Mirhosseini, H., Karjiban, R.A., Bakar, F.A., 2017. The use of innovative and efficient nanocomposite (magnetic graphene oxide) for the reduction on of fusarium mycotoxins in palm kernel cake. Sci. Rep. 7, 12453.

Prasad, R., Kumar, V., Prasad, K.S., 2014. Nanotechnology in sustainable agriculture: present concerns and future aspects. Afr. J. Biotechnol. 13 (6), 705–713.

Prasad, R., Bhattacharyya, A., Nguyen, Q.D., 2017a. Nanotechnology in sustainable agriculture: recent developments, challenges, and perspectives. Front. Microbiol. 8, 1014.

Prasad, R., Gupta, N., Kumar, M., Kumar, V., Wang, S., Abd-Elsalam, K.A., 2017b. Nanomaterials act as plant defense mechanism. In: Prasad, R., Kumar, V., Kumar, M. (Eds.), Nanotechnology. Springer, Singapore, pp. 253–269.

Puzyr, A.P., Purtov, K.V., Shenderova, O.A., Luo, M., Brenner, D.W., Bondar, V.S., 2007. The adsorption of aflatoxin B1 by detonation-synthesis nanodiamonds. Dokl. Biochem. Biophys. 417, 299–301.

Qin, Y., Zhao, P., Fan, S., Han, Y., Li, Y., Zou, N., Song, S., Zhang, Y., Li, F., Li, X., 2015. The comparison of dispersive solid phase extraction and multi-plug filtration cleanup method based on multi-walled carbon nanotubes for pesticides multi-residue analysis by liquid chromatography tandem mass spectrometry. J. Chromatogr. A 1385, 1–11.

Rai, M., Jogee, P.S., Ingle, A.P., 2015. Emerging nanotechnology for detection of mycotoxins in food and feed. Int. J. Food Sci. Nutr. 66, 363–370.

Rhouati, A., Bulbul, G., Latif, U., Hayat, A., Li, Z.H., Marty, J.L., 2017. Nano-aptasensing in mycotoxin analysis: recent updates and progress. Toxins 9, 349.

Sadowski, Z., 2010. Biosynthesis and application of silver and gold nanoparticles. In: Perez, D.P. (Ed.), Silver Nanoparticles. InTech.

Saito, R., Dresselhaus, G., Dresselhaus, M.S., 1998. Physical Properties of Carbon Nanotubes. Imperial College Press, London.

Sarlak, N., Taherifar, A., Salehi, F., 2014. Synthesis of nanopesticides by encapsulating pesticide nanoparticles using functionalized carbon nanotubes and application of new nanocomposite for plant disease treatment. J. Agric. Food Chem. 62, 4833–4838.

Schmitt, H., Creton, N., Prashantha, K., Soulestin, J., Lacrampe, M.F., Krawczak, P., 2015. Melt-blended halloysite nanotubes/wheat starch nanocomposites as drug delivery system. Polym. Eng. Sci. 55, 573–580.

Sekhon, B.S., 2014. Nanotechnology in agri-food production: an overview. Nanotechnol Sci Appl. 7, 31–53.

Serag, M.F., Kaji, N., Habuchi, S., Bianco, A., Baba, Y., 2013. Nanobiotechnology meets plant cell biology: carbon nanotubes as organelle targeting nanocarriers. RSC Adv. 3 (15), 4856–4862.

Shoala, T., 2018. Positive impacts of nanoparticles in plant resistance against different stimuli. In: Abd-Elsalam, K.A., Prasad, R. (Eds.), Nanobiotechnology Applications in Plant Protection. first ed. Springer International Publishing AG, Part of Springer Nature, pp. 267–279.

Simpson, D.R., Weston, G.E., Turner, J.A., Jennings, P., Nicholson, P., 2001. Differential control of head blight pathogens of wheat by fungicides and consequences for mycotoxin contamination of grain. Eur. J. Plant Pathol. 107, 421–431.

Suvarnaphaet, P., Pechprasarn, S., 2017. Graphene-based materials for biosensors: a review. Sensors 17 (10), 2161.

Tarigh, G.D., Shemirani, F., 2013. Magnetic multi-wall carbon nanotube nanocomposite as an adsorbent for preconcentration and determination of lead (II) and manganese (II) in various matrices. Talanta 115, 744–750.

Wang, J., Deo, R.P., Musameh, M., 2003a. Stable and sensitive electrochemical detection of phenolic compounds at carbon nanotube modified glassy carbon electrodes. Electroanalysis 15 (23–24), 1830–1834.

Wang, J., Liu, G., Jan, M.R., Zhu, Q., 2003b. Electrochemical detection of DNA hybridization based on carbon-nanotubes loaded with CdS tags. Electrochem. Commun. 5 (12), 1000–1004.

Wang, L., Fortner, J.D., Hou, L., Zhang, C., Kan, A.T., Tomson, M.B., Chen, W., 2013a. Contaminant-mobilizing capability of fullerene nanoparticles (nC60): effect of solvent-exchange process in nC60 formation. Environ. Toxicol. Chem. 32, 329–336.

Wang, X., Liu, X., Han, H., 2013b. Evaluation of antibacterial effects of carbon nanomaterials against copper-resistant *Ralstonia solanacearum*. Colloids Surf. B. Biointerfaces 103, 136–142.

Wang, H., Ma, H., Zheng, W., An, D., Na, C., 2014a. Multifunctional and recollectable carbon nanotube ponytails for water purification. ACS Appl. Mater. Interfaces 6, 9426–9434.

Wang, T., Zhao, D., Guo, X., Correa, J., Riehl, B.L., Heineman, W.R., 2014b. Carbon nanotube-loaded nafion film electrochemical sensor for metal ions: europium. Anal. Chem. 86, 4354–4361.

Wang, X., Liu, X., Chen, J., Han, H., Yuan, Z., 2014c. Evaluation and mechanism of antifungal effects of carbon nanomaterials in controlling plant fungal pathogen. Carbon 68, 798–806.

Wang, X., Zhou, Z., Chen, F., 2017. Surface modification of carbon nanotubes with an enhanced antifungal activity for the control of plant fungal pathogen. Materials 10, 1375.

Wu, K., Sun, Y., Hu, S., 2003. Development of an amperometric indole-3-acetic acid sensor based on carbon nanotubes film coated glassy carbon electrode. Sens. Actuators B 96 (3), 658–662.

Wurzinger, B., Mair, A., Pfister, B., Teige, M., 2011. Cross-talk of calcium-dependent protein kinase and MAP kinase signaling. Plant Signal. Behav. 6, 8–12.

Xu, Y., Ding, J., Chen, H., Zhao, Q., Hou, J., Yan, J., Wang, H., Ding, L., Ren, N., 2013. Fast determination of sulfonamides from egg samples using magnetic multiwalled carbon nanotubes as adsorbents followed by liquid chromatography–tandem mass spectrometry. Food Chem. 140 (1), 83–90.

Yadav, B.C., Kumar, R., 2008. Structure, properties and applications of fullerenes. Int. J. Nanotechnol. Appl. 2, 15–24.

Yu, X., Chattopadhyay, D., Galeska, I., Papadimitrakopoulos, F., Rusling, J.F., 2003. Peroxidase activity of enzymes bound to the ends of single-wall carbon nanotube forest electrodes. Electrochem. Commun. 5 (5), 408–411.

Zaytseva, O., Neumann, G., 2016. Carbon nanomaterials: production, impact on plant development, agricultural and environmental applications. Chem. Biol. Technol. Agric. 3, 17.

Zulkifli, H., Salam, F., Saad, S.M., Rahman, R.A., Rani, R.M., Karim, M.S.A., Ishak, Z., 2016. Preliminary study of electrochemical DNA sensor for cucumber mosaic virus. Procedia. Chem. 20, 98–101.

Carbon nanotubes: Plant gene delivery and genome editing

14

Rajkuberan Chandrasekaran[a], *P. Rajiv*[a], *Kamel A. Abd-Elsalam*[b]
[a]Department of Biotechnology, Karpagam Academy of Higher Education, Coimbatore, India, [b]Plant Pathology Research Institute, Agricultural Research Center (ARC), Giza, Egypt

1 Introduction

Richard Feynman, a physicist in 1959 gave a brief lecture about "There's Plenty of Room at the Bottom: An Invitation to Enter a New Field of Physics"; which set the basis for nanotechnology (Feynman, 1960). Until now, many research fraternity around the globe utilizing the nano concepts for their research and many novel intriguing scientific inventions have been achieved and a common man experiencing the ease of technology at present scenario (Baulcombe et al., 2009). The main thrust area of research in nanotechnology is electronics, energy, batteries, sensors, solar cells, the chemical industry, paints, cosmetics, etc. (Rajkuberan et al., 2015). Moreover, relevant to biological science, nanotechnology finds a niche in developing antibiotics, novel drug delivery systems, bioimaging, cancer treatment, and infectious disease management (Jahangirian et al., 2017).

Plant biotechnology is one of the promising approaches in the scientific discipline with the objective to ensure food productivity and security (Batlang et al., 2014). However, recently, plant biotechnology has faced hindrances in developing genetically modified plants. Gene delivery through conventional methods have major drawbacks like low efficiency, gene delivery mechanisms, narrow species range, limited cargo types and progeny regeneration (Cunningham et al., 2018). Consequently, the integration of nanotechnology with existing molecular tools will provide a platform to overcome the obstacles that lagged behind the progress in the development of genetically modified plants and biotransformation (Institute of Medicine (IOM), 2009). Specifically, the application of nanotechnology in agriculture is the focus of research, owing to its superior properties and the ability to facilitate the development of genetically modified crops, nano-based gene delivery, nanofertilizers, nanopesticides, plant protection, and production (Abd-Elsalam and Prasad, 2018; Prasad et al., 2017).

For the past two decades, plants have been utilized lavishly as a template for the synthesis of versatile metallic nanoparticles from the green chemistry perspective (Phougat et al., 2017). Hitherto, numerous reports have been published regarding the plant-mediated synthesis of metallic nanoparticles (Gour and Jain, 2019). The published reports clearly imply that plant-mediated synthesis of metallic nanoparticles will be a new modern entity to overcome the negative implications of the chemical-mediated synthesis of metallic nanoparticles. Moreover, these metallic nanoparticles

Carbon Nanomaterials for Agri-food and Environmental Applications. https://doi.org/10.1016/B978-0-12-819786-8.00014-1

have been proven to be superior materials for treating diseases as well as in antibiotics, bioimaging, textiles, pharmaceuticals, and agriculture (Duhan et al., 2017).

However, the utilization of nanomaterials for gene delivery to plants is limited, owing to the various factors that limit the application of nanoparticles in gene delivery. In spite of that, various dimensions of engineered nanoparticles such as carbon nanotubes (CNT), magnetic nanoparticles, titanium oxide, aluminum nanoparticles, zinc oxide, copper nanoparticle starch, and silver nanoparticles were used for gene delivery in plants (Duhan et al., 2017). These engineered nanomaterials have proven to be good carrier agents for gene delivery. The primary objective of nanotechnology in plant genetic engineering is to transfer the gene at the desired site with high efficiency. Other objectives include gene protection and reduced DNA damage; nano-based techniques can be easily amalgamated with existing techniques, leading to efficient transformation. In this chapter, we exclusively emphasize the role and function of carbon nanotubes in the delivery of the gene into the plants. Also, the future perspective for the application of carbon nanotubes as organic CRISPR device carriers for plant genome editing is described.

2 Nanomaterial for gene delivery in plants

After the success of transgenic plants in early 1980, researchers endeavored to develop new gene delivery techniques for plants (Darbani et al., 2008). In this pipeline, physical, chemical, and biological methods are available to deliver genes into the plant cells. The methods are particle bombardment (gene gun), electroporation, poly ethylene glycol (PEG), and *Agrobacterium*. Particle bombardment facilitates the transfer of the gene (>150kb DNA), but the concern is the integrity of the DNA after the transfer as well as low-level expression in plant cells due to the high pressure of gene machine plant cells; DNA damage may occur (Taylor and Fauquet, 2002). In electroporation method, production of strong magnetic field will damage the desired gene resulting in inaccurate translational end products and toxic to the plant cells. Though chemical-mediated gene transfer (PEG) allows for the successful transfection of protoplasts, the transfected protoplasts cannot be regenerated into whole plants. Unfortunately, this limits the use of PEG in gene delivery. *Agrobacterium*-mediated gene transfer has a high DNA delivery efficiency and integration of foreign DNA into the nuclear genome, but it is ineffective in numerous plant species, thus limiting the host range of this method of gene delivery (Demirer and Landry, 2017).

With the aforementioned pros and cons, it is indeed necessary to develop a novel delivery approach for gene delivery by using nanotechnology. The technology is inexpensive, facile, and robust. The plant scientific community can benefit from a delivery method by adopting nanoconcepts so that it can transfer genes to all phenotypes of plants with high efficiency and low toxicity.

Recently, with the advancement and progress made in the mammalian system by using nanoparticle-mediated gene delivery, researchers accelerated the research in plants by adopting NP-mediated gene delivery (Chen and Lai, 2015). For developing an appropriate gene transfer technology, the NPs should easily pass through the cell wall (which acts as a barrier for NPs translocation) and localize in the internal organelles (nuclear and cytosol). Preferably, cationic NPs is preferred for gene delivery technology because plant cell wall is negatively charged and cationic NPs bind with cell wall and proceed the transfer of gene effectively (Albanese et al., 2012; Torney et al., 2007). However, until now, nanocarrier delivery in plants has been accomplished with the help of external forces such as physical methods (biolistic, electroporation/sonoporation, magnetofection) and nonphysical methods (cationic carriers and infiltration). However, certain nanoparticles such as mesoporous silica nanoparticles (MSNs), DNA nanostructures and DNA origami, silicon carbide whiskers (SCWs), and layered double hydroxide (LDH) clay nanosheets have proved the efficacy of delivering a gene into the plant without any help of external sources (Chang et al., 2013).

Nanoparticles commonly used for gene delivery are comprised into four major categories: (i) metallic/magnetic NPs; (ii) carbon-based NPs; (iii) silicon-based NPs; and (iv) polymeric-based NPs. Metallic/magnetic NPs (gold, silver, iron oxide) will carry DNA only as a genetic cargo to deliver inside the plant/animal cells (Sandhu et al., 2002; Zhao et al., 2017). But carbon nanotubes (single-walled carbon nanotubes, multiwalled carbon nanotubes, fullerenes) carry both DNA and RNA for delivery while silicon-based NPs (silica spheres, mesoporous silica NPs, (MSN), silicon carbide) effectively deliver DNA and proteins. Polymeric NP (polyethylene-glycol (PEG), polyethylenimine (PEI), poly(lactic-*co*-glycolic acid) (PLGA)) vehicles encapsulate DNA, RNA, proteins, and ribonucleoprotein as a genetic cargo and deliver inside the cells (Demirer et al., 2018).

In Table 1, we have provided a comprehensive list of nanoparticles used in gene delivery. Various dimensions of nanomaterials have been used and the target gene is successfully transversed into the plant cell and integrated inside the host chromosome. Though in regard to plant gene delivery by nanocarriers, the plant cell wall acts as a barrier that can be overcome by the NPs with the help of chemical or mechanical aid. In Table 1, the NP gene transfer is achieved with the help of external sources. However, there is a limitation in NP-mediated gene transfer and progeny regeneration from the embryonic calli is still questionable, whereas the regeneration capacity varies among species, genotype, and plant depending on the developmental age of the source tissue (Cunningham et al., 2018). In some cases, NPs mediated passive gene delivery to leaves, roots, and protoplasts holds great promise for in vivo and in vitro genetic engineering of plants especially for transient expression in somatic tissues (Cunningham et al., 2018). Examples of NP-mediated passive delivery include *Nicotiana tabacum* protoplasts (Naqvi et al., 2012), *Arabidopsis thaliana* roots (Jiang et al., 2014), *Eruca sativa* leaves, and *Nicotiana benthamiana* leaves (Demirer et al., 2018).

NP-mediated gene delivery also in some case fails to transverse into the plant cell wall due to various factors. Therefore, adopting a suitable nanocarrier for delivery and regeneration is necessary to improve efficiency and to expand stable transformation to all plant species. Consistent research in nanoparticles mediated gene delivery is achieved so far and few NPs gene delivery methods were successfully delivered gene into various crop species.

Table 1 List of nanoparticles used in gene delivery.

Nanoparticle category	Nanoparticle type	Genetic material	Host	References
Metallic NPs	Gold	Plasmid DNA	*Nicotiana tabacum*	Vijayakumar et al. (2010)
	Iron oxide	Plasmid DNA	*Gossypium hirsutum*	Zhao et al. (2017)
	Calcium phosphate	pCambia 1301	*Brassica juncea*	Naqvi et al. (2012)
	ZnS nanoparticles	GUS-encoding plasmid DNA	*Nicotiana tabacum*	Fu et al. (2012)
	Magnetic gold NPs	β-Glucuronidase (GUS) plasmid	*Brassica napus*	Hao et al. (2013)
	Gold-plated MSNs	AmCyan1 and DsRed2 plasmids; Cre protein	*Zea mays*	Martin-Ortigosa et al. (2014)
Carbon NPs	Organically functionalized CNTs	YFP plasmid	*Nicotiana tabacum*	Burlaka et al. (2015a,b)
	Polymer functionalized CNTs	Green fluorescent protein (GFP) coding DNA	*Eruca sativa, N. benthamiana,* and *T. aestivum*	Demirer et al. (2018)
Dendrimer NPs	PAMAM dendrimer NPs	Double-stranded DNA for RNA interference	*Arabidopsis thaliana*	Jiang et al. (2014)
Silicon Based NPs	Mesoporous silica nanoparticles	Plasmid DNA	*Nicotiana tabacum*	Torney et al. (2007)
Polymers	Polymeric NPs	siRNA	*Nicotiana tabacum*	Silva et al. (2010)
Others	Starch nanoparticles	Plasmid DNA	*Dioscorea* sp.	Liu et al. (2008)

3 Types of carbon nanomaterials used in gene delivery

Carbon-based nanomaterials (CN) have attractive physicochemical properties such as surface morphology, small surface area, and increased chemical reactivity that are largely exploited in the engineering and chemical sciences (Cha et al., 2013). Some of the unique features of CNs are a stable molecular architecture and their uniform dispersive characteristic in the medium in which CNs are applied. CNs are classified based on the geometrical structure (tube shaped, horn shaped, spherical, or ellipsoidal). CNs in the shape of tubes are carbon nanotubes while horn-shaped particles are nanohorns and spheres or ellipsoids belong to the fullerene group (Zaytseva and Neumann, 2016).

Among the other carbon-based nanomaterials, CNTs are allotropes of carbon exhibiting stupendous properties such as dimension, structure, and topology as well as outstanding mechanical, electronic, and optical properties (Nasir et al., 2018). Structurally, carbon nanotubes vary in length, diameter, chirality (symmetry of the rolled graphite sheet), and the number of layers. Based on the structural pattern, CNTs may be classified into two main groups: single-walled carbon nanotubes (SWCNT) with a diameter around 1–3 nm and a length of a few micrometers, and multiwalled carbon nanotubes (MWCNTs) with a diameter of 5–40 nm and a length of around 10 μm (Aqel et al., 2012).

Despite the advantages of CNTs, their potential seemed paramount in nanoparticle production, electronics, solar cells, hydrogen storage, analytical devices, bioengineering, bioimaging, biosensors, drug delivery, and gene delivery (Cha et al., 2013). CNTs are produced by electric arc, laser ablation, chemical vapor deposition (CVD), and cyclic oxidation. Among the methods, CVD is the most preferred method of CNT production (Szabó et al., 2010). With the ever-growing merits of CNTs, their use in applications to living beings is hindered due to low biocompatibility. Pristine CNTs have highly hydrophobic surfaces that result in poor water dispersibility, making them unsuitable for biological applications. To improve biocompatibility, surface modification or functionalization is required to improve the solubility of CNTs. The surface functionalization of CNT is achieved by (i) covalent attachment of chemical groups, or (ii) noncovalent functionalization of CNT by supramolecular adsorption or wrapping of various functional molecules onto the tubes (Zhang et al., 2011). Functionalization influences modifying CNTs with desired properties exhibiting superior properties intended for biomedical, biotechnological, and pharmaceutical applications. Moreover, functionally modified CNTs will also effectively deliver genetic cargoes at the cellular or subcellular level and also exhibit controlled degradation and non toxic to the plant cells (Lombardo et al., 2019).

Covalent functionalization of CNTs involves a covalent bond formation between the CNT surface and the hydrophilic moiety. Functionalization of CNTs in the sidewall is carried out by agents such as aryl diazonium moieties, elemental metals, inorganic acids, esters, aldehydes, amines, aromatics, biomolecules, and polymers (Kharisov et al., 2008). In the tailoring of CNTs for genetic transformation methods, covalent bond formation between DNA and CNTs is not acceptable because it spoils DNA functions and prevents releasing DNA in cells (Burlaka et al., 2015a). To use CNTs in genetic transformation, it is preferable to form covalently grafted chains with noncovalent molecules (DNA). The DNA will immobilize on the side wall, which anchors to the covalently grafted chain. This kind of approach can be used to prepare CNT-based sensors and novel gene delivery systems (Sireesha et al., 2018). For anchoring, different residues (polymers polyallylamine (PAA), polyethyleneimine (PEI), or polyethylene glycol (PEG)) are grafted onto the side wall of the CNT for DNA attachment in the development of DNA delivery systems (Burlaka et al., 2015b). Such an approach will be effectively transferring DNA into mammalian cells with high expression and low toxicity. For gene transfer by CNTs in plant cells, it is amenable to covalently modify CNTs with enzymes that degrade the plant cell wall.

Noncovalent CNT functionalization is in the forefront of study for researchers, utilizing mild reaction conditions and nonhazardous chemicals. This approach will further develop the CNT-based nanotechnology approach involving green chemistry principles and practices. Noncovalent functionalization is based on the van der Waals and π-π interactions; adsorption or wrapping of biomolecules, surfactants, and polymers is also used (Tuncel, 2011). Specifically, for genetic transformation methods, noncovalent functionalization of CNTs with biomolecules such as DNA and protein should undergo sonication. During sonication, the protein or DNA will change its conformation with the opening of hydrophobic domains for interaction with the nanotube side wall (Witus et al., 2007). Such a process will facilitate the increased efficiency of noncovalent molecule binding (DNA, proteins) with CNTs and transform effectively in plant cells.

Protein, carbohydrate, and nucleic acid (DNA, RNA) functionalization with CNT improves water dispersibility, biocompatibility, and low toxicity, thus providing a new platform for green nanotechnology (Nandy et al., 2012). Proteins such as lysozyme, bovine serum albumin (BSA), hydrophobins, hemoglobin, myoglobin, ovalbumin, trypsin, glucose oxidase, synthetic oligopeptides, histone, streptavidin, and self-assembling peptide amphiphiles were functionalized with CNTs by π-π interactions between the aromatic residue of protein and the graphitic surface of CNTs (Nepal and Geckeler, 2007). Among the biomolecules, DNA has a well-defined length and sequence, high dispersion efficiency, and can be covalently or noncovalently functionalized with CNTs (Chen et al., 2007). In general, nucleic acids bind wrap on CNT by van der Waals and π-π stacking interactions. During the adsorption of DNA on SWCNTs, it appears to have to change in the internal structure of DNA. Similarly, siRNA strongly binds to the SWCNT surface via unzipping its base pairs; the propensity of unzipping is directly proportional to the increase in SWCNT diameter (Nakashima et al., 2003). Hydrophobic π-π stacking is not the only main factor for CNT-DNA complex formation. Other factors include electrostatic interaction between the phosphate backbone and the aqueous environment base stacking within the single-stranded polymer, and geometric and stereochemical limitations might also influence stability features (Becker et al., 2007).

Polysaccharides such as chitosan, schizophylla, hyaluronic acid, *N*-acetylgalactosamine phospholipids, phospholipid-dextran conjugates, and amino acids were used for conjugation of biomolecules with CNT (Patnaik et al., 2015). In a recent report, the dispersion efficiency and stability of functionalized CNTs are high when the following combination was used: dsDNA-deoxyribonucleoside triphosphates (dNTPs), bovine serum albumin (BSA), adenosine-triphosphate-Na (ATP-Na), and sodium humate (Burlaka et al., 2011). Thus, from the above information, it is obviously noted that CNT functionalization is an important phenomenon for gene delivery to microbial cells as well as plant and animal cells. Depending upon the host and genetic cargo, it is certain to modify the CNT with the appropriate agent for successful transformation and expression efficiency.

4 CNTs in genetic engineering

Among the other biotransformation methods, CNT-based methods of gene delivery are among the most robust, feasible, and convenient. Existing gene transfer techniques are very expensive, unspecific, and nonreproducible. In spite of this, CNTs will be an alternative that provides numerous benefits when utilized in genetic transformation. CNT-based delivery methods will offer the following advantages: making gene transfer fast and easy; protection of DNA by bonding or encapsulation to avoid the DNA damage from nucleases and others; and, in certain cases, energy-induced DNA damage-related gene mutation is strongly prohibited (Karimi et al., 2015; Rafsanjani et al., 2012). Certainly, the approach of nanoconcepts and principles with genetic transformation methods will provide novel gene transfer with high efficiency and expression.

Hitherto, CNTs have been a choice of many researchers for transferring DNA/RNA and protein to any kind of bacteria or mammalian cells. A recent report states that CNT surface modification with a suitable biomolecule and carrier effectively delivers DNA into the cells. Cai et al. (2005) reported the highly efficient delivery of plasmid DNA into ex vivo neurons and splenic B cells as well as transformed mouse B lymphocytes. Liu et al. (2005) developed carbon nanotubes by a grafting approach to prepare polyethyleneimine-graft multiwalled carbon nanotubes for transient delivery into COS7 and HepG2 cells. Mattos et al. (2011) postulated the action of MWCNTs against *Neisseria meningitidis* for the study of nanostructure action on the transformation process of *meningococci*. Pantarotto et al. (2004) utilized carbon nanotubes functionalized with supramolecular complexes with plasmid DNA through ionic interactions and delivered plasmid DNA effectively into CHO cells. Nunes et al. (2010) formulated cationic polymer-grafted multiwalled carbon nanotubes (polymer-g-MWNTs) to efficiently complex and transfer plasmid DNA (pCMV-β Gal) into lung cancer cells (AS549), and established their toxicological profile. Rojas-Chapana et al. (2005) demonstrated multiwalled carbon nanotubes for delivering plasmid DNA into *Escherichia coli* cells. These methods clearly show that functionalized CNT complexes are able to bind and penetrate within cells. Another important concern is functionalized CNT complexes displaying low toxicity and a high level of transgene expression rather than the conventional method of gene transfer (Mohajeri et al., 2019).

Another important phenomenon to address is the mode of entry of CNTs to mammalian cells. The uptake of CNTs in mammalian cells is suggested to occur by phagocytosis (Cheng et al., 2009; Cherukuri et al., 2004; Lacerda et al., 2012) or endocytosis (Liu et al., 2009). Imaging experiment electron microscopy and fluorescence microscopy studies affirm that uptake of CNTs inside the cells is plausible by either of the mechanisms (Kam et al., 2006). However, there is still no clarity regarding CNT uptake inside the cell and elucidation of the structural conformation inside the cells. Therefore, it is advocated to study the translocation mechanism of CNT entry in a particular biological system.

In a report, 5 nm of pristine and functionalized SWCNTs were studied in phospholipid bilayer models. This supports that nanotubes are directly translocated through the phospholipid membrane by a nanoneedle-like process (Kraszewski et al., 2012). Nanoneedle is a short and amino-functionalized nanotube to passively breach the cell membrane of mammalian cells. CNT uptake by the cells depends upon the length of the nanotube and the cell type (Lacerda et al., 2007). Various authors and their reports suggested that the cylindrical shape and high aspect ratio of CNTs as well as the length and functionalization moiety will influence the penetration through the plasma membrane.

5 CNTs for gene delivery in plants

Unlike mammalian cells, genetic transformation in a plant cell is difficult; plant cell walls act as a strong barrier to prevent the entry of any other external agents. The cell wall is made up of cross-linked polysaccharides, which discriminate between plant and animal cells. The plant cell wall architecture limits the use of any kind of nanomaterial for genetic cargo delivery. In an investigation, the uptake of fluorescent quantum dots directly into the plant cells was achieved after 24 h of starving, but unfortunately, the gene was not delivered (Etxeberria et al., 2006). Liu et al. (2009) reported SWNT penetration into walled plant cells, the *N. tabacum* L.cv. Bright Yellow (BY-2) cell, a popular plant cell model. They demonstrated effectively delivering DNA and small dye molecules into intact plant cells. Serag et al. (2011) showed the ability of MWCNTs to traverse across a plant cell membrane by escaping endocytosis. Further short MWCNTs were able to localize in the region of the nucleus, plastids, and vacuoles of plant cells.

Demirer et al. (2019) proved that pristine and chemically functionalized nanotubes efficiently deliver DNA and protein expression without transgene integration in mature plants of *N. benthamiana* (Nb), *E. sativa* (arugula), *Triticum aestivum* (wheat), and *Gossypium hirsutum* (cotton) leaves as well as arugula protoplasts. The CNTs not only facilitated the delivery of biomolecules into intact plant cells, but also protected the polynucleotides from nuclease degradation Fig. 1.

Recently, Kwak et al. (2019) attempted a novel strategy to deliver the transgene into the chloroplast specifically by nanotubes. In his study, he designed chitosan-complexed single-walled carbon nanotubes that selectively delivered plasmid DNA to chloroplasts of mature *E. sativa*, *Nasturtium officinale*, *N. tabacum*, and *Spinacia oleracea* plants and in isolated *A. thaliana* mesophyll protoplasts. The delivery was successful and exhibited transient expression. Kwak and his coworkers indicated gene delivery in plants by visualization of the transitive expression of a marker gene following infusion of the nanoparticle carriers to the leaf lamina. Kwak produced four groups of chitosan-complexed SWNTs by noncovalent coating and covalent amendment of the nanotube side wall to strengthen pDNA loading and delivery performance to the chloroplasts (Fig. 2).

Protoplast-based gene delivery methods result in a low succession rate because the viability of protoplasts and the capability of dividing are greatly reduced by the chemical treatment to disorganize the cell wall. To overcome this, Kaji et al. (2008) developed a new functionalized cup-stacked CNT (CSCNT-cellulase) and delivered this into the *A. thaliana* ecotype Columbia Col-0 and *Glycyrrhiza glabra*. The results showed that the CSCNT-cellulase was uptaken by 20% of *A. thaliana* cells and by 15% of *G. glabra* cells. This is the first to report the CNT penetration to cross the plant cell wall, the cell membrane, and the nuclear membrane, then localize inside the cell nucleus. Burlaka et al. (2015a,b) developed SWCNTs and MWCNTs with plasmid construct pGreen 0029 and delivered in the protoplasts, *N. tabacum* callus and leaf explants. The study exploits that the size of CNTs influences the penetration of tubes inside the cell wall, membrane, and inner membrane.

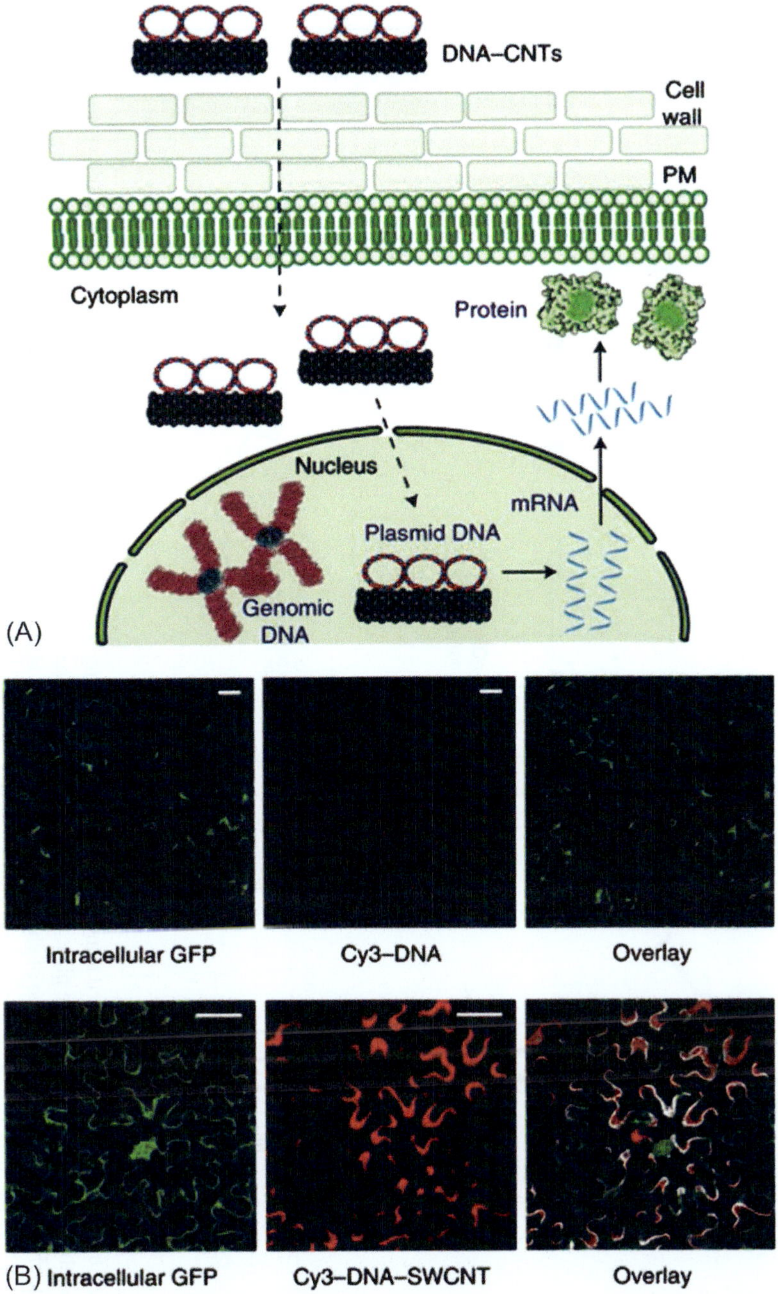

Fig. 1 (A) Diagram revealing DNA-CNT transfer in plant cells and following gene expression (*dotted lines* represent trafficking steps and *the rigid lines* represent gene expression steps). *PM*, plasma membrane. (B) Nanoparticle internalization into adult plant cells is demonstrated by imaging Cy3-tagged DNA-CNTs with confocal microscopy, in comparison to a control sample of Cy3-tagged DNA without CNTs, in a transgenic mGFP5 Nb plant. Republished from Demirer, Gozde S., et al., 2019. High aspect ratio nanomaterials enable delivery of functional genetic material without DNA integration in mature plants. Nat. Nanotechnol. with permission from Springer Nature.

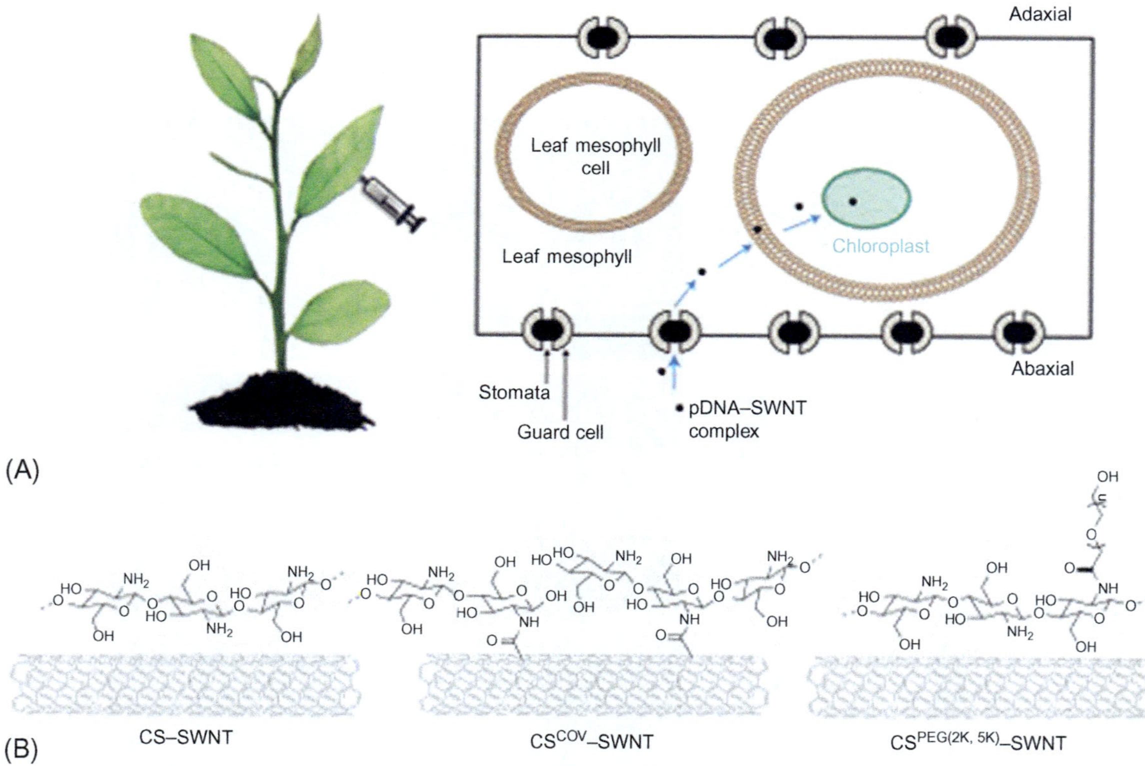

Fig. 2 Diagram representation of chloroplast-targeted transport of pDNA by employing SWNTs in a plant leaf. (A) pDNA-SWNT complexes enter the leaf mesophyll simply by stomata pores, traverse plant cell walls, plasma membranes, and, gradually chloroplast bilayers. Adversely charged pDNA is compacted to the positively charged surface of chitosan-complexed SWNTs via electrostatic interactions. (B) Chemical composition of chitosan-complexed SWNTs and their particular denotations; chitosan-SWNTs (CS-SWNTs), PEGylated chitosan-SWNTs (CSPEG-SWNTs with PEG5K and PEG2K), and covalently attached chitosan to the carboxylated SWNTs (CSCOV-SWNTs).
Republished from Kwak, S.Y., Lew, T.T.S., Sweeney, C.J., Koman, V.B., Wong, M.H., Bohmert-Tatarev, K., Snell, K.D., Seo, J.S., Chua, N.H., Strano, M.S., 2019. Chloroplast-selective gene delivery and expression in planta using chitosan-complexed single-walled carbon nanotube carriers. Nat. Nanotechol. https://doi.org/10.1038/s41565-019-0375-4 with permission from Springer Nature.

6 CNTs plant cell entering

In plants, the cell wall pore diameter ranging from 5 to 20 nm allows the entry of NPs, which are less than the cell wall pore diameter and can enter the plasma membrane of the cell. There is a shift in the pore cell wall or induction of new pores in the cell wall upon interaction with engineered NPs, which in turn uptake the NPs. In another viewpoint, nanoparticles enter through the cell membrane and internalize inside the cell organelles by endocytosis (Rai, 2015). Nanomaterials enter through the cell wall by energy-driven assisted transfer by a transporter protein (aquaporin) or through ion channels. Depending upon the type of location or part of the plant, NPs may enter inside the cells (Jia et al., 2005). For example, when NPs are applied on the leaf surface, they enter through the stomatal openings or through the bases of trichomes (Uzu et al., 2010). Another way of nanoparticle entry is through plasmodesmata, a specialized structure for transport between cells.

In regard to carbon-based nanomaterials, the only hurdle for translocation of CNTs in a plant cell is the cell wall. In the study, MWCNT escapes the endocytosis process and localizes in the protoplasts of *Catharanthus roseus* (Serag et al., 2011). In another study, SWCNTs were shown to enter the mesophyll protoplasts of *A. thaliana* by a nonenergy-dependent pathway. There are various factors that influence the penetration and localization of CNTs inside plant cells (Yuan et al., 2011). They are the size of NMs, genetic cargo, the functionalization agent, and the plant species. In some cases, some plants are easily and effectively uptaken by the NMs while in other cases, it is very difficult to transfer the gene. However, in such cases, CNTs with external aids (gene gun, electroporation, particle bombardment, or PEG) can be used to deliver the gene. But howsoever, in such case, CNTs with external aid (gene gun, electroporation, particle bombardment and PEG) can be used to deliver the gene but the success of transformation and regeneration efficiency is doubted one.

7 Gene entry mechanism

SWNTs may navigate the hard plant cell walls passing through the double lipid bilayers of chloroplasts and deliver genetic cargoes inside the chloroplasts (Giraldo et al., 2014). This type of nanoparticle uptake mode of action was defined by the lipid exchange envelope penetration (LEEP) model, wherein the potential of nanoparticles to permeate the cell membrane and the chloroplast envelope is influenced mainly by the nanoparticle size and surface charge (Lew et al., 2018; Wong et al., 2016). Depending on the previous mode of action, the tunable physical and optical characteristics of carbon nanotubes are usually leveraged to maximize the recurring delivery of biological cargo across many plant barriers that have hitherto been difficult to enter. Among various types of nanoparticles, SWNTs have attracted significant curiosity as nanocarriers for medication and gene delivery because of their large element ratio and significant surface area for substance alteration (Li et al., 2017). On the other hand, the offered functions of SWNTs in plants were mostly controlled to reports of the transfer of

SWNTs in plant cells (Liu et al., 2009), and nothing of the work investigated the probability of implementing SWNTs as nanocarriers for gene delivery inside particular targeted plant organelles.

8 Future prospective

A nanocarrier of any agent must precisely target the cells and transfer the gene effectively. In this regard, CNTs are attractive options, owing to their molecular properties and transfer efficiency. But we must solve some of the issues relating to CNTs to achieve a full functional delivering agent. Recently, surface modification with biomolecule moieties will spur the CNT to a great extent in a number of plant species. But depending on the type of plants, CNTs will vary such as in interactions with living cells, CNT toxicity, environmental fate, and biodegradation (Karousis et al., 2010). CNT production through chemical methods will result in various impurities as byproducts; in biological applications, such impurities will cause toxicity to the cells. Research should focus on the elimination of toxicity so that CNTs can be employed in a large number of biological applications (Lin et al., 2009).

Another important fundamental study is needed, and studies encouraging the interaction of CNTs with phospholipid layers at the molecular level are much needed (Pogodin and Baulin, 2010). Such studies will encourage others to develop novel hybrid CNs for gene delivery. Concerning the fate of CNs after entering plant cells, the mechanism of elimination is lacking, and only limited studies reported the defunctionalization of CNs (Yang et al., 2009). The report related to biodegradation of CNs should be encouraged.

In order to improve CNT-based gene delivery, some things must be noted for betterment. The surface charge of protoplasts and plasma membrane must vary from -10 to -30 mV (Reid et al., 2002). Raising the positive charge against the cell membrane will enhance the adhesion and transfer of CNTs across the membrane. The porosity of protoplasts is 0.1–5000 nm and the plasma membrane is 1–50 nm; the preparation of CNT size within the stipulated porosity and plasma membrane range will afford better transfer and integration of the gene into plant cells (Berestovsky et al., 2001).

Novel techniques should be adapted to degrade the plant cell wall because chemical treatment is toxic and protoplasts are unable to survive and divide rapidly. Techniques related to CNTs guided gene transfer relies on the types of CNT (SWCNT; MWCNT) used and functionalization with biomolecules (DNA/RNA/proteins) in regard to plant gene transformation. Further research is highly warranted in this aspect. Improvements in biocompatibility should be addressed; in general, raw carbon has good compatibility but nano-sized carbon has less compatibility. At the same time, nano-sized carbon has great functional features when compared with carbon. This contrasting feature (function and compatibility) must be tailored and developed in CNTs to lead to a promising approach in plant biotechnology. Finally, relentless efforts should be imposed by the researchers in order to acquire newer insights of CNTs in delivery of gene in plant cells with maximum efficiency of gene transfer and successful transformation and expression.

9 CNTs as a novel tool for plant genome editing

Some inorganic nanoparticles are natural potential CRISPR component carriers because they have already been used for similar purposes. Examples of these include AuNPs, CNTs, bare mesoporous silica nanoparticles (MSNPs), and dense silica nanoparticles (SiNPs). The use of AuNPs for CRISPR/Cas9 delivery was described above. While CNTs (Bates and Kostarelos, 2013), MSNPs (Luo et al., 2014), and SiNPs (Luo and Saltzman, 2000) have been used for many gene delivery applications, the use of these carriers for Cas9 delivery has yet to be reported. CRISPR-Cas systems are successfully implemented across a wide range of prokaryotic and eukaryotic species for efficient genome editing. The Innovative Genomics Institute (IGI) team wants to use this to deliver CRISPR-Cas9 genome editing tools in the future. The carbon nanotube would be coated in DNA that codes for the Cas9 protein and the guide RNA, which work together to edit a specific gene (Demirer et al., 2019). The genome editing tools are presented just long enough to edit the plant's DNA before being broken down by the plant, leaving no footprint. This means that, unlike some current plant genome editing methods, no foreign DNA would be inserted into the plant's genome.

10 Conclusion

Some nanomaterial delivery systems allow varied combined chemistries to obtain DNA transport without transgene utilization in the two model and crop plants, and in both dicot and monocot plants, with substantial performance that is devoid of toxicity or tissue destruction. To conclude, CNTs are an emerging nanocarrier for the delivery of a gene into the plant cells. The future will undoubtedly rely on carbon nano-based methods for gene delivery, sensors, imaging, etc. Significantly, plant biotechnology will be overwhelmed by the utilization of carbon nanotubes for delivering genes (DNA/RNA/siRNA and proteins) to develop and establish transient expression in plant cells in all kinds of species in comparison with existing methods. Carbon nanotubes may possibly grow to be an excellent system for editing and enhancing the chloroplast genome. Plant genome editing engineering, which includes the CRISPR/Cas9 strategy, enables particular changes of practically any crop genomic sequence; hence, they might have a significant function in the genetic enhancement of specific agronomic qualities. CRISPR/Cas9 concept continues to be effectively used to improve quality of plants producing disease resistance plants, increasing crop yields, and quality, increased stress tolerance in plants. Moreover, production of mutants by CRISPR/Cas9 will not hamper hybrid breeding, but also assists heterosis breeding in the possible future. Soon researchers will use this unique approach to respond promptly to pathogens, insect pests, salinity, and drought, all of which impact essential food crops such as soybeans, and to safeguard our most preferred plants, which can be under risk from plant pathogens and climate change.

References

Abd-Elsalam, K.A., Prasad, R., 2018. Nanobiotechnology Applications in Plant Protection. Springer International Publishing. https://doi.org/10.1007/978-3-319-91161-8. 394 pp.

Albanese, A., Tang, P.S., Chan, W.C., 2012. The effect of nanoparticle size, shape, and surface chemistry on biological systems. Annu. Rev. Biomed. Eng. 14, 1–16.

Aqel, A., El-Nour, K.M.A., Ammar, R.A., Al-Warthan, A., 2012. Carbon nanotubes, science and technology part (I) structure, synthesis and characterization. Arab. J. Chem. 5, 1–23.

Bates, K., Kostarelos, K., 2013. Carbon nanotubes as vectors for gene therapy: past achievements, present challenges and future goals. Adv. Drug Deliv. Rev. 65, 2023–2033.

Batlang, U., Tsurupe, G., Segwagwe, A., Obopile, M., 2014. Development and application of modern agricultural biotechnology in Botswana: the potentials, opportunities and challenges. GM Crops Food 5, 183–194.

Baulcombe, D., Crute, I., Davies, B., Dunwell, J., Gale, M., Jones, J., 2009. Reaping the Benefits: Science and the Sustainable Intensification of Global Agriculture. The Royal Society, London.

Becker, M.L., Fagan, J.A., Gallant, N.D., Bauer, B.J., Bajpai, V., Hobbie, E.K., Lacerda, S.H., Migler, K.B., Jakupciak, J.P., 2007. Length-dependent uptake of DNA-wrapped single-walled carbon nanotubes. Adv. Mater. 19 (7), 939–945.

Berestovsky, G., Ternovsky, V., Kataev, A., 2001. Through pore diameter in the cell wall of *Chara corallina*. J. Exp. Bot. 52, 1173–1177.

Burlaka, O.M., Pirko, Y.V., Yemets, A.I., Blume, Y.B., 2011. Carbon nanotubes and their use for genetic transformation of plants. Nanostruct. Mater. 2, 84–101.

Burlaka, O.M., Pirko, Y.V., Yemets, A.I., Blume, B.Y., 2015a. Plant genetic transformation using carbon nanotubes for DNA delivery. Cytol. Genet. 49, 349.

Burlaka, O., Pirko, Y., Yemets, A., Blume, Y., 2015b. Application of carbon nanotubes for plant genetic transformation. In: Fesenko, O., Yatsenko, L. (Eds.), Nanocomposites, Nanophotonics, Nanobiotechnology, and Applications. Springer Proceedings in Physicsvol. 156. Springer, Cham.

Cai, D., Mataraza, J., Qin, Z., 2005. Highly efficient molecular delivery into mammalian cells using carbon nanotube spearing. Nat. Methods 2, 449–454.

Cha, C., Shin, S.R., Annabi, N., Dokmeci, M.R., Khademhosseini, A., 2013. Carbon-based nanomaterials: multifunctional materials for biomedical engineering. ACS Nano 7, 2891–2897.

Chang, F.P., Kuang, L.Y., Huang, C.A., Jane, W.N., Hung, Y., Yue-ie, C.H., Mou, C.Y., 2013. A simple plant gene delivery system using mesoporous silica nanoparticles as carriers. J. Mater. Chem. B 1, 5279–5287.

Chen, Q., Lai, H., 2015. Gene delivery into plant cells for recombinant protein production. Biomed. Res. Int. 932161. https://doi.org/10.1155/2015/932161.

Chen, Y., Liu, H., Ye, T., Kim, J., Mao, C., 2007. DNA-directed assembly of single-wall carbon nanotubes. J. Am. Chem. Soc. 129, 8696–8697.

Cheng, C., Muller, K.H., Koziol, K.K., 2009. Toxicity and imaging of multi-walled carbon nanotubes in human macrophage cells. Biomaterials 30, 4152–4160.

Cherukuri, P., Bachilo, S.M., Litovsky, S.H., Weisman, R.B., 2004. Near-infrared fluorescence microscopy of single-walled carbon nanotubes in phagocytic cells. J. Am. Chem. Soc. 126, 15638–15639.

Cunningham, J., Natalie, S., Demirer, S., Matos, L., Landry, P., 2018. Nanoparticle-mediated delivery towards advancing plant genetic engineering. Trends Biotechnol. 36, 882–897.

Darbani, B., Farajnia, S., Toorchi, M., Zakerbostanabad, S., Noeparvar, S., Stewart, C.N., 2008. DNA-delivery methods to produce transgenic plants. Biotechnol. J. 7, 385–402.

Demirer, G.S., Landry, M.P., 2017. Delivering genes to plants. Chem. Eng. Prog. 113, 40–45.
Demirer, G.S., Zhang, H., Matos, J., Chang, R., Chio, L., Staskawicz, B., Landry, M.P., 2018. High aspect ratio nanomaterials enable biomolecule delivery and transgene expression or silencing in mature plants. Nat. Nanotechnol. https://doi.org/10.1038/s41565-019-0382-5.
Demirer, G.S., et al., 2019. High aspect ratio nanomaterials enable delivery of functional genetic material without DNA integration in mature plants. Nat. Nanotechnol. 14, 456–464.
Duhan, J.S., Kumar, R., Kumar, N., Kaur, P., Nehra, K., Duhan, S., 2017. Nanotechnology: the new perspective in precision agriculture. Biotechnol. Rep. 15, 11–23.
Etxeberria, E., Gonzalez, P., Baroja-Fernandez, E., Romero, J.P., 2006. Fluid phase endocytic uptake of artificial nano-spheres and fluorescent quantum dots by sycamore cultured cells: evidence for the distribution of solutes to different intracellular compartments. Plant Signal. Behav. 1, 196–200.
Feynman, R.P., 1960. There's plenty of room at the bottom. Eng. Sci. 23, 22–36.
Fouad, M., Kaji, N., Jabasini, M., Tokeshi, M., Baba, Y., 2008. Nanotechnology meets plant biotechnology: carbon nanotubes deliver DNA and incorporate into the plant cell structure. In: 12th International Conference on Miniaturized Systems for Chemistry and Life Sciences, San Diego, California, USA. pp. 12–16.
Fu, Y.Q., Li, L.H., Wang, P.W., Qu, J., Fu, Y.P., Wang, H., Sun, J.R., 2012. Delivering DNA into plant cell by gene carriers of ZnS nanoparticles. Chem. Res. Chin. Univ. 28, 672–676.
Giraldo, J.P., Landry, M.P., Faltermeier, S.M., McNicholas, T.P., Iverson, N.M., Boghossian, A.A., Reuel, N.F., Hilmer, A.J., Sen, F., Brew, J.A., Strano, M.S., 2014. Plant nanobionics approach to augment photosynthesis and biochemical sensing. Nat. Mater. 13, 400–408.
Gour, A., Jain, N.K., 2019. Advances in green synthesis of nanoparticles. Artif. Cells Nanomed. Biotechnol. 47, 844–851.
Hao, Y., Yang, X., Shi, Y., Song, S., Xing, J., Marowitch, J., Chen, J., Chen, J., 2013. Magnetic gold nanoparticles as a vehicle for fluorescein isothiocyanate and DNA delivery into plant cells. Botany 91, 457–466.
Institute of Medicine (IOM), 2009. Nanotechnology in Food Products: Workshop Summary. The National Academies Press, Washington, DC.
Jahangirian, H., Lemraski, E.G., Webster, T.J., Rafiee-Moghaddam, R., Abdollahi, Y.A., 2017. Review of drug delivery systems based on nanotechnology and green chemistry: green nanomedicine. Int. J. Nanomedicine 12, 2957–2978.
Jia, G., Wang, H., Yan, L., Wang, X., Pei, R., Yan, T., Zhao, Y., Xinbiao, G., 2005. Cytotoxicity of carbon nanomaterials: single-wall nanotube, multi-wall nanotube, and fullerene. Environ. Sci. Technol. 39, 1378–1383.
Jiang, L., Ding, L., He, B., Shen, J., Xu, Z., Yin, M., Zhang, X., 2014. Systemic gene silencing in plants triggered by fluorescent nanoparticle-delivered double-stranded RNA. Nanoscale 6, 9965.
Kam, N.W., Liu, Z.A., Dai, H.J., 2006. Carbon nanotubes as intracellular transporters for proteins and DNA: an investigation of the uptake mechanism and pathway. Angew. Chem. Int. Ed. 45, 577–581.
Karimi, M., Solati, N., Ghasemi, A., 2015. Carbon nanotubes part II: a remarkable carrier for drug and gene delivery. Expert Opin. Drug Deliv. 12, 1089–1105.
Karousis, N., Tagmatarchis, N., Tasis, D., 2010. Current progress on the chemical modification of carbon nanotubes. Chem. Rev. 110, 5366–5397.
Kharisov, B.I., Kharissova, O.V., Leija Gutierrez, H., Ortiz Méndez, U., 2008. Recent advances on the soluble carbon nanotubes. Ind. Eng. Chem. Res. 48, 10.

Kraszewski, S., Bianco, A., Tarek, M., Ramseyer, C., 2012. Insertion of short amino-functionalized single-walled carbon nanotubes into phospholipid bilayer occurs by passive diffusion. PLoS One 7, 40703.

Kwak, S.Y., Lew, T.T.S., Sweeney, C.J., Koman, V.B., Wong, M.H., Bohmert-Tatarev, K., Snell, K.D., Seo, J.S., Chua, N.H., Strano, M.S., 2019. Chloroplast-selective gene delivery and expression in planta using chitosan-complexed single-walled carbon nanotube carriers. Nat. Nanotechol. https://doi.org/10.1038/s41565-019-0375-4.

Lacerda, L., Raffa, S., Prato, M., 2007. Cell-penetrating CNTs for delivery of therapeutics. Nano Today 2, 38–43.

Lacerda, L., Russier, J., Pastorin, G., 2012. Translocation mechanisms of chemically functionalised carbon nanotubes across plasma membranes. Biomaterials 33, 3334–3343.

Lew, T.T.S., Wong, M.H., Kwak, S.Y., Sinclair, R., Koman, V.B., Strano, M.S., 2018. Rational design principles for the transport and subcellular distribution of nanomaterials into plant protoplasts. Small 14, 1802086.

Li, Z., de Barros, A.L.B., Soares, D.C.F., Moss, S.N., Alisaraie, L., 2017. Functionalized single-walled carbon nanotubes: cellular uptake, biodistribution and applications in drug delivery. Int. J. Pharm. 524, 41–54.

Lin, C., Fugetsu, B., Su, Y., Watari, F., 2009. Studies on toxicity of multiwalled carbon nanotubes on *Arabidopsis* T87 suspension cells. J. Hazard. Mater. 30, 578–583.

Liu, Y., Wu, D.C., Zhang, W.D., 2005. Polyethylenimine-grafted multiwalled carbon nanotubes for secure non-covalent immobilization and efficient delivery of DNA. Angew. Chem. Int. Ed. 44, 4782–4785.

Liu, J., Wang, F.H., Wang, L.L., Xiao, S.Y., Tong, C.Y., Tang, D.Y., Liu, X.M., 2008. Preparation of fluorescence starch-nanoparticle and its application as plant transgenic vehicle. J. Cent. S. Univ. Technol. 15, 768–773.

Liu, Q., Chen, B., Wang, Q., Shi, X., Xiao, Z., Lin, J., Fang, X., 2009. Carbon nanotubes as molecular transporters for walled plant cells. Nano Lett. 9, 1007–1010.

Lombardo, D., Kiselev, M.A., Caccamo, M.T., 2019. Smart nanoparticles for drug delivery application: development of versatile nanocarrier platforms in biotechnology and nanomedicine. J. Nanomater. 2019, 3702518. 26 pages.

Luo, D., Saltzman, W.M., 2000. Enhancement of transfection by physical concentration of DNA at the cell surface. Nat. Biotechnol. 18, 893–895.

Luo, G.F., Chen, W.H., Liu, Y., Lei, Q., Zhuo, R.-X., Zhang, X.-Z., 2014. Multifunctional enveloped mesoporous silica nanoparticles for subcellular co-delivery of drug and therapeutic peptide. Sci. Rep. 4, 6064.

Martin-Ortigosa, S., Peterson, D.J., Valenstein, J.S., Lin, V.S.Y., Trewyn, B.G., Lyznik, L.A., Wang, K., 2014. Mesoporous silica nanoparticle mediated intracellular Cre protein delivery for maize genome editing via loxP site excision. Plant Physiol. 164, 537–547.

Mattos, I.B., Alves, D.A., Hollanda, L.M., 2011. Effects of multi-walled carbon nanotubes (MWCNT) under *Neisseria meningitides* transformation process. J. Nanobiotechnol. 9, 53.

Mohajeri, M., Behnam, B., Sahebkar, A., 2019. Biomedical applications of carbon nanomaterials: drug and gene delivery potentials. J. Cell. Physiol. 234, 298–319.

Nakashima, N., Okuzono, S., Murakami, H., Nakai, T., Yoshikawa, K., 2003. DNA dissolves single-walled carbon nanotubes in water. Chem. Lett. 32, 456–457.

Nandy, B., Santosh, M., Maiti, P.K., 2012. Interaction of nucleic acids with carbon nanotubes and dendrimers. J. Biosci. 37, 457–474.

Naqvi, S., Maitra, A.N., Abdin, M.Z., Akmal, M.D., Arora, I., Samim, M.D., 2012. Calcium phosphate nanoparticle mediated genetic transformation in plants. J. Mater. Chem. 22, 3500.

Nasir, S., Hussein, M.Z., Zainal, Z., Yusof, N.A., 2018. Carbon-based nanomaterials/allotropes: a glimpse of their synthesis, properties and some applications. Materials 11, 295.

Nepal, D., Geckeler, K., 2007. Proteins and carbon nanotubes: close encounter in water. Small 3, 1259–1265. https://doi.org/10.1002/smll.200600511.

Nunes, A., Amsharov, N., Guo, C., 2010. Hybrid polymer-grafted multiwalled carbon nanotubes for in vitro gene delivery. Small 6, 2281–2291.

Pantarotto, D., Singh, R., McCarthy, D., 2004. Functionalized carbon nanotubes for plasmid DNA gene delivery. Angew. Chem. Int. Ed. 43, 5242–5246.

Patnaik, S., Bhatnagar, P., Bansal, R., Sarkar, J., Kumar, A., Kumar, P., C Gupta, K., 2015. Modified polysaccharides as carriers for biomolecules. Pharm. Nanotechnol. 3, 261–287.

Phougat, N., Kumar, M., Saini, R.V., Chhillar, A.K., 2017. Green chemistry approach towards nanoparticle synthesis. In: Kalia, V., Saini, A. (Eds.), Metabolic Engineering for Bioactive Compounds. Springer, Singapore.

Pogodin, S., Baulin, V.A., 2010. Can a carbon nanotube pierce through a phospholipid bilayer? ACS Nano 4, 5293–5300.

Prasad, R., Bhattacharyya, A., Nguyen, Q.D., 2017. Nanotechnology in sustainable agriculture: recent developments, challenges, and perspectives. Front. Microbiol. 8, 1014.

Rafsanjani, M.S.O., Alvari, A., Samim, M., 2012. Application of novel nanotechnology strategies in plant biotransformation: a contemporary overview. Recent Pat. Biotechnol. 6, 69–79.

Rai, M., 2015. Nanoparticles-based delivery systems in plant genetic transformation. In: Rai, M., Ribeiro, C., Mattoso, L., Duran, N. (Eds.), Nanotechnologies in Food and Agriculture. Springer, Cham.

Rajkuberan, C., Sudha, K., SathishKumar, G., Sivaramakrishnan, S., 2015. Antibacterial and cytotoxicity of silver nanoparticles synthesized using latex of *Calotropis gigantea*. Spectrochim. Acta A 136, 924–930.

Reid, W., Zhang, Q., Sekimoto, H., 2002. Influence of membrane surface charge on nutrient uptake by plants. Dev. Plant Soil Sci. 92, 198–199.

Rojas-Chapana, J., Troszczynska, J., Firkowska, I., 2005. Multi-walled carbon nanotubes for plasmid delivery into *E. coli* cells. Lab Chip 5, 536–539.

Sandhu, K.K., McIntosh, C.M., Simard, J.M., Smith, S.W., Rotello, V.M., 2002. Gold nanoparticle-mediated transfection of mammalian cells. Bioconjug. Chem. 13, 3–6.

Serag, M.F., Kaji, N., Gaillard, C., 2011. Trafficking and subcellular localization of multiwalled carbon nanotubes in plant cells. ACS Nano 5, 493–499.

Silva, A.T., Nguyen, A., Ye, C., Verchot, J., Moon, J.H., 2010. Conjugated polymer nanoparticles for effective siRNA delivery to tobacco BY-2 protoplasts. BMC Plant. Biol. 10, 291.

Sireesha, M., Jagadeesh Babu, V., Kranthi Kiran, A.S., Ramakrishna, S., 2018. A review on carbon nanotubes in biosensor devices and their applications in medicine. Nanocomposites 4 (2), 36–57.

Szabó, A., Perri, C., Csató, A., Giordano, G., Vuono, D., Nagy, J.B., 2010. Synthesis methods of carbon nanotubes and related materials. Materials 3, 3092–3140.

Taylor, N., Fauquet, C., 2002. Microparticle bombardment as a tool in plant science and agricultural biotechnology. DNA Cell Biol. https://doi.org/10.1089/104454902762053891.

Torney, F., Trewyn, B.G., Lin, V.S.Y., Wang, K., 2007. Mesoporous silica nanoparticles deliver DNA and chemicals into plants. Nat. Nanotechnol. 2, 295–300.

Tuncel, D., 2011. Non-covalent interactions between carbon nanotubes and conjugated polymers. Nanoscale 3, 3545–3554.

Uzu, G., Sobanska, S., Sarret, G., Munoz, M., Dumat, C., 2010. Foliar lead uptake by lettuce exposed to atmospheric pollution. Environ. Sci. Technol. 44, 1036–1042.

Vijayakumar, P.S., Abhilash, O.U., Khan, B.M., Prasad, B.L., 2010. Nanogold loaded sharp-edged carbon bullets as plant-gene carriers. Adv. Funct. Mater. 20, 2416–2423.
Witus, L.S., Rocha, J.D., Yuwono, V.M., 2007. Peptides that non-covalently functionalize single-walled carbon nanotubes to give controlled solubility characteristics. J. Mater. Chem. 17, 1909–1915.
Wong, M.H., Misra, R.P., Giraldo, J.P., Kwak, S.Y., Son, Y., Landry, M.P., Swan, J.W., Blankschtein, D., Strano, M.S., 2016. Lipid exchange envelope penetration (LEEP) of nanoparticles for plant engineering: a universal localization mechanism. Nano Lett. 16, 1161–1172.
Yang, S.T., Wang, H., Meziani, M.J., 2009. Bio-defunctionalization of functionalized single-walled carbon nanotubes in mice. Biomacromolecules 2009–2012.
Yuan, H., Hu, S., Huang, P., 2011. Single walled carbon nanotubes exhibit dual-phase regulation to exposed *Arabidopsis* mesophyll cells. Nanoscale Res. Lett. 6, 44.
Zaytseva, O., Neumann, G., 2016. Carbon nanomaterials: production, impact on plant development, agricultural and environmental applications. Chem. Biol. Technol. Agric. 3, 17.
Zhang, W., Zhang, Z., Zhang, Y., 2011. The application of carbon nanotubes in target drug delivery systems for cancer therapies. Nanoscale Res. Lett. 6, 555.
Zhao, X., Meng, Z., Wang, Y., Chen, W., Sun, C., Cui, B., Cui, H., 2017. Pollen magnetofection for genetic modification with magnetic nanoparticles as gene carriers. Nat. Plants 3, 956–964.

Further reading

Fernandez, V., Eichert, T., 2009. Uptake of hydrophilic solutes through plant leaves: current state of knowledge and perspectives of foliar fertilization. Crit. Rev. Plant Sci. 28, 36–68.
Simon, J., Flahaut, E., Golzio, M., 2019. Overview of carbon nanotubes for biomedical applications. Materials 12, 624.

Nanocarbon fertilizers: Implications of carbon nanomaterials in sustainable agriculture production

15

Ruchi Aacharya[a], *Hemraj Chhipa*[b]
[a]Department of Botany, Mohanlal Sukhadia University, Udaipur, India, [b]College of Horticulture and Forestry, Agriculture University, Kota, India

1 Introduction

Nanotechnology has shown its applicability in various fields and plays a significant role in the facilitation of processes in easy mode. Currently, nanotechnology has been applied in electronics, electrics, optics, solar, and the medical industry, and many commercial products have reached the market. Nanomaterials have specific optical, thermal, electrical, and chemical properties that give them specific characteristics compared to their bulk counterparts (Bennett et al., 2013; Hurt et al., 2006; Srivastava et al., 2015). Nanomaterials at this size scale provide a high volume-to-surface ratio, which improves their applicability in various sectors (Yin et al., 2018). Nanomaterials have shown applications as nanofertilizers, nanopesticides, and nanosensors to improve agriculture production (Chhipa, 2018). In agriculture, various metallic polymer-based oxide nanomaterials and nanocomposites have shown positive effects on plant growth. However, the phytotoxicity of metal nanomaterials, nanocomposites, and oxides has also been reported in many studies (Chhipa, 2017a,b; Chhipa and Joshi, 2016; Solanki et al., 2015), which narrows their applicability in agriculture. In recent years, carbon-based nanomaterials have shown wide applicability as plant growth regulators, pest control agents, nanosensors for pest detection, pesticide sorbents, and nutrient carriers in agriculture (Chhipa, 2019; Fraceto et al., 2016; Khadri et al., 2013; Ma and Wang, 2010; Morsy et al., 2014; Parizi et al., 2014; Samaj et al., 2004; Saxena et al., 2014; Sonkar et al., 2012; Wild and Jones, 2009).

Various carbon-based nanomaterials have been applied as agriculture inputs, such as carbon nanotubes, fullerenes, carbon nanodots, carbon nanofibers, graphene oxides, carbon nanohorns, and carbon nanoparticles (Mukherjee et al., 2016; Patel et al., 2019; Singh et al., 2018). Especially carbon nanotubes, graphene, and fullerene have been applied as nanocarriers for nutrients as well as active ingredients of pest control chemicals. Carbon nanomaterials provide ecofriendly and sustainable agents by reducing the amount of fertigation compounds and their controlled release, making them efficient agriculture inputs for next-generation fertilizers and pesticides (Georgakilas et al., 2015; Sonkar et al., 2012; Tiwari et al., 2013).

Carbon Nanomaterials for Agri-food and Environmental Applications. https://doi.org/10.1016/B978-0-12-819786-8.00015-3

The introduction of carbon nanomaterials to plants influenced plant physiology in many positive ways, such as enhancement in seed germination, increased seedling growth, raised shoot and root length, improved biochemical content, improved enzymatic activity, extended defense system by inducing defense-related genes, suppressed oxidative enzymes, improved photosynthesis rate, and many other metabolic activities that improved plant productivity (Cañas et al., 2008; Hao et al., 2018; Khodakovskaya et al., 2011; Lahiani et al., 2015; Lin and Xing, 2007; Liu et al., 2009; Mondal et al., 2011; Samaj et al., 2004; Saxena et al., 2014; Tiwari et al., 2013; Tripathi and Sarkar, 2014; Tripathi et al., 2011; Wang et al., 2012; Wild and Jones, 2009; Zhao et al., 2017). In addition to plant growth, carbon nanomaterials also assisted in maintaining soil structure by the slow release of nutrients, improved water retention capacity, enhanced nutrient retention capacity, and balanced soil pH (Mao et al., 2012; Sohi et al., 2009; Ueno et al., 2007). It has been reported in various manuscripts that carbon nanomaterials entered into plants from roots via sym-plastic or apo-plastic pathways and were transported in the plant body via xylem vessels, reaching different parts of the plant.

On the other side, the accumulation of carbon nanomaterials adversely affected plants by generating reactive oxygen species (ROS), electrolyte leakage, necrotic lesions in cells on the basis of the amount of accumulated carbon nanomaterials, and phytotoxicity (Anjum et al., 2013; Begum et al., 2011; Liu et al., 2013). The phytotoxicity is a concentration-dependent phenomena in the plant that varies with different types of carbon nanomaterials. Different research groups are working on the subject to understand plant-nanomaterial interaction and their toxicity level. Still, there is more research required to understand the tolerance level of various type of carbon nanomaterials in different crops for sustainable agriculture production. Detailed information on carbon nanomaterials in the agriculture sector has been reviewed by Singh et al. (2018) and Mukherjee et al. (2016).

In this chapter, we review various types of carbon nanomaterials used in agriculture, how they influence the plant system, their positive and negative effects on plant growth and production, and the future potential of carbon nanomaterials as fertilizers. We also discuss a crop-carbon nanomaterial model to understand the toxicity concentration for each system. Still, there is a gap to understand the long-term impact of carbon nanomaterials on the next progeny of crops as well as human and environmental health. Their impact on soil microbial communities also needs to be addressed to understand the impact on soil biological cycles.

2 Carbon nanomaterials

Carbon nanomaterials have been used in various industrial applications such as carriers in drug delivery systems and tissue engineering, optical devices, molecular switches, conducting materials, and quantum computers (Husen and Siddiqi, 2014). In agriculture, carbon-based nanofertilizers are more advantageous in plant growth and soil fertility by the controlled release of nutrients, fertilizing the soil in a sustainable manner. Their biocompatibility and ecofriendly nature as well as the easy surface modification of carbon-based nanomaterials have increased their application spectrum in agriculture. Carbon nanomaterials have shown specific mechanical, thermal, and chemical properties as well as high stability during their application, which improves their applicability as fertilizers, micronutrients, nanopesticides, nanoherbicides, and composite biofertilizers (Duhan et al., 2017; Zaytseva and Neumann, 2016). The application of carbon nanomaterials empowers agriculture through improved crop yield and production in a sustainable manner. In contrast, the adverse effect of carbon nanomaterials has also been reported, specifically in carbon nanotubes due to their hydrophobic nature as well as the surfactant and catalyst availability on their surface. Conversely, carbon nanotubes have been reported as a growth gene and reproductive gene stimulator. The entry of carbon nanotubes into the plant is inversely proportional to their size. Therefore, here we summarize various carbon nanomaterials used in agriculture.

3 Types and synthesis of nanocarbon

The carbon atom has the robust property to form covalent bonds via sp, sp^2, and sp^3 hybridization with organic and metallic atoms while also providing a wide range of structures from small to long-chain compounds (Mauter and Elimelech, 2008). Some of them have shown applications in agriculture and improved agriculture production. Various carbon-based nanomaterials such as fullerenes (C_{60}), carbon nanotubes, graphene oxides, carbon nanofibers, carbon nanodots, and carbon nanomaterials have been applied in agriculture as plant growth promoters, nanofertilizers, and pristine pest control agents, or they have been functionalized with other nanomaterials or chemical agriculture inputs (Fig. 1). Detailed information on carbon nanomaterials has been reviewed by Georgakilas et al. (2015).

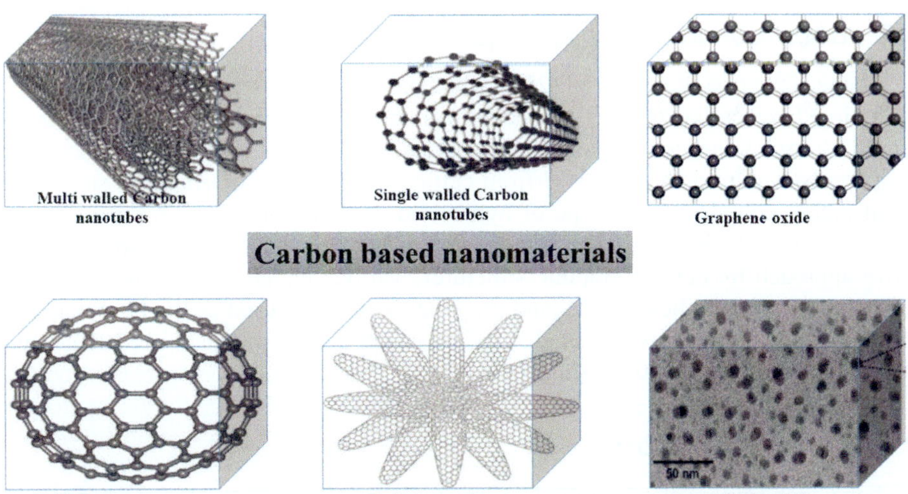

Fig. 1 Types of carbon-based nanomaterials used in agriculture as plant growth promoters.

3.1 Fullerenes (C_{60})

Fullerene (C_{60}) was the first carbon nanostructure, and it was discovered in 1985 (Kroto et al., 1985). In this structure, C_{60} sp^2 carbon atoms are arranged in chains of pentagons and hexagons and form a ball-like spherical structure. This represents the smallest carbon nanostructure and a zero-dimensional (0D) carbon nanoallotrope that has an external diameter of 0.71 nm (Georgakilas et al., 2015). Generally, fullerene is stable structure by at higher temperature about 1000°C it distracts. Fullerenes are produced by hydrocarbon combustion, naphthalene pyrolysis, and graphite vaporization using plasma discharge or laser irradiation (Dennis et al., 1991; Han et al., 2019; Howard et al., 1991; Krätschmer et al., 1990; Prikhod'ko, 2010; Richter et al., 1994).

3.2 Carbon nanotubes

Carbon nanotubes (CNTs) are a tubular structure that has sp^2 hybridized carbon atoms. In general, carbon nanotubes are formed by the fusion of edges of graphene sheets and developed cylindrical tube-like structures, which have a high aspect ratio. CNTs have shown their applicability in various sectors such as electronics, medicine, and energy harvesting (AbdulAlmohsin and Cui, 2012; Lacerda et al., 2006; Mink and Hussain, 2013). Further, carbon nanotubes are divided into single-cell carbon nanotubes (SWCNTs) and multiwall carbon nanotubes (MWCNTs). SWCNTs contain a single concentric layer of graphene while MWCNTs have several concentric cylinders. SWCNTs produced by arc discharge in the presence of a graphite electrode contain catalytic nanoparticles while MWCNTs are produced in the absence of a catalyst. CNTs have also been prepared by pulsed-laser ablation of a graphite target in a controlled atmosphere chamber, decomposition of carbon vapors by plasma irradiation, and then catalytic nucleation developed CNTs (Abbaslou et al., 2019; Fleming et al., 2019; Guo et al., 1995; Iijima and Ichihashi, 1993; Lim et al., 2010).

3.3 Graphene and graphene quantum dots

Graphene is the two-dimensional (2D) carbon nanostructure that has a sheet-like configuration. Each carbon atom is connected with three other carbon atoms with a sigma bond to form a honeycomb-like structure. Graphene is the strongest known material. Graphene quantum dots developed from cutting a monolayer of graphene sheet into small pieces contain the sp^2 type of hybridized carbon atom. They form a crystalline structure, which is 2–20 nm. Graphene dots are also synthesized through a top-down approach by cutting graphite structures. Various materials such as SWCNTs, MWCNTs, coal, graphene oxide, and graphite fibers were used as the base material in a top-down approach (Dong et al., 2012; Gong et al., 2014; Pan et al., 2010; Peng et al., 2012; Shen et al., 2012; Singh et al., 2019).

3.4 Carbon nanohorns

Carbon nanohorns are cone or tube-like structures, and are also developed from monolayer sheets of graphene. They are found in aggregates of 80–100 nm in diameter and look like the dahlia flower. Nanohorn's upper diameter is about 1–2 mm while at the bottom is wider with 4–5 mm diameter (Yudasaka et al., 2007). Microporosity and high surface area are the unique characteristics of carbon nanohorns that make them an efficient material as adsorbents and carriers.

3.5 Carbon dots (photoluminescent)

Carbon dots (CDs) are a combination of turbistratic and graphitic carbon with an sp^3 type of hybridization in covalent bonds; they form quasispherical structures with 2–10 nm diameter carbon nanoparticles. Carbon dots provide strong luminescence, depending on their size and excitation wavelength. Carbon dots are prepared by a top-down approach using graphite or cement as the initial base material. They are further reduced at the nanoscale by laser ablation and acid treatments (Li et al., 2012). Further, an electrochemical process was also investigated in the preparation of carbon dots using graphite electrodes, thermal decomposition, or carbonization of organic compounds, and microwave treatment of organic polymers (Lai et al., 2012; Liu et al., 2019; Milosavljevic et al., 2015; Wang et al., 2010).

3.6 Nanocarbon as a fertilizer

Nanocarbon materials are ecofriendly and have a high surface-to-volume ratio. Such characteristics make them proficient as nutrient carriers in the agricultural sector. Nanocarbon has shown the ability to retain water and nutrients for a long time and to release slowly as per requirements, which makes them smart nanofertilizers (Solanki et al., 2015). It has been found that nanomaterials are advantageous in the amount of reduction of synthetic fertilizers by their target-specific approach. They also provide nutrition to the cell, which helps in the excess leaching of chemicals in the environment. Reduction in soil and water pollution of synthetic chemicals by application of nanofertilizers has developed them as sustainable fertilizer agents. Therefore, various type of carbon nanomaterials have been used and studied as fertilizers on different crops. Saxena et al. (2014) found that the positive impact of water-soluble carbon nanoparticles (wsCNPs) was within the concentration range of 10–150 $mg L^{-1}$ on wheat plants; they achieved maximum growth at 50 $mg L^{-1}$. They derived wsCNPs from the chemical oxidation of raw carbon nanoparticles (rCNPs) in biochar. In the model experiment, they suggested that wsCNPs have the ability to release nutrients in a controlled release manner that boosts plant growth. Similarly, MWCNTs showed beneficial effects on the root, shoot, and seed germination in *Brassica juncea* (Mondal et al., 2011).

Functionalized MWCNTs also showed a beneficial effect on wheat, peanut, garlic, and corn (Srivastava and Rao, 2014). Plant biomass was significantly increased in the above crops at 0–50 ppm MWCNT concentration. Researchers hypothesized that MWCNTs entered into the seed coat by creating new pores while also enhancing water uptake, which would support improvement in plant biomass. Functionalization also affects the interactions in plants and influences the crop differently. Functionalization changed the surface charge on the nanomaterial and modified the mode of attraction.

Zhai et al. (2015) used three types of MWCNTs, such as positively charged MWCNT-NH_2, negatively charged MWCNT-COOH, and neutral MWCNT, to study their effect on corn and soybeans. They exposed corn and soybeans to MWCNTs at a concentration range from 10 to 50 mg L^{-1} for 18 days and observed increased growth in corn and suppressed growth in soybeans. Further, Lahiani et al. (2017) studied the long-term effect of carbon nanotubes on barley, soybean, and corn. They found that after continuous exposure of MWCNT for 20 weeks, several positive phenotypic changes were found in all crop systems without any toxicity. Previously, Lahiani et al. (2016) reported that carbon-based nanomaterials (CBNs) activate cell growth, germination, and crop development. CBNs induced genetic expression to stress-related factors in the tomato plant. MWCNTs functionalized with urea were also evaluated as nanocarbon fertilizers in rice plant growth (Yatim et al., 2019). The authors reported significant improvement in rice growth and localization of functionalized MWCNTs in root cells by TEM imaging. They also found that translocation occurred via phospholipid bilayers. Further, Lateef et al. (2019) synthesized nanocomposite-based biochar from agricultural corn cob waste and found it more suitable in providing micro- and macronutrients to plants. Lateef and coworkers found improved water retention, water absorbance, and swelling ration in nanocomposite biochar than corn cob biochar. They also observed that nanocomposite biochar indicated a slow release pattern of nutrients and provided nutrients to plants for a long time as sustainable fertilizers.

Similarly, Pandey et al. (2018) also reported that graphene and CNTs increased the seed germination rate in sorghum and switchgrass. They measured increased biomass in switchgrass by 28% at 200 ppm of graphene exposure. The size and shape of the carbon nanomaterial was also an important parameter by which CBNs showed variability in expressions. Lahiani et al. (2015) assessed the interaction of a carbon nanohorn with tomatoes, rice, corn, barley, soybeans, and switchgrass, and found that single-walled carbon nanohorn (SWCNH) enhanced seed germination and plant growth at various concentrations. The most effective concentration in corn and soybeans was found to be 100 ppm while in rice and tomatoes, it was 50 ppm. Wang et al. (2018) used carbon dots as nanofertilizer on the mung bean and observed a concentration-dependent effect on growth and photosynthesis. They found that the mung bean promoted root elongation, shoot elongation, increased biomass, increased root vigor, and the moisture level of the seed at an optimal concentration of carbon dots. Along with phenotypic characteristics, carbon dots also affected carbohydrate content, chlorophyll content, and rubisco activity in the mung bean while also significantly affecting the photosystem and photosynthesis reactions.

4 Role of nanocarbon in plant growth

Carbon nanomaterials have proved their role in plant growth by inducing various physiological and biochemical processes. Many researchers found various mechanistic outputs that were the result of carbon-based nanomaterials and plant interactions. Carbon nanotubes and fullerene C_{60} significantly affected the plant physiology by improving the water retention capacity and plant growth up to ~118%. This nanomaterial also induced the production of bioactive compounds in plants such as charantin (20%), cucurbitacin-B (74%), insulin (91%), and lycopene (82%), which were increased in bitter melon seeds after treatment with fullerene (Husen and Siddiqi, 2014). It has been reported that nanocarbon spread or dispersed in water was absorbed by the root system and transported via the xylem vessel channel in plants. Absorption and transportation of nanocarbon in plants depend on various factors, including size of the nanocarbon materials, the type and functionalization of nanocarbon, its solubility in water, and the plant behavior toward nanocarbons. Simultaneously, the effect of CNTs on seed germination was also studied by Haghighi and da Silva (2014). They exposed the seeds of *Lycopersicum esculentum* Mill. cv. Falcato, *Allium cepa* L. cv. Yellow Sweet Spanish, *Brassica rapa* L. cv. Toria, and *Raphanus sativus* L. cv. Small radish to CNT at 0–40 $mg\,L^{-1}$ concentration and found that *L. esculentum* and *A. cepa* showed improved germination by 8% and 95%, respectively. Further, in an experiment on seedling growth at the same concentration, Haghighi and coworkers also found that the dry weight of the *L. esculentum* and the *R. sativus* shoot also increased, but no effect on germination and growth was observed on *B. rapa*. It has also been reported that the variable effects of carbon nanomaterials on plants depend on the nature of the nanomaterial, the plant species, the nanomaterial concentration, the soil, and the time of exposure (Vera-Reyes et al., 2018). Similarly, it was also found that the charge on the carbon nanomaterial also affects its movement in the plant (Zhai et al., 2015). Recently, Fan et al. (2018) reported that MWCNTs affect the photosynthesis system of plants by stimulation of the electron transport rate and the photochemical quantum yield of the photosystem II up to 12% in comparison to control. They found that MWCNT exposure induces the photosynthesis-related proteins and suppressed the oxidative stress proteins in *Arabidopsis thaliana*.

Carbon nanomaterials also supported plants with stress tolerance. It has been reported that fullerenol helped the sugar beet plant in the alleviation of drought stress. Borišev et al. (2016) found that fullerenol penetrated into the root and plant leaves and supported additional water supply. They found that a foliar spray of fullerenol on the sugar beet inhibited the activity of stress-inducing enzymes. Similarly, Pandey et al. (2018) also found that CBN suppressed the toxic effect of salt stress in sorghum and switchgrass. They discovered that CBN restores the seed germination rate and the development of young seedlings. They found that CBN activates the expression of water channel-related genes in the stress condition and induced aquaporin protein production. Aquaporins are water channel-related proteins that regulate the plant-water relationship and express under various environmental stresses toward the plant, supporting plant growth. The authors found that the restoration of aquaporins in the sorghum plant after exposure to CBN, which have reduced in the presence of NaCl. Carbon-based nanomaterial supported the plant's tolerance under the salt stress condition by inducing the stress tolerance proteins in sorghum. Water-soluble carbon nanomaterials influenced more in comparison to nonfunctionalized carbon nanomaterials and enhanced more plant growth in gram and wheat (Saxena et al., 2014; Sonkar et al., 2012; Tripathi and Sarkar, 2014).

The size and shape of the carbon nanomaterials also influenced the plant growth by the rate of penetration in the cell and further transportation in the plant body.

Tripathi et al. (2016) studied the influence of various carbon nanomaterials on the growth of *Cicer arietinum*. The authors found that SWCNTs sized <10nm better promoted the growth of the gram plant in comparison to MWCNTs. They hypothesized that carbon nanomaterials small in size supported easy transportation of the nanomaterial into the plant. Small SWCNTs and carbon dots reached up to the root interior via the symplastic pathway while large carbon nanostructures preferred the apoplastic pathway. Similarly, Zhao et al. (2017) also evaluated the uptake of MWCNTs in rice, soybean, corn, and *A. thaliana* and found that the concentration of carbon nanomaterials significantly affected the biochemical parameters in comparison to the physiological parameters. They compared various physiological parameters such as pigment and protein content and water loss as well as biochemical parameters such as antioxidant enzyme activities in the above crops. The authors found that all crops accumulated MWCNTs in their root system, affecting both the biochemical and physiological parameters. Previously, it has also been reported that some plants such as cabbage, carrots, cucumbers, lettuce, onions, tomatoes, alfalfa, and wheat do not accumulate the MWCNTs in their root system (Cañas et al., 2008; Miralles et al., 2012a). Therefore, it has been concluded that the accumulation and transportation of nanomaterials are size-dependent phenomena that depend on the pore size of the cell.

On the other hand, Hao et al. (2018) applied various types of carbon-based nanomaterials on rice grown on loamy soil to understand the effects of carbon nanomaterials on plant physiology and soil bacterial community composition. They used fullerene C_{60}, reduced graphene oxide, and MWCNTs and found that at a higher concentration of carbon nanomaterials (CNM), various phytohormones including indoleacetic acid, gibberellic acid, and brassinosteroid were increased significantly. Similarly, antioxidant enzymes such as superoxide dismutase and peroxidase were also increased in roots at a higher amount of CNM. Auxins such as Indole acetic acid (IAA) played a significant role in cell elongation and division, a cell defense mechanism. On the other hand, gibberellic acid (GA) belongs to stem elongation, leaf expansion, fruit and flower development, and seed germination (Gopalan, 2008; Iqbal and Ashraf, 2013; Rodrigues et al., 2012; Song et al., 2014). Phytohormones regulate plant growth and provide tolerance to the plant during environmental stress. The effects of carbon nanomaterials on various crop species are given in Table 1.

5 The positive impact of nanocarbon on plant growth

Nanocarbons entered into the plants and affected seed and plant growth positively. Induction of the seed germination rate, seedling growth, and plant biomass improvement were major effects that have been reported in many studies (Fig. 2). Khodakovskaya et al. (2012) reported that multiwalled carbon nanotubes induced the cell division and water transport regulation genes in a tobacco cell culture. The effect of graphene oxide on *Vicia faba* L. at 100–1600ppm concentration was assessed by Anjum et al. (2013). They found that the germinated seedling appeared sensitive at 1600ppm and showed a decreased glutathione redox ratio and a reduced glutathione

Table 1 Effects of carbon nanomaterials on various crops: Carbon nanomaterials affected various crop systems in both positive and negative manners, which mainly depends on the concentration of carbon nanomaterial and the crop system.

S.·No	Crop/plant species	Carbon nanomaterial	Concentration range	Effect	Reference
1	Alfa-alfa and wheat	MWCNT	40–2560 $mg\,L^{-1}$	Seed germination rate increased	Miralles et al. (2012b)
2	Barley, corn, and soybean	MWCNT	100–200 $\mu g\,mL^{-1}$	Increased seed germination in all, shoot weight increased in corn, and root length in soybean	Lahiani et al. (2013)
3	Bitter melon	Fullerol	0.943–47.2 nM	Increased fruit weight, number, length	Kole et al. (2013)
4	Corn	MWCNT	500–5000 $\mu g\,mL^{-1}$	Plant growth and biomass increased	De La Torre-Roche et al. (2013)
5	Corn, soybean	Fullerenes	500–1000 $mg\,kg^{-1}$	Reduced biomass	De La Torre-Roche et al. (2013)
6	Gram	Water-soluble carbon nanotubes	3–6 $\mu g\,mL^{-1}$	Increased plant growth	Tripathi et al. (2011)
		Water-soluble carbon nanotubes	10–30 $\mu g\,mL^{-1}$	Increased plant growth	Sonkar et al. (2012)
7	Lettuce	PABS functionalized SWCNT	28–5000 $mg\,L^{-1}$	Root growth decreased	Cañas et al. (2008)
		MWCNT	2000 $mg\,kg^{-1}$	Reduced root length	Lin and Xing (2007)
8	Corn	MWCNT	10–40 $\mu g\,mL^{-1}$	Increment in plant biomass with Fe and Ca concentration	Tiwari et al. (2013)
		Carbon nanodots	250–2000 ppm	Phytotoxic	Chen et al. (2016)
9	Mustard	MWCNT	2.3–6.9 $\mu g\,mL^{-1}$	Seed germination, root and shoot length increased	Mondal et al. (2011)
10	Onion, cucumber	SWCNT	56–1750 $mg\,mL^{-1}$	Root growth increased	Cañas et al. (2008)

Continued

Table 1 Continued

S.·No	Crop/plant species	Carbon nanomaterial	Concentration range	Effect	Reference
11	Rape, corn, and cucumber	MWCNT	2000 mg L^{-1}	Root length increased	Lin and Xing (2007)
12	Rape, ryegrass, and radish	MWCNT	2000 mg L^{-1}	Seed germination and root length	Lin and Xing (2007)
13	Red spinach	MWCNT	125–1000 mg L^{-1}	Reduced plant growth	Begum and Fugetsu (2012)
14	Rice	SWCNT	50 μg mL^{-1}	Seed germination, root and shoot length increased	Nair et al. (2012)
		MWCNT	50 μg mL^{-1}	Increased seed germination, root and shoot length, and seedling growth	Nair et al. (2012)
		Fullerene	50 μg mL^{-1}	Seed germination, root and shoot length increased	Nair et al. (2012)
		MWCNT	10–80 μg mL^{-1}	Decrease in cell viability	Tan et al. (2009)
		MWCNT	400 mg L^{-1}	Reduced flowering	Lin et al. (2009)
15	Spinach	SWCNT	2.5–5 mg L^{-1}	Photosynthesis rate increased	Giraldo et al. (2014)
16	Tobacco	MWCNT	5–500 μg mL^{-1}	Cell growth and biomass increased	Khodakovskaya et al. (2012)
17	Tomato	SWCNT	50 μg mL^{-1}	Biomass and seedling growth increased	Alimohammadi et al. (2011)
		SWCNT	50 μg mL^{-1}	Biomass increased	Khodakovskaya et al. (2011)
		MWCNT	40 μg mL^{-1}	Seed germination and biomass increased	Villagarcia et al. (2012)
		MWCNT	50 μg mL^{-1}	Enhancement of heat shock protein and stress protein, enhancement of seed germination	Khodakovskaya et al. (2011)
		MWCNT	50–200 μg mL^{-1}	Number of seeds, fruits, and flowers increased	Khodakovskaya et al. (2013)
		SWCNT-QDs	50 μg mL^{-1}	Seed germination and root weight reduced	Alimohammadi et al. (2011)

18	Wheat	MWCNT	10–160 $\mu g\,mL^{-1}$	Increased root length, cell elongation, and dehydrogenase activity	Wang et al. (2012)
		wsCDs	150 $\mu g\,mL^{-1}$	Root and shoot length increased	Tripathi et al. (2011)
		Water-soluble carbon nanoparticle	10–150 $mg\,L^{-1}$	Increased plant growth	Saxena et al. (2014)
19	Zucchini	MWCNT	1000 $mg\,kg^{-1}$	Reduced biomass and root length	Stampoulis et al. (2009)
20	Wheat, corn, peanut, garlic	MWCNT	0–50 $\mu g\,mL^{-1}$	Increased plant growth and biomass	Srivastava and Rao (2014)
21	Tomato, onion, mustard, radish	CNT	0–40 $mg\,L^{-1}$	Improved seed germination	Haghighi and da Silva (2014)
22	Corn and soybean	MWCNTs	–	Enhanced growth in corn	Zhai et al. (2015)
3	Chinese cabbage	Liquid nano-carbon	–	Significantly increased the annual yield	Zhao et al. (2019)

Modified from Mohamed, M.A., Hashim, A.F., Alghuthaymi, M.A., Abd-Elsalam, K.A., 2018. Nano-carbon: plant growth promotion and protection. In: Abd-Elsalam, K.A., Prasad, R. (Eds.), Nanobiotechnology Applications in Plant Protection. Springer International Publishing, Cham, pp. 155–188 with Permission from Springer.

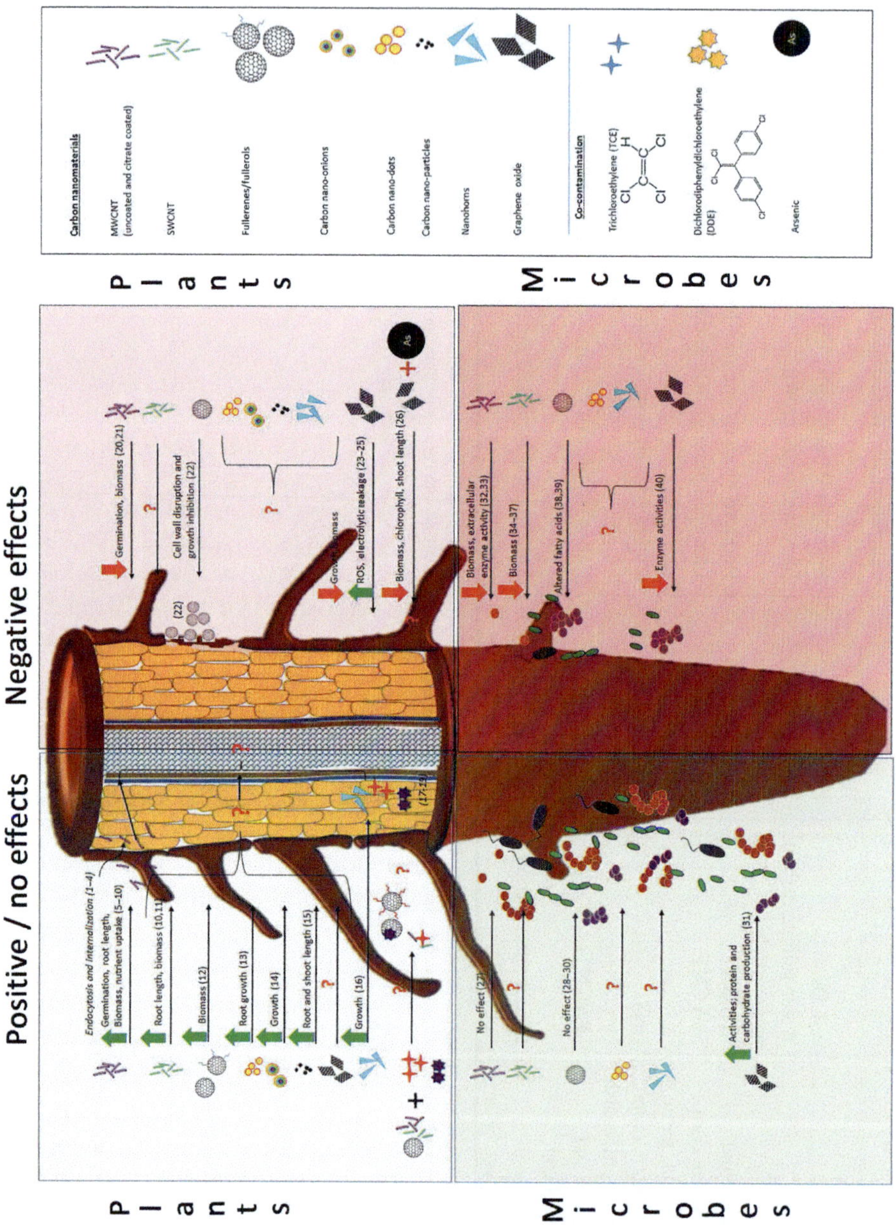

Fig. 2 See the legend on opposite page.

pool along with the increased activity of glutathione peroxidase, glutathione sulfotransferase, and glutathione reductase enzyme. Glutathione played an important role in plant growth via various metabolic processes such as biosynthesis, redox homeostasis, and detoxification (Noctor et al., 2012). The authors also found that at 800 and 400 ppm of graphene oxide glutathione redox ratio increased while glutathione peroxidase and glutathione sulfotransferase enzymes activity decreased. The authors inferred that the effect of graphene oxide concentration was directly dependent on the cellular redox state and the reduced glutathione pool. Tiwari et al. (2013) also observed the growth of corn seedlings at lower concentrations but inverse results at a higher concentration of commercial MWCNTs. It has been reported that MWCNTs affect mineral nutrition by opposing the inflow of minerals with water and retain in the medium by ion-CNT interactions. MWCNTs entered the seed through the seed coat and stimulated the seed coat (Mondal et al., 2011; Villagarcia et al., 2012).

Further, it has also been hypothesized that water-soluble carbon nanodots can be used as a growth promoter in cereal plants. It has been experimentally proved that water-soluble carbon nanodots (wsCNDs) increased the growth of the root and shoot of the wheat plant in the presence of light and dark conditions (Tripathi and Sarkar, 2014). It has been found that wsCNDs were absorbed on the cell and entered via pores on the cell wall and plasmodesmata into root cells through apoplastic and symplastic pathway and reached to the cortex of the cell. From the cortex it transported in various parts of plants like leaf, stem, fruit, and flower through xylem and provided nutrients to the cell. Recently, Verma et al. (2019) reviewed the application of carbon nanomaterials in the plant system in detail and explained that the effect on the carbon nanomaterial varies with the plant species, type of carbon nanomaterial, concentration, and time of exposure. In most cases, the yield and plant growth increased at a low concentration of CNM along with increased water transport, water uptake, water nutritional channel, nitrogenase activity, antioxidant activity, and nutrition absorption. Fullerene C_{60} also showed a positive impact on plant growth. Xiong et al. (2018) applied fullerol on *Brassica napus* seeds or leaves at 10–100 ppm and found improvement in seed germination, biomass, photosynthesis, and the antioxidation system of the plant under water stress conditions. Researchers reported that at 1–100 ppm concentration,

Fig. 2 Positive and negative effects of carbon-based nanomaterials on plants and microorganisms (Mukherjee et al., 2016), where 1–40 are (1) Samaj et al. (2004), (2) Liu et al. (2009), (3) Wild and Jones (2009), (4) Tripathi et al. (2011), (5) Lin and Xing (2007), (6) Mondal et al. (2011), (7) Wang et al. (2012), (8) Lahiani et al. (2015), (9) Tiwari et al. (2013), (10) Khodakovskaya et al. (2009, 2011), (11) Cañas et al. (2008), (12) Kole et al. (2013), (13) Tripathi and Sarkar (2014), (14) Sonkar et al. (2012), (15) Saxena et al. (2014), (16) Lahiani et al. (2015), (17) Ma and Wang (2010), (18) De La Torre-Roche et al. (2012), (19) Hamdi et al. (2015), (20) Stampoulis et al. (2009), (21) Mondal et al. (2011), (22) Liu et al. (2013), (23) Begum et al. (2011), (24) Anjum et al. (2013), (25) Anjum et al. (2014), (26) Hu et al. (2014), (27) Shrestha et al. (2013), (28) Tong et al. (2007), (29) Nyberg et al. (2008), (30) Cordeiro et al. (2014), (31) Wang et al. (2013), (32) Chung et al. (2011), (33) Kerfahi et al. (2015), (34) Liu et al. (2009), (35) Kang et al. (2007), (36) Rodrigues et al. (2013), (37) Jin et al. (2014), (38) Fang et al. (2007), (39) Johansen et al. (2008), and (40) Chung et al. (2015).

fullerol increased abscisic acid (ABA) biosynthesis and induced the expression of the downregulating gene of ABA biosynthesis CYP707A3 under stress conditions. Previously, Kole et al. (2013) also found enhancement in biomass and water content in a watermelon after treatment with fullerol. It has also been reported that under stress conditions, fullerol suppressed the stress-related chemicals such as H_2O_2 and MDA in corn and activate stress-related enzymes such as CAT, APX, and GPX in plants (Boriŝev et al., 2016; Liu et al., 2016).

Carbon-based nanomaterials not only alone but with other nanomaterials showed a synergistic effect on plant growth. Carbon nanomaterials along with copper nanoparticles (Cu NPs) in the form of CNF-CuNPs also increased the chlorophyll, protein content, root and shoot length, water uptake capacity, and germination rate in *C. arietinum* (Ashfaq et al., 2017). It was found that carbon nanofibers enhanced the effective translocation of Cu from the root to the shoot in the plant.

6 The adverse impact of nanocarbons on plant growth

In contrast with positive effects, carbon nanomaterials have also been shown to have a negative impact on plant growth by reducing seed germination, vigor index, and crop production. MWCNTs showed phytotoxicity to *B. juncea* and *Phaseolus mungo* in seed germination and seedling growth at 40 ppm concentration (Ghodake et al., 2010). Ghodake and coworkers also reported that the root growth of *B. juncea* was enhanced up to 202% at 30 ppm concentration, but further decreased and root growth remained at 135% at 40 ppm. Further, Hatami (2017) studied the effect of multiwalled carbon nanotubes on seeds of *Cucurbita pepo* L. The author applied various concentrations of multiwalled carbon nanotubes in the range of 125–1000 ppm at two water potential levels (well watered and water stress) for 14 days. He found decreased root and shoot length, biomass, vigor index, and germination percentage in the plant due to tissue electrolight leakage (ELI), and tissue injury by hydrogen peroxide and malondialdehyde oxidation. Previously, Begum et al. (2012) carried out the experiment of MWCNT in the concentration range of 0–2000 ppm on red spinach, lettuce, rice, cucumber, chili, lady's finger, and soybean seedlings in a hydroponic system. They found that after 15 days of exposure, red spinach, lettuce, and cucumber were negatively affected at a higher concentration of MWCNTs by a reduction in the root and length of seedlings at 1000 and 2000 ppm. They observed that the plant type also showed variation in cell death and sensitivity at higher concentrations of MWCNTs. Chili, lady's fingers, and soybeans were unaffected in comparison to rice and cucumbers while red spinach and lettuce showed cell death and electrolyte leakage at 1000–2000 ppm exposure of MWCNTs. The difference in xylem structures in different plants could be the possible reason for the variation in plant behavior and sensitivity at the same type of nanoparticle concentration. Further, Begum and Fugetsu (2012) reported the toxic impact of MWCNTs on red spinach, which was mediated by the generation of ROS. Generation of ROS induces oxidative stress in the cell, followed by damage of the cell membrane, cell organelles, and nucleic acid. They found broken mid-rid in spinach leaves as well as swelled epidermis and closed stomata after exposure of MWNCTs.

Previously, Begum et al. (2011) also reported phytotoxicity in the seedling stage of lettuce, tomatoes, red spinach, and cabbage at 500–2000ppm of graphene oxide after 20days of exposure. They found a reduction in the size and number of leaves after treatment of graphene oxide. The adverse effect of graphene oxide was observed due to the generation of ROS and necrotic lesion in red spinach and tomato. Larue et al. (2012) exposed the root of wheat and rapeseed to C^{14} radio-labeled MWCNTs in a hydroponic system and evaluated the impacts on seed germination, root elongation, evapotranspiration, chlorophyll, biomass, thiobarbituric acid reactive species, and H_2O_2 content. They found that only 0.005% of the nanomaterials was absorbed by the roots and transported up to the leaves. Such an absorbed amount was unable to affect the development and physiology of the plant. At this absorbed amount, no oxidative stress or modification in photosynthesis was measured by the authors. Therefore, the authors hypothesized that only a very small number of nanoparticles were absorbed by the plants and no harmful impact would be found during the flow of nanoparticles in the food chain.

Single-walled carbon nanotubes also negatively affected plant growth. Cañas et al. (2008) used functionalized and nonfunctionalized SWCNTs on tomatoes, cabbage, carrots, lettuce, and onions and found root length suppression in the tomato and root growth stimulation in the onion. While functionalized SWCNTs with poly-3-aminobenzene sulfonic acid showed adverse effects on root length in lettuce, carrots and cabbage which were not affected by unfunctionalized SWCNT. Further, Hatami et al. (2017) found that SWCNTs alleviated the drought stress in *Hyoscyamus niger* by increased water uptake, starch hydrolysis, and reduction in oxidation injury.

Chen et al. (2016) also assessed the phytotoxicity of carbon nanodots (C-dots) on corn in the range of 0–2000ppm concentration. The authors suggested that exposure of carbon nanodots induced the activity of ROS enzymes, lipid peroxidation, and H_2O_2 content after 500ppm of C-dots. They measured the reduction in the root and shoot's fresh weight up to 72% at 1000ppm concentration and increased enzyme activities of catalase, ascorbate peroxidase, guaiacol peroxidase, and superoxide dismutase at 2000ppm concentration; these were two times higher in comparison to control plants. They observed that C-dots absorbed and transported from root cap cells to the vascular bundle system and showed toxicity to corn plants. Further, Chen et al. (2018) also found phytotoxicity of graphene oxide to oats (*Avena sativa* L.) in a hydroponic system. They observed that graphene oxide inhibited the seedling length, biomass, and photosynthesis and distracted the root structure along with the generation of reactive oxygen species in the root. They compared the toxic effects of graphene oxide in a hydroponic system with soil and found that soil restricts the transportation of graphene oxide and reduces the adverse impact on the growth of *A. sativa*. On the other hand, the effect of fluorescent carbon dots on the mung bean was also measured. It was found that a concentration of carbon nanodots up to 100ppm did not show a negative impact on the mung bean, but a concentration at 400ppm affected the length of the root and shoot. Li and coworkers also observed through confocal microscopy that fluorescent carbon nanodots were transported in the mung bean through the root to the stem and further in leaves via the vascular system and the apoplastic pathway (Li et al., 2016).

Recently, Feng et al. (2019) found the adverse effect of graphene carbon dots (GCD) on *Vigna radiata* L. (mung bean) and *Solanum lycopersicum* L. (tomato) seedlings at a 1000–1500 ppm concentration. They observed that exposure of graphene carbon dots to *V. radiata* and *S. lycopersicum* seeds did not show a negative impact on their germination but the seedlings of *Vigna radiate* were more sensitive in comparison to *S. lycopersicum*. The chlorophyll content of both plants was increased at a limited concentration up to 500 ppm, but at a higher concentration of GCD, both types of plants showed an increment in stress-related enzymes such as catalase, H_2O_2, and glutathione reductase and stress-related molecules such as glutathione, malondialdehyde, proline, and glutathione.

7 Future prospects of nanocarbon as fertilizer

Nanocarbon materials have shown their potential as fertilizers by inducing plant growth and increasing crop production (Shukla et al., 2019). Many research groups observed that nanocarbon materials provided sufficient nutrients to seeds and growing seedlings in a controlled manner and induced their metabolic activities. Therefore, increased growth was observed in comparison to control (Hatami, 2017; Verma et al., 2019). Carbon nanomaterials have shown better stability and penetration ability in plants due to their nanoscale size. Carbon-based nanomaterials, especially carbon nanotubes and graphene, were the most favored materials that showed a positive effect on seed germination and plant growth (Tripathi et al., 2016). These nanomaterials showed increased water retention capacity and nutrient transport capacity in the plants (Pandey et al., 2018). Carbon nanomaterials were transported from the soil to the plant and then the food chains, where they interacted with various biological communities and environments (Hao et al., 2018; Wagay et al., 2019). Still, there are gaps in knowing the degradation patterns of carbon nanomaterials in the environment. More research is required to understand the fate of carbon nanomaterials in the environment. Further, an assessment of the carbon nanomaterial effect on soil microbial communities and soil health is required for commercial application as fertilizers.

8 Conclusion

Carbon-based nanomaterials showed significant potential in the field of agriculture. Various carbon nanomaterials such as carbon nanotubes, carbon dots, carbon nanohorns, fullerenes C_{60}, graphene oxide, and carbon nanoparticles have shown their applicability in seed germination, seedling growth, plant growth, and yield improvement. In contrast, some of the studies reported a negative impact on plant physiology. It has also been discovered how carbon nanomaterials entered the plant and were transported to various parts of the plant body. Carbon nanomaterials accumulated in the plants and affected the metabolic activity of plant cells. Carbon nanotubes have been used in most of the studies and were found to be a sustainable agent of next-generation fertilizers, among other carbon-based nanomaterials. The characteristics of carbon nanotubes such as being ecofriendly, a long period of stability in a pristine or functionalized state, a stable carrier of chemical agriculture input, and slow release profiles make them futuristic fertilizers that are capable of reducing the excess amounts of chemical fertilizers and pesticides. The above characteristics of carbon nanomaterials would be helpful to maintain ecological balance by the nondestruction of untargeted communities and the reduction of toxicity to untargeted microorganisms. Still, there are more studies required to understand the long-term impact of carbon nanomaterials on the environment and human health. We need more studies on the concentration optimization for each specific crop-carbon nanomaterial model for suppression of adverse effects in agriculture. Also, more data is required for each type of carbon nanomaterial and its interaction with specific crop types to implement them at a commercial level.et al.,

References

Abbaslou, R., Dalai, A., University of Saskatchewan, 2019. Promoted iron catalysts supported on carbon nanotubes for fischer-tropsch synthesis. U.S. Patent Application 10/213,770.

AbdulAlmohsin, S., Cui, J.B., 2012. Graphene-enriched P3HT and porphyrin-modified ZnO nanowire arrays for hybrid solar cell applications. J. Phys. Chem. C 116, 9433–9438.

Alimohammadi, M., Xu, Y., Wang, D.Y., Biris, A.S., Khodakovskaya, M.V., 2011. Physiological responses induced in tomato plants by a two-component nanostructural system composed of carbon nanotubes conjugated with quantum dots and its in vivo multimodal detection. Nanotechnology 22, 295101–295109.

Anjum, N.A., Singh, N., Singh, M.K., Shah, Z.A., Duarte, A.C., Pereira, E., Ahmad, I., 2013. Single-bilayer graphene oxide sheet tolerance and glutathione redox system significance assessment in faba bean (*Vicia faba* L.). J. Nanopart. Res. 15 (7), 1770.

Anjum, N.A., Singh, N., Singh, M.K., Sayeed, I., Duarte, A.C., Pereira, E., et al., 2014. Single-bilayer graphene oxide sheet impacts and underlying potential mechanism assessment in germinating faba bean (Vicia faba L.). Sci. Total Environ. 472, 834–841. https://doi.org/10.1016/j.scitotenv.2013.11.018.

Ashfaq, M., Verma, N., Khan, S., 2017. Carbon nanofibers as a micronutrient carrier in plants: efficient translocation and controlled release of Cu nanoparticles. Environ. Sci. Nano 4 (1), 138–148.591.

Begum, P., Fugetsu, B., 2012. Phytotoxicity of multi-walled carbon nanotubes on red spinach (*Amaranthus tricolor*) and the role of ascorbic acid as an antioxidant. J. Hazard. Mater. 243, 212–222.

Begum, P., Ikhtiari, R., Fugetsu, B., 2011. Graphene phytotoxicity in the seedling stage of cabbage, tomato, red spinach, and lettuce. Carbon 49 (12), 3907–3919.

Begum, P., Ikhtiari, R., Fugetsu, B., Matsuoka, M., Akasaka, T., Watari, F., 2012. Phytotoxicity of multi-walled carbon nanotubes assessed by selected plant species in the seedling stage. Appl. Surface Sci. 262, 120–124.

Bennett, S.W., Adeleye, A., Ji, Z., Keller, A.A., 2013. Stability, metal leaching, photoactivity and toxicity in freshwater systems of commercial single-wall carbon nanotubes. Water Res. 47, 4074–4085.

Borišev, M., Borišev, I., Župunski, M., Arsenov, D., Pajević, S., Ćurčić, Ž., Vasin, J., Djordjevic, A., 2016. Drought impact is alleviated in sugar beets (*Beta vulgaris* L.) by foliar application of fullerenol nanoparticles. PLoS One 11, e0166248.

Cañas, J.E., Long, M., Nations, S., Vadan, R., Dai, L., Luo, M., Ambikapathi, R., Lee, E.H., Olszyk, D., 2008. Effects of functionalized and nonfunctionalized single-walled carbon nanotubes on root elongation of select crop species. Environ. Toxicol. Chem. Int. J. 27 (9), 1922–1931.

Chen, J., Dou, R., Yang, Z., Wang, X., Mao, C., Gao, X., Wang, L., 2016. The effect and fate of water-soluble carbon nanodots in maize (*Zea mays* L.). Nanotoxicology 10 (6), 818–828.

Chen, L., Yang, S., Liu, Y., Mo, M., Guan, X., Huang, L., Sun, C., Yang, S.T., Chang, X.L., 2018. Toxicity of graphene oxide to naked oats (*Avena sativa* L.) in hydroponic and soil cultures. RSC Adv. 8 (28), 15336–15343.

Chhipa, H., 2017a. Nanofertilizers and nanopesticides for agriculture. Environ. Chem. Lett. 15 (1), 15–22.

Chhipa, H., 2017b. Nanopesticide: current status and future possibilities. Agri. Res. Technol. Open Access J. 5 (1), 555651.

Chhipa, H., 2018. Mycosynthesis of nanoparticles for smart agricultural practice: a green and eco-friendly approach. In: Shukla, A.K., Iravani, S. (Eds.), Green Synthesis, Characterization and Applications of Nanoparticles. Elsevier, Netherlands, pp. 87–110.

Chhipa, H., 2019. Applications of nanotechnology in agriculture. In: Ball, A., Soni, S., Gurtler, V. (Eds.), Nanotechnology. Methods in Microbiology-Nanotechnology, vol. 46. Academic Press.

Chhipa, H., Joshi, P., 2016. Nanofertilisers, nanopesticides and nanosensors in agriculture. In: Nanoscience in Food and Agriculture 1. Springer, Cham, pp. 247–282.

Chung, H., Son, Y., Yoon, T.K., Kim, S., Kim, W., 2011. The effect of multi-walled carbon nanotubes on soil microbial activity. Ecotoxicol. Environ. Saf. 74, 569–575. https://doi.org/10.1016/j.ecoenv.2011.01.004.

Chung, H., Kim, M.J., Ko, K., Kim, J.H., Kwon, H.A., Hong, I., et al., 2015. Effects of graphene oxides on soil enzyme activity and microbial biomass. Sci. Total Environ. 514, 307–313. https://doi.org/10.1016/j.scitotenv.2015.01.077.

Cordeiro, L.F., Marques, B.F., Kist, L.W., Bogo, M.R., Lopez, G., Pagano, G., et al., 2014. Toxicity of fullerene and nanosilver nanomaterials against bacteria associated to the body surface of the estuarine worm Laeonereis acuta (Polychaeta. Nereididae). Mar. Environ. Res. 99, 52–59. https://doi.org/10.1016/j.marenvres.2014.05.011.

De La Torre-Roche, R., Hawthorne, J., Deng, Y., Xing, B., Cai, W., Newman, L.A., et al., 2012. Fullerene-enhanced accumulation of p,p'-DDE in agricultural crop species. Environ. Sci. Technol. 46, 9315–9323. https://doi.org/10.1021/es301982w.

De La Torre-Roche, R., Hawthorne, J., Deng, Y., Xing, B., Cai, W., et al., 2013. Multiwalled carbon nanotubes and C60 fullerenes differentially impact the accumulation of weathered pesticides in four agricultural plants. Environ. Sci. Technol. 47, 12539–12547.

Dennis, T.J., Hare, J.P., Kroto, H.W., Taylor, R., Walton, D.R., Hendra, P.J., 1991. The vibrational Raman spectra of C60 and C70. Spectrochim. Acta A Mol. Spectrosc. 47 (9–10), 1289–1292.

Dong, Y., Chen, C., Zheng, X., Gao, L., Cui, Z., Yang, H., Guo, C., Chi, Y., Li, C.M., 2012. One-step and high yield simultaneous preparation of single-and multi-layer graphene quantum dots from CX-72 carbon black. J. Mater. Chem. 22 (18), 8764–8766.

Duhan, J.S., Kumar, R., Kumar, N., Kaur, P., Nehra, K., Duhan, S., 2017. Nanotechnology: the new perspective in precision agriculture. Biotechnol. Rep. 15, 11–23.

Fan, X., Xu, J., Lavoie, M., Peijnenburg, W.J.G.M., Zhu, Y., Lu, T., Fu, Z., Zhu, T., Qian, H., 2018. Multiwall carbon nanotubes modulate paraquat toxicity in *Arabidopsis thaliana*. Environ. Pollut. 233, 633–641.

Feng, P., Geng, B., Cheng, Z., Liao, X., Pan, D., Huang, J., 2019. Graphene quantum dots-induced physiological and biochemical responses in mung bean and tomato seedlings. Brazil J. Bot., 42 (1), 1–13.

Fleming, E., Du, F., Ou, E., Dai, L., Shi, L., 2019. Thermal conductivity of carbon nanotubes grown by catalyst-free chemical vapor deposition in nanopores. Carbon 145, 195–200.

Fang, J., Lyon, D.Y., Wiesner, M.R., Dong, J., Alvarez, P.J., 2007. Effect of a fullerene water suspension on bacterial phospholipids and membrane phase behavior. Environ. Sci. Technol. 41, 2636–2642. https://doi.org/10.1021/es 062181w.

Fraceto, L.F., Grillo, R., de Medeiros, G.A., Scognamiglio, V., Rea, G., Bartolucci, C., 2016. Nanotechnology in agriculture: which innovation potential does it have? Front. Environ. Sci. 4, 20.

Georgakilas, V., Perman, J.A., Tucek, J., Zboril, R., 2015. Broad family of carbon nanoallotropes: classification, chemistry, and applications of fullerenes, carbon dots, nanotubes, graphene, nanodiamonds, and combined superstructures. Chem. Rev. 115, 4744–4822.

Ghodake, G., Seo, Y.D., Park, D., Lee, D.S., 2010. Phytotoxicity of carbon nanotubes assessed by *Brassica juncea* and *Phaseolus mungo*. J. Nanoelectron. Optoelectron. 5 (2), 157–160.

Giraldo, J.P., Landry, M.P., Faltermeier, S.M., McNicholas, T.P., Iverson, N.M., Boghossian, A.A., Andrew, N.F.R., Sen, J.H.F., Brew, J.A., Strano, M.S., 2014. Plant nanobionics approach to augment photosynthesis and biochemical sensing. Nat. Mater. 13, 400–408.

Gong, X., Hu, Q., Paau, M.C., Zhang, Y., Shuang, S., Dong, C., Choi, M.M., 2014. Red-green-blue fluorescent hollow carbon nanoparticles isolated from chromatographic fractions for cellular imaging. Nanoscale 6 (14), 8162–8170.

Gopalan, S., 2008. Reversal of an immunity associated plant cell death program by the growth regulator auxin. BMC Res. Notes 1, 126.

Guo, T., Nikolaev, P., Thess, A., Colbert, D.T., Smalley, R.E., 1995. Catalytic growth of single-walled nanotubes by laser vaporization. Chem. Phys. Lett. 243 (1–2), 49.

Haghighi, M., da Silva, J.A.T., 2014. The effect of carbon nanotubes on the seed germination and seedling growth of four vegetable species. J. Crop. Sci. Biotechnol. 17 (4), 201–208.

Hamdi, H., De La Torre-Roche, R., Hawthorne, J., White, J.C., 2015. Impact of non-functionalized and amino-functionalized multiwall carbon nanotubes on pesticide uptake by lettuce (Lactuca sativa L.). Nanotoxicology 9, 172–180. https://doi.org/10.3109/17435390.2014.907456.

Han, F., Wang, R., Feng, Y., Wang, S., Liu, L., Li, X., Han, Y., Chen, H., 2019. On demand synthesis of hollow fullerene nanostructures. Nat. Commun. 10 (1), 1548.

Hao, Y., Ma, C., Zhang, Z., Song, Y., Cao, W., Guo, J., Zhou, G., Rui, Y., Liu, L., Xing, B., 2018. Carbon nanomaterials alter plant physiology and soil bacterial community composition in a rice-soil-bacterial ecosystem. Environ. Pollut. 232, 123–136.

Hatami, M., 2017. Toxicity assessment of multi-walled carbon nanotubes on *Cucurbita pepo* L. under well-watered and water-stressed conditions. Ecotoxicol. Environ. Saf. 142, 274–283.

Hatami, M., Hadian, J., Ghorbanpour, M., 2017. Mechanisms underlying toxicity and stimulatory role of single-walled carbon nanotubes in *Hyoscyamus niger* during drought stress simulated by polyethylene glycol. J. Hazard. Mater. 324, 306–320.

Howard, J.B., McKinnon, J.T., Makarovsky, Y., Lafleur, A.L., Johnson, M.E., 1991. Fullerenes C60 and C70 in flames. Nature 352 (6331), 139.

Hu, X., Kang, J., Lu, K., Zhou, R., Mu, L., Zhou, Q., 2014. Graphene oxide amplifies the phytotoxicity of arsenic in wheat. Sci. Rep. 4, 6122. https://doi.org/10.1038/srep06122.

Hurt, R.H., Monthioux, M., Kane, A., 2006. Toxicology of carbon nanomaterials: status, trends, and perspectives on the special issue. Carbon 44, 1028–1033.

Husen, A., Siddiqi, K.S., 2014. Carbon and fullerene nanomaterials in plant system. J. Nanobiotechnol. 12 (1), 16.

Iijima, S., Ichihashi, T., 1993. Single-shell carbon nanotubes of 1-nm diameter. Nature 363 (6430), 603.

Iqbal, M., Ashraf, M., 2013. Gibberellic acid mediated induction of salt tolerance in wheat plants: growth, ionic partitioning, photosynthesis, yield and hormonal homeostasis. Environ. Exp. Bot. 86, 76–85.

Jin, L., Son, Y., Deforest, J.L., Kang, Y.J., Kim, W., Chung, H., 2014. Single-walled carbon nanotubes alter soil microbial community composition. Sci. Total Environ. 466–467, 533–538. https://doi.org/10.1016/j.scitotenv.2013.07.035.

Johansen, A., Pedersen, A.L., Jensen, K.A., Karlson, U., Hansen, B.M., Scott-Fordsmand, J.J., et al., 2008. Effects of C60 fullerene nanoparticles on soil bacteria and protozoans. Environ. Toxicol. Chem. 27, 1895–1903. https://doi.org/10.1897/07-375.1.

Kang, S., Pinault, M., Pfefferle, L.D., Elimelech, M., 2007. Single-walled carbon nanotubes exhibit strong antimicrobial activity. Langmuir 23, 8670–8673. https://doi.org/10.1021/la701067r.

Kerfahi, D., Tripathi, B.M., Singh, D., Kim, H., Lee, S., Lee, J., et al., 2015. Effects of functionalized and raw multi-walled carbon nanotubes on soil bacterial community composition. PLoS ONE 10, e0123042. https://doi.org/10.1371/journal.pone.0123042.

Khadri, H., Alzohairy, M., Janardhan, A., Kumar, A.P., Narasimha, G., 2013. Green synthesis of silver nanoparticles with high fungicidal activity from olive seed extract. Adv. Nanopart. 2 (03), 241.

Khodakovskaya, M., Dervishi, E., Mahmood, M., Xu, Y., Li, Z., Watanabe, F., et al., 2009. Carbon nanotubes are able to penetrate plant seed coat and dramatically affect seed germination and plant growth. ACS Nano 3, 3221–3227. https://doi.org/10.1021/nn900887m.

Khodakovskaya, M.V., Silva, K.D., Nedosekin, D.A., Dervishi, E., Birisa, A.S., Shashkov, E.V., Galanzh, E.I., Zharov, V.P., 2011. Complex genetic, photothermal, and photoacoustic analysis of nanoparticle–plant interactions. Proc. Natl. Acad. Sci. 108, 1028–1033.

Khodakovskaya, M.V., de Silva, K., Biris, A.S., Dervishi, E., Villagarcia, H., 2012. Carbon nanotubes induce growth enhancement of tobacco cells. ACS Nano 6, 2128–2135.

Khodakovskaya, M.V., Kim, B.-S., Kim, J.N., Alimohammadi, M., Dervishi, E., Mustafa, T., Cernigla, C.E., 2013. Carbon nanotubes as plant growth regulators: effects on tomato growth, reproductive system, and soil microbial community. Small 9, 115–123.

Kole, C., Kole, P., Randunu, K.M., Choudhary, P., Podila, R., Ke, P.C., Rao, A.M., Marcus, R.K., 2013. Nanobiotechnology can boost crop production and quality: first evidence from increased plant biomass, fruit yield and phytomedicine content in bitter melon (*Momordica charantia*). BMC Biotechnol. 13, 37.

Krätschmer, W., Lamb, L.D., Fostiropoulos, K., Huffman, D.R., 1990. Solid C60: a new form of carbon. Nature 347 (6291), 354.

Kroto, H.W., Heath, J.R., O'Brien, S.C., Curl, R.F., Smalley, R.E., 1985. C_{60}: buckminsterfullerene. Nature 318 (6042), 162.

Lacerda, L., Bianco, A., Prato, M., Kostarelos, K., 2006. Carbon nanotubes as nanomedicines: from toxicology to pharmacology. Adv. Drug Deliv. Rev. 58 (14), 1460–1470.

Lahiani, M.H., Dervishi, E., Chen, J., Nima, Z., Gaume, A., Biris, A.S., Khodakovskaya, M.V., 2013. Impact of carbon nanotube exposure to seeds of valuable crops. ACS Appl. Mater. Interfaces 5, 7965–7973.

Lahiani, M.H., Chen, J., Irin, F., Puretzky, A.A., Green, M.J., Khodakovskaya, M.V., 2015. Interaction of carbon nanohorns with plants: uptake and biological effects. Carbon 81, 607–619.

Lahiani, M.H., Dervishi, E., Ivanov, I., Chen, J., Khodakovskaya, M., 2016. Comparative study of plant responses to carbon-based nanomaterials with different morphologies. Nanotechnology 27 (26), 265102.

Lahiani, M.H., Nima, Z.A., Villagarcia, H., Biris, A.S., Khodakovskaya, M.V., 2017. Assessment of effects of the long-term exposure of agricultural crops to carbon nanotubes. J. Agric. Food Chem. 66 (26), 6654–6662.

Lai, C.W., Hsiao, Y.H., Peng, Y.K., Chou, P.T., 2012. Facile synthesis of highly emissive carbon dots from pyrolysis of glycerol; gram scale production of carbon dots/mSiO2 for cell imaging and drug release. J. Mater. Chem. 22 (29), 14403–14409.

Larue, C., Pinault, M., Czarny, B., Georgin, D., Jaillard, D., Bendiab, N., Mayne-L'Hermite, M., Taran, F., Dive, V., Carrière, M., 2012. Quantitative evaluation of multi-walled carbon nanotube uptake in wheat and rapeseed. J. Hazard. Mater. 227, 155–163.

Lateef, A., Nazir, R., Jamil, N., Alam, S., Shah, R., Khan, M.N., Saleem, M., 2019. Synthesis and characterization of environmental friendly corncob biochar based nano-composite—a potential slow release nano-fertilizer for sustainable agriculture. Environ. Nanotechnol. Monit. Manage. 11, 100212.

Li, H., Kang, Z., Liu, Y., Lee, S.T., 2012. Carbon nanodots: synthesis, properties and applications. J. Mater. Chem. 22 (46), 24230–24253.

Li, W., Zheng, Y., Zhang, H., Liu, Z., Su, W., Chen, S., Liu, Y., Zhuang, J., Lei, B., 2016. Phytotoxicity, uptake, and translocation of fluorescent carbon dots in mung bean plants. ACS Appl. Mater. Interfaces 8 (31), 19939–19945.

Lim, S., Luo, Z., Shen, Z., Lin, J., 2010. Plasma-assisted synthesis of carbon nanotubes. Nanoscale Res. Lett. 5 (9), 1377.

Lin, D., Xing, B., 2007. Phytotoxicity of nanoparticles: inhibition of seed germination and root growth. Environ. Pollut. 150, 243–250.

Lin, S., Reppert, J., Hu, Q., Hudson, J.S., Reid, M.L., Ratnikova, T.A., Rao, A.M., Luo, H., Ke, P.C., 2009. Uptake, translocation, and transmission of carbon nanomaterials in rice plants. Small 5, 1128–1132.

Liu, Q., Chen, B., Wang, Q., Shi, X., Xiao, Z., Lin, J., et al., 2009. Carbon nanotubes as molecular transporters for walled plant cells. Nano Lett. 9, 1007–1010. https://doi.org/10.1021/nl803083u.

Liu, Q., Zhang, X., Zhao, Y., Lin, J., Shu, C., Wang, C., et al., 2013. Fullerene-induced increase of glycosyl residue on living plant cell wall. Environ. Sci. Technol. 47, 7490–7498. https://doi.org/10.1021/es4010224.

Liu, F.Y., Xiong, F.X., Fan, Y.K., Li, J., Wang, H.Z., Xing, G.M., Yan, F.M., Tai, F.J., He, R., 2016. Facile and scalable fabrication engineering of fullerenol nanoparticles by improved alkaline-oxidation approach and its antioxidant potential in maize. J. Nanopart. Res. 18, 338.

Liu, M.L., Chen, B.B., Li, C.M., Huang, C.Z., 2019. Carbon dots: synthesis, formation mechanism, fluorescence origin and sensing applications. Green Chem. 21 (3), 449–471.

Ma, X., Wang, C., 2010. Fullerene nanoparticles affect the fate and uptake of trichloroethylene in phytoremediation systems. Environ. Eng. Sci. 27, 989–992.

Mao, J.-D., Johnson, R.L., Lehmann, J., Olk, D.C., Neves, E.G., Thompson, M.L., Schmidt-Rohr, K., 2012. Abundant and stable char residues in soils: implications for soil fertility and carbon sequestration. Environ. Sci. Technol. 46, 9571–9576.

Mauter, M.S., Elimelech, M., 2008. Environmental applications of carbon-based nanomaterials. Environ. Sci. Technol. 42 (16), 5843–5859.

Milosavljevic, V., Nguyen, H.V., Michalek, P., Moulick, A., Kopel, P., Kizek, R., Adam, V., 2015. Synthesis of carbon quantum dots for DNA labeling and its electrochemical, fluorescent and electrophoretic characterization. Chem. Papers 69 (1), 192–201.

Mink, J.E., Hussain, M.M., 2013. Sustainable design of high-performance microsized microbial fuel cell with carbon nanotube anode and air cathode. ACS Nano 7 (8), 6921–6927.

Miralles, P., Church, T.L., Harris, A.T., 2012a. Toxicity, uptake, and translocation of engineered nanomaterials in vascular plants. Environ. Sci. technol. 46 (17), 9224–9239.

Miralles, P., Johnson, E., Church, T.L., Harris, A.T., 2012b. Multiwalled carbon nanotubes in alfalfa and wheat: toxicology and uptake. J. R. Soc. Interface 9 (77), 3514–3527.

Mondal, A., Basu, R., Das, S., Nandy, P., 2011. Beneficial role of carbon nanotubes on mustard plant growth: an agricultural prospect. J. Nanopart. Res. 13, 4519–4528.

Morsy, M.K., Khalaf, H.H., Sharoba, A.M., El-Tanahi, H.H., Cutter, C.N., 2014. Incorporation of essential oils and nanoparticles in pullulan films to control foodborne pathogens on meat and poultry products. J. Food Sci. 79 (4), M675–M684.

Mukherjee, A., Majumdar, S., Servin, A.D., Pagano, L., Dhankher, O.P., White, J.C., 2016. Carbon nanomaterials in agriculture: a critical review. Front. Plant Sci. 7, 172.

Nair, R., Mohamed, M.S., Gao, W., Maekawa, T., Yoshida, Y., Ajayan, P.M., Kumar, D.S., 2012. Effect of carbon nanomaterials on the germination and growth of rice plants. J. Nanosci. Nanotechnol. 12 (3), 2212–2220.

Noctor, G., Mhamdi, A., Chaouch, S., Han, Y.I., Neukermans, J., Marquez-Garcia, B.E.L.E.N., Queval, G., Foyer, C.H., 2012. Glutathione in plants: an integrated overview. Plant Cell Environ. 35 (2), 454–484.

Nyberg, L., Turco, R.F., Nies, L., 2008. Assessing the impact of nanomaterials on anaerobic microbial communities. Environ. Sci. Technol. 42, 1938–1943. https://doi.org/10.1021/es072018g.

Pan, D., Zhang, J., Li, Z., Wu, M., 2010. Hydrothermal route for cutting graphene sheets into blue-luminescent graphene quantum dots. Adv. Mater. 22 (6), 734–738.

Pandey, K., Lahiani, M.H., Hicks, V.K., Hudson, M.K., Green, M.J., Khodakovskaya, M., 2018. Effects of carbon-based nanomaterials on seed germination, biomass accumulation and salt stress response of bioenergy crops. PLoS One 13 (8), e0202274.

Parizi, M.A., Moradpour, Y., Roostaei, A., Khani, M., Negahdari, M., Rahimi, G., 2014. Evaluation of the antifungal effect of magnesium oxide nanoparticles on *Fusarium oxysporum* f. sp. *lycopersici*, pathogenic agent of tomato. Euro. J. Exp. Biol. 4 (3), 151–156.

Patel, A., Tiwari, S., Parihar, P., Singh, R., Prasad, S.M., 2019. (Chapter 2). Carbon nanotubes as plant growth regulators: impacts on growth, reproductive system, and soil microbial community. In: Tripathi, D.K., Ahmad, P., Sharma, S., Chauhan, D.K., Dubey, N.K. (Eds.), Nanomaterials in Plants, Algae and Microorganisms. Academic Press, pp. 23–42.

Peng, J., Gao, W., Gupta, B.K., Liu, Z., Romero-Aburto, R., Ge, L., Song, L., Alemany, L.B., Zhan, X., Gao, G., Vithayathil, S.A., 2012. Graphene quantum dots derived from carbon fibers. Nano Lett. 12 (2), 844–849.

Prikhod'ko, N.G., 2010. Synthesis of fullerenes in hydrocarbon flames assisted by a glow discharge. Russ. J. Phys. Chem. B 4 (3), 486–491.

Richter, H., Fonseca, A., Thiry, P., Gilles, J., Nagy, J., Lucas, A., 1994. Combustion synthesis of fullerenes. MRS Proc. 359, 17. https://doi.org/10.1557/PROC-359-17.

Rodrigues, C., de Souza Vandenberghe, L.P., de Oliveira, J., Soccol, C.R., 2012. New perspectives of gibberellic acid production: a review. Crit. Rev. Biotechnol. 32, 263–273.

Rodrigues, D.F., Jaisi, D.P., Elimelech, M., 2013. Toxicity of functionalized single-walled carbon nanotubes on soil microbial communities: implications for nutrient cycling in soil. Environ. Sci. Technol. 47, 625–633. https://doi.org/10.1021/es304002q.

Samaj, J., Baluska, F., Voigt, B., Schlicht, M., Volkmann, D., Menzel, D., 2004. Endocytosis, actin cytoskeleton, and signaling. Plant Physiol. 135, 1150–1161.

Saxena, M., Maity, S., Sarkar, S., 2014. Carbon nanoparticles in 'biochar' boost wheat (*Triticum aestivum*) plant growth. RSC Adv. 4, 39948–39954.

Shen, J., Zhu, Y., Yang, X., Li, C., 2012. Graphene quantum dots: emergent nanolights for bioimaging, sensors, catalysis and photovoltaic devices. Chem. Commun. 48 (31), 3686–3699.

Shrestha, B., Acosta-Martinez, V., Cox, S.B., Green, M.J., Li, S., Canas-Carrell, J.E., 2013. An evaluation of the impact of multiwalled carbon nanotubes on soil microbial community structure and functioning. J. Hazard. Mater. 261, 188–197. https://doi.org/10.1016/j.jhazmat.2013.07.031.

Shukla, P., Chaurasia, P., Younis, K., Qadri, O.S., Faridi, S.A., Srivastava, G., 2019. Nanotechnology in sustainable agriculture: studies from seed priming to post-harvest management. Nanotechnol. Environ. Eng. 4, 11.

Singh, A., Bhati, A., Tripathi, K.M., Sonkar, S.K., 2018. Nanocarbons in agricultural plants: can be a potential nanofertilizer? In: Hussain, C.M., Mishra, A.K. (Eds.), Nanotechnology in Environmental Science. Wiley Publisher, United States, pp. 153–190.

Singh, M., Sengupta, A., Zeller, K., Skoptsov, G., Vander Wal, R.L., 2019. Effect of hydrogen concentration on graphene synthesis using microwave-driven plasma-mediated methane cracking. Carbon 143, 802–813.

Sohi, S., Lopez-Capel, E., Krull, E., Bol, R., 2009. Biochar, climate change and soil: A review to guide future research. CSIRO Land and Water Science Report 05/09, pp.1–56.

Solanki, P., Bhargava, A., Chhipa, H., Jain, N., Panwar, J., 2015. Nano-fertilizers and their smart delivery system. In: Nanotechnologies in Food and Agriculture. Springer, Cham, pp. 81–101.

Song, J., Gong, X., Miao, W., Zheng, F., Song, C., Wang, M., 2014. Indole-3-acetic acid reverses the harpin-induced hypersensitive response and alters the expression of hypersensitive-response-related genes in tobacco. Biotechnol. Lett. 36, 1043–1048.

Sonkar, S.K., Roy, M., Babara, D.G., Sarkar, S., 2012. Water soluble carbon nano-onions from wood wool as growth promoters for gram plants. Nanoscale 4, 7670. https://doi.org/10.1039/c2nr32408c.

Srivastava, A., Rao, D.P., 2014. Enhancement of seed germination and plant growth of wheat, maize, peanut and garlic using multiwalled carbon nanotubes. Eur. Chem. Bull. 3 (5), 502–504.

Srivastava, V., Gusain, D., Sharma, Y.C., 2015. Critical review on the toxicity of some widely used engineered nanoparticles. Ind. Eng. Chem. Res. 54, 6209–6233.

Stampoulis, D., Sinha, S.K., White, J.C., 2009. Assay-dependent phytotoxicity of nanoparticles to plants. Environ. Sci. Technol. 43, 9473–9479. https://doi.org/10.1021/es901695c.

Tan, X., Lin, C., Fugetsu, B., 2009. Studies on toxicity of multi-walled carbon nanotubes on suspension rice cells. Carbon 47, 3479–3487.

Tiwari, D.K., Dasgupta-Schubert, N., Villaseñor Cendejas, L.M., Villegas, J., Carreto Montoya, L., Borjas García, S.E., 2013. Interfacing carbon nanotubes (CNT) with plants: enhancement of growth, water and ionic nutrient uptake in maize (*Zea mays*) and implications for nanoagriculture. Appl. Nanosci. 4, 577–591. https://doi.org/10.1007/s13204-013-0236-7.

Tong, Z., Bischoff, M., Nies, L., Applegate, B., Turco, R.F., 2007. Impact of fullerene (C60) on a soil microbial community. Environ. Sci. Technol. 41, 2985–2991. https://doi.org/10.1021/es061953l.

Tripathi, S., Sarkar, S., 2014. Influence of water soluble carbon dots on the growth of wheat plant. Appl. Nanosci. 5, 609–616. https://doi.org/10.1007/s13204-014-0355-9.

Tripathi, S., Sonkar, S.K., Sarkar, S., 2011. Growth stimulation of gram (*Cicer arietinum*) plant by water soluble carbon nanotubes. Nanoscale 3, 1176–1181.

Tripathi, S., Kapri, S., Datta, A., Bhattacharyya, S., 2016. Influence of the morphology of carbon nanostructures on the stimulated growth of gram plant. RSC Adv. 6 (50), 43864–43873.
Ueno, M., Kawamitsu, Y., Komiya, Y., Sun, L., 2007. Carbonisation and gasi fi cation of bagasse for effective utilisation of sugarcane biomass. Proc. Int. Soc. Sugar Cane Technol. 26, 1194–1201.
Vera-Reyes, I., Vázquez-Núñez, E., Lira-Saldivar, R.H., Méndez-Argüello, B., 2018. Effects of nanoparticles on germination, growth, and plant crop development. In: Agricultural Nanobiotechnology. Springer, Cham, pp. 77–110.
Verma, S.K., Das, A.K., Gantait, S., Kumar, V., Gurel, E., 2019. Applications of carbon nanomaterials in the plant system: a perspective view on the pros and cons. Sci. Total Environ. 667, 485–499.
Villagarcia, H., Dervishi, E., Silva, K., Biris, A.S., Khodakovskaya, M.V., 2012. Surface chemistry of carbon nanotubes impacts the growth and expression of water channel protein in tomato plants. Small 8, 2328–2334.
Wagay, J.A., Singh, S., Raffi, M., Rahman, Q.I., Husen, A., 2019. Effect of carbon-based nanomaterials on rhizosphere and plant functioning. In: Husen, A., Iqbal, M. (Eds.), Nanomaterials and Plant Potential. Springer International Publishing, Cham, pp. 553–575.
Wang, F., Pang, S., Wang, L., Li, Q., Kreiter, M., Liu, C.Y., 2010. One-step synthesis of highly luminescent carbon dots in noncoordinating solvents. Chem. Mater. 22 (16), 4528–4530.
Wang, X., Gu, X., Han, H., Chen, K., Liu, X., Lu, D., 2012. Multi-walled carbon nanotubes can enhance root elongation of wheat (*Triticum aestivum*) plants. J. Nanopart. Res. 14, 841–850.
Wang, D., Wang, G., Zhang, G., Xu, X., Yang, F., 2013. Using graphene oxide to enhance the activity of anammox bacteria for nitrogen removal. Bioresour. Technol. 131, 527–530. https://doi.org/10.1016/j.biortech.2013.01.099.
Wang, H., Zhang, M., Song, Y., Li, H., Huang, H., Shao, M., Liu, Y., Kang, Z., 2018. Carbon dots promote the growth and photosynthesis of mung bean sprouts. Carbon 136, 94–102.
Wild, E., Jones, K.C., 2009. Novel method for the direct visualization of in vivo nanomaterials and chemical interactions in plants. Environ. Sci. Technol. 43, 5290–5294.
Xiong, J.L., Li, J., Wang, H.C., Zhang, C.L., Naeem, M.S., 2018. Fullerol improves seed germination, biomass accumulation, photosynthesis and antioxidant system in *Brassica napus* L. under water stress. Plant Physiol. Biochem. 129, 130–140.
Yatim, N.M., Shaaban, A., Dimin, M.F., Mohamad, N., Yusof, F., 2019. Urea functionalized multiwalled carbon nanotubes as efficient nitrogen delivery system for rice. Adv. Nat. Sci. Nanosci. Nanotechnol. 10 (1), 015011.
Yin, J., Wang, Y., Gilbertson, L.M., 2018. Opportunities to advance sustainable design of nano-enabled agriculture identified through a literature review. Environ. Sci. Nano 5, 11–26.
Yudasaka, M., Iijima, S., Crespi, V.H., 2007. Single-wall carbon nanohorns and nanocones. In: Carbon nanotubes. Springer, Berlin, Heidelberg, pp. 605–629.
Zaytseva, O., Neumann, G., 2016. Phytotoxicity of carbon nanotubes is associated with disturbances of zinc homeostasis. Eur. Chem. Bull. 5 (6), 238–244.
Zhai, G., Gutowski, S.M., Walters, K.S., Yan, B., Schnoor, J.L., 2015. Charge, size, and cellular selectivity for multiwall carbon nanotubes by maize and soybean. Environ. Sci. Technol. 49 (12), 7380–7390.
Zhao, Q., Ma, C., White, J.C., Dhankher, O.P., Zhang, X., Zhang, S., Xing, B., 2017. Quantitative evaluation of multi-wall carbon nanotube uptake by terrestrial plants. Carbon 114, 661–670.
Zhao, Y., Zhao, P., Luo, J., Tian, L., Tian, Z., 2019. Effects of liquid nano-carbon biological fertilizer on physiological characters and yield of Chinese cabbage. J. Biobased Mater. Bioener. 13, 221–224.

Further reading

Kroto, H.W., 1987. The stability of the fullerenes Cn, with n= 24, 28, 32, 36, 50, 60 and 70. Nature 329 (6139), 529.

McKinnon, J.T., Bell, W.L., Barkley, R.M., 1992. Combustion synthesis of fullerenes. Combust. Flame 88, 102–112.

Mohamed, M.A., Hashim, A.F., Alghuthaymi, M.A., Abd-Elsalam, K.A., 2018. Nano-carbon: plant growth promotion and protection. In: Abd-Elsalam, K.A., Prasad, R. (Eds.), Nanobiotechnology Applications in Plant Protection. Springer International Publishing, Cham, pp. 155–188.

Rahman, M.S., Hossain, M.S., Ahmed, M.K., Akther, S., Jolly, Y.N., Akhter, S., Kabir, M.J., Choudhury, T.R., 2019. Assessment of heavy metals contamination in selected tropical marine fish species in Bangladesh and their impact on human health. Environ. Nanotechnol. Monit. Manag. 11, 100210.

Shojaei, T.R., Salleh, M.A.M., Tabatabaei, M., Mobli, H., Aghbashlo, M., et al., 2019. (Chapter 11). Applications of nanotechnology and carbon nanoparticles in agriculture. In: Rashid, S.A., Othman, R.N.I.R., Hussein, M.Z. (Eds.), Synthesis, Technology and Applications of Carbon Nanomaterials. Elsevier, pp. 247–277.

Wang, X., Liu, X., Chen, J., Han, H., Yuan, Z., 2014. Evaluation and mechanism of antifungal effects of carbon nanomaterials in controlling plant fungal pathogen. Carbon 68, 798–806.

Micro/nano biochar for sustainable plant health: Present status and future prospects

16

Mohamed M. Ramadan[a,b], *Asran-Amal*[a,b], *Kamel A. Abd-Elsalam*[a,b]
[a]Plant Pathology Research Institute, Agricultural Research Center (ARC), Giza, Egypt,
[b]Unit of Excellence in Nano-Molecular Plant Pathology Research Center—Plant Pathology Research Institute, Giza, Egypt

1 Introduction

Many of the distinctive procedures applied to reduction, recycling, composting, incineration, and removal into landfills have been practiced to manipulate compacted waste in different countries. The World Bank and Agency for International Development (USAID) stated that 20%–50% of the financial budget range of a municipal council in establishing nations is spent on solid waste reduction; these funds originate from less than 50% of the inner-city populace (Henry et al., 2006; Memon, 2010). World Bank survey data estimated that close to 1.3 billion tonnes of municipal solid waste are produced with the aid of world cities, which can be expected to generate as much as 2.2 billion tonnes by 2025. This is primarily because of population progress, increasing urbanization, and the sociofiscal progress of low- and core-earnings countries (Hoornweg and Bhada-Tata, 2012). Some agricultural actions influence waste formation, which may cause huge damage to human and animal health, ecosystems, biodiversity, and natural resources (Vergara and Tchobanoglous, 2012). At present, the handling of municipal solid waste creates some ecological problems such as using more energy and more heat or producing a large amount of recyclable material (Portugal-Pereira and Lee, 2016). The speedy boost of communal waste materials is the final result of the dramatic progress of the human and animal population, industrialization, and immigration of people from agriculture to cities through economic advancement (De Souza Melare et al., 2017).

Biochar is a pyrogenic black carbon derived from the thermal conversion of biomass feedstocks, including agricultural and forest residuals, in an inert atmosphere. Biochar technology has attracted great attention because of its potential to mitigate climate change and enhance soil fertility and health (Lehmann, 2007b). Moreover, many researchers have observed that biochar can be utilized instead of adsorbents to remove specific varieties of contaminants, including heavy metals, vitamins, minerals, and prescription drugs, from aqueous solutions (Inyang et al., 2012). Biochar usage could create soils that, in management and performance, start to resemble the fertile terra preta, which is a type of charcoal used to improve soil fertility (Sohi, 2012).

Carbon Nanomaterials for Agri-food and Environmental Applications. https://doi.org/10.1016/B978-0-12-819786-8.00016-5

Nevertheless, it has been recently proven that terra preta was produced through the unintended addition of inorganic (e.g., ash, fish bones) and natural (e.g., biomass wastes, manure, excrements, urine, biochar) substances to incredibly weathered and infertile Ferralsols (Glaser and Birk, 2012). Biochar production for soil application was initially recommended for long-term C storage with the aid of diverting waste biomass C from a quick to slow C-cycling pond in soil (Lehmann, 2007a, b). Additionally, to the long-term C potential of biochar, a tremendous study has been conducted in the last decade to comprehend other potential effects of biochars from their soil functions. For instance, biochar may decrease greenhouse gas emissions, increase agricultural commodity output, and improve microbial, fungal, and mycorrhizal growth and soil parameters (Chan et al., 2007; Lehmann et al., 2006). Despite their organic and natural origins, biochar is recalcitrant to decomposition as a result of its remarkably aromatic carbon structure. Therefore, biochar continues to be one of the main plant promotion elements in agriculture because of its ability to maximize cation exchange capacity (Chan et al., 2008), nutrient obtainability (Asai et al., 2009), and soil water-holding capacity (Masulili et al., 2010). This chapter details the trustworthy use of micro/nano biochar just for sustainable cultivation that might be helpful for wastewater purification, bioremediation of different types of pesticides, antimicrobial activity, fertilizers, controlled-launch fertilizers, plant growth promotion, fragments, and silage agents. Biochar's future is seriously bright because it is generally a low-cost option to improve plant health and crop yield.

2 What is biochar?

Biochar, formerly called "pyrochar," is the carbon-rich steady byproduct obtained from biomass carbonization incorporating timber, manure, or perhaps leaves, then heated to conditions between 300–1000°C under a low oxygen concentration. The principal confirmation of humans' deliberate use of charcoal shows that it was applied as a fuel at least since 5500 years ago in Southern Europe and the Middle East. With the commencement of the Bronze Age in Britain around 4000 years ago, the use of charcoal as a metallurgical fuel was commonplace. There is considerable evidence that pre-Columbian natives of the Amazon Basin used charcoal as a soil additive together with manure, bone, and pottery shards, turning otherwise unproductive soils into rich and fertile ones (Sombroek, 1993). Biochar is a solid material made through the thermochemical decomposition of biomass materials. This term is defined by the International Biochar Initiative (http://www.biochar-international.org/biochar); also, biochar is identified as "a solid material obtained from the carbonization of biomass." Within the last few years, biochar has received rising attention because of the prospect of a significant decrease in atmospheric greenhouse gas amounts (Lehmann, 2007a, b). Biochar states were recognized by Peter as a carbon-rich solid coproduction of pyrolysis, which may be the thermal degradation of biomass under oxygen-limited circumstances. The idea of biochar software to soil offers arisen predicated on observations produced in the Amazon Delta Dark Earth (terra preta), which includes higher organic C content material and better fertility compared to the surrounding organic (unamended) soils (Sombroek, 1993; Zech et al., 1990).

Biomass involves not only living organisms, which include crops and animals, but also the waste materials of various animals as well as sludge and waste wood (Tripathi et al., 2016). A few methods such as thermochemical decomposition procedures replace biomass components to syngas, and synthesis gas including both hydrogen, carbon monoxide and a few carbon dioxides, bio-oil, and biochar. They produce biochar that is known to us as terra preta, due to the partial burning of plant particles in pits. It was discovered that terra preta could keep soil fertile for many years (Marris, 2006; Read, 2009; Read and Parshotam, 2007). Biochar is generally constructed with long-term C storage in soil and offers additional possible advantages to soil, including physical, biological fertility and chemical. Biochar isn't a real carbon, as the component structures include carbon (C), hydrogen (H), oxygen (O), nitrogen (N), sulfur (S), potassium (K), magnesium (Mg), and ash. These elements can provide the most important nutrition in plant development (Fig. 1). The addition of biochar improves the quantity of organic matter in the soil (e.g., organic carbon), thus fixing soil's physicochemical and biological properties (Kavitha et al., 2018). Specifically, biochar includes a different high surface and porosity in comparison to other chemical substances or biological changes, enabling us to adsorb/preserve nutrients/water or even present habitats for supportive soil microorganisms (Lehmann et al., 2006), bioremediation of contaminated soil/drinking water (Beesley et al., 2010), and enhancement of soil health and crop yields (Steiner et al., 2008). Supplements of biochar to soil are generally determined to decrease emissions of greenhouse gases from agricultural soils, for instance, minimizing N2O emissions by up to 80% and totally reducing methane production (Lehmann et al., 2006; Rondon et al., 2007).

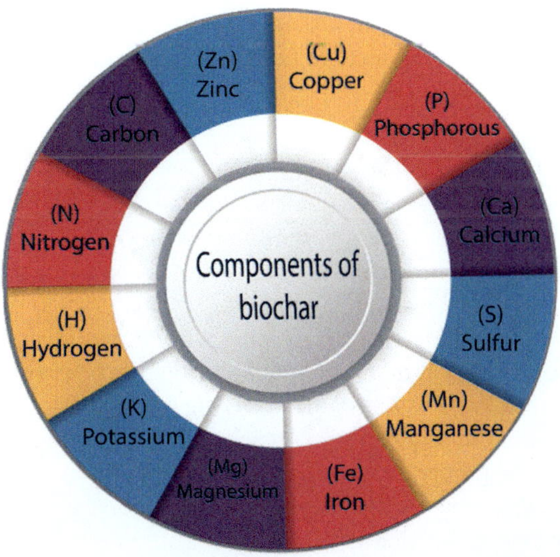

Fig. 1 The component structures of biochar, including plant macronutrient nitrogen (N), phosphorous (P), calcium (Ca), magnesium (Mg), potassium (K), and sulfur (S) and micronutrients copper (Cu), iron (Fe), manganese (Mn), zinc (Zn), and ash.

2.1 Biochar production by pyrolysis

Wood-based biochar is rough and extremely resistant to carbon content material close to 80% (Demirbas, 2004). Common charcoal production is offered using batch procedures that involve pits, earth mounds, and brick and metal kilns. Such methods are dependent on a basic method and these technologies are low cost to develop (Brown, 2009). Biochars have always been effectively generated from animal manures (Bird et al., 2011; Chan et al., 2008). In this approach, the green waste is exposed to pyrolysis, that is, heated to high temperatures in oxygen shortage (Martinez, 2006). The thermal decomposition of the wood commences when it is elevated to a temperature of about 300°C. The pyrolysis technique continues by itself and gives off extensive heat with an optimum temperature of around 500°C to produce highly proficient and quality biochar (FAO, 1983). Classic charcoal-producing methods release a significant amount of smoke, soot, and pollution and are energy ineffective (Brown, 2009). Biochar was manufactured in an old-style *earth* mound *kiln* where pyrolysis, gasification, and burning operations were performed in parallel (Brown, 2009; Duku et al., 2011). Quick pyrolysis of biomass generates a product that is used both as an energy source and as a feedstock for chemical production (Mohan et al., 2006). To really determine biochar composition and quality, various features consisting of density, particle size distribution, ash content material, water content, pH, character, and source of the feedstock should be characterized mutually with pyrolysis reaction conditions (Zhang et al., 2008). Electron microscope microstructure is shown in Fig. 2.

After only pyrolysis, the water and volatile organics from biomass may evaporate, raising the aromatic content. The pyrolysis temperature impacts the physiochemical properties of the obtained biochar. For instance, pyrolysis temperature may be the primary influence on atomic ratio and structural formula (Liu et al., 2010). The pyrolysis temperatures applied range between 300 and 1000°C. Biomass is produced from organic compounds that generally include cellulose, hemicellulose, and lignin (Fahmi

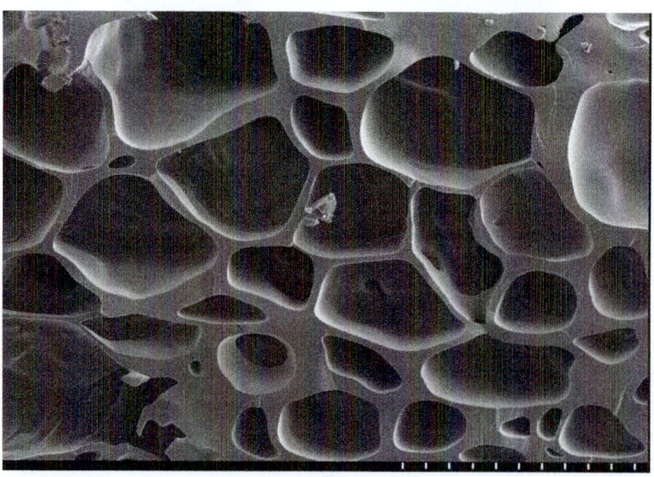

Fig. 2 Electron microscope image of biochar (freely provided by Biochar Project, 2011).

et al., 2008). Biochar is a carbonaceous material in a solid form that is produced when an organic mass is thermochemically deconstructed at an increased temperature, generally starting from 450 to 550°C in an oxygen-depleted atmosphere. This approach to managed thermochemical decomposition of organic mass is called pyrolysis (Hammond et al., 2011; Woolf et al., 2010). Biochar is mostly generated from crop residues, timber biomass, animal litter, and stable waste through various thermochemical processes, including slow pyrolysis, rapid pyrolysis, hydrothermal carbonization, flash carbonization, torrefaction, and gasification (Meyer et al., 2011). Lignin is one of the most stable of these compounds and is tolerant of destruction at even higher temperatures. In comparison, temperatures above 300°C can reduce the cellulose and hemicellulose items. The temperature and period of pyrolysis are identified based on the target purpose. In a few scenarios, catalytic additives such as K_3PO_4 and clinoptilolite are utilized to decrease the pyrolysis temperature (Mohamed et al., 2016). Graphene oxide covered with biochar composite was produced by slow pyrolysis of dip-coating water hyacinth biomass in the graphene oxide suspension (Shang et al., 2016). Commercial formation of biochar from urban waste materials may be helpful in minimizing the cost of biochar production (Fig. 3); however, because of the presence of potential contaminants, a proper regulatory platform is needed to prevent feasible hazards of soil contaminants. Toxic metal loadings in inner-city feedstocks can be transformed due to their selective reduction of ash (Hwang et al., 2008). Large-scale pyrolysis plants have been foiled in the past due to negative planning, but trials have demonstrated agronomically beneficial nutrient content material (Bridle and Pritchard, 2004; Wang et al., 2012). Learning the fate and impacts of biochar in the agricultural ecosystem is essential. Previous research has identified that biochar is usually a high-quality soil conditioner in agricultural soils. With regard to soil physical properties, the usage of biochar enhances the soil

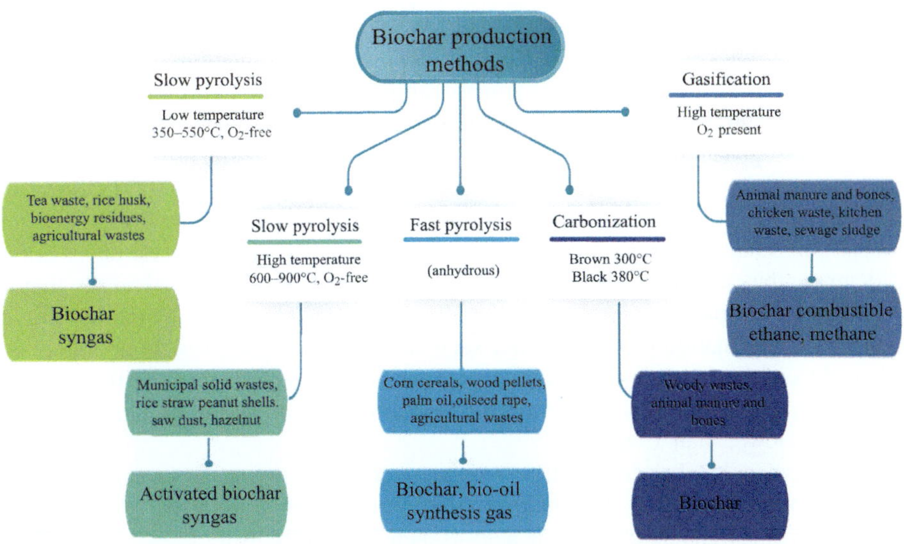

Fig. 3 Flowchart illustration of biochar production approaches.

blend stableness and water-maintaining capability via improving soil pore traits and water retention. The soil texture and type also provide a pivotal purpose. For example, some great benefits of biochar amendments will be explained more in a coarse-textured soil when compared to a fine-textured soil, and sandy soil is more reactive than clay-wealthy soils (Blanco-Canqui, 2017).

3 What is nanobiochar?

It has been shown that there is a comprehensive benefit in using nanoparticles with plants for agrifood utility (Torney et al., 2007). In fact, nanocarbons have received the interest of plant researchers as promising molecular transporters depending on the intrinsic capacity of carbon nanotubes to penetrate the cell membrane of many types of mammalian and plant cells to act as drug and gene delivery vehicles (Gao et al., 2006; Kam et al., 2005). Moreover, the thermochemical treatment of biomass can generate biofuels and syngas supported with biochar formation (Lehmann and Joseph, 2012). Consequently, producing biochar nanomaterials might succeed in four integrated goals, including pollutant removal, waste recycling, carbon sequestration, and energy production (Sohi, 2012; Tan et al., 2015).

All the benefits suggest that the newest nanocomposite could possibly be another solution and innovative adsorbent that may economically and efficiently remove Cr (VI) (an optimum adsorption ability of 150.02 mg g^{1}) from the aqueous solution. The modified biochar nanocomposites could possibly be fabricated by gradual pyrolysis of dip-coated hickory or bagasse biomass found in carbon nanotube (CNT) suspension systems (Inyang et al., 2015). Recent studies have also provided a manufactured useful composite using the impregnation of functional nanoparticles into biomass before pyrolysis (Table 1). For example, graphene improved the biochar composite by sluggish pyrolysis of graphene-pretreated wheat straw biomass. The total effects showed that graphene was covered in the surface of biochar, generating a larger surface area, greater thermal stability, many more functional groups, and higher removal effectiveness of phenanthrene and mercury compared to nontreated biochar (Tang et al., 2015).

Practically, producing biochar-based nanocomposites to mix the positive aspects of biochar with different types of nanomaterials, which is fairly significant. The forming composites commonly demonstrate impressive improvement in particular functional groups, pore characteristics, increased surface-active sites, and catalytic devastation proficiency; they are also fast to separate (Tan et al., 2016a, b) in a particularly major way. The biochar particle size is in the nanometer range determined by the technique employed for formation of nanobiochar such as fixing the pyrolyzing temperature or perhaps by exfoliation, as explored in several recent studies (Chen et al., 2017; Saxena et al., 2014; Zhang et al., 2013). Current researchers defined that the scale exfoliated biochar was lower than 10 nm.

Magnetic nanobiochar composites were produced using pyrolysis-carbonization of biomass and oil palm empty fruit bunches after pretreatment with $FeCl_3$. Magnetic nanobiochar was sulfonated by sulfuric acid to increase acidity using Oleic acid (MBC-SO_3H). Current research findings open new possibilities for the production of magnetic

Table 1 Recent studies on synthesis, characteristics, and applications of functional nanoparticle-coated biochar.

Feedstock	Temp. (°C) (during residence)	Time (h)	Product type of Nanomaterials	Adsorption temperature (°C)	Adsorption pH	Contaminants[a]	Kinetic mode[b]	Ref.
Cotton Wood	600	(1 h)	AlOOH particles/ flakes	22 ± 0.5		Methylene blue (Org)	pseudo-second-order	Zhang and Gao (2013)
Hickory chips	600	(1 h)	Hybrid multiwalled carbon nanotube (CNT)	22 ± 0.5		Methylene blue (Org)		Inyang et al. (2014)
Milled cotton wood	600	(1 h)	Graphene	22 ± 0.5		Methylene blue (Org)		Zhang et al. (2012b)
Sugarcane bagasse	600	(1 h)	Hybrid multiwalled carbon nanotube (CNT)	22 ± 0.5		Methylene blue (Org)	pseudo-second-order	Inyang et al. (2014)
Corn stover	600	(.5 h)	Magnetic iron oxide	25		Fluoride (Inorg)	pseudo-first-order	Mohan et al. (2014)
Sugar beet tailings	600	(1 h)	MgO	22 ± 0.5		Nitrate (Inorg)		Zhang et al. (2012a)
Tomato leaves	600	(1 h)	$Mg(OH)_2$ and MgO particles	22 ± 0.5		Phosphate (Inorg)		Yao et al. (2013)
Canola straw	350	4 h		25 ± 0.5		Methyl violet		Xu et al. (2011)
Eucalyptus	400	30 min		40		Methylene blue dye	pseudo-second-order	Sun et al. (2013)
Maize straw	300	1.5 h		25 ± 0.5		Oxytetracycline	pseudo-second-order	Jia et al. (2013)
Swine manure	400	1 h		25		Paraquat	pseudo-second-order	Tsai and Chen (2013)

Continued

Table 1 Continued

Feedstock	Temp. (°C) (during residence)	Time (h)	Product type of Nanomaterials	Adsorption temperature (°C)	Adsorption pH	Contaminants[a]	Kinetic mode[b]	Ref.
Cotton Wood	600	(1 h)	Hydrogel			Phenol (Org)	pseudo-second-order	Tang et al. (2015)
Hickory chips	600	(1 h)	Carbon nanotubes (CNTs)		6–7	Sulfapyridine (Org)	Elovich	Inyang et al. (2015)
Sugarcane bagasse	600	(1 h)	Carbon nanotubes (CNTs)		6–7	Sulfapyridine (Org)	Elovich	Inyang et al. (2015)
Marine macroalgae (*Laminaria japonica*)	600	(1 h)	Mg–Al assembled composites, MgO, spinel $MgAl_2O_4$, AlOOH, and Al_2O_3	20	6	Phosphate (Inorg)	pseudo-first-order	Jung et al. (2015)
Corn stalks	400	(2 h)	Fe_3O_4 nanoparticles	20	6	Crystal violet (Org)	pseudo-second-order	Sun et al. (2015)
Marine macroalgae (*Laminaria japonica*)	600	(1 h)	Nano-sized MgO particles with a highly crystalline structure	20 ± 2		Phosphate (Inorg)	pseudo-second-order	Jung and Ahn (2016)
Ramie stem	500	(2 h)	CaO	26 ± 0.5		Phosphate (Inorg)	pseudo-second-order	Liu et al. (2016a, b)
Maize straw	300	20 min	Nano-cerium oxide	20	Data not shown		pseudo-second-order	Feng et al. (2017)
Maize straw	600	100 min	Nano-cerium oxide	20	Data not shown		pseudo-second-order	Feng et al. (2017)

[a] Contaminants: organic contaminants (Org), inorganic contaminants (Inorg).
[b] Kinetic model: pseudo-first-order (PFO), pseudo-second-order (PSO)

Updated from Tan, X.F., Liu, Y.G., Gu, Y.L., Xu, Y., Zeng, G.M., Hu, X.J., Liu, S.B., Wang, X., Liu, S.M., Li, J., 2016a. Biochar-based nano-composites for the decontamination of wastewater: a review. Bioresour. Technol. 212, 318–333. Tan, X.F., Liu, Y.G., Gu, Y.L., Xu, Y., Zeng, G.M., Hu, X.J., Liu, S.B., Wang, X., Liu, S.M., Li, J., 2016b. Biochar–based nano–composites for the decontamination of wastewater: a review. Bioresour. Technol. 212, 318–333 with permission from Elsevier.

heterogenous acid catalysts from biomass waste (Jenie et al., 2017). Nanobiochar was generated from pine wood biochar using a planetary ball mill. Keeping biochar at −80°C before milling decreased the particle size up to 60 nm and raised the surface area of nanobiochar 15 times compared to the raw biochar. The created nanobiochar can be utilized for adsorption of carbamazepine pesticide from wastewater (Naghdi, et al., 2017). In contrast to various nanocomposite materials, varied specifications for using biochar as the substrate's material for nanocomposite formation are recognized. First of all, the feedstocks of biochar production are considerable and cheap, and are generally obtained from agricultural biomass and solid waste (Lehmann and Joseph, 2012; Sohi, 2012). Second, biochar production uses low energy supplies, which are often formed at quite low temperatures (<700°C) (Meyer et al., 2011). Therefore, micro/nano biochar can be used in various applications, including plant campaign, plant security, pesticide remediation, etc. This is confirmed by the considerable number of inspections (Alimohammadi et al., 2011; Ashfaq et al., 2017; Khodakovskaya et al., 2013; Tripathi et al., 2011) that are reliably generating significant and enriching results in plant systems.

4 Biochar applications in agriculture

A few biochar key features, which might be underpinned by its specific physicochemical architecture, are its capacity to sequester carbon and, consequently, reduce climate change; work as a soil transformer to boost agricultural output while handling fertilizer and irrigation water requirements; and assist in bioremediation efficiency. The addition of biochar to agricultural soil has been preferred to potentially reduce the amount and application frequencies of chemical (nitrogen-based) fertilizers to reduce nitrogen leaching to groundwater, the exhausts of greenhouse gases, and energy input (Orr and Ralebitso-Senior, 2014). Also, biochar has the ability to remediate pollution by immobilizing metals and adsorbing toxic chemicals (Cernansky, 2015) such as pesticides, which may cause environmental problems. Biochar has recently been proven to increase crop yield (Alburquerque et al., 2013), biomass (Soja et al., 2013), and total soil efficiency (Liu et al., 2013). In particular, rises in soil productivity are generally documented when biochar was employed in agronomic settings due to minimized soil acidity, upgraded cation exchange capacity, boosted water-holding capacity, and increased microbial population diversity (Shackley et al., 2009). Liu et al. (2015) and Manya (2012) evaluated the tremendous potential of biochar for sustainable production of carbon materials, demonstrating multifunctional applications as well as the agricultural sciences. Therefore, this material increases soil fertility (Sohi et al., 2009; Van Zwieten et al., 2009). Several of the applications involve soil remediation (Karami et al., 2011) CO2 capture, energy storage (Manya, 2012; Xiu et al., 2017) catalysis (Liu et al., 2015; Xiu et al., 2017), redox reactions, and environmental protection (Liu et al., 2015; Manya, 2012; Meyer et al., 2011). However, to the best of our knowledge, nanobiochar applications in agriculture have not been recorded. Here, considerable cases of the effective outcome of micro- and nanobiochar utility in sustainable agriculture are discussed (Fig. 4).

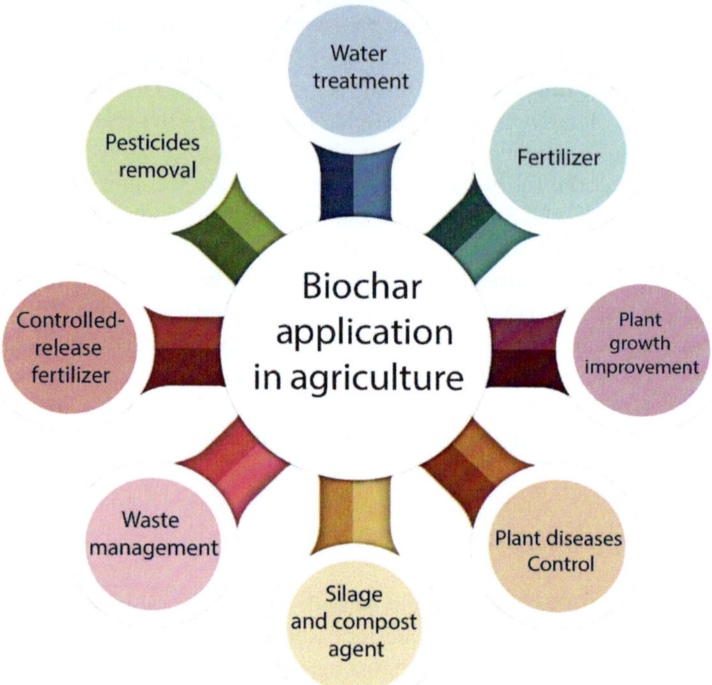

Fig. 4 Biochar applications in the agricultural sector indicate main areas of research such as water treatment, fertilizer, plant growth improvement, plant disease control, silage and compost agents, waste management, controlled-release fertilizer, pesticide removal, and more.

4.1 Silage and compost agents

The decrease in soil efficiency caused by rigorous cultivation and quick variations in agroclimatic growing conditions are crucial factors of food insecurity and poverty (Vanlauwe et al., 2010). To accomplish any soil control, it is essential to preserve an ideal level of soil organic situation and the biological cycle of critical nutrients. Com, mulch, manure, and cover crops have already been applied efficiently to promote fast cycling of soil nutrients by using microbial biomass activity and providing these kinds of nutrients to diverse plant species (Lehmann and Rondon, 2006). The benefits of such soil changes are, however, generally short-lived because decomposition rates are large and the supplied organic matter is commonly mineralized to carbon dioxide (Sohi et al., 2010). As a result of ever-increasing production of farm, livestock, and poultry products for individual usage, a significant quantity of organic and natural waste is produced from these industries (Li et al., 2011). These organic wastes are comprised of crop residues, animal manures, municipal solid wastes, biosolids, etc. The significant amounts generated should be used in an approach that adjusts to environmental regulations, integrating secure insertion onto agricultural land (Li et al., 2011). Farming waste materials, compost, and biochar combination amendments were applied to examine their particular role in managing the mobility of Cu (II) and Pb (II) and their resulting uptake in plants from the soil. As a result of biochar and compost addition, both Cu (II) and Pb (II) concentrations in ryegrass shoots were minimized, specifically for Pb (II) (Karami et al., 2011). A comprehensive study has been performed on the effect of biochar doses into fresh organic matter that is therefore composted, in a greenhouse test on sandy soil within moderate climate conditions, soil fertility completely elevated with raising amendment of biochar and compost. The impact was more described in sandy soil contrasted with loamy soil. Biochar application rates were commonly low with a maximum of 10 kg of biochar per ton of compost material (Fischer and Glaser, 2012). Recently, most research has mentioned the opportunity that biochar could increase soil quality features and maximize maize yields. A 27% increase in maize grain yield was noticed once fertilizer was combined with biochar, compared with a single fertilizer application (Chan et al., 2007). According to Sadegh-Zadeh et al. (2018), they discovered improved yield components of radish by the application of biochar and poultry manure on their own or in a combination the maximum yield resulted by 50 t ha^{-1} biochar and 5 t ha^{-1} poultry manure. Moreover, the combined application of produced biochars coming from rice hay waste and sugarcane bagasse, along with the fragment-improved feed yield of rice a lot more than separately making use of them. Many reports have represented excellent results of biochar (Butnan et al., 2015; Rogovska et al., 2014; Uzoma et al., 2011) or a negative impact (Deenik et al., 2010) or a neutral one (Jones et al., 2012).

4.2 Plant growth improvement

The risks of nutrients are broken-down soils that are connected with poor agricultural procedures, deforestation, overgrazing, and industrialization; this is becoming a global concern. Raising soil-nutrient destruction causing plant nutrient inadequacies have been recorded elsewhere. A lack of necessary plant nutrients and the following yield decline may seriously influence food security and economic development in the world. The most common approach to improving soil nutritional ingredients is the application of chemical fertilizers (Sanchez, 2002). Conversely, due to surface leaching and runoff, the nutrient concentration drops in soil, which motivates farmers to replicate the application of mineral fertilizers. Furthermore, the soil could have been overflowing with chemical fertilizers. Nevertheless, plant uptake is minimal due to low bioavailability. This may induce a hazardous impact on soil biology and plant ecosystems (Bah et al., 2014). Although a few possibilities are being proposed to manage the challenge of large application of nutrients, low resources, and nutrient decline, practical solutions are minimal (Eghball and Power, 1999; Withers et al., 2001). Biochar has developed most of the modern interest in soil nutrient management, especially in polluted soils; this is distinct from its application for environmental remediation and carbon sequestration (Atkinson et al., 2010; Sohi et al., 2010). Virtually, biochar efficiencies may potentially aid in the beneficial consumption of fertilizers by preservation and hence decrease nutrient leaching from the soil (Lehmann et al., 2003). Researchers have discovered an effective plant positive response to biochar amendment (Major et al., 2010; Steiner et al., 2007; Uzoma et al., 2011). Specifically, in which biochar can respond as a soil conditioner maximizing plant growth by delivering easily (Fig. 5), holding onto nutrients and by providing various solutions (Glaser et al., 2002). Biochar could positively impact plant nutrient uptake specifically due to its nutrient content and release characteristics, and also ultimately through better sorption of nutrients (Lehmann et al., 2003), improved soil cation exchange potential (Liang et al., 2006), improved soil physical properties (Chan et al., 2008), introducing an increase in water storage (Laird et al., 2010), and adjusting soil pH (Rondon et al., 2007). Modeling results show that biochar application to soil provides a promising tool to minimize greenhouse gas emissions on the order of 1.8 Pg CO_2-C every year; consequently, it may have a significant effect on climate change (Kammann et al., 2016; Sohi, 2012; Woolf et al., 2010). Modern investigations consider biochar to be more likely to be practical in fixing plant growth in acidic than alkaline soil types (Atkinson et al., 2010; Biederman and Harpole, 2013; Crane-Droesch et al., 2013; Jeffery et al., 2011). Biochar has been analyzed in the soil-less fertigation medium, in which vital benefits in pepper and tomato yields were discovered (Graber et al., 2010). There is universal awareness in lowering, or eradicating, the use of peat in the horticultural sector and in locating environmentally and cheaply sustainable growth media. Biochar may be a valuable alternative to peat (Tian et al., 2012). Different re-

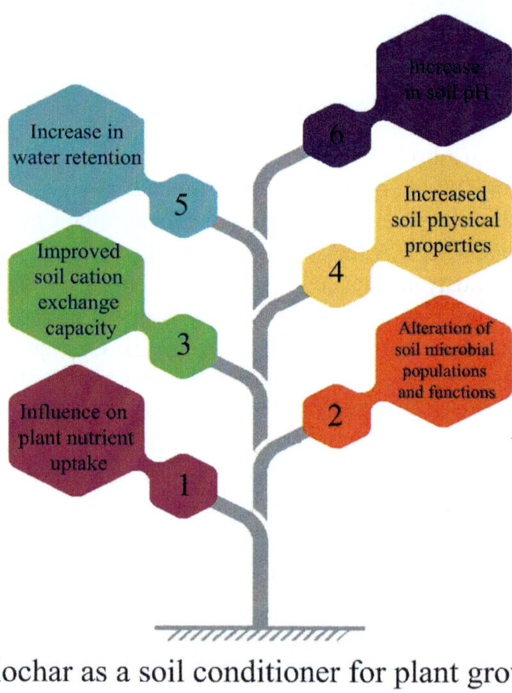

Fig. 5 The mechanism of biochar as a soil conditioner for plant growth includes influence on plant nutrient uptake, alteration of soil microbial populations and functions, improved soil cation exchange capacity, increased soil physical properties, increased water retention, and increased soil pH.

ports have proposed that the application of biochars in combo with selected chemicals and engineered nanomaterials may well possess extra positive aspects. Saxena et al. (2014) determined that biochar put with water-soluble carbon nanoparticles considerably increased wheat growth. Regarding root growth and development, root biomass can be elevated significantly once alkaline biochar is applied alternatively than acidic biochar (Bopp et al., 2016). When nonpyrogenic organic amendments (NPOA) are hybridized with biochar, the positive effects for plant growth promotion were investigated, but their particular shifting activities may well be a restricting factor for their intensive usage in agriculture (Bonanomi et al., 2017). In one other investigation, the application of bamboo biochars at concentrations of less than 10% developed nodulation and promoted soybean growth (Scheifele et al., 2017). The special effects of biochar addition on plant dry biomass and nutrition were based mostly on the biochar form and dose application rate. Plant soil amended with coconut husk biochar at an equal rate of 30 t ha− 1 triggered a 90% improvement in maize biomass and plant N and P concentrations of 0.88 and 0.12%, respectively. Therefore, the improvement in plant biomass and nutrition reveals the fineness of the coconut husk biochar as a soil amendment (Gonzaga et al., 2018).

4.3 Carbon fertilizer

In particular, total P and N content were observed to be excessive in biochar taken from a feedstock of animal sources such as sewage sludge and broiler liters than those from plants such as wood or green waste (Chan et al., 2008). Meanwhile biochars are extracted from biomass, contain *carbon*-rich product, and might include a high concentration of plant macronutrients such as nitrogen (N), phosphorous (P), calcium (Ca), magnesium (Mg), potassium (K), and sulfur (S)) and micronutrients such as copper (Cu), iron (Fe), manganese (Mn), zinc (Zn), and ash (Hossain et al., 2011). Nevertheless, nutrient elements in feedstock often mineralize, costabilize with carbon, or volatilize to form condensable solutions within pyrolysis. In particular, potassium and K are mainly conserved after transforming into inorganic forms while nitrogen is volatilized in proportion to carbon or involved with carbon in the resident fraction (Chan and Xu, 2009). Further optimization of biochar composites through the addition of minerals and small quantities of organic substances could produce more benefits in crop yield. Cover or intercalate urea (Manikandan and Subramanian, 2013). Field trials in China with composites constructed from biochar, clay, and chemical fertilizer have demonstrated a reliable increase in yield with reduced N input (Joseph et al., 2013; Qian et al., 2014). Likewise, the biological and chemical stability of biochar can make it an appropriate choice for long-term application as a promising fertilizer (Manya, 2012; Xiu et al., 2017). The maximum yield increases of 25%–30% were achieved with a composite consisting of wheat straw, municipal solid waste, and peanut hulls; N_2O and CH_4 exhausts were also reduced by 25%–50%.

Presently, there is a critical enhancement to plant growth and seed germination once the soil is modified with biochar and its influence is further boosted whenever combined with organic or inorganic fertilizers (Lehmann et al., 2011; Zhou et al., 2017). Studies noted the function of biochar in enhancing agronomical variables and environments for numerous plant species. Expansion of substantially lower than 5% biochar increased the germination, yield, and root progression of the halophytes consisting of sesbania and seashore mallow (Zheng et al., 2017). Alvarez-Campos et al. (2018) additionally showed that rice structure-based biochar applied at a rate of 2% to a sandy soil raised the biomass yield and sucrose percentage of sugarcane plants. Coconut husk biochar applied at a rate of 30 t ha^{-1} was linked with a 90% improvement of *Zea mays* biomass contrasted to the control without using any biochar amendments. The positive effects of the utilization of biochar and biochar-derived nanocarbons–its capability to maintain water and plant nutrients–are generally reviewed by many researchers (Manya, 2012; Xiu et al., 2017). Present results show the incredible outcome of nanocarbons on the growth of plants; they are simply utilized to enhance crop production (Khodakovskaya et al., 2013; Lin et al., 2009; Sonkar et al., 2012) in seed germination (Alimohammadi et al., 2011; Giraldo et al., 2014; Khodakovskaya et al., 2011, 2012; Tripathi et al., 2011; Wang et al., 2012), cell division (Khodakovskaya et al., 2012), protein expressions (Khodakovskaya et al., 2011, 2012; Wang et al., 2012), water uptake (Sonkar et al., 2012; Tripathi et al., 2011), and photosynthesis (Giraldo et al., 2014) as a fluorescent marker (Liu et al., 2009; Serag et al., 2013), DNA transporter (Liu et al., 2009), and in treatments for plant disease (Ocsoy et al., 2013). Khodakovskaya and his team confirmed that MWNTs in fairly low doses can enter thick seed coats, activate germination, and promote vegetative growth in tomato and tobacco plants. (Khodakovskaya et al., 2012). Nanocarbons are indubitably probable to provide sustained alternatives to the current difficulties in the field of agriculture (Serag et al., 2013; Singh et al., 2018; Zaytseva and Neumann, 2016) and can also easily be applied as nanofertilizers (Rose et al., 2011; Tiwari et al., 2014). The general findings show that monocot plants mainly display positive reactions in the plant to nanocarbon interactions (Lahiani et al., 2013; Saxena et al., 2014; Tripathi et al., 2016) in comparison to dicot plants, which usually indicate both positive and negative effects. (Sonkar et al., 2012). Furthermore, nanocarbon was integrated into slow-release fertilizer and the impact on rice yield and nitrogen loss in the surface water of paddy soil was practiced by field experiment. Rice grain yield and nitrogen use performance heightened considerably after employing slow-release fertilizers that added nanocarbon. These outcomes suggest that it is feasible that the nanocarbon is employed as a covering material for slow-release fertilizer and a benefit for minimizing water contamination (Wu, 2013). Saxena et al. (2014) evaluated the concentration-dependent effects of water-soluble carbon nanoparticles CNPs (ws-CNPs) on wheat. The ws-CNPs were separated from the naturally occurring raw CNPs present in biochar. When plants were treated with 10–150 mg/L ws-CNPs in the soil for up to 20 days, results showed perfect seedling growth at 50 mg/L treatment with root and shoot lengths elevated up to three times compared to those without treatment controls. Recently, the impact of nanocarbon water-retaining fertilizer on yield, quality of tuber mustard, and fertilizer application effectiveness was examined with field trials compared to the local tuber mustard fertilizer with the identical amount of practical composition. Nanocarbon water-retaining fertilizer and the treatment of 30% reduction may substantially increase the yield of tuber mustard and fertilizer consumption productivity (Wang et al., 2018). The results were investigated on liquid nanocarbon as a biological fertilizer, humic acid, and urea on Chinese cabbage and the potato. The liquid nanocarbon has a vital impact on protecting the growth and raising the yield of Chinese cabbage (Zhao et al., 2017).

4.4 Plant disease control

Plant feeder nematode suppression comes with an agronomic interest, both by its immediate role as a pest or as a vector of different types of plant diseases (Fry, 2012). Furthermore, to a few modest recommendations for biochar-induced plant protection against soil-borne diseases, the stimulation of systemic resistance against some foliar pathogens in three plant species has been confirmed. Plant resistance mechanisms of biochar depend on induced reactions along with both systemic acquired resistance (SAR) and induced systemic resistance (ISR) pathways, contributing to a broad-spectrum management ability in the canopy growth corps (Elad et al., 2011). Preparing biochar mixed with compost could strengthen pathogen resistance. This is risky but possible, considering biochar and compost may perhaps control different soil-borne pathogens. The disease suppressive issues of compost are extremely diverse. In particular, immature composts generate volatile compounds including sulfur, organic acids, and ammonia that are sensible for disease reduction. The suppressive outcome of mature compost is mostly biological, sometimes by using a microbial antagonism or elevated plant host resistance (Noble, 2011). It is possible that the growing impact of biochar on arbuscular mycorrhizal fungi will help to make the plants much more resistant to fungal pathogens (Elmer and Pignatello, 2011). Biochar and compost as a combination may possibly decrease the application of fungicides because of its adsorbent properties. However, it is known that biochar adsorbs organic molecules such as fungicides and pesticides (Zheng et al., 2010).

Multiple reports have demonstrated the management of plant pathogens by the utilization of biochar in farming soil, as exemplified in initial studies and important evaluations by Elad et al. (2010), Elmer and Pignatello (2011), Kolton et al. (2011), Harel et al. (2012a,b), Zwart and Kim (2012), Graber et al. (2014), and Bonanomi et al. (2015). Soil-applied biochar has effectively affected plant disease resistance in some foliar fungal pathogens *Botrytis cinerea* and *Leveillula taurica* in sweet pepper and tomato, and *Podosphaera aphanis* causes powdery mildew in strawberry plants. This induced systemic resistance phenomenon was caused by the interactions between specific soil microorganisms, such as *Bacillus, Pseudomonas*, and *Trichoderma* spp., and plant roots (Kolton et al., 2011). A few types of research have also outlined that biochar may lead to the elimination of plant diseases transferred by nematodes by controlling plant-feeding nematodes (Domene et al., 2015; Rahman et al., 2014; Zhang et al., 2013). Rahman et al. (2014) recorded a decline in plant-feeding nematodes in vineyard plots two years after the application of organic waste (compost and rice hull) and a poultry litter biochar to neutral soil, with the biochar treatment developing the most robust impact.

The potential of wood biochar and agricultural waste biochar to stimulate systemic resistance in strawberry crops against some fungal pathogens such as *Botrytis cinerea, Colletotrichum acutatum,* and *Podosphaera apahanis*, and to assay at the molecular mode of action of their effects on the level of plant defense was investigated. Three biochar treatments (1% GHW-450, 3% GHW-450, 3% CW) triggered an equal degree of minimized disease severity, as calculated by the area under the disease progression curves (AUDPC). The capability of biochar fertilizer to enhance transcriptional transformations combined with different plant protection pathways likely contributes to its broad-spectrum ability for disease reduction (Harel et al., 2012a, b). The addition of peat combined with 3% biochar led to considerably elevated root enhancement, fruit formation, and induced postharvest resistance of the fruits against *B. cinerea* (Fig. 6) while biochar had no influence on a leaf, petiole fresh, and dry weight (De Tender et al., 2016).

Biochar generally has excellent or simple impacts on plant growth contrasted with peat media once present in concentrations higher than 25%. Even so, studies of biochar effects on plant disease explain that although lower concentrations (≤1%) of biochar often reduced some diseases, higher concentrations (> 3%) were largely inadequate or activated plant disease (Frenkel et al., 2017). On the other hand, the capability of biochar to provide systemic resistance to plant pathogens is not totally known, and only some studies have dealt with this theme. Concomitant with many appropriate mechanisms by which biochar may well apply to disease control, these research groups determined important knowledge gaps and presented the demand for deeper studies.

4.5 Controlled-release fertilizer

Biochar has a good adsorbent capability, yet only a few reports have explored its application as a nutrient transporter in the development of slow-release fertilizers (Cai et al., 2016; Yao et al., 2013). These fertilizers owned high nitrogen-use

Fig. 6 Impact of biochar fertilizer on inhibition of grey mold on strawberry plant leaves. Plants were grown in a coconut peat potting medium amended with either agricultural waste (GHW-450) or citrus wood (CW) biochar for 25 days. *Botrytis cinere*a infection is seen as a rotten area after nine days. Images represent grey mold on strawberry leaves nine days after *B. cinerea* infection from treatments of no biochar (left), 1% GHW-450 (middle), and 3% GHW-450 (right).
Reprinted from Harel, Y.M., Elad, Y., Rav-David, D., Borenstein, M., Shulchani, R., Lew, B. and Graber, E.R., 2012a. Biochar mediates systemic response of strawberry to foliar fungal pathogens. Plant Soil, 357(1-2), 245-257; Harel, Y.M., Elad, Y., Rav-David, D., Borenstein, M., Shulchani, R., Lew, B., Graber, E.R., 2012b. Biochar mediates systemic response of strawberry to foliar fungal pathogens. Plant Soil, 357 (1–2), 245–257 with permission from Springer.

proficiency (64.27%), low nitrogen migrate-to-surface-loss amounts (7.4%), and low nitrogen-leaching-loss amounts (10.3%). Second, they successfully minimized the nitrogen-release rate (69.8% of nitrogen was released after 30 days), and therefore correctly elevated cotton plant growth. A controlled release fertilizer biochar can be regarded as an example of biochar-based EFFs (González et al., 2015).

A new investigation presented a polymer matrix constructed of cotton stalks (CSs), acrylic acid (AA), 2-acrylamide-2 methylpropane sulfonic acid (AMPS), and bentonite into NH^{4+}-loaded biochar (N-BC), and developed biochar-based slow-release nitrogen fertilizers (BSRFs) (Wen et al., 2016). A new biochar-based fertilizer consisting of rice husk biochar and urea-hydrogen peroxide (UHP) was produced and optimized, which could simultaneously slowly release nitrogen and immobilize cadmium (Cd). Increased crop growth was confirmed by pot tests, which revealed a vital improvement in the biomass of cabbage (Chen et al., 2018). To boost the efficiency of polyvinyl alcohol (PVA)-layered fertilizers, a new slow-release fertilizer was produced biochar and a waterborne copolymer of PVA and polyvinylpyrrolidone (PVP) as covering elements (Chen et al., 2018). A granular biochar-ammonium phosphate (BAP) as an uncoated-slow release fertilizer was formed by employing a reaction involving biochar (derived from rice husk), phosphoric acid, and ammonia gas for fertilization of sandy soil (El Sharkawi et al., 2018). A biochar nanocomposite produced from corn cob was designed to examine the viability of corn cob biochar (CB) for offering macro- and micronutrients to plants on a slow-release basis. This strategy will certainly not simply support waste elimination and control, but it will aid pollution protection by conversion of waste into value-applied products while fixing the leaching problems of traditional fertilizers (Lateef et al., 2019). Nevertheless, the efficiency of biochar-based fertilizer for simultaneously gradually delivering nitrogen and immobilizing heavy metals needs additional examination.

4.6 Biochar to remediate pesticides

A lot of diverse organic chemical compounds may possibly be turned into soil or water contaminants. Roughly 1100 distinct substances are generally and deliberately applied all over the world for the management of plant pests. Their commonly extensive persistent days and extremely high utility rates, mixed with flood or erosion events, might make every pesticide a soil or water pollutant. Carbon nanomaterials such as nanobiochars are recognized as having a great affinity for pesticides. Forty years ago, a couple of ash-C products caused a critical reduction in the diuron herbicide phytotoxicity recorded (Toth and Milham, 1975). Various studies demonstrated inhibition in the efficiency of pesticides in the existence of combustion residues in soil (Graber et al., 2012; Nag et al., 2011; Xu et al., 2008; Yang et al., 2006). Yang et al. (2006) also noted that the duplicity of the application rate of diuron failed to manage weed growth in the existence of 0.5% of wheat char in soil. Recently, nonetheless weed control and herbicide performance were inhibited in the reputation of biochars, the effect relied upon the specific surface area (SSA) of biochars, with higher SSA biochar resulting in poorer weed control (Graber et al., 2012). Some in vitro analyses revealed that there may be agronomic and/or economic effects, in consideration of the increased input cost of pesticides to the farmer, when herbicide application rates have to be improved for biochar-amended soils. Nag et al. (2011) observed a good impact of biochar application to soil on the usefulness of atrazine to control ryegrass weed and mentioned that the dose for biochar-amended (1% w/w) soil could need to be raised by 3–4 times to reach the needed weed control. Additionally, they documented that the level of biochar's effect on herbicide efficacy was based on the chemical structures of the herbicide molecule and its mechanism. Nevertheless, due to the determination that the aging of biochars may possibly decrease the sorption ability with time, which can be essential for herbicide efficiency control in biochar-amended soils (Martin et al., 2012). An updated summary of the remediation of pesticides using biochar, the know-how gaps, and the prospective research interest in this field to examine the influence of biochar addition on agroecosystems was described in detail. Increased biochar rates in pesticide-contaminated ecosystems provide these positive aspects: (a) boosts the soil water-keeping potential, (b) enhances air diffusion conditions in soil, and (c) supplies a natural environment for the best growth of microbes, therefore assisting the microbial population for metabolic actions and pesticide destruction (Varjani et al., 2019). The elevated sorption and reduced management of various pesticides in the occurrence of biochar may well reduce the concerns of environmental contamination and human exposure from the perspective of the ecosystem and human health. Second, the lowered bioavailability and plant uptake may maximize crop yield and lessen organic pesticide residues in agricultural crops. The elimination of dangerous pesticides and herbicides in agroecosystems by using biochar additions appears practical, but additional study is expected on the adsorbence of pesticides by biochar.

4.7 Micro- and nanobiochar for decontamination of wastewater

Inorganic and organic pollutants could be degraded but heavy metals are nondegradable and they also are persistent. A whole lot of heavy metals are available in contaminated water (Mielke et al., 2000). Lately, heavy metals in water have grown to be a great environmental issue in many countries worldwide because they are toxic to living organisms and in addition accumulate in biota. Specifically, Cd, Cu, Pb, and Zn are carcinogenic and harm disease-fighting capability (Cantor, 1997; Silva et al., 2012; Tapaswi et al., 2014). Many methods such as adsorption, ion exchange, chemical precipitation, etc., have already been applied to eliminate these heavy metals from the surroundings (Fu and Wang, 2011; Zhu et al., 2012). of these have disadvantages (e.g., the excessive time required, high costs, low efficiency, and high energy consumption), the adsorption process for aqueous effluents is becoming very attractive from an economic perspective (Gao et al., 2016; Kurniawan et al., 2006). Numerous absorbent components such as turned-on carbon, silica gel, flat iron oxide, carbonaceous nanofibers, zeolite and attapulgite (Ding et al., 2016). However, they will involve a few disadvantages incorporating their capability for oxidation process and assimilation, modest adsorbent capacity, cost-effective, and poor selectivity (Gan et al., 2015). New technology combining designed nanoparticles into biochar production devices could optimize the function of biochar for many uses and applications such as wastewater treatment. (Inyang et al., 2014). Therefore, we need to develop multifunctional adsorbents to take care of heavy metal pollution. Recently, biochar has been be used as an adsorbent. It really is carbon-rich and pyrolyzed in a closed container under anaerobic condition (Wardle et al., 2008). Lately, huge research efforts have used biochar adsorbents for the removal of aqueous contaminants, which may provide many benefits for both carbon sequestration and water pollution control (Ahmad et al., 2014; Tan et al., 2015). Hence, many methods have already been used to change biochar, especially surface modification, such as incorporating nanoparticles to get ready biochar nanocomposites, which enhance the removal ability for various water pollutants (Inyang et al., 2012; Zhang et al., 2012, 2013).

Biochar includes a great tendency to adsorb water contaminants because of its wide option of feedstocks, low price, and favorable physical/chemical surface characteristics (Tan et al., 2015). However, the most traditional biochar shows a limited capacity to absorb an ionic pollutant (Yao et al., 2013). Recent studies have synthesized different structures of nanobiochar to boost its capability of adsorbing aqueous an ionic pollutant. A novel nanocomposite has been treated with chitosan after pyrolysis, with the resulting composite containing amine functional groups on its surface to boost its adsorption convenience of toxic metals such as Pb (11), Cu (11), and Cd (11) (Zhou et al., 2013). The results also showed that by incorporating MgO nanoflakes on the biochar surface, the MgO nanobiochar showed efficient removal of phosphate and nitrate. The biochar/ALOOH nanocomposite could possibly be produced using a thermopyrolysis method of aluminum chloride pretreated biomass (Zhang and Gao, 2013).

Further benefits for coating nanoparticles on carbon structures could be that it might enhance the adsorption sites to entice organic pollutants from water (Sun et al., 2015; Zhang et al., 2013). Production of biochar-based nanocomposites isn't just a way to enhance the physical and chemical characteristics of biochar, but also to obtain an innovative composite and to incorporate positive aspects of biochar with other nanomaterials. Most recently, also nuclear wastewater and various substrate-based nanocomposite materials were formulated for purification of wastewater (Awual, 2016; Awual et al., 2014, 2016a, b). Also (Ce-MSB) nanocerium oxide-functionalized maize straw biochar was well prepared and valuable to eradicate P from agricultural wastewater. The response surface model was put to optimizing operating conditions. The response surface model (RSM) showed a higher material ratio and had a positive influence on PO_4 3 adsorption capacity while a higher pyrolysis temperature had a poor effect (Feng et al., 2017). According to Wang et al. (2018), a novel hydroxyapatite-biochar nanocomposite was successfully synthesized and used for removing Pb (11), Cu (11), and Zn (11) in both single and ternary mixed solutions. The hydroxylapatite biochar nanocomposite may be used specifically for adsorbing Pb (11). The isotherm studies of hydroxyapatite-biochar nanocomposite showed that Pb (11) adsorption followed the Langmuir model, as the two other ions followed the Freundlich in both single and ternary systems. Also, HCL solution may be used for recycling the adsorbed hydroxyapatite-biochar nanocomposite. The hydroxyapatite-biochar nanocomposite can be ecofriendly and a low-cost adsorbent for heavy metal removal from wastewater (Wang et al., 2018). The adsorption features of biochar-based nanocomposite hybrids with several types of heavy metals are offered in Table 2.

Table 2 Adsorption characteristics of various heavy metals with biochar-based nanocomposites.

Feedstock	Temp. (°C) (during a residence)	Time (h)	Product type of Nanomaterials	Adsorption temperature (°C)	Adsorption pH	Kinetic mode[a]	Ref.
Pine wood	600	(1 h)	Birnessite	22 ± 0.5	7	Elovich	Wang et al. (2015d)
Pine wood	600	(1 h)	Birnessite	22 ± 0.5	5.5	pseudo-second-order	Wang et al. (2015d)
Pine wood	600	(1 h)	Manganosite	22 ± 0.5	7	Elovich	Wang et al. (2015d)
Rice husks	550	(0.5 h)	Iron oxide		6.85 ± 0.23		Cope et al. (2014)
Rice husks	950	(0.5 h)	Iron oxide		6.85 ± 0.23		Cope et al. (2014)
Kans grass	500	(2 h)	Fe_3O_4	25	7	pseudo-second-order	Baig et al. (2014)
Kans grass	500	(4 h)	Fe_3O_4	25	7	pseudo-second-order	Baig et al. (2014)
Cotton Wood	600	(1 h)	γ-Fe_2O_3particles	22 ± 0.5		pseudo-second-order	Zhang et al. (2013)
Hickory chips	600	(1 h)	Iron hydroxide particles	20 ± 2	5.8 ± 0.2		Hu et al. (2015)
Loblolly pine wood	600	(1 h)	γ-Fe_2O_3particles	22 ± 0.5	7	Elovich	Wang et al. (2015e)
Hickory wood	600	(1 h)	MnOx	22 ± 0.5		Elovich	Wang et al. (2015a)
Pine bark waste	950	(1 h)	Magnetic $CoFe_2O_4$			pseudo-second-order	Reddy and Lee (2014)
Rice hull	400	(5 h)	Magnetic iron oxide and ZnS nanocrystals	20	6	pseudo-second-order	Yan et al. (2014)
Eichhornia crassipes	600	(1 h)	Chitosan and c-Fe_2O_3	20		pseudo-second-order	Zhang et al. (2015)
Sugarcane bagasse	450	(1 h)	ZnO	30		pseudo-second-order	Gan et al. (2015)
Corn straws	600	(3 h)	MnOx	25	6.0 ± 0.1		Song et al. (2014)
Wheat straw	600	(1 h)	Graphene	25	7	pseudo-second-order	Tang et al. (2015)
Bamboo	600	(2 h)	Chitosan	22 ± 0.5			Zhou et al. (2013)

Peeled pine wood	700	(3 h)	Hydrous-manganeseoxide	25		pseudo-second-order	Wang et al. (2015b)
Walnut shell	450	(1 h)	Chitosan	25		pseudo-second-order	Huang et al. (2016)
Banana peels	230	(2 h)	Latin plumbum		5	pseudo-second-order	Zhou et al. (2017b)
Rape straw	600	(3 h)	Latin plumbum		6	pseudo-second-order	Li et al. (2017)
Wetland reed	600	(2 h)	Magnetic iron nanoparticles			pseudo-second-order	Zhu et al. (2018)
Sugarcane leaf	550	(2 h)	Nanosize MgO flakes			pseudo-second-order	Li et al. (2018)
Red oak; Switchgrass	600	(3 h)	Magnetite			pseudo-second-order	Bakshi et al. (2018)
Camel's bones	500		$FeCl_3 \cdot H_2O$ $FeCl_2 \cdot 4H_2O$			pseudo-second-order	Alqadami et al. (2018)

[a] Kinetic model: pseudo-first-order (PFO), pseudo-second-order (PSO).

Updated from Tan, X.F., Liu, Y.G., Gu, Y.L., Xu, Y., Zeng, G.M., Hu, X.J., Liu, S.B., Wang, X., Liu, S.M., Li, J., 2016a. Biochar-based nano-composites for the decontamination of wastewater: a review. Bioresour. Technol. 212, 318–333. Tan, X.F., Liu, Y.G., Gu, Y.L., Xu, Y., Zeng, G.M., Hu, X.J., Liu, S.B., Wang, X., Liu, S.M., Li, J., 2016b. Biochar–based nano–composites for the decontamination of wastewater: a review. Bioresour. Technol. 212, 318–333 with permission from Elsevier.

4.8 Effect of biochar on soil microbial communities

Microbial communities are essential for plant soil ecosystem performance, which incorporates soil health, fertility, carbon and nutrient cycling, and crop yields (Thies et al., 2015). Biochars are generally revealed to alter many of the static soil physical properties consisting of surface area, bulk density, porosity, and tensile strength, which usually change soil biota (Lehmann et al., 2011). Biochar-induced abiotic processes (physical and chemical pathways) can have an impact on the soil microbial community function and diversity (Lehmann et al., 2011). Biochar may possibly have an effect on the microbe population profile simply by changing the soil physical environment, but it is also feasible for biochar-induced variations in microbial growth to modify soil physical characteristics. Biochar-induced physicochemical differences in soil can certainly influence the microbial community prosperity, diversity, and structure that eventually influence the nature of plant-microbe interactions in soil (Quilliam et al., 2013). The alterations in pH and oxidation potential considering that of the existence of biochar in the soil can influence the microbial activity and population structure (McCormack et al., 2013). The increase of biochars in the soil can potentially impact the adsorption dynamics of quickly mineralizable carbon and nutrient fertilizers that may change bacterial and fungal community patterns and actions (Muhammad et al., 2014). The positive influence of biochar incorporation of microbial activity and community is linked to its potential to provide conveniently degrading organic metabolites and nutrients for microbial growth (Watzinger et al., 2014). Biochar materials with high surface areas and varied charges assist in microorganism growth and sorption of nutrient ions (Mollinedo et al., 2015). The exceptionally porous composition of biochar can help in a preventative microhabitat for soil microbial populations and present important organic metabolites and minerals needed for microbial progress (Anderson et al., 2011). A comprehensive review of six probable biochar mechanisms may possibly determine soil microbial communities, with improvements in abiotic factors being explored (Thies et al., 2015). Caroline and Komang's book successfully gathered virtually all up-to-date research studies, introducing molecular-based DNA techniques on biochar mechanisms to enhance soil microbial ecology (Ralebitso-Senior and Orr, 2016).

Recognizing the effects of biochar on microbial population variations and functions may well present a crucial understanding basis for the production of biochar-based systems to boost degraded soils and preserve soil health and plant protection.

5 Conclusion and future prospective

Biochar application indicates vital environmental agroeconomic benefits, for example, enhancing soil quality, strengthening soil water, stimulating nutrient preservation, improving reduced CO2 exhaust, significantly raising plant growth and crop yield combined with the elimination of the pathogenic effect, and minimizing biomass waste from various crops. Because of its significant surface area and porosity, biochar has the promise to make improvements to the soil moisture holding capacity, minimize irrigation requirements, and therefore lessen agronomic carbon footprints.

Likewise, the amendment of biochar to soil rhizosphere has been proposed to potentially reduce the quantities and application frequencies of chemical fertilizers with the potential to then decrease nitrogen leaching to groundwater. Biochar efficiencies vary the microorganism population in soil by modulating soil pH, nitrogen levels, dissolved organic matter, and phosphorus quantity, which usually impacts pesticide bioremediation. Different know-how gaps with the reverence of the plant protection application of biochar included, what precisely is the influence of the continuous application of biochar on soil health? What are the positive and negative effects of biochar on diverse plant species and soil microbial communities? What are the biochar mechanisms used to control plant pathogens? Multiple future studies are expected to discover the mechanisms behind the detection of nonadditive impact and to develop biochar-organic amendment mixtures that improve plant yields in distinct agricultural systems.

Acknowledgments

Thanks to Biochar Industries/Biochar Projects and Friends of the Char that placed free hi-res versions (4 megs^{+}) of photos of biochar in the public domain (http://biocharproject.org/charmasters-log/biochar-electron-microscopeimages/#!prettyPhoto) for the benefit of the entire human race to feel free to use/share/reprint/distribute.

References

Ahmad, M., Rajapaksha, A.U., Lim, J.E., Zhang, M., Bolan, N., Mohan, D., Vithanage, M., Lee, S.S., Ok, Y.S., 2014. Biochar as a sorbent for contaminant management in soil and water: a review. Chemosphere 99, 19–33.

Alburquerque, J.A., Salazar, P., Barrón, V., Torrent, J., del Campillo, M.D.C., Gallardo, A., Villar, R., 2013. Enhanced wheat yield by biochar addition under different mineral fertilization levels. Agron. Sustain. Dev. 33 (3), 475–484.

Alimohammadi, M., Xu, Y., Wang, D., Biris, A.S., Khodakovskaya, M.V., 2011. Physiological responses induced in tomato plants by a two–component nanostructural system composed of carbon nanotubes conjugated with quantum dots and its in vivo multimodal detection. Nanotechnology 22 (29), 295101.

Alqadami, A., Khan, M., Otero, M., Siddiqui, M., Jeon, B., Batoo, K., 2018. A magnetic nanocomposites produced from camel bones for an efficient adsorption of toxic metals from water. J. Clean. Prod. 178, 293–304.

Alvarez-Campos, O., Lang, T.A., Bhadha, J.H., McCray, J.M., Glaz, B., Daroub, S.H., 2018. Biochar and mill ash improve yields of sugarcane on a sand soil in Florida. Agric. Ecosyst. Environ. 253, 122–130.

Anderson, C.R., Condron, L.M., Clough, T.J., Fiers, M., Stewart, A., Hill, R.A., Sherlock, R.R., 2011. Biochar induced soil microbial community change: implications for biogeochemical cycling of carbon, nitrogen and phosphorus. Pedobiologia 54, 309–320.

Asai, H., Samson, B.K., Stephan, H.M., Songyikhangsuthor, K., Homma, K., Kiyono, Y., Inoue, Y., Shiraiwa, T., Horie, T., 2009. Biochar amendment techniques for upland rice production in Northern Laos: 1. Soil physical properties, leaf SPAD and grain yield. Field Crop Res. 111 (1–2), 81–84.

Ashfaq, M., Verma, N., Khan, S., 2017. Carbon nanofibers as a micronutrient carrier in plants: efficient translocation and controlled release of Cu nanoparticles. Environ. Sci. Nano 4 (1), 138–148.

Atkinson, C.J., Fitzgerald, J.D., Hipps, N.A., 2010. Potential mechanisms for achieving agricultural benefits from biochar application to temperate soils: a review. Plant Soil 337 (1–2), 1–18.

Awual, M.R., 2016. Assessing of lead (III) capturing from contaminated wastewater using ligand doped conjugate adsorbent. Chem. Eng. J. 289, 65–73.

Awual, M.R., Yaita, T., Taguchi, T., Shiwaku, H., Suzuki, S., Okamoto, Y., 2014. Selective cesium removal from radioactive liquid waste by crown ether immobilized new class conjugate adsorbent. J. Hazard. Mater. 278, 227–235.

Awual, M.R., Miyazaki, Y., Taguchi, T., Shiwaku, H., Yaita, T., 2016a. Encapsulation of cesium from contaminated water with highly selective facial organic–inorganic mesoporous hybrid adsorbent. Chem. Eng. J. 291, 128–137.

Awual, M.R., Yaita, T., Miyazaki, Y., Matsumura, D., Shiwaku, H., Taguchi, T.A., 2016b. Reliable hybrid adsorbent for efficient radioactive cesium accumulation from contaminated wastewater. Sci. Rep. 6, 19937.

Bah, A., Husni, M.H.A., Teh, C.B.S., Rafii, M.Y., Syed Omar, S.R., Ahmed, O.H., 2014. Reducing runoff loss of applied nutrients in oil palm cultivation using controlled-release fertilizers. Adv. Agric., 2014.

Baig, S.A., Zhu, J., Muhammad, N., Sheng, T., Xu, X., 2014. Effect of synthesis methods on magnetic Kans grass biochar for enhanced As (III, V) adsorption from aqueous solutions. Biomass Bioenergy 71, 299–310.

Bakshi, S., Banik, C., Rathke, S., Laird, D., 2018. Arsenic sorption on zero-valent ironbiochar complexes. Water Res. 137, 153.

Beesley, L., Moreno-Jiménez, E., Gomez-Eyles, J.L., 2010. Effects of biochar and greenwaste compost amendments on mobility, bioavailability and toxicity of inorganic and organic contaminants in a multi–element polluted soil. Environ. Pollut. 158 (6), 2282–2287.

Biederman, L.A., Harpole, W.F., 2013. Biochar and its effects on plant productivity and nutrient cycling: a meta–analysis. Global Change Biol. Bioenergy 5, 202–214.

Bird, M.I., Wurster, C.M., de Paula Silva, P.H., Bass, A.M., De Nys, R., 2011. Algal biochar–production and properties. Bioresour. Technol. 102 (2), 1886–1891.

Blanco-Canqui, H., 2017. Biochar and soil physical properties. Soil Sci. Soc. Am. J. 81 (4), 687–711.

Bonanomi, G., Ippolito, F., Scala, F., 2015. A "black" future for plant pathology? Biochar as a new soil amendment for controlling plant diseases. J. Plant Pathol. 28 (2), 223–234.

Bonanomi, G., Ippolito, F., Cesarano, G., Nanni, B., Lombardi, N., Rita, A., Saracino, A., Scala, F., 2017. Biochar as plant growth promoter: better off alone or mixed with organic amendments? Front. Plant Sci. 8, 1570.

Bopp, C., Christl, I., Schulin, R., Evangelou, M.W., 2016. Biochar as possible long–term soil amendment for phytostabilisation of TE–contaminated soils. Environ. Sci. Pollut. Res. 23 (17), 17449–17458.

Bridle, T.R., Pritchard, D., 2004. Energy and nutrient recovery from sewage sludge via pyrolysis. Water Sci. Technol. 50, 169–175.

Brown, R., 2009. Biochar production technology. In: Lehmann, J., Joseph, S. (Eds.), Biochar for environmental management: science and technology. Earthscan, London.

Butnan, S., Deenik, J.L., Toomsan, B., Antal, M.J., Vityakon, P., 2015. Biochar characteristics and application rates affecting corn growth and properties of soils contrasting in texture and mineralogy. Geoderma 237, 105–116.

Cai, Y., Qi, H., Liu, Y., He, X., 2016. Sorption/desorption behavior and mechanism of NH4+ by biochar as a nitrogen fertilizer sustained–release material. J. Agric. Food Chem. 64, 4958–4964.

Cantor, K.P., 1997. Drinking water and cancer. Cancer Causes Control 8, 292–308.

Cernansky, R., 2015. Agriculture: state–of–the–art soil. Nature 517, 258–260.

Chan, K.Y., Xu, Z., 2009. Biochar: nutrient properties and their enhancement. Sci. Technol. 1 (1), 67–84.

Chan, K.Y., Van Zwieten, L., Meszaros, I., Downie, A., Joseph, S., 2007. Agronomic values of greenwaste biochar as a soil amendment. Aust. J. Soil Res. 45 (8), 629–634.

Chan, K.Y., Van Zwieten, L., Meszaros, I., Downie, A., Joseph, S., 2008. Using poultry litter biochars as soil amendments. Soil Res. 46 (5), 437–444.

Chen, L., Chen, X.L., Zhou, C.H., Yang, H.M., Ji, S.F., Tong, D.S., Zhong, Z.K., Yu, W.H., Chu, M.Q., 2017. Environmental-friendly montmorillonite-biochar composites: facile production and tunable adsorption-release of ammonium and phosphate. J. Clean. Prod. 156, 648–659.

Chen, S., Yang, M., Ba, C., Yu, S., Jiang, Y., Zou, H., Zhang, Y., 2018. Preparation and characterization of slow-release fertilizer encapsulated by biochar-based waterborne copolymers. Sci. Total Environ. 615, 431–437.

Cope, C.O., Webster, D.S., Sabatini, D.A., 2014. Arsenate adsorption onto iron oxide amended rice husk char. Sci. Total Environ. 488, 554–561.

Crane-Droesch, A., Abiven, S., Jeffery, S., Torn, M.S., 2013. Heterogeneous global crop yield response to biochar: a meta–regression analysis. Environ. Res. Lett. 8 (4), 044049.

De Souza Melare, A.V., González, S.M., Faceli, K., Casadei, V., 2017. Technologies and decision support systems to aid solid-waste management: a systematic review. Waste Manag. 59, 567–584.

De Tender, C., Haegeman, A., Vandecasteele, B., Clement, L., Cremelie, P., Dawyndt, P., Maes, M., Debode, J., 2016. Dynamics in the strawberry rhizosphere microbiome in response to biochar and *Botrytis cinerea* leaf infection. Front. Microbiol. 7, 2062.

Deenik, J.L., McClellan, T., Uehara, G., Antal, M.J., Campbell, S., 2010. Charcoal volatile matter content influences plant growth and soil nitrogen transformations. Soil Sci. Soc. Am. J. 74 (4), 1259–1270.

Demirbas, A., 2004. Effects of temperature and particle size on bio–char yield from pyrolysis of agricultural residues. J. Anal. Appl. Pyrolysis 72 (2), 243–248.

Ding, C., Cheng, W., Wang, X., Wu, Z.Y., Sun, Y., Chen, C., Wang, X., Yu, S.H., 2016. Competitive sorption of Pb (II), Cu (II) and Ni (II) on carbonaceous nanofibers: a spectroscopic and modeling approach. J. Hazard. Mater. 313, 253–261.

Domene, X., Mattana, S., Sanchez-Moreno, S., 2015. Soil nematode communities as indicators of medium-term ecological impacts of biochar application. In: 25th SETAC Europe Annual Meeting, 3–7 May, Barcelona, Spain.

Duku, M.H., Gu, S., Hagan, E.B., 2011. Biochar production potential in Ghana—a review. Renew. Sust. Energ. Rev. 15 (8), 3539–3551.

Eghball, B., Power, J.F., 1999. Composted and noncomposted manure application to conventional and no–tillage systems: corn yield and nitrogen uptake. Agron. J. 91 (5), 819–825.

El Sharkawi, H.M., Tojo, S., Chosa, T., Malhat, F.M., Youssef, A.M., 2018. Biochar-ammonium phosphate as an uncoated-slow release fertilizer in sandy soil. Biomass Bioenergy 117, 154–160.

Elad, Y., David, D.R., Harel, Y.M., Borenshtein, M., Kalifa, H.B., Silber, A., Graber, E.R., 2010. Induction of systemic resistance in plants by biochar, a soil-applied carbon sequestration agent. Phytopathology 100, 913–921.

Elad, Y., Cytryn, E., Harel, Y.M., Lew, B., Graber, E.R., 2011. The biochar effect: plant resistance to biotic stresses. Phytopathol. Mediterr. 50 (3), 335–349.

Elmer, W.H., Pignatello, J.J., 2011. Effect of biochar amendments on mycorrhizal associations and Fusarium crown and root rot of asparagus in replant soils. Plant Dis. 95, 960–966.

Fahmi, R., Bridgwater, A.V., Donnison, I., Yates, N., Jones, J.M., 2008. The effect of lignin and inorganic species in biomass on pyrolysis oil yields, quality and stability. Fuel 87 (7), 1230–1240.

FAO, Food and Agriculture Organisation of the United Nations, 1983. Simple Technologies for Charcoal Making. FAO Forestry Paper 41. FAO, Rome. www.fao.org/docrepS5328e.

Feng, Y., Lu, H., Liu, Y., Xue, L., Dionysiou, D.D., Yang, L., Xing, B., 2017. Nano–cerium oxide functionalized biochar for phosphate retention: preparation, optimization and rice paddy application. Chemosphere 185, 816–825.

Fischer, D., Glaser, B., 2012. Synergisms between compost and biochar for sustainable soil amelioration. In: Management of Organic Waste. IntechOpen.

Frenkel, O., Jaiswal, A.K., Elad, Y., Lew, B., Kammann, C., Graber, E.R., 2017. The effect of biochar on plant diseases: what should we learn while designing biochar substrates? J. Environ. Eng. Landsc. Manag. 25 (2), 105–113.

Fry, W.E., 2012. Principles of Plant Disease Management. Academic Press, London, UK.

Fu, F., Wang, Q., 2011. Removal of heavy metal ions from wastewaters: a review. J. Environ. Manag. 92 (3), 407–418.

Gan, C., Liu, Y., Tan, X., Wang, S., Zeng, G., Zheng, B., Li, T., Jiang, Z., Liu, W., 2015. Effect of porous zinc–biochar nanocomposites on Cr(VI) adsorption from aqueous solution. RSC Adv., 35107–35115.

Gao, L., Nie, L., Wang, T., Qin, Y., Guo, Z., Yang, D., Yan, X., 2006. Carbon nanotube delivery of the GFP gene into mammalian cells. Chem. Biol. Chem. 7 (2), 239–242.

Gao, X.M., Zhang, Y., Dai, Y., Fu, F., 2016. High-performance magnetic carbon materials in dye removal from aqueous solutions. J. Solid State Chem. 239, 265–273.

Giraldo, J.P., Landry, M.P., Faltermeier, S.M., McNicholas, T.P., Iverson, N.M., Boghossian, A.A., Reuel, N.F., Hilmer, A.J., Sen, F., Brew, J.A., Strano, M.S., 2014. Plant nanobionics approach to augment photosynthesis and biochemical sensing. Nat. Mater. 13 (4), 400.

Glaser, B., Birk, J.J., 2012. State of the scientific knowledge on properties and genesis of Anthropogenic Dark Earths in Central Amazonia (terra preta de Índio). Geochim. Cosmochim. Acta 82, 39–51.

Glaser, B., Lehmann, J., Zech, W., 2002. Ameliorating physical and chemical properties of highly weathered soils in the tropics with charcoal—a review. Biol. Fert. Soils 35, 219–230.

Gonzaga, M.I.S., Mackowiak, C., de Almeida, A.Q., de Carvalho Junior, J.I.T., Andrade, K.R., 2018. Positive and negative effects of biochar from coconut husks, orange bagasse and pine wood chips on maize (Zea mays L.) growth and nutrition. Catena 162, 414–420.

González, M.E., Cea, M., Medina, J., González, A., Diez, M.C., Cartes, P., Monreal, C., Navia, R., 2015. Evaluation of biodegradable polymers as encapsulating agents for the development of a urea controlled–release fertilizer using biochar as support material. Sci. Total Environ. 505, 446–453.

Graber, E.R., Harel, Y.M., Kolton, M., Cytryn, E., Silber, A., David, D.R., Tsechansky, L., Borenshtein, M., Elad, Y., 2010. Biochar impact on development and productivity of pepper and tomato grown in fertigated soilless media. Plant Soil 337 (1–2), 481–496.

Graber, E.R., Tsechansky, L., Gerstl, Z., Lew, B., 2012. High surface area biochar negatively impacts herbicide efficacy. Plant Soil 353 (1–2), 95–106.

Graber, E.R., Frenkel, O., Jaiswal, A.K., Elad, Y., 2014. How may biochar influence the severity of diseases caused by soilborne pathogens? Carbon Manag. 5 (2), 169–183.

Hammond, J., Shackley, S., Sohi, S., Brownsort, P., 2011. Prospective life cycle carbon abatement for pyrolysis biochar systems in the UK. Energy Policy 39 (5), 2646–2655.

Harel, Y.M., Elad, Y., Rav-David, D., Borenstein, M., Shulchani, R., Lew, B., Graber, E.R., 2012a. Biochar mediates systemic response of strawberry to foliar fungal pathogens. Plant Soil 357 (1-2), 245–257.

Harel, Y.M., Elad, Y., Rav-David, D., Borenstein, M., Shulchani, R., Lew, B., Graber, E.R., 2012b. Biochar mediates systemic response of strawberry to foliar fungal pathogens. Plant Soil 357 (1–2), 245–257.

Henry, R.K., Yongsheng, Z., Jun, D., 2006. Municipal solid waste management challenges in developing countries–Kenyan case study. Waste Manag. 26 (1), 92–100.

Hoornweg, D., Bhada-Tata, P., 2012. What a Waste: A Global Review of Solid Waste Management. vol. 15 World Bank, Washington, DC, p. 116.

Hossain, M.K., Strezov, V., Chan, K.Y., Ziolkowski, A., Nelson, P.F., 2011. Influence of pyrolysis temperature on production and nutrient properties of wastewater sludge biochar. J. Environ. Manag. 92 (1), 223–228.

Hu, X., Ding, Z., Zimmerman, A.R., Wang, S., Gao, B., 2015. Batch and column sorption of arsenic onto iron–impregnated biochar synthesized through hydrolysis. Water Res. 68, 206–216.

Huang, X., Liu, Y., Liu, S., Tan, X., Ding, Y., Zeng, G., Zhou, Y., Zhang, M., Wang, S., Zheng, B., 2016. Effective removal of Cr (VI) using β–cyclodextrin–chitosan modified biochars with adsorption/reduction bifuctional roles. RSC Adv. 6 (1), 94–104.

Hwang, I.H., Nakajima, D., Matsuto, T., Sugimoto, T., 2008. Improving the quality of waste–derived char by removing ash. Waste Manag. 28 (2), 424–434.

Inyang, M., Gao, B., Yao, Y., Xue, Y., Zimmerman, A.R., Pullammanappallil, P., Cao, X., 2012. Removal of heavy metals from aqueous solution by biochars derived from anaerobically digested biomass. Bioresour. Technol. 110, 50–56.

Inyang, M., Gao, B., Zimmerman, A., Zhang, M., Chen, H., 2014. Synthesis, characterization, and dye sorption ability of carbon nanotube–biochar nanocomposites. Chem. Eng. J. 236, 39–46.

Inyang, M., Gao, B., Zimmerman, A., Zhou, Y., Cao, X., 2015. Sorption and cosorption of lead and sulfapyridine on carbon nanotube–modified biochars. Environ. Sci. Pollut. Res. 22 (3), 1868–1876.

Jeffery, S., Verheijen, F.G., van der Velde, M., Bastos, A.C., 2011. A quantitative review of the effects of biochar application to soils on crop productivity using meta–analysis. Agric. Ecosyst. Environ. 144 (1), 175–187.

Jenie, S.A., Kristiani, A., Kustomo, Simanungkalit, S., Mansur, D., 2017. Preparation of nanobiochar as magnetic solid acid catalyst by pyrolysis–carbonization from oil palm empty fruit bunches. In: AIP Conference Proceedings. 1904. AIP Publishing, p. 020018, November.

Jia, M., Wang, F., Bian, Y., Jin, X., Song, Y., Kengara, F.O., Xu, R., Jiang, X., 2013. Effects of pH and metal ions on oxytetracycline sorption to maize-straw-derived biochar. Bioresour. Technol. 136, 87–93.

Jones, D.L., Rousk, J., Edwards-Jones, G., DeLuca, T.H., Murphy, D.V., 2012. Biochar-mediated changes in soil quality and plant growth in a three-year field trial. Soil Biol. Biochem. 45, 113–124.

Joseph, S., Graber, E.R., Chia, C., Munroe, P., Donne, S., Thomas, T., Nielsen, S., Marjo, C., Rutlidge, H., Pan, G.X., Li, L., 2013. Shifting paradigms: development of high–efficiency biochar fertilizers based on nano–structures and soluble components. Carbon Manag. 4 (3), 323–343.

Jung, K.W., Ahn, K.H., 2016. Fabrication of porosity-enhanced MgO/biochar for removal of phosphate from aqueous solution: application of a novel combined electrochemical modification method. Bioresour. Technol. 200, 1029–1032.

Jung, K.W., Jeong, T.U., Hwang, M.J., Kim, K., Ahn, K.H., 2015. Phosphate adsorption ability of biochar/Mg–Al assembled nanocomposites prepared by aluminum–electrode based electro–assisted modification method with MgCl2 as electrolyte. Bioresour. Technol. 198, 603–610.

Kam, N.W.S., O'Connell, M., Wisdom, J.A., Dai, H., 2005. Carbon nanotubes as multifunctional biological transporters and near–infrared agents for selective cancer cell destruction. Proc. Natl. Acad. Sci. 102 (33), 11600–11605.

Kammann, C., Glaser, B., Schmidt, H.P., 2016. Combining biochar and organic amendments. In: Biochar in European Soils and Agriculture. vol. 22, pp. 158–186.

Karami, N., Clemente, R., Jimenez, E., Lepp, N., Beesley, L., 2011. Efficiency of green waste compost and biochar soil amendments for reducing lead and copper mobility and uptake to ryegrass. J. Hazard. Mater. 191, 41–48.

Kavitha, B., Reddy, P.V.L., Kim, B., Lee, S.S., Pandey, S.K., Kim, K.H., 2018. Benefits and limitations of biochar amendment in agricultural soils: a review. J. Environ. Manag. 227, 146–154.

Khodakovskaya, M.V., de Silva, K., Nedosekin, D.A., Dervishi, E., Biris, A.S., Shashkov, E.V., Galanzha, E.I., Zharov, V.P., 2011. Complex genetic, photothermal, and photoacoustic analysis of nanoparticle–plant interactions. Proc. Natl. Acad. Sci. 108 (3), 1028–1033.

Khodakovskaya, M.V., De Silva, K., Biris, A.S., Dervishi, E., Villagarcia, H., 2012. Carbon nanotubes induce growth enhancement of tobacco cells. ACS Nano 6 (3), 2128–2135.

Khodakovskaya, M.V., Kim, B.S., Kim, J.N., Alimohammadi, M., Dervishi, E., Mustafa, T., Cernigla, C.E., 2013. Carbon nanotubes as plant growth regulators: effects on tomato growth, reproductive system, and soil microbial community. Small 9 (1), 115–123.

Kolton, M., Harel, Y.M., Pasternak, Z., Graber, E.R., Elad, Y., Cytryn, E., 2011. Impact of biochar application to soil on the root-associated bacterial community structure of fully developed greenhouse pepper plants. Appl. Environ. Microbiol. 14, 4924–4930.

Kurniawan, T.A., Chan, G.Y., Lo, W.H., Babel, S., 2006. Comparisons of low–cost adsorbents for treating wastewaters laden with heavy metals. Sci. Total Environ. 366 (2–3), 409–426.

Lahiani, M.H., Dervishi, E., Chen, J., Nima, Z., Gaume, A., Biris, A.S., Khodakovskaya, M.V., 2013. Impact of carbon nanotube exposure to seeds of valuable crops. ACS Appl. Mater. Interfaces 5 (16), 7965–7973.

Laird, D.A., Fleming, P.D., Davis, D.D., Wang, B., Horton, R., Karlen, D.L., 2010. Impact of biochar amendments on the quality of a typical Midwestern agricultural soil. Geoderma 158 (3–4), 443–449.

Lateef, A., Nazir, R., Jamil, N., Alam, S., Shah, R., Khan, M.N., Saleem, M., 2019. Synthesis and characterization of environmental friendly corncob biochar based nano-composite–A potential slow release nano-fertilizer for sustainable agriculture. Environ. Nanotechnol. Monitoring Manag. 11, 100212.

Lehmann, J., 2007a. Bio–energy in the black. Front. Ecol. Environ. 5, 381–387.

Lehmann, J., 2007b. A handful of carbon. Nature 447, 143–144.

Lehmann, J., Joseph, S., 2012. Biochar for Environmental Management: Science and Technology.

Lehmann, J., Rondon, M., 2006. Bio-char soil management on highly weathered soils in the humid tropics. In: Biological Approaches to Sustainable Soil Systems. vol. 113, pp. 517–530.

Lehmann, J., Kern, D., German, L., Mccann, J., Martins, G.C., Moreira, A., 2003. Soil fertility and production potential. In: Amazonian Dark Earths. Springer, Dordrecht, pp. 105–124.

Lehmann, J., Gaunt, J., Rondon, M., 2006. Bio–char sequestration in terrestrial ecosystems–a review. Mitig. Adapt. Strateg. Global Change 11 (2), 403–427.

Lehmann, J., Rillig, M.C., Thies, J., Masiello, C.A., Hockaday, W.C., Crowley, D., 2011. Biochar effects on soil biota–a review. Soil Biol. Biochem. 43, 1812–1836.

Li, Y., Park, S.Y., Zhu, J., 2011. Solid-state anaerobic digestion for methane production from organic waste. Renew. Sust. Energ. Rev. 15 (1), 821–826.

Li, B., Yang, L., Wang, C.Q., Zhang, Q.P., Liu, Q.C., Li, Y.D., Xiao, R., 2017. Adsorption of Cd(II) from aqueous solutions by rape straw biochar derived from different modification processes. Chemosphere 175, 332–340.

Li, R., Liang, W., Wang, J.J., Gaston, L.A., Huang, D., Huang, H., Xiao, R., 2018. Facilitativencapture of As(V), Pb(II) andmethylene blue fromaqueous solutions with MgO hybrid sponge-like carbonaceous composite derived from sugarcane leafy trash. J. Environ. Manag. 212, 77.

Liang, B., Lehmann, J., Solomon, D., Kinyangi, J., Grossman, J., O'neill, B., Skjemstad, J.O., Thies, J., Luizao, F.J., Petersen, J., Neves, E.G., 2006. Black carbon increases cation exchange capacity in soils. Soil Sci. Soc. Am. J. 70 (5), 1719–1730.

Lin, S., Reppert, J., Hu, Q., Hudson, J.S., Reid, M.L., Ratnikova, T.A., Ke, P.C., 2009. Uptake, translocation, and transmission of carbon nanomaterials in rice plants. Small 5 (10), 1128–1132.

Liu, Q., Chen, B., Wang, Q., Shi, X., Xiao, Z., Lin, J., Fang, X., 2009. Carbon nanotubes as molecular transporters for walled plant cells. Nano Lett. 9 (3), 1007–1010.

Liu, Z., Zhang, F.S., Wu, J., 2010. Characterization and application of chars produced from pinewood pyrolysis and hydrothermal treatment. Fuel 89 (2), 510–514.

Liu, X., Zhang, A., Ji, C., Joseph, S., Bian, R., Li, L., Pan, G., Paz-Ferreiro, J., 2013. Biochar'seffect on crop productivity and the dependence on experimental conditions—a metaanalysis of literature data. Plant Soil 373, 583–594.

Liu, W.J., Jiang, H., Yu, H.Q., 2015. Development of biochar–based functional materials: toward a sustainable platform carbon material. Chem. Rev. 115 (22), 12251–12285.

Liu, S.B., Tan, X.F., Liu, Y.G., Gu, Y.L., Zeng, G.M., Hu, X.J., Wang, H., Zhou, L., Jiang, L.H., Zhao, B.B., 2016a. Production of biochars from Ca impregnated ramie biomass (Boehmeria nivea (L.) Gaud.) and their phosphate removal potential. Res. Adv. 6 (7), 5871–5880.

Liu, S.B., Tan, X.F., Liu, Y.G., Gu, Y.L., Zeng, G.M., Hu, X.J., Wang, H., Zhou, L., Jiang, L.H., Zhao, B.B., 2016b. Production of biochars from Ca impregnated ramie biomass (Boehmeria nivea (L.) Gaud.) and their phosphate removal potential. RSC Adv. 6, 5871–5880.

Major, J., Rondon, M., Molina, D., Riha, S.J., Lehmann, J., 2010. Maize yield and nutrition during 4 years after biochar application to a Colombian savanna oxisol. Plant Soil 333 (1–2), 117–128.

Manikandan, A., Subramanian, K.S., 2013. Urea intercalated biochar—a slow release fertilizer production and characterisation. Indian J. Sci. Technol. 6 (12), 5579–5584.

Manya, J.J., 2012. Pyrolysis for biochar purposes: a review to establish current knowledge gaps and research needs. Environ. Sci. Technol. 46 (15), 7939–7954.

Marris, E., 2006. Putting the carbon back: Black is the new green.

Martin, S.M., Kookana, R.S., Van Zwieten, L., Krull, E., 2012. Marked changes in herbicide sorption–desorption upon ageing of biochars in soil. J. Hazard. Mater. 231, 70–78.

Martinez, J.A., 2006. Improvement of Kiln Design and Combustion/Carbonization Timing to Produce Charcoal From Agricultural Waste in Developing Countries. Doctoral dissertation, Massachusetts Institute of Technology.

Masulili, A., Utomo, W.H., Syechfani, M., 2010. Rice husk biochar for rice based cropping system in acid soil 1. The characteristics of rice husk biochar and its influence on the properties of acid sulfate soils and rice growth in West Kalimantan, Indonesia. J. Agric. Sci. 2 (1), 39–47.
McCormack, S.A., Ostle, N., Bardgett, R.D., Hopkins, D.W., Vanbergen, A.J., 2013. Biochar in bioenergy cropping systems: impacts on soil faunal communities and linked ecosystem processes. Global Change Biol. Bioenergy 5, 81–95.
Memon, M.A., 2010. Integrated solid waste management based on the 3R approach. J. Mater. Cycles Waste Manage. 12 (1), 30–40.
Meyer, S., Glaser, B., Quicker, P., 2011. Technical, economical, and climate–related aspects of biochar production technologies: a literature review Environ. Sci. Technol. 45 (22), 9473–9483.
Mielke, H.W., Gonzales, C.R., Smith, M.K., Mielke, P.W., 2000. Quantities and associations of lead, zinc, cadmium, manganese, chromium, nickel, vanadium, and copper in fresh Mississippi delta alluvium and New Orleans alluvial soils. Sci. Total Environ. 246 (2–3), 249–259.
Mohamed, B.A., Kim, C.S., Ellis, N., Bi, X., 2016. Microwave–assisted catalytic pyrolysis of switchgrass for improving bio–oil and biochar properties. Bioresour. Technol. 201, 121–132.
Mohan, D., Pittman, C.U., Steele, P.H., 2006. Pyrolysis of wood/biomass for bio–oil: a critical review. Energy Fuel 20 (3), 848–889.
Mohan, D., Kumar, S., Srivastava, A., 2014. Fluoride removal from ground water using magnetic and nonmagnetic corn stover biochars. Ecol. Eng. 73, 798–808.
Mollinedo, J., Schumacher, T.E., Chintala, R., 2015. Influence of feedstocks and pyrolysis on biochar's capacity to affect the soil water retention characteristics. J. Anal. Appl. Pyrolysis 114, 100–108.
Muhammad, N., Dia, Z., Xiao, K., Meng, J., Brookes, P.C., Liu, X., Wang, H., Wu, J., Xu, J., 2014. Changes in microbial community structure due to biochars generated from different feedstocks and their relationship with soil chemical properties. Geoderma 226, 270–278.
Nag, S.K., Kookana, R., Smith, L., Krull, E., Macdonald, L.M., Gill, G., 2011. Poor efficacy of herbicides in biochar–amended soils as affected by their chemistry and mode of action. Chemosphere 84 (11), 1572–1577.
Naghdi, M., Taheran, M., Brar, S.K., Rouissi, T., Verma, M., Surampalli, R.Y., Valero, J.R., 2017. A green method for production of nanobiochar by ball milling–optimization and characterization. J. Clean. Prod. 164, 1394–1405.
Noble, R., 2011. Risks and benefits of soil amendment with composts in relation to plant pathogens. Aust. Plant Pathol. 40 (2), 157–167.
Ocsoy, I., Paret, M.L., Ocsoy, M.A., Kunwar, S., Chen, T., You, M., Tan, W., 2013. Nanotechnology in plant disease management: DNA–directed silver nanoparticles on graphene oxide as an antibacterial against Xanthomonas perforans. ACS Nano 7 (10), 8972–8980.
Orr, C.H., Ralebitso-Senior, T.K., 2014. Tracking N–cycling genes in biochar–supplemented ecosystems: a perspective. OA Microbiol. 2 (1), 1.
Portugal-Pereira, J., Lee, L., 2016. Economic and environmental benefits of waste–to–energy technologies for debris recovery in disaster–hit Northeast Japan. J. Clean. Prod. 20 (112), 4419–4429.
Qian, L., Chen, L., Joseph, S., Pan, G., Li, L., Zheng, J., Zhang, X., Zheng, J., Yu, X., Wang, J., 2014. Biochar compound fertilizer as an option to reach high productivity but low carbon intensity in rice agriculture of China. Carbon Manage. 5 (2), 145–154.
Quilliam, R.S., Deluca, T.H., Jones, D.L., 2013. Biochar application reduces nodulation but increases nitrogenase activity in clover. Plant Soil 366, 83–92.

Rahman, L., Whitelaw-Weckert, M.A., Orchard, B., 2014. Impact of organic soil amendments, including poultry-litter biochar, on nematodes in a Riverina, New South Wales, vineyard. Soil Res. 52, 604–619.

Ralebitso-Senior, T.K., Orr, C.H., 2016. Biochar Application: Essential Soil Microbial Ecology. Elsevier, 332. pp. ISBN-13: 978-0128034330.

Read, P., 2009. A copenhagen initiative? Curing kyoto with a "Leaky bucket". UK Biochar Research Centre Working Paper 3.

Read, P., Parshotam, A., 2007. Holistic greenhouse gas management strategy. Institute of Policy Studies Working Paper 07/1.

Reddy, D.H.K., Lee, S.M., 2014. Magnetic biochar composite: facile synthesis, characterization, and application for heavy metal removal. Colloids Surf. A Physicochem. Eng. Asp. 454, 96–103.

Rogovska, N., Laird, D.A., Rathke, S.J., Karlen, D.L., 2014. Biochar impact on Midwestern Mollisols and maize nutrient availability. Geoderma 230, 340–347.

Rondon, M.A., Lehmann, J., Ramírez, J., Hurtado, M., 2007. Biological nitrogen fixation by common beans (*Phaseolus vulgaris* L.) increases with bio–char additions. Biol. Fertil. Soils 43 (6), 699–708.

Rose, P., Bhat, M., Vidhani, K., Ajmeri, N., Gole, A., Ghaisas, S., 2011. Intelligent informatics platform for nano–agriculture. In: 2011 11th IEEE International Conference on Nanotechnology 916–919 IEEE.

Sadegh-Zadeh, F., Tolekolai, S.F., Bahmanyar, M.A., Emadi, M., 2018. Application of biochar and compost for enhancement of rice (Oryza Sativa L.) grain yield in calcareous sandy soil. Commun. Soil Sci. Plant Anal. 49 (5), 552–566.

Sanchez, P.A., 2002. Soil fertility and hunger in Africa. Science 295 (5562), 2019–2020.

Saxena, M., Maity, S., Sarkar, S., 2014. Carbon nanoparticles in 'biochar'boost wheat (*Triticum aestivum*) plant growth. Res. Adv. 4 (75), 39948–39954.

Scheifele, M., Hobi, A., Buegger, F., Gattinger, A., Schulin, R., Boller, T., Mäder, P., 2017. Impact of pyrochar and hydrochar on soybean (*Glycine max* L.) root nodulation and biological nitrogen fixation. J. Plant Nutr. Soil Sci. 180 (2), 199–211.

Serag, M.F., Kaji, N., Habuchi, S., Bianco, A., Baba, Y., 2013. Nanobiotechnology meets plant cell biology: carbon nanotubes as organelle targeting nanocarriers. RSC Adv. 3 (15), 4856–4862.

Shackley, S., Sohi, S., Haszeldine, S., Manning, D., Mašek, O., 2009. Biochar, Reducing and Removing CO2 while Improving Soils: A Significant and Sustainable Response to Climate Change. . Evidence Submitted to the Royal Society Geo–Engineering Climate Enquiry. Available at: http://www.geos.ed.ac.uk/home/homes/sshackle/WP2.pdf (accessed 17.12.15).

Shang, M., Liu, Y., Liu, S., Zeng, G., Tan, X., Jiang, L., Huang, X., Ding, Y., Guo, Y.M., Wang, S., 2016. A novel graphene oxide coated biochar composite: synthesis, characterization and application for Cr(VI) removal. Res. Adv. 6, 85202–85212.

Silva, L.F.O., Sampaio, C.H., Guedes, A., Fdez-Ortiz de Vallejuelo, S., Madariaga, J.M., 2012. Multianalytical approaches to the characterisation of minerals associated with coals and the diagnosis of their potential risk by using combined instrumental microspectroscopic techniques and thermodynamic speciation. Fuel 94, 52–63.

Singh, A., Bhati, A., Gunture, Tripathi, K.M., Sonkar, S.K., 2018. In: Hussain, W.M., Mishra, A.K. (Eds.), Nanotechnology in Environmental Science.

Sohi, S.P., 2012. Carbon storage with benefits. Science 338 (6110), 1034–1035.

Sohi, S., Lopez-Capel, E., Krull, E., Bol, R., 2009. Biochar, climate change and soil: a review to guide future research. CSIRO Land Water Sci. Rep. 5 (09), 17–31.

Sohi, S.P., Krull, E., Lopez-Capel, E., Bol, R., 2010. A review of biochar and its use and function in soil. In: Advances in Agronomy. vol. 105. Academic Press, pp. 47–82.

Soja, G., Zehetner, F., Kitzler, B., 2013. Biochar as Yield Promoter for Agricultural Crops? Conjectures and Observations. http://www.oeaw.ac.at/forebiom/WS1lectures/SessionIII_Soja.pdf.

Sombroek, W., 1993. Amounts, dynamics and sequestering of carbon in tropical and subtropical soils. Ambio 22, 417–426.

Song, Z., Lian, F., Yu, Z., Zhu, L., Xing, B., Qiu, W., 2014. Synthesis and characterization of a novel MnOx–loaded biochar and its adsorption properties for Cu2+ in aqueous solution. Chem. Eng. J. 242, 36–42.

Sonkar, S.K., Roy, M., Babar, D.G., Sarkar, S., 2012. Water soluble carbon nano–onions from wood wool as growth promoters for gram plants. Nanoscale 4 (24), 7670–7675.

Steiner, C., Teixeira, W.G., Lehmann, J., Nehls, T., de Macêdo, J.L.V., Blum, W.E., Zech, W., 2007. Long term effects of manure, charcoal and mineral fertilization on crop production and fertility on a highly weathered Central Amazonian upland soil. Plant Soil 291 (1–2), 275–290.

Steiner, C., Glaser, B., Geraldes Teixeira, W., Lehmann, J., Blum, W.E., Zech, W., 2008. Nitrogen retention and plant uptake on a highly weathered central Amazonian Ferralsol amended with compost and charcoal. J. Plant Nutr. Soil Sci. 171 (6), 893–899.

Sun, L., Wan, S., Luo, W., 2013. Biochars prepared from anaerobic digestion residue, palm bark, and eucalyptus for adsorption of cationic methylene blue dye: characterization, equilibrium, and kinetic studies. Bioresour. Technol. 140, 406–413.

Sun, P., Hui, C., Khan, R.A., Du, J., Zhang, Q., Zhao, Y.H., 2015. Efficient removal of crystal violet using Fe3O4–coated biochar: the role of the Fe3O4 nanoparticles and modeling study their adsorption behavior. Sci. Rep. 5, 12638.

Tan, X., Liu, Y., Zeng, G., Wang, X., Hu, X., Gu, Y., Yang, Z., 2015. Application of biochar for the removal of pollutants from aqueous solutions. Chemosphere 125, 70–85.

Tan, X.F., Liu, Y.G., Gu, Y.L., Xu, Y., Zeng, G.M., Hu, X.J., Liu, S.B., Wang, X., Liu, S.M., Li, J., 2016a. Biochar-based nano-composites for the decontamination of wastewater: a review. Bioresour. Technol. 212, 318–333.

Tan, X.F., Liu, Y.G., Gu, Y.L., Xu, Y., Zeng, G.M., Hu, X.J., Liu, S.B., Wang, X., Liu, S.M., Li, J., 2016b. Biochar–based nano–composites for the decontamination of wastewater: a review. Bioresour. Technol. 212, 318–333.

Tang, J., Lv, H., Gong, Y., Huang, Y., 2015. Preparation and characterization of a novel graphene/biochar composite for aqueous phenanthrene and mercury removal. Bioresour. Technol. 196, 355–363.

Tapaswi, P.K., Moorthy, M.S., Park, S.S., Ha, C.S., 2014. Fast, selective adsorption of Cu2+ from aqueous mixed metal ions solution using 1, 4, 7–triazacyclononane modified SBA–15 silica adsorbent (SBA–TACN). J. Solid State Chem. 211, 191–199.

Thies, J.E., Rillig, M.C., Graber, E.R., 2015. Biochar effects on the abundance, activity and diversity of the soil biota. In: Lehmann, J., Joseph, S. (Eds.), Biochar for Environmental Management: Science, Technology and Implementation, second ed. Earthscan Publications Ltd., London, UK.

Tian, Y., Sun, X., Li, S., Wang, H., Wang, L., Cao, J., Zhang, L., 2012. Biochar made from green waste as peat substitute in growth media for Calathea rotundifola cv. Fasciata. Sci. Hort. 143, 15–18.

Tiwari, D.K., Dasgupta-Schubert, N., Cendejas, L.V., Villegas, J., Montoya, L.C., García, S.B., 2014. Interfacing carbon nanotubes (CNT) with plants: enhancement of growth, water and ionic nutrient uptake in maize (Zea mays) and implications for nanoagriculture. Appl. Nanosci. 4 (5), 577–591.

Torney, F., Trewyn, B.G., Lin, V.S.Y., Wang, K., 2007. Mesoporous silica nanoparticles deliver DNA and chemicals into plants. Nat. Nanotechnol. 2 (5), 295.

Toth, J., Milham, P.J., 1975. Activated-carbon and ash-carbon effects on the adsorption and phytotoxicity of diuron. Weed Res. 15 (3), 171–176.

Tripathi, S., Sonkar, S.K., Sarkar, S., 2011. Growth stimulation of gram (*Cicer arietinum*) plant by water soluble carbon nanotubes. Nanoscale 3 (3), 1176–1181.

Tripathi, S., Kapri, S., Datta, A., Bhattacharyya, S., 2016. Influence of the morphology of carbon nanostructures on the stimulated growth of gram plant. Res. Adv. 6 (50), 43864–43873.

Tsai, W.T., Chen, H.R., 2013. Adsorption kinetics of herbicide paraquat in aqueous solution onto a low-cost adsorbent, swine-manure-derived biochar. Int. J. Environ. Sci. Technol. 10, 1349–1356.

Uzoma, K.C., Inoue, M., Andry, H., Fujimaki, H., Zahoor, A., Nishihara, E., 2011. Effect of cow manure biochar on maize productivity under sandy soil condition. Soil Use Manag. 27 (2), 205–212.

Van Zwieten, L., Singh, B., Joseph, S., Kimber, S., Cowie, A., Chan, K.Y., 2009. Biochar and emissions of non–CO2 greenhouse gases from soil. In: Biochar for Environmental Management: Science and Technology. vol. 1, pp. 227–250.

Vanlauwe, B., Bationo, A., Chianu, J., Giller, K.E., Merckx, R., Mokwunye, U., Ohiokpehai, O., Pypers, P., Tabo, R., Shepherd, K.D., Smaling, E.M.A., 2010. Integrated soil fertility management: operational definition and consequences for implementation and dissemination. Outlook Agric. 39 (1), 17–24.

Varjani, S., Kumar, G., Rene, E.R., 2019. Developments in biochar application for pesticide remediation: current knowledge and future research directions. J. Environ. Manag. 232, 505–513.

Vergara, S.E., Tchobanoglous, G., 2012. Municipal solid waste and the environment: a global perspective. Annu. Rev. Environ. Resour. 37, 277–309.

Wang, X., Han, H., Liu, X., Gu, X., Chen, K., Lu, D., 2012. Multi–walled carbon nanotubes can enhance root elongation of wheat (Triticum aestivum) plants. J. Nanopart. Res. 14 (6), 841.

Wang, H., Gao, B., Wang, S., Fang, J., Xue, Y., Yang, K., 2015a. Removal of Pb (II), Cu (II), and Cd (II) from aqueous solutions by biochar derived from KMnO4 treated hickory wood. Bioresour. Technol. 197, 356–362.

Wang, M., Sheng, G., Qiu, Y., 2015b. A novel manganese–oxide/biochar composite for efficient removal of lead (II) from aqueous solutions. Int. J. Environ. Sci. Technol. 12, 1719–1726.

Wang, S., Gao, B., Li, Y., Mosa, A., Zimmerman, A.R., Ma, L.Q., Harris, W.G., Migliaccio, K.W., 2015d. Manganese oxide–modified biochars: preparation, characterization, and sorption of arsenate and lead. Bioresour. Technol. 181, 13–17.

Wang, S., Gao, B., Zimmerman, A.R., Li, Y., Ma, L., Harris, W.G., Migliaccio, K.W., 2015e. Removal of arsenic by magnetic biochar prepared from pinewood and natural hematite. Bioresour. Technol. 175, 391–395.

Wang, Y.Y., Liu, Y.X., Lu, H.H., Yang, R.Q., Yang, S.M., 2018. Competitive adsorption of Pb (II), Cu (II), and Zn (II) ions onto hydroxyapatite–biochar nanocomposite in aqueous solutions. J. Solid State Chem. 261, 53–61.

Wardle, D.A., Nilsson, M.C., Zackrisson, O., 2008. Fire–derived charcoal causes loss of forest humus. Science 320 (5876), 629.

Watzinger, A., Feichtmair, S., Kitzler, B., Zehetner, F., Kloss, S., Wimmer, B., Zechmeister Boltenstern, S., Soja, G., 2014. Soil microbial communities responded to biochar application in temperate soils and slowly metabolized 13C-labelled biochar as revealed by 13C PLFA analyses: results from a short-term incubation and pot experiment. Eur. J. Soil Sci. 65, 40–51.

Wen, P., Wu, Z., He, Y., Ye, B.C., Han, Y., Wang, J., Guan, X., 2016. Microwave–assisted synthesis of a semi–interpenetrating polymer network slow–release nitrogen fertilizer with water absorbency from cotton stalks. ACS Sustain. Chem. Eng. 4 (12), 6572–6579.

Withers, P.J., Clay, S.D., Breeze, V.G., 2001. Phosphorus transfer in runoff following application of fertilizer, manure, and sewage sludge. J. Environ. Qual. 30 (1), 180–188.

Woolf, D., Amonette, J.E., Street-Perrott, F.A., Lehmann, J., Joseph, S., 2010. Sustainable biochar to mitigate global climate change. Nat. Commun. 1, 56.

Wu, M.Y., 2013. Effects of incorporation of nano-carbon into slow-released fertilizer on rice yield and nitrogen loss in surface water of paddy soil. In: *2013 Third International Conference on Intelligent System Design and Engineering Applications*. IEEE, pp. 676–681.

Xiu, S., Shahbazi, A., Li, R., 2017. Characterization, modification and application of biochar for energy storage and catalysis: a review. Trends Renew. Energy 3 (1), 86–101.

Xu, C., Liu, W., Sheng, G.D., 2008. Burned rice straw reduces the availability of clomazone to barnyardgrass. Sci. Total Environ. 392 (2–3), 284–289.

Xu, R.K., Xiao, S.C., Yuan, J.H., Zhao, A.Z., 2011. Adsorption of methyl violet from aqueous solutions by the biochars derived from crop residues. Bioresour. Technol. 102, 10293–10298.

Yan, L., Kong, L., Qu, Z., Li, L., Shen, G., 2014. Magnetic biochar decorated with ZnS nanocrystals for Pb (II) removal. ACS Sustain. Chem. Eng. 3, 125–132.

Yang, Y., Sheng, G., Huang, M., 2006. Bioavailability of diuron in soil containing wheat–straw–derived char. Sci. Total Environ. 354 (2–3), 170–178.

Yao, Y., Gao, B., Chen, J., Yang, L., 2013. Engineered biochar reclaiming phosphate from aqueous solutions: mechanisms and potential application as a slow–release fertilizer. Environ. Sci. Technol. 47, 8700–8708.

Zaytseva, O., Neumann, G., 2016. Carbon nanomaterials: production, impact on plant development, agricultural and environmental applications. Chem. Biol. Technol. Agric. 3 (1), 17.

Zech, W., Haumaier, L., Reinhold, H., 1990. Ecological aspects of soil organic matter in tropical land use. In: Humic Substances in Soil and Crop Sciences: Selected Readings, (Humicsubstances), pp. 187–202.

Zhang, M., Gao, B., 2013. Removal of arsenic, methylene blue, and phosphate by biochar/AlOOH nanocomposite. Chem. Eng. J. 226, 286–292.

Zhang, Q., Yang, Z., Wu, W., 2008. Role of crop residue management in sustainable agricultural development in the North China Plain. J. Sustain. Agric. 32 (1), 137–148.

Zhang, M., Gao, B., Yao, Y., Xue, Y., Inyang, M., 2012. Synthesis of porous MgO–biochar nanocomposites for removal of phosphate and nitrate from aqueous solutions. Chem. Eng. J. 210, 26–32.

Zhang, M., Gao, B., Yao, Y., Xue, Y., Inyang, M., 2012a. Synthesis of porous MgO–biochar nanocomposites for removal of phosphate and nitrate from aqueous solutions. Chem. Eng. J. 210, 26–32.

Zhang, M., Gao, B., Yao, Y., Xue, Y., Inyang, M., 2012b. Synthesis, characterization, and environmental implications of graphene–coated biochar. Sci. Total Environ. 435, 567–572.

Zhang, M., Gao, B., Varnoosfaderani, S., Hebard, A., Yao, Y., Inyang, M., 2013. Preparation and characterization of a novel magnetic biochar for arsenic removal. Bioresour. Technol. 130, 457–462.

Zhang, M.M., Liu, Y.G., Li, T.T., Xu, W.H., Zheng, B.H., Tan, X.F., Wang, H., Guo, Y.M., Guo, F.Y., Wang, S.F., 2015. Chitosan modification of magnetic biochar produced from Eichhornia crassipes for enhanced sorption of Cr (vi) from aqueous solution. Res. Adv. 5, 46955–46964.

Zheng, W., Guo, M., Chow, T., Bennett, D.N., Rajagopalan, N., 2010. Sorption properties of greenwaste biochar for two triazine pesticides. J. Hazard. Mater. 181, 121–126.

Zheng, R., Li, C., Sun, G., Xie, Z., Chen, J., Wu, J., Wang, Q., 2017. The influence of particle size and feedstock of biochar on the accumulation of Cd, Zn, Pb, and As by Brassica chinensis L. Environ. Sci. Pollut. Res. 24 (28), 22340–22352.

Zhao, Q., Ma, C., White, J.C., Dhankher, O.P., Zhang, X., Zhang, S., Xing, B., 2017. Quantitative evaluation of multi-wall carbon nanotube uptake by terrestrial plants. Carbon 114, 661–670.

Zhou, Y., Gao, B., Zimmerman, A.R., Fang, J., Sun, Y., Cao, X., 2013. Sorption of heavy metals on chitosan–modified biochars and its biological effects. Chem. Eng. J. 231, 512–518.

Zhou, D., Liu, D., Gao, F., Li, M., Luo, X., 2017. Effects of biochar–derived sewage sludge on heavy metal adsorption and immobilization in soils. Int. J. Environ. Res. Public Health 14 (7), 681.

Zhou, N., Chen, H., Xi, J., Yao, D., Zhou, Z., Tian, Y., Lu, X., 2017b. Biochars with excellent Pb (II) adsorption property produced from fresh and dehydrated Banana peels via hydrothermal carbonization. Bioresour. Technol. 232, 204–210.

Zhu, Y., Hu, J., Wang, J., 2012. Competitive adsorption of Pb (II), Cu (II) and Zn (II) onto xanthate-modified magnetic chitosan. J. Hazard. Mater. 221, 155–161.

Zhu, S., Huang, X., Wang, D., Wang, L., Ma, F., 2018. Enhanced hexavalent chromium removal performance and stabilization by magnetic iron nanoparticles assisted biochar in aqueous solution: Mechanisms and application potential. Chemosphere 207.

Zwart, D.C., Kim, S.-H., 2012. Biochar amendment increases resistance to stem lesions caused by *Phytophthora* spp. in tree seedlings. HortScience 47 (12), 1736–1740.

Further reading

Awual, M.R., Yaita, T., Shiwaku, H., Suzuki, S., 2015. A sensitive ligand embedded nano–conjugate adsorbent for effective cobalt (II) ions capturing from contaminated water. Chem. Eng. J. 276, 1–10.

Chen, D., Suter, H., Islam, A., Edis, R., Freney, J.R., Walker, C.N., 2008. Prospects of improving efficiency of fertiliser nitrogen in Australian agriculture: a review of enhanced efficiency fertilisers. Soil Res. 46 (4), 289–301.

Lehmann, J., Joseph, S., 2015. Biochar for Environmental Management: Science, Technology and Implementation. Taylor & Francis, Routledge.

Masek O., n.d. Biochar Production Technologies, http://www.geos.ed.ac.uk/sccs/biochar/documents/BiocharLaunch-OMasek.pdf.

Rehman, M.Z.U., Rizwan, M., Ali, S., Fatima, N., Yousaf, B., Naeem, A., Sabir, M., Ahmad, H.R., Ok, Y.S., 2016. Contrasting effects of biochar, compost and farm manure on alleviation of nickel toxicity in maize (*Zea mays* L.) in relation to plant growth, photosynthesis and metal uptake. Ecotoxicol. Environ. Saf. 133, 218–225.

Sombroek, W.G., 1966. Amazon soils: A reconnaissance of the soils of the Brazilian Amazon region.' Verslagen van Landbouwkundige Onderzoekingen. Centre for Agricultural Publications and Documentation, Wageningen, the Netherlands.

Zwonitzer, J.C., Pierzynski, G.M., Hettiarachchi, G.M., 2003. Effects of phosphorus additions on lead, cadmium, and zinc bioavailabilities in a metal–contaminated soil. Water Air Soil Pollut. 143, 193–209.

Potential of nanoscale carbon-based materials for remediation of pesticide-contaminated environment

17

Josef Jampílek[a,b], *Katarína Kráľová*[c]
[a]Department of Analytical Chemistry, Faculty of Natural Sciences, Comenius University, Bratislava, Slovakia, [b]Regional Centre of Advanced Technologies and Materials, Faculty of Science, Palacky University, Olomouc, Czech Republic, [c]Institute of Chemistry, Faculty of Natural Sciences, Comenius University, Bratislava, Slovakia

1 Introduction

An increasing the human population needs an increasing amount of food (PotatoPro, 2016). While there is an undeniable trend in the increased use of plant life resources over livestock production (FAO, 2018; Röös et al., 2016; Thornton, 2010), food is still deficient, and higher yields are ensured by the use of artificial fertilizers (Ritchie, 2017; SFGate, 2019). Unfortunately, due to the changing climate of the planet, the number of different diseases and crop pests is growing, resulting in the increased application of a wide variety of herbicides, fungicides, and insecticides to control weeds and harmful fungi and insects. The overuse of fertilizers resulting in their accumulation in soil contributes to the impairment of soil quality (Aktar et al., 2009; Domene et al., 2008; Fargnoli et al., 2019). Although there has been a trend to use nanoparticles (NPs)/nanosystems for the delivery of nutrients, fertilizers, and pesticides to provide the benefits of targeted delivery of active substances (Jampílek and Kráľová, 2015, 2017a,b, 2018a,b), the accumulation of these agrochemicals in soil and water cannot be completely avoided. Thus, pesticides accumulated in soil and water can destroy nontarget organisms and cause much damage to biodiversity overall, resulting in the impairment of soil quality and the aquatic genofond (Bernardes et al., 2015; Carvalho, 2017). Moreover, the accumulation of increased levels of harmful pesticides in crops and fodder has a negative impact on the health of humans and farm animals (Al-Ahmadi, 2019; Gill and Garg, 2014). Therefore, it is desirable not only to reduce the applied pesticide amounts to achieve the most possible prevention of crops but also to secure the efficient removal of pesticides from the soil, water, and sediment surfaces by using powerful adsorbents.

Nanopesticides, including nanoscale biopesticides—agents of biological origin that allow achieving pest management in an environmentally friendly way—could significantly contribute to reducing the burden of pesticides on nature. As nanotechnology is able to provide new properties to old materials (Jampílek and Kráľová, 2015,

Carbon Nanomaterials for Agri-food and Environmental Applications. https://doi.org/10.1016/B978-0-12-819786-8.00017-7

2017a,b,c, 2019a,b; Jampílek et al., 2015, 2019; Pentak et al., 2016; Pisárčik et al., 2018), the increased attention of researchers has recently been focused on the use of nanomaterials (NMs) that are able to efficiently adsorb pesticides or their degradation products from soil and water and could also significantly contribute to the enhanced degradation of pesticides. For these purposes, carbon-based NMs have been preferably used. With the introduction of functional groups (e.g., carboxyl, hydroxyl, phenolic, and amine groups) on the surface of nanoscale carbon-based adsorbents, the removal of positively and negatively charged pesticides is possible.

A review paper summarizing older and newer techniques and materials used for the removal of a wide scale of pesticides using different processes, such as bioremediation with microorganisms or clay, activated carbon, and polymer materials to chemical treatment based on oxidation processes, was presented by Marican and Duran-Lara (2018). An overview of the advanced oxidation processes applied for the treatment of water pollutants defined in the recently launched Directive 2013/39/EU was presented by Ribeiro et al. (2015). Recent developments of electrochemical methods combined with nanotechnology for sensing and decontamination of pesticides were summarized by Viswanathan and Manisankar (2015). An up-to-date analysis related to NM application for the removal of toxic elements (e.g., detergents, pesticides, pharmaceuticals, and endocrine-disrupting chemicals) in water with a focus on inorganic NMs, organic polymer NMs, and organic polymer-supported nanocomposites (NCs) was presented by Lu and Astruc (2018). des Ligneris et al. (2018) published an up-to-date overview of the use of nanofibers for the treatment of wastewater contaminated by disinfectants and pesticides, highlighting the impact of adsorbent pore geometry and surface chemistry on the resulting adsorption performance against specific organic molecules.

It is important to note that for the monitoring of pesticides and their residues, various nanosensors, including those based on nanoscale carbon materials, are frequently used (Capoferri et al., 2018; Cui et al., 2018; Jampílek and Kráľová, 2015, 2018c; Kozitsina et al., 2018; Liu et al., 2014; Willner and Vikesland, 2018; Xu et al., 2018; Yang et al., 2014; Yu et al., 2015).

In the context of the above-mentioned facts, this chapter presents a comprehensive overview of recent research and applications of a wide spectrum of carbon-based materials used in the nanoscale for the removal of pesticides from the environment. Special attention is devoted to highly porous biochar and activated carbon possessing a broad range of nanoscale pore sizes as well as to NMs such as graphene, graphene oxide, reduced graphene oxide, fullerenes, single- and multiwalled carbon nanotubes, and various carbon-doped TiO_2 NCs showing superb photocatalytic activity.

2 Carbon-based materials used for pesticide removal

In a review paper, Sophia and Lima (2018) analyzed the adsorption process as an efficient method for the removal of emerging contaminants from the environment using different carbon-based materials while also discussing the mechanism of adsorption. Sorption of ionizable and ionic organic compounds to biochar, activated carbon, and other carbonaceous materials with a critical evaluation of related mechanisms and the role played by some environmental factors (e.g., pH, ionic strength) was reviewed by Kah et al. (2017). The application of graphene, graphene composites, electrospun membranes, and improved activated carbon water treatment systems focused mainly on the removal of heavy metals, pesticides, and harmful bacteria was reported by Rajapaksha et al. (2018). A magnetic NP-attached fluorographene-based sorbent was reported to show almost complete removal efficiency of the pesticides acetochlor and chlorantraniliprole from water with environmentally relevant concentrations within a short time (e.g., 10 min) (Wang et al., 2018).

Biochar and activated carbon show favorable adsorption ability to remove environmental contaminants. Therefore, they are applied for this purpose for a long time; however, recently, increasing interest has been devoted to nanoscale carbon-based materials (Cha et al., 2013; Ismail and Goh, 2018; Mauter and Elimelech, 2008; Photon, 2019; Wanekaya, 2011; Xia et al., 2015). Carbon-based nanomaterials (CNMs) are classified mostly according to their geometrical structure. They include graphene, which is a two-dimensional allotropic form of carbon formed by single layers of carbon atoms; graphene oxide; and reduced graphene oxide. They also include particles that are tube-shaped (single-walled (SWCNTs) and multiwalled carbon nanotubes (MWCNTs)), horn-shaped (nanohorns), spherical, or ellipsoidal (fullerenes). CNMs show immense diversity in their structures as well as size, which is not exclusively limited to the nanosize (<100 nm) in all dimensions (Mukherjee et al., 2016; Sharon and Sharon, 2010; Zaytseva and Neumann, 2016). A relatively small carbon amount in photocatalysts, such as C-doped TiO_2 NPs, was found to significantly improve the photocatalytic activity of TiO_2 NPs (Ameta and Ameta, 2018; Helmy et al., 2018; Liu et al., 2012). Various carbon-based materials used for pesticide removal are illustrated in Fig. 1.

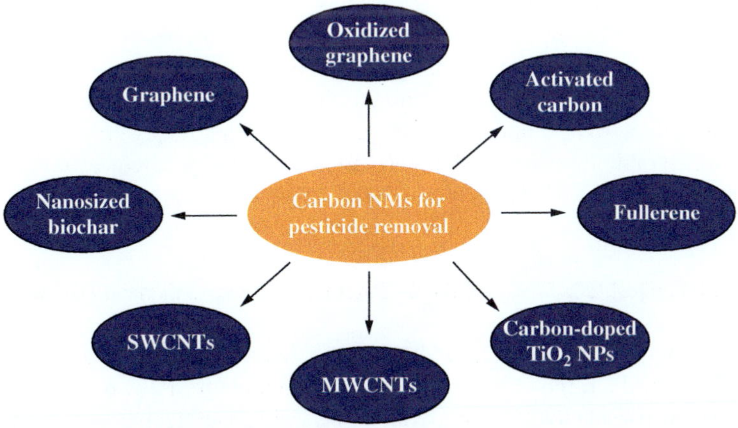

Fig. 1 Carbon-based materials used for pesticide removal.

2.1 Biochar

Biochar (BC) is produced from the pyrolysis of carbon-rich plant and animal residues under low-oxygen and high-temperature conditions. The presence of a highly porous structure and various functional groups (e.g., carboxyl, hydroxyl, and phenolic groups) in BC is responsible for its ability to adsorb and immobilize heavy metals and organic contaminants such as pesticide in the soil, whereby its specific physicochemical properties strongly depend on its feedstock (e.g, pinewood, wheat straw, rice husk, etc.) and the pyrolysis conditions (temperature, heating rate, and residence time). Moreover, amending the soil with BC could improve the crop yield (Githinji, 2013); it is also beneficial for the alleviation of adverse effects caused by climatic change (Woolf et al., 2010).

Biochar was found to increase the adsorption capacity for pesticides while also decreasing their desorption and mobility in soil layers as well as the bioavailability of pesticides in soil pore water. It can also improve soil physicochemical properties (e.g, pH, cation-exchange capacity, and water-holding capacity) and soil microbial activity by providing essential nutrients (e.g., Khorram et al., 2016).

The pores inside BC (intrapores) and the pores created between BC particles and soil particles (interpores) play fundamentally different roles in soil water retention when capillarity pressure is the main component of the soil water potential. Most BC intrapores have diameters (d) <0.01 mm. For example, the BC showing intraporosity of 0.6 m^3/m^3 can store in intrapores up to 0.6 m^3 water/m^3 BC. According to Liu et al. (2017), the increase of water retention of sandy soils amended by BC is caused by BC intraporosity at lower soil water potential while at higher soil water potential, it is connected with increasing interporosity due to the elongated BC particle shape increasing the interpore space. A microstructure of BC is shown in Fig. 2. Recently, great attention has been devoted to using BC as an effective pesticide sorbent from

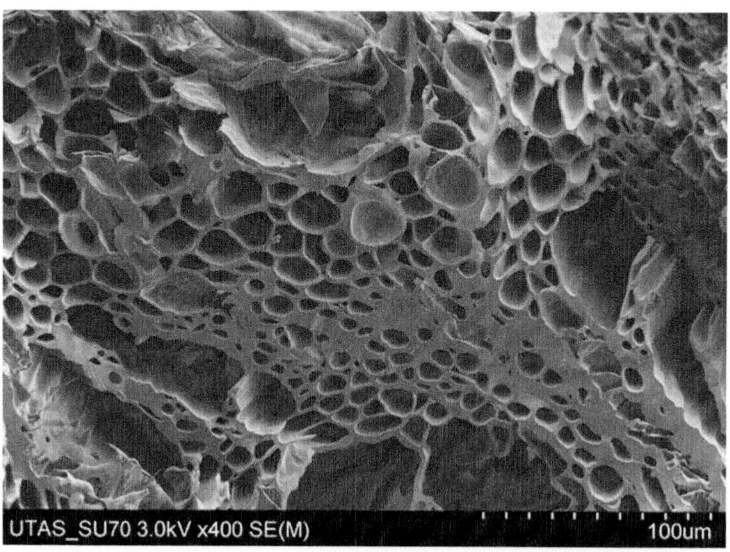

Fig. 2 Electron microscope image of biochar (freely provided by BiocharProject, 2011).

contaminated waters (Ahmad et al., 2014; Ameta and Ameta, 2018; Lehmann et al., 2011; Oliveira et al., 2017; Rajapaksha et al., 2016; Sizmur et al., 2017).

2.2 Activated carbon

Activated carbon (AC) is composed primarily of aromatic configurations of carbon atoms joined by random cross-linkages, its sheets or groups of atoms that are stacked unevenly in a disorganized manner; it is characterized by high porosity and high-surface-area reflected in the good adsorptive ability of this material. The spaces between the AC crystallites constitute the microporous structure with a large internal surface area, usually of 250–2500 m^2/g (Marsh and Rodriguez-Reinoso, 2006). The pores represent approx. 5% of the total surface area of the AC; their volume in the activated carbon ranges generally between 0.2 and 0.5 cm^3/g and their surface area from 0.5 to 2 m^2/g (Marsh and Rodriguez-Reinoso, 2006; McDougall, 1991).

AC powder products are fabricated by ground in roller, hammer, or jet mills to achieve a target particle-size distribution. For example, d_{50} (median particle size) can range from 5 to 20 μm. The highly porous nature of activated carbon could achieve surface areas even as large as 3000 m^2/g. While mesopores ranging in width from 2 to 50 nm could adsorb many contaminants, macropores with >50 nm in width exhibit minimal adsorption capacity, but are critical in determining adsorption kinetics within the particle. They also provide transport paths for contaminants to diffuse into the mesopore and micropores, where adsorption takes place (Koehlert, 2017).

2.3 Fullerene

Fullerenes are the carbon allotropes composed entirely of carbon and exist in the form of a hollow sphere (buckyballs), ellipsoid, or tube (buckytubes or carbon nanotubes). The cage-like structure of a fullerene consists of 12 five-member rings and a number of six-member rings, depending on the number of carbon atoms. Fullerene C_{60} (see Fig. 3), the most abundant molecule, is characterized with high symmetry. Its surface contains 20 hexagons and 12 pentagons, whereby all the rings are fused and the double bonds are conjugated; C_{60} fullerene is also a strong electron acceptor (Geckeler and Samal, 1999; Shanbogh and Sundaram, 2015). The size of the C_{60} fullerene is about 0.7 nm (Goel et al., 2004), and the sizes of various other fullerenes, both smaller and larger than C_{60}, range from 0.4–1.6 nm (Sloan et al., 2000).

Antioxidant effects, the mechanical properties and electrical properties of fullerene derivatives, and the catalytic properties of some hydrophobic fullerene derivatives were overviewed by Acquah et al. (2017). The use of fullerene for pesticide adsorption and in a composite with TiO_2, also for pesticide degradation, was described as well (Djordjevic et al., 2018; Srivastava et al., 2011).

2.4 Graphene, graphene oxide, and reduced graphene oxide

Graphene is a two-dimensional allotropic form of carbon, formed by single layers of sp^2-bonded carbon atoms arranged in a hexagonal lattice with a distance of

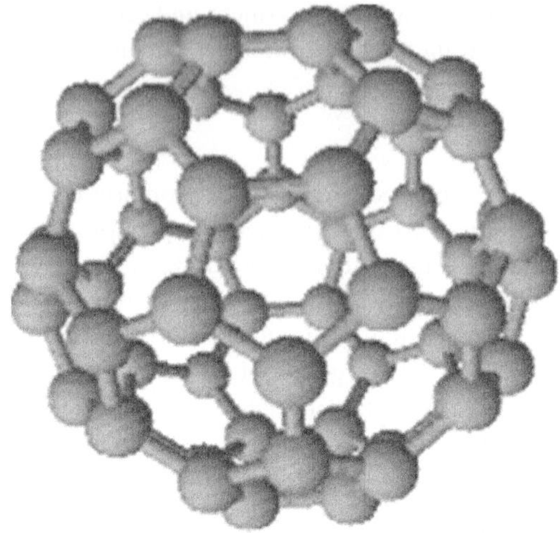

Fig. 3 Structure of C_{60} fullerene.

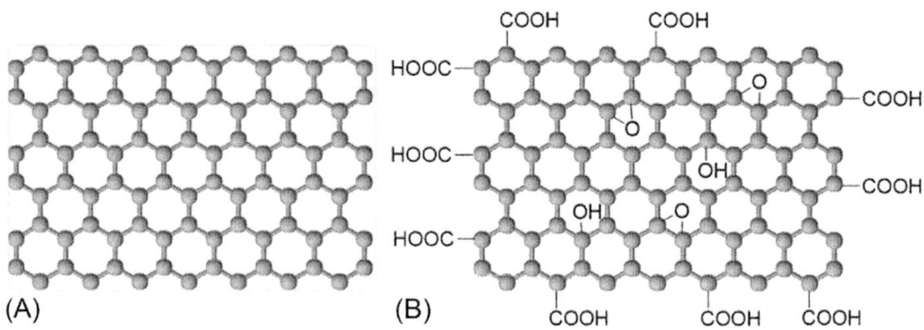

Fig. 4 Structures of graphene (A) and graphene oxide (B).

0.142 nm between neighboring atoms of carbon hexagons (Fig. 4A). It is the basic structural element of graphite, diamond, charcoal, carbon nanotubes, and fullerenes. The surface of graphene is hydrophobic. Graphene nanosheets are composed of >10 graphene sheets but are below 100 nm in thickness (Wick et al., 2014; Zaytseva and Neumann, 2016).

Graphene oxide (GO) is derivatized graphene with many oxygen functionalities through the introduction of carbonyl, hydroxy, and epoxy groups on the planar surfaces and edges of the carbon sheets during graphite oxide exfoliation (Fig. 4B). Coverage with oxygen atoms varies depending on the degree of oxidation during GO preparation. Hence, the GO surface consists of hydrophobic islands with hydrophilic regions showing various degrees of basal reactivity (Erickson et al., 2010; Wick et al., 2014).

Reduced graphene oxide (rGO) is obtained by the oxidation of graphite followed by reduction and exfoliation. rGO is the form of GO that is processed by chemical, thermal, and other methods to reduce the oxygen content, in which the C/O ratio is approximately 12:1. rGO consists of graphitic regions of up to 8 nm^2 and holes below 5 nm^2. The oxidized regions exhibit no order when they form a continuous network across the GO sheets (Erickson et al., 2010; Wick et al., 2014). The rGO sheets have properties that are similar to graphene, but they show deviation from the sp^2/planar character expected for the graphene monolayer. The presence of vacancies, heptagon and pentagon rings, edge effects, and functional groups also contributes to the formation of defects in the structure (Eda and Chhowalla, 2010; Morais et al., 2016).

2.5 Carbon nanotubes

Carbon nanotubes (CNTs) represent cylindrical structures with open or closed ends. They can be classified based on the number of concentric layers of rolled graphene sheets: single-walled CNTs (SWCNTs, see Fig. 5A) generally showing an outer diameter of 0.8–2 nm, and multiwalled CNTs (MWCNTs, see Fig. 5B) with outer diameters of 5–20 nm. Their lengths can range from 100 nm to several centimeters (Das, 2013; De Volder et al., 2013; Mukherjee et al., 2016; Yang et al., 2010).

Individual CNTs have closed caps, and even if the tubes have open ends, the smaller diameter of the tubes could not accommodate a typical contaminant of larger size. Therefore, a contaminant can be adsorbed by CNTs (i) at hollow interiors of nanotubes if they are open-ended, (ii) at interstitial pore spaces between the tube bundles, (iii) at grooves present at the boundary of the nanotube bundles, and (iv) at the external surface of the outermost CNTs (Agnihotri et al., 2005; Kang et al., 2008). Thus, potential adsorption sites are situated at the groove edges of nanotube bundles and the external surface areas of the outermost nanotubes are potential adsorption sites, whereby for the surface area and mesopore volume of pristine CNTs, values of 250 m^2/g and 0.85 cm^3/g were reported, respectively (Deng et al., 2008).

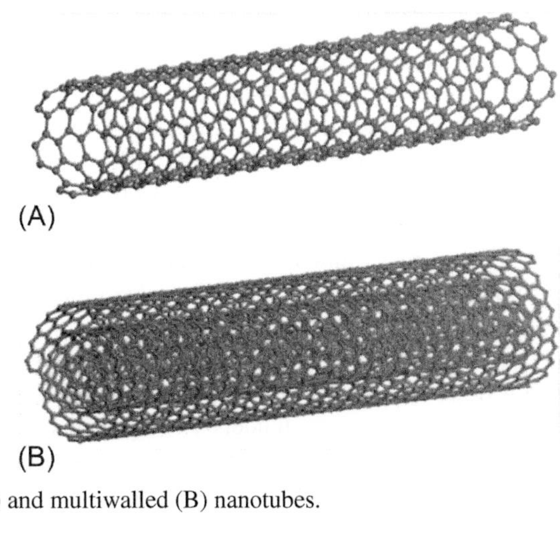

Fig. 5 Single- (A) and multiwalled (B) nanotubes.

2.6 Carbon-doped TiO_2 photocatalytic nanocomposites

In carbon-coated TiO_2, the carbon not only acts as a carrier but also can be used as a coating and pore agent (Palanivelu et al., 2007). These composite materials are favorable for the sorption of contaminants due to the increased effective surface. They are also more effective in photocatalytic oxidation as pristine TiO_2, which is effective only with UV light (Huang et al., 2016).

In C-doped TiO_2 that could be prepared by the addition of a small amount of carbon impurity (<1%) to TiO_2 in a controlled manner, the incorporation of carbonaceous species is responsible for the absorption tail in the visible-light region. Carbon can also be the lattice defect of TiO_2 to form interface states that effectively lower the band gap. With the carbon impurity, TiO_2 achieves an excess of conducting electrons or an excess of conducting holes, which is pivotal for making visible light active (Minnekhanov et al., 2012; Palanivelu et al., 2007). C-doped TiO_2 NPs are used as effective catalysts for the photodegradation of environmental pollutants, including pesticides (An et al., 2016; Cruz et al., 2017; Gómez-Avilés et al., 2019; Rasoulnezhad et al., 2017; Reddy et al., 2010).

3 Applications and benefits of individual carbon materials

3.1 Biochar

The role of BC in the adsorption-desorption and leaching of pesticides, its impact on the degradation and bioavailability of pesticides for living organisms, and its agricultural and environmental applications were summarized by Khorram et al. (2016). On the other hand, Kuppusamy et al. (2016) summarized the beneficial and possible negative impacts of BC on soil geochemistry, microflora and fauna, agrochemicals, greenhouse gas efflux, nutrients, crop yield, and organic and inorganic contaminants. Biochar addition in a pesticide-polluted environment contributes to the increase of soil water-holding capacity, improves aeration conditions in soil, and provides a habitat for the growth of microorganisms, which contribute to pesticide degradation (Varjani et al., 2018). Wu et al. (2017) summarized findings related to the impact of BC on the chemical characteristics (greenhouse gas emissions, nitrogen loss, decomposition, and humification of organic matter) and microbial community structure during composting of organic wastes, focusing also on the application of BC for remediation of soils contaminated with different environmental pollutants. A critical evaluation of the role of BC in sorption, desorption, and degradation of pesticides in the soil, including dominant properties of BC (e.g., porosity and surface area, pH, surface functional groups, carbon content and aromatic structure, and mineralogical composition) was presented by Liu et al. (2018a).

Nanobiochar with particle sizes of about 60 nm produced from pine wood BC using a planetary ball mill (milling time of 1.6 h, rotational speed 575 rpm, and ball-to-powder mass ratio of 4.5 g/g) had a great tendency to agglomerate, which could be inhibited by the conditioning of BC at cryogenic temperatures (24 h at −80°C) prior to milling. Such nanobiochar samples were able to remove 95% of carbamazepine from water (Naghdi et al., 2017). The amendments of BC prepared at a pyrolyzing temperature of 300°C promoted the biodegradation of thiacloprid in soil by increasing the microbe abundance and improving nitrite hydratase activity. On the other hand, BC prepared at higher pyrolyzing temperatures (500°C and 700°C, respectively) promoted chemical degradation mainly through elevated pH, active groups on the mineral surface, and generation of $^{\bullet}OH$ and other free radicals, suggesting that the impact of biochars on the fate of a pesticide in soil significantly depends on their type and properties (Zhang et al., 2018a). The application of slow-pyrolysis BC prepared from willows with a specific surface area (SSA) of $175\,m^2/g$ decreased the bioavailability of metsulfuron and sulfentrazone herbicides in the soil. This was in contrast to fast-pyrolysis BC with an SSA of $3\,m^2/g$, which was connected with increased herbicide adsorption on an adsorbent showing a greater surface area (Szmigielski et al., 2018). The evaluation of BC produced from cassava waste via pyrolyzation under oxygen-limiting conditions at 350°C, 550°C, and 750°C to sorb atrazine pesticide in tropical soil showed that the sorption amount at equilibrium increased with the increasing pyrolysis temperature (0.016, 0.025, and 0.050 mg/g); the equilibrium is reached more quickly with BC pyrolyzed at higher temperatures. The experimental data fitted well with the Freundlich isotherm. The sorption of atrazine in BC-added soil was a spontaneous and endothermic process primarily controlled by physisorption, whereby the lower pH was beneficial for the sorption of atrazine in BC-added soil (Deng et al., 2014). An investigation of the effects of BC properties on the fomesafen adsorption-desorption capacity of BC-amended soil showed that the adsorption capacity of BC was most likely controlled by SSA and dissolved organic carbon while the desorption and leaching processes were mainly affected by the micropore volume/total pore volume ratio. Hardwood BC applied at a dose of 2% with the highest SSA ($331.70\,m^2/g$) and lowest dissolved organic carbon (DOC) content (0.43%) showed a fomesafen adsorption coefficient of 20.67 while with the application of the same dose of rice straw BC (SSA of $63.10\,m^2/g$, DOC content of 3.67%), it reached only 16.64 (Khorram et al., 2018).

A study of interactions of triazine herbicides with hardwood biochars (300–700°C) showed that adsorption on BCs correlated poorly with pyrolysis temperature. The negative influence of solute molecular volume on the adsorption rate was estimated and the greatest mesoporosity-to-total-porosity ratio ensured a maximized adsorption rate for the BC, whereby the cationic forms of amines adsorbed more slowly than the neutral forms and the π-π electron donor-acceptor forces played a role in triazine polar interactions with BC (Xiao and Pignatello, 2015). The modification of potassium carbonate-activated porous biochar (KBC) by surface oxidation, surface amination, loading nanosized zero valent iron, and loading nanoiron oxyhydroxide affected the pore structure, especially the micropores. It also had a considerable impact on the functional groups. A specific surface area was found to be crucial for the adsorption of 2,4-dichlorophenoxyacetic acid (2,4-D). Slowing down of the 2,4-D adsorption rate was observed with the nitrogen-containing functional groups on the KBC surface. Although the surface oxidation destroyed the pore structure, the oxygen-containing functional groups enhanced the affinity of KBC for 2,4-D. On the other hand, nanosized zero valent iron and nanosized iron oxyhydroxide-modified KBC blocked the pores and decreased the specific surface area, which resulted in reduced adsorption capacity. Consequently, the beneficial impact on the adsorption of 2,4-D by BC showed surfaces rich in oxygen-containing functional groups and the adsorption was determined by the π-π interaction, chemisorption, and H-bonding (Zhu et al., 2018).

Carbonate-induced activated BC prepared from the waste of *Calligonum comosum* showed an SSA of 1473 m^2/g and a high amount of oxygenated surface functional groups, the maximum atrazine adsorption at the neutral water pH, and the increasing adsorption rate with the increase of solution temperature. The maximum adsorption at 40°C achieved 7143 mg/g, which considerably exceeded that observed with standard activated carbon. The adsorption equilibrium results fitted the Langmuir isotherm model and atrazine adsorption onto this adsorbent was found to be spontaneous chemisorption and an endothermic process (Alahabadi and Moussavi, 2017).

The dimethyl disulphide, an effective compound for fumigation in the soil, degraded faster in alkaline soil supplemented with BC amendment (Han et al., 2017). Although the aging process in BC-amended soil probably decreases the adsorption capacity of BC through changing its physicochemical properties, the adsorption capacity of rice hull BC used as a soil amendment for fomesan after six months of aging reached 2.5- to 5-fold higher values compared to the unamended soil (Khorram et al., 2017).

The amendment of biochars to soil enhanced the sorption and affected both the chemical and biological degradation of carbaryl. However, while BCs produced at 700°C and showing a strong liming effect more effectively enhanced the chemical hydrolysis of carbaryl in soil but reduced carbaryl biodegradation, BCs produced at 350°C promoted carbaryl biodegradation in soil (Ren et al., 2016).

The oil palm empty fruit bunch biochars produced at 300°C and applied to soils at a dose of 1% (w/w) were able to enhance the sorption of imazapic and imazapyr up to 59.90% and 69.94%, respectively (Yavari et al., 2016). The amendment of rape-straw derived BC in a paddy soil significantly enhanced pentachlorophenol sorption as well as its transformation due to an enhanced transferred extracellular electron in BC-amended soils; it also stimulated the growth and metabolism of soil microorganisms (Tong et al., 2014). The application of optimized biochars prepared using empty fruit bunches of the oil palm and rice husk biomasses as BC feedstock resulted in considerably higher biodegradation of imidazolinones in soil, with the greater effects from the rice husk BC. The polar herbicides were resistant to hydrolysis degradation and the photolysis rates of the herbicides were pronouncedly suppressed in the presence of biochars in the soil (Yavari et al., 2018).

A combined application of sepiolite (a soft white clay mineral known as meerschaum) and BC in soil contaminated by Cd and atrazine was better than individual applications of sepiolite or BC. Also, it was reported to be suitable for effective in situ remediations of the acid red soil of South China with combined pollution of Cd and atrazine (Qin et al., 2019).

The half-lives of 1,3-dichloropropene (1,3-D) fumigant in soil were found to be considerably affected by soil moisture, temperature, and amendment. An amendment of 1% BC resulted in 2.5- to 35-fold higher half-lives of 1,3-D in the soil, whereby significant correlation between the BC (air-dry) water content and the adsorption capacity of fumigants indicates that it could be used as a predictor for fumigant mitigation use (Wang et al., 2016).

Using BC supplementation of 0.50% (w/w), the BC-amended soil showed pronouncedly higher sorption capacity for chlorpyrifos (218.83 mg/kg) compared to compost-amended soils (111.85 mg/kg), suggesting that BC could be used to remediate chlorpyrifos polluted areas by immobilizing the pesticide and thus decreasing its bioavailability (Aziz et al., 2018).

3.2 Activated carbon

Using a small-scale granular activated carbon (GAC) (595–2380 μm bituminous coal) system to reduce pesticide concentrations in water, a contact time of 64 s resulted in the reduction of acephate, flurprimidol, and paclobutrazol below their minimum detection limits and caused the following reductions of other tested pesticides: 72.2% for bifenthrin, 89% for chlorpyrifos, 85.3% for imidacloprid, 99% for glyphosate, and 99.4% for triclopyr (Grant et al., 2019). Pumping neonicotinoid imidacloprid through columns packed with GAC under simulated flow conditions also removed all detectable imidacloprid from solution (Voorhees et al., 2017).

An investigation was done on wastewater containing herbicide glyphosate (200–3000 mg/L) treated via catalytic wet oxidation using as catalysts ACs modified by H_2O_2 oxidation and thermal treatment with ammonia or melamine related to degradation of herbicide to orthophosphate. AC modified by H_2O_2 and melamine had the highest catalytic activity, which was also characterized with superb stability in the continuous 55-day test, reaching 100% glyphosate removal and >93% orthophosphate removal using real wastewater. More pyrrolic, pyridinic, and graphitic nitrogen along with quinone oxygen functional groups on the AC surface contributed to the improved catalytic activity of the tested catalysts (Xing et al., 2018).

An investigation of the adsorption of chlordecone (CLD) on activated carbons prepared from sugar cane bagasse showed that the adsorption capacity increased with the amount of carbon and acidic groups at the AC surface while basic groups, hydroxyl, and ether groups had a negative impact on the adsorption. The adsorption capacity was maximized at a solution pH value of the point of zero charges and the CLD adsorption mechanism onto ACs was predominantly governed by interaction with carboxylic groups (Durimel et al., 2013). The adsorption efficiency of AC toward CLD was improved by in situ immobilized iron oxyhydroxide/iron oxide NPs generated on its surface, which was in the form of goethite (α-FeOOH) and hematite (α-Fe_2O_3). The calculated average monolayer capacity obtained for this NC was 79.8 mg/mg, which is 64% higher than that estimated for unmodified AC; also, the adsorption isotherms showed the best fit with the Langmuir model (Rana et al., 2017).

The iron oxide NP-doped ordered mesoporous carbon composite functionalized with carboxylate groups showing a high degree of order, a large SSA and pore volume, and good magnetic properties was found to exhibit rapid adsorption of 2,4-D from wastewater. The adsorption capacity reached 97% of the maximum within the first 5 min. The increase of the initial concentration of 2,4-D from 100 to 500 mg/L resulted in the increase of equilibrated adsorption capacity from 99.38 to 310.78 mg/g within 120 min. While enhancements in the initial 2,4-D concentration and contact time improved the adsorption capacity, increasing the pH and temperature resulted in reduced 2,4-D uptake (Tang et al., 2015). Also, granular activated carbon with an effective size of 1.5 mm modified with acid was reported as an appropriate adsorbent to reduce 2,4-D content in herbicide-contaminated waters; a pH of 3 and a contact time of 60 min were the optimal conditions (Dehghani et al., 2014).

An investigation of the remediation of lake sediment contaminated with organochlorine pesticide (hexachlorocyclohexane and DDT) using AC and CNTs (2 wt% of carbon-based material) for treatment of sediment or semipermeable membrane devices showed that AC exhibited pronouncedly higher remediation efficiency toward these pesticides in sediment than MWCNTs. This could be connected with a greater SSA of AC that is advantageous for binding contaminants, and thus AC could be considered as a suitable in situ sorbent amendment for sediment remediation. Moreover, the pesticide removal efficiencies could be improved by using increasing doses of AC and MWCNTs (Hua et al., 2017). By pyrolysis at 850°C of a composite made of commercially available synthetic clay Laponite and Cassava starch, followed by clay removal through acid digestion Pinto et al. (2016) prepared a porous carbonaceous material showing a partial graphitic structure containing predominantly sp^2 bounded C atoms, nearly double SSA, and up to 20-fold larger mesopore volume than the carbonaceous material obtained without the clay. This mesoporous carbonaceous material was found to effectively remove the organochlorine pesticide Dicamba from aqueous solution showing a maximum adsorption capacity of 251.9 mg/g.

The application of walnut (*Juglans regia*)-derived AC for the removal of adsorbed ethaboxam from soils containing this fungicide at concentrations of 5 and 7.5 ppm resulted in 99% and 89% ethaboxam removal, respectively (Ahmad, 2018a). It was found to also be effective for remediation of alachlor-contaminated soils, causing 84% and 72% alachlor removal at 5 and 7.5 ppm herbicide soil concentrations, respectively (Ahmad, 2018b). Charcoal-powdered AC was able to remove about 77% of the ethylene thiourea and 76% of the 1*H*-1,2,4-triazole from water, and the adsorption kinetics followed a pseudo second-order model (Amorim et al., 2013).

The γ-cyclodextrin (γ-CD)-functionalized AC showed a lower surface area and pore volume than AC alone (321 vs 637 m^2/g and 0.43 vs 0.69 cm^3/g, respectively), but a higher pore diameter (5.40 vs 4.33 nm). An enhancement of CLD removal from water when compared to AC alone could be achieved by a simple filtration of the insoluble inclusion complex of CLD with γ-CD from water with AC filters. Another method is by using a hybrid material produced by modification of γ-CD to form a bond/link to the surface of AC, whereby this γ-CD-grafted AC hybrid material was found to be an even better adsorbent for CLD from water (Rana et al., 2016).

The comparison of alpha-cypermethrin adsorption on carbon aerogel and xerogel prepared by polycondensation of resorcinol and formaldehyde in Na_2CO_3 catalysis, followed by a drying step in CO_2 supercritical or normal conditions and then a pyrolytic step, which was tested in batch conditions under magnetic stirring confirmed the higher removal of pyrethroid insecticide by carbon aerogel at optimized parameters (0.01 g adsorbent, $d < 0.025$ mm, and 45°C). The adsorption was an endothermic process, experimental data fitted well with the Langmuir isotherm, and the kinetics of the process can be described by the pseudo-second-order model (Cotet et al., 2013).

3.3 Fullerene

Srivastava et al. (2011) evaluated the interaction of fullerene with lindane and its role in the remediation of lindane from contaminated soil and water and compared the interaction of fullerene with lindane and trichloroethylene (TCE). They found that fullerene can be used as a potential NM for remediation of lindane due to stronger molecule-surface bonding of the fullerene-lindane complex than the fullerene-TCE complex.

Hasanzadeh et al. (2018) reported on photo-assisted electrochemical abatement of trifluralin using a cathode containing a C_{60}-CNT composite showing great efficiencies for electro-generating H_2O_2. This owes to the huge surface area and high conductivity, which was able to efficiently degrade trifluralin (20 mg/L) under its original pH, an Na_2SO_4 electrolyte concentration of 0.05 mol/L, an applied current intensity of 300 mA, and a flow rate of 12.5 L/h, resulting in 92.8% mineralization of the herbicide after 8 h.

The commercial TiO_2 Hombikat (TiO_2) was surface-modified using fullerene C_{60} NPs with tetrahydrofuran (THF-nC_{60}) as well as fullerenol $C_{60}(OH)_{24}$ NPs (FNPs). It was found that the NC prepared using TiO_2/FNPs increased the negative charge as well as the catalytic surface of TiO_2. The addition of electron acceptors H_2O_2 and $KBrO_3$ contributed to a higher mesotrione degradation efficiency compared to O_2 alone, with the most efficient system for mesotrione degradation and mineralization being TiO_2/7 mM $KBrO_3$ and TiO_2/7 mM $KBrO_3$/40 μL FNP, respectively. The researchers supposed that the substrate degradation in the first period takes place via $^{\bullet}OH$ and after 60 min of irradiation, the reaction mechanism proceeds mainly via holes (Djordjevic et al., 2018).

3.4 Graphene, graphene oxide and reduced graphene oxide

Xu et al. (2013) overviewed the preparation methods of graphene (GR) adsorbents and their applications in water purification as well as the adsorption of contaminants on GR, its influence by functional groups, and the interactions between GR and contaminants. Graphene was reported to exhibit more efficient removal of organochlorine (aldrin, dieldrin, heptachlor, heptachlor epoxide, methoxychlor) and organofluorine (trifluralin) pesticides than the GR-Ag composite but the latter was more efficient for remediation of water contaminated with different kinds of bacteria (El-Maali et al., 2018). Cost-effective few-layered GR nanosheets used as adsorbents for chlorophenoxyacetic acid herbicides from aqueous solutions showed the maximum removal efficiencies of 82%, 80%, 86%, 82%, and 70% for 2,4-D, 2,4,5-trichlorophenoxyacetic acid (2,4,5-T), 2-chloro-2-methylphenoxyacetic acid (MCPA) and 2-(2-methyl-4-chlorophenoxy) propionic acid (MCPP), respectively. The adsorption followed a pseudo second-order kinetics model and confirmed the Langmuir adsorption isotherm (K_a ranging from 1.327 L/g for 2,4-D to 23.971 L/g for 2,4,5-T), the process being endothermic and spontaneous (Kamaraj et al., 2017).

An investigation of GR quantum dots (QDs) as an adsorbent for the removal of the harmful pesticide oxamyl from aqueous solutions showed that maximum adsorption was observed at pH 8 (91.10 and 70.12 mg/g at an initial oxamyl concentration of 50 and 150 mg/L, respectively) at an optimized adsorbent dose of 0.6 g and a contact time <25 min. The experimental data fitted the Langmuir isotherm, and the pseudo second-order kinetic model, ionic interactions (chemical interactions) and electrostatic interactions (physical interactions) were predominant in the mechanism of adsorption. The temperature increase from 20 to 40°C led to the improvement of removal efficiency, which could be connected with the weakening of H-bonds and van der Waals interactions at higher temperatures, leading to the strengthening of physical interaction between the active sites of oxamyl and grapheme QDs (Agarwal et al., 2016).

A graphene sand composite prepared using a green-synthesized graphene material from cane sugar that was immobilized on sand without the need of any binder exhibited an adsorption capacity of 48 mg/g for chlorpyrifos, exceeding that of AC (Sen Gupta et al., 2012).

GR-coated cotton fibers as a sorbent for the extraction of multiclass pesticide residues from water on which the pesticides remain stable under certain storage conditions were recommended as suitable for environmental field studies, for example, the determination of pesticides in lake water samples (Montesinos et al., 2015).

Three-dimensional (3D) porous graphene hydrogels integrating the advantageous properties of both GR (e.g., fine chemical resistance and excellent mechanical property) and hydrogel (e.g., high water content, porous structure, and simple solution-based processability and scalability) showed excellent adsorption capacity for adsorbing and removing different environmental pollutants compared to conventional graphene aerogel, suggesting their promising potential to be used in environmental applications (Ma et al., 2017). Three-dimensional carboxymethyl chitosan-GR aerogels showing a high surface area and abundant hydroxyl and carboxyl groups were able to easily and rapidly extract glyphosate from the water. At the adsorption equilibrium, they reached the maximum adsorption capacity of 578.0 mg/g, being more effective in solutions with low pH. They could be regenerated for repeated use without pronounced loss of capacity, whereby the adsorbed glyphosate was effectively desorbed into the solution (Ding et al., 2018).

The GR-coated SiO_2 prepared by mixing exfoliated GO with acid-treated SiO_2 and reducing it with hydrazine hydrate showed higher levels of adsorption for the organophosphorus pesticides in water than graphite carbons, activated carbon, pure graphene, C18 silica, and silica sorbents. The adsorption mechanism was found to depend on the electron-donating abilities of the S, P, and N atoms and the strong π-bonding network of the benzene rings (Liu et al., 2013a).

A ZnO nanorod incorporated carboxylic GR/polyaniline (ZnR@CGR/PANI) composite showed the maximum adsorption of diuron at pH 4.0, reaching the maximum adsorption capacity of 13.40 mg/g, whereby the adsorption capacity showed an increase with increasing temperature, suggesting the endothermic nature of adsorption. The photocatalytic degradation of diuron under visible light irradiation using a ZnR@CGR/PANI NC photocatalyst followed first-order kinetics, with the optimum pH for degradation being 3.0. After regeneration (up to five cycles), the photocatalyst exhibited relative stability (Anirudhan et al., 2018). Magnetically recoverable Fe_3O_4/GR NC tested for the adsorption of five pesticides (ametryn, prometryn, simazine, simeton, and atrazine) in an aqueous medium showed the maximum adsorption capacity for ametryn of 54.8 mg/g at pH 5, which was enhanced in the presence of different ions (Mg^{2+}, Ca^{2+}, Na^{+}, and SO_4^{2-}), whereby in a seawater medium it reached 63.7 mg/g. The 93.61% adsorption of pesticides was connected with electrostatic, hydrophobic, and π-π interactions of the composite toward the heterocyclic conjugation of pesticide molecules. The advantages of this composite is that it could be easily and rapidly separated from an aqueous medium using an external magnet for reuse with a 88.66% adsorption efficiency up to seven cycles (Boruah et al., 2017). The phenyl-modified magnetic GR/mesoporous silica composites with a hierarchical bridge-pore structure exhibiting a large surface area (446.5 m^2/g), highly ordered mesopores with uniform pore size (2.8 nm) and pore volume (0.32 cm^3/g), and higher saturation magnetization (25 emu/g) showed the excellent adsorption capacity of pesticides exceeding that of AC, MWCNTs and SWCNTs and were recommended as low-cost and efficient adsorbents for the removal of toxic pesticides, being economically and technically feasible (Wang et al., 2017).

Quantitative characterization of the nanomechanical interaction mechanism between GO and a model aromatic organic compound (AOC) methylene blue and toluidine blue with the amino group to be end-tethered to polyethylene glycol showed that the formation of the AOC/GO complex is connected predominantly with the attraction between the cationic AOC molecule and the regions on GO with epoxy groups with an activation Gibbs energy for bond dissociation (ΔG) of −4.61 kcal/mol (Zhang et al., 2018b).

A fiber fabricated by mixing GO and OH-terminated polydimethylsiloxane coating the surface of etched stainless-steel wire could be used after aging by heating for direct solid phase microextraction of triazole fungicides from water samples with a high extraction efficiency (Li et al., 2016). GO-based composite ultrafiltration membranes fabricated on macroporous ceramic support tubes using chitosan (CS) as an intermediate matrix between the substrate and GO coating showed enhanced atrazine removal of >95% estimated in the side-stream membrane bioreactor treatment at the optimized conditions (Mukherjee et al., 2018). An AC derived from sieve-like cellulose/GO composites tested for the removal of several organophosphorus pesticides exhibited a higher adsorption capacity than MWCNTs, graphitized carbon black, AC, C18, and the primary secondary amine adsorbent, showing the maximum adsorption capacity for chlorpyrifos of 152.5 mg/g. The adsorption mechanism was dependent on the electron-donating abilities of the S and P atoms and this composite maintained an adsorption efficiency >80% after eight times of recycling (Suo et al., 2018).

A GO-mesoporous birnessite-type δ-MnO_2 NC was reported to be effective in the decomposition of parathion-methyl by cleavage of the P-O-aryl bond in the pesticide molecule (Tolasz et al., 2016). A metal organic framework/GO hybrid NC (UiO-67/GO), in which the abundant Zr-OH groups on the surface of GO as well as the large surface of available GO enabled excellent adsorption capacity toward glyphosate reaching approx. 2.855 mmol (482.69 mg/g) at pH 4, was recommended as a promising adsorbent for the efficient removal of organophosphorus pesticides from contaminated water (Yang et al., 2017).

At investigation of diazinon removal from aqueous solution by application of an ozone/magnetic GO hybrid pronounced acceleration of the organophosphorus pesticide reduction was estimated, which could be contributed to more effective generation of $^{\bullet}OH$ radicals mainly responsible for degradation of diazinon compared to sole ozonation and thus, the ozone/magnetic GO hybrid system could be considered to effectively remove harmful organic contaminants from aqueous solutions (Arfaeinia et al., 2018). In an investigation of the removal of endrin and dieldrin isomeric pesticides through stereoselective adsorption behavior on the GO-magnetic Fe_3O_4 NC, the higher removal percentage (86.0%) and higher slope value of the Langmuir adsorption isotherm were estimated for endrin compared to dieldrin (74.0%). This is connected with the endo-position of the oxygen atom in the endrine molecule, which is more beneficial for interaction with the NC than the exo-position of the oxygen present in dieldrin (Shrivas et al., 2017). A novel adsorbent composed of silica-coated magnetic microparticles (Fe_3O_4@SiO_2) and GO functionalized with phenylethylamine (PEA) characterized with the combined advantages of the excellent adsorption capability of modified GO and the magnetic separability of magnetite microparticles enabling both high adsorption capacity and easy isolation from sample solutions was found to be a superb nanoadsorbent for the extraction and enrichment of organophosphorus pesticides from fruit, vegetable, and water samples. In this adsorbent, the phenyl rings of PEA grafted on the GO nanosheets were accessible to interact effectively with organophosphorus pesticides via a delocalized π-electron system (Mahpishanian et al., 2015).

The magnetic rGO (rGO/Fe_3O_4) exhibited a considerable glyphosate adsorption capacity, mainly at the acid condition. The adsorption process was spontaneous, endothermic, and feasible and controlled by chemical adsorption and monolayer adsorption, suggesting that rGO/Fe_3O_4 could be considered a favorable adsorbent candidate for the treatment of glyphosate-contaminated water. A positive effect on herbicide adsorption was observed at high temperatures, and the H-bond interaction, electrostatic interaction, and coordination were reported to play a decisive role in the adsorption mechanism (Li et al., 2018). A hybrid composite of rGO decorated with $MnFe_2O_4$ prepared by immobilizing the $MnFe_2O_4$ microspheres on the GR nanosheets designed by Yamaguchi et al. (2016) was tested for glyphosate removal from water. It was found that the composite of single-layer rGO with $MnFe_2O_4$ microspheres showed adsorption properties for efficient removal of glyphosate from contaminated water. Also, the adsorption of herbicide onto the hybrid NC was spontaneous, exothermic, and feasible in the range of 5–45°C while the experimental data fitted the Freundlich model and the pseudo second-order kinetic model.

Enhanced photocatalytic degradation of 2,4-dichlorophenol (2,4-DCP) under infrared light irradiation was estimated using $Cu_2(OH)PO_4$/rGO NC compared to pure $Cu_2(OH)_4$; it was stable even after repeated cycling. With the introduction of H_2O_2, the photocatalytic and photo-Fenton effects were combined into one reaction system resulting in maximizing the infrared light photocatalytic efficiency, which was reflected at a 6.25-fold higher rate constant of $Cu_2(OH)PO_4$/rGO + H_2O_2 compared to $Cu_2(OH)PO_4$/rGO, and was even almost 10-fold greater than that of pure $Cu_2(OH)PO_4$ (Zhang et al., 2018c).

Rapid dehalogenation of pesticides and chlorinated and fluorinated alkyl halides at the interface of rGO-Ag NC showing high adsorption capacity (1534 mg/g) was observed. The dehalogenation of the target pollutants by AgNPs was followed by adsorption of the degraded compounds onto rGO. The adsorption kinetics could be described by a pseudo second-order rate equation, irrespective of tested pollutants (Koushik et al., 2016). Sen Gupta et al. (2015) described simultaneous dehalogenation and removal of persistent pesticide lindane from water using rGO@Ag NCs. Lindane adsorption was found to be physical in nature, but that of different isomers of trichlorobenzenes formed by its conversion was through π-π interactions. The conversion of lindane to trichlorobenzenes was observed only with the composite and did not occur on rGO and AgNPs separately.

Jo and Selvam (2017) designed a novel ternary hybrid photocatalyst (CdS/rGO/graphitic carbon nitride (g-C_3N_4)) enabling efficient H_2 generation and degradation of atrazine through Z-scheme electron transport due to improved charge separation and photostability. In this hybrid photocatalyst, the rGO as the electron mediator played a crucial role in accelerating electron transfer at the CdS/g-C_3N_4 interface via the Z-scheme electron transfer pathway.

The estimated adsorption capacities of chlorpyrifos, endosulfan, and malathion on GO and rGO from water reached values of about 1200, 1100, and 800 mg/g, respectively. Adsorption was found to be insensitive to pH or background ions. The adsorbent was reusable and the adsorption was mediated through water while direct interactions between graphene and the pesticides were found to be rather weak or unlikely (Maliyekkal et al., 2013).

3.5 Carbon nanotubes

An overview of the progress in CNT application at lindane monitoring and wastewater treatment for the detection and removal of contaminants and their use in energy conversion and storage was presented by Tan et al. (2012). A large surface area, a high aspect ratio, greater chemical reactivity, lower cost, less energy, less chemical mass, and a smaller impact on the environment predestine CNTs to be used successfully in water purification technologies such as adsorption, hybrid catalysis, desalination, and disinfection as well as the sensing and monitoring of organic, inorganic, and biological water pollutants (Das et al., 2014). In addition, in the review paper, Rawtani et al. (2018) focused on nanotechnology-based recent approaches for sensing and remediation of pesticides while attention was also devoted to CNTs.

An investigation of the CNT effects on atrazine biodegradation by *Arthrobacter* sp. showed that they enhanced or inhibited biodegradation through a balance of two effects: the toxic effects on microbial activity and the effects changing bioavailability resulting from adsorption and desorption. While at a low CNT concentration (25 mg/L) an increased biodegradation rate by up to 20% was observed due to enhanced bacterial growth and the overexpression of degradation genes, it decreased by up to 50% at a dose of 100 mg/L. It could be mentioned that the adsorption did not impact the biodegradation efficiency, and at CNT concentrations >25 mg/L, the preadsorbed atrazine on the CNTs was fully biodegraded (Zhang et al., 2015).

CNTs synthesized from waste polyethylene bottles used as an adsorbent for the removal of diuron from aqueous solution showed an adsorption capacity of approximately 40.37 mg/g at 303 K. The best fit with the experimental data showed that the pseudo second-order kinetic model of the adsorption process was exothermic (Deokar et al., 2017). Schwab et al. (2013) described that the toxic effect of diuron on *Chlorella vulgaris* algae was equal to or stronger in the presence of CNTs by a factor of up to 5, with the most significant increase in toxicity being estimated after a 24 h exposure of algae and CNTs. The progress in the application of CNT adsorbents and their composites for the removal of toxic organic pollutants from contaminated water was presented by Yu et al. (2014), who also analyzed the removal efficiencies, adsorption mechanisms, thermodynamics, and reaction kinetics.

Using CNTs impregnated with Ag loading of 10 wt% via a wet chemistry technique and then compacted at 200 MPa and sintered at 800°C for 3 h, a compact disk (membrane) was prepared showing high water permeate flux and strong antibacterial properties and being able to remove/kill 100% of the bacteria tested, suggesting that the use of such membranes could be beneficial for the removal of different contaminants from water in a continuous filtration system (Ihsanullah et al., 2015). Barbosa et al. (2017) reported on the application of versatile magnetic CNTs prepared by chemical deposition vapor for sampling and preconcentration of pesticides in environmental water samples, providing great dispersion at any water matrix and showing high efficiency of pesticide sorption.

Born-Oppenheimer molecular dynamics simulations performed by Bandyopadhyay et al. (2017) showed that transition metal atom-incorporated nitrogen-doped SWCNTs were able to efficiently trap the harmful pesticide molecules at room temperature due to the presence of strong η^6 bonding between the aromatic group of the pesticide and the transition metal adatom of the defective SWCNT. Ateia et al. (2017) refined the as received MWCNTs by applying a permanent magnet to a suspension of CNTs and separated the high-magnetic fraction with metallic nanoclusters of Fe and/or Ni located in the interior cavities of the nanotubes, which were responsible for their ferromagnetic character. The saturation magnetizations of such MWCNTs were up to 10-fold higher than that of the as-received materials, enabling their collection by magnetic attraction. The MWCNTs also showed high adsorption capacities (>40 mg atrazine/g), and good regeneration by ethanol.

The evaluation of SWCNTs used as a sorbent of 2,4-D from aqueous solutions showed a maximum adsorption capacity of 979.6 mg/g at pH 5, a contact time of 45 min, an initial concentration of 5000 μg/L, and 23 ± 2°C, whereby 97.96% of 2,4-D was removed, suggesting that SWCNTs could be considered effective adsorbents in the removal of 2,4-D from aqueous environments (Bazrafshan et al., 2013). Engel and Chefetz (2016) investigated the effects of solution chemistry on the adsorption and desorption of dissolved organic matter (DOM) by SWCNTs and found that the attachment of DOM molecules to the SWCNTs controlled the DOM adsorption and the solution parameters affected the initial adsorption rate. Limited adsorption of DOM at high ionic strength was supposed to be connected with alterations in SWCNT bundling, whereby elevated ionic strength and increased adsorbed amount of DOM reduced the oxidation temperature of the SWCNTs and DOM-coated SWCNTs at increased ionic strength provided fewer sites for atrazine adsorption. At high pH, the adsorbed DOM amount decreased while at low pH, an additional DOM adsorption was observed.

The investigation of sorption behaviors of the pesticides diuron, fluridone, and norflurazon by a SWCNT and three types of MWCNTs, including MWCNT10 (outer diameter <10 nm), MWCNT20 (10–20 nm), and MWCNT40 (20–40 nm), showed the linear relationships between the organic carbon-normalized saturated adsorption capacity and surface area (SA). This suggested that SA was likely responsible for the adsorption of diuron and norflurazon on CNTs and the adsorption of tested herbicides was primarily controlled by the hydrophobic interaction and hydrogen bonding. MWCNT10 application was recommended as an adsorbent suitable to reduce herbicide mobility in agricultural and environmental applications (Sun et al., 2012). SWCNTs and MWCNTs were used at concentrations of 0.058, 0.145, and 0.29 wt% to manage residual dichlorodiphenyltrichloroethanes (DDTs) and hexachlorocyclohexanes (HCHs) in Dong-ting Lake sediment. It was found that the 4,4′-DDT was desorbed less readily than its metabolites, and similarly, α-HCH was desorbed less easily than other HCH isomers from sediment. However, CNTs sprinkled on the surface of sediment as a cap and injected into the sediment as a mixture represented effective ways to prevent the release of DDTs and HCHs from sediment, suggesting their potential use as sorbents in in situ remediation. However, their effectiveness depended on the CNT type and dose as well as the sediment-sorbent contact time (Zhang et al., 2017a). Among four carbon-based nanoscale sorbents tested for removal of 4-chloro-2-methylphenoxyacetic acid from water, the SWCNTs were found to be more efficient that MWCNTs with outer diameters of 15, 30, and 50 nm. The herbicide was bound to the CNTs by a combination of electron donor-acceptor (π-π) interactions and hydrogen bonds, and the herbicide could be totally removed by only two sorption cycles on the SWCNT and MWCNTs (15 nm) (De Martino et al., 2012).

An investigation of commercial MWCNTs as an adsorbent to remove the insecticide triazophos showed that the adsorption kinetics belonged to the pseudo second-order, the adsorption rate constant was up to 1.7586 g/min, the maximum adsorption capacity was 42.02 mg/g, and the adsorption of insecticide onto MWCNTs was thermodynamically feasible and spontaneous (Xia et al., 2017). Chayawan (2016) analyzed the role of quantum-mechanical interactions between the electrons in the adsorption of probe compounds and biomolecules (steroid hormones and DNA bases) by MWCNTs in order to assess the risk of MWCNTs. As the major quantum-chemical factors contributing to the adsorption mean, the polarizability, electron-correlation energy, and electron-correlation contribution to the absolute electronegativity and lowest unoccupied molecular orbital (LUMO) energy were estimated. These electron correlation-based models also predict the strong adsorption of pesticides on CNTs and thus, the instantaneous interelectronic interactions could be used as a reliable predictor for MWCNT risk assessment.

MWCNTs with particle sizes of 10–40 nm and 9.0 m^2/g SA applied at a 2.0 g/L dose were able to remove 90% of fenuron pesticide from water (100.0 g/L; pH 7.0) at 25°C using 60 min contact time. The physical nature of the adsorption was estimated, and by using simulation studies, the binding energy and binding affinity of fenuron with MWCNTs was evaluated as −6.5 kcal/mol and 5.85×10^4/mol, respectively (Ali et al., 2018).

The optimized conditions estimated for the complete removal of malathion (100%) from water using MWCNTs were as follows: a malathion concentration of 6 mg/L, an initial MWCNT concentration of 0.5 g/L, and a contact time of 30 min (Dehghani et al., 2017). The MWCNTs used as solid-phase extraction sorbents for the extraction of carbamate insecticides from environmental waters showed a larger adsorption capacity than AC and could be applied to the analysis of tap and surface waters, showing mean recoveries of 92.2%–103.9% for the five tested carbamates (carbaryl, carbofuran, aminocarb, methiocarb, and zectra) (El Atrache et al., 2016). MWCNTs attenuated paraquat toxicity in *Arabidopsis thaliana* due to herbicide adsorption on MWCNTs. The beneficial effect of MWCNTs were reflected in improved photosynthesis, antioxidant protection, and a higher number and area of lateral roots, resulting in better nutrient assimilation (Fan et al., 2018). Using 4-nitrophenol (a hydrolysis product of insecticide paraoxon) as a template, 4-vinyl pyridine as the functional monomer, and divinylbenzene as the cross-linker, molecularly imprinted polymers grafted on the surface of MWCNTs were prepared to show superb adsorption ability compared to the MWCNT-based nonimprinted polymers. The adsorption of 4-nitrophenol contributed to the increased hydrolytic rate of paraoxon (an acetylcholinesterase inhibitor being the active metabolite of the insecticide parathion), suggesting the catalyzing effect of this NM on the hydrolysis of paraoxon, which was able to eliminate the harmful 4-nitrophenol as well (Chi et al., 2012). The maximum adsorption ability of vinyl-functionalized, MWCNT-based,molecular-imprinted polymers to adsorb chlorpyrifos from aqueous solutions evaluated using the Langmuir isotherm model was found to be 38.14 mg/g at 30°C (pH 7, saturation in 3 h); the kinetic followed a pseudo second-order model, indicating chemisorption and monolayer coverage of chlorpyrifos on the surface of the tested carbon-based sorbent (Anirudhan and Alexander, 2013).

Cross-linked NC films of polyvinyl alcohol (PVA) incorporated with functionalized MWCNTs showed the increase in capturing pesticides (diazinon, chlorpyrifos, pirimiphos-methyl, malathion) along with an increase of MWCNT concentration. The adsorption efficacy depended on the surface properties and porosity of the functionalized MWCNTs/PVA NCs and on the chemical contributions and geometry of organochlorine pesticides. A considerable antimicrobial activity of the functionalized MWCNTs/PVA NC films was estimated as well and was reflected in inhibition zones of approx. 30 mm, 28 mm, 28 mm, and 26 mm for *Staphylococcus aureus*, *Pseudomonas aeruginosa*, *Candidia albicans*, and *Aspargillus niger*, respectively, when the concentration of functionalized MWCNTs was 10% based on the weight of PVA. These findings suggest that such NC films could be used as wastewater treatment (Youssef et al., 2019).

MWCNT functionalization carried out by acidic oxidation, apart from increasing the wettability of the material, resulted in a better adsorption performance against seven triazine herbicides, which in spite of their common molecular skeleton, significantly differed each from other in terms of both solubilities in water and hydrophilicity/hydrophobicity. The prolongation of functionalization time (17–60 h) progressively enhanced the extraction of all tested pesticides, also producing a moderate increment of selectivity (D'Archivio et al., 2018).

The effect of pristine and nitrogen-doped carbon nanofibers fabricated both in powder (tested in a semibatch ozonation experiment) and in structured forms (used in continuous ozonation experiments) was investigated with respect to their catalytic activity for the ozonation of organic pollutants. It was found that N-containing functionalities on the surface of the carbon nanofibers improved their catalytical effectiveness (Restivo et al., 2016). On the other hand, the presence of strongly acidic *O*-containing groups in MWCNTs decreased their catalytic activity while the introduction of *S*-containing groups (mainly sulfonic acids) enhanced the removal rate of oxalic acid and phenol used as model compounds, particularly in the catalytic wet air oxidation of phenol (Restivo et al., 2014).

Ateia et al. (2018) described the ozone-assisted regeneration of magnetic MWCNTs used for removing atrazine from water and found that the atrazine removal capacity of the MWCNTs decreased from 57.8 to 27.6 mg/g in three cycles if only O_3 treatment was used. However, it increased significantly when MWCNTs were washed with ethanol following ozone treatment, when after 10 consecutive regeneration cycles, the MWCNTs retained 85%–93% of their original adsorption affinity. The researchers found that ozonation and/or washing with ethanol did not affect the π-π interactions of MWCNTs; however the degradation byproducts attached on the surface were removed. The catalytic ozonation of the herbicide metolachlor was tested using MWCNTs as photocatalysts in semibatch experiments and carbon nanofibers grown on a honeycomb cordierite monolith in continuous experiments. The carbon-based nanocatalysts improved the mineralization degree of metolachlor and decreased the toxicity of the solution subjected to ozonation. At application of the carbon-coated monolith subjected to the continuous ozonation process, improved total organic carbon from 5% to 35% was estimated (Restivo et al., 2012).

In the absence of algae, diuron sorption to MWCNTs (pristine and oxidized) was fast, strong, and nonlinear. However, at the addition of algae to equilibrated diuron-MWCNT mixtures, a 15%–20% diuron redissolution was observed, suggesting that diuron binds readily, but—particularly in the presence of algae—partially reversibly to MWCNTs, which could be considered at risk assessment (Schwab et al., 2014).

The adsorption of diuron and dichlobenil by MWCNTs in order to remove herbicides from contaminated water was closely correlated with the specific surface areas and micropore volumes of MWCNTs, whereby a rise in the oxygen content of MWCNTs with the same diameters and similar surface areas resulted in reduced adsorption of herbicides with increased adsorption of lead. Reduced adsorption of herbicides in the presence of lead could be connected with the hydration and complexation of the lead with –COOH groups, which may occupy part of the surface of MWCNTs-*O* (Chen et al., 2011). Using the carboxylic MWCNTs as an adsorbent for removing the insecticide fenvalerate from solution through batch experiments, a maximum adsorption capacity of 40.0 mg/g was estimated. The obtained experimental data could be described by pseudo first-order kinetics and fitted the Freundlich model, suggesting that sorption takes place on a heterogeneous material and this adsorbent at repeated use showed the stability and nonaggregatable attributes of MWCNTs (Naeimi et al., 2016). Carboxylated MWCNTs showed a superb catalytic effect on the degradation of triclosan, chlorofene, and dichlorophene disinfectants by a Fenton-like system Fe^{3+}/H_2O_2, whereby their structure and chemical composition did not show notable changes after the oxidative reaction process. In the presence of carboxylated MWCNTs, an enhanced reduction rate of Fe^{3+} ion to Fe^{2+} ion and the decomposition rate of H_2O_2 was observed, resulting in increased production of $^{\bullet}OH$ radicals predominantly responsible for triclosan oxidation (Peng et al., 2017).

Adsorption, dechlorination, and desorption of 2,4-DCP (used primarily as an intermediate in the preparation of the 2,4-D herbicide) investigated using Pd/Fe-Fe_3O_4@MWCNT nanohybrids showed that weak acidity (pH 5.0 and 6.5) supported the 2,4-DCP removal; the removal of 2,4-DCP in five consecutive runs ranged from 89.6% to 99.9% and the final product (phenol) was steadily desorbed to the liquid phase, leading to the recovery of active sites on the nanohybrids securing longer activity (Xu et al., 2016). The adsorption capacity of the covalently functionalized Fe_3O_4-coated SiO_2-decorated MWCNT (Fe_3O_4@SiO_2-MWCNT) core-shell magnetic microspheres tested for pentachlorophenol (PCP) adsorption from solution significantly depended on pH, reaching the maximum adsorption for PCP at pH 2.5 (96.4 mg/g). The equilibrium was achieved after 30 min, the adsorption kinetics could be described by the pseudo-second-order kinetic model, the experimental data fitted well with the Freundlich isotherm model, and the adsorption of pesticide onto functionalized MWCNTs within the temperature range of 15–45°C was spontaneous, endothermic, and random (Zhou et al., 2014).

An investigation of the removal of diquat dibromide herbicide from water by adsorption on oxidized MWCNTs (OMWCNTs) with the nonmagnetic property as well as magnetic NCs OMWCNT-Fe_3O_4 and OMWCNT-κ-carrageenan-Fe_3O_4 showed that a maximum adsorption capacity of OMWCNT at 25°C was approx. three- and fivefold higher than that of OMWCNT-Fe_3O_4 and OMWCNT-κ-carrageenan-Fe_3O_4, respectively, and could be regenerated and reused at least fivefold times. Herbicide adsorption onto MWCNT-based adsorbents was a spontaneous and endothermic process and they could be considered promising adsorbent materials for the efficient removal of herbicides from wastewaters (Duman et al., 2019).

Application of the functional absorbent metal-organic framework ZIF-8/MWCNTs, in which zeolitic imidazolate framework (ZIF-8) particles are deposited on the surface of the magnetic MWCNTs via coordination-polymerization strategy, used to remove the eight organophosphorus pesticides (triazophos, diazinon, phosalone, profenofos, methidathion, ethoprop, sulfotep, and isazofos) from environmental water and soil samples showed high static adsorption capacities for these pesticides due to its high SSA (127.95 m^2/g) and porous structure (pore volume of 0.564 mL/g). The data of static adsorption fitted the Freundlich adsorption model better than the Langmuir model while the maximum adsorption capacity of 3.89 mg/g was estimated for profenofos. Also, as a possible mechanism, valence-electron-driven adsorption by sharing or exchanging of electrons between the pesticide molecules and the vacant active sites of the hybrid adsorbent was assumed. This functional hybrid showing superparamagnetic properties used as an adsorbent was able to remove the tested organophosphorus pesticides from spiked tap water, underground water, and soil samples in $<$20 min (Liu et al., 2018b).

Investigation of in situ sequestrations of hexachlorocyclohexanes (HCHs) and DDT in sediments using MWCNTs as adsorbents showed that better sequestration effectiveness was observed with increasing dose, decreased MWCNT diameter, and prolonged contact time. The reduction of tested pesticides in aqueous equilibrium concentrations, semipermeable membrane device uptake, and quiescent flux to overlying water at MWCNT treatment exceeded that of control. Thus, a reduction of the amounts of toxic pesticides released from sediments due to their reduced bioavailability results in a decline of the risks to the ecosystem and humans (Guo et al., 2017). Using batch and saturated soil column experiments, Zhang et al. (2017b) studied sorption and cotransport of chlordecone and antibiotic sulfadiazine in the presence of functionalized MWCNTs in soils. The transport behaviors of contaminants and MWCNTs were tested at environmentally relevant concentrations (0.1–10 mg/L) and they were applied in the column studies at different times. The mobile MWCNTs facilitated remobilization of previously deposited chlordecone and its cotransport into deeper soil layers, suggesting that the high affinity and entrapment of contaminants to NPs could result in facilitated contaminant transport by NPs. On the other hand, the retained (immobile) MWCNTs caused enhanced sulfadiazine deposition in the topsoil layers, which was connected with the increased adsorption capacity of the soil.

Shen et al. (2014) observed the displacement of lindane presorbed on MWCNTs by atrazine, naphthalene, and phenanthrene, and the competitive sorption effect of these chemicals with lindane on MWCNTs decreased as follows: atrazine > naphthalene > phenanthrene. Thus, strong interactions between lindane and phenanthrene resulted in the lowest competition and displacement of lindane by phenanthrene while stronger displacement of lindane by atrazine than naphthalene and higher competitive sorption of lindane with atrazine than with naphthalene could be connected with the larger molecular volume of atrazine resulting in a higher steric hindrance effect compared to naphthalene. The use of MWCNT catalysts in the ozonation of atrazine in aqueous solutions was found to modify the distribution of atrazine degradation intermediates, resulting in reduced effluent toxicity compared to noncatalytic ozonation and an increase in the overall mineralization of atrazine. On the other hand, in the presence of MWCNTs, a partially reduced overall rate of atrazine conversion was observed, the apparent rate being 0.143/min compared to 0.233/min without the catalyst (Fan et al., 2014).

Using the oxidized treatment of as-prepared MWCNTs increased the surface area and pore volume, resulting in the enhanced adsorption of diuron. The adsorption of diuron herbicide was pH-dependent, with higher adsorption being estimated under neutral and basic conditions and the presence of Cu^{2+} ions pronouncedly decreased its adsorption onto oxidized MWCNTs (Deng et al., 2012).

CS/MWCNTs adsorbent containing 2.5% of MWCNTs to improve CS mechanical properties removed 82.5% of the diazinon from aqueous solution with an initial pesticide concentration 5 mg/L when used at a dose of 0.5 g/L, a contact time of 60 min, and pH 5.5. The maximum adsorption capacity of CS/MWCNT adsorbent for diazinon was 222.86 mg/g, the equilibrium adsorption isotherm data exhibited the best fit to the Sips isotherm model, and the experimental adsorption data fitted well to the pseudo second-order model. Due to the excellent performance of the CS/MWCNT adsorbent in adsorption and its simple and effective regeneration, it could be used for the removal of pesticides from contaminated streams (Firozjaee et al., 2017).

As the dominant mechanism for thiamethoxam molecules attaching to pristine and nitric acid-modified MWCNTs, aromatic π-π interactions (electron donor-acceptor interactions) were estimated while the adsorption was found to be spontaneous and of an exothermic nature. The surface modification of MWCNTs resulted in an increase of both the SSA (from 135 to 241 m^2/g) and the pore volume (from 0.6 to 0.8 cm^3/g) and was connected with an increase in the number of active sites and their accessibility to the adsorbate. However, the adsorption process most probably took place on the outer surface of the MWCNTs. After the oxidative treatment, the interstitial pores pronouncedly contributed to increases in SSA and pore volume and thus, the outer surface of the MWCNTs became more available to this neonicotinoid insecticide (Panic et al., 2017).

Yang et al. (2018) described the nonreductive dechlorination pathway (dehydrochlorination) of lindane in the presence of nitrogen-containing MWCNTs under environmentally relevant conditions, in which the amine and pyridinic nitrogen species were inferred as the main basic sites that were responsible for the dehydrochlorination of lindane because they could attack the hydrogen atom attached to the β carbon, resulting in lindane dehydrochlorination by a β-elimination mechanism. As the intermediate and final products, γ-1,3,4,5,6-pentachloro-cyclohexene and trichlorobenzene isomers were estimated.

3.6 Carbon-doped TiO_2 nanocomposites

Recent advances in the photocatalytic treatment of pollutants, including pesticides, in aqueous media introduced in order to upgrade photocatalytic performance with the modification of prototypical photocatalysts such as doping, dye sensitization, semiconductor coupling, mesoporous supports, single site, and nanobased catalysts were summarized by Reddy et al. (2016).

Carbon dot-decorated TiO_2 nanotube arrays were characterized with highly enhanced abilities from the aspect of photo/voltage-induced organic pollutant degradation and bacterial inactivation at the application of UV irradiation (365 nm) or an electrochemical potential >3 V (vs Ag/AgCl). This was connected with the efficient absorption of light by carbon dots, leading to multiple exciton generations and a reduction in the recombination of the e^-/h^+ pair produced in carbon dot/TiO_2 nanotube array film during photo/voltage-induced degradation. Moreover, the surface could be readily regenerated by UV light irradiation without damaging the structure of the whole film (Feng et al., 2016). TiO_2/GO NCs could be used as photocatalysts not only for the removal of emerging pollutants from water, but also in the photocatalytic disinfection of water contaminated with pathogenic microorganisms. Using TiO_2/GO composites under UV–Vis irradiation as the main reactive oxygen species involved in the disinfection process, H_2O_2, $^{\bullet}OH$, and $^1O_2$ were estimated while under application of visible irradiation, exclusively $^1O_2$ was observed (Cruz-Ortiz et al., 2017). Kumar and Mukherji (2018) assessed the feasibility of photocatalytic degradation of an organophosphorus pesticide dichlorvos (2,2-dichloroethenyl dimethyl phosphate) in aqueous solution using GO-TiO_2 NC under irradiation (UV and visible) with varying amounts of the photocatalyst. The photocatalysis of dichlorvos consisted of two steps, that is, adsorption of pesticide onto GO-TiO_2 NC followed by its degradation caused by the NC in the presence of an irradiation source.

Rasoulnezhad et al. (2017) reported on the visible light photocatalytic degradation of paraoxon and parathion pesticides on carbon-doped TiO_2 nanorod thin films. The pronouncedly improved optical properties of TiO_2 in the visible light region resulting in the good visible light photocatalytic activity of the thin films for degradation of these pesticides was connected with the presence of the carbon element and special nanorod morphology and the carbon-doped TiO_2 thin film with a 570 nm thickness showed a maximum photonic efficiency of 0.0312%, representing a 30% improvement with respect to standard reference TiO_2 films. A catalyst based on the composite GO-TiO_2 investigated in the photodegradation of a mixture of four pesticides (diuron, alachlor, isoproturon, and atrazine) under UV–Vis irradiation showed superb initial pesticide photodegradation rates in both ultrapure or natural water matrices in contrast to a bare TiO_2 photocatalyst exhibiting decreased degradation of pesticides in natural water, and under visible light illumination conditions, its photocatalytic activity was higher than that of commercial Evonik TiO_2 P25 catalyst (Cruz et al., 2017).

Nanohybrids consisting of green synthesized highly luminescent nitrogen-containing carbon dots prepared using bio-based citric acid and glycerol in the presence of cost-free cow urine and TiO_2 (anatase) showed promising photocatalytic degradation of organic pollutants (e.g., benzene, phenol) as well as an anthrogenic pesticide under sunlight (Hazarika and Karak, 2016). Zeolite-supported carbon-doped TiO_2 composite catalysts prepared using mordenite zeolites with SiO_2/Al_2O_3 ratios of 18 and 240 caused the effective photocatalytic degradation of 18 pharmaceuticals and pesticides in demineralized water under natural and simulated solar light irradiation and an enhanced removal rate of target pollutants was also observed in surface and hospital wastewaters. Although a dispersed layer of carbon-doped TiO_2 formed on the zeolite surface blocked the micropores of zeolites and reduced their surface area, the adsorption onto the zeolites was not pronouncedly affected (An et al., 2016).

Argentinean herb mate and the stevia plant were applied as the carbogenic source to fabricate the thermally induced formation of carbon nanodots with different levels of N and P doping and tunable photoluminescence response in the UV–Vis near-infrared ranges as TiO_2 sensitizers for a visible light-driven photocatalyst showing excellent activity in the photocatalytic degradation of 2,4-DCP, a persistent organic compound occurring in various pesticide formulations (Ortega-Liebana et al., 2017).

As an appropriate photocatalyst for enhanced degradation and detoxification of the emerging water contaminants, a Ga, N, F codoped TiO_2@rGO aerogel NC was recommended that was able to perform the complete degradation of 40 mg/L atrazine in the photocatalytic oxidation process within 60 min (Khavar et al., 2018a). The visible-light activated In, S-TiO_2@rGO NC consisting of rGO and TiO_2 co-doped with In and *S* showed notably increased photocatalytic activity and caused complete degradation and 95.5% mineralization of 20 mg/L atrazine within 20 min when the NC contained 3.0 mol% of In, 1 mol% of S, and 5 wt% of rGO, with the degradation of atrazine being predominantly caused by the reaction with $^{\bullet}OH$ radicals at the surface of the catalyst. This NC preserved its catalytic activity during four consecutive recycles and notable detoxification of the herbicide was observed during the removal process (Khavar et al., 2018b).

Bi-doped TiO_2 nanotube/graphene composites containing 2 wt% bismuth were tested for degradation of aqueous suspensions of the phenolic herbicide dinoseb. This phenolic herbicide showed outstanding photocatalytic activity under a visible light source due to the synergistic effect of graphene and Bi-doped TiO_2 nanotubes, resulting in supported interfacial charge transfer and improved efficiency of light harvesting in the visible region, whereby $^{\bullet}OH$ radicals solely contributed to degradation (Alam et al., 2017).

The use of rGO and $CuInS_2$ NPs codecorated TiO_2 nanotube (NT) arrays as photocatalysts, which were fabricated by consecutive electrodeposition of $CuInS_2$ and rGO film onto the top surface of TiO_2 nanotubes, for the removal of 2,4-D by photocatalytic degradation was reported by Liu et al. (2013b). Under simulated solar light (or visible light) irradiation, the photodegradation rate toward 2,4-D over this photocatalyst achieved 96.2% (or 87.7%) compared to 68.0% (or 58.4%) observed over $CuInS_2$-TiO_2 NTs, and 36.0% (or 26.3%) over bare TiO_2 NTs, suggesting that the synergetic effect of rGO and $CuInS_2$ notably contributed to the photocatalytic performance of TiO_2 NTs.

Electrosorption-assisted photoelectrocatalytic process under visible light irradiation using nitrogen-doped TiO_2 NPs and GR nanoplatelets deposited onto a titanium grid sheet by electrophoretic deposition showed 80% removal efficiency of diazinon compared to 28.3% for photocatalytic and 23.7% for electrosorption processes solely; this catalyst exhibited good activity even after five successive cycles (Ayoubi-Feiz et al., 2018).

4 Conclusion

In order to reduce the increasing contamination of environmental matrices with pesticides, it is necessary to secure their most efficient removal by using powerful and cheap adsorbents such as carbon-based nanomaterials characterized with high surface area or nanomaterials supporting the degradation of persistent pesticides. However, the complete disappearance of pesticide contamination from the environment could only be accomplished when the adsorbent is removed. Decontamination of water with pesticide residues could be successfully achieved using filtration or both filtration anda magnetic field for NCs consisting of a carbon-based material showing excellent adsorption capability and magnetic NPs exhibiting advantageous magnetic separability. However, very effective pesticide removal could also be achieved using carbon-doped TiO_2 NPs as photonanocatlysts, which generate harmful ROS such as H_2O_2, $^{\bullet}OH$, and $^1O_2$ under UV-Vis light, thus enabling not only the removal of emerging pollutants from water but also the photocatalytic disinfection of water contaminated with pathogenic microorganisms. It is desirable to develop optimized methods related to the removal of individual pesticides/pesticide classes with a focus on the physicochemical characteristics of the applied carbon-based nanosorbent/nanocatalyst and in order for the environmental matrix to be decontaminated. Economic costs need to be considered as well, and thus, the application of highly effective but cheap carbon-based sorbents/catalysts showing multiple recycling abilities is definitely preferred. However, it should be mentioned that a significant reduction of pesticide residues in environmental matrices could be achieved primarily by avoiding the application and overuse of harmful persistent pesticides, and priority must be given to the use of pesticides characterized with rapid degradation in environmental conditions and harmless degradation products. Such an approach, also considering the ethical aspects, requires responsible decisions of farmers as well as the interest of researchers to develop environmentally friendly agrochemicals and effective methods for removal of their residues from contaminated soils and water.

Acknowledgments

This study was supported by the Slovak Research and Development Agency (APVV-17-0373 and APVV-17-0318), and by the Ministry of Education of the Czech Republic (LO1305). The authors thank Biochar Industries, Biochar Projects, and Friends of the Char that placed free Hi-Res versions (4megs$^+$) of photos of biochar in the public domain (http://biocharproject.org/charmasters-log/biochar-electron-microscope-images) for the benefit of the entire human race to feel free to use/share/reproduce/disseminate or to do with them any other thing they might want.

References

Acquah, S.F.A., Penkova, A.V., Markelov, D.A., Semisalova, A.S., Leonhardt, B.E., Magi, J.M., 2017. Review—The beautiful molecule: 30 years of C_{60} and its derivatives. ECS J. Solid State Sci. Technol. 6 (6), M3155–M3162. https://doi.org/10.1149/2.0271706jss.

Agarwal, S., Sadeghi, N., Tyagi, I., Gupta, V.K., Fakhri, A., 2016. Adsorption of toxic carbamate pesticide oxamyl from liquid phase by newly synthesized and characterized graphene quantum dots nanomaterials. J. Colloid Interface Sci. 478, 430–438. https://doi.org/10.1016/j.jcis.2016.06.029.

Agnihotri, S., Mota, J.P.B., Rostam-Abadi, M., Rood, M.J., 2005. Structural characterization of single walled carbon nanotube bundles by experiment and molecular simulation. Langmuir 21 (3), 896–904. https://doi.org/10.1021/la047662c.

Ahmad, K.S., 2018a. Exploring the potential of *Juglans regia*-derived activated carbon for the removal of adsorbed fungicide Ethaboxam from soils. Environ. Monit. Assess. 190 (12), 737. https://doi.org/10.1007/s10661-018-7119-z.

Ahmad, K.S., 2018b. Sorption and *Juglans regia*-derived activated carbon-mediated removal of aniline-based herbicide Alachlor from contaminated soils. Environ. Earth Sci. 77 (12), 437. https://doi.org/10.1007/s12665-018-7633-x.

Ahmad, M., Rajapaksha, A.U., Lim, J.E., Zhang, M., Bolan, N., Mohan, D., Vithanage, M., Lee, S.S., Ok, Y.S., 2014. Biochar as a sorbent for contaminant management in soil and water: a review. Chemosphere 99, 19–33. https://doi.org/10.1016/j.chemosphere.2013.10.071.

Aktar, M.W., Sengupta, D., Chowdhury, A., 2009. Impact of pesticides use in agriculture: their benefits and hazards. Interdiscip. Toxicol. 2 (1), 1–12. https://doi.org/10.2478/v10102-009-0001-7.

Alahabadi, A., Moussavi, G., 2017. Parametric experiments and kinetics, equilibrium and thermodynamic modeling. J. Mol. Liq. 242, 40–52. https://doi.org/10.1016/j.molliq.2017.06.116.

Al-Ahmadi, M.S., 2019. Pesticides, Anthropogenic Activities, and the Health of Our Environment Safety. IntechOpen, Rijeka, Croatia1–24. https://doi.org/10.5772/intechopen.84161.

Alam, U., Fleisch, M., Kretschmer, I., Bahnemann, D., Muneer, M., 2017. One-step hydrothermal synthesis of Bi-TiO_2 nanotube/graphene composites: an efficient photocatalyst for spectacular degradation of organic pollutants under visible light irradiation. Appl. Catal. B-Environ. 218, 758–769. https://doi.org/10.1016/j.apcatb.2017.06.016.

Ali, I., Alharbi, O.M.L., Al-Othnam, Z.A., Al-Mohaimeed, A.M., Alwarthan, A., 2018. Modeling of fenuron pesticide adsorption on CNTs for mechanistic insight and removal in water. Environ. Res. 170, 389–397. https://doi.org/10.1016/j.envres.2018.12.066.

Ameta, S.C., Ameta, R., 2018. Advanced Oxidation Processes for Wastewater Treatment. Academic Press, ISBN: 9780128104996.

Amorim, C.C., Bottrel, S.E.C., Costa, E.P., Teixeira, A.P.C., Leao, M.M.D., 2013. Removal of ethylenthiourea and 1,2,4-triazole pesticide metabolites from water by adsorption in commercial activated carbons. J. Environ. Sci. Health B 48 (3), 183–190. https://doi.org/10.1080/03601234.2013.730287.

An, Y., de Ridder, D.J., Zhao, C., Schoutteten, K., Vanden Bussche, J., Zheng, H.L., Chen, G., Vanhaecke, L., 2016. Adsorption and photocatalytic degradation of pharmaceuticals and pesticides by carbon doped-TiO_2 coated on zeolites under solar light irradiation. Water Sci. Technol. 73 (12), 2868–2881. https://doi.org/10.2166/wst.2016.146.

Anirudhan, T.S., Alexander, S., 2013. Synthesis and characterization of vinyl-functionalized multiwalled carbon nanotubes based molecular imprinted polymer for the separation of chlorpyrifos from aqueous solutions. J. Chem. Technol. Biot. 88 (10), 1847–1858. https://doi.org/10.1002/jctb.4039.

Anirudhan, T.S., Shainy, F., Mohan, A.M., 2018. Fabrication of zinc oxide nanorod incorporated carboxylic graphene/polyaniline composite and its photocatalytic activity for the effective degradation of diuron from aqueous solutions. Sol. Energy 171, 534–546. https://doi.org/10.1016/j.solener.2018.06.111.

Arfaeinia, H., Rezaei, H., Sharafi, K., Moradi, M., Pasalari, H., Hashemi, S.E., 2018. Application of ozone/magnetic graphene oxide for degradation of diazinon pesticide from aqueous solutions. Desalin. Water Treat. 107, 127–135. https://doi.org/10.5004/dwt.2018.22158.

Ateia, M., Koch, C., Jelavic, S., Hirt, A., Quinson, J., Yoshimura, C., Johnson, M., 2017. Green and facile approach for enhancing the inherent magnetic properties of carbon nanotubes for water treatment applications. PLoS ONE 12 (7), e0180636. https://doi.org/10.1371/journal.pone.0180636.

Ateia, M., Ceccato, M., Budi, A., Ataman, E., Yoshimura, C., Johnson, M., 2018. Ozone-assisted regeneration of magnetic carbon nanotubes for removing organic water pollutants. Chem. Eng. J. 335, 384–391. https://doi.org/10.1016/j.cej.2017.10.166.

Ayoubi-Feiz, B., Mashhadizadeh, M.H., Sheydaei, M., 2018. Preparation of reusable nano N-TiO_2/graphene/titanium grid sheet for electrosorption-assisted visible light photoelectrocatalytic degradation of a pesticide: effect of parameters and neural network modeling. J. Electroanal. Chem. 823, 713–722. https://doi.org/10.1016/j.jelechem.2018.07.020.

Aziz, H., Murtaza, G., Usman, M., Basra, S.M.A., Niaz, A., 2018. Enhancing chlorpyrifos sorption potential of agricultural soil by biochar and compost addition. Pak. J. Agric. Sci. 55 (4), 833–841. https://doi.org/10.21162/PAKJAS/18.7463.

Bandyopadhyay, A., Ghosh, D., Pati, S.K., 2017. Trapping and sensing of hazardous insecticides by chemically modified single walled carbon nanotubes. Phys. Chem. Chem. Phys. 19 (35), 24059–24066. https://doi.org/10.1039/c7cp04980c.

Barbosa, F.H.F., Menezes, H.C., Teixeira, A.P.D., Serp, P., Antipoff, V., Cardeal, Z.D., 2017. Versatile magnetic carbon nanotubes for sampling and pre concentration of pesticides in environmental water. Talanta 167, 538–543. https://doi.org/10.1016/j.talanta.2017.02.054.

Bazrafshan, E., Mostafapour, F.K., Faridi, H., Farzadkia, M., Sargazi, S., Sohrabi, A., 2013. Removal of 2,4-dichlorophenoxyacetic acid (2,4-D) from aqueous environments using single-walled carbon nanotubes. Health Scope 2 (1), 39–46. https://doi.org/10.17795/jhealthscope-7710.

Bernardes, M.F.F., Pazin, M., Pereira, L.C., Dorta, D.J., 2015. Impact of pesticides on environmental and human health. In: Andreazza, A.C., Scola, G. (Eds.), Toxicology Studies—Cells, Drugs and Environment. IntechOpen, Rijeka, Croatia, pp. 195–233. https://doi.org/10.5772/59710.

BiocharProject, 2011. Spreading the Word About Biochar. http://biocharproject.org/charmasters-log/biochar-electron-microscope-images.

Boruah, P.K., Sharma, B., Hussain, N., Das, M.R., 2017. Magnetically recoverable Fe_3O_4/graphene nanocomposite towards efficient removal of triazine pesticides from aqueous solution: investigation of the adsorption phenomenon and specific ion effect. Chemosphere 168, 1058–1067. https://doi.org/10.1016/j.chemosphere.2016.10.103.

Capoferri, D., Della Pelle, F., Del Carlo, M., Compagnone, D., 2018. Affinity sensing strategies for the detection of pesticides in food. Foods 7 (9), E148. https://doi.org/10.3390/foods7090148.

Carvalho, F.P., 2017. Pesticides, environment, and food safety. Food Energ. Sec. 6 (2), 48–60. https://doi.org/10.1002/fes3.108.

Cha, C., Shin, S.R., Annabi, N., Dokmeci, M.R., Khademhosseini, A., 2013. Carbon-based nanomaterials: multi-functional materials for biomedical engineering. ACS Nano 7 (4), 2891–2897. https://doi.org/10.1021/nn401196a.

Chayawan, V., 2016. Quantum-mechanical parameters for the risk assessment of multi-walled carbon-nanotubes: a study using adsorption of probe compounds and its application to biomolecules. Environ. Pollut. 218, 615–624. https://doi.org/10.1016/j.envpol.2016.07.045.

Chen, G.C., Shan, X.Q., Pei, Z.G., Wang, H.H., Zheng, L.R., Zhang, J., Xie, Y.N., 2011. Adsorption of diuron and dichlobenil on multiwalled carbon nanotubes as affected by lead. J. Hazard. Mater. 188 (1–3), 156–163. https://doi.org/10.1016/j.jhazmat.2011.01.095.

Chi, W.H., Shi, H.G., Shi, W., Guo, Y., Guo, T.Y., 2012. 4-Nitrophenol surface molecularly imprinted polymers based on multiwalled carbon nanotubes for the elimination of paraoxon pollution. J. Hazard. Mater. 227, 243–249. https://doi.org/10.1016/j.jhazmat.2012.05.041.

Cotet, L.C., Maicaneanu, A., Fort, C.I., Danciu, V., 2013. Alpha-cypermethrin pesticide adsorption on carbon aerogel and xerogel. Sep. Sci. Technol. 48 (17), 2649–2658. https://doi.org/10.1080/01496395.2013.805782.

Cruz, M., Gomez, C., Duran-Valle, C.J., Pastrana-Martinez, L.M., Faria, J.L., Silva, A.M.T., Faraldos, M., Bahamonde, A., 2017. Bare TiO_2 and graphene oxide TiO_2 photocatalysts on the degradation of selected pesticides and influence of the water matrix. Appl. Surf. Sci. 416, 1013–1021. https://doi.org/10.1016/j.apsusc.2015.09.268.

Cruz-Ortiz, B.R., Hamilton, J.W.J., Pablos, C., Diaz-Jimenez, L., Cortes-Hernandez, D.A., Sharma, P.K., Castro-Alferez, M., Fernandez-Ibanez, P., Dunlop, P.S.M., Byrne, J.A., 2017. Mechanism of photocatalytic disinfection using titania-graphene composites under UV and visible irradiation. Chem. Eng. J. 316, 179–186. https://doi.org/10.1016/j.cej.2017.01.094.

Cui, H.F., Wu, W.W., Li, M.M., Song, X., Lv, Y., Zhang, T.T., 2018. A highly stable acetylcholinesterase biosensor based on chitosan-TiO_2-graphene nanocomposites for detection of organophosphate pesticides. Biosens. Bioelectron. 99, 223–229. https://doi.org/10.1016/j.bios.2017.07.068.

D'Archivio, A.A., Maggi, M.A., Odoardi, A., Santucci, S., Passacantando, M., 2018. Adsorption of triazine herbicides from aqueous solution by functionalized multiwall carbon nanotubes grown on silicon substrate. Nanotechnology 29 (6), 065701. https://doi.org/10.1088/1361-6528/aaa0a0.

Das, R., Hamid, S.B.A., Ali, M.E., Ismail, A.F., Annuar, M.S.M., Ramakrishna, S., 2014. Multifunctional carbon nanotubes in water treatment: the present, past and future. Desalination 354, 160–179. https://doi.org/10.1016/j.desal.2014.09.032.

Das, S., 2013. A review on carbon nano-tubes—a new era of nanotechnology. Int. J. Emerg. Technol. Adv. Eng. 3 (3), 774–783.

De Martino, A., Iorio, M., Xing, B.S., Capasso, R., 2012. Removal of 4-chloro-2-methylphenoxyacetic acid from water by sorption on carbon nanotubes and metal oxide nanoparticles. RSC Adv. 2 (13), 5693–5700. https://doi.org/10.1039/c2ra00038e.

De Volder, M.F., Tawfick, S.H., Baughman, R.H., Hart, A.J., 2013. Carbon nanotubes: present and future commercial applications. Science 339 (6119), 535–539. https://doi.org/10.1126/science.1222453.

Dehghani, M., Nasseri, S., Karamimanesh, M., 2014. Removal of 2,4-dichlorophenolyxacetic acid (2,4-D) herbicide in the aqueous phase using modified granular activated carbon. J. Environ. Health Sci. Eng. 12, 28. https://doi.org/10.1186/2052-336X-12-28.

Dehghani, M.H., Niasar, Z.S., Mehrnia, M.R., Shayeghi, M., Al-Ghouti, M.A., Heibati, B., McKay, G., Yetilmezsoy, K., 2017. Optimizing the removal of organophosphorus pesticide malathion from water using multi-walled carbon nanotubes. Chem. Eng. J. 310 (Pt 1), 22–32. https://doi.org/10.1016/j.cej.2016.10.057.

Deng, H., Yu, H.M., Chen, M., Ge, C.J., 2014. Sorption of atrazine in tropical soil by biochar prepared from cassava waste. Bioresources 9 (4), 6627–6643.

Deng, J., Shao, Y.S., Gao, N.Y., Deng, Y., Tan, C.Q., Zhou, S.Q., Hu, X.H., 2012. Multiwalled carbon nanotubes as adsorbents for removal of herbicide diuron from aqueous solution. Chem. Eng. J. 193, 339–347. https://doi.org/10.1016/j.cej.2012.04.051.

Deng, S., Upadhyayula, V.K.K., Smith, G.B., Mitchell, M.C., 2008. Adsorption equilibrium and kinetics of microorganisms on single walled carbon nanotubes. IEEE Sens. 8 (6), 954–962. https://doi.org/10.1109/JSEN.2008.923929.

Deokar, S.K., Bajad, G.S., Bhonde, P., Vijayakumar, R.P., Mandavgane, S.A., 2017. Adsorptive removal of diuron herbicide on carbon nanotubes synthesized from plastic waste. J. Polym. Environ. 25 (2), 165–175. https://doi.org/10.1007/s10924-016-0794-3.

des Ligneris, E., Dumee, L.F., Kong, L.X., 2018. Nanofiber-based materials for persistent organic pollutants in water remediation by adsorption. Appl. Sci.-Basel 8 (2), 166. https://doi.org/10.3390/app8020166.

Ding, C.F., Wang, X.Y., Liu, H., Li, Y., Sun, Y.L., Lin, Y.N., Sun, W.Y., Zhu, X.D., Dai, Y.X., Luo, C.N., 2018. Glyphosate removal from water by functional three-dimensional graphene aerogels. Environ. Chem. 15 (6), 325–335. https://doi.org/10.1071/EN18087.

Djordjevic, A., Merkulov, D.S., Lazarevic, M., Borisev, I., Medic, I., Pavlovic, V., Miljevic, B., Abramovic, B., 2018. Enhancement of nano titanium dioxide coatings by fullerene and polyhydroxy fullerene in the photocatalytic degradation of the herbicide mesotrione. Chemosphere 196, 145–152. https://doi.org/10.1016/j.chemosphere.2017.12.160.

Domene, X., Ramírez, W., Mattana, S., Alcañiz, J.M., Andrés, P., 2008. Ecological risk assessment of organic waste amendments using the species sensitivity distribution from a soil organisms test battery. Environ. Pollut. 155 (2), 227–236.

Duman, O., Ozcan, C., Polat, T.G., Tunc, S., 2019. Carbon nanotube-based magnetic and non-magnetic adsorbents for the high-efficiency removal of diquat dibromide herbicide from water: OMWCNT, OMWCNT-Fe_3O_4 and OMWCNT-κ-carrageenan-Fe_3O_4 nanocomposites. Environ. Pollut. 244, 723–732. https://doi.org/10.1016/j.envpol.2018.10.071.

Durimel, A., Altenor, S., Miranda-Quintana, R., Du Mesnil, P.C., Jauregui-Haza, U., Gadiou, R., Gaspard, S., 2013. pH dependence of chlordecone adsorption on activated carbons and role of adsorbent physico-chemical properties. Chem. Eng. J. 229, 239–249. https://doi.org/10.1016/j.cej.2013.03.036.

Eda, G., Chhowalla, M., 2010. Chemically derived graphene oxide: towards large-area thin-film electronics and optoelectronics. Adv. Mater. 22, 2392–2415.

El Atrache, L.L., Hachani, M., Kefi, B.B., 2016. Carbon nanotubes as solid-phase extraction sorbents for the extraction of carbamate insecticides from environmental waters. Int. J. Environ. Sci. Technol. 13 (1), 201–208. https://doi.org/10.1007/s13762-015-0861-7.

El-Maali, N.A., Nemr, A., Abd El-Wahab, M., Moharram, A., 2018. Monitoring of removal of organochlorine/fluorine pesticides from water using graphene composites by gas chromatography/mass spectrometry. J. Water Chem. Technol. 40 (4), 234–240. https://doi.org/10.3103/S1063455X18040094.

Engel, M., Chefetz, B., 2016. Removal of triazine-based pollutants from water by carbon nanotubes: impact of dissolved organic matter (DOM) and solution chemistry. Water Res. 106, 146–154. https://doi.org/10.1016/j.watres.2016.09.051.

Erickson, K., Erni, R., Lee, Z., Alem, N., Gannett, W., Zettl, A., 2010. Determination of the local chemical structure of graphene oxide and reduced graphene oxide. Adv. Mater. 22 (40), 4467–4472.

Fan, X.J., Xu, J.H., Lavoie, M., Peijnenburg, W.J.G.M., Zhu, Y.C., Lu, T., Fu, Z.W., Zhu, Z.T., Qian, H.F., 2018. Multiwall carbon nanotubes modulate paraquat toxicity in *Arabidopsis thaliana*. Environ. Pollut. 233, 633–641. https://doi.org/10.1016/j.envpol.2017.10.116.

Fan, X.L., Restivo, J., Orfao, J.J.M., Pereira, M.F.R., Lapkin, A.A., 2014. The role of multiwalled carbon nanotubes (MWCNTs) in the catalytic ozonation of atrazine. Chem. Eng. J. 241, 66–76. https://doi.org/10.1016/j.cej.2013.12.023.

FAO, 2018. The 10th Global Forum for Food and Agriculture. Food and Agriculture Organization of the United Nations. Shaping the Future of Livestock Sustainably, Responsibly, Efficiently, Berlin, January 18–20, 2018. http://www.fao.org/3/i8384en/I8384EN.pdf. (12 March 2019).

Fargnoli, M., Lombardi, M., Puri, D., Casorri, L., Masciarelli, E., Mandić-Rajčević, S., Colosio, C., 2019. The safe use of pesticides: a risk assessment procedure for the enhancement of occupational health and safety (OHS) management. Int. J. Environ. Res. Public Health 16 (3), 310. https://doi.org/10.3390/ijerph16030310.

Feng, L.Y., Sun, H.J., Ren, J.S., Qu, X.G., 2016. Carbon-dot-decorated TiO_2 nanotube arrays used for photo/voltage-induced organic pollutant degradation and the inactivation of bacteria. Nanotechnology 27 (11), 115301. https://doi.org/10.1088/0957-4484/27/11/115301.

Firozjaee, T.T., Mehrdadi, N., Baghdadi, M., Bidhendi, G.R.N., 2017. The removal of diazinon from aqueous solution by chitosan/carbon nanotube adsorbent. Desalin. Water Treat. 79, 291–300. https://doi.org/10.5004/dwt.2017.20794.

Geckeler, K.E., Samal, S., 1999. Syntheses and properties of macromolecular fullerenes, a review. Polym. Int. 48 (9), 743–757.

Gill, H.K., Garg, H., 2014. Pesticides: environmental impacts and management strategies. In: Larramendy, M.L., Soloneski, S. (Eds.), Pesticides—Toxic Aspects. IntechOpen, Rijeka, Croatia, pp. 187–230. https://doi.org/10.5772/57399.

Githinji, L., 2013. Effect of biochar application rate on soil physical and hydraulic properties of a sandy loam. Arch. Agron. Soil Sci. 60 (4), 457–470. https://doi.org/10.1080/03650340.2013.821698.

Goel, A., Howard, J.B., John, B. Vander Sande, J.B., 2004. Size analysis of single fullerene molecules by electron microscopy. Carbon 42 (10), 1907–1915. https://doi.org/10.1016/j.carbon.2004.03.022.

Gómez-Avilés, A., Peñas-Garzón, M., Bedia, J., Rodriguez, J.J., Belver, C., 2019. C-modified TiO_2 using lignin as carbon precursor for the solar photocatalytic degradation of acetaminophen. Chem. Eng. J. 358, 1574–1582.

Grant, G.A., Fisher, P.R., Barrett, J.E., Wilson, P.C., 2019. Removal of agrichemicals from water using granular activated carbon filtration. Water Air Soil Pollut. 230 (1), 7. https://doi.org/10.1007/s11270-018-4056-y.

Guo, Y.Y., Lai, C., Zeng, G.M., Gong, J.L., Su, C., Yang, C.P., Xu, P., 2017. Sequestration of HCHs and DDTs in sediments in Dongting Lake of China with multiwalled carbon nanotubes: implication for in situ sequestration. Environ. Sci. Pollut. Res. 24 (8), 7726–7739. https://doi.org/10.1007/s11356-017-8468-9.

Han, D.W., Yan, D.D., Cao, A.C., Fang, W.S., Liu, P.F., Li, Y., Ouyang, C.B., Wang, Q.X., 2017. Degradation of dimethyl disulphide in soil with or without biochar amendment. Pest Manag. Sci. 73 (9), 1830–1836. https://doi.org/10.1002/ps.4545.

Hasanzadeh, A., Khataee, A., Zarei, M., Joo, S.W., 2018. Photo-assisted electrochemical abatement of trifluralin using a cathode containing a C-60-carbon nanotubes composite. Chemosphere 199, 510–523. https://doi.org/10.1016/j.chemosphere.2018.02.061.

Hazarika, D., Karak, N., 2016. Photocatalytic degradation of organic contaminants under solar light using carbon dot/titanium dioxide nanohybrid, obtained through a facile approach. Appl. Surf. Sci. 376, 276–285. https://doi.org/10.1016/j.apsusc.2016.03.165.

Helmy, E.T., El Nemr, A., Mousa, M., Arafa, E., Eldafrawy, S., 2018. Photocatalytic degradation of organic dyes pollutants in the industrial textile wastewater by using synthesized TiO_2, C-doped TiO_2, S-doped TiO_2 and C,S co-doped TiO_2 nanoparticles. J. Water Environ. Nanotechnol. 3 (2), 116–127. https://doi.org/10.22090/jwent.2018.02.003.

Hua, S., Gong, J.L., Zeng, G.M., Yao, F.B., Guo, M., Ou, X.M., 2017. Remediation of organochlorine pesticides contaminated lake sediment using activated carbon and carbon nanotubes. Chemosphere 177, 65–76. https://doi.org/10.1016/j.chemosphere.2017.02.133.

Huang, F., Yan, A., Zhao, H., 2016. Influences of doping on photocatalytic properties of TiO_2 photocatalyst. In: Cao, W.B. (Ed.), Semiconductor Photocatalysis—Materials, Mechanisms and Applications. IntechOpen, Rijeka, Croatia, pp. 31–80. https://doi.org/10.5772/63234.

Ihsanullah, Laoui, T., Al-Amer, A.M., Khalil, A.B., Abbas, A., Khraisheh, M., Atieh, M.A., 2015. Novel anti-microbial membrane for desalination pretreatment: a silver nanoparticle-doped carbon nanotube membrane. Desalination 376, 82–93. https://doi.org/10.1016/j.desal.2015.08.017.

Ismail, A.F., Goh, P., 2018. Carbon-Based Polymer Nanocomposites for Environmental and Energy Applications. Elsevier, ISBN: 9780128135747.

Jampílek, J., Kráľová, K., 2015. Application of nanotechnology in agriculture and food industry, its prospects and risks. Ecol. Chem. Eng. S 22 (3), 321–361. https://doi.org/10.1515/eces-2015-0018.

Jampílek, J., Kráľová, K., 2017a. Nanopesticides: preparation, targeting and controlled release. In: Grumezescu, A.M. (Ed.), Nanotechnology in the Agri-Food Industry, Vol. 10—New Pesticides and Soil Sensors. Elsevier, London, UK, pp. 81–127. https://doi.org/10.1016/B978-0-12-804299-1.00004-7.

Jampílek, J., Kráľová, K., 2017b. Nanomaterials for delivery of nutrients and growth-promoting compounds to plants. In: Prasad, R., Kumar, M., Kumar, V. (Eds.), Nanotechnology: An Agricultural Paradigm. Springer, Singapore, pp. 177–226. https://doi.org/10.1007/978-981-10-4573-8_9.

Jampílek, J., Kráľová, K., 2017c. Nano-antimicrobials: activity, benefits and weaknesses. In: Ficai, A., Grumezescu, A.M. (Eds.), Nanostructures in Therapeutic Medicine, Vol. 2—Nanostructures for Antimicrobial Therapy. Elsevier, Amsterdam, Netherlands, pp. 23–54. https://doi.org/10.1016/B978-0-323-46152-8.00002-0.

Jampílek, J., Kráľová, K., 2018a. Application of nanobioformulations for controlled release and targeted biodistribution of drugs. In: Sharma, A.K., Keservani, R.K., Kesharwani, R.K. (Eds.), Nanobiomaterials: Applications in Drug Delivery. CRC Press, Warentown, NJ, USA, pp. 131–208.

Jampílek, J., Kráľová, K., 2018b. Benefits and potential risks of nanotechnology applications in crop protection. In: Abd-Elsalam, K., Prasad, R. (Eds.), Nanobiotechnology Applications in Plant Protection. Springer, Cham, Germany, pp. 189–246. https://doi.org/10.1007/978-3-319-91161-8_8.

Jampílek, J., Kráľová, K., 2018c. Nanomaterials applicable in food protection. In: Vittal, R.R., Bai, J.A. (Eds.), Nanotechnology Applications in the Food Industry. Taylor & Francis Group, Boca Raton, FL, USA, pp. 75–96.

Jampílek, J., Kráľová, K., 2019a. Nanotechnology based formulations for drug targeting to central nervous system. In: Keservani, R.K., Sharma, A.K. (Eds.), Nanoparticulate Drug Delivery Systems. Apple Academic Press & CRC Press, Warentown, NJ, USA, pp. 151–220.

Jampílek, J., Kráľová, K., 2019b. Natural biopolymeric nanoformulations for brain drug delivery. In: Keservani, R.K., Sharma, A.K., Kesharwani, R.K. (Eds.), Nanocariers for Brain Targeting: Principles and Applications. CRC Press, Warentown, NJ, USA, pp. 131–203. https://www.bokus.com/bok/9781771887304/nanocarriers-for-brain-targeting.

Jampílek, J., Záruba, K., Oravec, M., Kuneš, M., Babula, P., Ulbrich, P., Brezaniova, I., Triska, J., Suchy, P., 2015. Preparation of silica nanoparticles loaded with nootropics and their *in vivo* permeation through blood–brain barrier. Biomed. Res. Int. 2015, 812673. https://doi.org/10.1155/2015/812673.

Jampílek, J., Kos, J., Kráľová, K., 2019. Potential of nanomaterial applications in dietary supplements and foods for special medical purposes. Nanomaterials 9 (2), 296. https://doi.org/10.3390/nano9020296.

Jo, W.K., Selvam, N.C.S., 2017. Z-scheme CdS/g-C_3N_4 composites with RGO as an electron mediator for efficient photocatalytic H_2 production and pollutant degradation. Chem. Eng. J. 317, 913–924. https://doi.org/10.1016/j.cej.2017.02.129.

Kah, M., Sigmund, G., Xiao, F., Hofmann, T., 2017. Sorption of ionizable and ionic organic compounds to biochar, activated carbon and other carbonaceous materials. Water Res. 124, 673–692. https://doi.org/10.1016/j.watres.2017.07.070.

Kamaraj, R., Pandiarajan, A., Gandhi, M.R., Shibayama, A., Vasudevan, S., 2017. Eco-friendly and easily prepared graphene nanosheets for safe drinking water: removal of chlorophenoxyacetic acid herbicides. ChemistrySelect 2 (1), 342–355. https://doi.org/10.1002/slct.201601645.

Kang, S., Herzberg, M., Rodrigues, D.F., Elimelech, M., 2008. Antibacterial effects of carbon nanotubes: size does matter. Langmuir 24 (13), 6409–6413. https://doi.org/10.1021/la800951v.

Khavar, A.H.C., Moussavi, G., Mahjoub, A.R., Satari, M., 2018a. Facile preparation of multi-doped TiO_2/rGO cross-linked 3D aerogel (GaNF@TGA) nancomposite as an efficient visible-light activated catalyst for photocatalytic oxidation and detoxification of atrazine. Sol. Energy 173, 848–860. https://doi.org/10.1016/j.solener.2018.08.034.

Khavar, A.H.C., Moussavi, G., Mahjoub, A.R., Satari, M., Abdolmaleki, P., 2018b. Synthesis and visible-light photocatalytic activity of In, S-TiO_2@rGO nanocomposite for degradation and detoxification of pesticide atrazine in water. Chem. Eng. J. 345, 300–311. https://doi.org/10.1016/j.cej.2018.03.095.

Khorram, M.S., Zhang, Q., Lin, D.L., Zheng, Y., Fang, H., Yu, Y.L., 2016. Biochar: a review of its impact on pesticide behavior in soil environments and its potential applications. J. Environ. Sci. 44, 269–279. https://doi.org/10.1016/j.jes.2015.12.027.

Khorram, M.S., Lin, D.L., Zhang, Q., Zheng, Y., Fang, H., Yu, Y.L., 2017. Effects of aging process on adsorption-desorption and bioavailability of fomesafen in an agricultural soil amended with rice hull biochar. J. Environ. Sci. 56, 180–191. https://doi.org/10.1016/j.jes.2016.09.012.

Khorram, M.S., Sarmah, A.K., Yu, Y.L., 2018. The effects of biochar properties on fomesafen adsorption-desorption capacity of biohar-amended soil. Water Air Soil Pollut. 229 (3), 60. https://doi.org/10.1007/s11270-017-3603-2.

Koehlert, K., 2017. Activated carbon: fundamentals and new applications. Chem. Eng. 2017, 32–40.

Koushik, D., Sen Gupta, S., Maliyekkal, S.M., Pradeep, T., 2016. Rapid dehalogenation of pesticides and organics at the interface of reduced graphene oxide-silver nanocomposite. J. Hazard. Mater. 308, 192–198. https://doi.org/10.1016/j.jhazmat.2016.01.004.

Kozitsina, A.N., Svalova, T.S., Malysheva, N.N., Okhokhonin, A.V., Vidrevich, M.B., Brainina, K.Z., 2018. Sensors based on bio and biomimetic receptors in medical diagnostic, environment, and food analysis. Biosensors 8 (2), 35. https://doi.org/10.3390/bios8020035.

Kumar, R., Mukherji, S., 2018. Feasibility of photocatalytic degradation of an organophosphorus pesticide using graphene oxide-TiO_2 nanocomposite. J. Indian Chem. Soc. 95 (3), 217–222.

Kuppusamy, S., Thavamani, P., Megharaj, M., Venkateswarlu, K., Naidu, R., 2016. Agronomic and remedial benefits and risks of applying biochar to soil: current knowledge and future research directions. Environ. Int. 87, 1–12. https://doi.org/10.1016/j.envint.2015.10.018.

Lehmann, J., Rillig, M.C., Thies, J., Masiello, C.A., Hockaday, W.C., Crowley, D., 2011. Biochar effects on soil biota—a review. Soil Biol. Biochem. 43 (9), 1812–1836. https://doi.org/10.1016/j.soilbio.2011.04.022.

Li, Y.J., Zhao, C.Q, Wen, Y.J., Wang, Y.Y., Yang, Y.S., 2018. Adsorption performance and mechanism of magnetic reduced graphene oxide in glyphosate contaminated water. Environ. Sci. Pollut. Res. 25(21), 21036–21048. https://doi.org/10.1007/s11356-018-2282-x.

Li, Z., Ma, R.Y., Zhang, G.J., Zhang, S.H., Wang, C., Wu, Q.H., Wang, Z., 2016. Application of a solid-phase microextraction fiber coated with a graphene oxide-poly(dimethylsiloxane) composite for the extraction of triazoles from water. J. Sep. Sci. 39 (16), 3171–3177. https://doi.org/10.1002/jssc.201600485.

Liu, G., Han, C., Pelaez, M., Zhu, D., Liao, S., Likodimos, V., Ioannidis, N., Kontos, A.G., Falaras, P., Dunlop, P.S., Byrne, J.A., Dionysiou, D.D., 2012. Synthesis, characterization and photocatalytic evaluation of visible light activated C-doped TiO_2 nanoparticles. Nanotechnology 23 (29), 294003. https://doi.org/10.1088/0957-4484/23/29/294003.

Liu, G., Guo, W., Song, D., 2014. A multianalyte electrochemical immunosensor based on patterned carbon nanotubes modified substrates for detection of pesticides. Biosens. Bioelectron. 52, 360–366. https://doi.org/10.1016/j.bios.2013.09.009.

Liu, G.Y., Li, L.Y., Huang, X.D., Zheng, S.N., Xu, X.M., Liu, Z.X., Zhang, Y.G., Wang, J., Lin, H., Xu, D.H., 2018b. Adsorption and removal of organophosphorus pesticides from environmental water and soil samples by using magnetic multi-walled carbon nanotubes@organic framework ZIF-8. J. Mater. Sci. 53 (15), 10772–10783. https://doi.org/10.1007/s10853-018-2352-y.

Liu, X.N., Tang, Y.H., Luo, S.L., Wang, Y., Zhang, X.L., Chen, Y., Liu, C.B., 2013b. Reduced graphene oxide and $CuInS_2$ co-decorated TiO_2 nanotube arrays for efficient removal of herbicide 2,4-dichlorophenoxyacetic acid from water. J. Photochem. Photobiol. A 262, 22–27. https://doi.org/10.1016/j.jphotochem.2013.04.016.

Liu, X.T., Zhang, H.Y., Ma, Y.Q., Wu, X.L., Meng, L.X., Guo, Y.L., Yu, G., Liu, Y.Q., 2013a. Graphene-coated silica as a highly efficient sorbent for residual organophosphorus pesticides in water. J. Mater. Chem. A 1 (5), 1875–1884. https://doi.org/10.1039/c2ta00173j.

Liu, Y.X., Lonappan, B., Brar, S.K., Yang, S., 2018a. Impact of biochar amendment in agricultural soils on the sorption, desorption, and degradation of pesticides: a review. Sci. Total Environ. 645, 60–70. https://doi.org/10.1016/j.scitotenv.2018.07.099.

Liu, Z., Dugan, B., Masiello, C.A., Gonnermann, H.M., 2017. Biochar particle size, shape, and porosity act together to influence soil water properties. PLoS ONE 12 (6), e0179079. https://doi.org/10.1371/journal.pone.0179079.

Lu, F., Astruc, D., 2018. Nanomaterials for removal of toxic elements from water. Coord. Chem. Rev. 356, 147–164. https://doi.org/10.1016/j.ccr.2017.11.003.

Ma, J., Sun, Y.R., Zhang, M.Z., Yang, M.X., Gong, X., Yu, F., Zheng, J., 2017. Comparative study of graphene hydrogels and aerogels reveals the important role of buried water in pollutant adsorption. Environ. Sci. Technol. 51 (21), 12283–12292.

Mahpishanian, S., Sereshti, H., Baghdadi, M., 2015. Superparamagnetic core-shells anchored onto graphene oxide grafted with phenylethyl amine as a nano-adsorbent for extraction and enrichment of organophosphorus pesticides from fruit, vegetable and water samples. J. Chromatogr. A 1406, 48–58. https://doi.org/10.1016/j.chroma.2015.06.025.

Maliyekkal, S.M., Sreeprasad, T.S., Krishnan, D., Kouser, S., Mishra, A.K., Waghmare, U.V., Pradeep, T., 2013. Graphene: a reusable substrate for unprecedented adsorption of pesticides. Small 9 (2), 273–283. https://doi.org/10.1002/smll.201201125.

Marican, A., Duran-Lara, E.F., 2018. A review on pesticide removal through different processes. Environ. Sci. Pollut. Res. 25 (3), 2051–2064. https://doi.org/10.1007/s11356-017-0796-2.

Marsh, H., Rodriguez-Reinoso, F., 2006. Activated Carbon. Elsevier Science & Technology Books, Amsterdam, Netherlands.

Mauter, M.S., Elimelech, M., 2008. Environmental applications of carbon-based nanomaterials. Environ. Sci. Technol. 42 (16), 5843–5859. https://doi.org/10.1021/es8006904.

McDougall, G.J., 1991. The physical nature and manufacture of activated carbon. J. S. Afr. Inst. Min. Metal. 91, 109–120. https://www.saimm.co.za/Journal/v091n04p109.pdf.

Minnekhanov, A.A., Deygen, D.M., Konstantinova, E.A., Vorontsov, A.S., Kashkarov, P.K., 2012. Paramagnetic properties of carbon-doped titanium dioxide. Nanoscale Res. Lett. 7 (1), 333. https://doi.org/10.1186/1556-276X-7-333.

Montesinos, I., Sfakianaki, A., Gallego, M., Stalikas, C.D., 2015. Graphene-coated cotton fibers as a sorbent for the extraction of multiclass pesticide residues from water and their determination by gas chromatography with mass spectrometry. J. Sep. Sci. 38 (5), 836–843. https://doi.org/10.1002/jssc.201400957.

Morais, A., Carminati, S.D.A., Nogueira, A.F., 2016. Nanostructured hybrid materials based on reduced graphene oxide for solar energy conversion. Proc. SPIE 9923, 992314. https://doi.org/10.1117/12.2240029.

Mukherjee, A., Majumdar, S., Servin, A.D., Pagano, L., Dhankher, O.P., White, J.C., 2016. Carbon nanomaterials in agriculture: a critical review. Front. Plant Sci. 7, 172. https://doi.org/10.3389/fpls.2016.00172.

Mukherjee, D., Dewanjee, A., Ghosh, S., Majumdar, S., 2018. Development of graphene oxide/chitosan composite membrane on ceramic support for atrazine remediation by MBR process. Environ. Sci. Pollut. Res. Int. 25 (33), 33334–33352. https://doi.org/10.1007/s11356-018-3255-9.

Naeimi, A., Saeidi, M., Baroumand, N., 2016. Carboxylated carbon nanotubes as an efficient and cost-effective adsorbent for sustainable removal of insecticide fenvalerate from contaminated solutions. Int. Nano Lett. 6 (4), 265–271. https://doi.org/10.1007/s40089-016-0193-8.

Naghdi, M., Taheran, M., Brar, S.K., Rouissi, T., Verma, M., Surampalli, R.Y., Valero, J.R., 2017. A green method for production of nanobiochar by ball milling-optimization and characterization. J. Clean. Prod. 164, 1394–1405. https://doi.org/10.1016/j.jclepro.2017.07.084.

Oliveira, F.R., Patela, A.K., Jaisi, D.P., Adhikari, S., Lu, H., Khanal, S.K., 2017. Environmental application of biochar: current status and perspectives. Bioresour. Technol. 246, 110–122. https://doi.org/10.1016/j.biortech.2017.08.122.

Ortega-Liebana, M.C., Hueso, J.L., Ferdousi, S., Arenal, R., Irusta, S., Yeung, K., Santamaria, J., 2017. Extraordinary sensitizing effect of co-doped carbon nanodots derived from mate herb: application to enhanced photocatalytic degradation of chlorinated wastewater compounds under visible light. Appl. Catal. B-Environ. 218, 68–79. https://doi.org/10.1016/j.apcatb.2017.06.021.

Palanivelu, K., Im, J.S., Lee, Y.S., 2007. Carbon doping of TiO_2 for visible light photo catalysis—a review. Carbon Sci. 8 (3), 214–224. http://carbonlett.org/upload/files/carbonlett/214-224.pdf.

Panic, S., Guzsvany, V., Konya, Z., Kukovecz, A., Boskovic, G., 2017. Kinetic, equilibrium and thermodynamic studies of thiamethoxam adsorption by multi-walled carbon nanotubes. Int. J. Environ. Sci. Technol. 14 (6), 1297–1306. https://doi.org/10.1007/s13762-016-1237-3.

Peng, J., Xue, J., Li, J.H.H., Du, Z.K., Wang, Z.Y., Gao, S.X., 2017. Catalytic effect of low concentration carboxylated multi-walled carbon nanotubes on the oxidation of disinfectants with Cl-substituted structure by a Fenton-like system. Chem. Eng. J. 321, 325–334. https://doi.org/10.1016/j.cej.2017.03.133.

Pentak, D., Kozik, V., Bak, A., Dybal, P., Sochanik, A., Jampílek, J., 2016. Methotrexate and cytarabine-loaded nanocarriers for multidrug cancer therapy. Spectrosc. Study Mol. 21 (12), 1689. https://doi.org/10.3390/molecules21121689.

Photon, 2019. Carbon-Based Materials—Applications. Montreal, Canada. (5 March 2019) http://www.photonetc.com/carbon-based-materials.

Pinto, M.D.E., Goncalves, R.G.L., Santos, R.M.M., Araujo, E.A., Perotti, G.F., Macedo, R.D., Bizeto, M.A., Constantino, V.R.L., Pinto, F.G., Tronto, J., 2016. Mesoporous carbon derived from a biopolymer and a clay: preparation, characterization and application for an organochlorine pesticide adsorption. Micropor. Mesopor. Mater. 225, 342–354. https://doi.org/10.1016/j.micromeso.2016.01.012.

Pisárčik, M., Lukáč, M., Jampílek, J., Bilka, F., Bilková, A., Pašková, Ľ., Devínsky, F., Horáková, R., Opravil, T., 2018. Silver nanoparticles stabilised with cationic single-chain surfactants. Structure-physical properties-biological activity relationship study. J. Mol. Liq. 272, 60–72.

PotatoPro, 2016. The Effect of Population Growth on Efficiency in Food Production. Food Innovation Online Corp. https://www.potatopro.com/news/2016/effect-population-growth-efficiency-food-production. (12 March 2019).

Qin, X., Liu, Y.T., Huang, Q.Q., Liu, Y.Y., Zhao, L.J., Xu, Y.M., 2019. In-situ remediation of cadmium and atrazine contaminated acid red soil of South China using sepiolite and biochar. Bull. Environ. Contam. Toxicol. 102 (1), 128–133. https://doi.org/10.1007/s00128-018-2494-2.

Rajapaksha, AU., Chen, S.S., Tsang, D.C.W., Zhang, M., Vithanage, M., Mandal, S., Gao, B., Bolan, N.S., Ok, Y.S., 2016. Engineered/designer biochar for contaminant removal/immobilization from soil and water: potential and implication of biochar modification. Chemosphere 148, 276–291. https://doi.org/10.1016/j.chemosphere.2016.01.043.

Rajapaksha, P.P., Power, A., Chandra, S., Chapman, J., 2018. Graphene, electrospun membranes and granular activated carbon for eliminating heavy metals, pesticides and bacteria in water and wastewater treatment processes. Analyst 143 (23), 5629–5645. https://doi.org/10.1039/c8an00922h.

Rana, V.K., Kissner, R., Gaspard, S., Levalois-Grutzmacher, J., 2016. Cyclodextrin as a complexation agent in the removal of chlordecone from water. Chem. Eng. J. 293, 82–89. https://doi.org/10.1016/j.cej.2016.02.032.

Rana, V.K., Kissner, R., Jauregui-Haza, U., Gaspard, S., Levalois-Grutzmacher, J., 2017. Enhanced chlordecone (Kepone) removal by FeO-nanoparticles loaded on activated carbon. J. Environ. Chem. Eng. 5 (2), 1608–1617. https://doi.org/10.1016/j.jece.2017.01.039.

Rasoulnezhad, H., Kavei, G., Ahmadi, K., Rahimipour, M.R., 2017. Visible light photocatalytic degradation of paraoxon and parathion pesticides on carbon-doped TiO_2 nanorod thin films. J. Mater. Sci. Mater. Electr. 28 (24), 18337–18347. https://doi.org/10.1007/s10854-017-7780-y.

Rawtani, D., Khatri, N., Tyagi, S., Pandey, G., 2018. Nanotechnology-based recent approaches for sensing and remediation of pesticides. J. Environ. Manag. 206, 749–762. https://doi.org/10.1016/j.jenvman.2017.11.037.

Reddy, P.A.K., Reddy, P.V.L., Sharma, V.M., Srinivas, B., Kumari, V.D., Subrahmanyam, N., 2010. Photocatalytic degradation of isoproturon pesticide on C, N and S doped TiO_2. JWARP 2, 235–244. https://doi.org/10.4236/jwarp.2010.23027.

Reddy, P.A.K., Reddy, P.V.L., Kwon, E., Kim, K.H., Akter, T., Kalagara, S., 2016. Recent advances in photocatalytic treatment of pollutants in aqueous media. Environ. Int. 91, 94–103. https://doi.org/10.1016/j.envint.2016.02.012.

Ren, X.H., Zhang, P., Zhao, L.J., Sun, H.W., 2016. Sorption and degradation of carbaryl in soils amended with biochars: influence of biochar type and content. Environ. Sci. Pollut. Res. 23 (3), 2724–2734. https://doi.org/10.1007/s11356-015-5518-z.

Restivo, J., Orfao, J.J.M., Armenise, S., Garcia-Bordeje, E., Pereira, M.F.R., 2012. Catalytic ozonation of metolachlor under continuous operation using nanocarbon materials grown on a ceramic monolith. J. Hazard. Mater. 239, 249–256. https://doi.org/10.1016/j.jhazmat.2012.08.073.

Restivo, J., Rocha, R.P., Silva, A.M.T., Orfao, J.J.M., Pereira, M.F.R., Figueiredo, J.L., 2014. Catalytic performance of heteroatom-modified carbon nanotubes in advanced oxidation processes. Chinese J. Catal. 35 (6), 896–905. https://doi.org/10.1016/S1872-2067(14)60103-0.

Restivo, J., Garcia-Bordeje, E., Orfao, J.J.M., Pereira, M.F.R., 2016. Carbon nanofibers doped with nitrogen for the continuous catalytic ozonation of organic pollutants. Chem. Eng. J. 293, 102–111. https://doi.org/10.1016/j.cej.2016.02.055.

Ribeiro, A.R., Nunes, O.C., Pereira, M.F.R., Silva, A.M.T., 2015. An overview on the advanced oxidation processes applied for the treatment of water pollutants defined in the recently launched Directive 2013/39/EU. Environ. Int. 75, 33–51. https://doi.org/10.1016/j.envint.2014.10.027.

Ritchie, H., 2017. How Many People Does Synthetic Fertilizer Feed? Our World in Data. https://ourworldindata.org/how-many-people-does-synthetic-fertilizer-feed. (12 March 2019).

Röös, E., Patel, M., Spångberg, J., Carlsson, G., Rydhmer, L., 2016. Limiting livestock production to pasture and by-products in a search for sustainable diets. Food Policy 58, 1–13.

Schwab, F., Bucheli, T.D., Camenzuli, L., Magrez, A., Knauer, K., Sigg, L., Nowack, B., 2013. Diuron sorbed to carbon nanotubes exhibits enhanced toxicity to *Chlorella vulgaris*. Environ. Sci. Technol. 47 (13), 7012–7019. https://doi.org/10.1021/es304016u.

Schwab, F., Camenzuli, L., Knauer, K., Nowack, B., Magrez, A., Sigg, L., Bucheli, T.D., 2014. Sorption kinetics and equilibrium of the herbicide diuron to carbon nanotubes or soot in absence and presence of algae. Environ. Pollut. 192, 147–153. https://doi.org/10.1016/j.envpol.2014.05.018.

Sen Gupta, S., Sreeprasad, T.S., Maliyekkal, S.M., Das, S.K., Pradeep, T., 2012. Graphene from sugar and its application in water purification. ACS Appl. Mater. Interfaces 4 (8), 4156–4163. https://doi.org/10.1021/am300889u.

Sen Gupta, S., Chakraborty, I., Maliyekkal, S.M., Mark, T.A., Pandey, D.K., Das, S.K., Pradeep, T., 2015. Simultaneous dehalogenation and removal of persistent halocarbon pesticides from water using graphene nanocomposites: a case study of lindane. ACS Sustain. Chem. Eng. 3 (6), 1155–1163. https://doi.org/10.1021/acssuschemeng.5b00080.

SFGate, 2019. https://homeguides.sfgate.com/effects-synthetic-fertilizers-45466.html. (12 March 2019).

Shanbogh, P.P., Sundaram, N.G., 2015. Fullerenes revisited: materials chemistry and applications of C_{60} molecules. Resonance 20 (2), 123–135. https://doi.org/10.1007/s12045-015-0160-0.

Sharon, M., Sharon, M., 2010. Carbon Nano Forms and Applications. McGraw Hill Professional, New York, NY, USA.

Shen, X.F., Wang, X.L., Tao, S., Xing, B.S., 2014. Displacement and competitive sorption of organic pollutants on multiwalled carbon nanotubes. Environ. Sci. Pollut. Res. 21 (20), 11979–11986. https://doi.org/10.1007/s11356-014-3115-1.

Shrivas, K., Ghosale, A., Nirmalkar, N., Srivastava, A., Singh, S.K., Shinde, S.S., 2017. Removal of endrin and dieldrin isomeric pesticides through stereoselective adsorption behavior on the graphene oxide-magnetic nanoparticles. Environ. Sci. Pollut. Res. 24 (32), 24980–24988. https://doi.org/10.1007/s11356-017-0159-z.

Sizmur, T., Fresno, T., Akgül, G., Frost, H., Moreno-Jiménez, E., 2017. Biochar modification to enhance sorption of inorganics from water. Bioresour. Technol. 246, 34–47. https://doi.org/10.1016/j.biortech.2017.07.082.

Sloan, J., Dunin-Borkowski, R.E., Hutchison, J.L., Coleman, K.S., Williams, V.C., Claridge, J.B., York, A.P.E., Xu, C.G., Bailey, S.R., Brown, G., Friedichs, S., Green, M.K.H., 2000. The size distribution, imaging and obstructing properties of C_{60} and higher fullerenes formed within arc-grown single walled carbon nanotubes. Chem. Phys. Lett. 316 (3–4), 191–198. https://doi.org/10.1016/S0009-2614(99)01250-6.

Sophia, A.C., Lima, E.C., 2018. Removal of emerging contaminants from the environment by adsorption. Ecotoxicol. Environ. Saf. 150, 1–17. https://doi.org/10.1016/j.ecoenv.2017.12.026.

Srivastava, M., Abhilash, P.C., Singh, N., 2011. Remediation of lindane using engineered nanoparticles. J. Biomed. Nanotechnol. 7 (1), 172–174. https://doi.org/10.1166/jbn.2011.1255.

Sun, K., Zhang, Z.Y., Gao, B., Wang, Z.Y., Xu, D.Y., Jin, J., Liu, X.T., 2012. Adsorption of diuron, fluridone and norflurazon on single-walled and multi-walled carbon nanotubes. Sci. Total Environ. 439, 1–7. https://doi.org/10.1016/j.scitotenv.2012.08.022.

Suo, F.Y., Xie, G.X., Zhang, J., Li, J.Y., Li, C.S., Liu, X., Zhang, Y.P., Ma, Y.Q., Ji, M.S., 2018. A carbonised sieve-like corn straw cellulose-graphene oxide composite for organophosphorus pesticide removal. RSC Adv. 8 (14), 7735–7743. https://doi.org/10.1039/c7ra12898c.

Szmigielski, A.M., Hangs, R.D., Schoenau, J.J., 2018. Bioavailability of metsulfuron and sulfentrazone herbicides in soil as affected by amendment with two contrasting willow biochars. Bull. Environ. Contam. Toxicol. 100 (2), 298–302. https://doi.org/10.1007/s00128-017-2212-5.

Tan, C.W., Tan, K.H., Ong, Y.T., Mohamed, A.R., Zein, S.H.S., Tan, S.H., 2012. Energy and environmental applications of carbon nanotubes. Environ. Chem. Lett. 10 (3), 265–273. https://doi.org/10.1007/s10311-012-0356-4.

Tang, L., Zhang, S., Zeng, G.M., Zhang, Y., Yang, G.D., Chen, J., Wang, J.J., Wang, J.J., Zhou, Y.Y., Deng, Y.C., 2015. Rapid adsorption of 2,4-dichlorophenoxyacetic acid by iron oxide nanoparticles-doped carboxylic ordered mesoporous carbon. J. Colloid Interface Sci. 445, 1–8. https://doi.org/10.1016/j.jcis.2014.12.074.

Thornton, P.K., 2010. Livestock production: recent trends, future prospects. Philos. Trans. R. Soc. Lond. Ser. B Biol. Sci. 365 (1554), 2853–2867. https://doi.org/10.1098/rstb.2010.0134.

Tolasz, J., Stastny, M., Stengl, V., 2016. Graphene oxide-mesoporous δ-MnO_2 nanocomposite as a novel destructive sorbent. Mater. Today: Proc. 3 (8), 2795–2806. https://doi.org/10.1016/j.matpr.2016.06.029.

Tong, H., Hu, M., Li, F.B., Liu, C.S., Chen, M.J., 2014. Biochar enhances the microbial and chemical transformation of pentachlorophenol in paddy soil. Soil Biol. Biochem. 70, 142–150. https://doi.org/10.1016/j.soilbio.2013.12.012.

Varjani, S., Kumar, G., Rene, E.R., 2018. Developments in biochar application for pesticide remediation: current knowledge and future research directions. J. Environ. Manag. 232, 505–513. https://doi.org/10.1016/j.jenvman.2018.11.043.

Viswanathan, S., Manisankar, P., 2015. Nanomaterials for electrochemical sensing and decontamination of pesticides. J. Nanosci. Nanotechnol. 15 (9), 6914–6923. https://doi.org/10.1166/jnn.2015.10724.

Voorhees, J.P., Anderson, B.S., Phillips, B.M., Tjeerdema, R.S., 2017. Carbon treatment as a method to remove imidacloprid from agriculture runoff. Bull. Environ. Contam. Toxicol. 99 (2), 200–202. https://doi.org/10.1007/s00128-017-2074-x.

Wanekaya, A.K., 2011. Applications of nanoscale carbon-based materials in heavy metal sensing and detection. Analyst 136, 4383–4391. https://doi.org/10.1039/C1AN15574A.

Wang, Q.X., Gao, S.D., Wang, D., Spokas, K., Cao, A.C., Yan, D.D., 2016. Mechanisms for 1,3-dichloropropene dissipation in biochar-amended soils. J. Agric. Food Chem. 64 (12), 2531–2540. https://doi.org/10.1021/acs.jafc.5b04941.

Wang, W.J., Xu, Z.L., Zhang, X.X., Wimmer, A., Shi, E., Qin, Y., Zhao, X.P., Zhou, B.C., Li, L.X.Y., 2018. Rapid and efficient removal of organic micropollutants from environmental water using a magnetic nanoparticles-attached fluorographene-based sorbent. Chem. Eng. J. 343, 61–68. https://doi.org/10.1016/j.cej.2018.02.101.

Wang, X.M., Wang, H., Lu, M.X., Teng, R.J., Du, X.Z., 2017. Facile synthesis of phenyl-modified magnetic graphene/mesoporous silica with hierarchical bridge-pore structure for efficient adsorption of pesticides. Mater. Chem. Phys. 198, 393–400. https://doi.org/10.1016/j.matchemphys.2016.12.017.

Wick, P., Louw-Gaume, A.E., Kucki, M., Krug, H.F., Kostarelos, K., Fadeel, B., Dawson, K.A., Salvati, A., Vázquez, E., Ballerini, L., Tretiach, M., Benfenati, F., Flahaut, E., Gauthier, L., Prato, M., Bianco, A., 2014. Classification framework for graphene-based materials. Angew. Chem. Int. Ed. 53 (30), 7714–7718. https://doi.org/10.1002/anie.201403335.

Willner, M.R., Vikesland, P.J., 2018. Nanomaterial enabled sensors for environmental contaminants. J. Nanobiotechnol. 16 (1), 95. https://doi.org/10.1186/s12951-018-0419-1.

Woolf, D., Amonette, J.E., Street-Perrott, F.A., Lehmann, J., Joseph, S., 2010. Sustainable biochar to mitigate global climate change. Nat. Commun. 1 (56), 1–9. https://doi.org/10.1038/ncomms1053.

Wu, S.H., He, H.J., Inthapanya, X., Yang, C.P., Lu, L., Zeng, G.M., Han, Z.F., 2017. Role of biochar on composting of organic wastes and remediation of contaminated soils—a review. Environ. Sci. Pollut. Res. 24 (20), 16560–16577. https://doi.org/10.1007/s11356-017-9168-1.

Xia, D., Otyepka, M., Li, X., Liu, W., Zheng, Q., 2015. Carbon-based materials at nanoscale. J. Nanomater. 2015, 750242. https://doi.org/10.1155/2015/750242.

Xia, L.H., Luo, L.J., Li, Y.M., Zhao, T.T., Yang, W.R., Barrow, C.J., Yang, M., Wang, H.B., 2017. Study on triazophos adsorption behavior on the multi-walled carbon nanotubes. Desalin. Water Treat. 96, 97–103. https://doi.org/10.5004/dwt.2017.21531.

Xiao, F., Pignatello, J.J., 2015. Interactions of triazine herbicides with biochar: steric and electronic effects. Water Res. 80, 179–188. https://doi.org/10.1016/j.watres.2015.04.040.

Xing, B., Chen, H.L., Zhang, X.M., 2018. Efficient degradation of organic phosphorus in glyphosate wastewater by catalytic wet oxidation using modified activated carbon as a catalyst. Environ. Technol. 39 (6), 749–758. https://doi.org/10.1080/09593330.2017.1310935.

Xu, G., Huo, D., Hou, C., Zhao, Y., Bao, J., Yang, M., Fa, H., 2018. A regenerative and selective electrochemical aptasensor based on copper oxide nanoflowers-single walled carbon nanotubes nanocomposite for chlorpyrifos detection. Talanta 178, 1046–1052. https://doi.org/10.1016/j.talanta.2017.08.086.

Xu, J., Lv, H.D., Yang, S.T., Luo, J.B., 2013. Preparation of graphene adsorbents and their applications in water purification. Rev. Inorg. Chem. 33 (2–3), 139–160. https://doi.org/10.1515/revic-2013-0007.

Xu, J., Cao, Z., Liu, X., Zhao, H., Xiao, X., Wu, J.P., Xu, X.H., Zhou, J.L., 2016. Preparation of functionalized Pd/Fe-Fe_3O_4@MWCNTs nanomaterials for aqueous 2,4-dichlorophenol removal: interactions, influence factors, and kinetics. J. Hazard. Mater. 317, 656–666. https://doi.org/10.1016/j.jhazmat.2016.04.063.

Yamaguchi, N.U., Bergamasco, R., Hamoudi, S., 2016. Magnetic $MnFe_2O_4$-graphene hybrid composite for efficient removal of glyphosate from water. Chem. Eng. J. 295, 391–402. https://doi.org/10.1016/j.cej.2016.03.051.

Yang, J.R., Shen, F., Qiu, M., Qi, X.H., 2018. Catalytic dehydrochlorination of lindane by nitrogen-containing multiwalled carbon nanotubes (N-MWCNTs). Sci. Total Environ. 621, 1445–1452. https://doi.org/10.1016/j.scitotenv.2017.10.084.

Yang, Q.F., Wang, J., Zhang, W.T., Liu, F.B., Yue, X.Y., Liu, Y.N., Yang, M., Li, Z.H., Wang, J.L., 2017. Interface engineering of metal organic framework on graphene oxide with enhanced adsorption capacity for organophosphorus pesticide. Chem. Eng. J. 313, 19–26. https://doi.org/10.1016/j.cej.2016.12.041.

Yang, W., Ratinac, K.R., Ringer, S.P., Thordarson, P., Gooding, J.J., Braet, F., 2010. Carbon nanomaterials in biosensors: should you use nanotubes or graphene? Angew. Chem. Int. Ed. 49 (12), 2114–2138. https://doi.org/10.1002/anie.200903463.

Yang, Y., Asiri, A.M., Du, D., Lin, Y., 2014. Acetylcholinesterase biosensor based on a gold nanoparticle-polypyrrole-reduced graphene oxide nanocomposite modified electrode for the amperometric detection of organophosphorus pesticides. Analyst 139, 3055–3060. https://doi.org/10.1039/c4an00068d.

Yavari, S., Malakahmad, A., Sapari, N.B., Yavari, S., 2016. Sorption-desorption mechanisms of imazapic and imazapyr herbicides on biochars produced from agricultural wastes. J. Environ. Chem. Eng. 4 (4, Pt A), 3981–3989. https://doi.org/10.1016/j.jece.2016.09.003.

Yavari, S., Sapari, N.B., Malakahmad, A., Yavari, S., 2018. Degradation of imazapic and imazapyr herbicides in the presence of optimized oil palm empty fruit bunch and rice husk biochars in soil. J. Hazard. Mater. 366, 636–642. https://doi.org/10.1016/j.jhazmat.2018.12.022.

Youssef, A.M., El-Naggar, M.E., Malhat, F.M., El Sharkawi, H.M., 2019. Efficient removal of pesticides and heavy metals from wastewater and the antimicrobial activity of f-MWCNTs/PVA nanocomposite film. J. Clean. Prod. 206, 315–325. https://doi.org/10.1016/j.jclepro.2018.09.163.

Yu, G., Wu, W., Zhao, Q., Wei, X., Lu, Q., 2015. Efficient immobilization of acetylcholinesterase onto amino functionalized carbon nanotubes for the fabrication of high sensitive organophosphorus pesticides biosensors. Biosens. Bioelectron. 68, 288–294. https://doi.org/10.1016/j.bios.2015.01.005.

Yu, J.G., Zhao, X.H., Yang, H., Chen, X.H., Yang, Q., Yu, L.Y., Jiang, J.H., Chen, X.Q., 2014. Aqueous adsorption and removal of organic contaminants by carbon nanotubes. Sci. Total Environ. 482, 241–251. https://doi.org/10.1016/j.scitotenv.2014.02.129.

Zaytseva, O., Neumann, G., 2016. Carbon nanomaterials: production, impact on plant development, agricultural and environmental applications. Chem. Biol. Technol. Agric. 3, 17. https://doi.org/10.1186/s40538-016-0070-8.

Zhang, C.D., Li, M.Z., Xu, X., Liu, N., 2015. Effects of carbon nanotubes on atrazine biodegradation by *Arthrobacter* sp. J. Hazard. Mater. 287, 1–6. https://doi.org/10.1016/j.jhazmat.2015.01.039.

Zhang, C.Y., Du, Z., Zhou, R.Y., Xu, P., Dong, X.H., Fu, Y.Y., Wang, Q., Su, C.J., Yan, L., Gu, Z.J., 2018c. $Cu_2(OH)PO_4$/reduced graphene oxide nanocomposites for enhanced photocatalytic degradation of 2,4-dichlorophenol under infrared light irradiation. RSC Adv. 8 (7), 3611–3618. https://doi.org/10.1039/c7ra12684k.

Zhang, J., Gong, J.L., Zeng, G.M., Yang, H.C., Zhang, P., 2017a. Carbon nanotube amendment for treating dichlorodiphenyltrichloroethane and hexachlorocyclohexane remaining in Dong-ting Lake sediment—an implication for in-situ remediation. Sci. Total Environ. 579, 283–291. https://doi.org/10.1016/j.scitotenv.2016.11.105.

Zhang, J.W., Lu, X., Shi, C., Yan, B., Gong, L., Chen, J.S., Xiang, L., Xu, H.L., Liu, Q.X., Zeng, H.B., 2018b. Unraveling the molecular interaction mechanism between graphene oxide and aromatic organic compounds with implications on wastewater treatment. Chem. Eng. J. 358, 842–849. https://doi.org/10.1016/j.cej.2018.10.064.

Zhang, M.Y., Engelhardt, I., Simunek, J., Bradford, S.A., Kasel, D., Berns, A.E., Vereecken, H., Klumpp, E., 2017b. Co-transport of chlordecone and sulfadiazine in the presence of functionalized multi-walled carbon nanotubes in soils. Environ. Pollut. 221, 470–479. https://doi.org/10.1016/j.envpol.2016.12.018.

Zhang, P., Sun, H.W., Min, L.J., Ren, C., 2018a. Biochars change the sorption and degradation of thiacloprid in soil: insights into chemical and biological mechanisms. Environ. Pollut. 236, 158–167. https://doi.org/10.1016/j.envpol.2018.01.030.

Zhou, L.X., Pan, S.D., Chen, X.H., Zhao, Y.G., Zou, B.B., Jin, M.C., 2014. Kinetics and thermodynamics studies of pentachlorophenol adsorption on covalently functionalized Fe_3O_4@SiO_2-MWCNTs core-shell magnetic microspheres. Chem. Eng. J. 257, 10–19. https://doi.org/10.1016/j.cej.2014.07.060.

Zhu, L., Zhao, N., Tong, L.H., Lv, Y.Z., Li, G.J., 2018. Characterization and evaluation of surface modified materials based on porous biochar and its adsorption properties for 2,4-dichlorophenoxyacetic acid. Chemosphere 210, 734–744. https://doi.org/10.1016/j.chemosphere.2018.07.090.

Nanocarbon-based sensors for pesticide detection: Recent trends

18

Kamel A. Abd-Elsalam[a,b], *Mohamed M. Ramadan*[a], *Mona Kamal Hassanien*[c]
[a]Plant Pathology Research Institute, Agricultural Research Center (ARC), Giza, Egypt,
[b]Unit of Excellence in Nano-Molecular Plant Pathology Research Center—Plant Pathology Research Institute, Giza, Egypt, [c]Department of Microbiology, Faculty of Science, Ain Shams University, Cairo, Egypt

1 Introduction

Synthetic pesticides, which include insecticides, herbicides, and fungicides, are frequently utilized in greenhouses and agricultural fields. The unnecessary consumption of those agrochemicals has undesirable effects on the ecosystem, including a decreased population of beneficial insects as well as risks to vulnerable species and bird habitats. Pesticide pollution is becoming one of the most severe challenges of common public health around the globe because of their particular application in the agriculture sector to assure crop yield and productivity (Liu et al., 2019). As a result, it is necessary to design unique biosensors for the delicate detection of pesticide residues in food and water to prevent toxic impacts on humans and other organisms. Various nanomaterials are generally categorized into nanoparticles, nanotubes, and nanocomposites, which can be generally employed for the diagnosis, degradation, and adsorption of chemical pesticides. Carbon nanomaterials or nanoparticles (NPs) have specific characteristics, including a high surface-to-volume ratio, good electrical conductivity, catalytic action, beneficial biocompatibility, and can be simply modified with functional groups, which has made them be often used in pesticide biosensors to boost analytical efficiency (Xia and Gao, 2015). Carbon nanocomposites also obtain particular morphological constructions and features valuable for pesticides sensing (Qiu et al., 2018). The three-dimensional (3D) structures of graphene nanocomposites assist in the dissemination of distinct types of biomolecules and maintain their biocatalytic capabilities to enhance biosensing efficiency (Liu et al., 2013). Graphene-based combined with polymers and surface-covered with specific metal nanoparticles (Krishnan et al., 2019) have been examined for biosensing due to their good biocompatibility, large surface area, and site-selective conjugation with biomolecules (Chee et al., 2015; Wang et al., 2016a, b, c, d; Zhang and Chen, 2017).

Carbon nanotubes are actually applied in the construction of electrochemical sensors for the diagnosis of many contaminants, including drugs, chemical dyes, ions, phenols, and chemical pesticides (De Oliveira et al., 2015; Govindhan et al., 2015; Mani et al., 2015; Wang et al., 2014a, b; Wong et al., 2017). Consequently, there is an urgent need to

Carbon Nanomaterials for Agri-food and Environmental Applications. https://doi.org/10.1016/B978-0-12-819786-8.00018-9

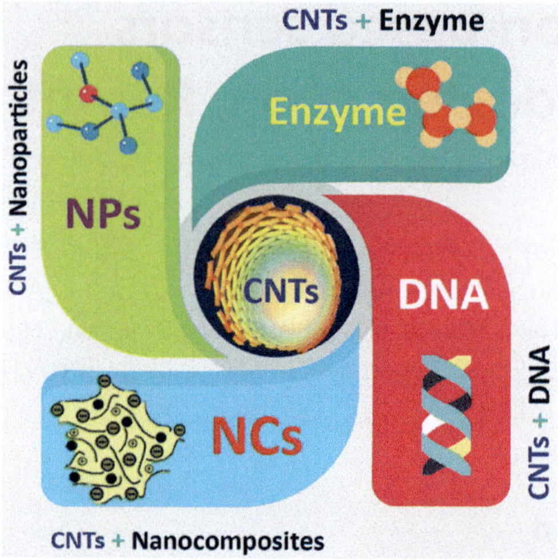

Fig. 1 Schematics of various types of materials such as nucleic acids, enzymes, inorganic and organic NPs, nanocomposites, nanosized polymers, molecularly imprinted polymers, and hybrid materials combined with CNMs to produce electrochemical biosensors.

produce a plain, sensitive, selective, low-cost, and mobile sensing system for pesticide detection. In contrast to the photometric and chromatographic techniques, electrochemical tests implementing improved electrodes offer ultrasensitive, portable, and easy-to-use systems combined with low-cost voltammetric equipment. Electrochemical sensors engineered with pristine carbon nanotubes or conjugated by carbon nanotubes with other modifiers may be identified within these. There are certainly transformed electrodes composed of CNTs and ionic liquids (ILs), porphyrin, phthalocyanines, metallic nanoparticles, and others (Wong et al., 2017). Zhao et al. (2018) discussed modern innovations and the latest trends in carbon-based nanosensors for the recognition of pesticides, incorporating the utility of graphene, carbon nanotubes, carbon dots, and other types of nanocarbons with two main biosensing approaches, that is, enzyme and enzyme-free biosensors. Likewise, a comprehensive reference source for experts in biosensors, graphene, materials science, nanotechnology, chemistry, and electrochemistry was observed by (Krishnan et al., 2019). All these improvements in nanocarbon biosensor exploration can provide unique and encouraging tools for the sensitive discovery of many pesticides (Fig. 1), which may present possible applications such as food, water, the agrifood sector, and ecosystem control. The current chapter is aimed to summarize the use of carbon-based nanomaterials for electrochemical and fluorescent for rapid sensing of chemical pesticides in various agroecosystem media.

2 Carbon-based pesticide nanosensors

2.1 CNTs as pesticides sensors

Presently, carbon nanomaterials may be used as transducers for pesticide detectors. Carbon nanomaterial-based pesticide biosensors primarily involve electrochemical and optical biosensors (Zhao et al., 2018). Sundari et al. (2010) detected carbendazim fungicide by applying a GCE improved with functionalized multiwalled carbon nanotubes (FMWCNTs/GCE). Through the use of an altered GCE with FMWCNTs, the analysis peak increased 15 times by choosing differential pulse adsorptive stripping voltammetry. The current sensor may start reading a large stepwise concentration range of 5.2×10^{-11} to 2.6×10^{-4} mol. Pesticide monitoring in soil and aquatic samples was achieved with high restoration rates of around 84.5% and 93.7%, demonstrating the good use of the recommended sensor. It was observed that producing a carbon paste electrode with tricresyl phosphate (TCP-CPE) as a liquefied binder is usually applied for the delicate assay of the Linuron herbicide 3-(3,4-dichlorophenyl)-1-methoxy-1-methylurea. They showed that the designed detection procedure supplied positive linearity results in the concentration range of 1.25–44.20 $\mu g\,mL^{-1}$ (Đorđević et al., 2012). The MWCNT paste electrode was applied to detect cyromazine insecticide obtained by square-wave adsorptive stripping voltammetry. The analytical curve data were linear between 2.5×10^{-6} and 5.0×10^{-4} mol L^{-1}. The current approach demonstrated a wonderful achievement in the immediate sensing of cyromazine in spiked normal water samples, with recoveries of $(102 \pm 2)\%$ and $(101 \pm 2)\%$ for river and tap water trials (Mercan et al., 2011). The MWCNT paste electrode was identified as methiocarb insecticide residue in both soil and river water samples, and the sensor and the provided findings were perfect, with recoveries of $(99 \pm 1)\%$ and $(98.5 \pm 0.3)\%$, respectively (İnam and Bilgin, 2013). Mani et al. (2015) offered a unique modification on the GCE surface with graphene oxide and MWCNTs for the fast and sensitive detection of two herbicides including diuron and fenuron. Salehzadeh et al. (2016) applied the electrochemical assay for the detection of fenitrothion (FT) and bifenox (BF) in river water samples using (FMWCNTs/GCE) with (SWV). The analytical sensitivity for FT and BF was 0.08 μM and the linear results were more than the range 0.2–60 μM. The present technique revealed restoration percentages of FT and BF in the range of 96.6%–104%. An electrochemical method for the numerical quantification of propham herbicide in spiked Bzura River water samples was assayed on (GCE/MWCNTs) by (SWAdSV). The limit of detection (LOD) and limit of quantification (LOQ) obtained from the calibration curve were 3.65×10^{-7} and 1.09×10^{-6} mol/L (Leniart et al., 2016), respectively. Modern studies on the application of carbon nanomaterials such as CNTs and graphene as pesticide nanosensors are described in Table 1.

2.2 *Electrochemical acetylcholinesterase (AChE, E.C.3.1.1.7)*

At the same time as the rapid production of biosensors, there are still several concerns regarding working with enzyme-based biosensors in an open field in contrast to the in vitro assay. The impact of distinct pesticide ingredients on the same enzyme is clearly not observed. Pesticide residue sensing by using enzyme systems is simply a successful approach because various pesticides are manufactured to prevent multifarious enzymes inside specific pests (Yan et al., 2015). Based on this theory, some enzymes such as acetylcholinesterase (AChE), butyrylcholinesterase (BChE), alkaline and acid phosphatase (ACP), tyrosinase, organophosphorus hydrolase, aldehyde dehydrogenase, and others may be employed to assess pesticide chemical residues in various ecosystem samples such as soil or water systems, and also in food or drinks (Lei et al., 2016;

Table 1 Application of CNTs and graphene in sensor platforms in the detection of pesticide residues.

Target analyte	Detection mechanism	Recognition element (receptor)	Transducer	Limit of detection	Ref.
Imidacloprid	CV	MIP (p-vinylbenzoic acid (VBA))	GCE/graphene	0.10 μM	Zhang et al. (2017)
Pyrimethanil	CV	β-Cyclodextrin	GCE/PANI/MWCNT-OOH	1.04 mM	Garrido et al. (2016)
6-Benzylaminopurine	CV and EIS	MIP (chitosan)	GCE/MWCNT@SnS_2	50 pM	Gan et al. (2016)
Carbofuran	DPV and EIS	MIP (o-phenylenediamine)	GCE/MWCNT/Pd-Ir/MB	1.7 pM	Li et al. (2016a, b, c)
Carbofuran	Amperometry	MIP (o-phenylenediamine)	CPE/MWCNT-COOH-Fe_3O_4@Au	3.8 nM	Amatatongchai et al. (2018)
Chlorpyrifos	DPV	AChE, Fe_3O_4	SPE/Fe_3O_4/graphene	0.02 $\mu g\,L^{-1}$	Wang et al. (2016a, b, c, d)
Organophosphorus pesticides	UHPLC-ESI/Q-Orbitrap MS	Neutral alumina, MWCNTs, diatomite, $MgSO_4$ (extraction step)	MS	0.08–0.29 $\mu g\,kg^{-1}$	Su et al. (2016)
Phthalates	HPLC	MWCNT (extraction step)	HPLC-MS	0.143–1.45 $\mu g\,L^{-1}$	González-Sálamo et al. (2018)
Fenitrothion	DPV	Surface adsorption	GCE/CeO_2/rGO	3 nM	Ensafi et al. (2017)
Malathion	ISFET	Au-gate, inhibition of functionalized enzyme acetylcholinesterase (AChE)	SWCNT-FET	Not determined	Bhatt et al. (2017)
Carbaryl	ISFET	Inhibition of urease enzyme	CVD graphene/Pt	10^{-8} $\mu g\,mL^{-1}$	Thanh et al. (2018)
Trichlorfon	DPV	Carboxymethyl chitosan	GCE/MWCNTs@TiO_2/	0.4 μM	Wei et al. (2016)

Abbreviations: *–COOH*, carboxylized; *CV*, cyclic voltammetry; *CVD*, chemical vapor deposition; *DPV*, differential pulse voltammetry; *EIS*, electrochemical impedance spectroscopy; *GCE*, glassy carbon electrode; *HPLC*, high performance liquid chromatography; *ISFET*, ion selective field effect transistor; *MIP*, molecular imprinted polymer; *MS*, mass spectrometry; *MWCNT*, multiwalled carbon nanotubes; *NP*, nanoparticles; *PANI*, polyaniline; *rGO*, reduced graphene oxide; *SWCNT*, single-walled carbon nanotubes; *UHPLC-ESI/Q-Orbitrap MS*, ultrahigh-performance liquid chromatograph with an electrospray ion source and a quadrupole orbitrap mass spectrometer.

Syshchyk et al., 2015). Presently, biosensors based upon enzyme-functionalized nanomaterials offer high performance and achievability. There will be a large possibility of application for these biosensors (Zhang et al., 2014; Zhou et al., 2014). A sensitive biosensor for chlorpyrifos (CPF), an organophosphorus pesticide, was formulated by inhibition (AChE) by using covalent binding to an oxidized exfoliated graphite nanoplatelet (xGnPs)-chitosan nanocomposite. The designated sensor may be used for the determination of organophosphorus pesticides (Ion et al., 2010). A fast and sensitive electrochemical (AChE) biosensor was effectively closed on a polyaniline (PANI) and (MWCNTs) core-shell improved glassy carbon electrode (GC), and employed to analyze carbamate pesticide residue in some fruits and vegetables, including apples, broccoli, and cabbage. The GC/MWCNT/PANI/AChE biosensor provides detection limits for carbaryl and methomyl that ranged from 1.4 to 0.95 $\mu mol\,L^{-1}$, respectively (Cesarino et al., 2012). A significant quantity of enzyme was frozen on a glassy carbon electrode throughout, particularly joining among functionalized ZnSe quantum dots and (AChE), and a graphene-chitosan nanocomposite was employed for the perseverance of methyl parathion insecticide in spiked water and soil samples (Dong et al., 2013). Yan and his research group recorded an innovative AChE biosensor by capturing multilayer films comprising AChE liposomes, MCNTs, and chitosan (Yan et al., 2013). AChE biosensors are among the most frequently utilized for pesticides that continue to be tested in ecosystem oversight as well as monitoring the quality and safety of food because of their particular high-sensitivity rampancy and offering of quick results (Ding et al., 2014a, b). Likewise, for diagnosis of organophosphorus pesticides (OPs), a unique amperometric biosensor based on executing polymer using an MWCNT-amended electrode is usually applied. AChE was properly frozen with covalent attachments on the altered different carbon nanomaterial electrodes. These groups of carbon composite electrodes are extremely promising for electroanalysis requirements, and potential electrochemical inquiries must be performed on the sensing and biosensing of different pesticides in environmental and agrifood sectors (Fig. 2).

The synthesized biosensor was screened for quick screening of pesticide residues in reinforced tap water samples (Kesik et al., 2014). The innovative electrochemical sensor based on the combined nanocomposite, which involves copper oxide nanowires (CuO NWs) combined with SWCNTs, was initially loaded for the detection of organophosphates. Malathion pesticide detection can be done under improved functional conditions; a differential pulse voltammetry technique was applied and the sensitivity of the detection limit was under 0.1 ppb. The current method offers great discrimination against several other prevalent pesticides, inorganic ions, and sugars, and has a positive stableness and repeatability (Huo et al., 2014). To be able to design a new photoelectrochemical AChE biosensor for two pesticides including paraoxon and dichlorvos, Li and his coworkers performed a nanocomposite based on CdSe@ZnS nanoparticle hybrid graphene onto an indium thin oxide glass electrode (Li et al., 2015). A sensitive electrochemical biosensor method was developed for the detection of carbaryl pesticide in some environmental samples based on freezing of the ACh enzyme on the (MWCNTs/GONRs) nanocomposite (Liu et al., 2015). Wei and Wang (2015) prepared a specific nanocomposite AChE sensor through binding honeycomb-like porous with carbon, gold nanoparticles, and ionic liquid loaded into

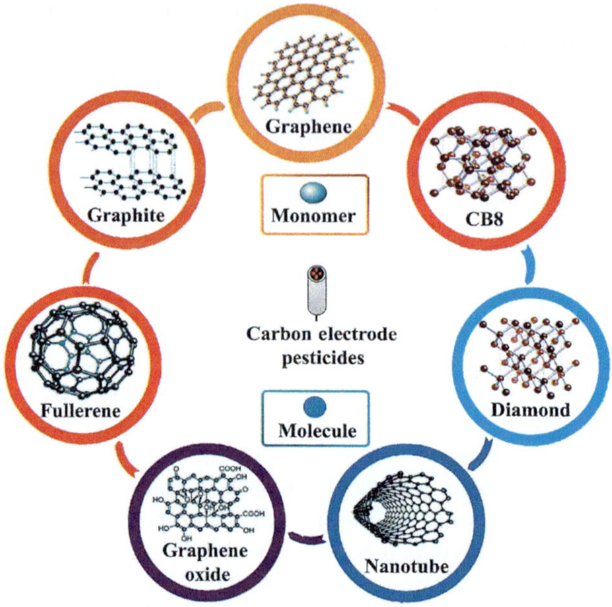

Fig. 2 Carbon nanomaterials such as graphene (Gr), carbon nanotubes (CNTs), and graphite are promising materials to produce the altered glassy carbon electrode (GCE) applied for pesticide detection due to its great sensitivity, low cost, low toxicity, good reproducibility, and high stability.

a boron-doped diamond electrode. The present procedure provided recoveries ranging from 80.8% to 93.1%. The manufactured biosensor was formulated based on a combination of AChE and MWCNT by joining with Pt/Au bimetallic NPs to generate the altered electrode and a unique sensor applied for the detection of chlorpyrifos, malathion, and methyl parathion pesticides (Miao et al., 2016). A simple electrochemical AChE biosensor that includes nafion (NA) and silver (Ag) nanoparticles jointly with amine-functionalized reduced graphene oxide (rGO-NH_2) to produce new pesticide nanosensor. The modified biosensor has a number of advantages such as it is very sensitive and steady with good reproducibility. It was employed to identify malathion, methidathion, and chlorpyrifos-ethyl with a broad linear range from 0.0063–0.077 mg/mL, 0.012–0.105 mg/mL, and 0.021–0.122 mg/mL, respectively (Guler et al., 2017). A new electrochemical biosensor has been produced by freezing AChE on electrochemically induced 3D graphene oxide network/MWCNT composites (e-GON-MWCNTs) under ideal conditions. The nanosensor displays a linear range of 0.03–0.81 ng mL^{-1} for sensitive detecting carbofuran, and two linear ranges of 0.05^{-1} ng mL^{-1} and 1–104 ng mL^{-1} for sensing paraoxon (Li et al., 2017a, b). An innovative electrochemical AChE biosensor consisting of 3DG-CuO NFs nanocomposites was employed for the detection of OPs pesticides. The created AChE biosensor was implemented below the most effective conditions such as the incubation time in the pesticide and the amount of AChE loading. The AChE-CS/3DG-CuO NFs/GCE biosensor presented a successful performance to malathion sensing with a large linear range, great sensitivity, excellent reproducibility, and high stability (Bao et al., 2015, 2019). The positive aspects in using enzymes, specifically AChE, in biosensors is because of enzyme-functionalized nanomaterials for the detection of (OPs). The high sensitivity, detection limit, recent challenges, disadvantages, and recommended solutions were observed in detail by Xiong et al. (2018). For instance, this includes the sensor limitations and general performance regarding nanomaterial type, and the capability of enzymes to distinguish among distinct pesticides. Because of its large surface area and exclusive electrocatalytic characteristics, graphene may significantly boost biosensing efficiency (Krishnan et al., 2019) (Table 2). In bi-enzyme systems, the strengths of multisignal amplification are generally regarded as a hypersensitive and selective procedure for the development of biosensors.

2.3 Molecularly imprinted polymers/CNTs sensors

Molecularly imprinted polymers (MIPs) are completely obtained throughout molecular imprinting for synthetic polymers that occur in the existence of the module molecule to be printed. An MIP is a large selective receptor with molecular adsorbent spots that assist large specificity and selectivity for targeted substances. This polymeric synthetic receptor is beneficial because of its technical, thermal, and chemical stabilities (Asman et al., 2015). Combining carbon nanomaterials such as graphene and CNT with polymers to increase the sensitivity of the MIP sensor results in a better particular surface area apart from enhanced electrical conductivity as a result of the synergistic impact in electrocatalytic functions. Furthermore, the utilization of CNTs enables greater stableness of GR sheets, which usually stack and aggregate through strong p-p stacking.

MIPs are prepared by combinatorial molecular imprinting on gold electrodes of quartz crystal microbalance (QCM, a piezoelectric transducer) for the selective detection of atrazine pesticide (Yaqub et al., 2011). Carbon nanomaterials such as MWCNTs-MIP were produced and employed as a molecular detector element to develop an electrochemical sensor for assaying parathion-methyl residues. The current sensor was also used in the quick detection of parathion-methyl in some fruits (pears and cucumbers) with average recoveries among 94.9% and 106.2% ($RSD < 5\%$) recorded (Zhang et al., 2012a, b). A new potentiometric sensor based upon an ion-imprinted polymer inclusion membrane (IPIM) was fabricated from the changes of MWCNT-based molecularly imprinted polymers for the detection of pesticide 2,4-D (2,4-dichlorophenoxyacetic acid) residues in water samples collected from the natural environment. The sensitivity detection limit was approximately 1.2×10^{-9} M. The membrane stability was determined to be three months without losing any sensitivity (Anirudhan and Alexander, 2014). A unique potentiometric sensor with large selectivity and sensitivity for pesticides was designed for residue detection of lindane, ($C_6H_6Cl_6$), based on the modification of γ-HCCH-imprinted polymer coated onto the surface of the Cu electrode. The MIP potentiometric sensor was obtained from the surface changed multiwalled carbon nanotube, with the percentage of recovery between 1×10^{-10} and 1×10^{-3} M (Anirudhan and Alexander, 2015).

The synthesized screen-imprinted SWCNT may boost the sensitivity of the electroanalytical assay due to its better binding ability for large concentrations of

Table 2 Graphene nanocomposite-modified glassy carbon electrode (GCE)-based acetylcholinesterase (AChE) electrochemical biosensors for detecting pesticides.

Graphene-based nanocomposites	Pesticides	Linear range	Limit of detection	Ref.
PtNPs-carboxylic GR-Na on	Methyl parathion	1×10^7 to 1 10^2 mM	5×10^5 nM	Yang et al. (2013a)
PtNPs-carboxylic GR-Na on	Carbofuran	1×10^6 to 1×10^2 mM	5×10^4 nM	Yang et al. (2013a)
NiONPs-carboxylic GR-Na on	Methyl parathion	1×10^7 to $1\times10^{\ 2}$ mM	5×10^5 nM	Yang et al. (2013b)
NiONPs-carboxylic GR-Na on	Chlorpyrifos	1×10^6 to 1×10^2 mM	5×10^4 nM	Yang et al. (2013b)
SnO_2NPs-carboxylic GR-Na on	Methyl parathion	1×10^7 to 1×10^2 mM	5×10^5 nM	Zhou et al. (2013)
SnNPs-carboxylic GR-Na on	Chlorpyrifos	1×10^6 to 1×10^2 mM	5×10^4 nM	Zhou et al. (2013)
MIPs/rGO-AuNPs	Carbofuran	5.0×10^2 to 0.2 mM	2.0×10^2 mM	Tan et al. (2015a, b)
MIPs-GR	Imidacloprid	0.5–15 mM	0.10 mM	Zhang et al. (2017)
EC-rGO/Na on	Methyl parathion	0.002–0.7 mM	1 nM	Jeyapragasam et al. (2017)
GR/GCE	Phoxim	20–20,000 nM	8 nM	Chao and Chen (2014)
Poly(3-methylthiophene)/N-doped GR	Phoxim	0.02–2.0 mM	6.4 nM	Wu et al. (2015)
Carboxylic chitosan/AgNCs-rGO composite	Phoxim	0.2–250 nM	81 pM	Zheng et al. (2015a, b, c)
TiO_2-GO/UiO-66 composite	Paraoxon	1.0–100 nM	0.2 nM	Karimian et al. (2019)
	Chlorpyrifos	5.0–300 nM	1.0 nM	
MWCNTs/GO nanoribbons	Carbaryl	5–5000 nM	1.7 nM	Liu et al. (2015)
AgNPs-carboxylic GR/Na on	Carbaryl	1.0×10^6 to 1×10^2 mM	5.45×10^4 nM	Liu et al. (2014)
AgNPs-carboxylic GR/Na on	Chlorpyrifos	1.0×10^7 to 1×10^2 mM	5.3×10^4 nM	Liu et al. (2014)
ZnONPs-carboxylic GR/Na on	Chlorpyrifos	1.0×10^7 to $1\times10^{\ 2}$ mM	5.0×10^5 nM	Wang et al. (2014a, b)
ZnONPs-carboxylic GR/Na on	Carbofuran	1.0×10^6 to 1×10^2 mM	$5.2\times10^{\ 4}$ nM	Wang et al. (2014a, b)
Il-GR-gelatin	Carbaryl	1.0×10^8 to 1×10^2 mM	5.3×10^6 nM	Zheng et al. (2015a, b, c)
Il-GR-gelatin	Monocrotophos	1.0×10^7 to 5×10^2 mM	4.6×10^5 nM	Zheng et al. (2015a, b, c)
PBNCs-rGO composite	Monocrotophos	1.0–600 ng mL^1	0.1 ng mL^1	Zhang et al. (2012a, b)
rGO-AuNPs/b-CD/PB-CS	Malathion	7.98 to 2.0 10^3 pg $mL^{\ 1}$	4.14 pg mL^1	Zhao et al. (2015)
rGO-AuNPs/b-CD/PB-CS	Carbaryl	4.3 to 1.0 10^3 pg mL^1	1.15 pg mL^1	Zhao et al. (2015)
CS@TiO_2-CS/rGO	Dichlorvos	0.036–22.6 mM	29 nM	Cui et al. (2018)
CeO_2-rGO composite	Fenitrothion	0.025–2.0 mg mL^1	3.0 ng mL^1	Ensafi et al. (2017)

rGO-AuNPs composite	Diuron	0.5–30.0 mg mL^{1}	0.125 mg mL^{1}	Shams et al. (2016)
rGO-AuNPs/PPy composite	Paraoxon-ethyl	1.0 nM to 5 mM	0.5 nM	Yang et al. (2014)
GR/PANI composite	Carbaryl	38–194 ng mL^{1}	20 ng mL^{1}	Li et al. (2016a, b, c)
Porous GO	Carbofuran	0.3–6.1 ng mL^{1}	0.15 ng mL^{1}	Li et al. (2017a, b)
PLaE-CS/AuNPs-GR	Methyl parathion	0.19–760 nM	0.19 nM	Bao et al. (2015)
PLaE-CS/AuNPs-GR	Malathion	1.5–1513.5 nM	1.51 nM	Bao et al. (2015)
EC-rGO-Na on	Dichlorvos	5–100 ng mL^{1}	2 ng mL^{1}	Wu et al. (2013)
3D GR/MWCNTs composite	Carbaryl paraoxon	0.3–0.81 ng mL^{1} 0.05–1 ng mL^{1}	0.015 ng mL^{1} 0.025 ng mL^{1}	Li et al. (2017a, b)
Ionic liquid-GR/CO_3O_4/CS	Dimethoate	5 10^3 nM to 0.1 mM	1×10^4 nM	Zheng et al. (2016)

Abbreviations: *AChE*, acetylcholinesterase; *AuNPs*, gold nanoparticles, carboxylic graphene (C-GR) Na on (NF); *b-CD*, b-cyclodextrin; *C-CS*, carboxylic chitosan; *CeO_2*, cerium oxide; *GCE*, glassy carbon electrode; *GNs*, graphene nanosheets; *Il-GR*, ionic liquid-functionalized graphene; *Methyl parathion*, *O,O*-dimethyl-*O*-(4-nitrophenyl) phosphorothioate; *MIPs*, molecularly imprinted polymers; *MWCNTs*, multiwalled carbon nanotubes; *NiONPs*, nickel oxide nanoparticles; *PANI*, polyaniline; *PBNCs*, Prussian blue nanocubes; *PLaE*, plant esterase; *PPy*, polypyrrole; *PtNPs*, platinum nanoparticles; *rGO*, reduced graphene oxide; *SnO_2NPs*, tin oxide nanoparticles.

Reprinted from Krishnan S.K., Singh E., Singh P., Meyyappan M., Nalwa H.S., 2019. A review on graphene-based nanocomposites for electrochemical and fluorescent biosensors. RSC Adv. 9 (16), 8778–8881, with permission from The Royal Society of Chemistry.

diuron. Moreover, researchers discovered the existence of catalytic characters and a rise in the oxidation process for the modified molecule by up to four times when compared to the normal electrode, which is credited to the great electrical properties of the SWCNT (Fernández-Cori et al., 2015). A unique MIP electrochemical sensor was manufactured from a glassy carbon electrode embellished with reduced graphene oxide and gold nanoparticles (rGO@Au) for the speedy detection of the carbofuran (CBF) pesticide. The developed sensor displayed substantial absorbent capability and positive selectivity for CBF. It was efficiently employed in an assay of CBF traces in vegetable samples collected from the market (Tan et al., 2015a). An MIP coating film was generated upon a graphene-modified electrode for the perseverance of phoxim, utilizing a free radical polymerization process. The current MIP electrochemical sensor was used to evaluate phoxim trace levels in cucumber samples with restoration starting from 98.06% to 101.13% (Tan et al., 2015b). An innovative selective MIP sensor was constructed for the assessment of hexazinone (HXZ) in distinct ecosystem samples such as river water samples (Toro et al., 2015). Wong et al. (2015) created an electrochemical sensor altered with an MIP and carboxyl-functionalized multiwalled carbon nanotubes (MWCNT-COOH). The new sensor was sensitive and selective to the fast detection of diuron in river water samples, and the obtained value of accuracy was 9.0×10^{-9} mol/L. MIP and MWCNT sensors were coated on a glassy carbon electrode for quick detection of 6-benzylaminopurine (BAP). The recommended sensor revealed a much better sensitivity and high selectivity compared with the imprinted chitosan, which is usually offered as the selective identification sites ($n=6$) as shown in Fig. 3. The engineered

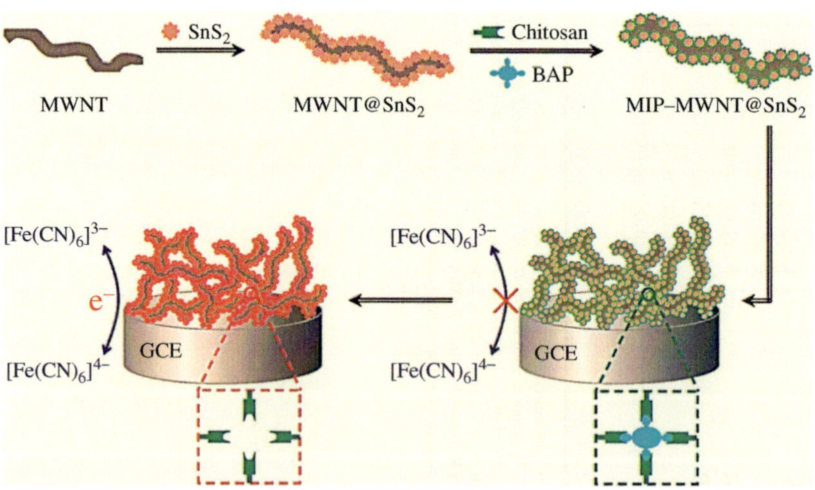

Fig. 3 Schematic diagram of the molecular imprinting mechanism on molecularly imprinted polymers (MIP) and MWCNT@SnS_2/GCE.
Reprinted from Gan, T., Lv, Z., Sun, Y., Shi, Z., Sun, J., Zhao, A., 2016. Highly sensitive and molecular selective electrochemical sensing of 6-benzylaminopurine with multiwall carbon nanotube@SnS_2-assisted signal amplification. J. Appl. Electrochem. 46 (3), 389–401, with permission from Springer.

sensor was displayed to be fast, sensitive, and economical for BAP detection in environmental samples (Gan et al., 2016).

The combined MWCNTs and MIP were employed as modifiers in the sensor composition. An MIP and a nonimprinted polymer (NIP) were fabricated and applied in the carbon paste electrode. The present sensor was used to determine the dicloran traces in human urine as well as tap and river water samples without a particular method for sample extraction prior to assays (Khadem et al., 2016, 2017). MIP nanoparticles were produced by the polymerization process and after that, utilized for adjustment of a carbon paste electrode (CPE) structure to produce a selective and sensitive sensor for quick detection of diazinon (DZN) residues. The provided findings demonstrated that the carbon paste electrode coated with MIP-NPs (nano-MIP-CP) offers high adsorption potential for diazinon compared to CPE-based nonimprinted polymer NPs (nano-NIP-CP) (Motaharian et al., 2016). A unique and highly selective sensor based on MIPs and carbon nanotubes altered carbon paste electrode was applied in detection of the dicloran pesticide (Shahtaheria et al., 2017). A new electrochemical sensor based on an MIP and platinum nanoparticle (Pt NPs)/carbon nitride nanotube (C_3N_4 NTs) nanocomposite was produced for atrazine (ATR) trace examination. The current voltammetric sensor was utilized to monitor ATR wastewater samples with great accuracy (Yola and Atar, 2017). As a result of the specific physical and chemical characteristics of the carbon-based nanomaterials and their excessive surface area, these kinds of products have an important function in the excellent analytical efficiency detected by MIP-based sensors.

2.4 CNT nanocomposites sensors

Carbon-based hybrid nanomaterial electrochemical sensors are effective techniques to perform this activity, mainly because of their high sensitivity and selectivity properties that provide successful detection of pesticide traces in ecosystem samples. Presently, there is an interest in using metal oxide-CNT nanocomposites and metal oxide-polymer composites in electrochemistry to boost the functionality of electrochemical sensors for the fast detection of biological and environmental pollutants, including pesticides. The synergistic impact from the hybrid use of metallic NPs and carbon nanotubes has continued to be investigated in recent years for sensing pesticide residues. Metallic NPs and carbon nanotubes increase the electrochemical efficiency of bare electrodes, offering a substantial electroactive surface area and, consequently, better analytical sensitivities and lower detection limits (Wong et al., 2017). The AuNP/CNT-altered GCE was fabricated. First, the GCE surface was improved with $5\,\mu L$ of a 0.5% m/m Nafion aqueous dispersion comprising CNTs. The designated voltammetric method was employed in the parathion detection of spiked water and vegetable samples with restoration rates starting from 98.3% to 104.3% (Zhang et al., 2009). Polymeric nanoparticles were integrated with a carbon paste electrode (CPE) to develop a fluoxetine selective sensor. The detection limit of the sensor was determined to be similar to $2.8\times10^{-9}\,mol\,L^{-1}$ (3S b/m). This unique sensor was applied effectively to fluoxetine monitoring in the spiked plasma samples in addition to fluoxetine ingredients (Alizadeh and Azizi, 2016).

Similarly, functionalized MWCNTs combined with @titanium dioxides/carboxymethyl chitosan to produce (MWCNTs@TiO_2/CMCS) nanocomposites were synthetized by in situ noncovalent covering of MWCNTs@TiO_2 with CMCS moieties while the MWCNTs@TiO_2 was fabricated throughout an in situ sol-gel reaction of tetrabutyl titanate in the presence of MWCNTs. Present MWCNTs@TiO_2/CMCS nanocomposites were designed to effectively determine trichlorfon pesticide traces, showing significant linear correlation among the various concentrations (Wei et al., 2016). Pd wormlike nanochains/graphitic carbon nitride (Pd WLNCs/g-C_3N_4) nanocomposites and AChE were developed for detection of both organophosphorus (OPs) and huperzine-A (hupA) pesticides. The fabricated electrochemical biosensor possessed appropriate reliability and a high potency for sensible functions (Ding et al. 2014a, b; Wang et al., 2016a, b, c, d). Graphene-conjugated nanocomposites were produced by reduction of graphene oxide in the existence of conductive polymers (PEDOT:PSS and polypyrrole) and AuNPs, which had been transferred by drop spreading onto gold interdigitated electrodes. The modified nanocomposite electrode offered sensing devices with a large precise surface area and substantial conductivity, yielding the production of a unique sensor with suited stability, positive repeatability, and significant sensitivity to pesticide samples; it was allowed to differentiate concentrations under 0.1 nmol L^{-1} (Facure et al., 2017). A new voltammetric sensor for the detection of diazinon pesticide residues was dependent on a glassy carbon electrode (GCE) surface altered with MWCNTs coated by TiO_2 nanoparticles (MWCNTs/TiO_2NPs). The sensors demonstrated a linear correlation range from 11 to 8360 nM and a detection limit of between 3 and 10 nM (Ghodsi and Rafati, 2017). A new selective and sensitive electrochemical sensor for the fast detection of an organophosphorus pesticide and methyl parathion (MP) was designed simply by using silver particle-supported graphene nanoribbons (Ag@GNRs). The current electrode nanocomposite is properly validated in the immediate MP assay in some food, vegetable, and fruit samples (Govindasamy et al., 2017).

On the other hand, an extremely consistent electrochemical AChE biosensor for monitoring of organophosphorus pesticides was formulated easily by adsorption of AChE on chitosan (CS), TiO_2 sol-gel, and reduced graphene oxide (rGO) to produce a CS @ TiO_2-CS/rGO detector. The biosensor has the capability to identify dichlorvos traces with range from 0.036 μM (7.9 ppb) to 22.6 μM, with a limit of detection of 29 nM (6.4 ppb) and a total time needed for assay of approximately 25 min. The present modified biosensor has advantages such as reproducibly and stability both in detection and storage, and it may effectively detect dichlorvos residues in cabbage juice samples, offering a reliable system for immobility of AChE (Cui et al., 2018). Other types of nanocomposites formulated with oxide nanoparticles (SnO_2) and carboxylated multiwalled carbon nanotubes (cMWCNTs) were employed to monitor methyl parathion (MP) pesticides in water samples. The sensor was significantly linear from a range of 1–160 μM, which is more efficient with a lesser diagnosis limit of 0.1 μM for pesticide detection assays in water samples. The current electrode was used 30 times and produced effective results (Dhull, 2018). Cu-doped TiO_2 hybrid nanocomposites were coated on the glass carbon electrode as the nonenzymatic electro-nanosensor for fast and selective detection of organophosphorous pesticides in groundwater.

The improved electrode may easily discover methyl parathion delicately on an extensive board from 0 to 2000 ppb with a lower limit of detection (LOD) of 1.21 ppb (Tian et al., 2018). A new modified carbon paste nanocomposite electrode (CPE) was employed as an electrochemical sensor for recognition of cypermethrin pesticide in various samples. The detection limit of the altered electrode was often down to the level of ~ 0.1 ppm, which may be under the optimum residual limits of cypermethrin in the ecosystem (~ 0.5 ppm) (Nurdin et al., 2019).

2.5 Ionic liquid (IL)/CNT sensors

Another type of carbon-based nanosensor, for instance ionic liquid (IL), is an organic salt that has specific physical and chemical characteristics that include great conductivity, substantial density, low toxicity, thermal and chemical stability, and nonvolatility. A fast, simplified, and delicate electroanalytical approach for monitoring pyrimethanil (PMT) applying a GCE-improved sensor with MWCNTs and 1-butyl-3-methylimidazolium hexafluorophosphate was created (Yang et al., 2015). Zheng et al. (2015a, b, c) modified an AChE biosensor to identify phorate insecticide utilizing a composite of ionic liquid functionalized graphene (IL-GR) and PVA (denoted as IL-GR-PVA). Within perfect variables, this amperometric biosensor displayed a linear correlation insecticide concentration of phorate was found in the range from 1.0×10^{-14} to 1.0×10^{-9} M and from 1.0×10^{-9} to 1.0×10^{-6} M, with a detection limit of 8.0×10^{-15} M (S/N = 3). The present modifiers allowed an improvement of the oxidation process, representing their particular electrochemical features. The modified sensor was applied effectively for monitoring pyrimethanil in fruit and water samples with recovery ranging from 96% to 110%. An easy method was used to produce a composite coating film comprising ionic liquid (IL), chitosan (CS), and electrochemically produced gold nanoparticles (AuNPs) on sole use pencil graphite electrodes (PGEs). The recommended nonenzymatic procedure was basic, effective, and economical. The confirmed technique was properly utilized for the detection of malathion residues in tomato and apple samples (Bolat and Abaci, 2018). Ionic liquid-assisted growth of a poly(3,4-ethylenedioxythiophene)/reduced graphene oxide-based electrode was prepared, three diverse OPs including chlorpyrifos (CP), malathion (ML) and methyl parathion (MP) were detected, the three organophosphorus pesticides were assayed with high sensitivity, selectivity and reproducibility (Rana et al., 2018).

2.6 Fullerene and quantum dot/CNT sensors

Fullerene-C60-MWCNT composite coated film is an ultrasensitive electrochemical sensing platform for the trace analysis of the endocrine disruptors vinclozolin and carbendaz. The sensor was employed for the detection of vinclozolin in wastewater samples with acceptable restoration percentages (Rather and De Wael, 2012). Feng et al. (2015) investigated the structure of an electrochemical sensor used for detection of pentachlorophenol in fish meat depending on a ZnSe quantum dot coated with MWCNTs. Fullerene (C60)-loaded gold nanoparticles (C60–Au) were utilized as the energy donor, and altered on a glassy carbon electrode. The detection accuracy of the current analysis was revealed to be 8.8×10^{-13} mol L $^{-1}$. Therefore, the sensing procedure explained in the following research could be modified for use in the recognition of numerous pesticide residues in environmental samples (Li et al., 2016a, b, c). Carbon quantum dots (CQDs) were prepared from commercialized chlorophyll and can be properly quenched by Au NPs via fluorescence resonance energy transfer (FRET). Depending on the FRET among CQDs and Au NPs, a fabricated, very sensitive, and practical nanosensor for organophosphorus pesticides was observed (Wu et al., 2017).

Özcan and Gürbüz (2018) designed a simplified and sensitive sensor for the voltammetric detection of the herbicide clopyralid in agroecosystem samples such as urine, river water, sugar beet, wheat, and herbicide formulation traces. This unique sensor was constructed by changes of a GCE with an MWCNT nanocomposite comprising oxygen-functionalized nanomaterials and fumed silica. A new fluorescent nanosensor was engineered for fast and sensitive monitoring of organophosphorus pesticides (OPs). Sipa et al. (2018) efficiently created an electrochemical sensor depending on a glassy carbon electrode altered with ß-cyclodextrins and MCNTs (ß-CDs/MWCNTs/GCE) for screening dichlorophen pesticide. The fluorescence of CDs is usually quenched by manganese dioxide (MnO_2) nanosheets through the Förster resonance energy transfer (FRET). A CD system is often applied to discover OPs measurably determined by the fluorescence turn "on-off-on" mechanism. Under the specific conditions, the current FRET-based procedure may easily identify paraoxon that ranges from 0.05 to 5 $ng\,mL^{-1}$ with a diagnosis limit of 0.015 $ng\,mL^{-1}$ (Yan et al., 2018). The agroecosystem screening and monitoring of various pesticides using electrochemical sensors based on carbon nanomaterials was explained in detail (Čović et al., 2018; Wong et al., 2017).

2.7 Aptasensors for pesticide detection

Aptamers are short RNA oligonucleotides or single-stranded DNA (ssDNA) that could particularly connect to a specific target consisting of tiny molecules such as amino acids, metal ions, and small-scale peptides, proteins, viruses, as well as some biological and bacterial cells (Sekhon et al., 2018; Song, et al., 2017). Aptamers are generally effectively applied in agrifood protection for fast detection and evaluation of antibiotics, biological toxins, pesticides, heavy metals, etc., in conjugation with various types of nanomaterials to develop electrochemistry and fluorescence nanosensor technology (Luan et al., 2016). The developed fluorescent aptasensors from some nanomaterials such as quantum dots (QDs), carbon dots (CDs), up-conversion nanoparticles (UCNPs), Au NPs, graphene oxide (GO), single-walled carbon nanotubes (SWNTs), etc., have been applied for quick detection of pesticide residues in environmental samples (Arvand and Mirroshandel, 2017; Bahreyni et al., 2018; Liu et al., 2019; Wang et al., 2018). These kinds of aptasensors may easily achieve a really sensitive and fast detection method for pesticide residues that never need to sample extraction techniques. A unique, sensitive, and selective electrochemical aptasensor based on mesoporous carbon (OMC) functionalized by chitosan (OMC-CS) and ferrocene hybrid chitosan (CS) dispersed multiwalled carbon nanotubes (Fc@MWCNTs-CS) was altered on the electrode surface. The modified electrode was practical for monitoring chlorpyrifos in a real vegetable sample. The chlorpyrifos concentration recoveries ranged from 98.5% to 107.2%, which definitely suggested that the aptasensor was suitable for the quick detection of pesticides in agroecosystem samples (Jiao et al., 2016). Fei et al. (2015) produced a sensitive electrochemical aptasensor for quick detection of acetamiprid traces at femtomole quantity using gold nanoparticles (Au NPs) coated with multiwalled carbon nanotube-reduced graphene oxide nanoribbon (Au/MWCNT-rGONR) composites, which were based on the variance of electron transfer resistance that was important to the development of the acetamiprid-aptamer complex. Furthermore, Li et al. (2016a) manufactured an electrogenerated chemiluminescence (ECL) sensor for carbofuran monitoring based upon ECL energy transfer and carbon dot (C-dot)-tagged aptamers as the identification element. The aptasensor displayed a large level of sensitivity to carbofuran with linear correlation results to carbofuran varying from 20 pM to 8.0 nM and a detection limit of 0.88 pM. A copper oxide nanoflower electrochemical aptasensor with a SWCNT nanocomposite for chlorpyrifos recognition exhibited a positive correlation for chlorpyrifos that ranged from 0.1 $ng\,mL^{-1}$ to 150 $ng\,mL^{-1}$, with a low detection limit of 70 pg/mL. The aptasensor also showed substantial selectivity and excellent reproducibility, and was efficiently employed for the detection of chlorpyrifos residues in spiked apples, celery, and cabbage with acceptable measures (Xu et al., 2018). Liu et al. (2019) evaluated distinct kinds of aptasensors utilized for fast monitoring of pesticide traces and residues in agroecosystem samples (Table 3).

2.8 Amperometric sensors for pesticide detection

Recently, some carbon nanomaterials have performed a critical role in the improved analytical efficiency of amperometric biosensors (Zhao et al., 2018). Wang et al. (2011) developed an ultrasensitive AChE amperometric biosensor based on the decoration of graphene with quantum dots made of CdS. The existence of CdS/graphene nanocomposites on glassy carbon electrodes supplied a level of sensitivity to organophosphates that was approximately three-fold greater than the unmodified glassy carbon electrodes.

Common chromatographic techniques such as high-performance liquid chromatography, HPLC, and an amperometric approach were employed for chlorpyriphos (CPF) monitoring in broccoli. The biosensor has a positive diagnosis limit of 10–10 M for chlorpyrifos in broccoli (Ion and Ion, 2012). Wu et al. (2013) formulated an amperometric electrochemical device consisting of a nanocomposite of rGO. Nafion was decorated on the electrode surface and used for the highly sensitive detection of dichlorvos in water and a few agrifood products. Bao et al. (2016) established an OPH-based biosensor for the detection of methyl parathion and parathion employing titanium dioxide nanofibers (TiO_2NFs) coated with carboxylic functionalized MWCNTs (cMWCNTs-TiO_2NFs). They demonstrated that a good connection of TiO_2NFs may result in the accumulation of OPs. Within the enhancing conditions, methyl parathion and parathion can be discovered with efficient results. The rGO-ZrO_2NPs nanocomposite was manufactured through a hydrothermal option and a final aerosol transferred on indium tin oxide (ITO) glass (Fig. 4). The recommended

Table 3 Aptasensors for the detection of pesticide residues.

Method Type	Target	Sample	Linear range	Detection limit	Reference
Electrochemical	Acetamiprid	Real water	50 fM to 10 μM	17 fM	Fei et al. (2015)
Electrochemical	Acetamiprid	Real water	10 pM to 100 nM	1 pM	Madianos et al. (2018)
Electrochemical	Acetamiprid	Fruit juice		0.086 μM	Rapini et al. (2016)
Electrochemical	Atrazine	Real water	100 pM to 1 μM	10 pM	Madianos et al. (2018)
Electrochemical	Chlorpyrifos	Apple, celery cabbage	0.1 to 150 ng mL^{-1}	70 pg/mL	Xu et al. (2018)
Fluorescent	Acetamiprid	River water and cabbage leaves	0–150 nM	0.7 nM	Lin et al. (2016)
Fluorescent	Acetamiprid	Tea	50 nM to 1 μM	3.2 nM	Hu et al. (2016)
Fluorescent	Acetamiprid	Tomatoes, cucumber, and cabbage	22.5 nM to 0.45 μM	4.85 nM	Wang et al. (2018)
Fluorescent	Acetamiprid	Tap water	5–50 nM	2.8 nM	Bahreyni et al. (2018)
Fluorescent	Chlorpyrifos	Vegetable		2.08 nM	Cheng et al. (2018)
Fluorescent	Diazinon			22.01 nM	Cheng et al. (2018)
Fluorescent	Malathion			2.24 nM	Cheng et al. (2018)
Fluorescent	Edifenphos	Water and rice	1.61–19.33 nM	0.42 nM	Arvand and Mirroshandel (2017)
Fluorescent	Acetamiprid	Vegetable	0.1 μM to 1.2 μM	9.12 nM	Liu et al. (2016)
Fluorescent	Malathion	Water, soil, and orange juice	10 pM to 1 μM	4 pM	Bala et al. (2018b)
Colorimetric	Omethoate	Soil	0.1 μM to 10 μM	0.1 μM	Wong et al. (2015)
Colorimetric	Malathion	Human serum	5 pM to 10 nM	1 pM	Abnous et al. (2018)
Colorimetric	Malathion	Apple		0.5 pM	Bala et al. (2018a)
Chemiluminescent	Acetamiprid	Waste water, soil, and cucumbers		62 pM	Qi et al. (2016)
Chemiluminescent	Carbofuran	Fruit and vegetable	20 pM to 8.0 nM	0.88 pM	Li et al. (2016a, b, c)

SERS	Isocarbophos	Apple juice		3.4 mM	Pang et al. (2014)
SERS	Omethoate			24 mM	Pang et al. (2014)
SERS	Phorate			0.4 mM	Pang et al. (2014)
SERS	Profenofos			14 mM	Pang et al. (2014)
SERS	Malathion	Tap water	500 nM to 10 μM		Nie et al. (2018)
Microfluidic chip	Omethoate	Vapor		4.8 nM	Fujii et al. (2017)

Updated from Liu, M., Khan, A., Wang, Z., Liu, Y., Yang, G., Deng, Y., He, N., 2019. Aptasensors for pesticide detection. Biosens. Bioelectron. 130, 174–184, with permission from Elsevier.

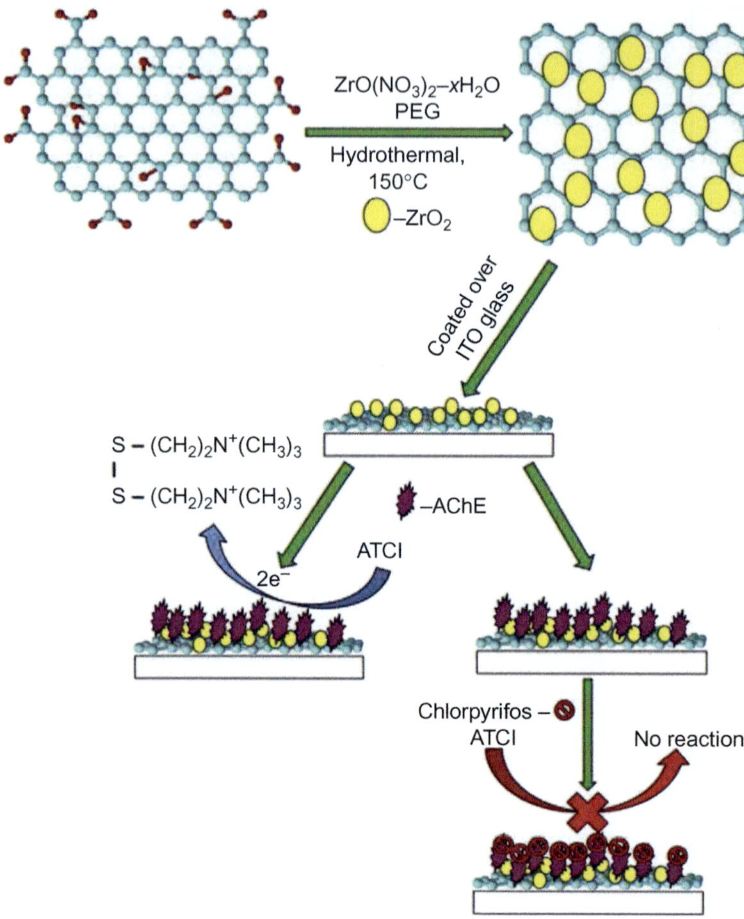

Fig. 4 Schematic illustration of the preparation of the rGO-ZrO_2 nanocomposite-based biosensor for chlorpyrifos detection.
Reproduced from Mogha N.K., Sahu V., Sharma, M., Sharma, R.K., Masram, D.T., 2016. Biocompatible ZrO_2-reduced graphene oxide immobilized AChE biosensor for chlorpyrifos detection. Mater. Des. 111, 312–320, with permission from Elsevier.

nanosensor could determine chlorpyrifos in the linear ranges from 0.1 to 103 pM (Mogha et al., 2016). An electrochemical biosensor depending on the assemblage of carbon nanotubes on a nanoporous gold electrode is explained for extremely sensitive monitoring of organophosphate pesticides. The nanoporous gold film (NPG) electrode is synthesized with an alloying/dealloying procedure, which usually has a large electroactive surface area and is a good substrate for sensor modifications. Carboxylated carbon nanotubes are extra conjugated with acetylcholinesterase (AChE) for amperometric detection of various pesticides (Ding et al., 2014a, b). Zhang et al. (2019) produced an amperometric AChE-biosensor applying hybrid polymer (furan-2-yl) benzothiadiazole (FBThF) and Ag-rGO-NH_2 nanocomposites. The modified sensor is reputable as it was examined in recognition of malathion and trichlorfon in tap water and apple samples, and the current findings advised that the biosensor is simple and fast in a specific sample (Zhang et al., 2019).

3 Conclusion and future perspectives

A number of environmental and human health challenges are related to the extensive consumption of chemical pesticides that have grown to be a severe concern due to the destruction of the ecosystem and human health. Therefore, the advancement of simple, quick, effective, and cheap detection strategies for pesticide discovery appears to be of high significance in scientific research and functional application. Various nanomaterials may easily offer a crucial advantage in the construction of biosensors in a hybrid with electrochemical sensors, as a result of their electrocatalytic process and large surface area. Carbon nanotube-based electrochemical sensors have been recommended for pesticide monitoring, combining distinct electrode modification approaches and electroanalytical methods. In this chapter, we present a review of the modern signs of progress in the utilization of carbon nanotubes for the improvement of electrochemical sensors devoted to the environmental verification of pesticides. Both electrochemical and fluorescent biosensors based on graphene nanocomposites, including graphene, GO, rGO/polymer nanocomposites, graphene/inorganic NP nanocomposites, molecularly imprinted polymer, aptasensors, and amperometric sensors, were employed for pesticide sensing. Carbon nanotubes have offered electrochemical sensors with the fairly positive rational efficiency needed for the detection of pesticides. Sometimes they were applied singly or in a hybrid with diverse choices of modifiers. The MWCNTs boost the sensor results due to elevating the electrode surface area and also enhancing the electron transmission around the electrode and supportive electrolyte. Most recently, nanomaterial/nanocomposites consisting of 3D graphene have become a unique and reliable procedure for onsite diagnosis of pesticides or herbicides with great functionality. Carbon-based nanomaterials are growing at an immediate pace, but as such, a complete evaluation of several nanomaterials under analytical performance conditions of the nano-engineered sensor is nonetheless a huge arena to be discovered.

References

Abnous, K., Danesh, N.M., Ramezani, M., Alibolandi, M., Emrani, A.S., Lavaee, P., Taghdisi, S.M., 2018. A colorimetric gold nanoparticle aggregation assay for malathion based on target-induced hairpin structure assembly of complementary strands of aptamer. Microchim. Acta 185 (4), 216.

Alizadeh, T., Azizi, S., 2016. Graphene/graphite paste electrode incorporated with molecularly imprinted polymer nanoparticles as a novel sensor for differential pulse voltammetry determination of fluoxetine. Biosens. Bioelectron. 81, 198–206.

Amatatongchai, M., Sroysee, W., Jarujamrus, P., Nacapricha, D., Lieberzeit, P.A., 2018. Selective amperometric flow-injection analysis of carbofuran using a molecularly-imprinted polymer and gold-coated-magnetite modified carbon nanotube-paste electrode. Talanta 179, 700–709.

Anirudhan, T.S., Alexander, S., 2014. Multiwalled carbon nanotube based molecular imprinted polymer for trace determination of 2, 4-dichlorophenoxyaceticacid in natural water samples using a potentiometric method. Appl. Surf. Sci. 303, 180–186.

Anirudhan, T.S., Alexander, S., 2015. Design and fabrication of molecularly imprinted polymer-based potentiometric sensor from the surface modified multiwalled carbon nanotube for the determination of lindane (γ-hexachlorocyclohexane), an organochlorine pesticide. Biosens. Bioelectron. 64, 586–593.

Arvand, M., Mirroshandel, A.A., 2017. Highly-sensitive aptasensor based on fluorescence resonance energy transfer between l-cysteine capped ZnS quantum dots and graphene oxide sheets for the determination of edifenphos fungicide. Biosens. Bioelectron. 96, 324–331.

Asman, S., Mohamad, S., Sarih, N., 2015. Effects of RAFT agent on the selective approach of molecularly imprinted polymers. Polymers 7 (3), 484–503.

Bahreyni, A., Yazdian-Robati, R., Ramezani, M., Abnous, K., Taghdisi, S.M., 2018. Fluorometric aptasensing of the neonicotinoid insecticide acetamiprid by using multiple complementary strands and gold nanoparticles. Microchim. Acta 185, 1–7.

Bala, R., Mittal, S., Sharma, R.K., Wangoo, N., 2018a. A supersensitive silver nanoprobe based aptasensor for low cost detection of malathion residues in water and food samples. Spectrochim. Acta A Mol. Biomol. Spectrosc. 196, 268–273.

Bala, R., Swami, A., Tabujew, I., Peneva, K., Wangoo, N., Sharma, R.K., 2018b. Ultra-sensitive detection of malathion using quantum dots-polymer based fluorescence aptasensor. Biosens. Bioelectron. 104, 45–49.

Bao, J., Hou, C., Chen, M., Li, J., Huo, D., Yang, M., Luo, X., Lei, Y., 2015. Plant esterase-chitosan/gold nanoparticles-graphene nanosheet composite-based biosensor for the ultrasensitive detection of organophosphate pesticides. J. Agric. Food Chem. 63, 10319–10326.

Bao, J., Hou, C., Dong, Q., Ma, X., Chen, J., Huo, D., Yang, M., El Galil, K.H.A., Chen, W., Lei, Y., 2016. ELP-OPH/BSA/TiO_2 nanofibers/c-MWCNTs based biosensor for sensitive and selective determination of p-nitrophenyl substituted organophosphate pesticides in aqueous system. Biosens. Bioelectron. 85, 935–942.

Bao, J., Huang, T., Wang, Z., Yang, H., Geng, X., Xu, G., Samalo, M., Sakinati, M., Huo, D., Hou, C., 2019. 3D graphene/copper oxide nano-flowers based acetylcholinesterase biosensor for sensitive detection of organophosphate pesticides. Sensors Actuat. B: Chem. 279, 95–101.

Bhatt, V., Joshi, S., Becherer, M., Lugli, P., 2017. Flexible, low-cost sensor based on electrolyte gated carbon nanotube field effect transistor for organo-phosphate detection. Sensors 17 (5), 1147.

Bolat, G., Abaci, S., 2018. Non-enzymatic electrochemical sensing of malathion pesticide in tomato and apple samples based on gold nanoparticles-chitosan-ionic liquid hybrid nanocomposite. Sensors 18 (3), 773.

Cesarino, I., Moraes, F.C., Lanza, M.R., Machado, S.A., 2012. Electrochemical detection of carbamate pesticides in fruit and vegetables with a biosensor based on acetylcholinesterase immobilised on a composite of polyaniline-carbon nanotubes. Food Chem. 135 (3), 873–879.

Chao, M., Chen, M., 2014. Electrochemical determination of phoxim in food samples employing a graphene-modified glassy carbon electrode. Food Anal. Methods. 7, 1729–1736.

Chee, W.K., Lim, H.N., Huang, N.M., Harrison, I., 2015. Nanocomposites of graphene/polymers: a review. RSC Adv. 5, 68014–68051.

Cheng, N., Song, Y., Fu, Q., Du, D., Luo, Y., Wang, Y., Xu, W., Lin, Y., 2018. Aptasensor based on fluorophore-quencher nano-pair and smartphone spectrum reader for on-site quantification of multi-pesticides. Biosens. Bioelectron. 117, 75–83.

Čović, J., Zarubica, A., Bojić, A., Ranđelović, M., 2018. Carbon nanotube-based electrochemical sensors for pesticide determination in aqueous solutions: a review. Chemia Naissensis 1, 130–149.

Cui, H.F., Wu, W.W., Li, M.M., Song, X., Lv, Y., Zhang, T.T., 2018. A highly stable acetylcholinesterase biosensor based on chitosan-TiO_2-graphene nanocomposites for detection of organophosphate pesticides. Biosens. Bioelectron. 99, 223–229.

De Oliveira, R., Hudari, F., Franco, J., Zanoni, M.V.B., 2015. Carbon nanotube-based electrochemical sensor for the determination of anthraquinone hair dyes in wastewaters. Chemosensors 3, 22–35.

Dhull, V., 2018. Fabrication of AChE/SnO_2-cMWCNTs/Cu nanocomposite-based sensor electrode for detection of methyl parathion in water. Int. J. Anal. Chem., 2018. 2874059.

Ding, H., Liao, S., Xiao, F., Shen, G., Yu, R., Wu, Z., 2014a. Acetylcholinesterase liquid crystal biosensor for identification of AChE inhibitors by a reactivator. Sci. China Chem. 57 (11), 1589–1595.

Ding, J., Zhang, H., Jia, F., Qin, W., Du, D., 2014b. Assembly of carbon nanotubes on a nanoporous gold electrode for acetylcholinesterase biosensor design. Sensors Actuat. B: Chem. 199, 284–290.

Dong, J., Zhao, H., Qiao, F., Liu, P., Wang, X., Ai, S., 2013. Quantum dot immobilized acetylcholinesterase for the determination of organophosphate pesticides using graphene-chitosan nanocomposite modified electrode. Anal. Methods 5 (11), 2866–2872.

Đorđević, J., Papp, Z., Guzsvány, V., Švancara, I., Trtić-Petrović, T., Purenović, M., Vytřas, K., 2012. Voltammetric determination of the herbicide linuron using a tricresyl phosphate-based carbon paste electrode. Sensors 12 (1), 148–161.

Ensafi, A.A., Noroozi, R., Zandi, N., Rezaei, B., 2017. Cerium (IV) oxide decorated on reduced graphene oxide, a selective and sensitive electrochemical sensor for fenitrothion determination. Sens. Actuators B 245, 980–987.

Facure, M.H., Mercante, L.A., Mattoso, L.H., Correa, D.S., 2017. Detection of trace levels of organophosphate pesticides using an electronic tongue based on graphene hybrid nanocomposites. Talanta 167, 59–66.

Fei, A., Liu, Q., Huan, J., Qian, J., Dong, X., Qiu, B., Mao, H., Wang, K., 2015. Label-free impedimetric aptasensor for detection of femtomole level acetamiprid using gold nanoparticles decorated multiwalled carbon nanotube-reduced graphene oxide nanoribbon composites. Biosens. Bioelectron. 70, 122–129.

Feng, S., Yang, R., Ding, X., Li, J., Guo, C., Qu, L., 2015. Sensitive electrochemical sensor for the determination of pentachlorophenol in fish meat based on ZnSe quantum dots decorated multiwall carbon nanotubes nanocomposite. Ionics 21 (12), 3257–3266.

Fernández-Cori, R., Gomero, J.C.M., Huayhuas-Chipana, B.C., Sotomayor, M.D.P.T., Montoya, J.G.R., 2015. Nanostructured sensors for determination of 3-(3, 4-dichlorophenyl)-1, 1-dimethylurea based in molecularly imprinted polymers (MIPs) deposited in screen printed carbon nanotubes. ECS Trans. 66 (37), 33–41.

Fujii, S., Nobukawa, A., Osaki, T., Morimoto, Y., Kamiya, K., Misawa, N., Takeuchi, S., 2017. Pesticide vapor sensing using an aptamer, nanopore, and agarose gel on a chip. Lab Chip 17 (14), 2421–2425.

Gan, T., Lv, Z., Sun, Y., Shi, Z., Sun, J., Zhao, A., 2016. Highly sensitive and molecular selective electrochemical sensing of 6-benzylaminopurine with multiwall carbon nanotube@SnS_2-assisted signal amplification. J. Appl. Electrochem. 46 (3), 389–401.

Garrido, J.M.P.J., Rahemi, V., Borges, F., Brett, C.M.A., Garrido, E.M.P.J., 2016. Carbon nanotube β-cyclodextrin modified electrode as enhanced sensing platform for the determination of fungicide pyrimethanil. Food Cont. 60, 7–11.

Ghodsi, J., Rafati, A.A., 2017. A voltammetric sensor for diazinon pesticide based on electrode modified with TiO_2 nanoparticles covered multi walled carbon nanotube nanocomposite. J. Appl. Electrochem. 807, 1–9.

González-Sálamo, J., Hernández-Borges, J., Afonso, M.D.M., Rodríguez-Delgado, M.Á., 2018. Determination of phthalates in beverages using multiwalled carbon nanotubes dispersive solid-phase extraction before HPLC-MS. J. Separ. Sci. 41 (12), 2613–2622.

Govindasamy, M., Mani, V., Chen, S.M., Chen, T.W., Sundramoorthy, A.K., 2017. Methyl parathion detection in vegetables and fruits using silver@ graphene nanoribbons nanocomposite modified screen printed electrode. Sci. Rep. 7, 46471.

Govindhan, M., Lafleur, T., Adhikari, B.R., Chen, A., 2015. Electrochemical sensor based on carbon nanotubes for the simultaneous detection of phenolic pollutants. Electroanalysis 27, 902–909.

Guler, M., Turkoglu, V., Basi, Z., 2017. Determination of malation, methidathion, and chlorpyrifos ethyl pesticides using acetylcholinesterase biosensor based on Nafion/Ag@ rGO-NH_2 nanocomposites. Electrochim. Acta 240, 129–135.

Hu, W., Chen, Q., Li, H., Ouyang, Q., Zhao, J., 2016. Fabricating a novel label-free aptasensor for acetamiprid by fluorescence resonance energy transfer between NH_2-NaYF4: Yb, Ho@ SiO_2 and Au nanoparticles. Biosens. Bioelectron. 80, 398–404.

Huo, D., Li, Q., Zhang, Y., Hou, C., Lei, Y., 2014. A highly efficient organophosphorus pesticides sensor based on CuO nanowires-SWCNTs hybrid nanocomposite. Sens. Actuators. B Chem. 199, 410–417.

İnam, R., Bilgin, C., 2013. Square wave voltammetric determination of methiocarb insecticide based on multiwall carbon nanotube paste electrode. J. Appl. Electrochem. 43 (4), 425–432.

Ion, A.C., Ion, I., Culetu, A., Gherase, D., Moldovan, C.A., Iosub, R., Dinescu, A., 2010. Acetylcholinesterase voltammetric biosensors based on carbon nanostructure-chitosan composite material for organophosphate pesticides. Mater. Sci. Eng. C 30 (6), 817–821.

Ion, I., Ion, A.C., 2012. Determination of chlorpyriphos in broccoli using a voltammetric acetylcholinesterase sensor based on carbon nanostructure-chitosan composite material. Mater. Sci. Eng. C 32 (4), 1001–1004.

Jeyapragasam, T., Saraswathi, R., Chen, S.M., Chen, T.W., 2017. Acetylcholinesterase biosensor for the detection of methyl parathion at an electrochemically reduced graphene oxide-nafion modified glassy carbon electrode. Int. J. Electrochem. Sci. 12. 4768–simultaneous 4781.

Jiao, Y., Jia, H., Guo, Y., Zhang, H., Wang, Z., Sun, X., Zhao, J., 2016. An ultrasensitive aptasensor for chlorpyrifos based on ordered mesoporous carbon/ferrocene hybrid multiwalled carbon nanotubes. RSC Adv. 6 (63), 58541–58548.

Karimian, N., Fakhri, H., Amidi, S., Hajian, A., Arduini, F., Bagheri, H., 2019. A novel sensing layer based on metal-organic framework UiO-66 modified with TiO_2-graphene oxide: application to rapid, sensitive and determination of paraoxon and chlorpyrifos. New J. Chem. 43, 2600–2609.

Kesik, M., Kanik, F.E., Turan, J., Kolb, M., Timur, S., Bahadir, M., Toppare, L., 2014. An acetylcholinesterase biosensor based on a conducting polymer using multiwalled carbon nanotubes for amperometric detection of organophosphorous pesticides. Sensors Actuat. B: Chem. 205, 39–49.

Khadem, M., Faridbod, F., Norouzi, P., Foroushani, A.R., Ganjali, M.R., Shahtaheri, S.J., 2016. Biomimetic electrochemical sensor based on molecularly imprinted polymer for dicloran pesticide determination in biological and environmental samples. J. Iran. Chem. Soc. 13 (11), 2077–2084.

Khadem, M., Faridbod, F., Norouzi, P., Rahimi Foroushani, A., Ganjali, M.R., Shahtaheri, S.J., Yarahmadi, R., 2017. Modification of carbon paste electrode based on molecularly imprinted polymer for electrochemical determination of diazinon in biological and environmental samples. Electroanalysis 29 (3), 70–715.

Krishnan, S.K., Singh, E., Singh, P., Meyyappan, M., Nalwa, H.S., 2019. A review on graphene-based nanocomposites for electrochemical and fluorescent biosensors. RSC Adv. 9 (16), 8778–8881.

Lei, Y., Sun, R., Zhang, X., Feng, X., Jiang, L., 2016. Oxygen-rich enzyme biosensor based on superhydrophobic electrode. Adv. Mater. 28, 1477–1481.

Leniart, A., Brycht, M., Burnat, B., Skrzypek, S., 2016. Voltammetric determination of the herbicide propham on glassy carbon electrode modified with multi-walled carbon nanotubes. Sens. Actuators B Chem. 231, 54–63.

Li, S., Wu, X., Liu, C., Yin, G., Luo, J., Xu, Z., 2016a. Application of DNA aptamers as sensing layers for detection of carbofuran by electrogenerated chemiluminescence energy transfer. Anal. Chim. Acta 941, 94–100.

Li, S., Yin, G., Wu, X., Liu, C., Luo, J., 2016b. Supramolecular imprinted sensor for carbofuran detection based on a functionalized multiwalled carbon nanotube-supported Pd-Ir composite and methylene blue as catalyst. Electrochim. Acta 188, 294–300.

Li, X., Zheng, Z., Liu, X., Zhao, S., Liu, S., 2015. Nanostructured photoelectrochemical biosensor for highly sensitive detection of organophosphorous pesticides. Biosens. Bioelectron. 64, 1–5.

Li, Y., Zhang, Y., Han, G., Xiao, Y., Li, M., Zhou, W., 2016c. An acetylcholinesterase biosensor based on graphene/polyaniline composite film for detection of pesticides. Chin. J. Chem. 34, 82–88.

Li, Y., Shi, L., Han, G., Xiao, Y., Zhou, W., 2017a. Electrochemical biosensing of carbaryl based on acetylcholinesterase immobilized onto electrochemically inducing porous graphene oxide network. Sens. Actuators B 238, 945–953.

Li, Y., Zhao, R., Shi, L., Han, G., Xiao, Y., 2017b. Acetylcholinesterase biosensor based on electrochemically inducing 3D graphene oxide network/multi-walled carbon nanotube composites for detection of pesticides. RSC Adv. 7, 53570–53577.

Lin, B., Yu, Y., Li, R., Cao, Y., Guo, M., 2016. Turn-on sensor for quantification and imaging of acetamiprid residues based on quantum dots functionalized with aptamer. Sens. Actuators B Chem. 229, 100–109.

Liu, F., Piao, Y., Choi, J.S., Seo, T.S., 2013. Three-dimensional graphene micropillar based electrochemical sensor for phenol detection. Biosens. Bioelectron. 50, 387–392.

Liu, M., Khan, A., Wang, Z., Liu, Y., Yang, G., Deng, Y., He, N., 2019. Aptasensors for pesticide detection. Biosens. Bioelectron. 130, 174–184.

Liu, Q., Fei, A., Huan, J., Mao, H., Wang, K., 2015. Effective amperometric biosensor for carbaryl detection based on covalent immobilization acetylcholinesterase on multiwall carbon nanotubes/graphene oxide nanoribbons nanostructure. J. Electroanal. Chem. 740, 8–13.

Liu, X., Li, Y., Liang, J., Zhu, W., Xu, J., Su, R., Yuan, L., Sun, C., 2016. Aptamer contained triple-helix molecular switch for rapid fluorescent sensing of acetamiprid. Talanta 160, 99–105.

Liu, Y., Wang, G., Li, C., Zhou, Q., Wang, M., Yang, L., 2014. A novel acetylcholinesterase biosensor based on carboxylic graphene coated with silver nanoparticles for pesticide detection. Mater. Sci. Eng. C. 35, 253–258.

Luan, Y., Lu, A., Chen, J., Fu, H., Xu, L., 2016. A label-free aptamer-based fluorescent assay for cadmium detection. Appl. Sci. 6 (12), 432.

Madianos, L., Tsekenis, G., Skotadis, E., Patsiouras, L., Tsoukalas, D., 2018. A highly sensitive impedimetric aptasensor for the selective detection of acetamiprid and atrazine based on microwires formed by platinum nanoparticles. Biosens. Bioelectron. 101, 268–274.

Mani, V., Devasenathipathy, R., Chen, S.M., Wu, T.Y., Kohilarani, K., 2015. High-performance electrochemical amperometric sensors for the sensitive determination of phenyl urea herbicides diuron and fenuron. Ionics 21 (9), 2675–2683.

Mercan, H., Inam, R., Aboul-Enein, H.Y., 2011. Square wave adsorptive stripping voltammetric determination of cyromazine insecticide with multi-walled carbon nanotube paste electrode. Anal. Lett. 44 (8), 1392–1404.

Miao, S.S., Wu, M.S., Ma, L.Y., He, X.J., Yang, H., 2016. Electrochemiluminescence biosensor for determination of organophosphorous pesticides based on bimetallic Pt-Au/multi-walled carbon nanotubes modified electrode. Talanta 158, 142–151.

Mogha, N.K., Sahu, V., Sharma, M., Sharma, R.K., Masram, D.T., 2016. Biocompatible ZrO_2-reduced graphene oxide immobilized AChE biosensor for chlorpyrifos detection. Mater. Des 111, 312–320.

Motaharian, A., Motaharian, F., Abnous, K., Hosseini, M.R.M., Hassanzadeh-Khayyat, M., 2016. Molecularly imprinted polymer nanoparticles-based electrochemical sensor for determination of diazinon pesticide in well water and apple fruit samples. Anal. Bioanal. Chem. 408 (24), 6769–6779.

Nie, Y., Teng, Y., Li, P., Liu, W., Shi, Q., Zhang, Y., 2018. Label-free aptamer-based sensor for specific detection of malathion residues by surface-enhanced Raman scattering. Spectrochim. Acta A. Mol. Biomol. Spectrosc. 191, 271–276.

Nurdin, M., Maulidiyah, M., Muzakkar, M.Z., Umar, A.A., 2019. High performance cypermethrin pesticide detection using anatase TiO_2-carbon paste nanocomposites electrode. Microchem. J. 145, 756–761.

Özcan, A., Gürbüz, M., 2018. Development of a modified electrode by using a nanocomposite containing acid-activated multi-walled carbon nanotube and fumed silica for the voltammetric determination of clopyralid. Sens. Actuators B Chem. 255, 262–267.

Pang, S., Labuza, T.P., He, L., 2014. Development of a single aptamer-based surface enhanced Raman scattering method for rapid detection of multiple pesticides. Analyst 139 (8), 1895–1901.

Qi, Y., Xiu, F.R., Zheng, M., Li, B., 2016. A simple and rapid chemiluminescence aptasensor for acetamiprid in contaminated samples: sensitivity, selectivity and mechanism. Biosens. Bioelectron. 83, 243–249.

Qiu, L., Li, D., Cheng, H.-M., 2018. Structural control of graphene-based materials for unprecedented performance. ACS Nano 12, 5085–5092.

Rana, S., Kaur, R., Jain, R., Prabhakar, N., 2018. Ionic liquid assisted growth of poly (3, 4-ethylenedioxythiophene)/reduced graphene oxide-based electrode: An improved electro-catalytic performance for the detection of organophosphorus pesticides in beverages. Arab. J. Chem. https://doi.org/10.1016/j.arabjc.2018.08.008.

Rapini, R., Cincinelli, A., Marrazza, G., 2016. Acetamiprid multidetection by disposable electrochemical DNA aptasensor. Talanta 161, 15–21.

Rather, J.A., De Wael, K., 2012. C_{60}-functionalized MWCNT based sensor for sensitive detection of endocrine disruptor vinclozolin in solubilized system and wastewater. Sens. Actuators B Chem. 171, 907–915.

Salehzadeh, H., Ebrahimi, M., Nematollahi, D., Salarian, A.A., 2016. Electrochemical study of fenitrothion and bifenox and their simultaneous determination using multiwalled carbon nanotube modified glassy carbon electrode. J. Electroanal. Chem. 767, 188–194.

Sekhon, S.S., Park, G.Y., Park, D.Y., Kim, S.Y., Wee, J.H., Ahn, J.Y., Kim, Y.H., 2018. Aptasensors for pesticide detection. Toxicol. Environ. Health Sci. 10 (5), 229–236.

Shahtaheria, S.J., Faridbod, F., Khadem, M., 2017. Highly selective voltammetric sensor based on molecularly imprinted polymer and carbon nanotubes to determine the dicloran pesticide in biological and environmental samples. Procedia Technol. 27, 96–97.

Shams, N., Lim, H.N., Hajian, R., Yusof, N.A., Abdullah, J., Sulaiman, Y., Ibrahim, I., Huang, N.M., Pandikumar, A., 2016. A promising electrochemical sensor based on Au nanoparticles decorated reduced graphene oxide for selective detection of herbicide diuron in natural waters. J. Appl. Electrochem. 46, 655–666.

Sipa, K., Brycht, M., Leniart, A., Urbaniak, P., Nosal-Wiercińska, A., Pałecz, B., Skrzypek, S., 2018. β-Cyclodextrins incorporated multi-walled carbon nanotubes modified electrode for the voltammetric determination of the pesticide dichlorophen. Talanta 176, 625–634.

Song, M.S., et al., 2017. Detecting and discriminating shigella sonnei using an aptamer-based fluorescent biosensor platform. Molecules 22 (5), 825.

Su, R., Li, D., Wang, X., Yang, H., Shi, X., Liu, S., 2016. Determination of organophosphorus pesticides in ginseng by carbon nanotube envelope-based solvent extraction combined with ultrahigh-performance liquid chromatography mass spectrometry. J. Chromatogr. B 1022, 141–152.

Sundari, P.A., Palaniappan, S.P., Manisankar, P., 2010. Enhanced sensing of carbendazim, a fungicide on functionalized multiwalled carbon nanotube modified glassy carbon electrode and its determination in real samples. Analyt. Lett. 43 (9), 1457–1470.

Syshchyk, O., Skryshevsky, V.A., Soldatkin, O.O., Soldatkin, A.P., 2015. Enzyme biosensor systems based on porous silicon photoluminescence for detection of glucose, urea and heavy metals. Biosens. Bioelectron. 66, 89–94.

Tan, X., Hu, Q., Wu, J., Li, X., Li, P., Yu, H., Li, X., Lei, F., 2015a. Electrochemical sensor based on molecularly imprinted polymer reduced graphene oxide and gold nanoparticles modified electrode for detection of carbofuran. Sens. Actuators B Chem. 220, 216–221.

Tan, X., Wu, J., Hu, Q., Li, X., Li, P., Yu, H., Li, X., Lei, F., 2015b. An electrochemical sensor for the determination of phoxim based on a graphene modified electrode and molecularly imprinted polymer. Analyt. Methods 7 (11), 4786–4792.

Thanh, C.T., Binh, N.H., Van Tu, N., Thu, V.T., Bayle, M., Paillet, M., Sauvajol, J.L., Thang, P.B., Dai Lam, T., Minh, P.N., Van Chuc, N., 2018. An interdigitated ISFET-type sensor based on LPCVD grown graphene for ultrasensitive detection of carbaryl. Sens. Actuators B Chem. 260, 78–85.

Tian, X., Liu, L., Li, Y., Yang, C., Zhou, Z., Nie, Y., Wang, Y., 2018. Nonenzymatic electrochemical sensor based on CuO-TiO_2 for sensitive and selective detection of methyl parathion pesticide in ground water. Sens. Actuators B Chem. 256, 135–142.

Toro, M.J.U., Marestoni, L.D., Sotomayor, M.D.P.T., 2015. A new biomimetic sensor based on molecularly imprinted polymers for highly sensitive and selective determination of hexazinone herbicide. Sens. Actuators B: Chem. 208, 299–306.

Wang, M., Duan, X., Xu, Y., Duan, X., 2016a. Functional three-dimensional graphene/polymer composites. ACS Nano 10, 7231–7247.

Wang, B., Ye, C., Zhong, X., Chai, Y., Chen, S., Yuan, R., 2016b. Electrochemical biosensor for organophosphate pesticides and huperzine-a detection based on pd wormlike nanochains/graphitic carbon nitride nanocomposites and acetylcholinesterase. Electroanalysis 28 (2), 304–311.

Wang, G., Tan, X., Zhou, Q., Liu, Y., Wang, M., Yang, L., 2014a. Synthesis of highly dispersed zinc oxide nanoparticles on carboxylic graphene for development a sensitive acetylcholinesterase biosensor. Sens. Actuators B 190, 730–736.

Wang, H., Zhao, G., Chen, D., Wang, Z., Liu, G., 2016c. A sensitive acetylcholinesterase biosensor based on screen printed electrode modified with Fe3O4 nanoparticle and graphene for chlorpyrifos determination. Int. J. Electrochem. Sci. 11, 10906–10918.

Wang, J., Wu, Y., Zhou, P., Yang, W., Tao, H., Qiu, S., Feng, C., 2018. A novel fluorescent aptasensor for ultrasensitive and selective detection of acetamiprid pesticide based on the inner filter effect between gold nanoparticles and carbon dots. Analyst 143 (21), 5151–5160.

Wang, K., Liu, Q., Dai, L., Yan, J., Ju, C., Qiu, B., Wu, X., 2011. A highly sensitive and rapid organophosphate biosensor based on enhancement of CdS-decorated graphene nanocomposite. Anal. Chim. Acta 695 (1–2), 84–88.

Wang, P., Wan, Y., Ali, A., Deng, S., Su, Y., Fan, C., Yang, S., 2016d. Aptamer-wrapped gold nanoparticles for the colorimetric detection of omethoate. Sci. China Chem. 59 (2), 237–242.

Wang, T., Zhao, D., Guo, X., Correa, J., Riehl, B.L., Heineman, W.R., 2014b. Carbon nanotube-loaded nafion film electrochemical sensor for metal ions: Europium. Anal. Chem. 86, 4354–4361.

Wei, M., Wang, J., 2015. A novel acetylcholinesterase biosensor based on ionic liquids-AuNPs-porous carbon composite matrix for detection of organophosphate pesticides. Sens. Actuators B: Chem. 211, 290–296.

Wei, X.P., Luo, Y.L., Xu, F., Chen, Y.S., Yang, L.H., 2016. In-situ non-covalent dressing of multi-walled carbon nanotubes@ titanium dioxides with carboxymethyl chitosan nanocomposite electrochemical sensors for detection of pesticide residues. Mater. Design 111, 445–452.

Wong, A., Foguel, M.V., Khan, S., de Oliveira, F.M., Tarley, C.R.T., Sotomayor, M.D., 2015. Development of an electrochemical sensor modified with MWCNT-COOH and MIP for detection of diuron. Electrochim. Acta 182, 122–130.

Wong, A., Silva, T., Caetano, F., Bergamini, M., Marcolino-Junior, L., Fatibello-Filho, O., Janegitz, B., 2017. An overview of pesticide monitoring at environmental samples using carbon nanotubes-based electrochemical sensors. C 3 (1), 8.

Wu, L., Lei, W., Han, Z., Zhang, Y., Xia, M., Hao, Q., 2015. A novel non-enzyme amperometric platform based on poly (3-methylthiophene)/nitrogen doped graphene modified electrode for determination of trace amounts of pesticide phoxim. Sens. Actuators B 206, 495–501.

Wu, S., Huang, F., Lan, X., Wang, X., Wang, J., Meng, C., 2013. Electrochemically reduced graphene oxide and Nafion nanocomposite for ultralow potential detection of organophosphate pesticide. Sens. Actuators B Chem. 177, 724–729.

Wu, X., Song, Y., Yan, X., Zhu, C., Ma, Y., Du, D., Lin, Y., 2017. Carbon quantum dots as fluorescence resonance energy transfer sensors for organophosphate pesticides determination. Biosens. Bioelectron. 94, 292–297.

Xia, N., Gao, Y.P., 2015. Carbon nanostructures for development of acetylcholinesterase electrochemical biosensors for determination of pesticides. Int. J. Electrochem. Sci. 10, 713–724.

Xiong, S., Deng, Y., Zhou, Y., Gong, D., Xu, Y., Yang, L., Chen, H., Chen, L., Song, T., Luo, A., Deng, X., 2018. Current progress in biosensors for organophosphorus pesticides based on enzyme functionalized nanostructures: a review. Anal. Methods 10 (46), 5468–5479.

Xu, G., Huo, D., Hou, C., Zhao, Y., Bao, J., Yang, M., Fa, H., 2018. A regenerative and selective electrochemical aptasensor based on copper oxide nanoflowers-single walled carbon nanotubes nanocomposite for chlorpyrifos detection. Talanta 178, 1046–1052.

Yan, J.X., Guan, H.N., Yu, J., Chi, D.F., 2013. Acetylcholinesterase biosensor based on assembly of multiwall carbon nanotubes onto liposome bioreactors for detection of organophosphates pesticides. Pestic. Biochem. Physiol. 105, 197–202.

Yan, X., Li, H., Han, X., Su, X., 2015. A ratiometric fluorescent quantum dots-based biosensor for organophosphorus pesticides detection by inner-alter effect. Biosens. Bioelectron. 74, 277–283.

Yan, X., Song, Y., Zhu, C., Li, H., Du, D., Su, X., Lin, Y., 2018. MnO_2 nanosheet-carbon dots sensing platform for sensitive detection of organophosphorus pesticides. Anal. Chem. 90 (4), 2618–2624.

Yang, J., Wang, Q., Zhang, M., Zhang, S., Zhang, L., 2015. An electrochemical fungicide pyrimethanil sensor based on carbon nanotubes/ionic-liquid construction modified electrode. Food Chem. 187, 1–6.

Yang, L., Wang, G., Liu, Y., 2013a. An acetylcholinesterase biosensor based on platinum nanoparticles-carboxylic graphene-nafion-modified electrode for detection of pesticides. Anal. Biochem. 437, 144–149.

Yang, L., Wang, G., Liu, Y., Wang, M., 2013b. Development of a biosensor based on immobilization of acetylcholinesterase on NiO nanoparticles-carboxylic graphene-nafion modified electrode for detection of pesticides. Talanta 113, 135–141.

Yang, Y., Asiri, A.M., Du, D., Lin, Y., 2014. Acetylcholinesterase biosensor based on a gold nanoparticle-polypyrrole-reduced graphene oxide nanocomposite modified electrode for the amperometric detection of organophosphorus pesticides. Analyst 139, 3055–3060.

Yaqub, S., Latif, U., Dickert, F.L., 2011. Plastic antibodies as chemical sensor material for atrazine detection. Sens. Actuators B: Chem. 160 (1), 227–233.

Yola, M.L., Atar, N., 2017. Electrochemical detection of atrazine by platinum nanoparticles/carbon nitride nanotubes with molecularly imprinted polymer. Ind. Eng. Chem. Res. 56 (27), 7631–7639.

Zhang, R., Chen, W., 2017. Recent advances in graphene-based nanomaterials for fabricating electrochemical hydrogen peroxide sensors. Biosens. Bioelectron. 89, 249–268.

Zhang, W.Y., Asiri, A.M., Liu, D.L., Du, D., Lin, Y.H., 2014. Nanomaterial-based biosensors for environmental and biological monitoring of organophosphorus pesticides and nerve agents. Trends Anal. Chem. 54, 1–10.

Zhang, D., Yu, D., Zhao, W., Yang, Q., Kajiura, H., Li, Y., Zhou, T., Shi, G., 2012a. A molecularly imprinted polymer based on functionalized multiwalled carbon nanotubes for the electrochemical detection of parathion-methyl. Analyst 137 (11), 2629–2636.

Zhang, L., Zhang, A., Du, D., Lin, Y., 2012b. Biosensor based on Prussian blue nanocubes/reduced graphene oxide nanocomposite for detection of organophosphorus pesticides. Nanoscale 4, 4674–4679.

Zhang, M., Zhao, H.T., Xie, T.J., Yang, X., Dong, A.J., Zhang, H., Wang, J., Wang, Z.Y., 2017. Molecularly imprinted polymer on graphene surface for selective and sensitive electrochemical sensing imidacloprid. Sens. Actuators B 252, 991–1002.

Zhang, P., Sun, T., Rong, S., Zeng, D., Yu, H., Zhang, Z., Chang, D., Pan, H., 2019. A sensitive amperometric AChE-biosensor for organophosphate pesticides detection based on conjugated polymer and Ag–rGO–NH_2 nanocomposite. Bioelectrochemistry 127, 163–170.

Zhang, Y., Kang, T.F., Wan, Y.W., Chen, S.Y., 2009. Gold nanoparticles–carbon nanotubes modified sensor for electrochemical determination of organophosphate pesticides. Microchim. Acta 165 (3–4), 307–311.

Zhao, F., Wu, J., Ying, Y., She, Y., Wang, J., Ping, J., 2018. Carbon nanomaterial-enabled pesticide biosensors: design strategy, biosensing mechanism, and practical application. TrAC Trends Anal. Chem.

Zhao, H., Ji, X., Wang, B., Wang, N., Li, X., Ni, R., Ren, J., 2015. An ultra-sensitive acetylcholinesterase biosensor based on reduced graphene oxide–Au nanoparticles–β-cyclodextrin/Prussian blue-chitosan nanocomposites for organophosphorus pesticides detection. Biosens. Bioelectron. 65, 23–30.

Zheng, Y.Y., Liu, Z.M., Zhan, H.J., Li, J., Zhang, C.C., 2015a. Development of a sensitive acetylcholinesterase biosensor based on a functionalized graphene-polyvinyl alcohol nanocomposite for organophosphorous pesticide detection. Anal. Methods 7, 9977–9983.

Zheng, Y., Liu, Z., Jing, Y., Li, J., Zhan, H., 2015b. An acetylcholinesterase biosensor based on ionic liquid functionalized graphene–gelatin-modified electrode for sensitive detection of pesticides. Sens. Actuators B 210, 389–397.

Zheng, Y., Wang, A., Lin, H., Fu, L., Cai, W., 2015c. A sensitive electrochemical sensor for direct phoxim detection based on an electrodeposited reduced graphene oxide–gold nanocomposite. RSC Adv. 5, 15425–15430.

Zheng, Y., Liu, Z., Zhan, H., Li, J., Zhang, C., 2016. Studies on electrochemical organophosphate pesticide (OP) biosensor design based on ionic liquid functionalized graphene and a Co3O4 nanoparticle modified electrode. Anal. Methods 8, 5288–5295.

Zhou, Q., Yang, L., Wang, G., Yang, Y., 2013. Acetylcholinesterase biosensor based on SnO_2 nanoparticles–carboxylic graphene–nafion modified electrode for detection of pesticides. Biosens. Bioelectron. 49, 25–31.

Zhou, Y., Tang, L., Zeng, G., Chen, J., Cai, Y., Zhang, Y., et al., 2014. Mesoporous carbon nitride-based biosensor for highly sensitive and selective analysis of phenol and catechol in compost bioremediation. Biosens. Bioelectron. 61, 519–525.

Further reading

Bobrinetskiy, I.I., Knezevic, N.Z., 2018. Graphene-based biosensors for on-site detection of contaminants in food. Anal. Methods 10 (42), 5061–5070.

Sundari, P.A., Manisankar, P., 2011. Development of nano poly (3-methyl thiophene)/multi-walled carbon nanotubes sensor for the efficient detection of some pesticides. J. Braz. Chem. Soc. 22 (4), 746–755.

Carbon nanotubes: An efficient sorbent for herbicide sensing and remediation

19

Mousa Alghuthaymi[a], *Asran-Amal*[b], *Manal Mostafa*[b], *Kamel A. Abd-Elsalam*[b,c]
[a]Department of Biology, Science and Humanities College, Shaqra University, Alquwayiyah, Saudi Arabia, [b]Plant Pathology Research Institute, Agricultural Research Center (ARC), Giza, Egypt, [c]CIHEAM IAMB, Mediterranean Agronomic Institute of Bari, Valenzano, Italy

1 Introduction

In recent years, environmental pollution from the use of phytosanitary products has received much attention because of the negative impact of those products on natural resources and living organisms. Mainly, phytosanitary products such as pesticides, fertilizers, and herbicides accumulate in the ecosystem and may cause an apocalypse later. Therefore, finding a clean technique to remove the components of phytosanitary products is a must, and one of the most promising techniques is absorption using nanoparticles. It is an established technique utilized widely, due to the ability of nanoparticles and their unique properties. Taghizade et al. (2018) explained that nanostructures are a recommended candidate and reactive media for phytosanitary product removal because of their wide range of physicochemical properties. In another aspect, solid-phase extraction is an established technique used for remediation. Moreover, different substances have been proposed and used as solid-phase extraction sorbents, a chelating resin, modified silica, activated carbon, polyurethane foam, cellulose, biological substances, and carbon nanotubes (Liang et al., 2005). Carbon nanotubes (CNTs) are among the most widely utilized kind of nanomaterial; they are considered materials with sizes below 100 nm and at least two dimensions (Hochella, 2002). Iijima (1991) illustrated the structure of CNTs, which are a group of carbon-based nanomaterials that can be imagined as one or more micrometer-scale graphene sheets rolled into a nanoscale-cylinder. They can also be imagined as a graphite sheet that has been rolled into a tube and divided into multiwalled carbon nanotubes. Since their discovery in 1991, CNTs have attracted great attention because of their unique properties (Liang et al., 2005), such as electrical conductivity, optical activity, structure, and mechanical strength. This unique new material class has proved promising. Additionally, due to their increasing use, CNTs are spreading rapidly in the environment (Mauter and Elimelech, 2008; Nowack and Bucheli, 2007). Single-walled carbon nanotubes (SWCNTs) in the wall of the nanotube depending on the carbon atom layers. The hexagonal array of carbon atoms in CNT surface graphene sheets has a strong interaction with other molecules or atoms, making CNTs in many ways promising adsorbent materials that replace activated carbon (Liang et al., 2005). Recently, CNTs have been

Carbon Nanomaterials for Agri-food and Environmental Applications. https://doi.org/10.1016/B978-0-12-819786-8.00019-0

among the record promising groups of materials with some attractive properties, such as lightness, rigidity, high surface area, high mechanical strength in tension, good thermal conductivity, and mechanical damage resistance. These interesting properties can make CNTs an inexpensive alternative to traditional sorbents applied in analytical chemistry, especially in purification, identification, and degradation procedures. In this chapter, the potential ability of carbon nanotubes as a sorbent factor for diverse forms of herbicides and their traces as well as the remediation mechanisms will be illustrated.

2 Types of carbon nanotubes

Conceptually, carbon nanotubes are usually deemed a graphene sheet rolled into a nanoscale micron cylinder (Cao, 2004). These materials have driven a broad range of applications due to their extreme properties such as high surface areas, large aspect ratios, remarkably high mechanical strength, and electrical and thermal conductivities. Currently, however, carbon nanotubes (CNTs) are the hottest nanocarbon material. There are two structural forms of CNTs as primarily divided: single-walled carbon nanotubes (SWCNTs) that consist of one layer of graphene and multiwalled carbon nanotubes (MWCNTs) that consist of multiple layers of graphene, with lengths that can be as short as a few hundred nanometers or as long as several microns. The SWCNT diameters range from 1 to 10 nm and are normally at the end. MWCNT diameters, on the other hand, are much larger (from 5 nm to a few hundred nanometers) because their structure is made up of many concentrated cylinders retained together by van der Waals forces (Wepasnick et al., 2010) (Fig. 1). Mainly, one type of MWCNT, which is double-walled carbon nanotubes (DWNTs); DWNTs use increases rapidly

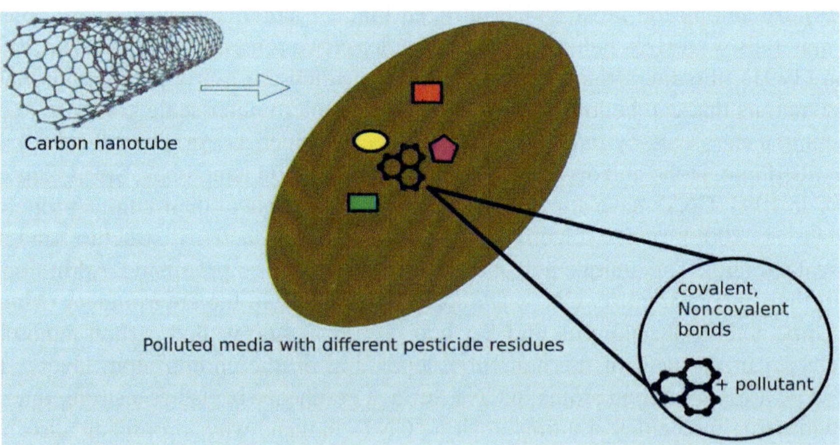

Fig. 1 The ability of carbon nanotubes to adsorb the residues of pesticides using covalent and noncovalent forces (Van der Waals, π-effect).

because of their ability to resist chemicals. Hydrophilic groups such as hydroxyl and carboxyl groups can be used for both SWCNTs and MWCNTs, to be functionalized.

3 The agricultural importance of carbon nanotubes

With a continually increasing world population, increasing and maximizing agricultural production on limited areas of arable land with minimal negative environmental impacts is a significant challenge for the future (Hong et al., 2013; Khot et al., 2012; Parisi et al., 2014). Controlled released nanofertilizers improve plant growth, yield, and productivity (Zaytseva and Neumann, 2016); A few studies indicate germination rate impacts of SWCNTs. Seed germination was stimulated in response to SWCNT (10–40 $mg\,L^{-1}$) treatments, potentially induced by seed coat perforation, and has been reported for salvia, pepper, and high fescue (Pourkhaloee et al., 2011). Similar to SWCNTs, the stimulating effects of MWCNTs for a wider range of different crops have been reported (Zaytseva and Neumann, 2016). Additionally, further application of various nanomaterials in plant protection sector including purification, sensing, and degradation of pesticides and herbicides were observed (Gogos et al., 2012; Pereira et al., 2014; Sarlak et al., 2014; Suresh Kumar et al., 2013). Moreover, the general reduction of applied agrochemicals is one of the challenges that needs to be overcome. Nanencapsulated plant protection products are used as slow-release fertilizers for that purpose (Zaytseva and Neumann, 2016). According to Gogos et al. (2012), carbon-based nanomaterials acting as additives as well as active components will be used by 40% of all nanotechnology contributions to agriculture. In addition, due to the novel chemical, physical, and mechanical properties of carbon-based nanostructures, they are utilized to develop highly sensitive sensors and diagnostic devices for many agricultural and environmental applications (Zaytseva and Neumann, 2016). In that regard, CNT sensors have been successfully used as stress indicators for in vivo monitoring of ROS formation in plant tissues (Ren et al., 2013). Furthermore, an electronic device that can be placed into insects or plants to detect toxic gases or plant diseases in real time can be applied in remote sensing systems. Covering a field site with multiple sensors to have a potentially complete picture of the disease severity spatial distribution at a field scale allows hot spot detection, that requires special treatment, as the main contribution to decreasing the agrochemicals input (Zaytseva and Neumann, 2016). The most important agricultural application for carbon nanotubes is described in Fig. 2.

4 Herbicides

Herbicides are usually chemical substances that are weed killers used to control unwanted plants that compete with crops for light, water, and nutrients. They kill all the plant materials with which they come into contact. Hence, they can be used to clear waste ground, industrial and construction sites, railways, and railway embankments. Other important distinctions, apart from selective/nonselective, include persistence (also known as residual action: how long the product remains in place and remains

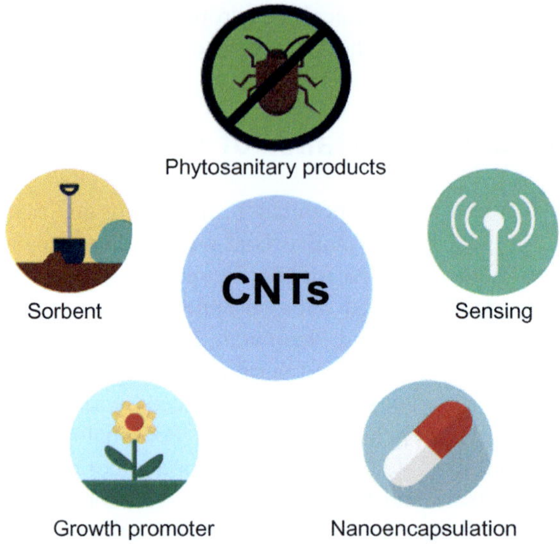

Fig. 2 The agricultural importance of carbon nanotubes: soil contamination sorbent, growth promoter such as fertilizers, nanoencapsulation for the agroproducts for faster release and efficacy, smart sensors for disease detection and weather forecasts, and phytosanitary products such as pesticides, fungicides, and herbicides.

active) and means of absorption (whether absorbed by foliage only above ground, through the roots, or by other means) (Vats, 2015). Herbicides are widely used in weeds and other vegetation. Even at low concentrations, they are relatively cheap and very powerful. Most herbicides are applied directly to the soil or sprayed over crop fields, meaning they are released directly into the environment as a result of large production and high stability. Thus, herbicides may contaminate streams, rivers, or lakes directly from the drainage of agricultural areas (Pyrzyńska, 2011). The dicamba herbicide (3,6-dichloro-2-methoxy benzoic acid) is one of the most widely used agrochemicals for plant protection in the world (Hatterman-Valenti and Mayland, 2005); other phenoxyalkanoic acid herbicides, it imitates auxins, a type of plant hormone, it affects cell division and causes abnormal growth. Dicamba has been listed for restricted utilization because of its high potential for soil leaching and groundwater persistence (Magdalena and Krystyna, 2006). Aryloxyphenoxy-propionate herbicide (AOPP) is a type of selective post—the emergence of herbicide that has been registered to be used in many crops to control annual and perennial grassy weeds (Lucini and Molinari, 2011). Nevertheless, as mentioned previously, living organisms experience adverse impacts from herbicides. Additionally, resistance develops from excessive use of the herbicide. Therefore, because of the unique properties of nanoherbicices, researchers recently suggested their use as an alternative solution. The nanostructured formulation performs by controlling the release mechanism. Nanoherbicides contain a broad range of entities, including polymeric and metallic nanoparticles. Nanoherbicides need a glance to place nanotechnology at the top level (Abgail and Chidambaram, 2017). The nanoherbicide high penetration efficacy helps eliminate weeds before

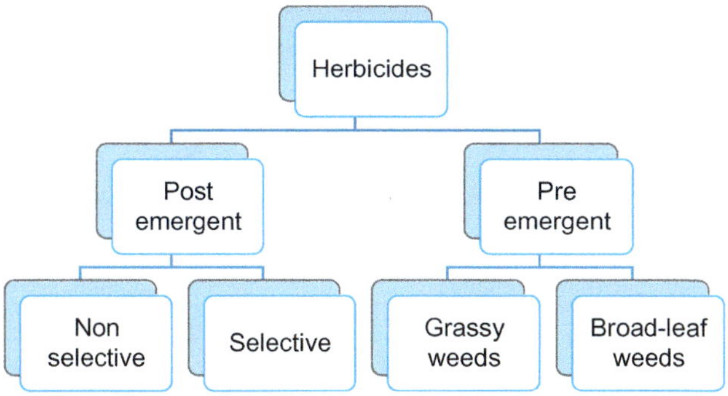

Fig. 3 Herbicide classification based on their time application.

resistance develops. The nanocarriers required to prepare nanoherbicides supply short—and long—residual herbicides on the basis of their need by preventing a lethal dose at which the plant could develop herbicide resistance. Preparing a nanoformulation with suitable carriers would provide a basis for sustainable and economic farming. Nanoherbicides will begin a high localization of the active substances within the target plants to avoid resistance development to specific herbicides at the basic level. Consequently, it is good to apply this nanotechnology-based miracle worker, nanoherbicide, to overcome herbicide resistance (Abgail and Chidambaram, 2017). Herbicides are classified as preemergence or postemergence based on the application time (Wagner and Nadasy, 2006), as shown in Fig. 3.

They could possibly be practical against grassy weeds or broad-leaf weeds when preemergent is applied (Grossmann and Ehrhardt, 2007; Wright et al., 2010). In contrast, the application of postemergent can be selective (specific target) or nonselective (broad target) (Grossmann and Ehrhardt, 2007). The herbicide can be added to the soil at the preemergent stage or the seeds could even be treated with it. The seedlings are treated with certain herbicides and postemergent applications to eliminate weeds. Crafts (1946) defined selectivity as the ability of a herbicide to destroy a target plant without damaging or eliminating the nontarget plants. The Global Herbicide Resistance Action Committee (HRAC) group has established a classification system for the Weed Science Society of America (WSSA) based on the herbicide mode and site of action (Table 1 and Fig. 4).

4.1 *Types of herbicides*

The various herbicide types are prepared to kill tissues in plants, and there are two main types: contact herbicides and systemic herbicides.

4.1.1 Contact herbicides

Principally, the contact method means the chemical will kill the parts of the plant it contacts in that particular type of herbicide. This means that the aboveground leafy parts of the plants will be killed specifically for broad-leaf weeds. However, the belowground plant parts, such as roots, bulbs, tubers, or rhizomes, will be killed indirectly. The popularity of contact herbicides is due to their rapid effectiveness as

Table 1 Herbicide classification according to HRAC and WSSA classification system and their sites of action.

Groups of classification system for herbicides	Chemical structural herbicides	Inhibitors effect of herbicides	References
HRAC Group: A: ACCase inhibitors (WSSA Group: 1 Aussie Group: A)	Aryloxyphenoxypropionate (FOPs) and cyclohexanedione (DIMs)	Inhibit the enzyme acetylCoAcarboxylase (ACCase)	Stoltenberg et al. (1989)
HRAC Group: B: ALS inhibitors (WSSA Group: 2 Aussie Group: B)	Imidazolinones, pyrimidinylthiobenzoates, sulfonylaminocarbonyltriazolinones, sulfonylureas, and triazolopyrimidines	Inhibition of acetolactate synthase ALS (acetohydroxyacid synthase AHAS)	LaRossa and Schloss (1984)
HRAC Group: C1: Photosystem II inhibitors (WSSA Group: 5 Aussie Group: C)	Phenylcarbamates, pyridazinones, triazines, triazinones, and uracilses,	Inhibition of photosynthesis at photosystem II	Devine et al. (1993)
HRAC Group: C2: PSII inhibitor (Ureas and amides) (WSSA Group: 7 Aussie Group: C)	Ureas and amides	Inhibition of photosynthesis at photosystem II	Devine et al. (1993)
HRAC Group: C3: PSII inhibitors (Nitriles) (WSSA Group: 6 Aussie Group: C)	Benzothiadiazinones, nitriles, and phenylpyridazines	Inhibition of photosynthesis at photosystem II	Devine et al. (1993)
HRAC Group: D: PSI Electron Diverter (WSSA Group: 22 Aussie Group: L)	Bipyridyliums	Photosystem-I-electron diversion	Dodge (1982)
HRAC Group: E: PPOinhibitors(WSSA Group: 14 Aussie Group: G)	Diphenylethers, *N*-phenylphthalimides, oxadiazoles, oxazolidinediones, phenylpyrazoles, pyrimidindiones, thiadiazoles, and triazolinones	Inhibition of protoporphyrinogen oxidase (PPO)	Duke et al. (1991)
HRAC Group: F1: Carotenoid biosynthesis inhibitors (WSSA Group: 12 Aussie Group: F)	Amides, anilidex, furanones, phenoxybutan-amides, pyridiazinones, and pyridines	Inhibition ocarotenoid biosynthesis at the phytoene desaturase step (PDS)	Sandmann and Böger (1989)

HRAC Group: F2: HPPD inhibitors (WSSA Group: 27 Aussie Group: H)	Callistemones, isoxazoles, pyrazoles, and triketones	Inhibition of 4 hydroxyphenyl-pyruvate-dioxygenase (4-HPPD)	
HRAC Group: F3: Carotenoid biosynthesis (WSSA Group: 11 Aussie Group: Q)	Clomazone—5-keto form of clomazone	Inhibition of carotenoid biosynthesis (unknown target)	Barry and Pallett (1990)
HRAC Group: F4: DOXP inhibitors (WSSA Group: 13 Aussie Group: Q)	Glycines (glyphosate)	Inhibitor of 1-deoxy-d-xylulose 5-phosphate synthase	Amrhein et al. (1980)
HRAC Group: H: Glutamine synthase inhibitors (WSSA Group: 10 Aussie Group: N)	Phosphinic acids (glufosinate and bialophos)	Inhibition of glutamine synthetase	Tachibana et al. (1986)
HRAC Group: I: DHP synthase inhibitors (WSSA Group: 18 Aussie Group: R)	Carbamate	Inhibition of DHP (dihydropteroate) synthase	Fedtke (1982)
HRAC Group: K1: Microtubule inhibitors (WSSA Group: 3 Aussie Group: D)	Benzamide, benzoic acid (DCPA), dinitroaniline, phosphoramidate, and pyridine	Microtubule assembly inhibition	Vaughn and Lehnen (1991)
HRAC Group: K2: Mitosis inhibitors (WSSA Group: 23 Aussie Group: E)	Carbamate, carbetamide, chlorpropham, and propham	Inhibition of mitosis/microtubule polymerization	
HRAC Group: K3: Long chain fatty acid inhibitors (WSSA Group: 15 Aussie Group: K)	Acetamide, chloroacetamide, oxyacetamide, and tetrazolinone	Inhibition of cell division (inhibition of very long chain fatty acids)	Husted et al. (1966)
HRAC Group: L: Cellulose inhibitors (WSSA Group: 20, 21, 26, 29 Aussie Group: I, O, Z)	Benzamides and nitriles	Inhibition of cell wall (cellulose) synthesis	Heim et al. (1990)
HRAC Group: M:Uncouplers (WSSA Group: 24 Aussie Group: Z)	Oxidative Phosphorylation	Uncoupling (membrane disruption)	
HRAC Group: N: Lipid Inhibitors (WSSA Group: 16, 26, 8 Aussie Grou: J)	Benzofuranes, chlorocarbonic acids, phosphorodithioates and thiocarbamates	Inhibition of lipid synthesis—not ACCase inhibition	Gronwald (1991)

Continued

Table 1 Continued

Groups of classification system for herbicides	Chemical structural herbicides	Inhibitors effect of herbicides	References
HRAC Group: O: Synthetic Auxins (WSSA Group: 4 Aussie Group: I)	Benzoic acids, phenoxycarboxylic acids, pyridine carboxylic acids, and quinoline carboxylic acids	Synthetic auxins (action like indoleacetic acid)	
HRAC Group: P: Auxin transport inhibitors (WSSA Group: 19 Aussie Group: P)	Phthalamates (naptalam) and semicarbazones (diflufenzopyr)	Inhibition of auxin transport	
HRAC Group: Z: Unknown (WSSA Group: 27 Aussie Group: Z)	Unknown	Unknown	
HRAC Group: Z: Cell elongation inhibitors (WSSA Group: 8 Aussie Group: Z)	Unknown	Unknown	
HRAC Group: Z: Antimicrotubule mitotic disrupter (WSSA Group: 25 Aussie Group: Z)	Unknown	Unknown	
HRAC Group: Z: Nucleic acid inhibitors (WSSA Group: 17 Aussie Group: Z)	Unknown	Unknown	

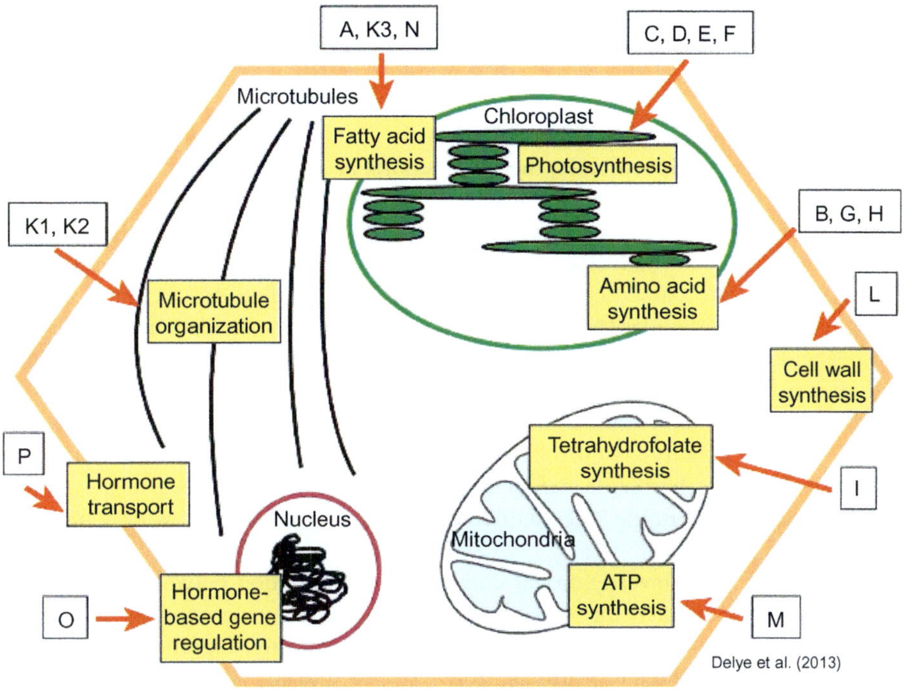

Fig. 4 Cellular targets of herbicide action and herbicide classification by mode of action according to the Herbicide Resistance Action Committee (HRAC). Herbicides target only a few proteins or processes among the tremendous range present in plants.
Reprinted from Délye, C., Jasieniuk, M., Le Corre, V., 2013. Deciphering the evolution of herbicide resistance in weeds. Trends Genet. 29 (11), 649–658.

one day by killing the tissue. For faster effect, some herbicides integrate contact with systemic chemicals. For faster control, the new Round-Up weed and grass killers have combined for both. Others did the same thing.

Killing sections aboveground will not be sufficient for some plants to completely destroy the plant. Yet most plants will regrow plant tissue and the reapplication of the herbicide will be a must. On the other hand, every time the plant uses energy to start growing again, the plant will weaken and eventually die.

4.1.2 Systemic herbicides

Systemic means the plant absorbs via the leaves or stems for systemic types of herbicides and carries it internally throughout the plant. The chemical moves with the plant fluid, so the quick "knockdown" effect usually doesn't occur. A systemic type of herbicide has the greatest advantage of killing the entire plant, roots, and everything. Nonetheless, the temperature of the soil and air almost controls the speed of chemical movement. For instance, a chemical sprayed in mid-summer takes a couple of weeks less to work than the same chemical sprayed in early spring. Moreover, the "mode of action" counts for the speed of kill.

5 Negative effects of herbicides

5.1 Humans

Herbicides are produced and utilized worldwide, and many reports have studied the impact of herbicides. Specifically, there are reports about the negative impact of herbicide residues on crops, soil, and water. Therefore, herbicide residues in environmental resources have attracted much attention because they can accumulate through the food chain and cause adverse effects on the environment and human health. Thus, many countries have passed correlative environmental regulations to control the potential toxicity of herbicides (Nogueira et al., 2004). Women who feed on food sprayed with pesticides pass the chemical to their babies and toddlers through breastfeeding. Moreover, a woman can pass the chemical to her fetus during pregnancy (Jurewicz and Hanke, 2008). Many studies have shown that environmental exposure and neurodevelopmental toxicity are correlated (Grandjean, 2013; Grandjean and Landrigan, 2006). Although the herbicide paraquat dichloride (PQ) is restricted in the United States and banned in Europe (Cicchetti et al., 2009), it is still utilized widely in developing countries.

Even though some investigational data report a correlation between PQ and developmental neurotoxicity, the impact of neuronal marker gene expression, altering basal amino acid neurotransmitters levels or impairing learning, cognition and memory (Benitez-Diaz and Miranda-Contreras, 2009; Hogberg et al., 2009; Li et al., 2016). Different studies propose that during brain development, exposure to PQ may result in permanent changes in brain functions for adults, or increase vulnerability to this pesticide if exposure occurs again in adulthood (Cory-Slechta et al., 2005a,b). Furthermore, PQ is suspected of being associated with Parkinson's disease by inducing mechanisms that perform a pathophysiological role of diseases (Tanner et al., 2011). This is reactive oxidative stress that is related to mitochondrial dysfunction and even neuroinflammation and deterioration of the nigrostriatal pathway (Castello et al., 2007; Cicchetti et al., 2009; Drechsel and Patel, 2009; Mitra et al., 2011; Purisai et al., 2007; Sandström von Tobel et al., 2014). Sandström von Tobel et al. (2017) using the in vitro system consists of major types of brain cell neurons, astrocytes, oligodendrocytes and microglia, and over time, are differentiated and matured into a tissue. PQ was repeatedly applied in the submicromolar range over 10 days and the effects on neurons and glial cells were evaluated. The observation showed that PQ induced significant negative effects on glutamatergic, GABAergic, and dopaminergic neurons, which were assessed by gene expression and enzymatic activity, despite a higher uptake in mature cultures. The attack path may result from direct or indirect consumption, improper use resulting in the herbicide coming into direct contact with humans or wildlife, inhalation of aerial sprays, or food consumption prior to the preharvest interval labeled.

5.2 Soil and water

Regarding the soil microbiome, the application of herbicides may disrupt the soil ecosystem functional balance by changing soil enzymatic activity as well as the structure and diversity of microorganism groups (Ratcliff et al., 2006). Although Glyphosate is a herbicide that is environmentally friendly, many studies indicate that glyphosate may have extensive indirect effects on disease severity (Harper, 2007; Larson et al., 2006) and the efficacy of nutrients (Gordon, 2007; Reichenberger, 2007). In addition, by direct spraying, surface runoff, and infiltration, glyphosate can enter the aquatic environment, and it has recently been frequently detected in surface water (Coupe et al., 2012; Kolpin et al., 2006; Struger et al., 2008; Tsui and Chu, 2008). For instance, according to Peruzzo et al. (2008), 0.7 mg/L glyphosate was found in a surface water system in Argentina. Despite the fact that using pesticides and herbicides may help to increase crop yields, the existence of their residues in soil, water, and air provides potential environmental and public health risks (Celis et al., 2008). In general, conditions promoting herbicide transportation include intense storm events (especially shortly after application) and the limited capacity of the soil to absorb or retain herbicides.

Herbicide properties that raise transportation probability include persistence (degradation resistance) and high water solubility (Andreotti et al., 2018). In particular, the movement of herbicides is possible under certain conditions and contaminant sources of groundwater or remote surface water through leaching or surface runoff. Many countries have improved their regulations. The maximum acceptable herbicide concentration in drinking water has been severely restricted as has that in vegetable foods, either local or imported, particularly with the increase of public awareness and concern for agrochemicals and their potential movement and impact in the ecosystem. Some agricultural practices, including the widespread use of herbicides to enhance yields and improve product quality and quantity, may lead to soil and water contamination. Due to the importance in the ecosystem as well as in the supply of water for drinking and convenience, the quality of soil and surface water deserves special attention. Therefore, their protection and ultimately their remediation are primary necessities.

5.3 Microorganisms

Pesce et al. (2009) found that glyphosate can affect the algae composition community. Wong (2000) observed that glyphosate at low concentrations could possibly boost the growth of *Scenedesmus quadricauda*. Studies of the effects of glyphosate on cyanobacteria, however, are insufficient. Many studies have recently found that several industrial, agricultural, and residual contaminants, including antibiotics and herbicides, can affect cyanobacterial growth (Perron and Juneau, 2011; Phlips et al., 1992). Wu et al. (2016) probed cyanobacteria's physiological characteristics *Microcystis aeruginosa (M. aeruginosa)* after glyphosate exposure and the results indicated reduced changes in the production of cell density, chlorophyll a, and protein content. In addition, in *M. aeruginosa*, oxidative glyphosate stress showed that two days of exposure increased malondialdehyde (MDA) concentration and increased the activity of superoxide dismutase (SOD), catalase (CAT), and peroxidase (POD) while further triggering toxin release. By affecting the food chain, it can affect the entire ecosystem. Causative cyanobacteria can produce microcystins (MCs) or other cyclical hepatotoxins that are toxic to domestic animals and wildlife worldwide (Wu et al., 2016). Kucharski et al. (2016) evaluated the Boreal 58 WG herbicide, the dose of which influenced dehydrogenase activity in only 0.84%, urease activity in 2.04%, b-glucosidase activity in 8.62%, catalase activity in 12.40%, arylsulfatase activity in 12.54%, acid phosphatase

Fig. 5 The negative effects following *herbicide application* include air pollution, soil pollution, water pollution, animal health, plant health, and farmer health.

activity in 42.11%, organotrophic bacteria counts in 18.29%, and actinomyces. Hence, accidental soil contamination with Boreal 58 WG leads to relatively minor microbiological and biochemical disturbances in the soil (Kucharski et al., 2016). The adverse impacts of herbicide applications are investigated in Fig. 5.

6 Nanomaterials for herbicides remediation

Maintaining and restoring air, water, and soil quality are some of the major challenges of our time; most countries are facing serious environmental problems such as drinking water availability, waste and wastewater treatment, air pollution, and soil and groundwater contamination (Mueller and Nowack, 2010). Conventional remediation and treatment technologies have in many cases shown only limited efficacy in reducing pollutant levels, especially in soil and water (Rickerby and Morrison, 2007). As remediation approaches, nanotechnology may cause a potential revolution. Due to their much larger surface area per unit of mass, nanomaterials can be significantly more reactive than larger particles (Rickerby and Morrison, 2007). New nanomaterials with properties unique to the nanometer (nm) scale have increased their use in analytical sciences in recent years (Costa-Fernandez et al., 2006; Merkoçi, 2006; Valcarcel et al., 2007). The combined promise of their unique nm-scale behavior with their obvious miniaturization utility provides the main drive for their use, particularly in environmental remediation. Nanoparticles can generally be used in two configurations as sorbent materials: chemically bonded to microparticles in general through a covalent bond or directly used as raw material.

Research into new types of sorbents offers a focus on selectivity and capacity improvement. Mixed-setting polymer sorbents, molecularly amazing polymers (MIPs), and nanocarbons involve the unique types of sorbents that could be useful in herbicide evaluation for enrichment and clean up. MIPs tend to be more selective than mixed-mode sorbents; nevertheless, they have greater capability than MIPs. Carbon nanomaterials possess a good affinity for adsorption for a large selection of organic substances, including pesticides, and also have a higher sorption surface area. Carbon-encapsulated magnetic nanoparticle utilization prevents time-consuming column passage and filtration and will offer great analytical potential in the preconcentration of sizeable volumes of actual samples of drinking water (Zhao et al., 2008). Improvements within their production provided the facilitated application of carbon nanostructures, as the cost was a major factor in limiting marketing. However, it is broadly suspected that if formation volumes were enhanced, costs would be minimized significantly, thus considerably strengthening the usage of the superb properties of nanostructured carbon. New solvent-free procedures for the creation of CNTs from utilized polymers through thermal dissociation in the shut reactor in the inert or air flow atmosphere have been recently planned (Pol and Thiyagarajan, 2010). And in the real environment, organic matter is usually frequently evaluated as among the factors regularly studied because of its improvement of Cu (II) adsorption on carbon nanotubes, which may be the real impact among carbon nanomaterials and heavy metal ions (Hyung and Kim, 2008).

7 Carbon nanotubes for pesticide remediation

In agricultural production around the world, pesticides are widely used to protect plants from pests, fungi, and weeds. Consequently, pesticide residues are distributed heavily in drinking water, groundwater, and soil as contaminants (Nasrabadi et al., 2011). This results in an adverse impact on living organisms, ecosystems, and services (Joo and Cheng, 2006; Maddah and Hasanzadeh, 2017). The use of carbon nanostructures started in 1985 when Kroto et al. (1985) discovered Buckminsterfullerene C60. Since then, the number of structures discovered has increased rapidly, such as nanotubes. For the first time, carbon nanotubes were used to extract five *N*-methylcarbamate insecticides (i.e., carbaryl, carbofuran, aminocarb, methiocarb, and zectran) from different surface water as solid-phase extraction sorbents. For some contaminants, such as chlorobenzenes and dioxin, pesticides are considered the main source. With respect to dioxin, Long and Yang (2001) used MWCNTs as a dioxin-removal sorbent due to the strong interaction between dioxin and carbon. The removal effectiveness owes to the unique structure and electronic properties of carbon nanotubes. The carbon nanotube surfaces, consisting of hexagonal carbon atom arrays of graphene sheets surrounding the tube axis, interact strongly with the two dioxin benzene rings. Additionally, chlorobenzenes were used in large quantities in the manufacture of other chemical products such as solvents, lubricants, pesticides, and deodorants, or as intermediates

(Beurskens et al., 1994). However, chlorobenzenes have been common in the environment as a result of their widespread use over several decades, for example in water, soils, sediments, sewage sludge, and aquatic biota (Doong and Liao, 2001; Murena and Gioia, 2002; Van Noort et al., 2003). Regarding human beings, some chlorobenzenes are known carcinogens (Gustafson et al., 2000). MWCNTs were utilized by Liu et al. (2004) as chlorobenzene–sorbent solid–phase extraction (SPE) compared to two different sorbents, silica-bonded C18 and activated carbon. The results demonstrated that for chlorobenzenes, MWCNTs had higher extraction recoveries. Organochlorine pesticides (OCPs) such as dichlorodiphenyltrichloroethanes (DDTs) and hexachlorocyclohexane (HCH) are the most typical persistent organic pollutants (POPs), with high toxicity and persistence as well as half volatile and bioaccumulation characteristics that can potentially have adverse effects on human health (Huang et al., 2008; Zeng et al., 2013). Mainly because of the strong bioaccumulation and the difficult degradation of OCPs, residual DDTs and HCHs have remained in the environment for decades; they are even sometimes prohibited for use on a large scale as they can cause major damage to ecosystems (Qian et al., 2006). A study by Zhang et al. (2017) showed the prominent effects of sequestering DDTs and HCHs for both SWCNTs and MWCNTs in samples from contaminated lake sediment. In addition, due to SWCNTs' larger specific surface area (SSA), they showed more efficacy. Generally, the efficiency of carbon nanotube modification was dependent on type, dose, and sediment sorbent contact time, enhancing with increasing dose and contact time. It could be concluded that carbon nanotubes have better effects than those of HCHs on the reduction of DDT aqueous concentration of equilibrium as well as DDT uptake in SPMDs. Under optimal conditions (sediment treated with 0.29 wt% SWCNTs for four months), carbon nanotubes showed high efficacy in treating DDTs and HCHs. Furthermore, sufficient amounts of carbon nanotubes (0.29 wt%) achieved significant reductions in DDT and HCH aqueous equilibrium concentration as well as DDT and HCH intake in SPMDs even for one month, which was a relatively short period. In addition, small amounts (0.058 wt%) of carbon nanotubes could also have similar effects on the reduction of aqueous equilibrium concentrations of DDTs and HCHs as well as on the uptake of DDTs and HCHs in SPMDs due to sufficient sediment-sorbent contact time (4 months). This may have implications for in situ remediation by adding carbon nanotubes to the overall cost of the modification, efficiency, and time effect (Zhang et al., 2017). Pyrethroid pesticides are used more in public health and animal breeding because they are less toxic to living organisms compared to, for example, organochlorine pesticides (Mahmood et al., 2016). El-Sheikh et al. (2007) conducted comparative research to select the most appropriate type of carbon nanotube. These studies indicated that even short MWCNTs of 40–60 nm (long or short MWCNTs of different external diameters: 10–20, 10–30, 20–40, 40–60, 60–100 nm) were the most appropriate among the tested types of nanomaterials and ensured the highest efficiency of extraction (El-Sheikh et al., 2007). Taking this information into account, in contrast, for selected pesticide residues, López-Feria et al. (2009) demonstrated that single-walled carbon nanotubes (i.e., 1–2 nm) with carboxylic groups have significantly better adsorption capacity than multilayered carbon nanotubes. The application of 5.3 mg CNTs per 100 mL of a sample solution containing seven pesticides (pirimicarb, pyrifenox, penconazole, carbendazim, cyromazine, pyrimethanil, and cyprodinil) was optimal in the work by Asensio-Ramos et al. (2008). So, as illustrated, even using this small sorbent amount showed the ability to ensure recoveries in the range between 53% and 94% for Milli-Q water, which is satisfactory in regard to the relatively large number of analytes belonging to the different groups of plant protection products.

8 Carbon nanotubes for herbicide sensing

8.1 Herbicide purification

Hundreds of herbicides possessing different chemical structures are widely used in agriculture because of their persistence, polar nature, and water solubility. They are dispersed in the environment and present in several environmental matrices through their residues and transformation products. Hence, the availability of sensitive, selective, precise, and rapid methods of residue analysis is essential. This generally requires several steps such as sample extraction, removal of interfering coextractives, enrichment of analysis, and quantification of their content (Pyrzyńska, 2011). The better dispersion of the extracting directly enhances the contact area between the sample and the stationary phase, which will consequently improve the extraction efficiency of the method. This was evident in the research of Lasarte-Aragonés et al. (2013), which described the effectiveness of CNT dispersion in the micro-solid-phase extraction (μ-SPE) of certain herbicides from environmental water samples (104). In addition, researchers incorporated methanol to eluted retained compounds from the analyte-enriched MWCNTs and optimized the μ-SPE technique for the determination of herbicide numbers in waters through using ultraperformance liquid chromatography (UPLC) combined with ultraviolet detection (UV). The MWCNT-modified dispersive solid-phase extraction (*d*-SPE) technique is fast, effective, cheap, rugged, and safe. It was validated using gas chromatography combined with tandem mass spectrometry (GC-MS/MS) (Liang et al., 2005). The analytical potential of MWCNTs as an SPE adsorbent for the preconcentration of copper traces was assessed using the column method; a microcolumn packed with MWCNTs acting as a sorbent has been developed for the preconcentration in water samples prior to its determination by flame atomic absorption spectrometry (FAAS). CNTs, which are a new form of carbon-based sorbent, represent another promising material in the SPE of herbicides (Pyrzynska, 2008).

MWCNTs were first used to extract five N-methylcarbamate insecticides (i.e., carbofuran, methiocarb, carbaryl, aminocarb, and zectran) from different surface water samples as solid-phase extraction sorbents (Latrous El Atrache et al., 2016). MWCNTs were also used to determine the four chloroacetanilide herbicides (alachlor, acetochlor, metolachlor, and butachlor) in water as solid-phase extraction (SPE) adsorbents. The main factors affecting the effectiveness of the SPE are the quantity of the adsorbent, the eluent solvent, the pH, and the sample mass (Maofeng et al., 2009). For a wide variety of organic compounds, including pesticides, CNTs have a strong adsorption affinity, and they are also characterized by their high sorption surface. These interesting properties were exploited in the extraction of solid phases (Krystyna, 2011). Among the most important applications of these materials in analytical science is the use of CNTs in solid-phase extraction (SPE) (Asensio-Ramos et al., 2011). Additionally, Hou et al. (2014) reported a multiresidue technique for the quantization of pesticides in tea samples using MWCNTs as a dispersive solid-phase extraction sorbent. Using their methods, it was possible to analyze 78 pesticide residues in tea (Hou et al., 2014).

A solid-phase extraction microcartridge containing a nonpolar polystyrene absorbent matrix was combined with an electrochemical immunoassay analyzer (EIA) and used in real samples to detect ultrasensitive phenyl urea herbicide diuron. The h-CNT modification of the SPE surface, alongside the highly specific diuron antibody, exhibited an excellent detection limit of 0.1 pg mL^{-1} for diuron-spiked water samples. As for the modification of the sensor surface, good stability was observed due to the hydrophobic nature of the hapten, which tends to form aggregates of high aspect ratio h-CNT disordered ensembles (Sharma et al., 2012a, b). The application of the solid-liquid-solid dispersive extraction (SLSDE) on herbicides from tobacco samples using MWCNTs as clean-up adsorbents is a feasible method that permitted the discovery of numerous herbicide residues (Liao et al., 2014). For the simultaneous separation and determination of six types of sulfonylurea herbicides (SUs) in environmental water samples, magnetic solid-phase extraction (MSPE) using magnetic multiwalled carbon nanotubes (mag-MWCNTs) as adsorbents was combined with a high-performance liquid chromatography-diode-array detector (HPLC-DAD) (Ma et al., 2016).

8.2 Herbicide detection

Recently, CNTs have gained interest for their use in electrochemical immunoassays due to their remarkable tensile strength, surface area, flexibility, and other unique structural, mechanical, electrical, and physicochemical properties that reflect increased signal currents (Chen et al., 2008). The development of a metal ion-derivatized electrochemical immunoassay based on gold-iron (Au/Fe) nanobioprobes for the detection of commonly used phenylurea herbicide diuron on reduced graphene oxide-carbon nanotubes (rGO/CNT) modified biosensing platform using specific anti-diuron antibodies. Moreover, magnetic nanoparticles offer a rapid immunocomplex formation on magneto-microtiter plates and their further electrochemical bursting into a large number of Fe^{2+} ions presents ultrahigh sensitivity for diuron detection on SPE (Sharma et al., 2012a, b). The chlorophenoxy herbicide MCPA (4-chloro-2-methyl-phenoxy-acetic acid) was determined by cyclic voltammetry using an electrochemical sensor based on GCE modified with MWCNTs, β-CD, and polyaniline (PANI) film. The method was successfully applied to the MCPA quantification of river water samples, providing similar results to those found by high-performance liquid chromatography (HPLC) (Rahemi et al., 2013). Mani et al. (2015) modified a GCE surface with graphene oxide and multiwalled carbon nanotubes (GO-MWCNT) for the sensitive detection of diuron and fenuron herbicides. GO-MWCNT film-modified GCE exhibited excellent electrocatalytic effects on the oxidation process of diuron and fenuron in terms of lower overpotential and highly enhanced peak currents.

9 Carbon nanotubes for herbicide remediation

CNTs were evaluated as novel sorbents for removing the highly polar and acidic herbicide "dicamba" from aqueous samples for environmental protection (Magdalena and Krystyna, 2006). They can be described as a graphite sheet rolled up into a nanoscale tube; they have lengths up to several centimeters with both ends normally capped with fullerene-like structures. Nanotubes have the simplest chemical composition and atomic bond configuration. However, they exhibit the most extreme diversity in structure and structure-dependent properties among nanochemicals (Dai, 2002). Chen et al. (2011) were able to remove two herbicides (diuron and dichlobenil) from contaminated water using MWCNTs.

Magnetic ionic liquid-modified multiwalled carbon nanotubes (m-IL-MWCNTs) were prepared by the spontaneous assembly of magnetic nanoparticles and imidazolium-modified carbon nanotubes, used as a d-MSPE sorbent for the simultaneous extraction of aryloxyphenoxypropionate herbicides (AOPPs) and their polar acid metabolites as a result of excellent π electron donor acceptance or interactions and anion exchange ability (Mai Luo et al., 2014). Diuron adsorption on MWCNTs indicated that MWCNT oxidation treatment increased the surface area and volume of the pores and subsequently the capacity for adsorption. In addition, diuron adsorption was spontaneous and exothermic on MWCNTs (Deng et al., 2012). SWCNTs have been shown to have a higher phenoxy acid herbicide adsorption capacity "4-chloro-2-methyl phenoxy acetic acid" (MCPA) than three types of MWCNTs with different outer diameters and several nanoscale metal oxides (Al_2O_3, TiO_2, and ZnO). The adsorption kinetics usually follow pseudo-second-order kinetics, with the adsorption process being spontaneous and exothermic (De Martino et al., 2012). Triazine, a powerful herbicide used for weed control, has often been used in agriculture. However, it is also listed as a chemical pollutant that needs to be heavily monitored due to its environmental effects such as toxicity, persistence, and accumulation as well as its effect on human health (Hernándea et al., 1998). Atrazine and simazine are typical candidates of fundamental importance and improved weed control efficiency, hence their high rate of consumption. The use of MWCNTs as absorbents provided excellent atrazine and simazine enrichment efficiency (Zhou et al., 2006). Sulfonylurea is another type of herbicide that can hinder the susceptible plant's growth under specific conditions such as high pH levels, low temperatures, reduced rainfall, and poor microbial activity (Fahl et al., 1995). Additionally, their accumulation can lead to severe environmental pollution (Zhou et al., 2006). However, MWCNTs presented good suitability for the preconcentration of sulfonylurea herbicides in complex water samples, therefore yielding better recoveries (Zhou et al., 2006). This was deduced after achieving analytical performances from real-world water samples such as river water, reservoir water, tap water, and wastewater after primary pretreatment (Zhou et al., 2006). According to Liang et al. (2005), MWCNTs possess a higher adsorption capacity toward copper, and the analyte retained on MWCNTs can be easily disorbed while no carryover is observed. In other words, MWCNTs have promising potential as an adsorbent for copper preconcentration. According to DQ herbicide adsorption, the maximum adsorption capacity values were not determined for single-walled carbon nanotubes, single-walled hydroxyl-functional carbon nanotubes, and multiwalled carbon nanotubes (Dichiara et al., 2015). Therefore, additional investigation is required to assess and understand the adsorption performances and mechanisms of diverse forms of carbon nanotubes. Previous studies have shown that magnetic MWCNTs have been modified with κ-carrageenan-Fe_3O_4 nanocomposite can be applied as a magnetic adsorbent to eradicate the cationic dyes from water solutions (Duman et al., 2016a, b). However, it is unclear whether such modifications can increase the herbicide adsorption capacity of carbon nanotubes. Evaluations were conducted on CNTs synthesized from waste polyethene bottles and used as adsorbents for the removal of diuron herbicide from aqueous solutions. The adsorption capacity for diuron removal was approximately 40.37 mg/g at 303 K, determined to use the Hill isotherm (Deokar et al., 2017). Among agricultural pesticides such as an aromatic hydrocarbon, 2,4-dichlorophenoxyacetic acid (2,4-D) is a dangerous and toxic organic pollutant. The biochar made from rice husk (BRH), granular activated carbon (GAC), and MWCNTs was investigated for adsorption of 2,4-D in a fixed-bed column system. Results confirmed that the BRH is an affordable and sustainable material that can be a viable alternative to GAC and MWCNTs for remediation and treatment strategies (Bahrami et al., 2018). Simazine and propazine are selective triazine herbicides currently used to control broad-leaved weeds and annual grasses. The adsorption of the two triazines and Bisphenol A on the surface of NPS of iron (III) oxide, NPS of carbon, bulk iron (III) oxide, and aluminium oxide at pH 6 and pH 8 using UV-visible spectroscopy. The NPS of carbon has shown the highest affinity for all three compounds (Eshete et al., 2018). The adsorption efficiency of MWCNTs grown on silicon support for seven triazine herbicides dissolved in water was investigated on a small volume scale. Results showed that MWCNTs performed with acidic oxidation increased the material's wettability and resulted in improved adsorption performance (D'Archivio et al., 2018). Duman et al. (2019) compared the adsorption performances of three types of multiwalled carbon nanotubes (OMWCNT, OMWCNT-Fe_3O_4 nanocomposite, and OMWCNT-κ-carrageenan-Fe_3O_4 nanocomposite) and magnetic Fe_3O_4 nanoparticles for the removal of DQ from the water. The effects of initial herbicide concentration, contact time, and temperature on the adsorption process were also examined (Fig. 6).

10 Remediation mechanism

Environmental remediation is primarily based on the use of various technologies (e.g., adsorption, absorption, chemical reactions, photocatalysis, and filtration) to remove contaminants from various environmental media (e.g., soil, water, and air). Chlorophenols have enhanced the application of carbon nanotubes to solve environmental analytical problems (Cai et al., 2005), herbicides (Zhou et al., 2006), and organic chemical adsorption mechanisms for CNTs. Heterogeneity and hysteresis adsorption are two widely recognized characteristics of organic chemical–CNT interactions. However, because different mechanisms can operate at the same time (mainly hydrophobic interactions, π-π bonds, electrostatic interactions, and hydrogen bonds), it is not easy to predict organic chemical adsorption on CNTs. The dominant

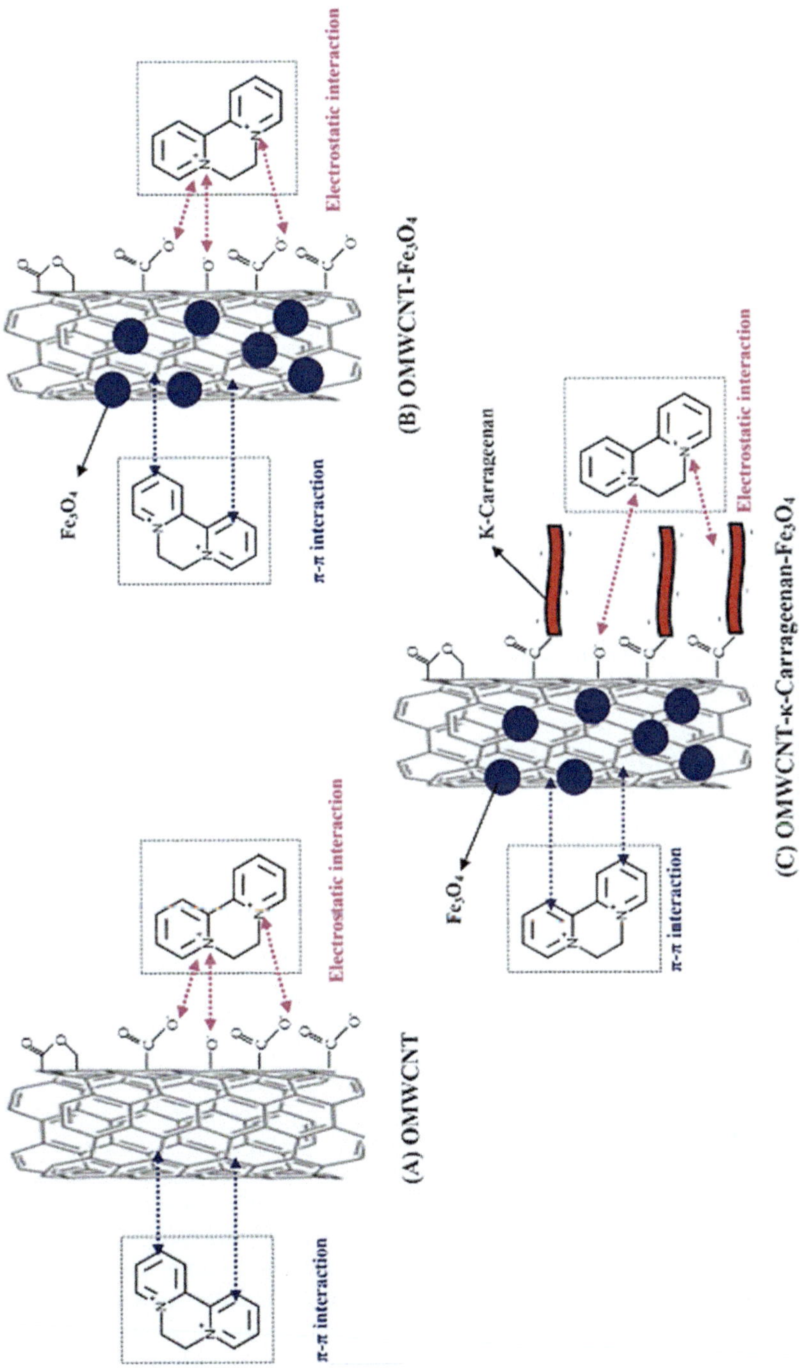

Fig. 6 Schematic illustration of the possible interactions between adsorbents and DQ. (A) OMWCNT and DQ; (B) OMWCNT-Fe3O4 and DQ; and (C) OMWCNT-κ Carrageenan-Fe_3O_4 and DQ.
Reprinted from Duman, O., Özcan, C., Polat, T.G., Tunç, S., 2019. Carbon nanotube-based magnetic and non-magnetic adsorbents for the high-efficiency removal of diquat dibromide herbicide from water: OMWCNT, OMWCNT-Fe_3O_4 and OMWCNT-κ-carrageenan-Fe_3O_4 nanocomposites. Environ. Pollut. 244, 723–732.

mechanism of adsorption differs from one type of organic chemical to another (for instance, polar and nonpolar). Hence, to predict organic chemical-CNT interactions, different models may be needed. Nevertheless, mechanisms for adsorption will be clearer by investigating the effects of both CNTs and organic chemical properties influenced by environmental conditions (Pan and Xing, 2008).

11 Three general mechanisms of organic chemical adsorption on CNTs

11.1 Heterogeneous adsorption

Heterogeneous adsorption indicates that organic chemical adsorption on CNTs could not be described using a single adsorption coefficient. If a single coefficient is used, a significant error will occur when predicting the organic chemical–CNT interaction, which will consequently lead to a wrong conclusion regarding the environmental risk of both organic chemicals and CNTs (Wang et al., 2007).

11.2 Hysteresis

An adsorption/desorption hysteresis on CNTs was observed for small molecules such as organic vapors of methane, ethylene, and benzene as well as polymers such as poly(aryleneethynylene)s (Chen et al., 2002).

11.3 Multiple mechanisms acting simultaneously

The outer surface of individual CNTs provides evenly distributed hydrophobic sites for organic chemicals. Several studies have highlighted hydrophobic interactions with protein naphthalene acid herbicides (Pyrzynska et al., 2007). MWCNT oxidation leads to surface functionalization with groups containing oxygen (e.g., carboxylic, carbonyl, and hydroxyl), allowing cations to be retained. pH performs a major role as it affects the CNT surface charge. When the pH is higher than the isoelectric point of the CNTs, the adsorbent surface has a negative charge that allows electrostatic interactions with the solution's cations (Valcàrcel et al., 2008).

12 Conclusions and upcoming views

Carbon nanotubes (CNTs) are advanced materials that have been proposed as components for enzymatic biosensors, DNA samples, and solid-phase extraction due to their high surface area, good thermal stability, and resistance. This chapter investigates that in pesticide determination, carbon nanotubes provide a relatively good analytical performance of electrochemical sensors. Herbicides in different classes were quantified electrochemically using electrochemical sensors based on carbon nanotubes. Their use has been investigated for the sorption of herbicides, particularly with regard to environmental technologies. In some aspects, using carbon nanotubes as a herbicide remediation can cause an environmental hazard as organic contaminants adsorbed on nanotubes can form during water treatment and then enter the environment (water, sediment, and soil) as contamination. Also, this process can happen in both groundwater and surface water where these contaminants may already be present. If the pollutants in significant concentrations were not released from the adsorbent, they would ostensibly remain nontoxic to living organisms. On the other hand, there is the risk that contaminants released from CNTs may adversely affect the environment and organisms with significant desorption. This potential threat makes understanding interactions between CNMs and organic pollutants vital, especially the desorption range of pollutants as a potential source of secondary water contamination. There is also a need to focus on understanding the organic contaminant adsorption mechanisms on CNMs. In the near future, we need to focus more on herbicide degradation/transformation products (from hydrolysis, oxidation, biodegradation, or photolysis) as they may be present in the environment at higher levels than the parent herbicide, and they may be just as toxic or even more toxic at times. New compounds (such as glyphosate and organophosphorus herbicides) have also been introduced on the market, and studies are being conducted to understand their fate and transport in the environment.

References

Abgail, E.A., Chidambaram, R., 2017. Nanotechnology in herbicide resistance. Open Sci. https://doi.org/10.5772/intechopen.68355.

Amrhein, N., Deus, B., Gehrke, P., Steinrücken, H.C., 1980. The site of the Inhibition of the shikimate pathway by glyphosate: II. Interference of glyphosate with chorismate formation in vivo and in vitro. Plant Physiol. 66, 830.

Andreotti, G., Koutros, S., Hofmann, J.N., Sandler, D.P., Lubin, J.H., Lynch, C.F., Lerro, C.C., De Roos, A.J., Parks, C.G., Alavanja, M.C., Silverman, D.T., Beane Freeman, L.E., 2018. Glyphosate use and cancer incidence in the agricultural health study. J. Natl. Cancer Inst. 110 (5), 509–516. https://doi.org/10.1093/jnci/djx233. PMID 29136183.

Asensio-Ramos, M., Hernández-Borges, J., Ravelo-Pérez, L.M., Rodríguez-Delgado, M.A., 2008. Simultaneous determination of seven pesticides in waters using multi-walled carbon nanotube SPE and NACE. Electrophoresis 29 (21), 4412–4421.

Asensio-Ramos, M., D'Orazio, G., Hernandez-Borges, J., Rocco, A., Fanali, S., 2011. Multiwalled carbon nanotubes–dispersive solid-phase extraction combined with nano-liquid chromatography for the analysis of pesticides in water samples. Anal. Bioanal. Chem. 400 (4), 1113–1123.

Bahrami, M., Amiri, M.J., Beigzadeh, B., 2018. Adsorption of 2, 4-dichlorophenoxyacetic acid using rice husk biochar, granular activated carbon, and multi-walled carbon nanotubes in a fixed bed column system. Water Sci. Technol. 78 (8), 1812–1821.

Barry, P., Pallett, K.E., 1990. Herbicides inhibition of care lenogenesis detected by HPLC. Z. Naturforsch. 45C, 492–497.

Benitez-Diaz, P., Miranda-Contreras, L., 2009. Effects of prenatal exposure to paraquat on the development of amino acid synaptic transmission in mouse cerebral parietal cortex. Investig. Clin. 50, 465–478.

Beurskens, J.E.M., Dekker, C.G.C., Vandenheuvel, H., Swart, M., Dewolf, J., 1994. Dechlorination of chlorinated benzenes by an anaerobic microbial consortium that selectively mediates the thermodynamic most favorable reactions. Environ. Sci. Technol. 28 (4), 701–706.

Cai, Y., Mou, S., Lu, Y., 2005. Multi-walled carbon nanotubes as a solid-phase extraction adsorbent for the determination of chlorophenols in environmental water samples. J. Chromatogr. A 1081, 245–247. ISSN: 0021-9673.

Cao, G.Z., 2004. Nanostructures and Nanomaterials Synthesis, Properties and Applications. Imperial College Press [Orig.-Prod.], London.

Castello, P.R., Drechsel, D.A., Patel, M., 2007. Mitochondria are a major source of paraquat-induced reactive oxygen species production in the brain. J. Biol. Chem. 282, 14186–14193.

Celis, E., Elefsiniotis, P., Singhal, N., 2008. Biodegradation of agricultural herbicides in sequencing batch reactors under aerobic or anaerobic conditions. Water Res. 42, 3218–3224.

Chen, J., Liu, H.Y., Weimer, W.A., Halls, M.D., Waldeck, D.H., Walker, G.C., 2002. Noncovalent engineering of carbon nanotube surfaces by rigid, functional conjugated polymers. J. Am. Chem. Soc. 124, 9034–9035.

Chen, H., Zuo, X., Su, S., Tang, Z., Wu, A., Song, S., Zhang, D., Fan, C., 2008. An electrochemical sensor for pesticide assays based on carbon nanotube-enhanced acetycholinesterase activity. Analyst 133 (9), 1182–1186.

Chen, G.C., Shan, X.Q., Pei, Z.G., Wang, H.H., Zheng, L.R., Zhang, J., 2011. Adsorption of diuron and dichlobenil on multiwalled carbon nanotubes as affected by lead. J. Hazard. Mater. 188, 156–163.

Cicchetti, F., Drouin-Ouellet, J., Gross, R.E., 2009. Environmental toxins and Parkinson's disease: what have we learned from pesticide-induced animal models? Trends Pharmacol. Sci. 30, 475–483.

Cory-Slechta, D.A., Thiruchelvam, M., Barlow, B.K., Richfield, E.K., 2005a. Developmental pesticide models of the Parkinson disease phenotype. Environ. Health Perspect. 113, 1263–1270.

Cory-Slechta, D.A., Thiruchelvam, M., Richfield, E.K., Barlow, B.K., Brooks, A.I., 2005b. Developmental pesticide exposures and the Parkinson's disease phenotype. Birth Defects Res. A Clin. Mol. Teratol. 73, 136–139.

Costa-Fernandez, J.M., Pereiro, R., Sanz-Medel, A., 2006. The use of luminescent quantum dots for optical sensing. Trends Anal. Chem. 25 (3), 207–218.

Coupe, R.H., Kalkhoff, S.J., Capel, P.D., Gregoire, C., 2012. Fate and transport of glyphosate and aminomethylphosphonic acid in surface waters of agricultural basins. Pest Manag. Sci. 68, 16–30.

Crafts, A.S., 1946. Selectivity of herbicides. Plant Physiol. 21, 345–361.

D'Archivio, A.A., Maggi, M.A., Odoardi, A., Santucci, S., Passacantando, M., 2018. Adsorption of triazine herbicides from aqueous solution by functionalized multiwall carbon nanotubes grown on silicon substrate. Nanotechnology 29 (6), 065701.

Dai, H., 2002. Carbon nanotubes: synthesis, integration, and properties. Acc. Chem. Res. 35 (12), 1035–1044.

De Martino, A., Iorio, M., Xing, B., Capasso, R., 2012. Removal of 4-chloro-2-methylphenoxyacetic acid from water by sorption on carbon nanotubes and metal oxide nanoparticles. RSC Adv. 2, 5693–5700.

Deng, J., Shao, Y., Gao, N., Deng, Y., Tan, C., Zhou, S., Hu, X., 2012. Multiwalled carbon nanotubes as adsorbents for removal of herbicide diuron from aqueous solution. J. Chem. Eng. 193, 339–347.

Deokar, S.K., Bajad, G.S., Bhonde, P., Vijayakumar, R.P., Mandavgane, S.A., 2017. Adsorptive removal of diuron herbicide on carbon nanotubes synthesized from plastic waste. J. Polym. Environ. 25 (2), 165–175.

Devine, M., Duke, S.O., Fedtke, C., 1993. Physiology of Herbicide Action. Prentice Hall, New Jersey.

Dichiara, A.B., Harlander, S.F., Rogers, R.E., 2015. Fixed bed adsorption of diquat dibromide from aqueous solution using carbon nanotubes. RSC Adv. 5 (76), 61508–61512.

Dodge, A.D., 1982. In: Moreland, D.E., St. John, J.B., Hess, F.D. (Eds.), Biochemical Responses Induced by Herbicides, pp. 57–77.

Doong, R.A., Liao, P.L., 2001. Determination of organochlorine pesticides and their metabolites in soil samples using headspace solid-phase microextraction. J. Chromatogr. A 918 (1), 177–188.

Drechsel, D.A., Patel, M., 2009. Chapter 21. Paraquat-induced production of reactive oxygen species in brain mitochondria. Methods Enzymol. 456, 381–393.

Duke, S.O., John, L., José, M.B., Timothy, D.S., Larry, P.L.J.R., Hroshi, M., 1991. Protoporphyrinogen oxidase-inhibiting herbicides. Weed Sci. 39, 465–473.

Duman, O., Tunç, S., Polat, T.G., Bozoğlan, B.K., 2016a. Synthesis of magnetic oxidized multiwalled carbon nanotube-κ-carrageenan-Fe_3O_4 nanocomposite adsorbent and its application in cationic Methylene Blue dye adsorption. Carbohydr. Polym. 147, 79–88.

Duman, O., Tunç, S., Polat, T.G., Bozoğlan, B.K., 2016b. Removal of triphenylmethane and reactive azo dyes from aqueous solution by magnetic carbon nanotube-κ-carrageenan-Fe_3O_4 nanocomposite. J. Alloys Compd. 68 (7), 370–383.

Duman, O., Özcan, C., Polat, T.G., Tunç, S., 2019. Carbon nanotube-based magnetic and non-magnetic adsorbents for the high-efficiency removal of diquat dibromide herbicide from water: OMWCNT, OMWCNT-Fe_3O_4 and OMWCNT-κ-carrageenan-Fe_3O_4 nanocomposites. Environ. Pollut. 244, 723–732.

El-Sheikh, A.H., Insisi, A.A., Sweileh, J.A., 2007. Effect of oxidation and dimensions of multiwalled carbon nanotubes on solid phase extraction and enrichment of some pesticides from environmental waters prior to their simultaneous determination by high performance liquid chromatography. J. Chromatogr. A 1164 (1-2), 25–32. https://doi.org/10.1016/j.chroma.2007.07.009.

Eshete, M., Bowleg, J., Perales, S.G., Okunrobo, M., Watkins, D., Spencer, H., 2018. Adsorption of propazine, simazine and bisphenol a on the surface of nanoparticles of iron oxide nanoparticles of carbon and metallic oxides. J. Environ. Prot. 9 (01), 13.

Fahl, G.M., Kreft, L., Altenburger, R., Faust, M., Boedeker, W., Grimme, L.H., 1995. pH-Dependent sorption, bioconcentration and algal toxicity of sulfonylurea herbicides. Aquat. Toxicol. 31 (2), 175–187.

Fedtke, C., 1982. Biochemistry and Physiology of Herbicide Action. Springer-Verlag, Berlin, 148 pp.

Gogos, A., Knauer, K., Bucheli, T.D., 2012. Nanomaterials in plant protection and fertilization: current state, foreseen applications, and research priorities. J. Agric. Food Chem. 60 (39), 9781–9792.

Gordon, B., 2007. Manganese nutrition of glyphosate-resistant and conventional soybeans. Bett. Crop. Plant. Food. 91, 12–13.

Grandjean, P., 2013. How Environmental Pollution Impairs Brain Development and How to Protect the Brains of the Next Generation. Oxford University press, New York.

Grandjean, P., Landrigan, P.J., 2006. Developmental neurotoxicity of industrial chemicals. Lancet 368 (9553), 2167–2178.

Gronwald, J.W., 1991. Lipid biosynthesis inhibitors. Weed Sci 39 (3), 435–449.

Grossmann, K., Ehrhardt, T., 2007. On the mechanism of action and selectivity of the corn herbicide topramezone: a new inhibitor of 4-hydroxyphenylpyruvate dioxygenase. Pest Manage. Sci.: formerly. Pestic. Sci. 63 (5), 429–439.

Gustafson, D.L., Long, M.E., Thomas, R.S., Benjamin, S.A., Yang, R.S.H., 2000. Comparative hepatocarcinogenicity of hexachlorobenzene, pentachlorobenzene, 1,2,4,5-tetrachlorobenzene, and 1,4-dichlorobenzene: application of a medium-term liver focus bioassay and molecular and cellular indices. Toxicol. Sci. 53 (2), 245–252.

Harper, M., 2007. The Review of the Moratorium on GM Canola. Australia.

Hatterman-Valenti, H., Mayland, P., 2005. Annual flower injury from sublethal rates of dicamba, 2,4-D, and premixed 2,4-D + mecoprop plus dicamba. Hortiscience 40, 680–684.

Heim, D.R., Skomp, J.R., Tschabold, E.E., Larrinua, I., 1990. Isoxaben inhibits the synthesis of acid-insoluble cell wall materials in *Arabidopsis thaliana*. Plant Physiol. 93, 695–700.

Hernándea, F., Hidalgo, C., Sancho, J.C., López, F.J., 1998. Coupled-column liquid chromatography applied to the trace-level determination of triazine herbicides and some of their metabolites in water samples. Anal. Chem. 70 (15), 3322–3328.

Hochella, M.F., 2002. Nanoscience and technology: the next revolution in the Earth sciences. Earth Planet. Sci. Lett. 203, 593–605.

Hogberg, H.T., Kinsner-Ovaskainen, A., Hartung, T., Coecke, S., Bal-Price, A.K., 2009. Gene expression as a sensitive endpoint to evaluate cell differentiation and maturation of the developing central nervous system in primary cultures of rat cerebellar granule cells (CGCs) exposed to pesticides. Toxicol. Appl. Pharmacol. 235, 268–286.

Hong, J., Peralta-Videa, J.R., Gardea-Torresdey, J.L., 2013. Nanomaterials in agricultural production: benefits and possible threats? In: Shamim, N., Sharma, V.K. (Eds.), Sustainable Nanotechnology and the Environment: Advances and Achievements. American Chemical Society, Washington, DC, pp. 73–90.

Hou, X., Lei, S., Qiu, S., Guo, L., Yi, S., Liu, W., 2014. A multi-residue method for the determination of pesticides in tea using multi-walled carbon nanotubes as a dispersive solid-phase extraction absorbent. Food Chem. 153, 121–129.

Huang, D.L., Zeng, G.M., Feng, C.L., Hu, S., Jiang, X.Y., Tang, L., Su, F.F., Zhang, Y., Zeng, W., Liu, H.L., 2008. Degradation of lead-contaminated lignocellulosic waste by phanerochaete chrysosporium and the reduction of lead toxicity. Environ. Sci. Technol. 42 (13), 4946–4951.

Husted, R.F., Olin, J.F., Upchurch, R.P., 1966. A new selective preemergence herbicide: CP-50144. In: Proceedings of the Northern Central Weed Control Conference. vol. 21, pp. 44.

Hyung, H., Kim, J.H., 2008. Natural organic matter (NOM) adsorption to multiwalled carbon nanotubes: effect of NOM characteristics and water quality parameters. Environ. Sci. Technol. 42, 4416–4421.

Iijima, S., 1991. Synthesis of carbon nanotubes. Nature 354, 56–58. https://doi.org/10.1038/354056a0.

Joo, S.H., Cheng, F., 2006. Nanotechnology for Environmental Remediation. Modern Inorganic Chemistry. Springer, New York.

Jurewicz, J., Hanke, W., 2008. Parental and childhood exposure to pesticides and neurobehavioral development. Review of epidemiological studies. Int. J. Occup. Med. Environ. Health 21 (2), 211–232.

Khot, L.R., Sankaran, S., Maja, J.M., Ehsani, R., Schuster, E.W., 2012. Applications of nanomaterials in agricultural production and crop protection: a review. Crop Prot. 35, 64–70.

Kolpin, D.W., Thurman, E.M., Lee, E.A., Meyer, M.T., Furlong, E.T., Glassmeyer, S.T., 2006. Urban contribution of glyphosate and its degradate AMPA to streams in the United States. Sci. Total Environ. 354, 191–197.

Kroto, H.W., Heath, J.H., O'Brian, S.C., Carl, R.F., Smalley, R.E., 1985. C60: Buckminsterfullerene. Nature 318, 162–163.

Krystyna, P., 2011. Carbon nanotubes as sorbents in the analysis of pesticides. Chemosphere 83, 1407–1413.

Kucharski, J., Tomkiel, M., Baćmaga, M., Borowik, A., Wyszkowska, J., 2016. Enzyme activity and microorganisms diversity in soil contaminated with the Boreal 58 WG herbicide. J. Environ. Sci. Heal. B 51 (7), 446–454. https://doi.org/10.1080/03601234.2016.1159456.

LaRossa, R.A., Schloss, J.V., 1984. The sulfonylurea herbicide sulfometuron methyl is an extremely potent and selective inhibitor of acetolactate synthase in Salmonella typhimurium. J. Biol. Chem. 259 (14), 8753–8757.

Larson, R.L., Hill, A.L., Fenwick, A., Kniss, A.R., Hanson, L.E., Miller, S.D., 2006. Influence of glyphosate on Rhizoctonia and Fusarium root rot in sugar beet. Pest Manag. Sci. 62, 1182–1192.

Lasarte-Aragonés, G., Lucena, R., Cárdenas, S., Valcárcel, M., 2013. Effervescence-assisted carbon nanotubes dispersion for the micro-solid-phase extraction of triazine herbicides from environmental waters. Anal. Bioanal. Chem. 405 (10), 3269–3277.

Latrous El Atrache, L., Hachani, M., Kefi, B.B., 2016. Carbon nanotubes as solid-phase extraction sorbents for the extraction of carbamate insecticides from environmental waters. Int. J. Environ. Sci. Technol. 13, 201–208.

Li, B., He, X., Sun, Y., Li, B., 2016. Developmental exposure to paraquat and maneb can impair cognition, learning and memory in Sprague-Dawley rats. Mol. Biosyst. 12, 3088–3097.

Liang, P., Ding, Q., Song, F., 2005. Application of multiwalled carbon nanotubes as solid phase extraction sorbent for preconcentration of trace copper in water samples. J. Sep. Sci. 28, 2339–2343.

Liao, Q.G., Zhou, Y.M., Luo, L.G., Wang, L.B., Feng, X.H., 2014. Determination of twelve herbicides in tobacco by a combination of solid–liquid–solid dispersive extraction using multi-walled carbon nanotubes, dispersive liquid-liquid micro-extraction, and detection by GC with triple quadrupole mass spectrometry. Microchim. Acta 181 (1-2), 163–169.

Liu, G., Wang, J., Zhu, Y., Zhang, X., 2004. Application of multiwalled carbon nanotubes as a solid-phase extraction sorbent for chlorobenzenes. Anal. Lett. 37 (14), 3085–3104. https://doi.org/10.1081/AL-200035912.

Long, R.Q., Yang, R.T., 2001. Carbon nanotubes as superior sorbent for dioxin removal. J. Am. Chem. Soc. 123 (9), 2058–2059.

López-Feria, S., Cárdenas, S., Valcárcel, M., 2009. One step carbon nanotubes-based solid-phase extraction for the gas chromatographic-mass spectrometric multiclass pesticide control in virgin olive oils. J. Chromatogr. A 1216 (43), 7346–7350.

Lucini, L., Molinari, G.P., 2011. Detection of the herbicide fenoxaprop-p-ethyl, its agronomic safener isoxadifen ethyl and their metabolites residue in rice. Qual. Assur. Saf. Crops Foods. 3, 63–68.

Luo, M., Liu, D., Zhao, L., Han, J., Liang, Y., Wang, P., Zhou, Z., 2014. A novel magnetic ionic liquid modified carbon nanotube for the simultaneous determination of aryloxyphenoxy-propionate herbicides and their metabolites in water. Anal. Chim. Acta 852, 88–96. https://doi.org/10.1016/j.aca.2014.09.024.

Ma, J., Jiang, L., Wu, G., Xia, Y., Lu, W., Li, J., Chen, L., 2016. Determination of six sulfonylurea herbicides in environmental water samples by magnetic solid-phase extraction using multi-walled carbon nanotubes as adsorbents coupled with high-performance liquid chromatography. J. Chromatography A 1466, 12–20.

Maddah, B., Hasanzadeh, M., 2017. Fe_3O_4/CNT magnetic nanocomposites as adsorbents to remove organophosphorus pesticides from environmental water. Int. J. Nanosci. Nanotechnol. 13, 139–149.

Magdalena, B., Krystyna, P., 2006. The evaluation of carbon nanotubes as a sorbent for dicamba herbicide. J. Sep. Sci. 29, 2241–2244.
Mahmood, I., Imadi, S.R., Shazadi, K., Gul, A., Hakeem, K.R., 2016. Effects of pesticides on environment. In: Plant, Soil and Microbes. Springer, Cham, pp. 253–269.
Mani, S., Cheemalapati, S., Chen, S.-M., Devadas, B., 2015. Anti-tuberculosis drug pyrazinamide determination at multiwalled carbon nanotubes/graphene oxide hybrid composite fabricated electrode. Int. J. Electrochem. Sci. 10, 7049–7062.
Maofeng, D., Yongqiang, M., Ercheng, Z., Chuanfan, Q., Lijun, H., Shuren, J., 2009. Using multiwalled carbon nanotubes as solid phase extraction adsorbents for determination of chloroacetanilide herbicides in water. Microchim. Acta 165, 123–128.
Mauter, M.S., Elimelech, M., 2008. Environmental applications of carbon-based nanomaterials. Environ. Sci. Technol. 42, 5843–5859.
Merkoçi, A., 2006. Carbon nanotubes in analytical sciences. Microchim. Acta 152 (3–4), 157–174.
Mitra, S., Chakrabarti, N., Bhattacharyya, A., 2011. Differential regional expression patterns of alpha-synuclein, TNF-alpha, and IL-1beta; and variable status of dopaminergic neurotoxicity in mouse brain after Paraquat treatment. J. Neuroinflammation 8, 163.
Mueller, N.C., Nowack, B., 2010. Nanoparticles for remediation: solving big problems with little particles. Elements 6, 395–400. https://doi.org/10.2113/gselements.6.6.395.
Murena, F., Gioia, F., 2002. Catalytic hydrotreatment of water contaminated by chlorinated aromatics. Catal. Today 75 (1–4), 57–61.
Nasrabadi, T., Bidhendi, G.N., Karbassi, A., Grathwohl, P., Mehrdadi, N., 2011. Impact of major organophosphate pesticides used in agriculture to surface water and sediment quality (Southern Caspian Sea basin, Haraz River). Environ. Earth Sci. 63, 873–883.
Nogueira, J.M.F., Sandrab, T., Sandra, P., 2004. Multiresidue screening of neutral pesticides in water samples by high performance liquid chromatography–electrospray mass spectrometry. Anal. Chim. Acta 505 (2), 209–215. https://doi.org/10.1016/j.aca.2003.10.065.
Nowack, B., Bucheli, T.D., 2007. Occurrence, behavior and effects of nanoparticles in the environment. Environ. Pollut. 150, 5–22.
Pan, B., Xing, B., 2008. Adsorption mechanisms of organic chemicals on carbon nanotubes. Environ. Sci. Technol. 42 (24), 9005–9013.
ParisiC.ViganiM.Rodríguez-CerezoE., editors. Proceedings of a Workshop on "Nanotechnology for the Agricultural Sector: From Research to the Field," Luxembourg, 2014.
Pereira, A., Grillo, R., Mello, N.F., Rosa, A.H., Fraceto, L.F., 2014. Application of poly(epsilon-caprolactone) nanoparticles containing atrazine herbicide as an alternative technique to control weeds and reduce damage to the environment. J. Hazard. Mater. 268, 207–215.
Perron, M.C., Juneau, P., 2011. Effect of endocrine disrupters on photosystem II energy fluxes of green algae and cyanobacteria. Environ. Res. 111, 520–529.
Peruzzo, P.J., Porta, A.A., Ronco, A.E., 2008. Levels of glyphosate in surface waters, sediments and soils associated with direct sowing soybean cultivation in north pampasic region of Argentina. Environ. Pollut. 156, 61–66.
Pesce, S., Batisson, I., Bardot, C., Fajon, C., Portelli, C., Montuelle, B., Bohatier, J., 2009. Response of spring and summer riverine microbial communities following glyphosate exposure. Ecotoxicol. Environ. Saf. 72, 1905–1912.
Phlips, E.J., Hansen, P., Velardi, T., 1992. Effect of the herbicide diquat on the growth of microalgae and cyanobacteria. Bull. Environ. Contam. Toxicol. 49, 750–756.
Pol, V.G., Thiyagarajan, P., 2010. Remediating plastic waste into carbon nanotubes. J. Environ. Monit. 12 (2), 455–459.
Pourkhaloee, A., Haghighi, M., Saharkhiz, M.J., Jouzi, H., Doroodmand, M.M., 2011. Carbon nanotubes can promote seed germination via seed coat penetration. Seed Technol. 33 (2), 155–169.

Purisai, M.G., McCormack, A.L., Cumine, S., Li, J., Isla, M.Z., Di Monte, D.A., 2007. Microglial activation as a priming event leading to paraquat-induced dopaminergic cell degeneration. Neurobiol. Dis. 25, 392–400.

Pyrzynska, K., 2008. Carbon nanotubes as a new solid-phase extraction material for removal and enrichment of organic pollutants in water. Sep. Purif. Rev. 37 (4), 375–392.

Pyrzyńska, K., 2011. Solid-phase extraction for enrichment and separation of herbicides. In: Soloneski, S., Larramendy, M. (Eds.), Herbicides. Theory and Application. INTECH, pp. 325–344.

Pyrzynska, K., Stafiej, A., Biesaga, M., 2007. Sorption behavior of acidic herbicides on carbon nanotubes. Microchim. Acta 159, 293–298.

Qian, Y., Zheng, M., Zhang, B., Gao, L., Liu, W., 2006. Determination and assessment of HCHs and DDTs residues in sediments from Lake Dongting, China. Environ. Monit. Assess. 116, 157–167.

Rahemi, V., Garrido, J.M.P.J., Borges, F., Brett, C.M.A., Garrido, E.M.P.J., 2013. Electrochemical determination of the herbicide bentazone using a carbon nanotube β-cyclodextrin modified electrode. Electroanalysis 25, 2360–2366.

Ratcliff, A.W., Busse, M.D., Shestak, K.J., 2006. Changes in microbial community structure following herbicide (glyphosate) additions to forest soils. Appl. Soil Ecol. 34 (2–3), 114–124.

Reichenberger, L., 2007. Missing micronutrients: using glyphosate is complicating the uptake of some minor nutrients. Furrow 22–23, .

Ren, Q.Q., Yuan, X.J., Huang, X.R., Wen, W., Zhao, Y.D., Chen, W., 2013. In vivo monitoring of oxidative burst on aloe under salinity stress using hemoglobin and single-walled carbon nanotubes modified carbon fiber ultramicroelectrode. Biosens. Bioelectron. 50, 318–324.

Rickerby, D., Morrison, M., 2007 Report From the Workshop on Nanotechnologies for Environmental Remediation, JRC Ispra. Available at www.nanowerk.com/nanotechnology/reports/reportpdf/report101.pdf.

Sandmann, G., Böger, P., 1989. Inhibition of carotenoid biosynthesis by herbicides. In: Boger, P., Sandmann, G. (Eds.), Target Sites for Herbicide Action. CRC Press, Boca Raton, FL, pp. 25–44.

Sandström von Tobel, J., Zoia, D., Althaus, J., Antinori, P., Mermoud, J., Pak, H.S., Scherl, A., Monnet-Tschudi, F., 2014. Immediate and delayed effects of subchronic Paraquat exposure during an early differentiation stage in 3D-rat brain cell cultures. Toxicol. Lett. 230, 188–197.

Sandström von Tobel, J., Broyer, A., Zoia, D., Schilt, C., Greggio, C., Fournier, M., Do, K.Q., Monnet-Tschudi, F., 2017. Potential mechanisms of development-dependent adverse effects of the herbicide paraquat in 3D rat brain cell cultures. Neurotoxicology 60, 116–124. https://doi.org/10.1016/j.neuro.2017.04.010.

Sarlak, N., Taherifar, A., Salehi, F., 2014. Synthesis of nanopesticides by encapsulating pesticide nanoparticles using functionalized carbon nanotubes and application of new nanocomposite for plant disease treatment. J. Agric. Food Chem. 62 (21), 4833–4838.

Sharma, P., Bhalla, V., Dravid, V., Shekhawat, G., Prasad, E.S., Suri, C.R., 2012a. Enhancing electrochemical detection on graphene oxide-CNT nanostructured electrodes using magneto-nanobioprobes. Sci. Rep. 2, 877.

Sharma, P., Bhalla, V., Tuteja, S., Kukkar, M., Suri, C.R., 2012b. Rapid extraction and quantitative detection of the herbicide diuron in surface water by a hapten-functionalized carbon nanotubes based electrochemical analyzer. Analyst 137 (10), 2495–2502.

Stoltenberg, D.E., Gronwald, J.W., Wyse, D.L., Burton, J.D., 1989. Effect of sethoxydim and haloxyfop on acetyl-CoA carboxylase activity in Festuca species. Weed Sci. 37, 512–516.

Struger, J., Thompson, D., Staznik, B., Martin, P., McDaniel, T., Marvin, C., 2008. Occurrence of glyphosate in surface waters of Southern Ontario. Bull. Environ. Contam. Toxicol. 80, 378–384.

Suresh Kumar, R.S., Shiny, P.J., Anjali, C.H., Jerobin, J., Goshen, K.M., Magdassi, S., Mukherjee, A., Chandrasekaran, N., 2013. Distinctive effects of nanosized permethrin in the environment. Environ. Sci. Pollut. Res. Int. 20 (4), 2593–2602.

Tachibana, K., Watanabe, T., Sekizawa, Y., Takematsu, T., 1986. Herbicide-resistant crops: agricultural, economic, environmental. J. Pest. Sci. 11, 33–37.

Taghizade, F.T., Mehrdadi, N., Baghdadi, M., Nabi, B.G.R., 2018. Application of nanotechnology in pesticides removal from aqueous solutions—a review. Int. J. Nanosci. Nanotechnol. 14 (1), 43–56.

Tanner, C.M., Kamel, F., Ross, G.W., Hoppin, J.A., Goldman, S.M., Korell, M., Marras, C., Bhudhikanok, G.S., Kasten, M., Chade, A.R., Comyns, K., Richards, M.B., Meng, C., Priestley, B., Fernandez, H.H., Cambi, F., Umbach, D.M., Blair, A., Sandler, D.P., Langston, J.W., 2011. Rotenone, paraquat, and Parkinson's disease. Environ. Health Perspect. 119, 866–872.

Tsui, M.T.K., Chu, L.M., 2008. Environmental fate and non-target impact of glyphosate-based herbicide (Roundup) in a subtropical wetland. Chemosphere 71, 439–446.

Valcarcel, M., Cardenas, S., Simonet, B.M., 2007. Role of carbon nanotubes in analytical science. Anal. Chem. 79 (13), 4788–4797.

Valcàrcel, M., Càrdenas, S., Simonet, B.M., Moliner-Martínez, Y., Lucena, R., 2008. Carbon nanostructures as sorbent materials in analytical processes. Trends Anal. Chem. 27, 34–43.

Van Noort, P.C.M., Cornelissen, G., Ten Hulscher, T.E.M., Vrind, B.A., Rigterink, H., Belfroid, A., 2003. Slow and very slow desorption of organic compounds from sediment: influence of sorbate planarity. Water Res. 37 (10), 2317–2322.

Vats, S., 2015. Herbicides: history, classification and genetic manipulation of plants for herbicide resistance. In: Lichtfouse, E. (Ed.), Sustainable Agriculture Reviews. vol. 15. Springer International Publishing, pp. 153–192.

Vaughn, K.C., Lehnen, J.R., 1991. Mitotic disrupter herbicides. Weed Sci. 39, 450–457.

Wagner, G., Nadasy, E., 2006. Effect of pre-emergence herbicides on growth parameters of green pea. Commun. Agric. Appl. Biol. Sci. 71 (3 Pt A), 809–813.

Wang, S.G., Liu, X.W., Gong, W.X., Nie, W., Gao, B.Y., Yue, Q.Y., 2007. Adsorption of fulvic acids from aqueous solutions by carbon nanotubes. J. Chem. Technol. Biotechnol. 82, 698–704.

Wepasnick, K.A., Smith, B.A., Bitter, J.L., Fairbrorther, D.H., 2010. Chemical and structural, characterization of carbon nanotube surfaces. Anal. Bioanal. Chem. 396, 1003–1014.

Wong, P.K., 2000. Effect of 2, 4-D, glyphosate and paraquat on growth, photosynthesis and cholorophyll-a synthesis of Scenedesmus quadricauda Berb 614. Chemosphere 41, 177–182.

Wright, T.R., Shan, G., Walsh, T.A., Lira, J.M., Cui, C., Song, P., Zhuang, M., Arnold, N.L., Lin, G., Yau, K., Russell, S.M., 2010. Robust crop resistance to broadleaf and grass herbicides provided by aryloxyalkanoate dioxygenase transgenes. Proc. Natl. Acad. Sci. 107 (47), 20240–20245.

Wu, L., Qiu, Z., Zhou, Y., Du, Y., Liu, C., Ye, J., Hu, X., 2016. Physiological effects of the herbicide glyphosate on the cyanobacterium *Microcystis aeruginosa*. Aquat. Toxicol. 178, 72–79.

Zaytseva, O., Neumann, G., 2016. Carbon nanomaterials: production, impact on plant development, agricultural and environmental applications. Chem. Biol. Technol. Agric. 3 (1), 17.

Zeng, G., Chen, M., Zeng, Z., 2013. Risks of neonicotinoid pesticides. Science 340 (6139), 1403.

Zhang, J., Gong, J.L., Zeng, G.M., Yang, H.C., Zhang, P., 2017. Carbon nanotube amendment for treating dichlorodiphenyltrichloroethane and hexachlorocyclohexane remaining in Dong-ting Lake sediment—an implication for in-situ remediation. Sci. Total Environ. 579, 283–291. https://doi.org/10.1016/j.scitotenv.2016.11.105.

Zhao, X., Shi, Y., Cai, Y., Mou, S., 2008. Cetyltrimethylammonium bromide-coated magnetic nanoparticles for the preconcentration of phenolic compounds from environmental water samples. Environ. Sci. Technol. 42 (4), 1201–1206.

Zhou, Q., Xiao, J., Wang, W., Liu, G., Shi, Q., Wang, J., 2006. Determination of atrazine and simazine in environmental water samples using multiwalled carbon nanotubes as the adsorbents for preconcentration prior to high performance liquid chromatography with diode array detector. Talanta 68 (4), 1309–1315.

Further reading

Aschner, M., Ceccatelli, S., Daneshian, M., Fritsche, E., Hasiwa, N., Hartung, T., Hogberg, H.T., Leist, M., Li, A., Mundi, W.R., Padilla, S., Piersma, A.H., Bal-Price, A., Seiler, A., Westerink, R.H., Zimmer, B., Lein, P.J., 2017. Reference compounds for alternative test methods to indicate developmental neurotoxicity (DNT) potential of chemicals: example lists and criteria for their selection and use. ALTEX 34 (1), 49–74. https://doi.org/10.14573/altex.1604201.

Blewetta, T.C., Robertsa, D.W., Brintona, W.F., 2005. Phytotoxicity factors and herbicide contamination in relation to compost quality management practices. Renew. Agri. Food Syst. 20, 67.

Délye, C., Jasieniuk, M., Le Corre, V., 2013. Deciphering the evolution of herbicide resistance in weeds. Trends Genet. 29 (11), 649–658.

Ding, S., Zhao, L., Qi, Y., Lv, Q., 2014. Preparation and characterization of lecithin-nano Ni/Fe for effective removal of PCB77. J. Nanomater. 2014, 1–7. Article No. 5.

Guerra, F.D., Attia, M.F., Whitehead, D.C., Alexis, F., 2018. Nanotechnology for environmental remediation: materials and applications. Molecules 23, 1760. https://doi.org/10.3390/molecules23071760.

Jaworski, E.G., 1972. Mode of action of N-phosphonomethyl-glycine: inhibition of aromatic amino acid biosynthesis. J. Agric. Food Chem. 20, 1195–1198.

Li, S., 2012. A Study of Environmental Fate and Application of Commercially Available Carbon Nanotubes. Texas Tech University, p.152.

Madden, J.C., Enoch, S.J., Hewitt, M., Cronin, M.T.D., 2009. Pharmaceuticals in the environment: good practice in acute ecotoxicological effect. Toxicol. Lett. 185 (2), 85–101.

Pyrzynska, K., Trojanowicz, M., 1999. Functionalized cellulose sorbents for preconcentration of trace metals in environmental analysis. Anal. Chem. 29 (4), 313–321.

Zhao, H., Wang, L., Qiu, Y., Zhou, Z., Zhong, W., Li, X., 2007. Multiwalled carbon nanotubes as a solid-phase extraction adsorbent for the determination of three barbiturates in pork by ion trap gas chromatography–tandem mass spectrometry (GC/MS/MS) following microwave assisted derivatization. Anal. Chim. Acta 586, 399–406. ISSN: 0003-2670.

Effect of nanocarbons on physical and mechanical properties of soils

20

Jamal M.A. Alsharef[a,d], *Mohd Raihan Taha*[d], *Panbarasi Govindasamy*[b], *Ali Akbar Firoozi*[d], *Ramez A. Al-Mansob*[c]
[a]Department of Civil Engineering, Faculty of Engineering Science and Technology, Sebha University, Sebha, Libya, [b]Civil Engineering Dept, Politeknik Ungku Omar, Ipoh, Malaysia, [c]Faculty of Engineering, Technology and Built Environment, UCSI University, Kuala Lumpur, Malaysia, [d]Department of Civil and Structural Engineering, Universiti Kebangsaan Malaysia (UKM), Bangi, Malaysia

1 Introduction

Construction of infrastructure over a weak or soft soil is common in Malaysia, especially along the west coast of Peninsular Malaysia where the soil has poor shear and tensile strength and undergoes excessive and differential settlements to many constructions of infrastructure (Kassim et al., 2005). Further, most of the slopes in Malaysia which cut through the residual soil expose various weathering grades of the soil (Huat et al., 2007). The infiltration of the slopes varies at different parts of the slope as the complexity of the weathering grades of the residual soil, namely the difference in the size of the particles, mineral content, density, cohesion, and void ratios. Generally, the type of foundation chosen is influenced by the soil strata and cost involved.

Occasionally, when there is a need to erect a high-rise building over weak soil, it is vital to improve the load-bearing capacity of the soil. Various methods such as soil reinforcement are utilized to improve the load-bearing capacity of the soil. The traditional method of using fibers changes the physical properties of soil without affecting the environment. Therefore, there is a growing interest in the mechanical characteristics of fiber-reinforced soils. One of the newly developed methods to enhance the mechanical characteristics of soil is discrete fiber reinforcement (Ahmad et al., 2010). Subsequently, the tensile characteristic of soil plays a vital role in numerous reinforcement engineering applications. Compacted soils utilized in geotechnical construction, for example, in a dam, backfill for retaining walls, highway base materials, and construction of a cover barrier, can crack because of tensile failure (Tang et al., 2011).

Moreover, fiber reinforcement in sandy soils is noted to have increased the peak shear strength and limited the postpeak reductions in shear resistance of dry sand that has been subjected to direct shear tests (Maher and Gray, 1990). Nataraj and Mcmanis (1997) conducted a study on the effect of fibers on the characteristic of clay soils. They have concluded that the usage of fibers in clay soils has improved the strength and

Carbon Nanomaterials for Agri-food and Environmental Applications. https://doi.org/10.1016/B978-0-12-819786-8.00020-7

ductility of clay soil and increased the splitting tensile strength, hydraulic conductivity, and flexural toughness index.

The addition of carbon filament in a cement matrix will increase the tensile strength but significantly decrease the compressive strength of the cement matrix composites (Metaxa et al., 2013). Microfibers will be able to delay the growth of cracks and nucleation at the microscale level. Nano-reinforcements can further delay the growth of cracks and nucleation at a nanoscale and prevent them from spreading to the microlevel, thus resulting in improved mechanical characteristics of soil. Nevertheless, nanomaterial can extraordinarily improve the mechanical characteristics of structural composites, for instance, soil-cement or concrete (Figueiredo et al., 2015). The main problem faced when a nanomaterial is utilized in composites is dispersion whereby nanocarbon (NCs) fibers attract each other strongly because of van der Waals forces (Kang et al., 2006). In fact, one of the main features that strongly influences the properties of nanocomposites is the dispersion of NCs (Montazeri and Chitsazzadeh, 2014). This research concluded that nanocarbons (NCs), namely carbon nanofibers (CNFs) and carbon nanotubes (CNTs), are superior complements or alternatives to traditional fibers and have the potential to be the next-generation multifunctional and high-performance soil-based materials and structures.

Recently, many studies have been done on the effects of nanomaterials on plant germination and growth with the aim of promoting their usage for agricultural applications. Recently, several carbon-based NMs, for instance CNMs such as fullerenes, fullerol, carbon NPs, and MWCNTs/WCNTs have gained interest because of their possible applications in regulating plant growth (Khot et al., 2012). A literature review showed that carbon-based NMs have both positive and negative influences on terrestrial species of plants based on CNM concentration and type, plant species, and growth conditions.

Recent studies on CNMs revealed their stimulatory and beneficial impact on plants in vitro or cultured conditions. Thus, these findings have enhanced the possibility of applications in food production and the agricultural sector (Serag et al., 2015), even though the findings are not consistent. MWCNTs are noted to be internalized in plant roots (Lin et al., 2009). In addition, SWCNTs have been verified to efficiently cross the cell walls and membranes of tobacco cells upon in vitro exposure, with subsequent transport to specific cellular organelles. This provides a proof for possible usage as nanotransporters (Liu et al., 2009). Plants are reported to have undertaken suppression of organic contaminants in the presence of select CNMs (De La Torre-Roche et al., 2013).

Although there is a wide scope of CNM applications in the agricultural sector, there are still a number of limiting factors that impede their extensive usage. First, in the artificial plant cultures selected, CNMs have produced toxicity at the physiological, genetic, and cellular levels (Ghosh et al., 2015). However, there are very few studies that have been carried out to evaluate the chronic phytotoxicity of CNMs under environmentally relevant conditions. Evaluation of CNM toxicity in soil is influenced by an array of factors such as low solubility, reactivity with organic matter in the soil, and a large background from the residuals in soil, which limits detection and complicates a mechanistic understanding of cellular interactions. Second, there is not much information on the nutritional benefits and potential genetic damage or modifications in CNM-exposed plants, including trans generational effects (Lin et al., 2009). Third, the stability of CNMs in a complex media and residual leaching of attached metals or contaminants from CNMs (Bennett et al., 2013) is not understood well. Finally, CNM shifts from soil to plants and then to higher herbivore and carnivore trophic levels are also completely unknown. Hence, direct application in the agricultural sector is still in the experimental level because there is limited knowledge about CNMs and their influence on agroecosystems.

Nevertheless, despite the limitations stated above, CNTs are one of the 10 most-produced ENMs (Keller et al., 2013), and along with graphene, CNTs are considered viable to be developed for commercial applications (De Volder et al., 2013). Hence, with the increase in application and production, their release into the environment with subsequent human exposure is unavoidable. Petersen et al. (2011) has given a comprehensive review on the environmental exposure, challenges, and risks linked to the usage of CNTs. A comprehensive knowledge of the mechanistic interactions of CNMs with plants and associated microbial communities is essential to assess the risk related to widespread usage in the agricultural sector. The current review aims to provide a comprehensive assessment of the existing literature on CNM effects on the soil, including its influence on plants and related microorganisms in the soil. The limitations in the current literature will be evaluated and topics worthy of future studies will also be highlighted.

2 Soil reinforcement

There are two types of reinforcements, which are divided according to high or low modulus reinforcements (Mcgown et al., 1978). The first type includes inextensible reinforcement such as metal bars, strips, and bars. The second type includes extensible fibers, either natural or synthetic fibers such as plant roots and polymeric fabrics. Current geotechnical practice requires the design of the fiber-reinforcement project to test the specific fiber-reinforced specimens, which may be costly or time consuming. In order to overcome the need for project-specific testing, even for small projects, a discrete methodology that depends on independent characterization of the properties of soil and refibers has been developed by Zornberg (2002). Short discrete fibers are mixed uniformly with the soil mass to provide an isotropic increase in the strength of the soil composite without introducing continuous planes of weakness. Besides, fiber-reinforcement solutions need not take into consideration anchorage in their design as required by planar reinforcements. The fiber-reinforcement method is increasingly gaining popularity in geotechnical projects involving repairing of failed slopes and stabilizing thin soil veneers (Brown and Sheu, 1975).

3 Nano-reinforcement

The development of nanomaterials for soil reinforcement has many benefits, particularly in concrete technology. Carbon nanofibers and carbon nanotubes are popular nanomaterials due to their unique mechanical features. The nanoscale and high ratio of CNFs and CNTs denote that they can be distributed on a lesser scale than the microreinforced fibers (Alsharef et al., 2016; Nochaiya and Chaipanich, 2011). Therefore, microcracks can be treated faster during their proliferation using a nano-reinforced matrix. Hence, much smaller cracks are formed at the point of first contact between the moving crack front and the reinforcement. During the last few decades, studies have been widely done to investigate the use of macro to microfibers in controlling the formation of cracks in cemented materials such as cement paste, concrete, and cement grout (Al-Rub et al., 2011; Wang et al., 2008). Scanning electron microscopy is used to analyze the microstructure of cementatious materials reinforced with both treated NCs and untreated NCs. It is discovered that nanofibers enhanced the mechanical characteristics of the cementations material. Nevertheless, enhancements indicate the existence of interfacial bonds between the cement and NCs, which helped to improve the mechanical characteristics of the cement paste. These enhancements were noted only for untreated NCs. Based on the results seen in concrete, NCs can be used as excellent alternatives to microfibers to provide nano-reinforcement in high-performance soil (Taha and Alsharef, 2018a, b). The effectiveness of NCs depends on two major characteristics: (a) the distribution of NCs within the material, and (b) the bonding energy and strength between the material and the NC surface (Nochaiya and Chaipanich, 2011). NCs are strongly attracted to each other because of their high van der Waals forces. Furthermore, CNTs are more influenced by van der Waals forces than CNFs because their surface-to-volume ratio is higher (Taha and Alsharef, 2018a, b). CNTs are more susceptible to bundling and agglomeration because of this higher attraction (Yazdanbakhsh et al., 2010).

4 Materials

4.1 Soil

This study made used of residual soil from within Bangi Malaysia. The sample soil was obtained from a depth of $1\,m \pm 20\,cm$ below the ground surface. The soil was mixed with different ratios of bentonite, namely S1 = 0%, S2 = 10%, and S3 = 20%, to create three kinds of soil samples with different PIs. The chemical and physical properties of each soil sample are stated in Table 1. According to the Unified Soil Classification System, the soil (S1 native soil) has low plasticity index clay (CL). However, when the quantity of bentonite was raised to 10% and 20%, respectively, the soil has high plasticity index clay (CH). The S3 soil content has more clay content or percent of fines than S2 soil samples and S2 soil sample have more than S1 soil sample because of the increase in bentonite content. The chemical and physical properties of soil samples S1, S2, and S3 are stated in Table 1.

5 Nanocarbons

In this study, the two types of nanocarbons used are carbon nanotubes and carbon nanofibers.

Table 1 Chemical and physical features of soil samples.

Characteristics	Value and descriptions		
	S1	S2	S3
Specific gravity	2.60	2.61	2.61
Plasticity index (%)	13.6	31.4	64.5
Linear shrinkage (%)	8.9	14.0	19.8
Passing no 200 sieve (wt%)	45.4	55.1	64.9
Unified soil classification system (USCS)	CL	CH	CH
Optimum water content (%)	14.3	15.7	18
Chemical composition			
SiO_2 (%)	62.1	55.2	53.1
Al_2O_3 (%)	29.5	21.3	23.5
Fe_2O_3 (%)	5.7	5.2	5.2
MgO (%)	0.58	1.5	1.1
CaO (%)	0.03	1.23	1.0
TiO_2 (%)	1.17	0.9	0.9
Na_2O (%)	–	1.2	0.8
K_2O (%)	0.8	0.7	0.7
Other	0.1	0.4	0.3
Heat loss	0.2	11.1	12.3
Organic matter (%)	4.2	–	–

5.1 Carbon nanotubes (CNTs)

The multiwalled CNT supplied by France Company, Arkema Incorporation, marketed under the trade name Graphistrength is used in this study. The dimensions are 1–10 μm in length and 10–15 nm in diameter (5–15 walls). The properties and other relevant information about the CNT provided by the company are as stated in Table 2.

5.2 Carbon nanofibers (CNFs)

The nanofiber material PR-24-XT-LHT is a CNF supplied by Pyrograf (United States). The properties and other relevant information about CNF as provided by the company are provided in Table 3.

Table 2 Properties of CNT–(Graphistrength C100).

Property	Value
The average diameter (nm)	10–15
Average length (μm)	1–10
Carbon purity, %	>95
Apparent density (kg/mc)	50–150
Relative density (g/mL) at 25°C	2.1
Aspect ratio	600–700
Applications	Reinforcements

Table 3 Properties of CNF -PR-19-XT-LHT.

Property	Value
Average diameter, nm	200
Average length (μm)	50–200
Carbon purity, %	>98
Apparent density (kg/mc)	30–300
CVD carbon overcoat present on the fiber	No
Nanofiber wall density, g/cc	2–2.1
Iron, ppm	12,466
Aspect ratio	1300–1500
Applications	Mechanical and electrical

6 Preparation of nanocarbon-soil mixtures

The quantity of NCs mixed into the soil was 0%, 0.05%, 0.075%, 0.1%, and 0.2% of the total dry weight of the soil. If the concentration was >0.2%, NC would be less effective in enhancing the mechanical characteristics, for instance, flexural strength, because of poor dispersion of NC (Kowald and Trettin, 2009). First, NCs were added to the water-soil solution and an ultrasonic tip was placed in the beaker. Next, the mixture was sonicated for 4 min to deagglomerate the NCs and avoid possible tube fragmentation (Vaisman et al., 2006). The 4 min time frame was also enough for the dispersal of soil in the aqueous solution (Alsharef et al., 2017) and at the same time prevent the suspensions from overheating (Lee et al., 2007). After sonication, the suspension was thoroughly mixed with the dry soil to obtain uniformity. Next, this suspension was mixed with dry soil and allowed to stand for 24 h as recommended by ASTM D698-7 to allow the water in the suspension to be thoroughly absorbed by the dry soil particles.

7 Experimental laboratory test program

7.1 pH test

An electrometric technique (BS 1377: part 3: 1990: Clause 9) was utilized to determine the pH value of the soil suspension in water because it was considered to be the most accurate method. This technique needs a soil-water ratio of 1:2.5, which can be acquired by passing 30 g soil through a 2 mm sieve and diluted with 75 mL of distilled water for at least 8 h.

7.2 X-ray diffraction (XRD)

XRD is a very prominent and reliable method utilized to identify minerals in clay and rocks (Viennet et al., 2015). XRD analysis was carried out to determine the chemical reactions and development of new products. A particular diffraction pattern would be produced if X-rays were in contact with a crystalline substance. Each crystalline substance would have a specific pattern and the same substance would produce the same pattern. In a mixture of substances, every substance would produce its independent pattern (Hull, 1919).

7.3 Atterberg limits test

The Atterberg limits test, namely the liquid limit and plastic limit of each of the natural and treated soil samples, was identified according to BS 1377, part 2, 1990. The liquid limit tests were carried out using a penetration cone assembly that was made of a stainless steel cone with a cone angle of 30 ± 1°C. The plastic limit tests were carried out manually. Each sample was rolled at a suitable pressure on a glass plate to form a thread with a uniform diameter of 3.2 mm along the full length of the sample. The liquid limit is the empirically established moisture content whereby the soil changes from the plastic state to the liquid state. This would pave the way to classify the soil, particularly if the plastic limit was known. The plasticity index could be obtained from the difference between the water content at the liquid and plastic limits.

7.4 Unconfined compressive strength (UCS) test

An axial compressive load was gradually applied in an unconfined compressive strength (UCS) test to a cylindrical soil sample until it failed. Lateral confinement was not provided and the load was applied quickly so undrained conditions prevailed in this test.

After ultrasonication, the suspension was mixed thoroughly with the dry soil to obtain a uniform and homogeneous mixture. The mixture was permitted to stand for 24 h, which was the standing time suggested by ASTM D698-7 to enable the water to be absorbed thoroughly by the soil particles. After that, the quantity of water needed to achieve optimum moisture content (OMC) was added to the mixture. Next, the amount of material needed was placed inside the mold and compressed in three steps (Tang et al., 2007) by using Harvard Miniature Compaction to obtain uniform compaction. This technique could produce homogeneous samples because uniform color could be obtained after compaction. Nevertheless, homogeneous mixtures must be prepared at each stage.

Compacted cylindrical specimens of 70 mm height and 34 mm diameter were used to quickly obtain the strength of the soil samples; the average value was recorded from three measurements of three samples. The loading rate of 1 mm/min was applied to the specimens until the specimens experienced failure in the test. The unconfined compressive strength was taken as the maximum load attained per unit area. The unconfined compression test (UCS) was carried out according to ASTM D2166-66. A strain-controlled machine was used as it was required by the ASTM D2166 test procedure.

7.5 Indirect tensile test

An indirect tensile test (ITS) was carried out to observe the characteristics of soil-NC mixtures. The samples were compacted cylindrical in shape at 115 mm height and 105 mm diameter. The samples were subjected to a loading rate of 1 mm/min until the samples experienced test failure. The ITS test was carried out by applying the load along the cores in between two flat parallel plates according to the indirect Brazilian test proposed by Dexter and Kroesbergen (1985).

The ITS value was calculated by using a modified equation suggested by Frydman (1964).

$$\text{ITS} = \frac{2Fg_{(x)}}{\pi dl} \tag{1}$$

whereby F is the applied force, d is the specimen diameter, and l represents length. (Frydman, 1964) has proposed a flattening coefficient $g_{(x)}$ as represented by the equation below:

$$g_{(x)} = \left(-\frac{d}{2a}\right)\left\{2x - \sin 2x - \left(\frac{2y}{d}\right)\ln\left(\frac{\pi}{4} + \frac{x}{2}\right)\right\} \tag{2}$$

where x is the flattening ratio whereby $x = a/y$, and a represents the width of the flattened portion and y represents the distance between the flattened portion at failure.

A strain-controlled machine, which was a prerequisite for the ITS test, was also utilized. The tensile strength was noted at the end of the indirect tensile test.

7.6 Direct shear test

A direct shear device has been utilized in numerous practical and theoretical studies in many laboratories worldwide because of its many benefits and simplicity. ASTM D3080 was utilized to run the direct shear tests. This test utilized a shear box measuring 60 * 60 mm in the plane and 20 mm in depth. The parameters of shear strength were thoroughly determined at vertical normal stresses of 100, 200, and 300 kN/m^2. A constant shear force at a rate of 1.2 mm/min has been used on all test samples and the data was gathered every 25 s. The tests were continued until the applied force reached a constant value or 20% strain of the sample length. The maximum shear stress for every sample was chosen and utilized to plot the shear stress at failure against the corresponding normal stress. The standard Proctor compaction equipment was utilized to compact the soil samples at OMC according to ASTM D698-70.

Design of Experiments (DOE) is a very important program used in engineering application. This program is utilized to assess the results acquired from actual data and compare them with the predicted data obtained from the program through statistical analysis. Furthermore, DOE can reduce the cost and time used for running the number of tests required. This program is able to decrease the number of direct shear tests needed for each type of NCs from 45 to 17.

8 Impact of the physicochemical and mechanical of nanocarbon reinforcement

8.1 The pH test

The impact of NCs on pH has been studied to assess chemical reactions on the soil. In the current study, samples that have been treated with 0.05 to 0.2% NCs have increased the pH from 4.14 to 4.8, respectively, for native soil. The test results revealed that the pH of the soil samples would increase as the nanocarbon percentage for both NCs increases (Fig. 1).

When UKM soil plus 5% bentonite was treated with the same dosage of nanoalumina content, the pH increased from 6.81 to 7.32. Taha (2014), also discovered that for UKM soil plus 0% and 5% bentonite, 0.7% nanocopper would increase the pH from 4.57 to 5.98 and 6.81 to 7.68. Hence, the pH results of the current study are less with the findings by Taha (2014). The NCs would not have a chemical reaction with the soil because they are a superhydrophobic and chemically inert material (Wang et al., 2008).

8.2 X-ray diffraction (XRD)

The XRD results for soil specimens with NCs and native soil cured for 84 days are displayed in Fig. 2. The XRD data was analyzed using Bruker's EVA software. Sample S1 with 0.2% CNT and CNF have produced identical curves. The peaks were matched with possible compounds found in the soil. Matching the peaks with the database revealed that SiO_2, Al_2O_3, and Fe_2O_3 as the three main compounds found in the native soil. The XRD results for NC soil samples and native soil are displayed in Fig. 2. There were no changes to the chemical composition of the soil added with any of the NCs. The NCs utilized in this research were insoluble materials and would not react with soil when mixed with water. The findings were similar to Wang et al. (2007).

8.3 Atterberg limits test

Nanoparticles interact very strongly with other soil components, namely the liquid phase, organic matter, clay minerals, and cations, because of their extraordinary high specific surface (SSA) and reactive surfaces to changes. Hence, even in a small quantity nanoparticles influence the microstructure as well as the physical, chemical, and

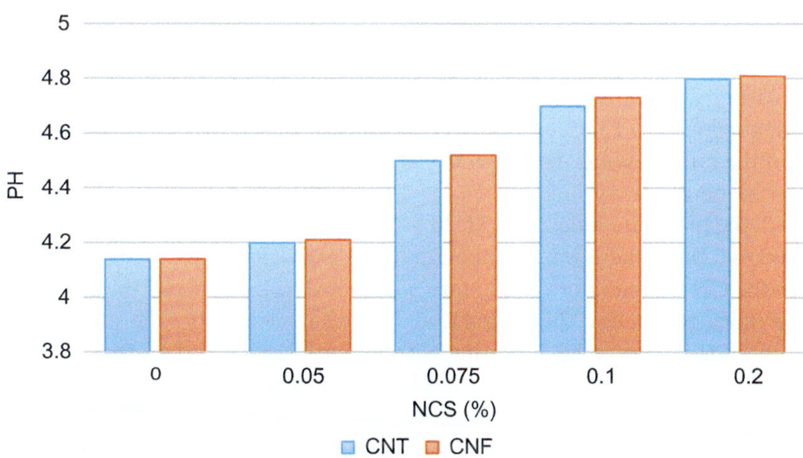

Fig. 1 Effect of CNT and CNFs on the pH.

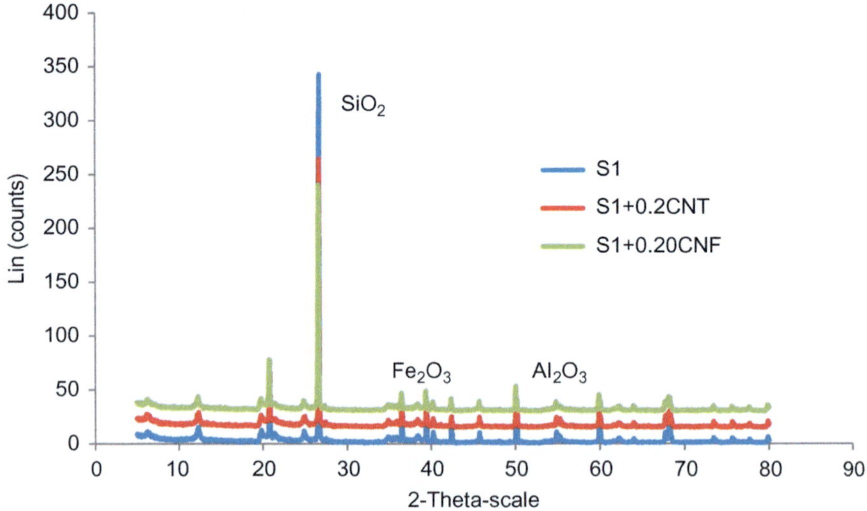

Fig. 2 X-ray diffraction (XRD) patterns.

engineering features of soils. Soils which have high PI will be difficult to work within construction because they are stick and instable when wet. Soils with high PI most likely will have damaging volume changes while wet and dry. Hence, there is a need to add nanomaterial, even in a small quantity, to the soil to enhance its geotechnical features.

Generally, the plasticity index is the mathematical difference between the liquid limit (LL) and the plastic limit (PL) of a particular material. The plasticity index defines the range of moisture content over which the soil remains as plastic. It is a measurement of the cohesive qualities of the binder resulting from the clay content. The results of LL and PL for both with and without NC soils (S1, S2, and S3) are displayed in Table 4. The decrease in PI for soil treated with CNF and CNT, as in Table 4, is an indication of soil improvement because the PI denotes the range of water content whereby the soil is in a plastic state (Rajasekaran, 2005).

The plasticity index was 13.61 for soil sample S1 at 0% CNT content. The plasticity index reduced to 13.95 when CNT content was raised to 0.05%. When the CNT content of the mixtures was increased to 0.075%, 0.1%, and 0.2%, there was a further reduction in the plasticity index to 13.48, 13.34, and 13.31, respectively. As for soil sample S2, the plasticity index was 31.38 at 0% CNT content. When the CNT content was raised to 0.05%, there was a reduction in the plasticity index to 30.1.

When the CNT content of the mixture was raised to 0.075%, 0.1%, and 0.2%, the plasticity index reduced to 29.8, 29.41, and 28.62 respectively. The plasticity index was 63.97 for soil sample S3 when the NT content was 0%. When the CNT content of the mixture was raised to 0.05%, the plasticity index dropped to 63.45. When the CNT content of the mixture was raised to 0.075%, 0.1%, and 0.2%, the plasticity index dropped to 63.13, 62.98, and 61.48, respectively. When the CNF content of soil sample S1 was raised from 0% to 0.05%, 0.075%, 0.1%, and 0.2%, the plasticity index

Table 4 Values of liquid limit (LL), plastic limit (PL), and plasticity index (PI) of the soil-nanocarbon mixture.

Sample symbol	NC (%)	CNT			CNF		
		LL (%)	PL (%)	PI (%)	LL (%)	PL (%)	PI (%)
S1	0.0	34.34	20.73	13.61	34.34	20.73	13.61
	0.05	33.54	19.95	13.59	33.02	20.5	12.52
	0.075	32.53	19.05	13.48	31.96	19.58	12.38
	0.10	31.59	18.25	13.34	30.1	18.61	11.49
	0.20	30.54	17.23	13.31	29.14	17.98	11.16
S2	0.0	57.38	26	31.38	57.38	26	31.38
	0.05	54.21	24.11	30.1	52.03	23.11	28.92
	0.075	53.11	23.31	29.8	49.81	22.63	27.18
	0.10	51.89	22.48	29.41	48.69	21.32	27.37
	0.20	50.34	21.72	28.62	45.94	20.22	25.72
S3	0.0	92.51	28.54	63.97	92.51	28.54	63.97
	0.05	91.23	27.78	63.45	90.34	27.5	62.84
	0.075	90.28	27.12	63.16	88.65	27.12	61.53
	0.10	89.88	26.9	62.98	86.98	26.96	60.02
	0.20	88.14	26.3	61.84	86.19	26.5	59.69

dropped from 13.61 to 12.52, 12.38, 11.49, and 11.16, respectively. In soil sample C2, the plasticity index dropped from 31.38 to 28.92, 27.18, 27.37 and 25.72 as the CNF content raised from 0% to 0.05%, 0.075%, 0.1%, and 0.2%, respectively. In soil sample S3, the plasticity index dropped from 63.97 to 62.84, 61.53, 60.02, and 59.69 as the CNF content raised from 0% to 0.05%, 0.075%, 0.1%, and 0.2%, respectively.

The maximum drop in the plasticity index was seen at 0.2% CNT and CNF content for all types of soil. The reduction in the plasticity index for soil sample S1 was 0.3 and 2.45 after adding 0.2% CNT and 0.2% CNF, respectively. The reduction in the plasticity index for soil sample S2 was 2.67 and 5.66 after adding 0.2% CNT and 0.2% CNF. The plasticity index of soil sample S3 dropped to 2.13 and 4.28 after 0.2% CNT and 0.2% CNF were added.

8.4 Unconfined compressive strength (UCS) test

The effect of including CNT on the unconfined compressive strength and deformation was noted as a function of CNT content on all soil types. The stress-deformation correlations for CNT are displayed in Table 5.

The reinforced sample S1 with CNT content of 0.05%, 0.075%, 0.1%, and 0.2% showed compressive strength as there was an increase from 83, 127, and 178 to 187, 232, and 284, respectively, from the original soil strength of 102 to 307 kPa.

For S2 samples with 0.05%, 0.075%, 0.1%, and 0.2%, the CNT content revealed compressive strengths of 213, 222, 239, and 245 kPa, respectively, with an increase of 21%, 26%, 35%, and 39%, respectively, from the original soil strength of 176 kPa. The compressive strength for S3 with 0.05%, 0.075%, 0.1%, and 0.2% CNT content

Table 5 Values of unconfined compressive strength (UCS) and strain peak of soil-nanocarbon mixture.

Sample symbol	NC (%)	UCS (kPa)		Strain (%)	
		CNT	CNF	CNT	CNF
S1	0.0	102	102	4.2	4.2
	0.05	187	249	4.2	4.2
	0.075	232	298	4.2	4.9
	0.10	284	319	3.5	4.2
	0.20	307	345	4.2	4.9
S2	0.0	176	176	4.2	4.2
	0.05	213	220	1.4	2.1
	0.075	222	240	2.1	2.8
	0.10	239	234	2.1	1.4
	0.20	245	262	2.1	2.1
S3	0.0	216	216	4.2	4.2
	0.05	259	235	2.1	1.4
	0.075	274	271	2.8	2.8
	0.10	283	286	2.1	2.1
	0.20	286	332	2.1	3.5

revealed compressive strengths of 259, 274, 283, and 286 kPa, respectively, resulting in a strength increase of 19%, 26%, 31%, and 32%, respectively, from the original soil strength of 216 kPa.

The stress–strain correlation for CNT is displayed in Table 5. The CNT content influenced the overall characteristic of the CNT-reinforced soil. Adding CNT increased the strength of the content. Table 5 displays the stress–strain correlation of soil sample S1 with the addition of different percentages of CNT. Apparently peak strength increased significantly as the CNT was increased. The rate of increase in strength can rise to 0.10%. After that, the rate of increase in strength is lower but it can still increase from 0.10% to 0.20%. The maximum increase in strength was obtained with 0.20% CNT content, whereby the value increased from 102 kPa when the soil was without CNT to 307 kPa. Hence, there was an increase of 201% when 0.20% CNT was added Nevertheless, the characteristics of soil with and without CNT did not change in terms of peak strength as it occurred at the same strain value of between 4% and 5%. Table 5 displays the results of the stress–strain correlations for S2 and S3 soils, respectively. There was an increase in strength for S2 and S3 soils similar to S1, but the raise was not as significant as for S1 soil. The maximum increase in strength for S2 soil was from 176 to 245 kPa, and for S3 soil from 216 to 286 kPa. However, the rise in strength after 0.10% was very small. Nevertheless, unlike soil sample S1, the characteristics of soil samples S2 and S3 changed from ductile to stiff when the peak strength of reinforced soil happened at a lower strain compared to the original soil. This proved that CNT did not change the brittle or ductile characteristics of soil sample S1 and the strain continued to be the same for different CNT content. Nevertheless, CNT caused the characteristics of soil samples S2 and S3 to be a little brittle. Moreover, both the soil samples have reduced strain at maximum strength for different CNT content.

The bridge-connecting effect of CNT verified by FSEM analysis in Fig. 3 served as a bridge across voids and cracks, besides assuring load transfer in the case of tension (Nochaiya and Chaipanich, 2011). At the same time, the compressive strength was also improved.

The effect of including CNF on the unconfined compressive strength and deformation correlation was noted as a function of CNF content for all types of soil. Stress-deformation correlation for CNF is displayed in Table 5.

Soil sample S1 with 0.05%, 0.075%, 0.1%, and 0.2% CNF content showed compressive strengths of 249, 289, 319, and 345 kPa, respectively, resulting in a strength increase of 144%, 183%, 212%, and 238%, respectively, from the original soil strength of 102 kPa. Soil sample S2 with 0.05%, 0.075%, 0.1%, and 0.2% CNT content showed compressive strengths of 220, 240, 234, and 262 kPa, respectively which revealed an increase in strength of 25%, 36%, 34%, and 48%, respectively, from the original strength of 176 kPa. The compressive strength for S3 with 0.05%, 0.075%, 0.1%, and 0.2% CNT content revealed compressive strengths of 235, 271, 286, and 332 kPa, respectively, which showed a strength increase of 9%, 25%, 32%, and 53%, respectively, from the original strength of 216 kPa.

The stress–strain correlation for CNF is provided in Table 5. Similar to CNT, CNF raised the peak strength for all three soil samples. In soil sample S1 as shown in Table 5, the strength rose from 102 to 345 kPa, which was an increase of 240%. Table 5 displays the results of stress–strain correlations for soil sample S2 and S3.

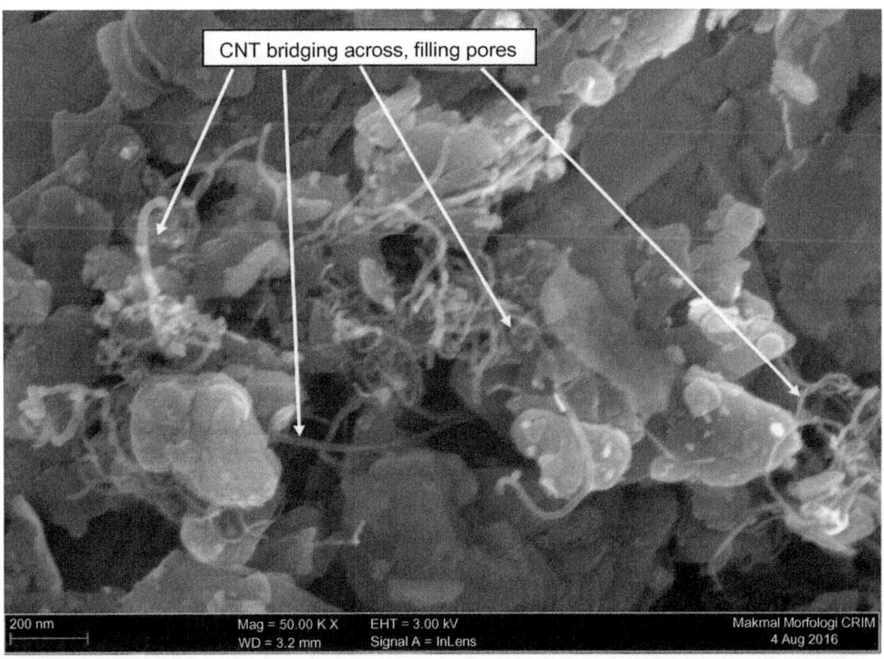

Fig. 3 Typical FESEM images of microcrack bridging and voids by CNT for soil mixture composites @ Mag = 50.00 K X.

Similar to the S1 soil, there was an increase in strength for soil samples S2 and S3, but the increase is not as large as for the S1 soil. The maximum increase in strength in sample soil S2 was from 176 to 262 kPa, and for soil sample S3 from 208 to 332 kPa. Similar to CNT, the CNF was more effective in increasing the strength of soil sample S1, which was without bentonite, than soil samples S2 and S3. Moreover, CNF did not change the characteristics (ductile or brittle) of sample soil S1 and the strain remained the same for different CNF content. Nevertheless, CNF altered the characteristics of sample soil S2 and S3 to being a little brittle. The strain at maximum strength for different CNFs decreased in both soils. As mentioned earlier, the bridge-connecting influence of CNF as verified by FSEM analysis in Fig. 4 produced bridges across voids and cracks. This enabled load transfer in the case of tension.

8.5 *Indirect tensile test*

The influence of CNT inclusion on the deflection and tensile load relationship was ascertained as a function of CNT content for all types of soils. Figs. 5–7 provide the deflection-tensile load correlation for CNT. As illustrated in the figures, there is a linear increase of the load with a deflection to peak load for all samples.

The CNT content influences the overall characteristics of the CNT-soil mix. The tensile load is raised by the presence of CNT with rising CNT content. The S1 soil's deflection-tensile load correlation for varying percentages of CNT is shown in Fig. 5.

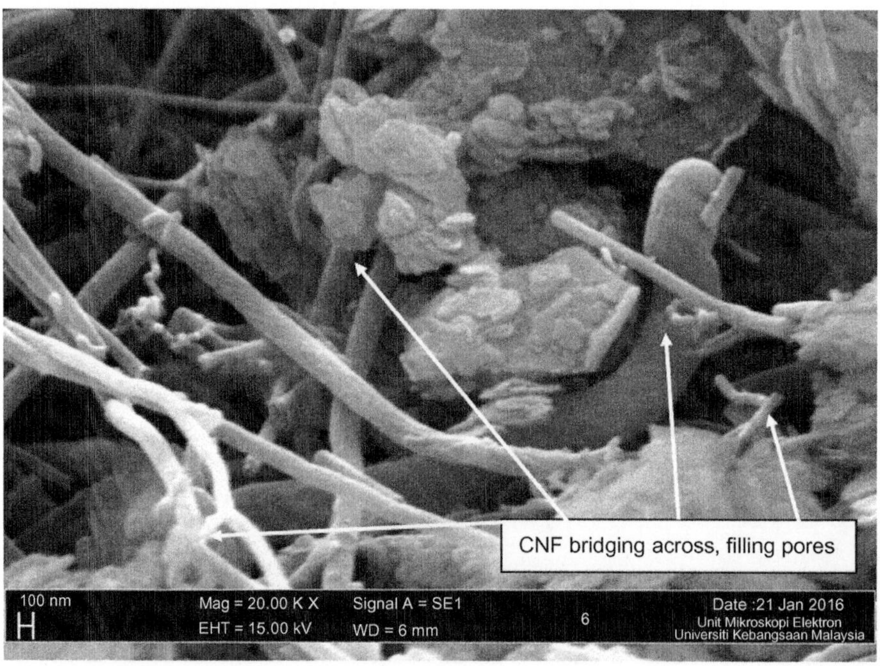

Fig. 4 Typical FESEM images of microcrack bridging and voids by CNF for soil mixture composites @ Mag = 20.00 K X.

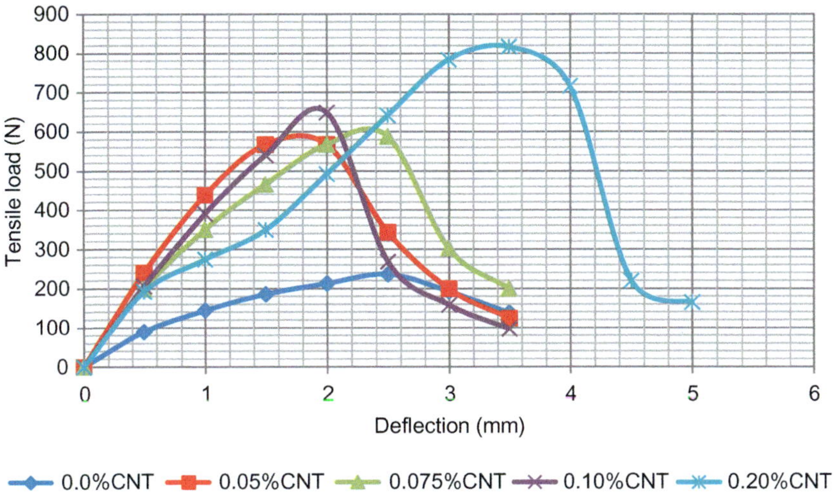

Fig. 5 Tensile load-deflection relationship for S1 with various CNT contents.

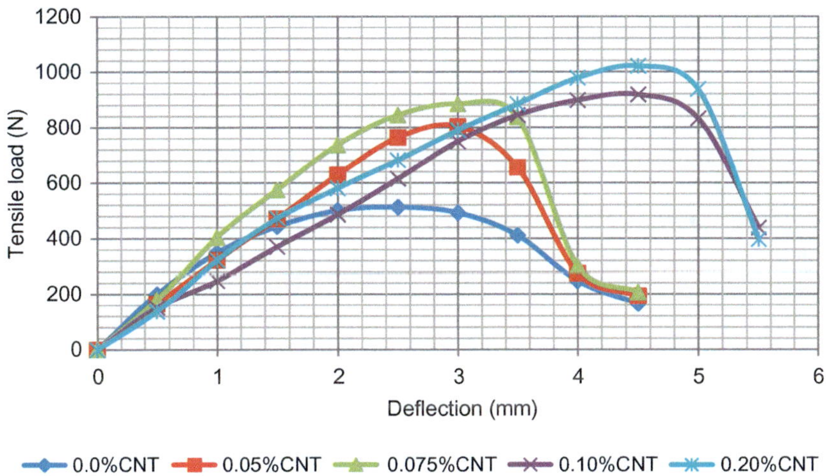

Fig. 6 Tensile load-deflection relationship for S2 with various CNT contents.

It is undeniable that there is a considerable rise in peak tensile load, which rises with rising CNT content. The rate of rise in the tensile load is as high as 0.10%. Next, despite a reduced rate of increase, there is still a rise in the tensile load from 0.10% to 0.20%. The 0.20% CNT content showed a maximum rise in the tensile load in which the value rose from 237 N (soil without CNT) to 817 N–an increase of 245%. Nonetheless, there were no changes in soil behavior without and with CNT, whereby the peak tensile load took place at a deflection value between 4% and 2%. The results of deflection-tensile load correlation for S3 and S2 soils are illustrated in Figs. 6 and 7, respectively. There is a rise in tensile load for S3 and S2 soil that is similar to S1

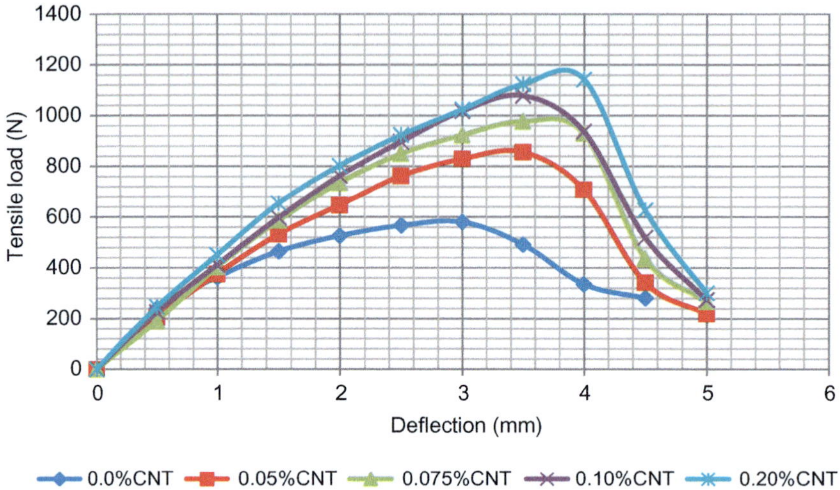

Fig. 7 Tensile load-deflection relationship for S3 with various CNT contents.

soil, despite not being as significant a rise as for S1 soil. The maximum rise in the tensile load for S3 soil was from 581 N to 1144 N, resulting in a 96% increase, while the maximum rise in the tensile load in S2 soil was from 513 to 1021 N, resulting in a 99% increase. It is obvious that the rise in tensile load following 0.10% is marginally low when compared with S1. In contrast to the S1 soil, the behavior of S3 and S2 soils were altered from ductile to stiff, that is, the soil with the CNT content's peak tensile load took place at a lower deflection in comparison to unreinforced soil. The values of ITS against varying CNT percentages are presented in Fig. 8. Results highlight the fact that the presence of CNT raises the three soils' tensile strength. At the beginning,

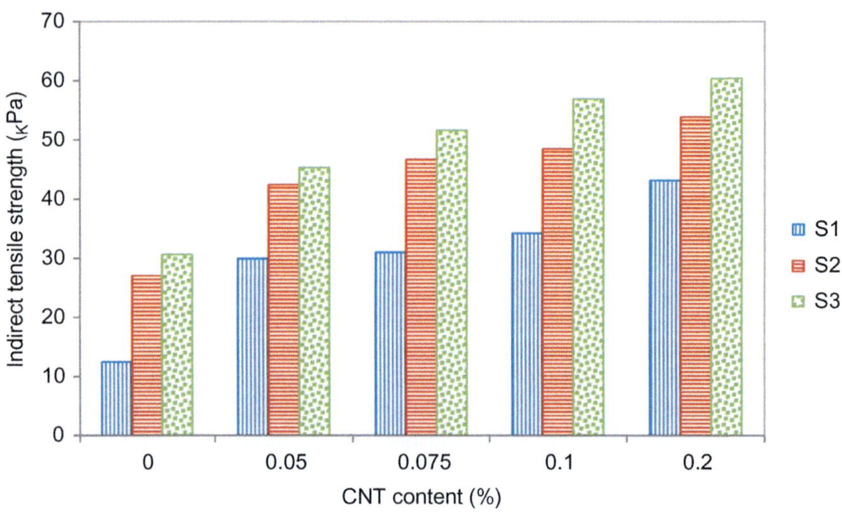

Fig. 8 The values of indirect tensile strength with various CNT contents.

with 0.2% CNT, the tensile strength rose to 96%, 99%, and 245% for S3, S2, and S1, respectively.

Figs. 9–11 show the way the deflection-tensile load link for CNF is affected by the inclusion of CNF. Quite similar to CNT, there was a linear increase of the load with a deflection to the peak load for all samples.

The CNF content also influenced the overall behavior of the CNF-reinforced soil. The incorporation of CNF raised the tensile load to 0.1%. Subsequently, there was a marginal drop as the CNF reached 0.2% content. The deflection-tensile load link of S1 soil for varying percentages of CNG is shown in Fig. 9. It is clear that there is a

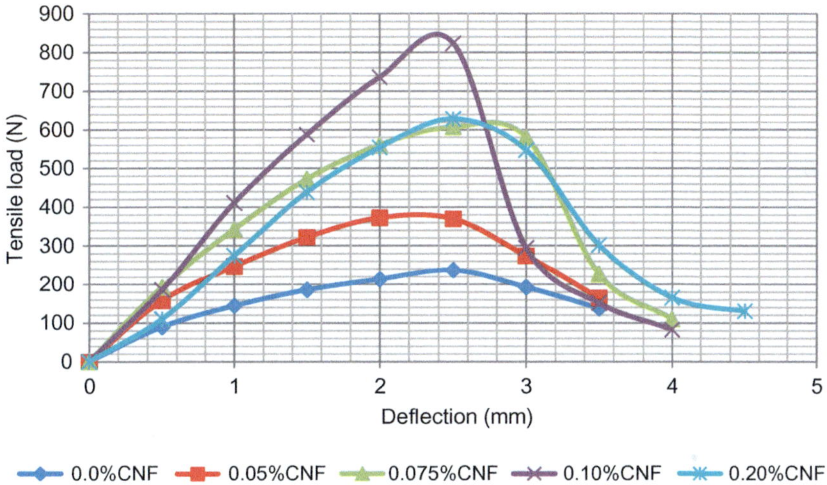

Fig. 9 Tensile load-deflection relationship for S1 with various CNF contents.

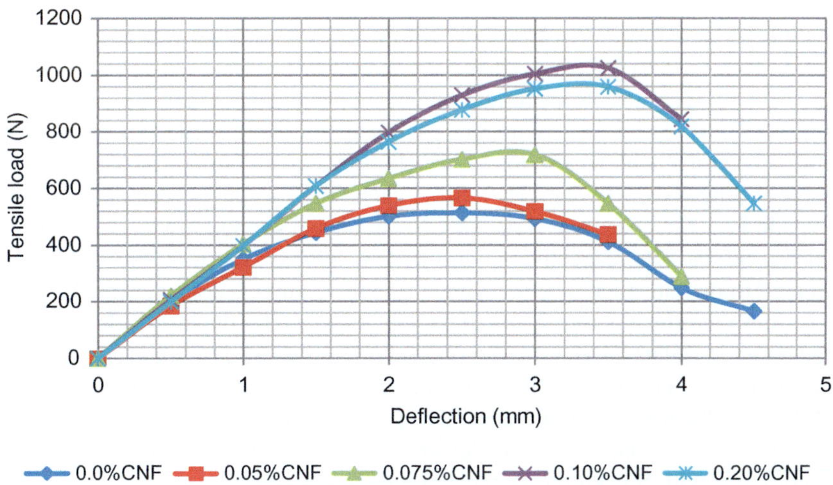

Fig. 10 Tensile load-deflection relationship for S2 with various CNF contents.

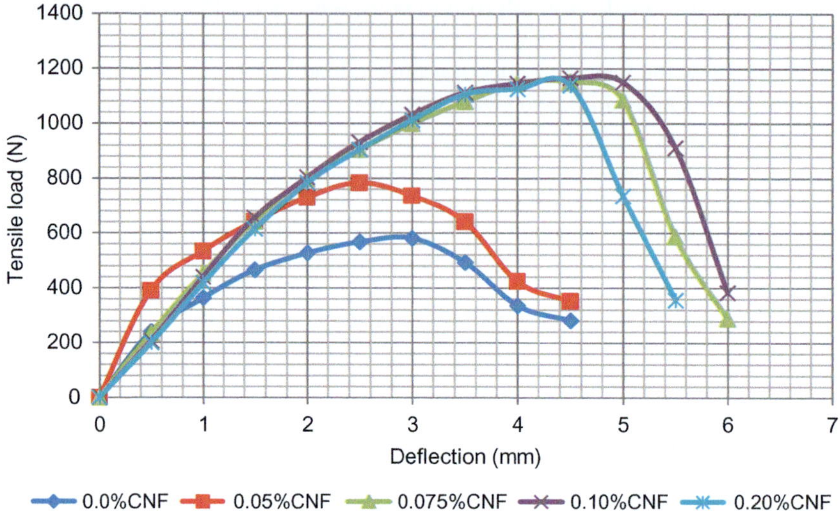

Fig. 11 Tensile load-deflection relationship for S3 with various CNF contents.

significant rise in peak tensile load, which rises with rising CNF content. The rate of rise in the tensile load is high up to 0.10%. Subsequently, the tensile load fell from 0.10% to 0.20%. The tensile load showed a maximum increase at 0.10% CNF content, in which the value rose from 237 N (soil in the absence of CNF) to 824 N, thus causing a rise of 247%. Nonetheless, there was no alteration of behavior with or without CNF, that is, the maximum tensile load took place at the deflection value (between 3% and 2%). The results of the deflection-tensile load link for S3 and S2 soils are illustrated in Figs. 11 and 10, respectively. There was a rise in tensile load for S3 and S2 soils up to 0.1% CNF, similar to S1 soil, followed by a marginal drop at 0.2% CNF content. Nonetheless, the rise is not as significant as for the S1 soil. The maximum rise in tensile load for S3 soil was from 581 N to 1165 N, causing a rise of 100.5%, and for S2 soil from 513 to 1025 N, causing a rise of 100%. It is obvious that the rise in tensile load following 0.10% is comparatively less significant than S1. In contrast to S1 soil, the S3 and S2 soil characteristics were altered from stiff to ductile, whereby the reinforced soil's peak tensile load took place at a lower deflection as compared to soil in the absence of CNF. The ITS values against the varying percentage of CNF are displayed in Fig. 12. There was a rise of CNF in tensile strength up to 100.5, 100, and 247 for S3, S2 and S1, respectively, with 0.10% content. The possibility of this trait is attributed to the interlock of CNF with soil particles that is considerably better than CNT. A strong interface between soil and CNF is formed by the interlock, thus raising the load transfer. Enlarging the surface contact area between fibers and soil also raised the cohesion force. This is possible by raising the NC content in the soil specimens. As such, following the formation of cracks, the specimens that have NC retained their tendency to absorb more loads to extend deformation in a much more ductile trend, which is in accordance with Sivakumar Babu et al. (2008). In addition, the soil peak strength and postpeak (residual) strength were significantly raised by the inclusion of NCs.

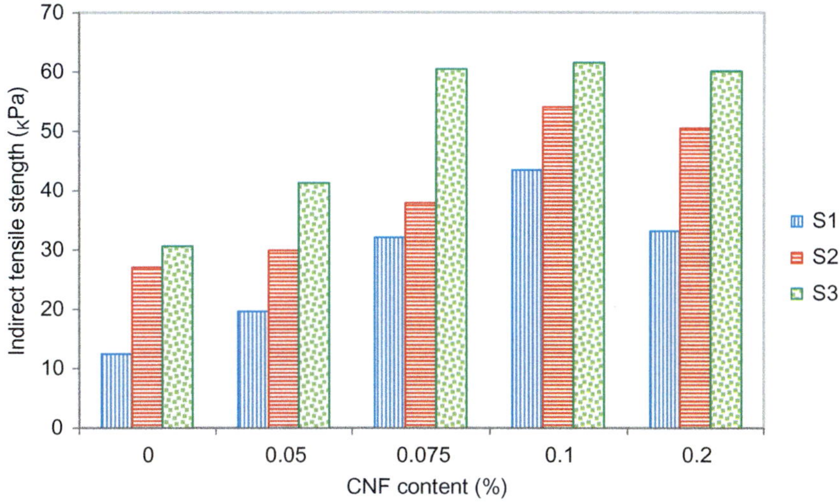

Fig. 12 The values of indirect tensile strength with various CNF contents.

When the tension cracks due to loading began to appear, the nanofibers operated as connections (Figs. 3 and 4) and efficiently blocked the further opening and growth of cracks. Hence, this can prevent the samples from complete failure.

In the current study, soil with CNF and CNT at 0.1% nanomaterial content improved ITS by 247% and 245%, respectively. Shojaei (2016) also concluded that adding 3% bitumen emulsion, 7% Rovene@4045, and 8% Tylac@4190 content to the soil-cement mixtures that had 4% cement improved the ITS by 128%, 184%, and 168%, respectively. The results obtained from the current study showed that more improvements have been made compared to Shojaei (2016), possibly because of the usage of higher-density nanomaterials.

8.6 *Direct shear test*

Globally, most laboratories utilize the direct shear strength (DSS) via direct shear box device for various theoretical and practical research projects. The direct shear tests are run via ASTM D 3080. In this study, the results collected from actual data were evaluated using the DOE program. The results were then compared with estimated data obtained from the program via statistical analysis.

A contour two-dimensional (2D) graph form of DOE and a three-dimensional (3D) response surface were formed by employing a second-order polynomial model to assess the impact of each factor on shear stress. In order to obtain the results of the distinctive interactions among variables, two variables within the experimental range were diversed. Figs. 13–15 and present the results. These influences were independently discussed with additional evidence in the way the results occurred within the varying range of factors based on statistical values. It is observed in Fig. 13 that the shear stress increases with rising CNT of 0.05% to 0.10%. The steady reduction of shear stress was observed at higher CNT content (0.2%). Conversely,

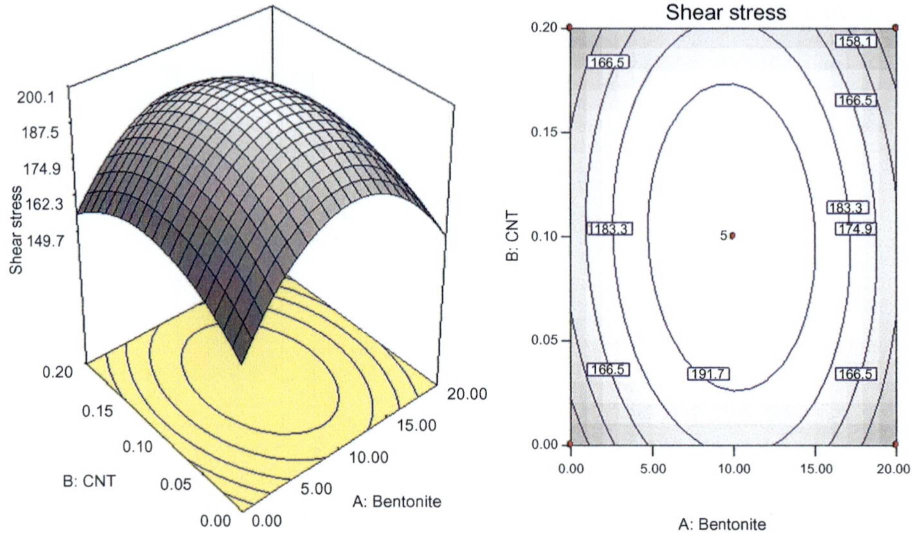

Fig. 13 Effect of interaction between the bentonite ratio and CNT content on the shear stress as three-dimensional response surface and two-dimensional contour plot.

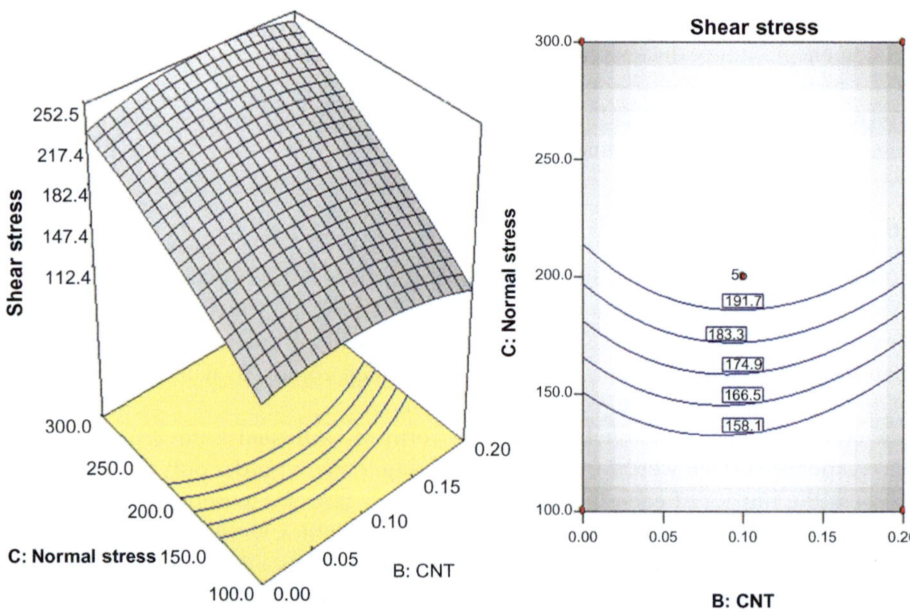

Fig. 14 Effect of interaction between the CNT content and normal stress on the shear stress as three-dimensional response surface and two-dimensional contour plot.

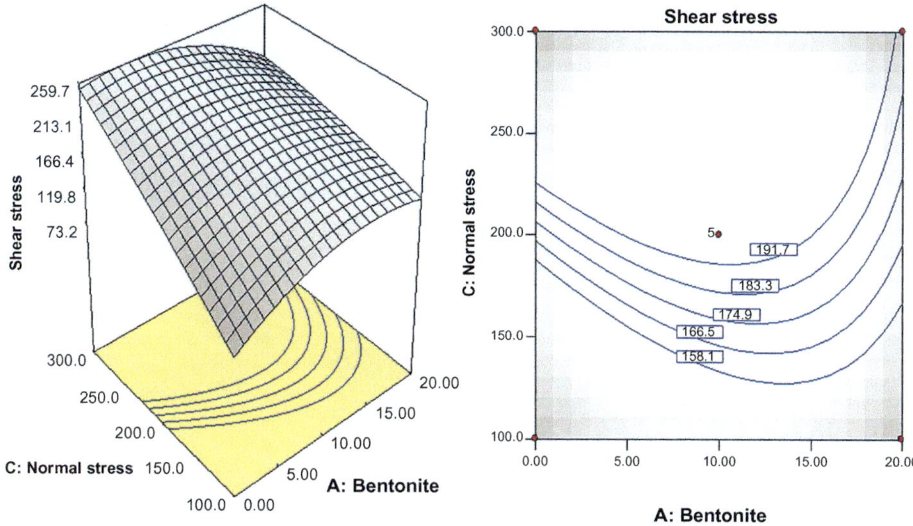

Fig. 15 Effect of interaction between the bentonite ratio and normal stress on the shear stress as three-dimensional response surface and two-dimensional contour plot.

the shear stress slightly upturned with rising bentonite content to 10% (S2) and then lowered with 20% content (S3). As obtained from direct shear tests, the shear stress behavior of soil that is incorporated with altered CNT content with distinct normal stresses is illustrated in Fig. 14. This figure highlights that the CNT-reinforced soil's shear stress increases with rising CNT content and normal stresses of 100–300 kPa, followed by a stagnant trend of shear stress following CNT content of 0.11%. These results may be explained by the interface between the CNT and soil. In a scenario where the normal stresses are low, the particle of the interface between the CNT and soil is not within proximity and thus the shear stress rises only marginally, or stays constant (Qu et al., 2013).

A steady increase in shear stress occurs as the CNT and interface soil become closer with rising normal stress. As illustrated in Fig. 15, the maximum shear stress takes place at roughly 259.7 at 0% bentonite content (S1) and normal stress of 300 kPa. Subsequently, they fall with rising bentonite content. This may be attributed to overwhelming bentonite content within the soil, which may reduce the interface between soil particles and the CNT section. Generally, the incorporation of CNT causes the reduction of soil pore size and porosity. In addition, when compared to all soil mixes in the absence of CNT, the CNT functions as a filler that causes a denser microstructure and greater strength (shear strength) (Chaipanich et al., 2010). The CNT bridge-connecting influence, as shown by FSEM analysis (Fig. 16), enables it to act as a bridge across voids and cracks, and thus ensures load transfer in the case of shear strength. Definitely, the shear stress is improved. Similar traits were shown when nanocarbons were incorporated into the cement matrix (Nochaiya and Chaipanich, 2011).

The result of each factor on the shear stress was assessed by using the polynomial model for CNF in order to form a contour (2D) and 3D response surface. In order

Fig. 16 Microcrack bridging by the CNT.

to obtain results of distinct interactions between all the variables, the two variables within the experiment's variety were diverse. Results are illustrated in Figs. 17–19. Statistical values were used to individually discuss these effects with more evidence on the way the outcomes occurred within the varying range of factors. The shear stress rises with rising CNF content of 0.05% to 0.17%, as illustrated in Fig. 17. A steady decrease of shear stress was noted at higher CNF content, that is, at 0.2%. Conversely, the shear stress had a marginal increase with rising bentonite content to 10% (S2) and then fell to 20% bentonite (S3).

The shear stress trait of soil that is incorporated with distinct CNF content with distinct normal stress is illustrated in Fig. 18, as obtained from direct shear tests. The figure shows that the shear stress of soil incorporated with CNF rises with increasing CNF content while for normal stresses of 100–300 kPa, followed by a stagnant level of shear stress beyond 0.12% CNF content. These results may be explained by the interface between CNF and the soil. In a scenario where the normal stresses are considerably low, the particle of the interface between the CNF and soil is not in close proximity, causing the shear stress to have a marginal increase or a constant level (Qu et al., 2013). As normal stress increases, the CNF and interface soil gets closer, thus causing a steady increase in shear stress. The maximum shear stress of 280 kPa takes place at 0% bentonite content (S1) and a normal stress of 300 kPa, as shown in Fig. 19. Subsequently, they show a lowering trend as the bentonite content rises. This may be due to the fact that overwhelming bentonite in the soil may have resulted in lowered interface between soil particles and the CNF section. In general, with the incorporation

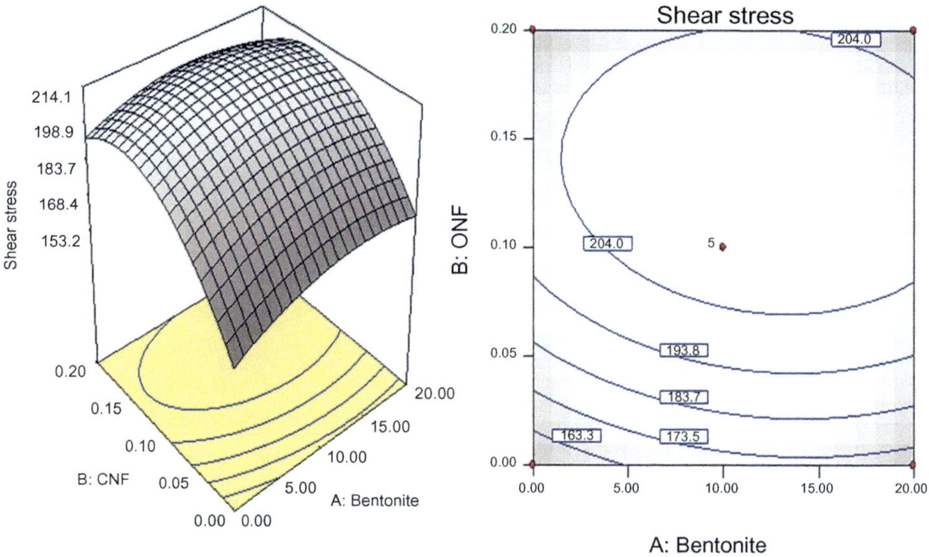

Fig. 17 Effect of interaction between the bentonite content and CNF content on the shear stress as three-dimensional response surface and two-dimensional contour plot.

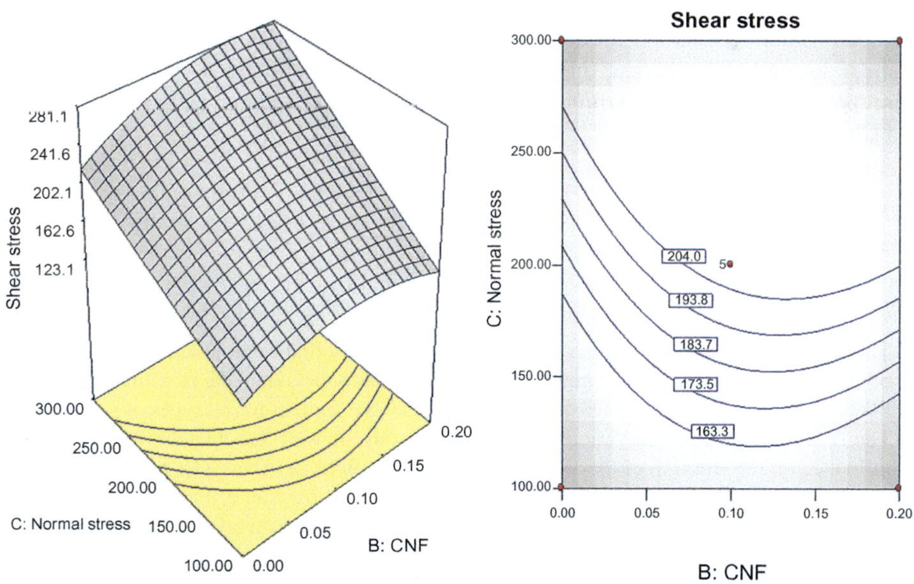

Fig. 18 Effect of interaction between the CNF content and normal stress on the shear stress as three-dimensional response surface and two-dimensional contour plot.

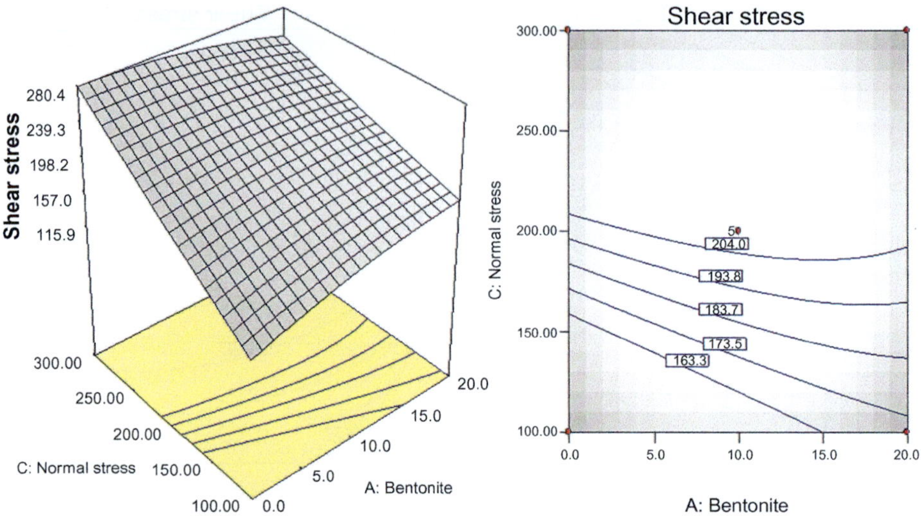

Fig. 19 Effect of interaction between the bentonite content and normal stress on the shear stress as three-dimensional response surface and two-dimensional contour plot.

of CNF, the soil's porosity and pore size decreased. In addition, in comparison to all soils mixed in the absence of CNF, the CNF functions as a filler, thus causing a denser microstructure and a greater strength (shear strength) (Chaipanich et al., 2010). The CNF's bridge-connecting influence is verified by FSEM analysis, as shown in Fig. 20. The CNF functions as a bridge between voids and cracks, thus ensuring load transfer at shear strength. For certain, there is enhancement of shear stress, and similar traits were displayed when nanocarbon was incorporated into the cement matrix (Nochaiya and Chaipanich, 2011).

9 Conclusion

The following conclusions are derived based on the results and discussion in Chapter 4:

1. The effect of various percentages of nanocarbons showed that for soil samples S1, S2, and S3, the liquid limit decreases with increasing NC. It was noticed that the maximum reduction in the plasticity index was noted at 0.2% CNT and CNF content for all types of soil.
2. The results obtained from various tests done on soils showed that NCs have a significant influence on the strength of the soil. Interfacial interactions between NCs and soil particles can generate a high bonding strength and increase the load-transfer efficiency from soil particles to NCs. The bond and friction strength at the interface appeared to be the main feature controlling the benefit of NCs.
3. The NCs and interface soil will get closer as the normal stress is increased and this will lead to a steady increase in shear stress. The maximum shear stress occurred at about 259.7 kPa for CNT and 280.44 kPa for CNF at 0% bentonite content and a normal stress of 300 kPa, which will reduce if the bentonite content is increased.

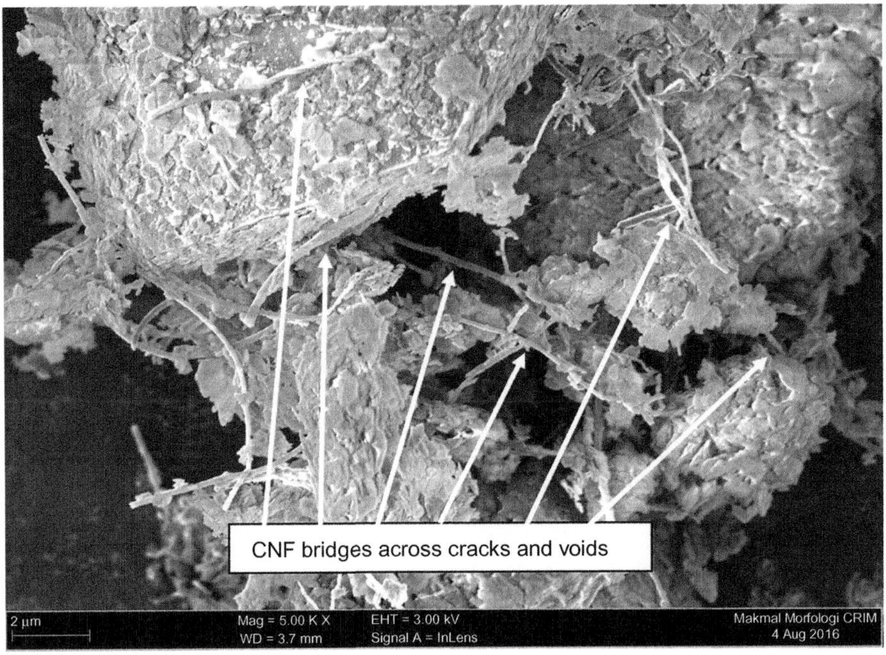

Fig. 20 Microcrack bridging by the CNF.

References

Ahmad, F., Bateni, F., Azmi, M., 2010. Performance evaluation of silty sand reinforced with fibers. Geotext. Geomembr. 28 (1), 93–99.

Al-Rub, R.K.A., Tyson, B.M., Yazdanbakhsh, A., Grasley, Z., 2011. Mechanical properties of nanocomposite cement incorporating surface-treated and untreated carbon nanotubes and carbon nanofibers. J. Nanomech. Micromech. 2 (1), 1–6.

Alsharef, J., Taha, M.R., Firoozi, A.A., Govindasamy, P., 2016. Potential of using nanocarbons to stabilize weak soils. Appl. Environ. Soil Sci. 2016.

Alsharef, J.M., Taha, M.R., Khan, T.A., 2017. Physical dispersion of nanocarbons in composites–a review. J. Teknol. 79 (5), 69–81.

Bennett, S.W., Adeleye, A., Ji, Z., Keller, A.A., 2013. Stability, metal leaching, photoactivity and toxicity in freshwater systems of commercial single wall carbon nanotubes. Water Res. 47 (12), 4074–4085.

Brown, C.B., Sheu, M.S., 1975. Effects of deforestation of slopes. J. Soil Mech. Found. Div. 101 (2), 147–165.

Chaipanich, A., Nochaiya, T., Wongkeo, W., Torkittikul, P., 2010. Compressive strength and microstructure of carbon nanotubes–fly ash cement composites. Mater. Sci. Eng. A 527 (4–5), 1063–1067.

De La Torre-Roche, R., Hawthorne, J., Deng, Y., Xing, B., Cai, W., Newman, L.A., Wang, Q., Ma, X., Hamdi, H., White, J.C., 2013. Multiwalled carbon nanotubes and C60 fullerenes differentially impact the accumulation of weathered pesticides in four agricultural plants. Environ. Sci. Technol. 47 (21), 12539–12547.

De Volder, M.F., Tawfick, S.H., Baughman, R.H., Hart, A.J.J.S., 2013. Carbon nanotubes: present and future commercial applications. Science 339 (6119), 535–539.
Dexter, A., Kroesbergen, B., 1985. Methodology for determination of tensile strength of soil aggregates. J. Agric. Eng. Res. 31 (2), 139–147.
Figueiredo, D.T.R., Correia, A.A.S., Hunkeler, D., Rasteiro, M.G.B.V., 2015. Surfactants for dispersion of carbon nanotubes applied in soil stabilization. Colloids Surf. A Physicochem. Eng. Asp. 480, 405–412.
Frydman, S., 1964. The applicability of the brazilian (indirect iension) test to soils. Aust. J. Appl. Sci. 15 (4), 335–343.
Ghosh, M., Bhadra, S., Adegoke, A., Bandyopadhyay, M., Mukherjee, A.J.M.R.F., Mutagenesis, M.M.O., 2015. MWCNT uptake in allium cepa root cells induces cytotoxic and genotoxic responses and results in DNA hyper-methylation. Mutat. Res. 774 (2015), 49–58.
Huat, B.B., Gue, S.S., Ali, F.H., 2007. Tropical Residual Soils Engineering. CRC Press.
Hull, A., 1919. A new method of chemical analysis. J. Am. Chem. Soc. 41 (8), 1168–1175.
Kang, I., Heung, Y.Y., Kim, J.H., Lee, J.W., Gollapudi, R., Subramaniam, S., Narasimhadevara, S., Hurd, D., Kirikera, G.R., Shanov, V., Schulz, M.J., Shi, D., Boerio, J., Mall, S., Ruggles-Wren, M., 2006. Introduction to carbon nanotube and nanofiber smart materials. Compos. B Eng. 37 (6), 382–394.
Kassim, K.A., Hamir, R., Kok, K., 2005. Modification and stabilization of Malaysian cohesive soils with lime. Geotech. Eng. 36 (2), 123–132.
Keller, A.A., Mcferran, S., Lazareva, A., Suh, S.J.J.O.N.R., 2013. Global life cycle releases of engineered nanomaterials. J. Nanopart. Res. 15 (6), 1692.
Khot, L.R., Sankaran, S., Maja, J.M., Ehsani, R., Schuster, E.W.J.C.P., 2012. Applications of nanomaterials in agricultural production and crop protection: a review. Crop Prot. 35, 64–70.
Kowald, T., Trettin, R., 2009. Improvement of cementitious binders by multi-walled carbon nanotubes. In: Nanotechnology in Construction. vol. 3. Springer, pp. 261–266. 2009.
Lee, J., Kim, M., Hong, C.K., Shim, S.E., 2007. Measurement of the dispersion stability of pristine and surface-modified multiwalled carbon nanotubes in various nonpolar and polar solvents. Meas. Sci. Technol. 18 (12), 3707–3712.
Lin, S., Reppert, J., Hu, Q., Hudson, J.S., Reid, M.L., Ratnikova, T.A., Rao, A.M., Luo, H., Ke, P.C.J.S., 2009. Uptake, translocation, and transmission of carbon nanomaterials in rice plants. Small 5 (10), 1128–1132.
Liu, Q., Chen, B., Wang, Q., Shi, X., Xiao, Z., Lin, J., Fang, X.J.N.L., 2009. Carbon nanotubes as molecular transporters for walled plant cells. Nano Lett. 9 (3), 1007–1010.
Maher, M.H., Gray, D.H., 1990. Static response of sands reinforced with randomly distributed fibers. J. Geotech. Eng. 116 (11), 1661–1677.
Mcgown, A., Andrawes, K., Al-Hasani, M., 1978. Effect of inclusion properties on the behaviour of sand. Géotechnique 28 (3), 327–346.
Metaxa, Z.S., Konsta-Gdoutos, M.S., Shah, S.P., 2013. Carbon nanofiber cementitious composites: effect of debulking procedure on dispersion and reinforcing efficiency. Cem. Concr. Compos. 36, 25–32.
Montazeri, A., Chitsazzadeh, M., 2014. Effect of sonication parameters on the mechanical properties of multi-walled carbon nanotube/epoxy composites. Mater. Design 56, 500–508.
Nataraj, M., Mcmanis, K., 1997. Strength and deformation properties of soils reinforced with fibrillated fibers. Geosynth. Int. 4 (1), 65–79.
Nochaiya, T., Chaipanich, A., 2011. Behavior of multi-walled carbon nanotubes on the porosity and microstructure of cement-based materials. Appl. Surf. Sci. 257 (6), 1941–1945.
Petersen, E.J., Zhang, L., Mattison, N.T., O'carroll, D.M., Whelton, A.J., Uddin, N., Nguyen, T., Huang, Q., Henry, T.B., Holbrook, R.D., 2011. Potential release pathways, environmental fate, and ecological risks of carbon nanotubes. Environ. Sci. Technol. 45 (23), 9837–9856.

Qu, J., Li, C., Liu, B., Chen, X., Li, M., Yao, Z., 2013. Effect of random inclusion of wheat straw fibers on shear strength characteristics of shanghai cohesive soil. Geotech. Geol. Eng. 31 (2), 511–518.

Rajasekaran, G., 2005. Sulphate attack and ettringite formation in the lime and cement stabilized marine clays. Ocean Eng. 32 (8–9), 1133–1159.

Serag, M.F., Kaji, N., Tokeshi, M., Baba, Y., 2015. Carbon Nanotubes and Modern Nanoagriculture. Nanotechnol. Plant Sci. 183–201. Springer.

Shojaei, M., 2016. Investigation of engineering characteristics on cement-treated. Road base using bitumen emulsion, rovene® 4045 and tylac® 4190 additives. Thesis PhD, Civil and Structural Engineering, Universiti Kebangsaan.

Sivakumar Babu, G.L., Vasudevan, A.K., Sayida, M.K., 2008. Use of coir fibers for improving the engineering properties of expansive soils. J. Nat. Fibers 5 (1), 61–75.

Taha, O.M.E., 2014. Effect of Nanomaterials on the Shrinkage and Expansive Behaviour of Soil. Civil and Structural Engineering, Universiti Kebangsaan Malaysia, Thesis PhD.

Taha, M.R., Alsharef, J.M., 2018a. Performance of soil stabilized with carbon nanomaterials. Chem. Eng. Trans. 757–762. Italian Association of Chemical Engineering-AIDIC.

Taha, M.R., Alsharef, J.M.A., 2018b. Effect of dispersion of nanomaterial on adhesion properties of soil using atomic force microscopy. Int. J. Adv. Mech. Civil Eng. 5 (2), 8–12.

Tang, C., Shi, B., Gao, W., Chen, F., Cai, Y., 2007. Strength and mechanical behavior of short polypropylene fiber reinforced and cement stabilized clayey soil. Geotext. Geomembr. 25 (3), 194–202.

Tang, C.-S., Shi, B., Liu, C., Suo, W.-B., Gao, L., 2011. Experimental characterization of shrinkage and desiccation cracking in thin clay layer. Appl. Clay Sci. 52 (1–2), 69–77.

Vaisman, L., Marom, G., Wagner, H.D., 2006. Dispersions of surface-modified carbon nanotubes in water-soluble and water-insoluble polymers. Adv. Funct. Mater. 16 (3), 357–363.

Viennet, J.-C., Hubert, F., Ferrage, E., Tertre, E., Legout, A., Turpault, M.-P., 2015. Investigation of clay mineralogy in a temperate acidic soil of a forest using X-ray diffraction profile modeling: Beyond the HIS and HIV description. Geoderma 241, 75–86.

Wang, Z., Ci, L., Chen, L., Nayak, S., Ajayan, P.M., Koratkar, N., 2007. Polarity-dependent electrochemically controlled transport of water through carbon nanotube membranes. Nano Lett. 7 (3), 697–702.

Wang, C., Li, K.-Z., Li, H.-J., Jiao, G.-S., Lu, J., Hou, D.-S., 2008. Effect of carbon fiber dispersion on the mechanical properties of carbon fiber-reinforced cement-based composites. Mater. Sci. Eng. A 487 (1), 52–57.

Yazdanbakhsh, A., Grasley, Z., Tyson, B., Abu Al-Rub, R., 2010. Distribution of carbon nanofibers and nanotubes in cementitious composites. Transportation research record: J. Transport. Res. Board 2142 (1), 89–95.

Zornberg, J., 2002. Discrete framework for limit equilibrium analysis of fibre-reinforced soil. Géotechnique 52 (8), 593–604.

Further Reading

Jawad, I.T., 2014. Nano Magnesium Oxide for Treatment of Residual Soil. Civil and Structural Engineering, Universiti Kebangsaan Malaysia, Thesis PhD.

Interaction of carbon nanotubes with rhizosphere microbial communities

21

Asma Irshad[a], *Nadeem Sarwar*[b], *Maira Zahid*[a], *Irfan Irshad*[c]
[a]National Centre of Excellence in Molecular Biology, University of the Punjab, Lahore, Pakistan, [b]Department of Computer Science, Bahria University Lahore Campus, Lahore, Pakistan, [c]Department of Pathology, University of Veterinary and Animal Sciences, Pattoki Campus, Pakistan

1 Introduction

Nanotechnology is a quickly increasing field that includes the application of nanomaterials and nanodevices in different fields such as biology, health sciences, physics, chemistry, materials science, therapeutics, and even engineering (Krishna et al., 2018; Tripathi et al., 2017). It mainly deals with properties in the nanoscale dimension and has the potential to transform each field (Satvekar et al., 2014). The prefix "nano" is a Greek word that means "dwarf," and it is used to represent one billion parts. The design, characteristics, and applications of nanomaterials and nanosystems of scale in the nanometer range (which ranges from 1 to 100 nm) are included in this field. Nanomaterials are generally classified as nanorods, nanotubes, nanospheres, nanofibers, nanocages, nanotubes, and nanogels (Volkova et al., 2017).

Several recent inventions of various types of carbon nanostructures have revolutionized the study of various useful applications in different fields. They have many promising applications in areas such as bioscience, gene delivery, medicine, and drug delivery. Several techniques for nanomaterial production have been reported while their functionalization and various chemical modifications have been accomplished. Further, the characterization, separation, and utilization of nanomaterials are promising and probable. Different factors such as chemistry, charge, size distribution, structure, surface area, and aggregation state as well as purification of samples possess significant impact for carbon nanotubes' (CNTs) reactivity. Otherwise, CNTs have the potential to be utilized in controlling other nanoscale structures because of their flexibility and strength, which proposes a significant role in nanotechnology engineering.

The utilization of innovative nanomaterials such as CNTs is becoming progressively cooperative in various research fields. CNTs reflect many extraordinary conductive, mechanical, and thermal properties. Such exceptional nanomaterials have one or sometimes more rolled-up graphene sheets that have high mechanical strength and surface area, electronic properties, highly stable and chemical properties, and low weight. Because of these properties, CNTs are favorable in various technological fields (Fabbro et al., 2012). This chapter has detailed information about CNTs, their

Carbon Nanomaterials for Agri-food and Environmental Applications. https://doi.org/10.1016/B978-0-12-819786-8.00021-9

structure, and advantages. This chapter details the properties, applications, and effects of various types of CNT and their concentrations levels on soil microbial communities.

2 Carbon nanotubes

CNTs are basically allotropes and nanostructures of carbon are more than 1,000,000 length-to-diameter ratio. Sumio Iijima discovered CNTs in 1991 by using HREM (high-resolution electron microscopy). CNTs generally have a cylindrical shape but can also be presented as diamond, graphite, or tubes, which are the different allotropes of carbon. Generally, the confirmation of CNTs exists in the powder form but they are classified as single-walled carbon nanotubes (SWCNTs) and multiwalled carbon nanotubes (MWCNTs) (Ahmad et al., 2012).

Basically, CNTs nanostructures are made from planes of rolled graphene. They have several notable chemical as well as physical properties, and for this this reason, they have a wide use in biomedicine, environmental, and agri-food applications. Many hypothetical studies have represented exciting electronic properties for CNTs. The most likely application of CNTs to synthesize nanowires has been reported in one study. *HREM* is a vital approach to characterize the microstructure of CNT and this approach is very suitable in various nanotube studies. It should be noted that the image gained is a projection of two dimensions of an object having three dimensions. CNTs have various biofunctionalities and on this basis, they can be conjugated with different biological molecules such as drugs, proteins, and even nucleic acid (Parijat and Mandeep, 2016).

Fullerenes are another allotrope of carbon that was discovered by scientists in 1985. In various studies, CNTs are alternatively called cylindrical fullerenes or buckytubes. Fullerenes and graphite have the same structures that are made up of sheets of hexagonal rings that are linked with each other. However, they possess pentagonal or heptagonal rings that avert the structure from being planar (Ahmad et al., 2012). Their diameters sized in nanometers while the lengths range in micrometers. The arrangement of carbon atoms in CNTs (fullerenes and graphite) is hexagonal. CNTs are considered the new substance in carbon allotropes; they are present between fullerenes and graphite (Fig. 1).

Tanaka et al. (2015) demonstrated the rolling of graphene sheets to form CNTs with a cylindrical confirmation and a 1.32×108:1 length-to-diameter ratio, which is an expressively large value compared to other materials. Sometimes the length of CNTs can reach 18 cm. Researchers reported CNTs as the most noticeable nanomaterial because of their remarkable use in various fields, including biomedicine (Kroto et al., 1985).

3 Properties of CNTs

The diameters of MWCNTs range from 20 to 200 nm with lengths that range from 106 to 1000 nm. They have a straight shape and are partially rigid, but sometimes they can be bent or curled because of their flexibility. They can exist as a single particle or present in the form of clusters or maybe ropes. According to the WHO definition,

Fig. 1 The powder form of CNTs.

MWCNTs as single particles with a length of 5 nm are considered to be fiber, which has an aspect ratio of more than 3:1. PCOM cannot detect particles with a diameter less than 200 nm because of the limits in optical detection. It is well acknowledged that fibers such as asbestos have a dispersal of diameters (Ken et al., 2006).

Carbon is an essential chemical that has six atomic numbers, which means it possesses six electrons and the electronic configuration is $1s^2$, $2s^2$, and $2p^2$ in an atomic orbital. Hybridization can form in sp., sp^2, or sp^3. The discoveries of very persistent nanometer-sized sp^2 carbon-bonded substances, that is, graphene, fullerenes, and CNTs, became the reason to make inquiries in nanotechnology. CNTs derive their physical properties from graphene, as they are composed of rolls of graphene sheets. In the graphene structure, carbon atoms are compactly arranged in a regular sp^2-bonded atomic scale just like a honeycomb (hexagonal) pattern. This arrangement is an elementary structure for other sp2 carbon-bonded substances (allotropes), that is, fullerenes and CNTs. Theoretical studies defined CNTs as different cylindrical structures obtained from rolled-up graphene sheets (Ali et al., 2014). CNTs with single wells are called SWCNTS and were first described in 1993. However, those with more than one well are considered MWCNTs, and they were first discovered by Iijima (1991).

4 Structure of CNTs

In order to form different nanostructures, carbon has diverse ways to bond and show completely different properties. Carbon has sp2 hybridization that forms a layered construction because of van der Waals forces (weak and out of the plane) and strong in-plane

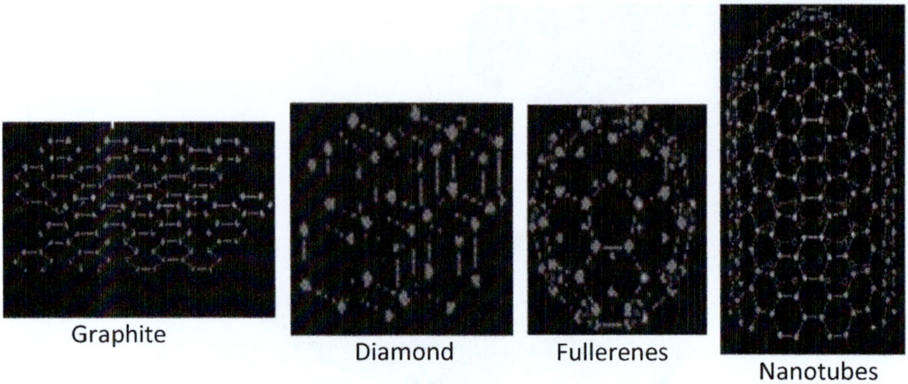

Fig. 2 Types of carbon allotropes. Hexagonal arrangement of carbon atoms in CNTs.

forces (bonds). MWCNT concentric cylinders, whose numbers range from a few to a few tens, are arranged around a fundamental hollow tube having a regular periodic layer. In different studies, it has been reported that the real-space examination of MWCNTs has a range of 0.34–0.39 nm interlayer spacing (Ajayan and Ebbesen, 1997). On the basis of layers, the MWCNTs have internal diameters that range from 0.4 nm to several nm, but possess an outer diameter of 2–30 nm. MWCNTs typically have both ends closed and they are covered with the help of a half fullerene molecule that is dome-shaped, which is defective pentagonally, and the axial size varies from one micrometer to a few cm. In closing the tube from both ends, half of the fullerene molecule, which has a pentagonal ring defect, plays a crucial role. On the other hand, SWCNTs possess fewer diameters that range from 0.4 to 3 nm, and the lengths vary in the range of micrometers. Usually, these SWCNTs tend to come together and close to form bundles, which appear as ropes to form a bundle or rope-like structure. SWCNTs are organized hexagonally, as that makes a crystal-like construction (Chico et al., 1996) (Fig. 2).

5 Types of CNTs

CNTs are categorized into two types:

Single-walled carbon nanotubes (SWCNTs).
Multiwalled carbon nanotubes (MWCNTs).

5.1 Single-walled carbon nanotubes

SWCNTs have only one carbon layer in a cylindrical shape that has a diameter of 0.4–2 nm, which totally depends on the synthesis temperature. It is reported that upon increasing the synthesis temperature, the diameter of the SWCNT can be enhanced (Klumpp et al., 2006). Depending on the synthesis temperature, SWCNTs can have many forms such as chiral, armchair, zigzag, or sometimes even helical arrangements. SWCNTs have an excessive surface area as large as 1300 m^2/g, and that represents enough space to load the drug and even to conjugate them with biomolecules that are in bioconjugation. In comparison with MWCNTs, SWCNTs are more suitable for drug delivery because of their enhanced surface area and proficient capacity to load the drug. In one study, it was reported that SWCNTs have more retention time for the drug in the anticancer drug complex. The tumor cells have increased permeability and a high retention effect. The functionalized SWCNTs are expelled gradually by the biliary path and feces after the release of drugs to the targeted region. All these characteristics make SWCNTs appropriate candidate tools for delivering drugs and a promising nanoplatform, especially in therapeutics (Saito et al., 2014).

5.2 *Multiple walled carbon nanotubes*

MWCNTs have multiple cylindrical structures and each cylinder is made from a single sheet of graphene that surrounds the core. MWCNTs have an outer diameter of 2–100 nm and the internal diameter ranges from 1 to 3 nm while the length ranges from 1 to a few micrometers (Rastogi et al., 2014). Depending on the arrangement of graphite layers, MWCNTs are categorized into two categories: one is a parchment-like structure that has a graphene sheet that is folded around it and the second is the Russian doll model where different layers of graphene sheets are organized as a concentric structure (Madani et al., 2011). MWCNTs have depositing materials on the walls or ends, which are bonded by physical interaction that can be potentially applied in catalysis, electronic devices, biomedical, magnetic data storage, and biosensors. The different methods used for this purpose are chemical decomposition, hydrolysis at high temperatures, or precipitation of a metal precursor (Uc-Cayetano et al., 2014) (Fig. 3).

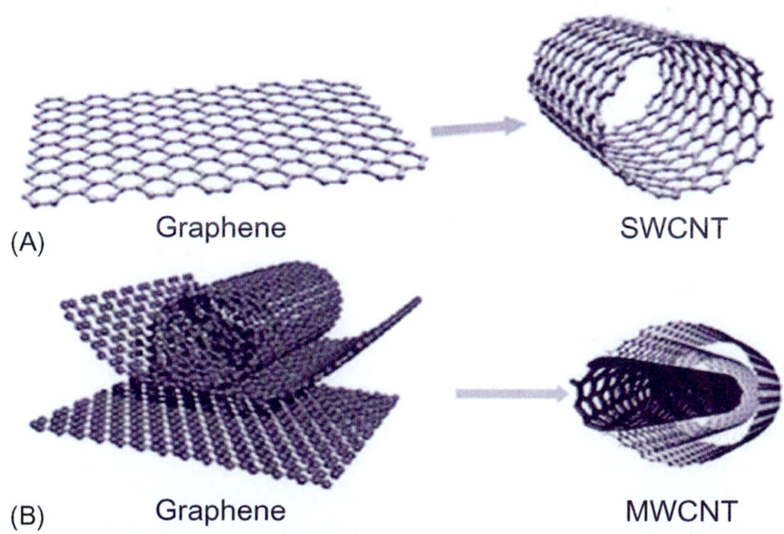

Fig. 3 Graphene and CNTs as (A) SWCNT and (B) MWCNT structures.
Reprinted from Vidu, R., Rahman, M., Mahmoudi, M., Enachescu, M., Poteca, T.D. and Opris, I., 2014. Nanostructures: a platform for brain repair and augmentation. Front. Systems Neurosci. 8, 91, Open Access Journal.

Table 1 Comparison of SWCNTs and MWCNTs.

SWCNTs	MWCNTs
They have a single layer of graphene in their structure.	They possess numerous layers of graphene in their structure.
For SWCNT synthesis, the catalyst is compulsory.	They can be synthesized without a catalyst.
SWCNT synthesis in bulk is tough because an appropriate atmospheric condition and overall growth conditions are needed.	MWCNT synthesis in bulk is not very difficult.
They have the least purity.	They show high purity.
During functionalization, there is a chance of defects.	Defect during functionalization is the least but when it occurs, it is difficult to recover.
Their accumulation is the least in the body.	They can accumulate more in the body.
Their characterization and evaluation methods are easy.	They have complex characterization and evaluation methods.
They are easy to twist and they are more pliable.	It is not easy to twist them.

Reprinted from Ali, E., Hadis, D., Hamzeh, K., Mohammad, K., Nosratollah, Z., Abolfazl, A., Mozhgan, A., Younes, H., Sang, W. J. 2014. Carbon nanotubes: properties, synthesis, purification, and medical applications. Nanoscale Res. Lett. 9, 393.

The Russian doll and parchment models are two different structural models for MWCNTs. If a CNT consists of more than one nanotube within it and the diameter of an outer nanotube is larger than the inner nanotube, then it is referred to as the Russian doll model. But if only one graphene sheet is rolled around it many times, just like a piece of paper folded many times on itself, it is referred to as the parchment model. MWCNTs and SWCNTs possess the same properties. MWCNTs possess multilayers and the outer walls play a role to protect the internal CNTs from exposure to various interactions (chemical) with externally present materials. MWCNTs have properties such as high tensile strength that are lacking in SWCNTs (or exist only partially) (WanderWal, 2003) (Table 1).

5.3 Properties of MWCNTs and SWCNTs

On the basis of folding the sheet in a cylinder, we can categorize the SWCNTs into three forms: armchair, chiral, and zigzag. SWCNT structure is categorized on the basis of a pair of indices that can be represented as n and m, which define chiral vector and have a direct influence on the electrical properties of nanotubes. These integers n and m are used to find the number of unit vectors in graphene, which is just like a honeycomb-like structure. It is reported by various researchers that when the m integer is equal to 0, then the CNTs are categorized as zigzag nanotubes; when the n integer is equal to the m integer, the nanotubes are categorized as armchair nanotubes while other forms are called chiral. The chiral vector $C = na_1 + ma_2$ (a_1 and a_2 both are the base cell vectors of graphite) also defines the diameter of tube d (Ali et al., 2014), and this vector determines the direction of folding a graphene sheet. There is a mathematical expression to calculate the diameter of a CNT:

$$a = 1.42 \times \sqrt{3}\,\text{Å} \qquad (1)$$

corresponds to the lattice constant in the graphite sheet.

When $n - m$ is a multiple of 3, then the nanotube is described as "metallic" or a highly conducting nanotube; if not, then the nanotube is semimetallic or a semiconductor.

At all times, the armchair form is metallic, whereas other forms can make the nanotube a semiconductor.

5.4 Factors affecting the structure

The following are factors that can affect the structure of nanotubes (Wenkai et al., 2018):

1. Translational vector $= \mathrm{T} = t_1a_1 + t_2a_2$ » (t_1, t_2).
2. Chiral vector $= \mathrm{Ch} = na_1 + na_2$ » (n, m).
3. Length of chiral vector $= L = a\ \sqrt{(n_2 + m_2 + n \times m)}$, where a is the lattice constant.
4. Chiral angle $= \cos\theta = (2n + m)/(2 \times \sqrt{(n_2 + m_2 + n \times m)})$.
5. Number of hexagons in the unit cell $= N = (2 \times (n_2 + m_2 + n \times m)/dR)$.
6. Diameter $= dt = L/\pi$.
7. Rotation angle of the symmetry vector $= \psi = 2\pi/N$ (in radians).
8. Symmetry vector $= \mathrm{R} = pa_1 + qa_2$ » (p, q).
9. Pitch of the symmetry vector $= \tau = ((m \times p - n \times q) \times T)/N$.

6 Synthesis of CNTs

Arc discharge or laser ablation are the high-temperature approaches that were first utilized in CNT synthesis. Recently, these approaches have been changed with low-temperature methods such as chemical vapor deposition (CVD), which is less than 800°C. In this method, the diameter, length, density, alignment, and purity of CNTs can be accurately assayed (He et al., 2010).

Most of the synthesis methods involve supporting gases and vacuum, although the gas-phase methods involve volume. Because of this, they are considered appropriate for implementations, for instance, composite materials that are required in large quantities of nanotubes and industrial level production approaches to make them cost-effective. On the other hand, the drawbacks of gas-phase synthesis methods include low catalyst yield, which has a minute percentage of catalysts to form nanotubes; the short lifetime of the catalyst; and the low number density of the catalyst (Unrau et al., 2010).

In the existence of metal catalysts, mechanisms of CNT formation are considered where carbon is accumulated on a catalyst nanoparticle. From that, the network of C is finally extended to make a graphite tube and its growth has been detected using in situ electron microscopy. The tube's diameter is evaluated with the size of the metal particle; metal particles having diameters of a few nanometers produce SWCNTs while particles having large diameters produce wider MWCNTs (Vinciguerra et al., 2003).

6.1 Arc discharge method

Arc discharge is the synthesis method that needs higher temperatures (more than 1700°C) to synthesize CNTs, which usually results in CNT growth with fewer structural defects in comparison with other techniques.

6.2 Laser ablation method

Pulsed laser deposition (PLD) is considered the best method for SWCNT production and it was first introduced in 1995 by Smalley's group. The features of CNTs synthesized by PLD are mainly reliant on parameters that are as follows: energy fluency, repetition rate, power of peak, cw versus pulse, wavelength oscillation, the composition of the target material chemically and structurally, the pressure of the chamber and the chemical composition, the flow along with the pressure of the buffer gas, and the distance between the target and the substrates. High-quality and high-purity SWCNT growth is critical to the PLD approach.

However, the equipment requirement and the huge energy consumption by these methods make them the least suitable for nanotube production. With the arc and laser approaches, only samples in powder form with nanotubes that are tangled in bulk can be formed (Guo et al., 1995; Prášek et al., 2011).

6.3 Flame pyrolysis

Liu and his coworkers presented this technique for the mass production of CNTs with simple conditions and equipment. This technique is also known as V-type pyrolysis flame. Researchers used carbon monoxide as a carbon source in this method and synthesized the high-yield CNTs with fewer impurities. Combustion provided the acetylene/air premixed gas, pentacarbonyl was used as the catalyst, and the hydrogen/helium premixed gas acted as the diluted and protection gas. CNTs with a 10–20 nm diameter and length ranges to dozens of microns were obtained by this method (Liu et al., 2011).

7 Purification of CNTs

After the production of CNTs, they are purified by removing soot (the amorphous carbon), the residues of the free and left metal catalyst, and other materials. This is typically done by washing or ultrasonicating with dilute acid. After high purification of tubes, they will undergo oxidation for further purification. Removal of other materials such as silica and alumina requires a stronger treatment in the presence of concentrated acid. Because purification also destroys CNTs, the removal of impurities must be balanced against the introduction of defects into tubes (Borowiak-Palen et al., 2002).

7.1 Advantages

CNTs are considered biocompatible and nonbiodegradable with a nonimmunogenic nature.
CNTs have an increased elastic nature that makes them usable for intracellular delivery.
They have shown minimum cytotoxicity.
In experimentation, 96% of CNTs are excreted by urine and the remaining 4% through feces.
They are extremely lightweight and that cannot be further broken during processing.
Both sides have open ends, which makes the inner surface available for easy subsequent integration of species within nanotubes.
Nanotubes possess a larger inner volume in relation to their diameter for casing.
CNTs have the ability to enter into cells by a spontaneous mechanism because of their shape, which is nano as well as tubular.
CNTs have a discrete inner and outer surface, which can be modified differently for chemical as well as biochemical functionalization (He et al., 2013; Mehra et al., 2008).

8 Antimicrobial applications of CNTs

In various studies related to biology, different types of CNTs are being efficiently used as drug delivery systems, sensing materials, antibacterial substances, and biomaterials. There is a huge amount of data and literature, as well as many related publications that are rapidly emerging, showing that CNTs are efficient for use in various applications such as biology, in which they can be used as bactericidal agents (Bitounis et al., 2013).

The first investigation was published in 2007 by Kang et al. It reported the antimicrobial property of SWCNTs against the pathogenic bacteria *Escherichia coli*. Various research has shown that both SWCNTs and MWCNTs have strong repressive properties for microbial growth, even exposed for a short period of time, which reflects that carbon-based nanostructures have different biomedical applications. On the basis of toxicity, SWCNTs have shown significantly higher antibacterial performance when compared with MWCNTs or with the fullerene-C60 structure. Despite the material having unbelievable properties and a huge number of publications on their antibacterial effects, the bactericidal mechanism of CNTs has yet to be fully understood (Ahmed et al., 2017).

8.1 Factors affecting the antimicrobial activates of CNTs

Microbial toxicity mechanism of CNTs is affected by various parameters such as a functional group on the surface, nanoparticles diameter, residues of catalyst, electronic structure, coating and chemistry of the surface. Among all these parameters, the length of C-based nanotubes is very important because it plays a role in the interaction with the bacterial cell membrane. Shorter tubes are found to have higher bactericidal activity compared to longer tubes. CNTs having shorter lengths may boost the probabilities of interaction among open ends of nanotubes and microorganisms, which can lead to increased damage to cell membranes. It has been reported that MWCNTs having the length of 50 μm enfolded around the surface of the microbial cell and results in the osmotic lysis of the microorganism. Long nanotubes have revealed very high antibacterial activity as compared to short nanotubes.

8.2 Effects of solid/liquid interaction with CNTs

When compared with solid surfaces, CNTs have a quite different interaction with cells in liquids. In a liquid medium, the CNTs with short lengths tend to form aggregates without comprising the microbial cells in a large number while the nanotubes with long lengths form aggregates that are more bioeffective because they directly affect a very large number of cells because of these aggregates. It is well recognized that CNT aggregates are certain because of their distinctive configuration and strong van der Waals interactions (Johnston et al., 2010; Yang et al., 2010).

In the inactivation process of a microbial cell, the diameter of the tube has a critical role. CNTs having smaller diameters cause damage to the cell membrane of the microorganism by the interaction between cell and surface. A CNT of 1.5 nm diameter that is attached at the end of the microbial cell is bulging out from the cell and has a needle-like appearance at the other end; this has been observed through direct microscopy. Further, it is reported that the small nanotubes made close interactions with bacterial cells. Nanotubes with a larger diameter ranged from 15 to 30 nm tend to interact with the bacterial cell, where the cells are just present on top of the nanotubes instead of interacting with CNTs aggregates tightly (Bitounis et al., 2013). CNTs have shown the least bactericidal activity toward bacteria having a rod shape when compared with bacteria of a spherical shape.

8.3 Effect of electronic structure on antimicrobial activity

Another significant feature that affects the CNT antimicrobial property is raised from the electronic structure of nanotubes. Among other parameters, the diameter and orientation of the tubes as well as the wrapping angle that can highly influence the electronic conductivity of single-walled CNTs and small differences in these features could alter the tubes from a metallic to semiconducting state (Saifuddin et al., 2013). Various reports have revealed that the electronic structure is a very important element in determining the antimicrobial property of SWCNTs. The ratio of SWCNTs having the same diameter, defective spots, and length in a soil sample can directly affect the viability of bacteria. CNTs are considered good electrical conductors that possess high dielectric breakdown strength and even electron emission properties. When CNTs are employed in the corona plasma system or the discharge electrode, the threshold voltage is reduced to a great extent in the plasma field. A small operational voltage can be generated that results in a low concentration of ozone in the discharge system by the CNT-based corona that is ultimately used in bioaerosol eradication. It was reported in one study that 97% of *E. coli* bioaerosols can be inactivated by using the CNT-based corona discharge system when the discharge voltages are −7.5 kV, which is more than using electrodes of stainless steel. By incorporating CNTs with various dielectric polymers such as poly(dimethylsiloxane) (PDMS)/poly(methyl methacrylate) (PMMA) fibers, the behavior of electrodes can be enhanced and further applied in many applications such as artificial muscles, sensors, nanoscale aerosol filters, biomedical implants, and robots (Lolla et al., 2016).

8.4 Lethal effect of CNTs on biofilm formation

CNT toxicity against biofilm formation is also studied by different researchers to determine the ability of CNTs to hinder microbial attachment to the surfaces and their proliferation at different steps of microbial colonization. Biofilm is a structure that is very important for the protection of bacterial cells. It renders them highly resistant to detachment by physical forces and harmful nanoparticles. CNTs have depicted strong bactericidal actions against bacterial cells in biofilms. After microscopic observation of the bottom layer of the biofilms, it was reported that bacterial species such as *E. coli* and *Bacillus subtilis* were in direct contact with coatings having SWCNTs, showing that approximately 80%–90% of the bacterial cells were dead. Yet, the interface of CNTs with biofilm is totally dependent on different stages of biofilm formation, but the efficiency of CNTs is more noticeable in the initial stages of biofilm formation. Microbial cells are the least susceptible or remain unaffected by the effect of CNTs when the biofilm became mature to provide protection to the microbial cells when compared with the effects in the initial or early stages of biofilm (Dong and Yang, 2014; Rodrigues and Elimelech, 2010).

8.5 Effect of CNTs on soil microbial communities

Different studies also revealed that CNTs do not always affect the community of soil microbes on exposure because various microbes are able to develop resistance even at exposure with a high concentration of CNTs. Khodakovskaya et al. (2013) demonstrated that MWCNTs are not efficient enough to affect the species diversity and richness of microbial communities while they are quite efficient to influence the abundance of each microbial group by treatment. In one study, researchers compared the influence of two types of CNTs: raw MWCNTs and functionalized MWCNTs in a community of different soil microbes. They used different concentrations of MWCNTs: 0 μg/g, 50 μg/g, 500 μg/g, and 5000 μg/g of soil. After exposure, they extracted the DNA and then amplified it, revealing that the soil bacteria were not influenced by the raw MWCNTs but the composition of the soil bacterial community was influenced by different concentrations of functionalized MWCNTs. However, Shrestha et al. (2013) determined the influence of MWCNTs at the concentration of 10–10,000 mg/g on the soil microbes (sandy and loam) after 90 days of exposure. Two effects were noted after the exposure: (i) reduction in the soil bacterial genera such as Waddlia, Derxia, Opitutus, and Holophaga, and (ii) an increase in the number of soil bacterial genera such as Pseudomonas, Rhodococcus, Nocardioides, and Cellulomonas, which are the strong degraders of pollutants such as polycyclic aromatic hydrocarbons. It can be concluded that on the exposure of a high concentration of MWCNTs, there is a variation in the soil microbial community composition to more tolerant microbial populations (Kerfahi et al., 2015).

9 Impacts of CNTs on living organisms and plants

In today's society, because of great use in industries, nanomaterials are in high demand. Because of the increased applications, nanoparticles were released into the environment. Kerfahi et al. (2015) reported that the production of CNTs has increased by approximately 25% each year and has been extended to 9300 tons with a production cost of $1.3 billion. A lack of rules and regulations in the synthesis, use, and discard of CNTs has allowed nanotube residues to bind with contaminants such as metals. When released into the environment, CNTs are quite capable of affecting human health. Very few studies have addressed the effects of CNTs on living organisms in the environment. CNTs have demonstrated toxicity to several organisms, such as protozoa, copepods, mice, and rainbow trout. However, very little is known about the toxicity of these nanomaterials to plants and plant parts (Shrestha et al., 2013).

MWCNTs have been extensively used by researchers because of their cost effectiveness and the ease in their bulk production when compared with different CNTs. Researchers used the bush beans known as *Phaseolus vulgaris* (which has economic as well as nutritional importance) in one study. These are among the oldest crops in the world. They determined the influence of CNTs on the bush beans under hydroponic conditions. In the world for human consumption, this plant is considered an important grain of legumes. Total worldwide production of bush beans is more than 23 million metric tons, out of which seven million metric tons are grown in Latin America and Africa. These beans are among the low-calorie plants and are considered to be rich in dietary fibers. The high amount of minerals such as Fe, Ca, K, Mn, Mg, vitamin A promoting flavonoid, vitamin B such as B1 and B6, polyphenolic antioxidants such as lutein, zeaxanthin and α/β-carotene and vitamin C (Kappen et al., 2009).

Different studies have been done to evaluate CNT influence on the microbial community present in the soil and even on plants. It has been reported that there is a reduction in the enzymatic actions of soil and also in the biomass of microbes when the different soils are treated with CNTs. In 2011, Chung et al. treated two kinds of soils with 5000 μg of MWCNTs: one with grass and the other with coniferous trees having different physicochemical properties of soil. Researchers evaluated the enzymatic actions as well as biomass of microbes after exposure of 20 days and reported that the activities of different enzymes such as xylosidase, 1,4-b-glucosidase, 1,4-b-*N*-acetylglucosaminidase, phosphatase, and cellobiohydrolase was decreased along with microbial biomass C and N at the concentration of 5000 μg MWCNTs per gram of soil (Chung et al., 2011). In another study, different soil samples were exposed for 3 weeks with both suspended as well as powder SWCNTs at different concentrations ranging from 0 to 1000 μg/g of soil. They also determined that SWCNTs of concentration 300–1000 μg/g of soil are efficient enough to decrease the enzyme actions such as cellobiohydrolase, β-1,4-glucosidase, β-1,4-xylosidase, β-1,4-*N*-acetylglucosaminidase, L-leucine aminopeptidase, acid phosphatase, and microbial biomass. In 2014, Jin et al. reported that SWCNTs on the exposure of 25 reduced profiles of phospholipid-derived fatty acids (PLFA), which measures phospholipid fatty acid content, used a concentration of 0, 0.03, 0.1, 0.3, 0.6, and 1 mg/g of soil in the study. They also showed that SWCNTs have an important negative impact on the biomass of a broad range of bacterial groups, including both gram-positive and gram-negative as well as on fungi. During one study in 2007, Kang et al. revealed that SWCNTs with high purification showed strong antimicrobial actions. When *E. coli* K12 was exposed for 1 h with SWCNTs (1–50 µg/mL of aqueous solution), survival was determined using a fluorescence-based assay. The results showed that cells exposed to SWCNTs presented a considerable decline in viability. It proved that damage in the cell membrane occurs with direct contact with SWCNT aggregates and this is the main mechanism behind the bactericidal action of CNTs (Chung et al., 2011; Jin et al., 2014).

Studies have demonstrated that sometimes the exposure of CNTs to soil microbial communities may generate resistance when exposed, even at the high concentrations of these nanotubes.

The metabolic system of plants can be changed after absorbing CNTs by altering the biomass, roots, and fruits of the plants. In 2014, researchers showed in one study that CNTs can easily enter into the seed coat and even into the cell wall of a plant cell. After entering the plant, CNTs make different variations in the metabolic system that will result in the enhancement of biomass of plants and the yield of fruits. It is also reported that SWCNTs can enter through passive transport and then be confined in the lipid envelope that is extracted from the plant chloroplast. That results in an increase in the photosynthetic process and also the rate of electron transport (Jin et al., 2014). Miralles et al. (2012) treated alfalfa and wheat with MWCNTs having concentrations of 40–2560 mg/L of deionized water. The results showed that both crops tolerated this high concentration of MWCNTs, and in the alfalfa plant, this led to an increase in the elongation of plant roots while in wheat it increased germination. Researchers reported that tomato plants produce flowers and fruits twice but with the same amount of leaves when grown in CNTs with the treated soil having a concentration of 50 µg m/L and even 200 µg m/L of CNTs, as compared to that of tomato plants grown in regular soil (Husen and Siddiqi, 2014) (Fig. 4).

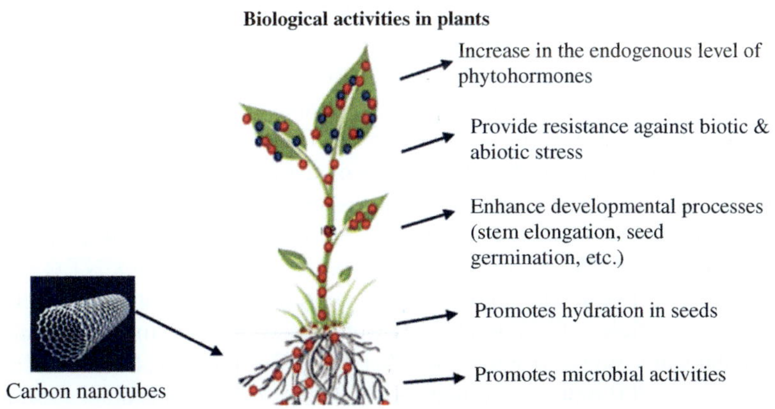

Fig. 4 Impacts of carbon nanotubes on the endogenous levels of phytohormones, biotech and biotech stress, developmental processes, hydration of seeds, and microbial activities in plants.

In 2014, Srivastava and Rao also proved that MWCNTs significantly increase the growth of the plant as well as the biomass of garlic, wheat, maize, and peanuts. They reported in their study that seeds tend to be germinated 3–4 times faster when treated with MWCNTs compared to the seeds of the control group. The roots of peanut plants showed increased delivery of water after exposure to CNTs. Ratnikova et al. (2015) conducted one study using tomatoes in which they tested the permeability of the tomato seed coat, which contrasts the above finding. The seeds of the tomato plant were soaked in water having 50 mg/L of MWCNTs for 0–60 min. The CNT penetration of the seed coats to the embryo affects the seed germination as well as the seedling growth. A study by Ratnikova et al. (2015) found that CNTs were not capable of penetration into tomato seed coats.

It has been observed by different researchers that the growth of some plants has been enhanced when treated with CNTs. On the other hand, other plants display inhibition, or sometimes the CNTs do not affect the plant's growth. Khodakovskaya et al. (2012) studied the different effects of MWCNTs of 5–500 μg m/L concentration on the germination of tobacco seeds and reported that MWCNTs are able to increase the development of the tobacco cell culture by 55%–64%. However, researchers showed that activated carbon is able to stimulate cell growth only at low concentrations of 5 μg m/L and significantly repressed the growth of plant cells when a high concentration of 100–500 μg m/L was used. During one study, Cañas et al. (2008) showed that cucumbers, tomatoes, onions, and plants grow effectively when treated with both kinds of CNTs that are functionalized and nonfunctionalized. It was observed in another study that lettuce showed repression of root elongation and CNTs had no special effects on the root elongation of cabbage and carrots. In 2009, Stampoulis et al. determined the effect of MWCNTs on zucchini in hydroponic surroundings for 15 days. They reported a 60% reduction in plant biomass when compared with control plants and even with bulk carbon at a 1000 mg/mL concentration (Ratnikova et al., 2015; Srivastava and Rao, 2014).

There is a direct interaction of CNTs with the soil microbial community and environment when they are introduced into the soil while watering the plants. By analyzing the soil microbial community and the phenotype, researchers assessed the effect of CNTs on the soil environment. The denaturing gradient gel electrophoresis (DGGE) method was reported as the initial screening method to observe the difference in the microbial community in soil samples. This method can be used to compare the bacterial community among different soil samples having treatment with CNTs. Researchers used the DGGE profiles very frequently to compare bacterial communities in different soils with different treatments of CNTs. In one study, DGGE profiles reflected the similarity. These correspondences showed that one profile of triplicate CNT 0 and 200 samples clustered with the CNT 50 sample group. It was reported that there were 94.3% and 83.02% correspondences of profiles within CNT 0 and CNT 200 samples, respectively. The homologies of microbial communities in different soil samples (having different CNT treatment) were shown by the similarities in DGGE profiles. In the study, higher resemblances between CNT 0 (control) and CNT 50 than CNT 50 and CNT 200 represented that the differences in the microbial communities were dependent on the CNT concentration. Therefore, it can be concluded that the microbial populations could be affected by introducing high concentrations of CNTs as compared to the low treated levels in the soil (Khodakovskaya et al., 2012).

10 Toxicity of CNTs

It was reported in one study that *E. coli* may have been killed after exposure to CNTs. Further, there is disagreement about the toxicity of CNTs because of various experimental approaches. Researchers studied the CNT's toxicity in vivo as well as in vitro and the parameters of CNTs related to toxicity such as length, concentration, functionalization, method of exposure, duration of exposure, and the used surfactant. Mariya et al. stated that the length, concentration, and exposure conditions of CNTs are responsible for the toxicity against the plant species (Mariya et al., 2013). The main reason for the harmful effects of CNTs solid particles depends on two main factors that act in harmony, these factors including surface area and surface reactivity. The small particles tend to have increased surface area per unit mass; consequently, the particle's intrinsic property is associated with the surface area. Thus, when the particle size reduced, the cytotoxic effects of CNTs was increased (Kerfahi et al., 2015).

There are different concerns related to CNT fortification due to their three properties that are directly linked to the pathogenicity of the particles:

1. They tend to be more toxic when compared with larger particles.
2. Having a fiber shape, they can behave as asbestos or any other pathogenic fibers and those that are considered toxic because of their needle shape.
3. They are considered graphitic and they tend to be biologically biopersistent.

11 Conclusion

This chapter deals with the general properties, structural features, and several factors affecting functional efficiencies of CNTs. It also deals with fundamental types of CNTs, including SWCNTs and MWCNTs as well as their properties, explaining the strength, purity, and twisting ability to describe their potential antimicrobial characteristics. Moreover, it also elaborates the arc discharge method, the laser ablation method, and flame pyrolysis for synthesis and the universal method for the purification of CNTs. Furthermore, the main applications of CNTs dealt with in this chapter are related to the antibacterial root system significantly on soil microbial communities and biofilm formation along with its positive and negative impacts on living organisms and plants.

References

Ahmad, A., Kholoud, M.M., Reda, A.A., Abdulrahman Al-Warthan, 2012. Carbon nanotubes, science and technology part (I) structure, synthesis and characterization. Arab. J. Chem. 5, 1–23.

Ahmed, A.J., Surjith, A., Kateryna, B., Mohan, V., 2017. Review on the antimicrobial properties of carbon nanostructures. Jacob Mater. 10, 1066. https://doi.org/10.3390/ma10091066.

Ajayan, P.M., Ebbesen, T.W., 1997. Nanometre-size tubes of carbon. Rep. Prog. Phys. 60, 1025.

Ali, E., Hadis, D., Hamzeh, K., Mohammad, K., Nosratollah, Z., Abolfazl, A., Mozhgan, A., Younes, H., Sang, W.J., 2014. Carbon nanotubes: properties, synthesis, purification, and medical applications. Nanoscale Res. Lett. 9, 393.

Bitounis, D., Ali-Boucetta, H., Hong, B.H., Min, D.H., Kostarelos, K., 2013. Prospects and challenges of graphene in biomedical applications. Adv. Mater. 25, 2258–2268.

Borowiak-Palen, E., Pichler, T., Liu, X., Knupfer, M., Graff, A., Jost, O., Pompe, W., Kalenczuk, R.J., Fink, J., 2002. Reduced diameter distribution of single-wall carbon nanotubes by selective oxidation. Chem. Phys. Lett. 363, 567–572.

Cañas, J.E., Long, M., Nations, S., Vadan, R., Dai, L., Luo, M., Ambikapathi, R., Lee, E.H., Olszyk, D., 2008. Effects of functionalized and nonfunctionalized single-walled carbon nanotubes on root elongation of select crop species. Environ. Toxicol. Chem. 27 (9), 1922–1931.

Chico, L., Crespi, V.H., Benedict, L.X., Louie, S.G., Cohen, M.L., 1996. Pure carbon nanoscale devices: nanotube heterojunctions. Phys. Rev. Lett. 76, 971–974.

Chung, H., Son, Y., Yoon, T.K., Kim, S., Kim, W., 2011. The effect of multi-walled carbon nanotubes on soil microbial activity. Ecotoxicol. Environ. Saf. 74, 569–575.

Dong, X., Yang, L., 2014. Inhibitory effects of single-walled carbon nanotubes on biofilm formation from *Bacillus anthracis* spores. Biofouling 30, 1165–1174.

Fabbro, A., Bosi, S., Ballerini, L., Prato, M., 2012. Carbon nanotubes: artificial nanomaterials to engineer single neurons and neuronal networks. ACS Chem. Nerosci. 3 (8), 611–618.

Guo, T., Nikolaev, P., Thess, A., Colbert, D.T., Smalley, R.E., 1995. Catalytic growth of single-walled nanotubes by laser vaporization. Chem. Phys. Lett. 243, 49–54.

He, H., Pham-Huy, L.A., Dramou, P., Xiao, D., Zuo, P., PhamHuy, C., 2013. Carbon nanotubes: applications in pharmacy and medicine. Biomed. Res. Int. 12, 1456.

He, Z.B., Maurice, J.L., Lee, C.S., Cojocaru, C.S., Pribat, D., 2010. Nickel catalyst faceting in plasma–enhanced direct current chemical vapor deposition of carbon nanofibers Arabian. J. Sci. Eng. Sect. B. 35, 19–28.

Husen, A., Siddiqi, K.S., 2014. Carbon and fullerene nanomaterials in plant system. J. Nanobiotechol. 12, 16.

Iijima, S., 1991. Helical microtubules of graphitic carbon. Nature 354, 56–58.

Jin, L., Son, Y., De Forest, J.L., Kang, Y.J., Kim, W., 2014. Single-walled carbon nanotubes alter soil microbial community composition. Sci. Total Environ. 466, 533–538.

Johnston, H.J., Hutchison, G.R., Christensen, F.M., Peters, S., Hankin, S., Aschberger, K., Stone, V., 2010. A critical review of the biological mechanisms underlying the in vivo and in vitro toxicity of carbon nanotubes: the contribution of physico–chemical characteristics. Nanotoxicology 4, 207–246.

Kappen, P., Halstead, B., Rider, A., Pigram, P.J., Brack, N., 2009. Multi-walled carbon nanotubes grown from chemical vapor: links between atomic near range order and growth parameters. J. Phys. Chem. C 113, 4307–4314.

Ken, D., Robert, A., Lang, T., Vicki, S., Rodger, D., Gavin, F., Andrew, A., 2006. Carbon nanotubes: a review of their properties in relation to pulmonary toxicology and workplace safety. Toxicol. Sci. 92, 5–22.

Kerfahi, D., Tripathi, B.M., Singh, D., Kim, H., Lee, S., 2015. Effects of functionalized and raw multi-walled carbon nanotubes on soil bacterial community composition. PLoS One 10, 1–18.

Khodakovskaya, M.V., De Silva, K., Biris, A.S., Dervishi, E., Villagarcia, H., 2012. Carbon nanotubes induce growth enhancement of tobacco cells. ACS Nano 6, 2128–2135.

Khodakovskaya, M.V., Kim, B.S., Kim, J.N., Alimohammadi, M., Dervishi, E., Mustafa, T., Cernigla, C.E., 2013. Carbon nanotubes as plant growth regulators: effects on tomato growth, reproductive system, and soil microbial community. Small 9 (1), 115–123.

Klumpp, C., Kostarelos, K., Prato, M., Bianco, A., 2006. Functionalized carbon nanotubes as emerging nanovectors for the delivery of therapeutics. Biochem. Biophys. Acta 9, 404–412.

Krishna, V.D., Wu, K., Su, D., Cheeran, M.C., Wang, J., Perez, A., 2018. Nanotechnology: review of concepts and potential application of sensing platforms in food safety. Food Microbiol. 75, 47–54.

Kroto, H.W., Heath, J.R., O'Brien, S.C., Curl, R.F., Smalley, R.E., 1985. Nature 318, 162.

Liu, Y.C., Zheng, N.N., Huang, J.D., Sun, B.M., 2011. In: Wang, C.H., Ma, L.X., Yang, W. (Eds.), Advanced Polymer Science and Engineering. Trans Tech Publications Ltd, Stafa-Zurich, pp. 99–103.

Lolla, D., Lolla, M., Abutaleb, A., Shin, H.U., Reneker, D.H., Chase, G.G., 2016. Fabrication, polarization of electrospun polyvinylidene fluoride electret fibers and effect on capturing nanoscale solid aerosols. Materials 9, 671.

Madani, S.Y., Naderi, N., Dissanayake, O., Tan, A., Seifalian, A.M., 2011. A new era of cancer treatment: carbon nanotubes as drug delivery tools. Int. J. Nanomedicine 6, 29632979.

Mariya, V., Khodakovskaya, B.K., Jong, K., Mohammad, A.M., Enkeleda, D., Thikra, M., Carl, E.C., 2013. Carbon nanotubes as plant growth regulators: effects on tomato growth, reproductive system, and soil microbial community. Small 9, 4765–4772.

Mehra, N.K., Jain, A.K., Lodhi, N., Dubey, V., Mishra, D., Raj, R., 2008. Challenges in the use of carbon nanotubes in biomedical applications. Crit. Rev. Ther. Drug Carr. Syst. 25, 169–176.

Miralles, P., Church, T.L., Harris, A.T., 2012. Toxicity, uptake, and translocation of engineered nanomaterials in vascular plants. Environ. Sci. Technol. 46 (17), 9224–9239.

Parijat, P., Mandeep, D., 2016. Carbon nanotubes: types, methods of preparation and applications. Inter. J. Pharm. Sci. Res. 1, 2455–4685.

Prášek, J., Drbohlavova, J., Chomoucka, J., Hubalek, J., Ondrej, J., Adam, V., Kizek, R., 2011. Methods for carbon nanotubes synthesis—review. J. Mater. Chem. 21, 15872–15884.

Rastogi, V., Yadav, P., Bhattacharya, S.S., Mishra, A.K., Verma, N., Verma, A., 2014. Carbon nanotubes: an emerging drug carrier for targeting cancer cells. J. Drug Del. 23.

Ratnikova, T.A., Podila, R., Rao, A.M., Taylor, A.G., 2015. Tomato seed coat permeability to selected carbon nanomaterials and enhancement of germination and seedling growth. Sci. World J 2015, 419215.

Rodrigues, D.F., Elimelech, M., 2010. Toxic effects of single-walled carbon nanotubes in the development of *E. coli* biofilm. Environ. Sci. Technol. 44, 4583–4589.

Saifuddin, N., Raziah, A.Z., Junizah, A.R., 2013. Carbon nanotubes: a review on structure and their interaction with proteins. J. Chem. 2013, 1–18.

Saito, N., Haniu, H., Usui, Y., Aoki, K., Hara, K., Takanashi, S., 2014. Safe clinical use of carbon nanotubes as innovative biomaterials. Chem. Rev. 114, 6040–6079.

Satvekar, R.K., Tiwale, B., Pawar, S., 2014. Emerging trends in medical diagnosis: a thrust on nanotechnology. Med. Chem. 4, 407–416.

Shrestha, B., Acosta-Martinez, V., Cox, S.B., Green, M.J., Li, S., 2013. An evaluation of the impact of multiwalled carbon nanotubes on soil microbial community structure and functioning. J. Hazard. Mater. 261, 188–197.

Srivastava, A., Rao, D.P., 2014. Enhancement of seed germination and plant growth of wheat, maize, peanut and garlic using. Euro. Chem. Bull. 3, 502–504.

Tanaka, H., Arima, R., Fukumori, M., Tanaka, D., Negishi, R., Kobayashi, Y., et al., 2015. Method for controlling electrical properties of single-layer graphene nanoribbons via adsorbed planar molecular nanoparticles. Sci. Rep. 5, 12341.

Tripathi, D.K., Singh, S., Singh, S., Pandey, R., Singh, V.P., Sharma, N.C., Prasad, S.M., Dubey, N.K., Chauhan, D.K., 2017. An overview on manufactured nanoparticles in plants: uptake, translocation, accumulation and phytotoxicity. Plant Physiol. Biochem. 110, 2–12.

Uc-Cayetano, E.G., Aviles, F., Cauich-Rodriguez, J.V., Schonfelder, R., Bachmatiuk, A., Rummeli, M.H., 2014. Influence of nanotube physicochemical properties on the decoration of multiwall carbon nanotubes with magnetic particles. J. Nanopart. Res. 16, 2192.
Unrau, C.J., Axelbaum, R.L., Lo, C.S., 2010. An overview of nano–particles effect on mechanical properties of composites. J. Phys. Chem. C 114, 10430–10435.
Vinciguerra, V., Buonocore, F., Panzera, G., Occhipinti, L., 2003. Growth mechanisms in chemical vapour deposited carbon nanotubes. Nanotechology 14, 655–660.
Volkova, N., Pavlovich, O., Fesenko, O., Budnyk, O., Kovalchuk, S., Goltsev, A., 2017. Studies of the influence of gold nanoparticles on characteristics of mesenchymal stem cells. J. Nanomater. 2017.
WanderWal, R.L., 2003. Carbon nanotube synthesis in a flame using laser ablation for in situ catalyst generation. Ref Type: Generic. 77, 885–889.
Wenkai, X., Xujun, L., Pengfei, M., Xian, Z., Taotao, F., Xiaotuo, L., 2018. Structure factors of carbon nanotubes on the thermal conductivity of carbon nanotube/epoxy composites. AIP Adv. 8, 035107.
Yang, C., Mamouni, J., Tang, Y., Yang, L., 2010. Antimicrobial activity of single-walled carbon nanotubes: length effect. Langmuir 26, 16013–16019.

Further reading

Anuradha, P., Sanjesh, T., Parul, P., Rachana, S., Sheo, M.P., 2019. Carbon nanotubes as plant growth regulators: impacts on growth, reproductive system, and soil microbial community. Nanomater. Plants Algae Microorganisms 13, 23–42.
Broughton, W.J., Hernandez, G., Blair, M., Beebe, S., Gepts, P., 2003. Beans (*Phaseolus* spp.) model food legumes. Plant and Soil 252, 55–128.
Jin, L., Son, Y., Yoon, T.K., Kang, Y.J., Kim, W., 2012. High concentrations of single-walled carbon nanotubes lower soil enzyme activity and microbial biomass. Ecotoxicol. Environ. Saf. 88, 9–15.
Kang, S., Mauter, M.S., Elimelech, M., 2009. Microbial cytotoxicity of carbon-based nanomaterials: implications for river water and wastewater effluent. Environ. Sci. Technol. 43, 2648–2653.
Vidu, R., Rahman, M., Mahmoudi, M., Enachescu, M., Poteca, T.D., Opris, I., 2014. Nanostructures: a platform for brain repair and augmentation. Front. Syst. Neurosci. 8, 91.
Yunhe, Z., Jing, F., Jian, L., Yanling, G., Qingwen, W., 2017. The effect of carbon nanotubes on the mechanical properties of wood plastic composites by selective laser sintering. Polymers 9, 728.

Nanocarbons: Antibacterial, antifungal, and antiviral activity and the underlying mechanism

22

Mohamed Amine Gacem[a,b], *Hiba Gacem*[c], *Aminata Ould-El-Hadj-Khelil*[a]
[a]Laboratory of Ecosystems Protection in Arid and Semi-Arid Area, University of Kasdi Merbah, Ouargla, Algeria, [b]Department of Biology, Faculty of Science, University of Amar Telidji, Laghouat, Algeria, [c]Epidemiology Service and Preventive Medicine, Hassani Abdelkader University Hospital Center, Faculty of Medicine, University of Djillali Liabes, Sidi-Bel-Abbes, Algeria

1 Introduction

Beginning in 1990, the discovery of carbon nanostructures (CNs) launched development in physics, electronics, optics, mechanics, biology, and medicine (Yang, 2015). The emergence and exponential growth in the development of chronic diseases, such as high blood pressure, cancers, neurological diseases, and diabetes, has pushed current research toward the pursuit of a new nanometric platform that could pave the way to new pharmacological, biotechnological, and medicinal applications. This new nanotechnological approach associated with nanoscience consists of delivering innovations and specific techniques for synthesizing substances that have therapeutic potential with dimensions between 0.001 and 0.1 μm (Nagaich, 2014). A wide variety of CN forms such as fullerenes, carbon nanotubes, carbon nanofibers, graphene sheets, and nanodiamonds (Fig. 1) reveal many interesting and promising properties as new nanomaterials (Aguilar, 2012).

The electrical, chemical, mechanical, and thermal properties of CNs also give them a promising potential for biotechnological and biomedical applications (Bekyarova et al., 2005). The application of these nanoparticles in medical microbiology as an antimicrobial agent delivers fundamental data about the toxicity of these materials against microorganisms. Moreover, the progress of new CNs with a multitude of structures and features is important for the uncontrolled growth of multidrug-resistant microbial strains (Maas, 2016). Other authors have revealed that these nanomaterials can be used for water disinfection and microbial control (Hossain et al., 2014; Lee et al., 2009; Li et al., 2008a), the manufacture of graphene-based paper with antimicrobial properties (Hu et al., 2010a), the manufacture of antibacterial cotton fabrics (Zhao et al., 2013), and the manufacture of beneficial products for healing (Lu et al., 2012). This activity strongly depends on the chemical nature of the surface, which will determine their toxic potential (Maas, 2016), such as a hydrophobic nature that permits these CNs to cross cell membranes (Monteiro-Riviere et al., 2005) and produce reactive oxygen species (ROS) (Pulskamp et al., 2007). According to published reviews,

Carbon Nanomaterials for Agri-food and Environmental Applications. https://doi.org/10.1016/B978-0-12-819786-8.00022-0

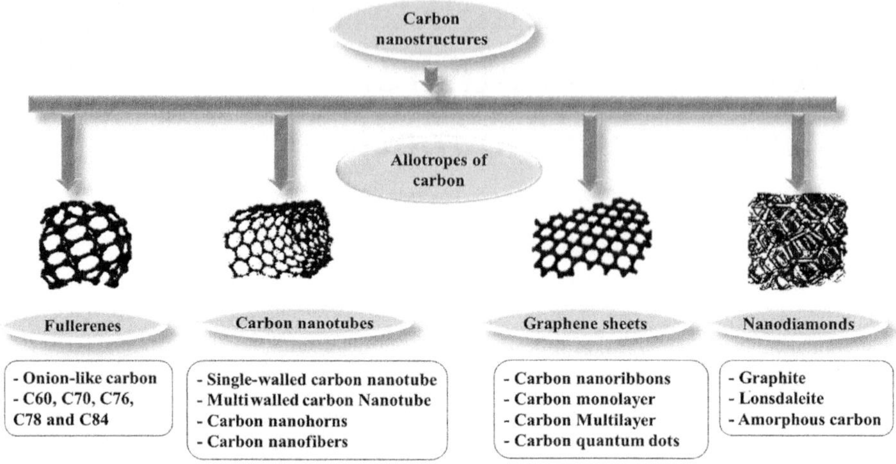

Fig. 1 Different forms of carbon nanostructures.

all CNs have moderately toxic activity for most prokaryotic cells (Jia et al., 2005; Yang et al., 2013). This chapter discusses the interest and use of CNs in medical and agricultural microbiology as microbiocidal agents, the influence of various intrinsic and extrinsic factors on the activity of these nanocomposites, and finally, the mechanism of action and toxicity of these nanoparticles against pathogenic microorganisms.

2 Antimicrobial activity of carbon nanostructures

2.1 Antibacterial, antifungal, and antiviral activities of fullerene

Since its discovery in 1985 (Spohn et al., 2009) and the development of a synthesis protocol in 1990, fullerene has been widely used due to its flexibility. This carbon compound with a spherical cage structure and an electronic cloud can act as a transfer complex. These elements produce oxygen species following their photosensation, and they can also be doped with other molecules to adapt them to the desired objectives (Lyon et al., 2005). In FE1-Mutatrade mark Mouse lung epithelial cells, fullerene is less toxic than carbon nanotubes (Jacobsen et al., 2008); these nanocomposites are attractive because of their antioxidant and antiradical properties. Furthermore, the current published data do not allow a conclusion with a definitive decision concerning fullerene toxicity (Aschberger et al., 2010). The most-studied fullerene is the buckminsterfullerene (C_{60}), which contains 60 carbon atoms with a hydrophobic character and can disperse in water as a colloidal aggregate (Maas, 2016).

The methods of preparing fullerenes have great consequence on their toxicity toward bacteria. According to Lyon and his contributors, the fullerene parts containing the smallest aggregates had greater antibacterial activity (Lyon et al., 2006), and the manufacture of fullerene by solvent also plays a role in its toxicity (Markovic et al., 2007).

Zhang and his team demonstrated that the toxicity of nC_{60} during the synthesis of unwashed THF/nC_{60} is due to a THF peroxide, that is, tetrahydro-2-(tetrahydrofuran-2-ylperoxy)furan, without ignoring the reactivity and toxicity of the nC_{60} (Zhang et al., 2009). In another work testing diverse suspensions of fullerenes (nC_{60}) with aggregates of 150 nm diameter against *Escherichia coli* AMC198, THF/nC_{60} was the most toxic, Tol/nC_{60} presented as a cryoprotectant, and Aqu-N(2)/nC_{60} and Aqu-O(2)/nC_{60} improved the metabolism of the bacteria. The toxicity of THF/nC_{60} is due to THF and THF peroxide without eliminating the nC_{60} (Dai et al., 2012). The stability of fullerene should be studied in media containing biological molecules to verify the stability of nanocomposites because their biological action can vanish with their interaction with biomolecules (Fortner et al., 2005).

The synthesis of nano-C_{60} (nC_{60}) revealed its bactericidal damage against *E. coli* and *Bacillus subtilis* with MICs of 0.5–3 $mg\,L^{-1}$ (Lyon et al., 2005). The antibacterial effect is due to an oxidative stress that is independent of ROS with oxidation of proteins and changes in the membrane potential accompanied by an interruption of cellular respiration (Lyon and Alvarez, 2008; Lyon et al., 2008). Fullerenes cause changes in the bacterial membrane but do not lead to lipid peroxidation (Lyon et al., 2008). Another study indicated that the colloidal suspension of nano-C_{60} disrupts cell function by lipid peroxidation (Sayes et al., 2005). An experimental study demonstrated that nC_{60} induced a change in the lipid membrane composition; in the presence of nC_{60}, *Pseudomonas putida* decreased unsaturated fatty acid levels and increased the proportions of cyclopropane fatty acids (Fang et al., 2007).

Decacationic fullerene C_{70} monoadducts produces $^1O_2$ and HO•; moreover, it can diffuse more easily into the walls of gram-positive bacteria, whereas for gram-negative bacteria, it is less permeable and requires greater levels of HO• to damage the wall (Huang et al., 2013). The high doses of fullerene derivatives induce an increase in O_2 uptake by the bacterial cell, which subsequently converts O_2 into H_2O_2, with the latter compound interacting with the respiratory chain (Mashino et al., 2003). The electrostatic charge plays a role in the antimicrobial dynamics of fullerenes; a study testing C_{60}, C_{60}-OH, C_{60}-COOH, and C_{60}-NH_2 against *E. coli* W3110 and *Shewanella oneidensis* MR-1 demonstrated that positively charged (C_{60}-NH_2) inhibited the development of bacteria with cell structure damage and that neutrally charged (C_{60} and C_{60}-OH) caused slight effects, whereas negatively charged (C_{60}-COOH) did not affect the microorganism growth (Tang et al., 2007). The previous publications clearly show that positively charged fullerenes are more toxic against bacterial membranes than negatively charged ones.

Fullerene is a good photocatalyst. A study indicates that ROS can be generated through fullerol under UV and polychromatic visible light (Pickering and Wiesner, 2005). Fulleropyrrolidine, with its photodynamic capacity, generates singlet oxygen and superoxide anion, causes photodamage in the *Staphylococcus aureus* envelope, and can induce a rupture to the cell membrane (Grinholc et al., 2015). The nanocomposite structure is the main cause of facilitating the absorption of light and the generation of ROS (Al-Jumaili et al., 2017). As was described previously, positively charged fullerenes possess good antimicrobial activity. Therefore, the structural change of C_{60} by iodide intensifies the antibacterial potential of fullerenes, and the mechanism may include a photo-induced reduction of electrons of (1)(C_{60}>)* or (3)(C_{60}>)* by an iodide producing I or I_2, followed by subsequent intermolecular electron transfer events of (C_{60}>)-• to produce reactive radicals (Zhang et al., 2015a).

PVP/C_{60} (C_{60} encapsulated with poly(N-vinylpyrrolidone)) and fullerol (hydroxylated C_{60}) produce $^1O_2$ and O_2^{-}* but do not produce OH*, which suggests that both nanocomposites target microorganisms that are most sensitive to superoxide or singlet oxygen (Brunet et al., 2009). Novel fullerene derivatives bearing s-triazine demonstrated good bactericidal activity against gram-negative and gram-positive bacteria (Kumar and Menon, 2009). The determination of C_{60} and C_{60}-COOH toxicity against *Phanerochaete chrysosporium* demonstrated that the carboxylated form is more toxic, as it inhibits weight gain, which destroys the fibrous structure of the mycelium, blocks enzymatic activity, and decreases the degradation capacity of wood (Ming et al., 2018a). The study of the antimycotoxigenic effect of C_{60} fullerenes at concentrations ranging from 0 to 100 ng ml^{-1} against *Aspergillus flavus* producing aflatoxin revealed that fullerene causes an antiaflatoxigenic effect in fungal cells, with an unexpected increase in aflatoxin production at 50 ng ml^{-1}. The reduction of aflatoxin B1 production is due to the antioxidant activity of the nanocomposite. The production of AFB1 at 50 ng mL^{-1} is due to the prooxidant effects of fullerene, which exceeds cellular defenses (Kovač et al., 2018). Novel cationic fullerene derivatives bearing a substituted-quinazolin-4(3H)-one showed significant activity against *Mycobacterium tuberculosis* (H37RV), *S. aureus, S. pyogenes, P. aeruginosa, K. pneumonia, E. coli, Candida albicans, Aspergillus clavatus,* and *A. niger* (Patel et al., 2013). Fullerene established good antifungal action against *Botrytis cinerea* on rose petals (Hao et al., 2017). Other conjugate fullerenes doped with peptide fragments have been synthesized but not yet tested (Barron, 2016).

The inactivation of viruses by fullerene has been suggested in several studies. Badireddy and his team demonstrated that under UV, fullerol produces singlet oxygen and superoxide; these two ROS inactivate bacteriophage MS2 (Badireddy et al., 2007). Hotze and his group confirmed the toxicity mechanism of fullerol against viruses; fullerol aggregates associate with the viruses and allow for the efficient transfer of $^1O_2$ to the capsids, which causes significant oxidative modifications to the capsid proteins and minor modifications of DNA and phospholipids. These modifications decrease the ability of bacteriophage to bind to the surface receptors of their bacterial hosts (Hotze et al., 2009); the inactivation of the bacteriophage may be due to the loss of capsid structural integrity (localized deformation or capsid rupture) and reduced ability to eject genomic RNA in the bacterium (Badireddy et al., 2012).

2.2 Antibacterial, antifungal, and antiviral activities of carbon nanotubes

Carbon nanotubes (CNTs) are hollow structures with a very precise shape; these structures are obtained by winding laminated graphene sheets. CNTs may contain a single wall (SWCNT) or multiple walls (MWCNT), and those with multiple walls are composed of several nanotubes that are maintained in a single tube; these two different structures characterize the antimicrobial activity of the CNTs (Maas, 2016).

Nanostructural films made from CNTs can be used in various biomedical applications (Haung et al., 2013). The study of the mutagenic effect of MWCNTs was revealed to be negative in *E. coli* and *S. typhimurium* (Di Sotto et al., 2009); however, high-aspect-ratio MWCNTs were found to be more lethal with indirect genotoxicity by the bias of oxidative stress or inflammation (Kim et al., 2011).

The first evidence of the CNT bactericidal effect was revealed by Kang and his team in 2007. SWCNTs show strong antimicrobial activity by direct contact with the cell membrane and cause damage that results in the death of bacterial cells (Kang et al., 2007). In 2008, the same team demonstrated that SWCNTs are much more lethal to bacteria than MWCNTs (Kang et al., 2008a). The SWCNT mechanism of action is described in the study of Vecitis in three steps: (1) the contact of SWCNTs with the membrane; (2) the disruption of the bacterial membrane; and (3) the oxidation dependence on the electronic structure (Vecitis et al., 2010). The CNT mechanism of action is also detailed in the study conducted by Chen and his team on human intestinal microbes. In their study, NTCs (SWCNT, MWCNT) and functionalized MWCNTs demonstrated antimicrobial effects. The activity is dependent on the diameters and lengths of the CNTs, which cause a lysis of walls, a release of intracellular components, and the loss of bacterial membrane potential. Moreover, thin and rigid SWCNTs show a more efficient piercing of spherical bacterial walls than MWCNTs (Chen et al., 2013). Another study testing three SWCNTs of different sizes ($<1\,\mu m$, $1–5\,\mu m$, and $\sim 5\,\mu m$) has stated that longer SWCNTs agglomerate more efficiently and show a stronger bactericidal effect while shorter SWCNTs aggregate without involving many bacterial cells (Yang et al., 2010). Another factor influencing CNT antimicrobial activity is the electronic structure, which is related to other factors such as length, diameter, and winding angle. These factors characterize the nanotube to be metallic or a semiconductor; in addition, CNTs with a metallic electronic structure are more harmful to microbe viability (Al-Jumaili et al., 2017).

The study of SWCNT action against the growth of *Bacillus anthracis* cells as well as the formation and germination of the spores demonstrated that at a concentration of $100\,\mu g.ml^{-1}$, the nanoparticles inhibit the development of *B. anthracis* by direct contact and induce damage to the cell membrane. The nanoparticles did not reduce the formation of spores, which was explained by the absence of contact with SWCNT (Aferchich et al., 2012). The treatment of *B. anthracis* biofilms by SWCNT in the early stages of their formation was shown to be more effective than treatment at later stages, as mature biofilms are highly resistant to treatment (Dong and Yang, 2014). The high adsorption and hydrophobicity of MWCNTs make these nanomaterials ideal for nisin coating (NIS); MWCNT-NIS reduced the germination and adhesiveness of *B. anthracis* spores as well as the formation of biofilms (Dong et al., 2014) (Fig. 2). Another study suggested that ordered sequential treatment with SWCNTs and H_2O_2 or NaOCl reveals effective sporicidal activity by creating more permeable spores (Lilly et al., 2012). The study by Arias and Yang tested the antibacterial activity of SWCNTs and MWCNTs with different functional groups on their surface against different gram-positive and gram-negative bacterial forms. This demonstrated that SWCNT activity is dependent on the buffer and concentration while functional groups are not a critical factor, toxicity is explained by membrane damage, and oxidative stress is

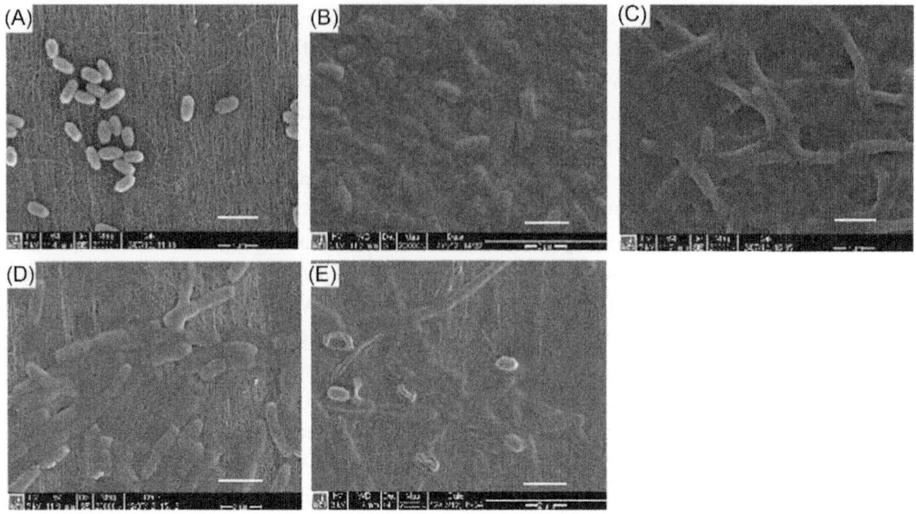

Fig. 2 Scanning electron microscopy (SEM) images of *B. anthracis* spores and biofilms on various surfaces. Spore attachment on MWCNT sheets on poly(methyl methacrylate) (PMMA-MWCNT) surface (A), The biofilm was formed by tight cells and appeared as small sticks (PMMA surfaces (B)). The biofilm was formed by tightly packed chain-forming rod cells; nisin could be the factor to cause morphology change (PMMA-nisin surface (C)). The biofilm did not completely cover the surface, and the cells appeared as chain-forming rods longer (PMMA-MWCNT surface (D)). The biofilm did not completely cover the surface (PMMA-MWCNT-nisin surface (E)). Magnification 20,000×. Scale bar: 2 μm.
Reprinted from Dong, X., McCoy, E., Zhang, M., Yang, L. 2014. Inhibitory effects of nisin-coated multi-walled carbon nanotube sheet on biofilm formation from Bacillus anthracis spores. J. Environ. Sci. 26, 2526–34.

proven by the expression of several genes related to the oxidative stress response. MWCNTs with –OH, –COOH, and $-NH_2$ surface groups did not exert any antibacterial activity (Arias and Yang, 2009).

(CNTs) are widely used in many applications because of their structure and excellent mechanical and electrical properties. The physicochemical modification of MWNTs affects their cytotoxicity against microbes (Kang et al., 2008b); for example, modified SWNTs with Ag nanoparticles presented strong bactericidal activity against *E. coli* and *S. aureus* (Kosaka et al., 2009. The antibacterial activity of freestanding polymer mats with incorporated SWCNTs demonstrated that the loss of *E. coli* correlated with the increase of SWCNT incorporation into the carpet (Schiffman and Elimelech, 2011). MWCNT-ZnO demonstrated excellent photoinactivation of *E. coli* under UV–visible light illumination. That activity is attributed to charge transfer through Zn–O–C bands formed between the oxygen atoms of the carboxylic functional groups of the CNT and the Zn atoms of the ZnO film (Akhavan et al., 2011).

The water suspension of a nanocomposite containing an electroactive polymer, polyvinyl-N-carbazole (PVK), and SWNT demonstrated good activity against the adhesion of *E. coli* and *B. subtilis* and a remarkable eliminatory power against biofilm formation (Ahmed et al., 2012). In another study, the antibiofilm effect of CNTs on *K. oxytoca, P. aeruginosa,* and *S. epidermidis* was effective (Malek et al., 2016). Improvement in CNT antimicrobial activity may be achieved by near-infrared (NIR) laser treatments. The first application of laser-activated CNTs as antimicrobial agents was performed by Kim and his team. Multipulse lasers at 532 and 1064 nm with a pulse duration of 12 ns are used for irradiation of *E. coli* K12 and photothermal SWCNT and MWCNT mixtures. The study demonstrated that irradiated CNTs have the ability to self-assemble as clusters on the surface of bacteria and cause irreparable damage (Kim et al., 2007). Previous results are confirmed by other studies that demonstrate that the bactericidal effect of NIR-treated SWCNTs tested against *Salmonella* has improved (Mamouni et al., 2011). Furthermore, the treatment of $10\,\mu g\,mL^{-1}$ SWCNT with an NIR for 20 min improved the antimicrobial effect by reducing the number of *B. anthracis* spores. In addition, the SWCNT-NIR treatment activated the germination of surviving spores with the release of dipicolinic acid (Dong et al., 2013).

CNTs are also photochemical under UV irradiation with ROS production (Chae et al., 2011). Several researchers have tried to improve CNT photoactivity by the junction of various metallic elements. TiO_2/MWNTs/Si presented excellent photocatalyst activity by destroying *E. coli* (Akhavan et al., 2009). On the other hand, branched TiO_2/CNTs revealed good photocatalytic behavior in the reduction of *C. albicans* growth; this activity is due to the generation of electron–hole pairs under visible light excitation (Darbari et al., 2011).

Concerning antifungal activity, the development of nanomaterials pioneered new drug delivery systems. Amphotericin B-doped conjugate CNTs (CNT-AMB) demonstrated very good activity against several AMBD-resistant *Candida* strains (Benincasa et al., 2011). Modified MWCNTs by functional groups in their surfaces (OH, COOH, and NH_2) exhibited an increased inhibition of the elongation and germination of *Fusarium graminearum* spores by an accumulation of MWCNTs around the spores, which became an unfavorable factor for spore growth and development (Wang et al., 2017). Another result indicated that the preexposure of nematodes to MWCNTs increased the colony formation of *C. albicans* in their bodies and suppressed innate immunity by decreasing the expression of certain antimicrobial genes (Shakoor et al., 2016).

A preliminary study has shown that CNT-based vaccines can be used in infectious disease settings (Scheinberg et al., 2013). Regarding the antiviral activity of CNTs, one study modeled three CNTs: armchair, chiral, and zigzag, which are used as ligands for three target proteins involved in HIV-mediated infection (HIV-Vpr, Nef, and Gag); demonstrated a strong binding affinity with CNTs, confirming the antagonistic effect of CNTs to the disease targets (Krishnaraj et al., 2014).

2.3 Antibacterial and antifungal activities of graphene

Graphene is a two-dimensional material that has demonstrated high crystalline and electronic quality since its sudden discovery in 2004 (Chen et al., 2010). Graphene is very hydrophobic (Tadyszak et al., 2018) while the chemical conversion or modification allowed for the development of a new approach to stabilizing these materials through electrostatic stabilization by forming aqueous colloids without the need for surfactants or stabilizing polymers (Li et al., 2008b). The exfoliation mechanism by pyrenecarboxylic acid makes it possible to obtain multilayer flakes of stable graphene in aqueous dispersions (An et al., 2010). Another type of chemically modified graphene, such as graphene oxide (GO) prepared by oxidation incorporating different oxygenated functional groups (Ibarra-Hernández et al., 2018), demonstrates more hydrophilic properties, which can be explained by the strong interaction of hydrogen bonds in the interface (Xu et al., 2018). The choice of these compounds as potential candidates in biomedicine is influenced by their physicochemical properties such as surface functional groups and structural defects (Chng and Pumera 2013). Other chemical, physical, and microbiological treatments can be performed on GO to reduce its oxygen content and to convert it to reduced graphene oxide (rGO) (Tadyszak et al., 2018).

The antibacterial and antifungal activities of graphene or other graphene-based materials have been studied widely against different bacterial and fungal strains (Athinarayanan et al., 2018; Farzanegan et al., 2018; Li et al., 2013; Ye et al., 2017). The prediction of their antimicrobial activities and mechanisms is very difficult because of their divergent properties such as the size, number of layers, morphology, oxidation, presence of other nanocomposites, hydrophilicity, etc. To understand the interaction mechanisms of these nanomaterials with microorganisms, several publications have been reviewed to provide an overview of the microbicidal activities. The comparison of the antimicrobial activity of four graphene-based materials–graphite (Gt), graphite oxide (GtO), graphene oxide (GO), and reduced graphene oxide (rGO)–under the same experimental conditions against *E. coli* demonstrated that GO has the highest antibacterial activity, followed sequentially by rGO, Gt, and GtO. The mechanism is well explained by SEM photos (Fig. 3), which demonstrated that the nanosheets of GO disrupt the cell membrane with the absence of ROS production. Moreover, the rGO and Gt nanosheets possess higher antioxidant activity compared to GO and GtO. The same study showed that the bactericidal action engendered by these three types of materials passes through three stages: (1) the deposition of microbial cells on nanomaterials, (2) the disruption of membranes, and (3) oxidation (Liu et al., 2011). Another previous study demonstrated a superior antibacterial effect of GO against *E. coli* (Hu et al., 2010b).

A study explained this underlying molecular mechanism following strong lipid-dispersion interactions with graphene nanosheets, which can penetrate cell membranes and extract large quantities of phospholipids (Tu et al., 2013). These interactions between GOs and bacteria are explained by the strong presence of radicals •C (Li et al., 2014). The nanosheets of graphene pierce the phospholipid membrane only if an appropriate geometric orientation is respected; moreover, the larger nanosheets are deposited on the membrane bilayer by damaging it, which creates a patch of returning phospholipids (Dallavalle et al., 2015). The same author demonstrated in 2018 that graphene oxide GO nanosheets penetrate the phospholipid bilayer only if the hydrophilic groups are dispersed randomly in the structure; moreover, the major reorganization of the lipids is triggered only if oxidation is located at the edge of the sheets. This lipid reorganization is characterized by lipid chains that form dynamic structures

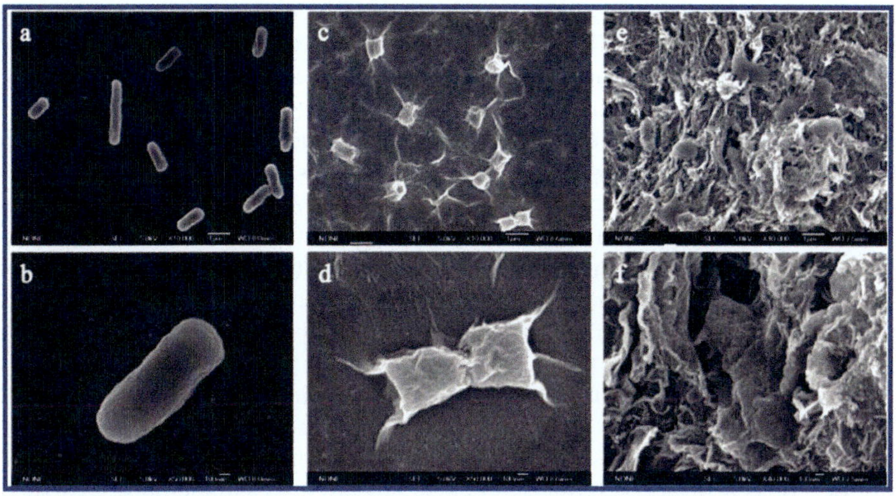

Fig. 3 SEM images of (A, B) *E. coli* after incubation with saline solution for 2 h without graphene-based materials, (C, D) *E. coli* cells after incubation with GO dispersion ($40\,\mu g\,mL^{-1}$) for 2 h, (E, F) *E. coli* cells after incubation with rGO dispersion ($40\,\mu g\,mL^{-1}$) for 2 h. *E. coli* cells become flattened and lose their cellular integrity after exposure to GO or rGO dispersions.
Reprinted from Liu, S., Zeng, T.H., Hofmann, M., Burcombe, E., Wei, J., Jiang, R., Kong, J., Chen, Y. 2011. Antibacterial activity of graphite, graphite oxide, graphene oxide, and reduced graphene oxide: membrane and oxidative stress. ACS Nano 5, 6971–80.

that resemble a swirl of water. Small GOs do not affect the overall density distribution of the hydrophobic and hydrophilic moieties of phospholipids, whereas large GOs adhere to certain phospholipids, which induces an incorrect reorganization of the membrane (Dallavalle et al., 2018). Oxidative stress and oxidation are other mechanisms of graphene toxicity (Castrillón et al., 2015; Krishnamoorthy et al., 2012). In *P. aeruginosa*, the production of $O_2^{\bullet-}$ induces oxidative stress by DNA fragmentation and the loss of cell viability (Gurunathan et al., 2012). When using graphene oxide nanosheets, the antimicrobial activity increases when the surface of the nanocomposites decreases from 0.65 to 0.01 μm^2 (Perreault et al., 2015).

The experimental bioinformatic tools and computer simulations of graphene nanosheets demonstrated that the antimicrobial activity results from the formation of pores in the bacterial wall, thus leading to an osmotic imbalance that is related to the density of graphene edges (Pham et al., 2015). The rGO is more toxic to bacteria than the unreduced-GO nanowalls, as it is attributed to the best charge transfer during the interaction between bacteria and the edges (Akhavan and Ghaderi, 2010). This load transfer was explained by Li and his team, who investigated the antibacterial activity of a large surface monolayer graphene film on a Cu conductor, a Ge semiconductor, and an SiO_2 insulator. The results demonstrated that the antimicrobial activity was surprising on Cu and Ge, whereas on SiO_2 it was limited; this activity is attributed to electron transfer away from the cell membrane (Li et al., 2014). Other treatments such as irradiation revealed that GO nanosheets irradiated under UV have more pronounced antibacterial behavior than GO nanosheets and kanamycin; UV causes functional changes in oxygen radicals in the GO surface and causes significant changes in the GO electronic structure (Veerapandian et al., 2013). Following a study on *E. coli*, GO nanosheet alignment control improves their antibacterial activities; in this study, larger GOs have higher antibacterial activity than smaller ones. These larger nanoparticles more easily cover the cells that inhibit their development once they are completely covered and cause the loss of cell viability (Liu et al., 2012). The photocatalytic activity of nanocomposites is a known phenomenon; multifunctional nanocomposites (GO-TiO_2-Ag) have demonstrated their photocatalytic activities for the degradation of phenol under solar irradiation and their photobiocidal capacity against *E. coli* relative to GO-TiO_2 and GO-Ag (Liu et al., 2013).

Resistance to natural antimicrobial agents increased in recent years following the massive use of antibiotics It is unlikely that this microorganism resistance will develop toward nanomaterials, as this last point influenced researchers to develop and improve the graphene nanocomposite as a new strategy to fight antimicrobial resistance. Graphene-based materials have attracted researchers to develop innovative hybrid systems with enhanced antimicrobial activities. Graphene-based hybrid structures proved their antibacterial activities as the graphene silver oxide nanocomposites (GO-Ag), with an average size of 9.4 ± 2.8 nm, revealed excellent antibacterial activity against methicillin-resistant strains (*S. aureus, Acinetobacter baumannii, Enterococcus faecalis,* and *E. coli*) without any detected toxicities (de Moraes et al., 2015). These results are confirmed in *E. coli* and *S. aureus*, and the activity is due to a synergistic mechanism, including the destruction of cell membranes and oxidative stress (Shao et al., 2015; Song et al., 2016). The nanocomposite GO-Ag tested against *P. aeruginosa* demonstrated good antimicrobial activity (MIC was 2.5–5.0 $\mu g\, mL^{-1}$) and antiadherence against biofilm formation (de Faria et al., 2014). Other nanocomposites such as GO-Fe_3O_4 were tested against different pathogenic strains. In particular, *Klebsiella pneumoniae, Proteus hauseri, S. aureus,* and *S. pyogenes* revealed good antimicrobial activity (Singh et al., 2018). The nanostructures synthesized from GO combined with N-alkylated poly (4-vinyl pyridine) (NPVP) and Fe_3O_4, forming nanocomposites of GO-Fe_3O_4 @ NPVP-Ag, demonstrated an excellent antibacterial property against *E. coli* and *S. aureus* (Li et al., 2018). Another nanocomposite, the graphene-tin oxide nanocomposite synthesized from the coconut shell, demonstrated good antimicrobial activity against *P. aeruginosa* compared with those synthesized from carbonized wood; this bactericidal performance is strongly related to nanomorphology and the influence of the precursor materials used (Mohan et al., 2019).

Silver nanoparticle-decorated magnetic graphene oxide (MGO–Ag) synthesized from silver nanoparticles and Fe_3O_4 demonstrated excellent antibacterial activity against *E. coli* and *S. aureus*; their mechanisms of action are essentially based on a large amount of oxygenated functional groups (the epoxy, carboxyl, and hydroxyl groups). Once the bacteria are anchored to GO surfaces, membrane integrity can be destroyed by physical pressure, oxidative stress, and contact with nano-Fe_3O_4 and nano-Ag, which leads to lipopolysaccharide degradation and increased permeability. This therefore results in the leakage of intracellular materials as well as DNA destruction. These nanoparticles can be easily separated after antibacterial treatment

(Zhang et al., 2016). The Ag-$CoFe_2O_4$-graphene oxide (Ag-$CoFe_2O_4$-GO) nanocomposite also exhibited an excellent antimicrobial activity in comparison with the $CoFe_2$O4, Ag-$CoFe_2O_4$, and $CoFe_2O_4$-GB composites due to the loss of division and cell membrane integrity (Ma et al., 2015). Ag/rGO prepared by a microwave irradiation process revealed good antimicrobial activity against *E. coli, P. aeruginosa, S. aureus, Enterococcus,* and the white rot fungus (Fan et al., 2018).

Graphene nanoplatelets decorated with zinc oxide nanorods (ZNGs) are capable of inducing significant mortality in *C. albicans* cells and affecting the main virulence factors. ROS formation is one of the mortality mechanisms (Ficociello et al., 2018), and the same result is observed with indolicidin graphene oxide (GO-IN) (Farzanegan et al., 2018). ZnGOs are also effective against the cariogenic bacterium, *Streptococcus mutans* (Zanni et al., 2016). The graphene oxide-borneol (GOB) composite demonstrates excellent antifungal activity (Li et al., 2017). GO-L-cys-Ag nanocomposites demonstrated good bactericidal activity against *E. coli* MTCC1687 and *S. aureus* MTCC3160 (Chandraker et al., 2017). GO AgNPs (TOB-GO-Ag) based on silver nanoparticles, graphene oxide, and tobramycin demonstrated excellent bactericidal activity against multidrug-resistant *E. coli* (BL21 DE3); the positively charged TOB-GO-Ag interacts with the cell membrane (negatively charged) and inhibits bacterial growth by the loss of the membrane potential from -52 ± 2 mV (control) to -2 ± 1 mV (treated) (Ullah et al., 2018). The titanium dioxide-graphene oxide (TiO_2-GO) nanocomposite possessed good photocatalytic activity for rainwater disinfection; this nanocomposite deactivated 99.5% of *E. coli* in 90 min, for which a slight decrease in photocatalytic activity is noted after 10 cycles of treatment (Wang et al., 2016). Reduced graphene oxide wrapped Pd/TiO_2 (rGO@Pd/TiO_2) showed high photocatalytic activity under visible light; this activity has not yet been tested as a photodiode against microbes (Dao et al., 2016). Other improvements that were performed in GO did not yield any significant effects, such as the combination of GO with Zn^{2+} and alginates (Frígols et al., 2019).

According to published studies, gram-positive bacteria are more sensitive to graphene sheet damage than gram-negative bacteria due to the presence of the LPS layer in the structure of gram-negative bacteria (Akhavan and Ghaderi 2010). Otherwise, some gram-negative bacteria are able to grow in the presence of graphene and are reduced under ambient conditions. In an anaerobic mixed fermentation, *E. coli* was able to reduce GO sheets, which showed inhibition of bacterial growth on their surfaces (Akhavan and Ghaderi 2012). *Shewanella* is also able to grow in the presence of graphene and to reduce it under ambient conditions (Wang et al., 2011). *P. chrysosporium* is able to oxidize rGOs by adding oxygen and decreasing carbon accordingly (Yang et al., 2019). Following these contradictory results, it is difficult to compare the results of published studies.

2.4 Antibacterial and antifungal activity of Nanodiamond, diamond-like carbon, and diamond thin films

Nanodiamonds (NDs) (charcoal, lampblack, and diamond particles) are composed of sp^3 hybridized carbon, which allows them to create dangling bonds on the surface. Nanodiamonds are compounds that are rich in oxygen species and residual graphene.

The synthesis of these compounds is based on detonation, which gives nanodiamonds a diameter of approximately 5 nm, and the grinding of large diamonds produces larger nanodiamonds; these nanocomposites form very stable aqueous dispersions with high zeta potential (Maas, 2016). Research done on nanocomposites showed that they are not toxic to eukaryotic cells (Moche et al., 2017; Paget et al., 2014; Schrand et al., 2007). In addition to this important character, NDs have the highest biocompatibility compared with that of all CNs while their surface can be derivatized easily with various functional groups (Vaijayanthimala and Chang, 2009). Their strong and stable photoluminescence, small size, large surface area, and ability to receive biomolecules make NDs attractive for various biomedical applications (Xing and Dai, 2009).

The study of the antimicrobial activity of NDs against *S. aureus* and *C. albicans* demonstrated that NDs bind closely to the surface of microorganisms without causing visible damage and cell destruction (Chwalibog et al., 2010). Another study showed that NDs can attach to *Salmonella* flagella by creating links (Sawosz et al., 2011). The size of the surface terminations has a significant influence on ND activity. The study carried out by Beranová and his group showed that *E. coli* has a higher sensitivity than *B. subtilis*. The activity is related to the concentration, size, and shape of NDs; NDs of 5 nm were the most effective against *E. coli* while NDs of 18–50 nm were effective against *B. subtilis* (Beranová et al., 2014). Regarding the chemical nature of the surfaces, NDs contain partially oxidized surfaces (carboxyl anhydride groups) and are negatively charged; this characteristic is the most active and controls the bactericidal effects by binding to proteins, which elucidates the biocompatibility of nanodiamonds (Wehling et al., 2014). On solid culture media containing *P. mirabilis*, NDs cause an appreciable reduction in the colony area and act as a barrier to lateral growth (Lišková et al., 2018); the same result was observed with *E. coli* on an agar medium (Beranová et al., 2012). Another study with NDs (reduced and oxidized) against *E. coli* in Mueller-Hinton broth (MH) demonstrated that the reduced form decreases cell viability, whereas for the oxidized form, the activity disappeared after contact with biomolecules in the nutrient media (Kromka et al., 2016).

Detonated NDs exhibit remarkable photocatalytic activity (Jang et al., 2012). The insertion of TiO_2 on the ND matrices demonstrated degradation by oxidation of two pollutants (diphenhydramine and amoxicillin) under irradiation close to UV (Pastrana-Martínez et al., 2018). However, no antibacterial action associated with the photocatalytic activity of NDs has been reported.

Diamond-like carbon (DLC) is a class of carbon coating (Maas, 2016) having a disordered assembly with a mixture of sp^2 and sp^3 hybridized carbon (Yan et al., 2002). Ultrahard DLC films are characterized by low internal compression stress, high hardness, and high deposition rates, and these DLCs possess antimicrobial activity against *E. coli* (Marciano et al., 2009a). Due to their exceptional biocompatibility, chemical inertness, and larger mechanical properties, DLCs were tested against *E. coli* ATCC 25922, *P. aeruginosa* ATCC 27853, *S. typhimurium* ATCC 14028, and *S. aureus* ATCC 25923 with a good bactericidal effect; the adhesion of bacteria to the DLC surface is influenced by bacterial species, DLC structure, chemical bonds, and hydrogen content (Marciano et al., 2011). In addition, pure DLCs cause irreparable physical damage to the outer membrane of the cells, which causes the release of intracellular content

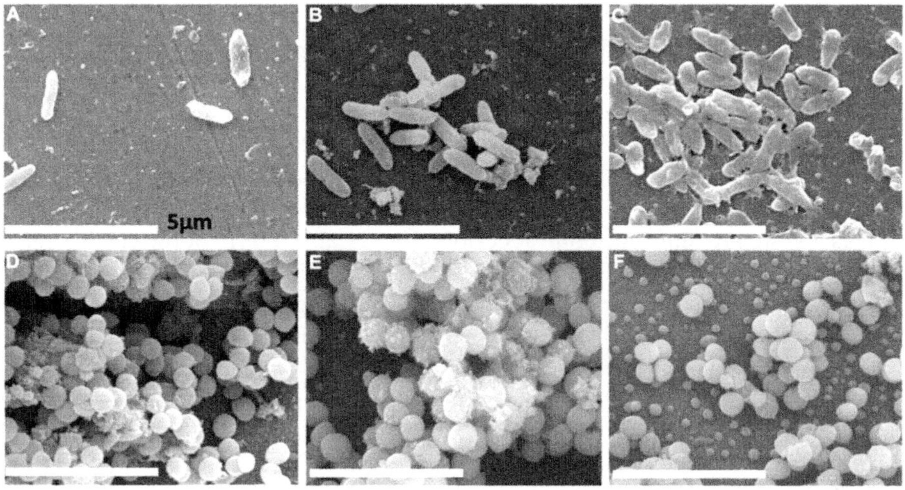

Fig. 4 SEM analysis of microbial biofilms on substrates. (A): *P. aeruginosa* on SS316, (B): *P. aeruginosa* on DLC, (C): *P. aeruginosa* on Ge-DLC, (D): *S. aureus* on SS316, (E): *S. aureus* on DLC, (F): *S. aureus* on Ge-DLC. Scale bar = 5 μm.
Reprinted from Robertson, S.N., Gibson, D., MacKay, W.G., Reid, S., Williams, C., Birney, R. 2017. Investigation of the antimicrobial properties of modified multilayer diamond-like carbon coatings on 316 stainless steel. Surf. Coat. Technol. 314, 72–78.

(Marciano et al., 2009b). A study testing the activity of DLC and DLC/germanium-doped coatings demonstrated that the last one has a bactericidal effect against gram-negative and gram-positive bacteria. Cell wall rupture and the leakage of cellular components are observed under SEM (Fig. 4) (Robertson et al., 2017). DLC and its hybrids revealed good antibiofilm property against *S. epidermidis* and *S. aureus* (Myllymaa et al., 2013); the decrease of bacterial adhesion can be explained by the free energy of the bond, which predicts whether microbial adhesion is energetically favorable (Wang et al., 2004).

Surface modification by oxygen plasma treatment causes an increase in the sp^3/sp^2 ratio due to the generation of unstable carbon bonds, which rounds the surface to make it rougher and superhydrophilic (Marciano et al., 2009c). Other studies have demonstrated that surface-free energy is an important factor in the control of antibacterial activities, as the initial fixation of bacteria is associated with the total surface energy, which suggests that the number of adhered cells decreases with the decrease in the total surface energy (Su et al., 2010). Several elements can be introduced in DLC films, which form an innovative approach to add unique functional characteristics to DLC coatings. For example, the addition of N or Si on DLC revealed a significant effect on bacterial adhesion by inhibiting the bond of *P. aeruginosa* (Liu et al., 2008). The same result is detected with silicon- and nitrogen-doped diamond-like carbon coatings (Zhao et al., 2009), and the addition of fluorine to DLC increases the adhesion of *P. aeruginosa* films to DLC surfaces (Bendavid et al., 2009). Diamond-like carbon (aC:H) with a copper dopant offered superior antibacterial activity against

E. coli (Chan et al., 2011). The copper added to the DLC gives it the capacity to effectively resist the establishment of the protein-associated conditioning layer and the further destruction of microorganisms by the prolonged release of copper ions (Liu et al., 2016). The incorporation of silver in DLC enhanced the bactericidal properties (Bociąga et al., 2016).

Si/Ag-incorporated DLC layers possess increased bactericidal activity (Swiatek et al., 2016). Si- or N-doped DLC coatings confirmed better results than pure DLC coatings by suppressing bacterial adhesion (Liu et al., 2008). Boron-incorporated diamond-like carbon (B-DLC) causes an increase in sp^3-bound carbon in films, increased hardness, and reduced surface roughness (Zhang et al., 2015b). The energy of the surface coating and the interaction energy between coatings and bacteria influences the bacterial bond (Shao et al., 2010).

Nanocrystalline diamond (NCD) coatings also have good adhesion between the coating and the underlying substrate (Narayan et al., 2006). NCD surfaces have the greatest resistance to the bacteria-binding process in comparison to titanium (Jakubowski et al., 2004) and greater antimicrobial activity than microcrystalline diamond (MCD) with mechanical hardness, oxidation and corrosion resistance, and biological compatibility. NCD surfaces may interact with bacterial membranes, which causes detorsion, the inhibition of adhesion, and the colonization of surfaces (Medina et al., 2012). This miraculous nature makes the medical implants covered by NCD more resistant with a barrier between the implant and the human environment (Mitura et al., 2006). In addition to NCD, ultrananocrystalline diamonds (UNCD) also have a remarkable antimicrobial property; this activity is attributed to surface properties such as their wettability and surface energy dominated by the dispersive component (Jelinek et al., 2013). According to Zhou's team, hydrophobicity and surface stability play important roles in antimicrobial activity (Zhou et al., 2008). It is noted that NCDs are photocatalytic; however, no bactericidal effects associated to the photocatalytic potential of NCDs have been reported thus far.

3 Management of plant diseases

The increase in agricultural production with limited land use is a challenge for the future where the world's population continues to grow. The incorporation of nanobiotechnology in this context lies in the improvement of agricultural yield through the incorporation of plant growth factors and the application of plant protection products based on nanomaterials (Fig. 5) (Zaytseva and Neumann, 2016). Based on the results of the studies presented above, carbon-based nanomaterials have an immense antimicrobial potential for the control of plant pathogens by offering an effective and unique application methodology in nanobiotechnology with less harmful impacts against the environment. These nanostructures can play a key role as insecticides against pests while their application in food production is an area under exploitation for future use (Khota et al., 2012). Due to the antifungal power of carbon-based nanomaterials, in particular, carbon nanotubes, fullerene, and graphene, these CNs have become a promising raw material for the development and innovation of nanoantifungal pesticides

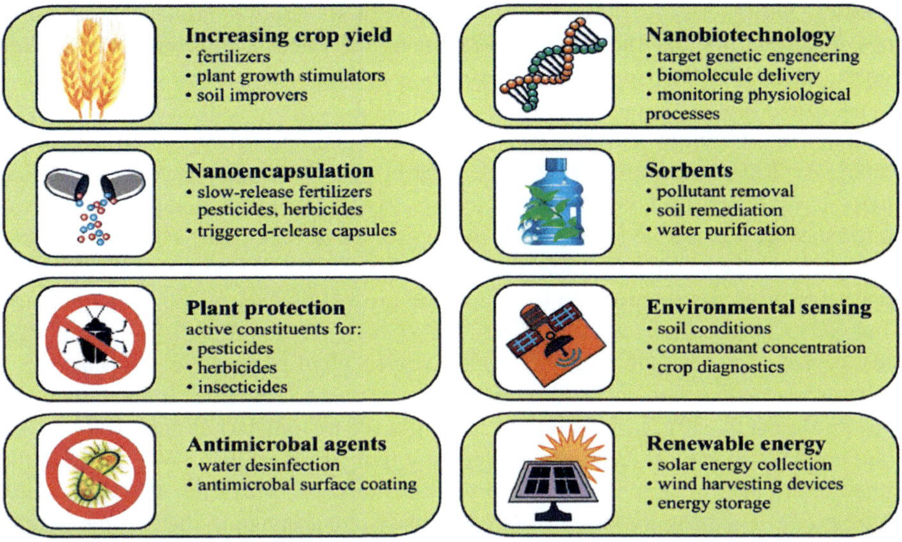

Fig. 5 Potential applications of carbon-based nanomaterials in environmental and agricultural sectors.
Reprinted from Zaytseva, O., Neumann, G. 2016. Carbon nanomaterials: production, impact on plant development, agricultural and environmental applications. Chem. Biol. Technol. Agric. 3, 17. Open Access journal.

and suitable disinfectants for agricultural purposes in order to develop sustainable agricultural conditions (Raliya et al., 2013). CNs are also cleaner, as the study by Camilli and his colleagues showed that CNs containing sulfur possess a high absorption of o-dichlorobenzene in water (Camilli et al., 2014). In general, CNs and their derivatives are materials currently known to be highly effective in the retention of contaminants such as pesticides, agricultural fertilizers, and pharmaceutical products. However, on the other hand, CNs can alter the vitality of human cells in a dose-dependent manner by penetrating the cytoplasm and locating in the nucleus (Porter et al., 2007).

The study by Wang and his team on several carbon-based nanostructures against two molds–*F. graminearum* and *F. poae*, known to be phytotoxic to agricultural crops–demonstrated that SWCNTs show the highest antifungal activities. This activity is due to the plasmolysis induction by CN deposition on the surface of the spores and inhibition of water absorption (Wang et al., 2014). Commercial MWCNTs are not highly cytotoxic for the growth and sporification of mycelium and fungal conidia of *Paecilomyces fumosoroseus* (Gorczyca et al., 2009). A study on Mazafati fresh dates demonstrated that the packaging of dates in nanocomposite low-density polyethylene films with CNT prepared by solution casting from boiling xylene prolongs shelf life, with the loss of certain sensory characteristics (Asgari et al., 2014).

Hybrid nanocarbons such as oxidized multilayer carbon nanotubes decorated with citric acid (MWCNT-*graft*-poly(citric acid)) are soluble in water and able to trap ions and pesticides such as zineb and mancozeb by forming encapsulated pesticides; these last formed compounds are very toxic to *Alternaria alternata* compared to traditional pesticides (Sarlak et al., 2014). Otherwise, between six nanomaterials tested on rose petals, fullerene demonstrated the highest fungicidal effect against *Botrytis cinerea* (Hao et al., 2017). *Podosphaera pannosa,* is a pathogen that causes powdery mildew on roses, four nanomaterials (MWCNTs, rGO, copper oxide (CuO) nanoparticles (NPs), and titanium dioxide (TiO_2) NPs) are tested to determine their antifungal potency. At 200 mg L^{-1} all NPs reduced the growth of fungi, the most important activity is exercised by CuO-NPs. The results suggest that nanomaterials increased plant resistance against *P. pannosa* by increasing the levels of the endogenous hormone, particularly zeatin riboside (Hao et al., 2019). MWCNTs decorated with zinc, copper, and silver oxide are able to inhibit the growth of *Aspergillus fumigatus* and *A. ochraceus* (Fosso-Kankeu et al., 2016). MWCNTs decorated with lysine and arginine revealed good antifungal activity against several phytopathogenic molds: *A. niger, A. fumigatus, Penicillium chrysogenum, Purpureocillium lilacinum,* and *F. culmorum* (Zare-Zardini et al., 2012). The study of carbon-based nanomaterials against *Ralstonia solanacearum*, known for its phytopathogenic activity, showed that SWCNTs have the strongest antibacterial activity, followed in descending order by GO, MWCNT, rGO, and fullerene. The primary target of CN toxicity is the damage to the cell membrane and the release of cytoplasmic material (Wang et al., 2013).

Mycotoxins are secondary metabolites produced by toxigenic molds, known in the literature by their toxicities. These metabolites are produced in fields on agricultural products and even in food processing chains, causing severe damage (Gacem and Ould El Hadj-Khelil, 2016). CNs are nanomaterials that are very attractive by their high adsorbent potential and can become a new scavenger of mycotoxins (Chen et al., 2007). Modified nanodiamonds in powder exhibited high colloidal stability in hydrosols and demonstrated rapid aflatoxin B1 adsorption (Puzyr et al., 2007). Another study confirmed the previous result and reported that positively charged NDs are effective at adsorbing ochratoxin A by electrostatic interaction while AFB_1 adsorption is directly related to aggregate size (Gibson et al., 2011). According to LC–MS/MS graphs, magnetic graphene oxide nanocomposites were shown to be able to adsorb mycotoxins (deoxynivalenol, HT-2, zearalenone, and T-2 toxin) produced by *Fusarium* and naturally present in the palm kernel cake (Pirouz et al., 2017).

Nanomaterials can be applied in other ways in plant protection and can be used as a depolluting agent in irrigation water to minimize microbial and toxic loads. The literature search conducted by Upadhyayula and his collaborators concluded that CN-based adsorption filters are effective for water treatment and pathogen removal (Upadhyayula et al., 2009), such as the adsorption of trihalomethanes by CNTs (Lu et al., 2005).

P. chrysosporium is a species that breaks down lignin. The study done by Ming and his collaborators demonstrated that pristine MWCNTs (p-MWCNTs) and oxidized MWCNTs (o-MWCNTs) are disruptors and induce an inhibition of laccase by p-MWCNT accompanied by an inhibition of wood degradation (Ming et al., 2018b). Fig. 6 demonstrates the effects of p-MWCNTs and o-MWCNTsv in *P. chrysosporium.* The implications of GO for ecological safety are also demonstrated in the study of Xie. Researchers showed that a low dose does not affect the activity of *P. chrysosporium* while high doses cause a complete inhibition (Xie et al., 2016). MWCNTs are

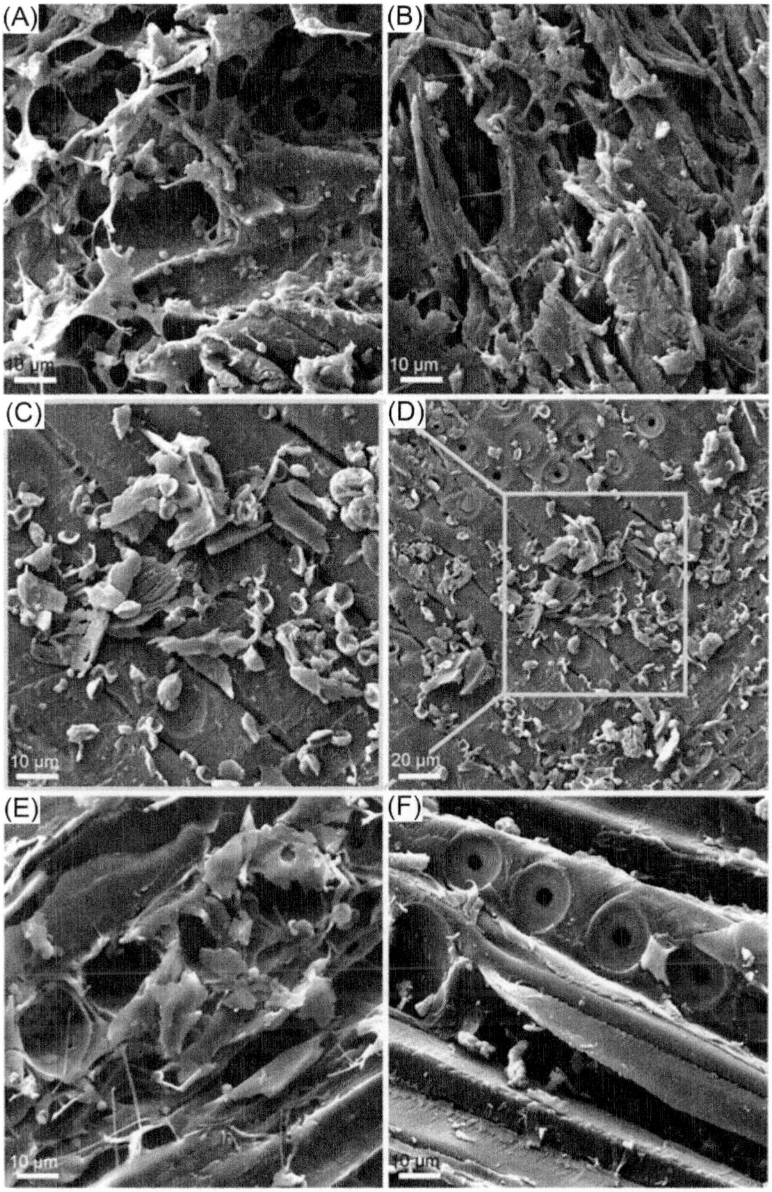

Fig. 6 Representative SEM images of sawdust before and after the degradation by *P. chrysosporium* selected from three specimens per treatment. (A) control; (B) 0.3 $\text{mg}\,\text{mL}^{-1}$ of p-MWCNTs; (C, D) 1.0 $\text{mg}\,\text{mL}^{-1}$ of p-MWCNTs; (E) 1.0 $\text{mg}\,\text{mL}^{-1}$ of o-MWCNTs; (F) nondegraded sawdust.
Reprinted from Ming, Z., Feng, S., Yilihamu, A., Yang, S., Ma, Q., Yang, H., Bai, Y., Yang, S.T. 2018b. Toxicity of carbon nanotubes to white rot fungus Phanerochaete chrysosporium. Ecotoxicol. Environ. Saf. 162, 225–234.

a stressor and modify the metabolism of *Beauveria bassiana* cells; some spores of this species placed in contact with MWCNTs undergo growth stimulation (Gorczyca et al., 2014). However, due to the unknown fate of CNs in complex environmental matrices, it is quite difficult to know the manifestation of in vitro properties of CNs already described and their impacts when they are released into the environment. This requires much deeper research in real field conditions with a wide range of crops and pathogens (Zaytseva and Neumann, 2016).

4 Conclusion and perspectives

The biotechnological development of CNs has developed in recent years as a result of their importance in biomedical medicine. This progress opens opportunities and creates a new strategy for the fight against antibiotic-resistant microbes, and recent research suggests that CNs have shown an excellent microbiocidal property against methicillin-resistant bacteria. This chapter clearly demonstrates that the potential of the antimicrobial activity of carbon-based nanomaterials depends strictly on the pretreatment of nanocomposites while their toxicity actions are based on several physical and chemical pathways. The production of ROSs generating oxidative stress against microbes is among the cytotoxic mechanisms of these nanomaterials; other toxic mechanisms are also responsible for reducing cell viability, such as cell wall damage and cytoplasmic separation. In the photocatalytic activity that characterizes certain nanomaterials, the size, shape, functionality, number of nanosheet layers, electronic configuration, and the nature and the form of the target microorganisms are all factors that influence antibacterial activity.

The dispersion of nanocomposites is a crucial parameter in the determination of antimicrobial activity and its potential under laboratory conditions. This parameter is measured in simple solutions, whereas in biological fluids, this parameter is perhaps strongly altered by the presence of biomolecules that can decrease the microbiocidal power of nanocomposites. This fundamental parameter in cytotoxicity must be tested in eukaryotic cells or even in vivo in the presence of biomolecules to determine the possible interactions between biomolecules and to evaluate toxicity. To understand how these nanostructures can prevent bacterial adhesion, biofilm formation, and the inhibition of microorganism growth, theoretical and experimental investigations and bioinformatics simulations by molecular docking need to be put in place to clearly elucidate their exact mechanisms. Following these restricted researches, the antifungal activity must be tested on a wide range of fungal strains of biomedical and agricultural interest. These antimicrobial activities must even be expanded to be tested against the most important human, animal, and plant diseases. To conclude this chapter, the promising capacity of CNs to destroy microorganisms will make these structures encouraging candidates for a wide range of applications in medical and agricultural microbiology.

References

Aferchich, K., Lilly, M., Yang, L., 2012. Effect of single-walled carbon nanotubes on *Bacillus anthracis* cell growth, sporulation, and spore germination. J. Nanosci. Nanotechnol. 12, 3821–3830.

Aguilar, Z., 2012. Nanomedical devices. In: Nanomaterials for Medical Applications. Elsevier, pp. 235–292.
Ahmed, F., Santos, C.M., Vergara, R.A., Tria, M.C., Advincula, R., Rodrigues, D.F., 2012. Antimicrobial applications of electroactive PVK-SWNT nanocomposites. Environ. Sci. Technol. 46, 1804–1810.
Akhavan, O., Abdolahad, M., Abdi, Y., Mohajerzadeh, S., 2009. Synthesis of titania/carbon nanotube heterojunction arrays for photoinactivation of *E. coli* in visible light irradiation. Carbon 47, 3280–3287.
Akhavan, O., Azimirad, R., Safa, S., 2011. Functionalized carbon nanotubes in ZnO thin films for photoinactivation of bacteria. Mater. Chem. Phys. 130, 598–602.
Akhavan, O., Ghaderi, E., 2010. Toxicity of graphene and graphene oxide nanowalls against bacteria. ACS Nano 4, 5731–5736.
Akhavan, O., Ghaderi, E., 2012. *Escherichia coli* bacteria reduce graphene oxide to bactericidal graphene in a self-limiting manner. Carbon 50, 1853–1860.
Al-Jumaili, A., Alancherry, S., Bazaka, K., Jacob, M.V., 2017. Review on the antimicrobial properties of carbon nanostructures. Materials 10, 1066.
An, X., Simmons, T., Shah, R., Wolfe, C., Lewis, K.M., Washington, M., Nayak, S.K., Talapatra, S., Kar, S., 2010. Stable aqueous dispersions of noncovalently functionalized graphene from graphite and their multifunctional high-performance applications. Nano. Lett. 10, 4295–4301.
Arias, L.R., Yang, L., 2009. Inactivation of bacterial pathogens by carbon nanotubes in suspensions. Langmuir 25, 3003–3012.
Aschberger, K., Johnston, H.J., Stone, V., Aitken, R.J., Tran, C.L., Hankin, S.M., Peters, S.A., Christensen, F.M., 2010. Review of fullerene toxicity and exposure--appraisal of a human health risk assessment, based on open literature. Regul. Toxicol. Pharmacol. 58, 455–473.
Asgari, P., Moradi, O., Tajeddin, B., 2014. The effect of nanocomposite packaging carbon nanotube base on organoleptic and fungal growth of Mazafati brand dates. Int. Nano Lett. 4, 98.
Athinarayanan, J., Periasamy, V.S., Krishnamoorthy, R., Alshatwi, A.A., 2018. Evaluation of antibacterial and cytotoxic properties of green synthesized Cu2O/Graphene nanosheets. Mater. Sci. Eng. C Mater. Biol. Appl. 93, 242–253.
Badireddy, A.R., Budarz, J.F., Chellam, S., Wiesner, M.R., 2012. Bacteriophage inactivation by UV-A illuminated fullerenes: role of nanoparticle-virus association and biological targets. Environ. Sci. Technol. 46, 5963–5970.
Badireddy, A.R., Hotze, E.M., Chellam, S., Alvarez, P., Wiesner, M.R., 2007. Inactivation of bacteriophages via photosensitization of fullerol nanoparticles. Environ. Sci. Technol. 41, 6627–6632.
Barron, A.R., 2016. Fullerene-peptides: bio-nano conjugates with structural and chemical diversity. J. Enzyme Inhib. Med. Chem. 31 (sup1), 164–176.
Bekyarova, E., Ni, Y., Malarkey, E.B., Montana, V., McWilliams, J.L., Haddon, R.C., Parpura, V., 2005. Applications of carbon nanotubes in biotechnology and biomedicine. J. Biomed. Nanotechnol. 1, 3–17.
Bendavid, A., Martin, P., Randeniya, L., Amin, M., 2009. The properties of fluorine containing diamond-like carbon films prepared by plasma-enhanced chemical vapour deposition. Diam. Relat. Mater. 18, 66–71.
Benincasa, M., Pacor, S., Wu, W., Prato, M., Bianco, A., Gennaro, R., 2011. Antifungal activity of amphotericin B conjugated to carbon nanotubes. ACS Nano 5, 199–208.
Beranová, J., Seydlová, G., Kozak, H., Benada, O., Fišer, R., Artemenko, A., Konopásek, I., Kromka, A., 2014. Sensitivity of bacteria to diamond nanoparticles of various size differs in gram-positive and gram-negative cells. FEMS Microbiol. Lett. 351, 179–186.

Beranová, J., Seydlová, G., Kozak, H., Potocký, Š., Konopásek, I., Kromka, A., 2012. Antibacterial behavior of diamond nanoparticles against *Escherichia coli*. Phys. Status Solidi B 249, 2581–2584.

Bociąga, D., Jakubowski, W., Komorowski, P., Sobczyk-Guzenda, A., Jędrzejczak, A., Batory, D., Olejnik, A., 2016. Surface characterization and biological evaluation of silver-incorporated DLC coatings fabricated by hybrid RF PACVD/MS method. Mater. Sci. Eng C. Mater. Biol. Appl. 63, 462–474.

Brunet, L., Lyon, D.Y., Hotze, E.M., Alvarez, P.J., Wiesner, M.R., 2009. Comparative photo-activity and antibacterial properties of C60 fullerenes and titanium dioxide nanoparticles. Environ. Sci. Technol. 43, 4355–4360.

Camilli, L., Pisani, C., Gautron, E., Scarselli, M., Castrucci, P., D'Orazio, F., Passacantando, M., Moscone, D., De Crescenzi, M., 2014. A three-dimensional carbon nanotube network for water treatment. Nanotechnology 25, 065701.

Castrillón, S.R.V., Perreault, F.O., De Faria, A.F., Elimelech, M., 2015. Interaction of graphene oxide with bacterial cell membranes: Insights from force spectroscopy. Environ. Sci. Technol. Lett. 2, 112–117.

Chae, S.R., Watanabe, Y., Wiesner, M.R., 2011. Comparative photochemical reactivity of spherical and tubular fullerene nanoparticles in water under ultraviolet (UV) irradiation. Water. Res. 45, 308–314.

Chan, Y.H., Huang, C.F., Ou, K.L., Peng, P.W., 2011. Mechanical properties and antibacterial activity of copper doped diamond-like carbon films. Surf. Coat. Technol. 206, 1037–1040.

Chandraker, K., Nagwanshi, R., Jadhav, S.K., Ghosh, K.K., Satnami, M.L., 2017. Antibacterial properties of amino acid functionalized silver nanoparticles decorated on graphene oxide sheets. Spectrochim. Acta A Mol. Biomol. Spectrosc. 181, 47–54.

Chen, D., Tang, L., Li, J., 2010. Graphene-based materials in electrochemistry. Chem. Soc. Rev. 39, 3157–3180.

Chen, H., Wang, B., Gao, D., Guan, M., Zheng, L., Ouyang, H., Chai, Z., Zhao, Y., Feng, W., 2013. Broad-spectrum antibacterial activity of carbon nanotubes to human gut bacteria. Small 9, 2735–2746.

Chen, W., Duan, L., Zhu, D., 2007. Adsorption of polar and nonpolar organic chemicals to carbon nanotubes. Environ. Sci. Technol. 41, 8295–8300.

Chng, E.L., Pumera, M., 2013. The toxicity of graphene oxides: dependence on the oxidative methods used. Chemistry 19, 8227–8235.

Chwalibog, A., Sawosz, E., Hotowy, A., Szeliga, J., Mitura, S., Mitura, K., Grodzik, M., Orlowski, P., Sokolowska, A., 2010. Visualization of interaction between inorganic nanoparticles and bacteria or fungi. Int. J. Nanomedicine 5, 1085–1094.

Dai, J., Wang, C., Shang, C., Graham, N., Chen, G.H., 2012. Comparison of the cytotoxic responses of *Escherichia coli* (*E. coli*) AMC 198 to different fullerene suspensions (nC60). Chemosphere 87, 362–368.

Dallavalle, M., Bottoni, A., Calvaresi, M., Zerbetto, F., 2018. Functionalization pattern of graphene oxide sheets controls entry or produces lipid turmoil in phospholipid membranes. ACS Appl. Mater. Interfaces 10, 15487–15493.

Dallavalle, M., Calvaresi, M., Bottoni, A., Melle-Franco, M., Zerbetto, F., 2015. Graphene can wreak havoc with cell membranes. ACS Appl. Mater. Interfaces 7, 4406–4414.

Dao, A.Q., Zheng, B., Liu, H., Dong, S., Thi, T.T., Fu, C., Liu, H., 2016. Facile synthesis of r-GO@Pd/TiO2 nanocomposites and its photocatalytic activity under visible light. J. Nanosci. Nanotechnol. 16, 3557–3563.

Darbari, S., Abdi, Y., Haghighi, F., Mohajerzadeh, S., Haghighi, N., 2011. Investigating the antifungal activity of TiO2 nanoparticles deposited on branched carbon nanotube arrays. J. Phys. D Appl. Phys. 44, 245401.

de Faria, A.F., Martinez, D.S., Meira, S.M., de Moraes, A.C., Brandelli, A., Filho, A.G., Alves, O.L., 2014. Anti-adhesion and antibacterial activity of silver nanoparticles supported on graphene oxide sheets. Colloids Surf. B Biointerfaces 113, 115–124.

de Moraes, A.C., Lima, B.A., de Faria, A.F., Brocchi, M., Alves, O.L., 2015. Graphene oxide-silver nanocomposite as a promising biocidal agent against methicillin-resistant *Staphylococcus aureus*. Int. J. Nanomedicine 10, 6847–6861.

Di Sotto, A., Chiaretti, M., Carru, G.A., Bellucci, S., Mazzanti, G., 2009. Multi-walled carbon nanotubes: Lack of mutagenic activity in the bacterial reverse mutation assay. Toxicol. Lett. 184, 192–197.

Dong, X., McCoy, E., Zhang, M., Yang, L., 2014. Inhibitory effects of nisin-coated multi-walled carbon nanotube sheet on biofilm formation from *Bacillus anthracis* spores. J. Environ. Sci. 26, 2526–2534.

Dong, X., Tang, Y., Wu, M., Vlahovic, B., Yang, L., 2013. Dual effects of single-walled carbon nanotubes coupled with near-infrared radiation on *Bacillus anthracis* spores: inactivates spores and stimulates the germination of surviving spores. J. Biol. Eng. 7, 19.

Dong, X., Yang, L., 2014. Inhibitory effects of single-walled carbon nanotubes on biofilm formation from *Bacillus anthracis* spores. Biofouling 30, 1165–1174.

Fan, B., Li, Y., Han, F., Su, T., Li, J., Zhang, R., 2018. Synthesis of Ag/rGO composite materials with antibacterial activities using facile and rapid microwave-assisted green route. J. Mater. Sci. Mater. Med. 29, 69.

Fang, J., Lyon, D.Y., Wiesner, M.R., Dong, J., Alvarez, P.J., 2007. Effect of a fullerene water suspension on bacterial phospholipids and membrane phase behavior. Environ. Sci. Technol. 41, 2636–2642.

Farzanegan, A., Roudbary, M., Falahati, M., Khoobi, M., Gholibegloo, E., Farahyar, S., Karimi, P., 2018. Synthesis, characterization and antifungal activity of a novel formulated nanocomposite containing Indolicidin and graphene oxide against disseminated candidiasis. J. Mycol. Med. 28, 628–636.

Ficociello, G., De Caris, M.G., Trillò, G., Cavallini, D., Sarto, M.S., Uccelletti, D., Mancini, P., 2018. Anti-candidal activity and *in vitro* cytotoxicity assessment of graphene nanoplatelets decorated with zinc oxide nanorods. Nanomaterials 8, 752.

Fortner, J.D., Lyon, D.Y., Sayes, C.M., Boyd, A.M., Falkner, J.C., Hotze, E.M., Alemany, L.B., Tao, Y.J., Guo, W., Ausman, K.D., Colvin, V.L., Hughes, J.B., 2005. C60 in water: nanocrystal formation and microbial response. Environ. Sci. Technol. 39, 4307–4316.

Fosso-Kankeu, E., De Klerk, C.M., Botha, T.A., Waanders, F., Phoku, J., Pandey, S., 2016. The antifungal activities of multi-walled carbon nanotubes decorated with silver, copper and zinc oxide particles. In: International Conference on Advances in Science, Engineering, Technology and Natural Resources (ICASETNR-16), Parys (South Africa), pp. 55–59.

Frígols, B., Martí, M., Salesa, B., Hernández-Oliver, C., Aarstad, O., Teialeret Ulset, A.S., Inger Sætrom, G., Aachmann, F.L., Serrano-Aroca, Á., 2019. Graphene oxide in zinc alginate films: Antibacterial activity, cytotoxicity, zinc release, water sorption/diffusion, wettability and opacity. PLoS ONE 14, e0212819.

Gacem, M.A., Ould El Hadj-Khelil, A., 2016. Toxicology, biosynthesis, bio-control of aflatoxin and new methods of detection. Asian Pac. J. Trop. Biomed. 6, 808–814.

Gibson, N.M., Luo, T.J., Brenner, D.W., Shenderova, O., 2011. Immobilization of mycotoxins on modified nanodiamond substrates. Biointerphases 6, 210–217.

Gorczyca, A., Kasprowicz, M.J., Lemek, T., 2009. Physiological effect of multi-walled carbon nanotubes (MWCNTs) on conidia of the entomopathogenic fungus, *Paecilomyces fumosoroseus* (*Deuteromycotina*: *Hyphomycetes*). J. Environ. Sci. Health A Tox. Hazard. Subst. Environ. Eng. 44, 1592–1597.

Gorczyca, A., Kasprowicz, M.J., Lemek, T., 2014. The physiological effects of multi-walled carbon nanotubes (MWCNTs) on conidia and the development of the entomopathogenic fungus, *Metarhizium anisopliae* (Metsch.) Sorok. J. Environ. Sci. Health A Tox. Hazard. Subst. Environ. Eng. 49, 741–752.

Grinholc, M., Nakonieczna, J., Fila, G., Taraszkiewicz, A., Kawiak, A., Szewczyk, G., Sarna, T., Lilge, L., Bielawski, K.P., 2015. Antimicrobial photodynamic therapy with fulleropyrrolidine: photoinactivation mechanism of *Staphylococcus aureus*, *in vitro* and *in vivo* studies. Appl. Microbiol. Biotechnol. 99, 4031–4043.

Gurunathan, S., Han, J.W., Dayem, A.A., Eppakayala, V., Kim, J.H., 2012. Oxidative stress-mediated antibacterial activity of graphene oxide and reduced graphene oxide in *Pseudomonas aeruginosa*. Int. J. Nanomedicine 7, 5901–5914.

Hao, Y., Cao, X., Ma, C., Zhang, Z., Zhao, N., Ali, A., Hou, T., Xiang, Z., Zhuang, J., Wu, S., Xing, B., Zhang, Z., Rui, Y., 2017. Potential Applications and antifungal activities of engineered nanomaterials against gray mold disease agent *Botrytis cinerea* on rose petals. Front. Plant Sci. 8, 1332.

Hao, Y., Fang, P., Ma, C., White, J.C., Xiang, Z., Wang, H., Zhang, Z., Rui, Y., Xing, B., 2019. Engineered nanomaterials inhibit *Podosphaera pannosa* infection on rose leaves by regulating phytohormones. Environ. Res. 170, 1–6.

Haung, C.F., Chan, Y.H., Chen, L.K., Liu, C.M., Huang, W.C., Ou, S.F., Ou, K.L., Wang, D.J., 2013. Preparation, characterization, and properties of anticoagulation and antibacterial films of carbon-based nanowires fabricated on surfaces of Ti implants. J. Electrochem. Soc. 160, H392–H397.

Hossain, F., Perales-Perez, O.J., Hwang, S., Román, F., 2014. Antimicrobial nanomaterials as water disinfectant: applications, limitations and future perspectives. Sci. Total. Environ. 466–467, 1047–1059.

Hotze, E.M., Badireddy, A.R., Chellam, S., Wiesner, M.R., 2009. Mechanisms of bacteriophage inactivation via singlet oxygen generation in UV illuminated fullerol suspensions. Environ. Sci. Technol. 43, 6639–6645.

Hu, P., Zhang, J., Li, L., Wang, Z., O'Neill, W., Estrela, P., 2010b. Carbon nanostructure-based field-effect transistors for label-free chemical/biological sensors. Sensors 10, 5133–5159.

Hu, W., Peng, C., Luo, W., Lv, M., Li, X., Li, D., Huang, Q., Fan, C., 2010a. Graphene-based antibacterial paper. ACS Nano 4, 4317–4323.

Huang, L., Wang, M., Sharma, S.K., Sperandio, F.F., Maragani, S., Nayka, S., Chang, J., Hamblin, M.R., Chiang, L.Y., 2013. Decacationic [70] Fullerene approach for efficient photokilling of infectious bacteria and cancer cells. ECS Trans. 45.

Ibarra-Hernández, A., Vega-Rios, A., Osuna, V., 2018. Synthesis of graphite oxide with different surface oxygen contents assisted microwave radiation. Nanomaterials 8, 106.

Jacobsen, N.R., Pojana, G., White, P., Møller, P., Cohn, C.A., Korsholm, K.S., Vogel, U., Marcomini, A., Loft, S., Wallin, H., 2008. Genotoxicity, cytotoxicity, and reactive oxygen species induced by single-walled carbon nanotubes and C(60) fullerenes in the FE1-Mutatrade mark Mouse lung epithelial cells. Environ. Mol. Mutagen. 49, 476–487.

Jakubowski, W., Bartosz, G., Niedzielski, P., Szymanski, W., Walkowiak, B., 2004. Nanocrystalline diamond surface is resistant to bacterial colonization. Diam. Relat. Mater. 13, 1761–1763.

Jang, D.M., Myung, Y., Im, H.S., Seo, Y.S., Cho, Y.J., Lee, C.W., Park, J., Jee, A.Y., Lee, M., 2012. Nanodiamonds as photocatalysts for reduction of water and graphene oxide. Chem. Commun. 48, 696–698.

Jelinek, M., Voss, A., Kocourek, T., Mozafari, M., Vymetalova, V., Zezulova, M., Pisarik, P., Kotzianova, A., Popov, C., Miksovsky, J., 2013. Comparison of the surface properties of DLC and ultrananocrystalline diamond films with respect to their bio-applications. Phys. Status. Solidi Appl. Mater. Sci. 210, 2106–2110.

Jia, G., Wang, H., Yan, L., Wang, X., Pei, R., Yan, T., Zhao, Y., Guo, X., 2005. Cytotoxicity of carbon nanomaterials: single-wall nanotube, multi-wall nanotube, and fullerene. Environ. Sci. Technol. 39, 1378–1383.

Kang, S., Herzberg, M., Rodrigues, D.F., Elimelech, M., 2008a. Antibacterial effects of carbon nanotubes: size does matter! Langmuir 24, 6409–6413.

Kang, S., Mauter, M.S., Elimelech, M., 2008b. Physicochemical determinants of multiwalled carbon nanotube bacterial cytotoxicity. Environ. Sci. Technol. 42, 7528–7534.

Kang, S., Pinault, M., Pfefferle, L.D., Elimelech, M., 2007. Single-walled carbon nanotubes exhibit strong antimicrobial activity. Langmuir 23, 8670–8673.

Khota, L.R., Sankarana, S., Majaa, J.M., Ehsania, R., Schuster, E.W., 2012. Applications of nanomaterials in agricultural production and crop protection: a review. Crop. Prot. 35, 64–70.

Kim, J.S., Lee, K., Lee, Y.H., Cho, H.S., Kim, K.H., Choi, K.H., Lee, S.H., Song, K.S., Kang, C.S., Yu, I.J., 2011. Aspect ratio has no effect on genotoxicity of multi-wall carbon nanotubes. Arch. Toxicol. 85, 775–786.

Kim, J.W., Shashkov, E.V., Galanzha, E.I., Kotagiri, N., Zharov, V.P., 2007. Photothermal antimicrobial nanotherapy and nanodiagnostics with self-assembling carbon nanotube clusters. Lasers Surg. Med. 39, 622–634.

Kosaka, T., Ohgami, A., Nakamura, T., Ohana, T., Ishihara, M., 2009. The preparation of Ag nanoparticle-modified single-walled carbon nanotubes and their antibacterial activity. Biocontrol Sci. 14, 133–138.

Kovač, T., Šarkanj, B., Klapec, T., Borišev, I., Kovač, M., Nevistić, A., Strelec, I., 2018. Antiaflatoxigenic effect of fullerene C60 nanoparticles at environmentally plausible concentrations. AMB Express 8, 14.

Krishnamoorthy, K., Veerapandian, M., Zhang, L.H., Yun, K., Kim, S.J., 2012. Antibacterial efficiency of graphene nanosheets against pathogenic bacteria via lipid peroxidation. J. Phys. Chem. C 116, 17280–17287.

Krishnaraj, R.N., Chandran, S., Pal, P., Berchmans, S., 2014. Investigations on the antiretroviral activity of carbon nanotubes using computational molecular approach. Comb. Chem. High Throughput Screen. 17, 531–535.

Kromka, A., Jira, J., Stenclova, P., Kriha, V., Kozak, H., Beranova, J., Vretenar, V., Skakalova, V., Rezek, B., 2016. Bacterial response to nanodiamonds and graphene oxide sheets. Phys. Status Solidi B 253, 2481–2485.

Kumar, A., Menon, S.K., 2009. Fullerene derivatized s-triazine analogues as antimicrobial agents. Eur. J. Med. Chem. 44, 2178–2183.

Lee, I., Mackeyev, Y., Cho, M., Li, D., Kim, J.H., Wilson, L.J., Alvarez, P.J., 2009. Photochemical and antimicrobial properties of novel C60 derivatives in aqueous systems. Environ. Sci. Technol. 43, 6604–6610.

Li, C., Wang, X., Chen, F., Zhang, C., Zhi, X., Wang, K., Cui, D., 2013. The antifungal activity of graphene oxide-silver nanocomposites. Biomaterials 34, 3882–3890.

Li, D., Müller, M.B., Gilje, S., Kaner, R.B., Wallace, G.G., 2008b. Processable aqueous dispersions of graphene nanosheets. Nat. Nanotechnol. 3, 101–105.

Li, G., Zhao, H., Hong, J., Quan, K., Yuan, Q., Wang, X., 2017. Antifungal graphene oxide-borneol composite. Colloids Surf. B Biointerfaces 160, 220–227.

Li, J., Wang, G., Zhu, H., Zhang, M., Zheng, X., Di, Z., Liu, X., Wang, X., 2014. Antibacterial activity of large-area monolayer graphene film manipulated by charge transfer. Sci. Rep. 4, 4359.

Li, Q., Mahendra, S., Lyon, D.Y., Brunet, L., Liga, M.V., Li, D., Alvarez, P.J., 2008a. Antimicrobial nanomaterials for water disinfection and microbial control: potential applications and implications. Water. Res. 42, 4591–4602.

Li, Q., Yong, C., Cao, W., Wang, X., Wang, L., Zhou, J., Xing, X., 2018. Fabrication of charge reversible graphene oxide-based nanocomposite with multiple antibacterial modes and magnetic recyclability. J. Colloid Interface Sci. 511, 285–295.

Lilly, M., Dong, X., McCoy, E., Yang, L., 2012. Inactivation of *Bacillus anthracis* spores by single-walled carbon nanotubes coupled with oxidizing antimicrobial chemicals. Environ. Sci. Technol. 46, 13417–13424.

Lišková, P., Beranová, J., Ukraintsev, E., Fišer, R., Kofroňová, O., Benada, O., Konopásek, I., Kromka, A., 2018. Diamond nanoparticles suppress lateral growth of bacterial colonies. Colloids Surf. B Biointerfaces 170, 544–552.

Liu, C., Zhao, Q., Liu, Y., Wang, S., Abel, E.W., 2008. Reduction of bacterial adhesion on modified DLC coatings. Colloids Surf. B Biointerfaces 61, 182–187.

Liu, L., Bai, H., Liu, J., Sun, D.D., 2013. Multifunctional graphene oxide-TiO_2-Ag nanocomposites for high performance water disinfection and decontamination under solar irradiation. J. Hazard. Mater. 261, 214–223.

Liu, S., Hu, M., Zeng, T.H., Wu, R., Jiang, R., Wei, J., Wang, L., Kong, J., Chen, Y., 2012. Lateral dimension-dependent antibacterial activity of graphene oxide sheets. Langmuir 28, 12364–12372.

Liu, S., Zeng, T.H., Hofmann, M., Burcombe, E., Wei, J., Jiang, R., Kong, J., Chen, Y., 2011. Antibacterial activity of graphite, graphite oxide, graphene oxide, and reduced graphene oxide: membrane and oxidative stress. ACS Nano 5, 6971–6980.

Liu, Y., Guo, P., He, X., Li, L., Wang, A., Li, H., 2016. Developing transparent copper-doped diamond-like carbon films for marine antifouling applications. Diam. Relat. Mater. 69, 144–151.

Lu, B., Li, T., Zhao, H., Li, X., Gao, C., Zhang, S., Xie, E., 2012. Graphene-based composite materials beneficial to wound healing. Nanoscale 4, 2978–2982.

Lu, C., Chung, Y.L., Chang, K.F., 2005. Adsorption of trihalomethanes from water with carbon nanotubes. Water. Res. 39, 1183–1189.

Lyon, D.Y., Adams, L.K., Falkner, J.C., Alvarezt, P.J., 2006. Antibacterial activity of fullerene water suspensions: effects of preparation method and particle size. Environ. Sci. Technol. 40, 4360–4366.

Lyon, D.Y., Alvarez, P.J., 2008. Fullerene water suspension (nC60) exerts antibacterial effects via ROS-independent protein oxidation. Environ. Sci. Technol. 42, 8127–8132.

Lyon, D.Y., Brunet, L., Hinkal, G.W., Wiesner, M.R., Alvarez, P.J., 2008. Antibacterial activity of fullerene water suspensions (nC60) is not due to ROS-mediated damage. Nano Lett. 8, 1539–1543.

Lyon, D.Y., Fortner, J.D., Sayes, C.M., Colvin, V.L., Hughe, J.B., 2005. Bacterial cell association and antimicrobial activity of a C60 water suspension. Environ. Toxicol. Chem. 24, 2757–2762.

Ma, S., Zhan, S., Jia, Y., Zhou, Q., 2015. Highly Efficient antibacterial and Pb(II) removal effects of Ag-CoFe2O4-GO nanocomposite. ACS Appl. Mater. Interfaces 7, 10576–10586.

Maas, M., 2016. Carbon nanomaterials as antibacterial colloids. Materials 9, 617.

Malek, I., Schaber, C., Heinlein, T., Schneider, J., Gorb, S., Schmitz, R., 2016. Vertically aligned multi walled carbon nanotubes prevent biofilm formation of medically relevant bacteria. J. Mater. Chem. B 4, 5228–5235.

Mamouni, J., Tang, Y., Wu, M., Vlahovic, B., Yang, L., 2011. Single-walled carbon nanotubes coupled with near-infrared laser for inactivation of bacterial cells. J. Nanosci. Nanotechnol. 11, 4708–4716.

Marciano, F., Bonetti, L., Santos, L., Da-Silva, N., Corat, E., Trava-Airoldi, V., 2009b. Antibacterial activity of DLC and Ag–DLC films produced by PECVD technique. Diam. Relat. Mater 18, 1010–1014.

Marciano, F.R., Bonetti, L.F., Da-Silva, N.S., Corat, E.J., Trava-Airoldi, V.J., 2009a. Diamond-like carbon films produced from high deposition rates exhibit antibacterial activity. Synth. Met. 159, 2167–2169.

Marciano, F.R., Bonetti, L.F., Da-Silva, N.S., Corat, E.J., Trava-Airoldi, V.J., 2009c. Wettability and antibacterial activity of modified diamond-like carbon films. Appl. Surf. Sci. 255, 8377–8382.

Marciano, F.R., Bonetti, L.F., Mangolin, J.F., Da-Silva, N.S., Corat, E.J., Trava-Airoldi, V.J., 2011. Investigation into the antibacterial property and bacterial adhesion of diamond-like carbon films. Vacuum 85, 662–666.

Markovic, Z., Todorovic-Markovic, B., Kleut, D., Nikolic, N., Vranjes-Djuric, S., Misirkic, M., Vucicevic, L., Janjetovic, K., Isakovic, A., Harhaji, L., Babic-Stojic, B., Dramicanin, M., Trajkovic, V., 2007. The mechanism of cell-damaging reactive oxygen generation by colloidal fullerenes. Biomaterials 28, 5437–5448.

Mashino, T., Usui, N., Okuda, K., Hirota, T., Mochizuki, M., 2003. Respiratory chain inhibition by fullerene derivatives: hydrogen peroxide production caused by fullerene derivatives and a respiratory chain system. Bioorg. Med. Chem. 11, 1433–1438.

Medina, O., Nocua, J., Mendoza, F., Gómez-Moreno, R., Ávalos, J., Rodríguez, C., Morell, G., 2012. Bactericide and bacterial anti-adhesive properties of the nanocrystalline diamond surface. Diam. Relat. Mater. 22, 77–81.

Ming, Z., Feng, S., Yilihamu, A., Ma, Q., Yang, S., Yang, S.T., 2018a. Toxicity of pristine and chemically functionalized fullerenes to white rot fungus *Phanerochaete chrysosporium*. Nanomaterials 8, 120.

Ming, Z., Feng, S., Yilihamu, A., Yang, S., Ma, Q., Yang, H., Bai, Y., Yang, S.T., 2018b. Toxicity of carbon nanotubes to white rot fungus *Phanerochaete chrysosporium*. Ecotoxicol. Environ. Saf. 162, 225–234.

Mitura, K., Niedzielski, P., Bartosz, G., Moll, J., Walkowiak, B., Pawłowska, Z., Louda, P., Kieć-Świerczyńska, M., Mitura, S., 2006. Interactions between carbon coatings and tissue. Surf. Coat. Technol. 201, 2117–2123.

Moche, H., Paget, V., Chevalier, D., Lorge, E., Claude, N., Girard, H.A., Arnault, J.C., Chevillard, S., Nesslany, F., 2017. Carboxylated nanodiamonds can be used as negative reference in *in vitro* nanogenotoxicity studies. J. Appl. Toxicol. 37, 954–961.

Mohan, A.N., Manoj, B., Panicker, S., 2019. Facile synthesis of graphene-tin oxide nanocomposite derived from agricultural waste for enhanced antibacterial activity against *Pseudomonas aeruginosa*. Sci. Rep. 9, 4170.

Monteiro-Riviere, N.A., Nemanich, R.J., Inman, A.O., Wang, Y.Y., Riviere, J.E., 2005. Multi-walled carbon nanotube interactions with human epidermal keratinocytes. Toxicol. Lett. 155, 377–384.

Myllymaa, K., Levon, J., Tiainen, V.M., Myllymaa, S., Soininen, A., Korhonen, H., Kaivosoja, E., Lappalainen, R., Konttinen, Y.T., 2013. Formation and retention of staphylococcal biofilms on DLC and its hybrids compared to metals used as biomaterials. Colloids Surf. B Biointerfaces 101, 290–297.

Nagaich, U., 2014. Nanotechnology: the vision of 2025. J. Adv. Pharm. Technol. Res. 5, 05–06.

Narayan, R.J., Wei, W., Jin, C., Andara, M., Agarwal, A., Gerhardt, R.A., Shih, C.C., Shih, C.M., Lin, S.J., Su, Y.Y., 2006. Microstructural and biological properties of nanocrystalline diamond coatings. Diam. Relat. Mater. 15, 1935–1940.

Paget, V., Sergent, J.A., Grall, R., Altmeyer-Morel, S., Girard, H.A., Petit, T., Gesset, C., Mermoux, M., Bergonzo, P., Arnault, J.C., Chevillard, S., 2014. Carboxylated nanodiamonds are neither cytotoxic nor genotoxic on liver, kidney, intestine and lung human cell lines. Nanotoxicology 1, 46–56.

Pastrana-Martínez, L.M., Morales-Torres, S., Carabineiro, S.A.C., Buijnsters, J.G., Figueiredo, J.L., Silva, A.M.T., Faria, J.L., 2018. Photocatalytic activity of functionalized nanodiamond-TiO2 composites towards water pollutants degradation under UV/Vis irradiation. Appl. Surf. Sci. 458, 839–848.

Patel, M.B., Harikrishnan, U., Valand, N.N., Modi, N.R., Menon, S.K., 2013. Novel cationic quinazolin-4(3H)-one conjugated fullerene nanoparticles as antimycobacterial and antimicrobial agents. Arch. Pharm. 346, 210–220.

Perreault, F., de Faria, A.F., Nejati, S., Elimelech, M., 2015. Antimicrobial properties of graphene oxide nanosheets: Why size matters. ACS Nano 9, 7226–7236.

Pham, V.T., Truong, V.K., Quinn, M.D., Notley, S.M., Guo, Y., Baulin, V.A., Al Kobaisi, M., Crawford, R.J., Ivanova, E.P., 2015. Graphene induces formation of pores that kill spherical and rod-shaped bacteria. ACS Nano 9, 8458–8467.

Pickering, K.D., Wiesner, M.R., 2005. Fullerol-sensitized production of reactive oxygen species in aqueous solution. Environ. Sci. Technol. 39, 1359–1365.

Pirouz, A.A., Selamat, J., Iqbal, S.Z., Mirhosseini, H., Karjiban, R.A., Bakar, F.A., 2017. The use of innovative and efficient nanocomposite (magnetic graphene oxide) for the reduction on of *Fusarium* mycotoxins in palm kernel cake. Sci. Rep. 7, 12453.

Porter, A.E., Gass, M., Muller, K., Skepper, J.N., Midgley, P.A., Welland, M., 2007. Direct imaging of single-walled carbon nanotubes in cells. Nat. Nanotechnol. 2, 713–717.

Pulskamp, K., Diabaté, S., Krug, H.F., 2007. Carbon nanotubes show no sign of acute toxicity but induce intracellular reactive oxygen species in dependence on contaminants. Toxicol. Lett. 168, 58–74.

Puzyr, A.P., Purtov, K.V., Shenderova, O.A., Luo, M., Brenner, D.W., Bondar, V.S., 2007. The adsorption of aflatoxin B1 by detonation-synthesis nanodiamonds. Dokl. Biochem. Biophys. 417, 299–301.

Raliya, R., Tarafdar, J.C., Gulecha, K., Choudhary, K., Ram, R., Mal, P., Sarna, R.P., 2013. Review article; scope of nanoscience and nanotechnology in agriculture. J. Appl. Biol. Biotechnol. 1, 041–044.

Robertson, S.N., Gibson, D., MacKay, W.G., Reid, S., Williams, C., Birney, R., 2017. Investigation of the antimicrobial properties of modified multilayer diamond-like carbon coatings on 316 stainless steel. Surf. Coat. Technol. 314, 72–78.

Sarlak, N., Taherifar, A., Salehi, F., 2014. Synthesis of nanopesticides by encapsulating pesticide nanoparticles using functionalized carbon nanotubes and application of new nanocomposite for plant disease treatment. J. Agric. Food Chem. 62, 4833–4838.

Sawosz, E., Chwalibog, A., Mitura, K., Mitura, S., Szeliga, J., Niemiec, T., Rupiewicz, M., Grodzik, M., Sokołowska, A., 2011. Visualisation of morphological interaction of diamond and silver nanoparticles with *Salmonella Enteritidis* and *Listeria monocytogenes*. J. Nanosci. Nanotechnol. 11, 7635–7641.

Sayes, C.M., Gobin, A.M., Ausman, K.D., Mendez, J., West, J.L., Colvin, V.L., 2005. Nano-C60 cytotoxicity is due to lipid peroxidation. Biomaterials 26, 7587–7595.

Scheinberg, D.A., McDevitt, M.R., Dao, T., Mulvey, J.J., Feinberg, E., Alidori, S., 2013. Carbon nanotubes as vaccine scaffolds. Adv. Drug Deliv. Rev. 65, 2016–2022.

Schiffman, J.D., Elimelech, M., 2011. Antibacterial activity of electrospun polymer mats with incorporated narrow diameter single-walled carbon nanotubes. ACS Appl. Mater. Interfaces 3, 462–468.

Schrand, A.M., Huang, H., Carlson, C., Schlager, J.J., Omacr Sawa, E., Hussain, S.M., Dai, L., 2007. Are diamond nanoparticles cytotoxic? J. Phys. Chem. B 111, 2–7.

Shakoor, S., Sun, L., Wang, D., 2016. Multi-walled carbon nanotubes enhanced fungal colonization and suppressed innate immune response to fungal infection in nematodes. Toxicol. Res. 5, 492–499.

Shao, W., Liu, X., Min, H., Dong, G., Feng, Q., Zuo, S., 2015. Preparation, characterization, and antibacterial activity of silver nanoparticle-decorated graphene oxide nanocomposite. ACS Appl. Mater. Interfaces 7, 6966–6973.

Shao, W., Zhao, Q., Abel, E.W., Bendavid, A., 2010. Influence of interaction energy between Si-doped diamond-like carbon films and bacteria on bacterial adhesion under flow conditions. J. Biomed. Mater. Res. A 93, 133–139.

Singh, R., Smitha, M.S., Karuppiah, S., Singh, S.P., 2018. Enhanced bioactivity of a GO-Fe3O4 nanocomposite against pathogenic bacterial strains. Int. J. Nanomedicine 13, 63–66.

Song, B., Zhang, C., Zeng, G., Gong, J., Chang, Y., Jiang, Y., 2016. Antibacterial properties and mechanism of graphene oxide-silver nanocomposites as bactericidal agents for water disinfection. Arch. Biochem. Biophys. 604, 167–176.

Spohn, P., Hirsch, C., Hasler, F., Bruinink, A., Krug, H.F., Wick, P., 2009. C60 fullerene: a powerful antioxidant or a damaging agent? The importance of an in-depth material characterization prior to toxicity assays. Environ. Pollut. 157, 1134–1139.

Su, X., Zhao, Q., Wang, S., Bendavid, A., 2010. Modification of diamond-like carbon coatings with fluorine to reduce biofouling adhesion. Surf. Coat. Technol. 204, 2454–2458.

Swiatek, L., Olejnik, A., Grabarczyk, J., Jedrzejczak, A., Sobczyk-Guzenda, A., Kaminska, M., Jakubowski, W., Szymanski, W., Bociaga, D., 2016. Multi-doped diamond like-carbon coatings (DLC-Si/Ag) for biomedical applications fabricated using the modified chemical vapour deposition method. Diam. Relat. Mater. 67, 54–62.

Tadyszak, K., Wychowaniec, J.K., Litowczenko, J., 2018. Biomedical applications of graphene-based structures. Nanomaterials 8, 944.

Tang, Y.J., Ashcroft, J.M., Chen, D., Min, G., Kim, C.H., Murkhejee, B., Larabell, C., Keasling, J.D., Chen, F.F., 2007. Charge-associated effects of fullerene derivatives on microbial structural integrity and central metabolism. Nano Lett. 7, 754–760.

Tu, Y., Lv, M., Xiu, P., Huynh, T., Zhang, M., Castelli, M., Liu, Z., Huang, Q., Fan, C., Fang, H., Zhou, R., 2013. Destructive extraction of phospholipids from *Escherichia coli* membranes by graphene nanosheets. Nat. Nanotechnol. 8, 594–601.

Ullah, S., Ahmad, A., Subhan, F., Jan, A., Raza, M., Khan, A.U., Rahman, A.U., Khan, U.A., Tariq, M., Yuan, Q., 2018. Tobramycin mediated silver nanospheres/graphene oxide composite for synergistic therapy of bacterial infection. J. Photochem. Photobiol. B 183, 342–348.

Upadhyayula, V.K., Deng, S., Mitchell, M.C., Smith, G.B., 2009. Application of carbon nanotube technology for removal of contaminants in drinking water: a review. Sci. Total. Environ. 408, 1–13.

Vaijayanthimala, V., Chang, H.C., 2009. Functionalized fluorescent nanodiamonds for biomedical applications. Nanomedicine 4, 47–55.

Vecitis, C.D., Zodrow, K.R., Kang, S., Elimelech, M., 2010. Electronic-structure-dependent bacterial cytotoxicity of single-walled carbon nanotubes. ACS Nano 4, 5471–5479.

Veerapandian, M., Zhang, L., Krishnamoorthy, K., Yun, K., 2013. Surface activation of graphene oxide nanosheets by ultraviolet irradiation for highly efficient anti-bacterials. Nanotechnology 24, 395706.

Wang, G., Feng, W., Zeng, X., Wang, Z., Feng, C., McCarthy, D.T., Deletic, A., Zhang, X., 2016. Highly recoverable TiO2-GO nanocomposites for stormwater disinfection. Water Res. 94, 363–370.

Wang, G., Qian, F., Saltikov, C.W., Jiao, Y., Li, Y., 2011. Microbial reduction of graphene oxide by *Shewanella*. Nano Res. 4, 563–570.

Wang, J., Huang, N., Pan, C., Kwok, S., Yang, P., Leng, Y., Chen, J., Sun, H., Wan, G., Liu, Z., 2004. Bacterial repellence from polyethylene terephthalate surface modified by acetylene plasma immersion ion implantation–deposition. Surf. Coat. Technol. 186, 299–304.

Wang, X., Liu, X., Chen, J., Han, H., Yuan, Z., 2014. Evaluation and mechanism of antifungal effects of carbon nanomaterials in controlling plant fungal pathogen. Carbon 68, 798–806.

Wang, X., Liu, X., Han, H., 2013. Evaluation of antibacterial effects of carbon nanomaterials against copper-resistant *Ralstonia solanacearum*. Colloids Surf. B Biointerfaces 103, 136–142.

Wang, X., Zhou, Z., Chen, F., 2017. Surface modification of carbon nanotubes with an enhanced antifungal activity for the control of plant fungal pathogen. Materials 10, 1375.

Wehling, J., Dringen, R., Zare, R.N., Maas, M., Rezwan, K., 2014. Bactericidal activity of partially oxidized nanodiamonds. ACS Nano 8, 6475–6483.

Xie, J., Ming, Z., Li, H., Yang, H., Yu, B., Wu, R., Liu, X., Bai, Y., Yang, S.T., 2016. Toxicity of graphene oxide to white rot fungus *Phanerochaete chrysosporium*. Chemosphere 151, 324–331.

Xing, Y., Dai, L., 2009. Nanodiamants pour la nanomédecine. Nanomedicine 4, 207–218.

Xu, K., Zhang, J., Hao, X., Zhang, C., Wei, N., Zhang, C., 2018. Wetting properties of defective graphene oxide: a molecular simulation study. Molecules 23. pii: E1439.

Yan, J.Y., Chen, F.R., Kai, J.J., 2002. Mapping of sp^2/sp^3 in DLC thin film by signal processed ESI series energy-loss image. J. Electron Microsc. 51, 391–400.

Yang, C., Mamouni, J., Tang, Y., Yang, L., 2010. Antimicrobial activity of single-walled carbon nanotubes: length effect. Langmuir 26, 16013–16019.

Yang, H., Wu, X., Ma, Q., Yilihamu, A., Yang, S., Zhang, Q., Feng, S., Yang, S.T., 2019. Fungal transformation of graphene by white rot fungus *Phanerochaete chrysosporium*. Chemosphere 216, 9–18.

Yang, K., Li, Y., Tan, X., Peng, R., Liu, Z., 2013. Behavior and toxicity of graphene and its functionalized derivatives in biological systems. Small 9, 1492–1503.

Yang, L., 2015. Carbon nanostructures: new materials for orthopedic applications. In: Yang, L. (Ed.), Nanotechnology-Enhanced Orthopedic Materials, Fabrications, Applications and Future Trends. Woodhead Publishing, pp. 97–120.

Ye, X., Feng, J., Zhang, J., Yang, X., Liao, X., Shi, Q., Tan, S., 2017. Controlled release and long-term antibacterial activity of reduced graphene oxide/quaternary ammonium salt nanocomposites prepared by noncovalent modification. Colloids Surf. B Biointerfaces 149, 322–329.

Zanni, E., Chandraiahgari, C.R., De Bellis, G., Montereali, M.R., Armiento, G., Ballirano, P., Polimeni, A., Sarto, M.S., Uccelletti, D., 2016. Zinc oxide nanorods-decorated graphene nanoplatelets: a promising antimicrobial agent against the cariogenic bacterium *Streptococcus mutans*. Nanomaterials 6, 179.

Zare-Zardini, H., Amiri, A., Shanbedi, M., Memarpoor-Yazdi, M., Asoodeh, A., 2012. Studying of antifungal activity of functionalized multiwalled carbon nanotubes by microwave-assisted technique. Surf. Interface Anal. 45, 751–755.

Zaytseva, O., Neumann, G., 2016. Carbon nanomaterials: production, impact on plant development, agricultural and environmental applications. Chem. Biol. Technol. Agric. 3, 17.

Zhang, B., Cho, M., Fortner, J.D., Lee, J., Huang, C.H., Hughes, J.B., Kim, J.H., 2009. Delineating oxidative processes of aqueous C60 preparations: role of THF peroxide. Environ. Sci. Technol. 43, 108–113.

Zhang, H.Z., Zhang, C., Zeng, G.M., Gong, J., Ou, X.M., Huan, S.Y., 2016. Easily separated silver nanoparticle-decorated magnetic graphene oxide: synthesis and high antibacterial activity. J. Colloid Interface Sci. 471, 94–102.

Zhang, L.L., Yang, Q., Tang, Y., Yang, L., Zhang, C., Hu, Y., Cui, X., 2015b. Synthesis and characterization of boron incorporated diamond-like carbon thin films. Thin Solid Films 589, 457–464.

Zhang, Y., Dai, T., Wang, M., Vecchio, D., Chiang, L.Y., Hamblin, M.R., 2015a. Potentiation of antimicrobial photodynamic inactivation mediated by a cationic fullerene by added iodide: *in vitro* and *in vivo* studies. Nanomedicine 10, 603–614.

Zhao, J., Deng, B., Lv, M., Li, J., Zhang, Y., Jiang, H., Peng, C., Li, J., Shi, J., Huang, Q., Fan, C., 2013. Graphene oxide-based antibacterial cotton fabrics. Adv. Healthc. Mater. 2, 1259–1266.

Zhao, Q., Su, X., Wang, S., Zhang, X., Navabpour, P., Teer, D., 2009. Bacterial attachment and removal properties of silicon- and nitrogen-doped diamond-like carbon coatings. Biofouling 25, 377–385.

Zhou, H., Xu, L., Ogino, A., Nagatsu, M., 2008. Investigation into the antibacterial property of carbon films. Diam. Relat. Mater. 17, 1416–1419.

Toxic and beneficial effects of carbon nanomaterials on human and animal health

23

Atef A. Hassan[a], Mogda K. Mansour[b], Rasha M.H. Sayed El Ahl[a], Ahmed M.A. El Hamaky[a], Noha H. Oraby[a]
[a]Department of Mycology and Mycotoxins, Animal Health Research Institute, Agriculture Research Center, Dokki, Egypt [b]Department of Biochemistry, Animal Health Research Institute, Agriculture Research Center, Dokki, Egypt

1 Introduction

Until now, nanotechnology has numerous benefits to human and animal health such as increasing animal productivity, which is helpful for human welfare. In addition, the worldwide application of nanotechnology in several fields related to human and animal health increases awareness about their toxicity and the adverse effects of their use. The particle size of nanomaterials ranges from 1 to 100 nm and this smaller size can help in the delivery of effective drugs. This new nanomedicine technology increases the efficacy of drug delivery and its potential in cancer treatment by increasing drug exposure to target cells (Da Silva et al., 2016; Duncan and Gaspar, 2011). In addition, nanoparticle-mediated drug delivery provides longer circulation half-lives and reduced undesirable side effects (Davis et al., 2008; Hu and Zhang, 2010; Zhang et al., 2008). Nanoparticle-based therapy can improve the balance between the efficacy and the toxicity of systemic therapeutic interventions (Peer et al., 2007).

Carbon-based nanomaterials (CBNs) such as carbon nanotubes (CNTs), graphene (G), carbon fibers, and carbon nanoparticles have huge potential in biomedicine, nanoelectronics, and mechanical engineering (De Volder et al., 2013; Gu et al., 2007; Le Croy et al., 2016; Osawa et al., 2009; Yamashita et al., 2011). Other studies reported that carbon nanomaterials can be used in lymph node imaging, disease diagnosis, drug delivery, gene therapy, physiological treatments (Li et al., 2017; Luo et al., 2014; Son et al., 2016; Wang et al., 2017; Zhang et al., 2017), and carbon-based drug delivery systems in cancer treatment (Reina et al., 2017).

In addition, carbon nanotubes are airborne and sometimes aggregate in air streams of fuel gas, contaminating the atmospheric environment and potentially reaching the lungs (Murr et al., 2004). The significance application of carbon-based nanomaterials in biological activities as DNA, protein synthesis and drug delivery were evaluated (Wang et al., 2004; Wu et al., 2004).

However, the potential toxicity of carbon nanomaterials was evaluated before their application in the environment; some studies detected the lower toxicity of carbon nanomaterials compared to carbon black or filaments (Magrez et al., 2006). The toxicity of

Carbon Nanomaterials for Agri-food and Environmental Applications. https://doi.org/10.1016/B978-0-12-819786-8.00023-2

CNTs to animal cells was suggested to be due to interruption of the membrane electron transfer, penetration of the cell envelope, oxidation of cell components, and production of secondary products such as dissolved heavy metal ions or reactive oxygen species (ROS). Several studies indicated that well-functionalized CNTs are safe for animal cells while raw CNTs or CNTs without functionalization showed severe toxicity to animal or human cells at even a moderate dosage (Khalid et al., 2016).

The progressive widespread application of carbon-based nanomaterials resulted in an elevation in their presence in the aquatic environment and increase their toxic effects and microbial pollution in aquatic animals (Jang and Hwang, 2018). Recently, several studies reported that carbon nanotubes without any additive were toxic to the lungs of mice (Lam et al., 2004; Shvedova et al., 2005; Warheit et al., 2004). The functionalization of carbon nanotubes was nontoxic to animals through its conjunction with other beneficial materials (Dumortier et al., 2006; Sayes et al., 2006; Schipper et al., 2008; Wu et al., 2008). Hence, nonfunctionalized long multiwalled carbon nanotubes (MWCNTs) may be carcinogenic to mice (Ding et al., 2005). In general, nanomedicine used several materials such as drug delivery vehicles, including liposomes, lipid micelles, polymer drug conjugates, dendrimers, protein carriers, biologically synthesized nanoparticles, and inorganic nanoparticles (Bai et al., 2018).

Therefore, this chapter was undertaken to spotlight on the effects of CBNs on humans and animals as well as ways to ameliorate their toxicity. In addition, the benefits of these materials for disease diagnosis, human health, animal reproduction, and drug delivery were fully discussed.

2 Types and synthesis of carbon-based nanomaterials

Several studies reported that carbon materials are divided into graphene (Ouyang et al., 2001), fullerenes (Kim et al., 2003), and carbon nanotubes (Chico et al., 1996); these are utilized for several purposes in the medical field. The single-walled carbon nanotubes (SWCNTs) have a single wall (Iijima and Ichihashi, 1993) while the MWCNTs have more than one wall and were detected for the first time (Iijima, 1991). Recently, different forms of carbon nanomaterials were employed to be used in human and animal health. The more widely used carbon nanomaterials include:

2.1 Fullerenes

Fullerenes are pure carbon molecules composed of at least 60 atoms of carbon. They are inert and nontoxic while, because of their size, they can interact easily with cells, proteins, and viruses. In addition, they are hollow inside so it is very easy to insert drugs into them, and they can enter efficiently into the insides of cells (Feneque, 2003).

2.2 Nanotubes (crystal form, graphite, and diamonds)

Nanotubes are a sequence of nanoscale C_{60} atoms arranged in a long thin cylindrical structure. They have potential applications in drug delivery. The antibacterial activity of carbon nanotubes is due to its ability to inter the bacterial cells membrane (Chico et al., 1996).

2.2.1 Carbon nanotubes

During the past decades, various studies illustrated carbon and its derivatives, which resulted from changes of the structures. While, Ajayan and Ebbesen (1997) and WanderWal (2003) showed that the graphene sheet can be rolled up forming concentric cylinders in one or multilayers leading to the formation of SWCNTs and MWCNTs, respectively. Also, WanderWal (2003) detected the similarity of SWCNTs and MWCNTs in properties and the latter have high tensile actions (Fig. 1; Merkoçi, 2006).

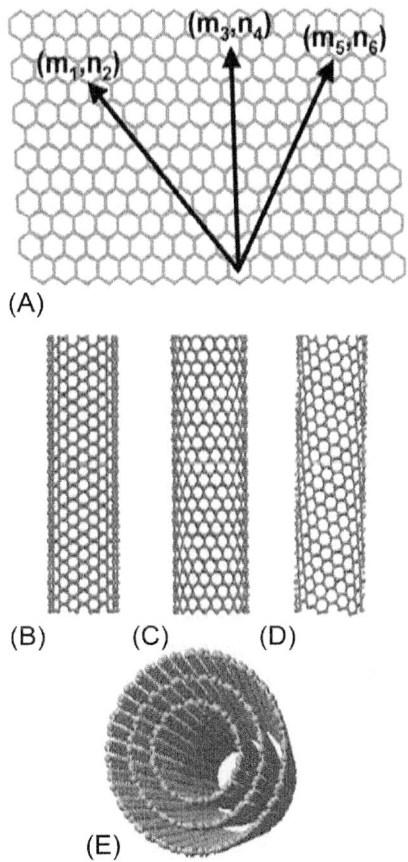

Fig. 1 Schematic of the honeycomb structure of a graphene sheet (A). SWCNTs can be formed by folding the sheet along the shown lattice vectors, leading to armchair (B), zigzag (C), and chiral (D) tubes, respectively. The graphene sheets rolled up into concentric cylinders form MWCNT (E).
From Merkoçi, A., 2006. Review: carbon nanotubes in analytical sciences. Microchim. Acta 152, 157. https://doi.org/10.1007/s00604-005-0439-z. Springer-Verlag 2005.

2.3 Synthesis of carbon nanotube

Regarding the synthesis of CNTs, various methods are reported that mainly involve gas-phase processes. These procedures included three methods. The first is chemical vapor deposition (CVD) as reported by Jose-Yacaman et al. (1993) and Abbasi et al. (2014), the second is laser ablation (Ajayan and Ebbesen, 1997; Chico et al., 1996), and the third is carbon arc discharge (Hirlekar et al., 2009; Hou et al., 2002; Thess et al., 1996).

Currently, CNTs synthesis by laser ablation methods used high temperatures, which affected control of the purity and other properties of products. However, CVD used a low temperature in synthesis, which is preferred because the nanotube length, purity, and density of CNTs can be controlled (Ganesh, 2013). Hence, the CNTs synthesis method produced CNTs that differed in impurities, and the type and amount depended on the technique used.

Recently, Eatemadi et al. (2014) evaluated different methods of synthesis CNTs to produce powders containing different fractions of CNTs. They added that other carbonaceous particles were found in prepared powder such as nanocrystalline graphite, amorphous carbon, fullerenes, and different metals that act as catalysts during synthesis. Similarly, the impurities of synthesized CNTs affected the desired properties and caused several unwanted changes in their characteristics and applications. Hence, efficient purification is urgently required during CNTs synthesis (Kruusenberg et al., 2011) and the common purification methods are based on acid treatment of synthesized CNTs (Mubarak et al., 2011).

3 Toxicity of carbon-based nanomaterials for human and animals cells

CBNs such as nanodiamonds, graphite, graphene oxide, fullerenes, carbon nanotubes, and carbon nanohorns have been widely studied and utilized for biomedical applications, such as diagnostic and therapeutic tools, due to their biocompatibility and low toxicity. Nowadays, the exposure of lung that had tumors to CBNs treated with acid enhanced severe inhibition of tumor cells proliferation resulted in cell death (Magrez et al., 2006). However, the potential health hazards related to CNT exposure were rare and require more study (Ding et al., 2005; Hurt et al., 2006).

On the other hand, graphene-based materials can provide a larger surface area and better dispensability in most solvents than carbon nanotubes. They also have attractive electronic, thermal, optical, and mechanical properties (Novoselov, 2004; Sanchez et al., 2012). Recently, graphene oxide (GO) was found to be toxic to mice by formation of a macrophage nodule in the lungs after intravenous injection at 2.1 mg/kg body weight (Liu et al., 2015). Although dextran functionalization of graphene caused a decrease in toxicity, this elevated the tolerable dose to 125 mg/kg and the toxicity signs were observed only at 250 mg/kg or higher (Kanakia et al., 2014). Parvongnukul and Lumb (1978) used a porous polytetrafluoroethylene-graphite coating for the stabilization of total hip prostheses in goats. In addition, graphite-furnace atomic absorption spectroscopy was used to determine the concentration of heavy metals and blood selenium concentrations of goats and cattle (Bischoff et al., 2010; McComb et al., 2010) as well as lead and cadmium concentrations in goat, cow, sheep, and buffalo milk (Rahimi, 2013).

Furthermore, nanographene was used for cellular imaging and drug delivery (Sun et al., 2008). Reduced graphene oxide-coated polydimethylsiloxane film was used as a good selection for hearing transmitter for high-pressure and high-frequency ultrasound generation, which is used in human and veterinary medicine to treat and diagnose diseases (Lee et al., 2015). In addition, the exposure of laboratory animals to raw nonfunctionalized CNTs via intratracheal instillation caused pulmonary toxicity, but the use of functionalized CNTs showed lower toxicity and safe effects for animals (Lam et al., 2004).

However, the acute toxicity of CNTs decreased the lead metal toxicity to *Daphnia magna* by the reduced bioavailability of the free form of lead through its adsorption on the surface of the CNTs (Jang and Hwang, 2018). Hepatotoxicity was detected in mice after intravenous injection with CNTs (Zhang et al., 2010). CNTs can be accumulated inside the animal body after long administration, but caused toxicity and oxidative stress (Yang et al., 2008). The viability of macrophages was reduced in a cell-cultured medium when CNTs or MWCNTs were added, but caused damage to cell membranes with dose- and time-dependent oxidative stress. In addition, the use of SWCNTs caused a significant decrease in viability and induction of oxidative stress. It triggers oxidative stress and is cytotoxic in cultured cell lines (Fadeel and Garcia-Bennett, 2010); toxicity is significant in clinical application (Lamprecht et al., 2009; Surrendering et al., 2009). However, the bronchial epithelial cells appeared to be much less susceptible to MWCNTs (Khaliullin et al., 2015). In another study, Lam et al. (2004) concluded that the CNTs are much more toxic than carbon black and can be more toxic than quartz for lungs, which is considered a serious occupational health hazard in chronic inhalation exposures.

Several reports detected that the carbon nanomaterial had induced apoptosis and DNA damage during testing under laboratory conditions (Asakura et al., 2010; Bottini et al., 2006). Other studies observed the occurrence of inflammation and granuloma in the lungs of laboratory animals as a result of carbon nanomaterial injection (Ma-Hock et al., 2009; Shvedova et al., 2005). MWCNT inhalation for long periods resulted in general toxicity, especially for respiratory systems (Lan et al., 2009). In addition, CNS could cause atherosclerotic lesions in the brachiocephalic artery and result in tumor formation in the abdominal cavity (Li et al., 2007; Sakamoto et al., 2009).

During the past decades, healthy and safe reproduction in animals represents huge national economic importance in animal wealth. The increased worldwide requirements of animal meats, milks, leathers, and eggs initiated several workers to evaluate the reproductive toxicity of nanomaterial (Pietroiusti et al., 2011; Yang et al., 2012a,b). However, Bai et al. (2010) reported that the administrations of CNTs caused reversible testicular damage without affecting fertility in male mice. Wei et al. (2014) detected that the intravenous injection of MWCNTs into pregnant mice could cross the placental barrier into the fetus and mainly accumulate in the fetal animal's liver, lungs, and heart. MWCNTs decreased progesterone levels and increased estradiol levels in mice serum.

The occurrence of damaged and placental dysfunctions caused delayed fetal growth and further harmed the fetal heart and brain; abortion may occur. Currently, Muller et al. (2005) evaluated MWCNTs in rats by intratracheal instillation to female rats (0.5, 2, and 5 mg per animal) and observed dose-dependent inflammation and granuloma formation. They added that the dose-related pulmonary fibrosis was diagnosed by measuring the hydroxyl proline in the lung tissue. Moreover, Mitchell et al. (2007) detected immunosuppression changes after inhalation exposure of mice to 0.3–5 mg/m^3 MWCNT for 14 days. Various studies indicated that after the injection of carbon nanomaterial in mice, it was trapped in the liver, spleen, and lungs (Wang et al., 2013; Yang et al., 2012a,b) and combined with proteins in the blood of animals (Chang et al., 2014).

On the other hand, Xie et al. (2017a) studied the toxicity of carbon nanoparticle suspension injection (CNSI) after intravenous exposure to mice. They revealed that the body weight of animals increases with no changes in hematological and biochemical parameters. The induction of higher apoptosis levels was not observed in the reticuloendothelial system (RES) organs. They added that no inflammatory cell infiltration occurred in the lung sections after injection of the CNSI, which was widely observed for CNTs and graphene (Bengtson et al., 2016; Snyder-Talkington et al., 2016; Yang et al., 2008). This was due to the high dispensability of CNSI and small particle sizes. Nevertheless, apoptosis was widely observed in the toxicity of carbon nanomaterials in vitro rather than in vivo after intravenous injection (Yang et al., 2008; Zhang et al., 2014). Furthermore, Ding et al. (2005) detected that the necrosis of human skin fibroblasts was increased by MWCNTs while T-lymphocyte apoptosis was observed by oxidized MWCNTs (Bottini et al., 2006). In addition, the oxidative stress of carbon nanomaterial was a significant toxic effect in laboratory animals and caused animal disease hazards (Shvedova et al., 2005). However, the metal catalyst used during the synthesis of CNTs should also be considered an important factor when the toxicity of carbon nanotubes is studied (Plata et al., 2008). The effects of concentration-dependent cytotoxicity of SWCNTs on a lung carcinoma cell was reported (Davoren et al., 2007). Also, Sharma et al. (2007) detected that the CNTs caused oxidative stress in rat lung cells. Furthermore, it was reported that the toxicity of CNTs depends on the density of functionalization, where lower toxicity was observed after use of functionalization with the high density of phenyl animals (Sayes et al., 2006). This action may be due to aggregation of CNTs that bind with biological materials as proteins induce less cell toxicity. When purified SWCNTs were compared with commercial CNTs, it was revealed that all the biological effects were associated with metal traces.

The significant toxicity of carbon nanoparticles to animal cells, particularly to lungs, urgently requires intensive studies for measuring the safe nontoxic doses before their field use in animal therapy.

4 Antimicrobial potential of carbon nanomaterials against bacterial and fungal cells

Several studies evaluated the effects of environmental contamination with CNTs against the activity of bacteria and fungi (Boor, 2006). There are many reasons regarding the role of toxicity by CNTs to microorganisms, including their penetration of the microbial cell envelope, the oxidation of cell components, interruption of the cell membrane, and formation of ROS or dissolved heavy metal ions (Kang et al., 2007). The antimicrobial potential of SWCNTs was investigated against bacteria, which caused the failure of bacterial film formation and caused cell death (Kang et al., 2007). Also, the SWCNTs are toxic to bacterial cells that have the ability to interrupt bacterial membranes (Liu et al., 2009). However, the toxicity of CNTs depends on their composition along with their geometry and surface functionalization. Hence, the CNT-bacteria interaction is influenced by the functionalization and length of CNTs. Numerous studies have indicated that well-functionalized, serum-stable CNTs are safe to animal cells, whereas CNTs without functionalization appeared severely toxic to human or animal cell lines at moderate dosages (Dumortier et al., 2006). In addition, the negatively charged or neutral SWCNT-functionalized aggregates were efficient in decreasing bacteria viability as compared to the positively charged SWCNTs (Arias and Yang, 2009).

Regarding the length of SWCNTs, it was detected that as length decreased, the concentration and time-dependent toxicity to bacteria increased (Yang et al., 2010). The suspension of SWCNTs in saline decreases their toxicity and interactions with bacterial cells and hence reduced cells aggregation (Kang et al., 2009). In addition, the pure SWCNTs caused lower toxicity of bacteria than SWCNTs with higher metal content due to glutathione oxidation (Vecitis et al., 2010). The higher rigidity of MWCNTs showed lower toxicity to microorganisms than SWCNT; this may be due to fewer interactions between bacteria and MWCNTs (Kang et al., 2008). In the first report about the toxicity of SWCNTs to rainbow trout, Catherine et al. (2007) exposed SWCNTs to a freshwater control and solvent control (0.1, 0.25, or 0.5 mg/L SWCNT), and administered for up to 10 days. This revealed that SWCNTs caused significant increases in Na^+/K^+-ATPase activity in the gills and intestine, but not in the brain and the total glutathione levels in the gills (28%) and livers (18%), compared to the solvent control. They added that the SWCNTs are toxic to respiratory system in trout and cellular pathological changes, which caused cell cycle defects, neurotoxicity. However, Hussain et al. (2009) studied the effect of carbon black on the bronchial epithelial cell and detected that it produced reactive oxygen species under abiotic conditions. Increased granulocyte macrophage colony stimulating factor mRNA expression and secretion were also detected. These effects of carbon nanoparticles are dependent on the surface area and amount. On the other hand, the microbial contamination of the environment and food, including mycotoxins producing fungi, has carcinogenic effects on humans and animals (Refai and Hassan, 2013). Gibson et al. (2011) used modified nanodiamond substrates in carboxylation, hydrogenation, and hydroxylation for immobilization of mycotoxins such as aflatoxin B_1 (AFB_1) and ochratoxin A (OTA). Carbon nanotubes and graphene have antibacterial potential throughout penetration of bacterial cell membranes (Fernandez-Lopez et al., 2001) as they inhibited the growth of *Listeria monocytogenes* and Salmonella species (Ye et al., 2015). In addition, the functionalized (SWCNTs) were selected as a drug carrier to carry an antiviral drug for the treatment of viral diseases. Particularly, the SWCNTs have potential antiviral activity against the reovirus (Gurunathan et al., 2013). Similarly, Ye et al. (2015) detected the antiviral activity of GO against pseudo rabies virus DNA and the porcine epidemic diarrhea virus.

Furthermore, other CNTs (fullerenes) and their derivatives showed antiviral activity against HIV (Friedman et al., 1993; Sijbesma et al., 1993), Semliki forest virus (SFV, Togaviridae), or vesicular stomatitis virus (VSV, Rhabdoviridae) (Kaesermann and Kempf, 1997). Recently, Gunawardana et al. (2015) regarded that the unformulated oligodeoxynucleotides (ODN) containing CpG motifs (CpG-ODN) stimulated the innate immune system of animals against a variety of bacterial, viral, and protozoan infections. They added that this material was able to significantly protect neonatal broiler chickens against *Escherichia coli* or Salmonella Typhimurium infections. In another study, Khaliullin et al. (2015) investigated the toxic effects of SWCNTs and MWCNTs on the cultures of macrophages and bronchial epithelial cells. The results revealed that macrophage viability was not affected after the addition of MWCNTs and caused damage to cell membranes with dose- and time-dependent oxidative stress. SWCNTs caused a significant decrease in cell viability as well as induction of oxidative stress and reduction in glutathione. Bronchial epithelial cells appeared to be much less susceptible to MWCNTs. Recently, Meena et al. (2018) evaluated the application of nanotechnology in human and veterinary medicine and revealed major benefits for animal health and production. Carbon nanotubes could be used as sensors, antimicrobial agents, cancer therapy, and drug delivery. As the first step in this direction, the authors of this chapter evaluated the antimicrobial potential of CNTs against *Candida albicans*, *Aspergillus flavus*, *E. coli*, and *Staphylococcus aureus* and prevention of aflatoxin production in feed by toxigenic *A. flavus*. The results revealed that the viability of microbial cells decreased as the concentration of used CNTs increased until the final death of microbial cells and complete prevention of aflatoxin production at a level of 125 μg/mL of CNTs. The scanning electron microscope images of the treated microbial cells showed destruction of the constituents of cell membranes and cytoplasm and death of cells (Figs. 2–4).

5 Beneficial application of carbon nanomaterial for animal health and production

Until now, nanotechnology yielded huge benefits for human and animal health as a synthesis of new materials and tools that helps increase animal production (Bai et al., 2018). Several applications of material nanoparticles were detected in various studies such as human and animal disease diagnosis; treatment; drug delivery; animal nutrition, breeding, and reproduction; valuable additives to animal products; and food safety. Recently, nanotechnology potentiated the production of nontoxic antimicrobial agents against various bacterial and fungal pathogens that caused chronic infections in livestock, including fungi such as *A. flavus*, *A. ochraceus* and bacteria as *S. aureus*, and *E. coli* (Hassan et al., 2015, 2016a) as well as *Brucella*, *Mycobacterium bovis*, *S. aureus*, and *Rhodococcus equi* (Muktar et al., 2015). As the first step in the degradation of mycotoxins by metal nanomaterials, Hassan et al. (2016b, 2017) successfully eliminated the toxic carcinogenic effects of aflatoxins in rabbits by the addition of zinc nanoparticles in feed. The serious changes due to aflatoxicosis such as chemical parameters and carcinogenic changes on liver cells of rabbits were improved by

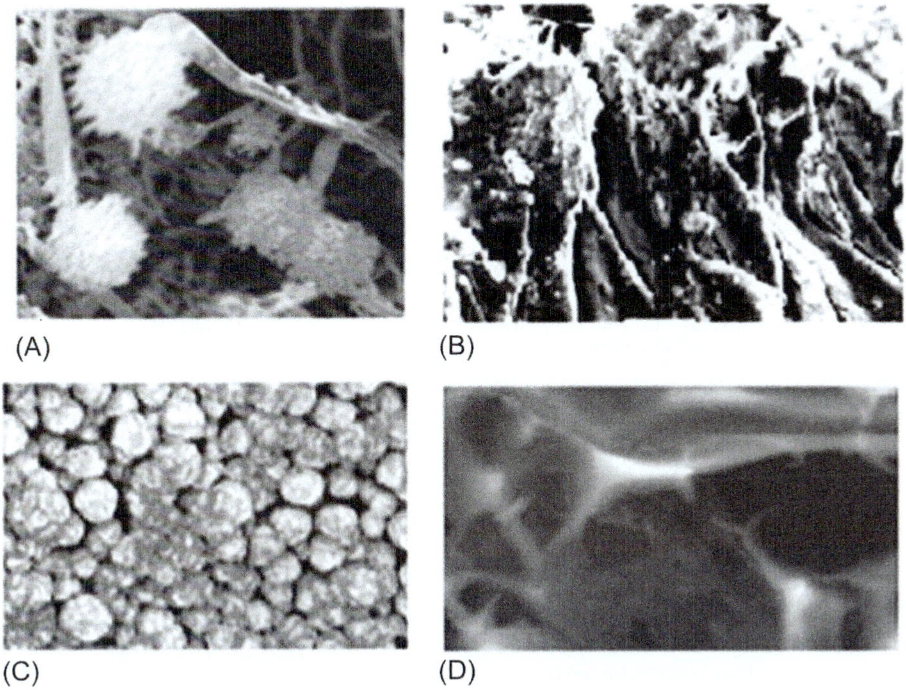

Fig. 2 Scanning electron microscope of fungal spores treated with CNTs: (1) *A. flavus* (A) before and (B) after treatment (2) *C. albicans* and (C) before (D) after treatment.

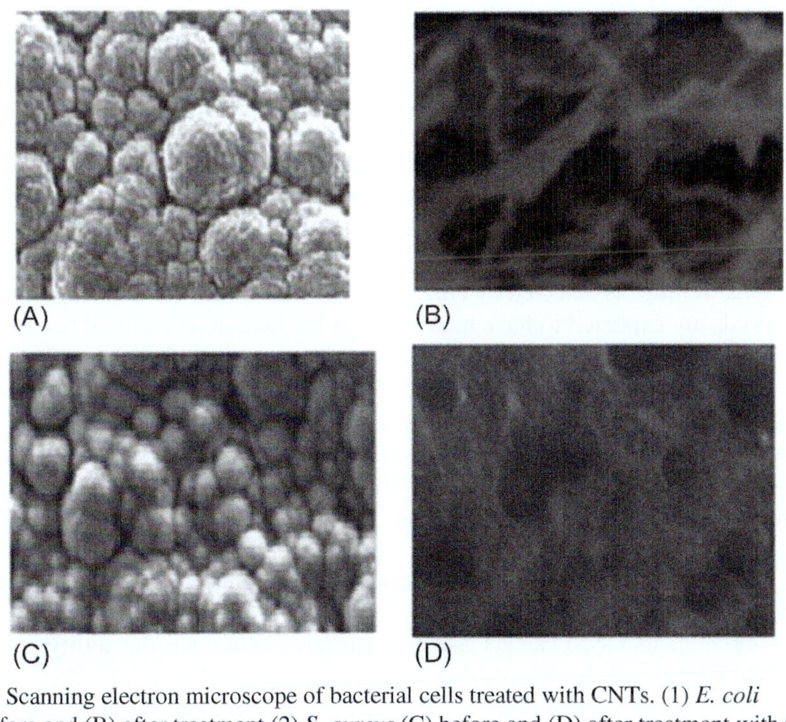

Fig. 3 Scanning electron microscope of bacterial cells treated with CNTs. (1) *E. coli* (A) before and (B) after treatment (2) *S. aureus* (C) before and (D) after treatment with CNTs.

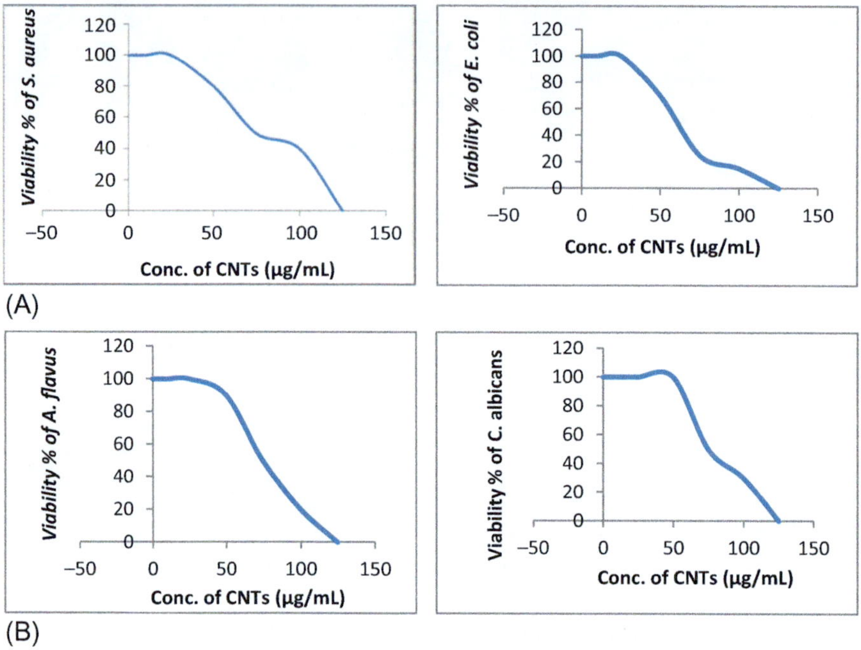

Fig. 4 (A) Changes in viability % of bacterial cell at different conc. of CNTs (*S. aureus* and *E. coli*). (B) Changes in viability % of fungal cell at different conc. of CNTs (*A. flavus* and *C. albicans*).

the addition of zinc nanomaterials in the feed. In addition, the nanobiosensors could be used for reproductive management, such as the detection of estrus, hormone levels, and metabolite profiles (Monerris et al., 2012; Sagadevan and Periasamy, 2014). Also, nanomaterials have the ability to preserve gonadal tissues, sperm, oocytes, and embryos (Saragusty and Arav, 2011). Quantum dots improved the understanding of mammalian spermatozoon and oocyte movement and their interactions in a different physiological setting, as detected by Hill and Li (2017). On the other hand, the carbon nanomaterials are expected to have huge benefits for human and animal health in the future by their functionalization with other nontoxic materials. Currently, CNSI were applied in clinical treatments as an injection during oncological surgery (Xie et al., 2017b). In addition, the CNSI could be used for lymphatic mapping and parathyroid gland and lymph node tumor imaging, such as in gastric cancer (Gu et al., 2015; Li et al., 2016; Wu et al., 2015; Zhu et al., 2016) and breast cancer (Wu et al., 2015). Zhu et al. (2016) successfully used CNSI in lymph node dissection in the diagnosis of thyroid carcinoma. In addition, the use of CNSI in dissection surgery of thyroid carcinoma caused their staining which potentiated avoiding false resection of this gland (Gu et al., 2015).

On the other hand, CNSI can act as a drug carrier, where it has the ability to adsorb drugs as in the case of epirubicin and doxorubicin drugs (Xie et al., 2016), which are used in lymphatic chemotherapy (Yang et al., 2012a,b). Furthermore, CNSI might include photothermal therapy, gene delivery, and use as an immune adjuvant (Liu et al., 2013; Xie et al., 2017b,c). The quantum dots of functionalized carbon nanoparticles with polyethylene glucose (PEG) were nontoxic to mice after intravenous exposure (Yang et al., 2009). Another study detected that the toxicity of CNTs depends on the type of functionalization, aggregation behavior, and the presence of metal catalyst particles leftover during synthesis (Khalid et al., 2016). Therefore, it is suggested that SWCNTs and MWCNTs could be functionalized and combined into bioactive molecules such as carbohydrates, amino acids, peptides, nucleic acids, and proteins to be beneficial in diverse biological applications (Endo et al., 2008). The carbohydrate-functionalized carbon nanotubes have already been used for the recognition of pathogenic microorganisms such as *E. coli*, *C. albicans*, *A. flavus* (Elkin et al., 2005), and *Bacillus anthracis* (Wang et al., 2008). Cells were viable upon internalization of an individually encapsulated DNA-wrapped SWNT complex (Kam et al., 2005). Also, CNTs have the ability to detect particular DNA sequences (Tu et al., 2009), cancer biomarkers (Thakare et al., 2010), and viruses (Patolsky et al., 2004). In addition, they have huge potential in energy generation and storage (Kim et al., 2006), Li-ion batteries (Leroux et al., 1999), solar cells (Landi et al., 2005), and fuel cells (Wang et al., 2004). Carbon nanomaterial toxicity is low after intravenous injection and the properties of carbon nanomaterials regulate their biosafety (Chang et al., 2014; Yang et al., 2012a,b). Therefore, the advanced future benefits of functionalized carbon nanomaterials are expected for clinical applications in human and animal health. In addition, nanotechnology could help in the production of meat and meat products free from contamination (Lee et al., 2011). Recently, Rakesh et al. (2018) illustrated the application of CNTs functionalized with magnetic nanoparticles including SWCNs, endohedral with C_{60}, covalent side wall, noncovalent exohedral with surfactants or polymers, and metal plating of CNTs (Fig. 5) (Rakesh et al., 2018).

Until now, Hassan et al. (2016b, 2017) reported the efficacy of zinc nanomaterials in the degradation of mycotoxins in animal feed, particularly aflatoxins. Today, the authors investigated the potential of an inhibition of aflatoxin production by mycotoxigenic *A. flavus* by the addition of CNTs in feed. The effective doses of CNTs for degradation of Afs contamination in feeds ranged from 60 to 125 μg/g of feed while complete elimination of AFs in feed occurred at 100–125 μg/g of feed. Currently, nanosensors can diagnose biological material or tissue samples and measure blood gases, ions, metabolites, blood pressure, and blood flow (Rolfe, 2012). Furthermore, the activity of nanosensors is due to catalytic and affinity sensing, where catalytic sensors use enzymes, cells, tissues/organelles, and microorganisms as the recognition agent. The affinity sensors use antibodies, antibody fragments, nucleic acid, phages, and synthetic protein binding agents as recognition agents (Akkoyun et al., 2000). While, Lu and Bowles (2013) used nanosensors in the detection of cations, anions, and organic compounds in food; the detection of pesticides, antibiotics, heavy metals, microbial cells and toxins, and in feed and nutrient components and quick detection systems. However, Neethirajan et al. (2017) detected that some sensors can be implanted in animals to measure body temperature, antibiotic detection, observation of behavior and movement, and stress. In a recent study, sensors were used for the early diagnosis of subclinical ketosis in dairy cattle (Bai et al., 2018). In addition, nanosensors have

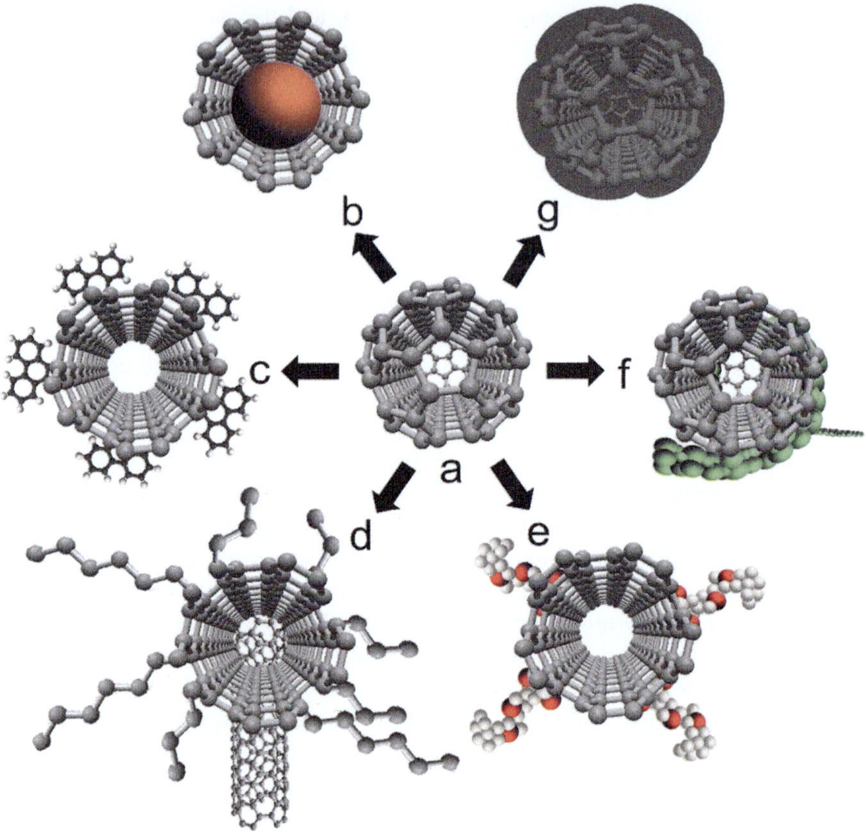

Fig. 5 Schematic of different methods of functionalizing SWNTs: (a) Single-walled carbon nanotube, (b) endohedral functionalization with, for example, C_{60}, (c) covalent side wall functionalization, (d) defect-group functionalization, (e) noncovalent exohedral functionalization with surfactants, (f) noncovalent exohedral functionalization with polymers, and (g) metal plating of carbon nanotubes.
From Rakesh, P.S., Abdalla, A.M., Abdel Fattah, A., Ghosh, S., Puri, I.K., 2018. Synthesis, characterization, and applications of carbon nanotubes functionalized with magnetic nanoparticles. In: G. Balasubramanian (Ed.), Advances in Nanomaterials (Chapter 2). © Springer International Publishing AG, p. 37. doi:10.1007/978-3-319-64717-3_2. https://www.springer.com/cda/content/document/cda_downloaddocument/9783319647159-c1.pdf?SGWID=0-0-45-1623598-p181071311.

a major role in veterinary sciences; they enable eliminating chemical contaminants, viruses, or bacteria, which is helpful for agriculture and food systems that in return improve animal health and production. In addition, this can be used in animal health care to diagnose some diseases, the spreading of viruses, and microbial pathogens of animals. Hence, all activities of the nanomaterial products can help in elevating the efficacy of animal health and production that is associated with a country's economic, social, and environmental status.

6 Conclusions and future perspective

The increased worldwide requirements for animal meats, milks, leathers, and eggs initiated several workers to evaluate all factors related to nanomaterials, including their toxicity. As no toxicology report is available on the emerging applications of nanotechnology in medical technology, therefore the practical data required addressing the toxicological risks of the application of nanotechnology in humans and animals. Also, the significant toxicity of carbon nanoparticles to animal cells, particularly lungs, urgently required intensive studies for the measurement of safe nontoxic doses before their field use in animal therapy. The toxicity of carbon nanomaterials to animal cells, bacteria, and fungi is suggested to be due to interruption of membrane electron transfer, penetration of the cell envelope, and oxidation of cell components. The well-functionalized carbon nanomaterials are safe while raw carbon nanomaterials or carbon nanomaterials without functionalization show severe toxicity to animal or human cells at even moderate dosages. In addition, carbon nanomaterials have no significant effects on the growth and biomass production of fungi and bacteria and the antimicrobial effects are improved by functionalization, which causes cell death. In general, carbon nanomaterial toxicity is low after intravenous injection and the properties of carbon nanomaterials regulate their biosafety. After functionalization or injection, these materials can be used for lymphatic mapping and parathyroid gland and lymph node tumor imaging, such as in gastric cancer and breast cancer and the diagnosis of papillary thyroid carcinoma. In addition, it can be a drug carrier due to its ability to adsorb drugs, such as in chemotherapy. Therefore, carbon nanomaterials are expected to have huge benefits for human and animal health in the future by their functionalization with other nontoxic materials or their use as injected materials with physiological reagents. Hence, all activity of the nanomaterial products and processes elevated the efficacy of animal health and production that is associated with a country's economic, social, and environmental status.

Acknowledgment

Great appreciation to our Prof. Dr. M.K. Refai, the Worldwide Professor of Microbiology, Cairo University, for his continuous initiation and advice to achieve this work.

References

Abbasi, E., Sedigheh, F.A., Abolfazl, A., Morteza, M., Hamid, T.N., Younes, H., Kazem, N.K., Roghiyeh, P.A., 2014. Dendrimers: synthesis, applications, and properties. Nanoscale Res. Lett. 9 (1), 247–255.

Ajayan, P.M., Ebbesen, T.W., 1997. Nanometre-size tubes of carbon. Rep. Prog. Phys. 60 (10), 1025–1030.

Akkoyun, A., Kohen, V., Bilitewski, U., 2000. Detection of sulphamethazine with an optical biosensor and antiidiotypic antibodies. Sensors Actuators B Chem. 70, 12–18.

Arias, L.R., Yang, L., 2009. Inactivation of bacterial pathogens by carbon nanotubes in suspensions. Langmuir 25 (5), 3003–3012.

Asakura, M., Sasaki, T., Sugiyama, T., Takaya, M., Koda, S., Nagano, K., Fukushima, S., 2010. Genotoxicity and cytotoxicity of multi-wall carbon nanotubes in cultured Chinese hamster lung cells in comparison with chrysotile a fibers. J. Occup. Health 52 (3), 155–166.

Bai, Y., Zhang, Y., Zhang, J., Mu, Q., Zhang, W., Butch, E.R., Yan, B., 2010. Repeated administrations of carbon nanotubes in male mice cause reversible testis damage without affecting fertility. Nat. Nanotechnol. 5 (9), 683–689.

Bai, D.P., Xin-Yu, L., Yi-Fan, H., Xi-Feng, Z., 2018. Review theranostics aspects of various nanoparticles in veterinary medicine. Int. J. Mol. Sci. 19 (11), 3299–3338.

Bengtson, S., Knudsen, K.B., Kyjovska, Z.O., Berthing, T., Skaug, V., Levin, M., Koponen, I.K., Shivayogimath, A., Booth, T.J., Alonso, T., 2016. Differences in inflammation and acute phase response but similar genotoxicity in mice following pulmonary exposure to graphene oxide and reduced graphene oxide. PLoS One 12, e0178355–e0178362.

Bischoff, K., Gaskill, C., Erb, H.N., Ebel, J.G., Hillebrandt, J., 2010. Comparison of two methods for blood lead analysis in cattle: graphite-furnace atomic absorption spectrometry and Lead Care(R) II system. J. Vet. Diagn. Investig. 22, 729–733.

Boor, K.J., 2006. Bacterial stress responses: what does not kill them can make them stronger. PLoS Biol. 4 (1), e23–e29.

Bottini, M., Bruckner, S., Nika, K., Bottini, N., Bellucci, N., Magrini, A., Mustelin, T., 2006. Multiwalled carbon nanotubes induce T lymphocyte apoptosis. Toxicol. Lett. 160 (2), 121–126.

Catherine, J.S., Benjamin, J.S., Richard, D.H., 2007. Toxicity of single walled carbon nanotubes to rainbow trout (*Oncorhynchus mykiss*): respiratory toxicity, organ pathologies, and other physiological effects. Aquatic Toxicol. 82 (2), 94–109.

Chang, X., Yang, S.T., Xing, G., 2014. Molecular toxicity of nanomaterials. J. Biomed. Nanotechnol. 10, 2828–2851.

Chico, L., Crespi, V.H., Benedict, L.X., Louie, S.G., Cohen, M.L., 1996. Pure carbon nanoscale devices: nanotube heterojunctions. Phys. Rev. Lett. 76 (6), 971–974.

Da Silva, C.G., Rueda, F., Löwik, C.W., Ossendorp, F., Cruz, L.J., 2016. Combinatorial prospects of nano-targeted chemoimmunotherapy. Biomaterials 83, 308–320.

Davis, M.E., Chen, Z.G., Shin, D.M., 2008. Nanoparticle therapeutics: an emerging treatment modality for cancer. Nat. Rev. Drug Discov. 7, 771–782.

Davoren, M., Herzog, E., Casey, A., Cottineau, B., Chambers, G., Byrne, H., Lyng, F.M., 2007. In vitro toxicity evaluation of single walled carbon nanotubes on human A549 lung cells. Toxicol. in Vitro 21 (3), 438–448.

De Volder, M.F., Tawfick, S.H., Baughman, R.H., Hart, A.G., 2013. Carbon nanotubes: present and future commercial applications. Science 339 (6119), 535–539.

Ding, L.H., Stilwell, J., Zhang, T.T., Elboudwarej, O., Jiang, H.G., Selegue, J.P., Cooke, P.A., Gray, J.A., Chen, F.Q.F., 2005. Molecular characterization of the cytotoxic mechanism of multiwall carbon nanotubes and Nano-onions on human skin fibroblast. Nano Lett. 5 (12), 2448–2464.

Dumortier, H., Lacotte, S., Pastorin, G., Marega, R., Wu, W., Bonifazi, D., Briand, J.P., Prato, M., Muller, S., Bianco, A., 2006. Functionalized carbon nanotubes are non-cytotoxic and preserve the functionality of primary immune cells. Nano Lett. 6 (7), 1522–1528.

Duncan, R., Gaspar, R., 2011. Nanomedicine(s) under the microscope. Mol. Pharm. 8, 2101–2141.

Eatemadi, H., Daraee, H., Karimkhanloo, M., Kouhi, N., Zarghami, A., Akbarzadeh, M., Abasi, Y., Hanifehpour, S., Joo, S.W., 2014. Carbon nanotubes: properties, synthesis, purification, and medical applications. Nanoscale Res. Lett. 9, 1–13.

Elkin, T., Jiang, X., Taylor, S., Lin, W., Gu, L., Yang, H., Brown, J., Collins, S., Sun, S., 2005. Immuno carbon nanotubes and recognition of pathogens. ChemBioChem 6 (4), 640–643.

Endo, M., Strano, M.S., Ajayan, P.M., 2008. Potential applications of carbon nanotubes. Appl. Phys. 111, 13–61.

Fadeel, B., Garcia-Bennett, A.E., 2010. Better safe than sorry: understanding the toxicological properties of inorganic nanoparticles manufactured for biomedical applications. Adv. Drug Deliv. Rev. 62, 362–374.

Feneque, J., 2003. Brief Introduction to the Veterinary Applications of Nanotechnology. http://www.nanotechnow.com/JoseFeneque/VeterinaryApplicationsNanotechnology.htm.

Fernandez-Lopez, S., Kim, H.S., Choi, E.C., Delgado, M., Granja, J.R., Khasanov, A., Kraehenbuehl, K., Long, G., Weinberger, D.A., Wilcoxen, K.M., 2001. Antibacterial agents based on the cyclic DL-alpha-peptide architecture. Nature 412, 452–455.

Friedman, S.H., De Camp, D.L., Sijbesma, R.P., 1993. Inhibition of the HIV-1 protease by fullerene derivatives: model building studies and experimental verification. J. Am. Chem. Soc. 115, 6506–6509.

Ganesh, E.N., 2013. Single walled and multi walled carbon nanotube structure, synthesis and applications. Int. J. Innov. Technol. Explor. Eng. 2 (4), 311–320. ISSN 2278-3075.

Gibson, N.M., Luo, T.J., Brenner, D.W., Shenderova, O., 2011. Immobilization of mycotoxins on modified nanodiamond substrates. Biointerphases 6, 210–217.

Gu, G., Nie, S., Feenstra, R.M., Devaty, R.P., Choyke, W.J., Chan, W.K., Kane, M.G., 2007. Field effect in epitaxial graphene on a silicon carbide substrate. Appl. Phys. Lett. 90 (25), 253507–2535115.

Gu, J., Wang, X., Nie, W., Wang, W., Shang, J., 2015. Potential role for carbon nanoparticles identification and preservation in situ of parathyroid glands during total thyroidectomy and central compartment node dissection. Int. J. Clin. Exp. Med. 8, 9640–9648.

Gunawardana, T., Foldvari, M., Zachar, T., Popowich, S., Chow-Lockerbie, B., Ivanova, M.V., Tikoo, S., Kurukulasuriya, S., Willson, P., Gomis, S., 2015. Protection of neonatal broiler chickens following in ovum delivery of oligodeoxynucleotides containing CpG motifs (CpG-ODN) formulated with carbon nanotubes or liposomes. Avian Dis. 59, 31–37.

Gurunathan, S., Han, J.W., Kim, J.H., 2013. Green chemistry approach for the synthesis of biocompatible graphene. Int. J. Nanomedicine 8, 2719–2732.

Hassan, A.A., Mahmoud, H.K., Hesham, T., El-Ahl, R.M.H.S., Mahmoud, H.H., 2015. Herbal biosynthesis of zinc nanoparticles and evaluation of their antifungal and antibacterial effect for buffaloes skin affections. Int. J. Curr. Res. 7 (12), 24338–24349.

Hassan, A.A., Mansour, M.K., Noha, H.O., Aliaa, A.E., 2016a. The efficiency of using silver nanoparticles singly and in combination with traditional antimicrobial agents in control of some fungal and bacterial affection of buffaloes. Int. J. Curr. Res. 8 (04), 29758–29770.

Hassan, A.A., Mogda, K.M., Ibrahim, M., Sayed El-Ahl, R.M., Al-Kalamawey, N., El Kattan, Y., Ali, M.A., 2016b. Efficacy of zinc oxide nanoparticles and curcumin in amelioration the toxic effects in aflatoxicated rabbits. Int. J. Cur. Microbiol. App. Sci. 5 (12), 795–818.

Hassan, A.A., El-Mokhtar, N.M., El-Hamaky, A.M., 2017. Evaluation of the efficacy of Ozone fumigation and zinc oxide nanoparticles in control of aflatoxins contamination in cattle feeds. 1st International Conference, Animal Health Research Institute, Egypt 1, 9-13, Anim. Health Res. J. 5, 165–180.

Hill, E.K., Li, J., 2017. Current and future prospects for nanotechnology in animal production. J. Anim. Sci. Biotechnol. 8, 26–30.

Hirlekar, R., Yamagar, M., Garse, H., Vij, M., Kadam, V., 2009. Carbon nanotubes and its applications: a review. Asian J. Pharm. Clin. Res. 2 (4), 17–27.

Hou, P.X., Bai, S., Yang, Q.H., Liu, C., Cheng, H.M., 2002. Multi-step purification of carbon nanotubes. Carbon 40 (1), 81–85.

Hu, S., Zhang, Y., 2010. Endostar-loaded PEG-PLGA nanoparticles. In vitro and in vivo evaluation. Int. J. Nanomedicine 5, 1039–1048.

Hurt, R.H., Monthioux, M., Kane, A., 2006. Toxicology of carbon nanomaterial: status, trends, and perspectives on the special issue. Carbon 44 (6), 1028–1033.

Hussain, S., Sonja, B., Armelle, B., Rodolphe, H., Leen, C.J., Johan, A.M., Marie, A., Jocelyne, F., Frédéric, M., Jean-Claude, P., Francelyne, M., 2009. Oxidative stress and pro-inflammatory effects of carbon black and titanium dioxide nanoparticles: role of particle surface area and internalized amount. Toxicology 260, 142–149.

Iijima, S., 1991. Helical microtubules of graphitic carbon. Nature 354, 56–58.0.

Iijima, S., Ichihashi, T., 1993. Single-shell carbon nanotubes of 1-nm diameter. Nature 363, 603–605.

Jang, M.H., Hwang, Y.S., 2018. Effects of functionalized multi-walled carbon nanotubes on toxicity and bioaccumulation of lead in *Daphnia magna*. PLoS One 13 (3), e0194935–e0194943.

Jose-Yacaman, M., Miki-Yoshida, M., Rendon, L., Santiesteban, J.G., 1993. Catalytic growth of carbon microtubules with fullerene structure. Appl. Phys. Lett. 62 (2), 202–204.

Kaesermann, F., Kempf, C., 1997. Photodynamic inactivation of enveloped viruses by buckminsterfullerene. Antivir. Res. 34, 65–70.

Kam, N.W.S., O'Connell, M., Wisdom, J.A., Dai, H., 2005. Carbon nanotubes as multifunctional biological transporters and near-infrared agents for selective cancer cell destruction. Proc. Natl. Acad. Sci. U. S. A. 102 (33), 11600–11605.

Kanakia, S.J.D., Toussaint, S.M., Chowdhury, T., Tembulkar, S., Lee, Y., Jiang, R.Z., Lin, K.R., Shroyer, S., Moore, S., Sitharaman, B., 2014. Dose ranging, expanded acute toxicity and safety pharmacology studies for intravenously administered functionalized graphene nanoparticle formulations. Biomaterials 35, 7022–7031.

Kang, S., Pinault, M., Pfefferle, L.D., Elimelech, M., 2007. Single walled carbon nanotubes exhibit strong antimicrobial activity. Langmuir 23 (17), 8670–8673.

Kang, S., Herzberg, M., Rodrigues, D.F., Elimelech, M., 2008. Antibacterial effects of carbon nanotubes: size does matter. Langmuir 24 (13), 6409–6413.

Kang, S., Mauter, M.S., Elimelech, M., 2009. Microbial cytotoxicity of carbon based nanomaterials: implications for river water and wastewater effluent. Environ. Sci. Technol. 43 (7), 2648–2653.

Khalid, P., Hussain, S.A., Suman, A.V., Arun, A.B., 2016. Toxicology of carbon nanotubes—a review. Int. J. Appl. Eng. Res. 11 (1), 148–157.

Khaliullin, T.O., Fatkhutdinova, L.M., Zalyalov, R.R., Kisin, E.R., Murray, A.R., Shvedova, A.A., 2015. *In vitro* toxic effects of different types of carbon nanotubes. Nanobiotech. IOP Conf. Ser. Nanobiotech. Mater. Sci. Eng. 98, 1–6. https://doi.org/10.1088/1757-899X/98/1/012021. 012021.

Kim, H., Lee, J., Kahng, S.J., Son, Y.W., Lee, S.B., Lee, C.K., Ihm, J., Kuk, Y., 2003. Direct observation of localized defect states in semiconductor nanotube junctions. Phys. Rev. Lett. 90 (21), 216107–216115.

Kim, Y.J., Kim, Y.A., Chino, T., Suezaki, H., Endo, M., Dresselhaus, M.S., 2006. Chemically modified multi walled carbon nanotubes as an additive for supercapacitors. Small 2 (3), 339–345.

Kruusenberg, I., Alexeyeva, S., Tammeveski, K., Kozlova, J., Matisen, L., Sammelselg, V., Solla-Gullon, J., Feliu, J.M., 2011. Effect of purification of carbon nanotubes on their electrocatalytic properties for oxygen reduction in acid solution. Carbon 49 (12), 4031–4039.

Lam, C.W., James, J.T., McCluskey, R., Hunter, R.L., 2004. Pulmonary toxicity of single wall carbon nanotubes in mice 7 and 90 days after intratracheal instillation. Toxicol. Lett. 77 (1), 126–134.

Lamprecht, C., Liashkovich, I., Neves, V., Danzberger, J., Heister, E., Rangl, M., Coley, H.M., McFadden, J., Flahaut, E., Gruber, H.J., 2009. AFM imaging of functionalized carbon nanotubes on biological membranes. Nanotechnology 20, 434001.

Lan, M., Silk, T., Volker, S., Sandra, B., Michael, M., 2009. Inhalation of multiwall carbon nanotubes in rates exposed for 3 months. Toxicol. Sci. 112 (2), 468–481.

Landi, B.J., Castro, S.L., Ruf, H.J., Evans, C.M., Bailey, S.G., Raffaelle, R.P., 2005. CdSe quantum dot single wall carbon nanotube complexes for polymeric solar cells. Sol. Energy Mater. Sol. Cells 87 (1), 733–746.

Le Croy, G.E., Yang, S.T., Yang, F., Liu, Y., Fernando, K.S., Bunker, C.E., Sun, Y.P., 2016. Functionalized carbon nanoparticles: syntheses and applications in optical bio imaging and energy conversion. Coord. Chem. Rev. 320, 66–81.

Lee, J., Jo, M., Kim, T.H., Ahn, J.Y., Lee, D.K., Kim, S., Hong, S., 2011. Aptamer sandwich-based carbon nanotube sensors for single-carbon-atomic-resolution detection of non-polar small molecular species. Lab Chip 11, 52–56.

Lee, C., Kim, J., Zhang, Y., Jeon, M., Liu, C., Song, L., Lovell, J.F., Kim, C., 2015. Dual-color photoacoustic lymph node imaging using nanoformulated naphthalocyanines. Biomaterials 73, 142–148.

Leroux, F., Metenier, K., Gautier, S., Frackowiak, E., Bonnamy, S., Beguin, F., 1999. Electrochemical insertion of lithium in catalytic multi walled carbon nanotubes. J. Power Sources 8, 317–322.

Li, Z., Hulderman, T., Salmen, R., Chapman, R., Leonard, S.S., Young, S.H., Shvedova, A., Luster, M.I., Simeonova, P.P.M., 2007. Cardiovascular effects of pulmonary exposure to single-wall carbon nanotubes. Environ. Health Perspect. 115, 377–392.

Li, Z., Ao, S., Bu, Z., Wu, A., Wu, X., Shan, F., Ji, X., Zhang, Y., Xing, Z., Ji, J., 2016. Clinical study of harvesting lymph nodes with carbon nanoparticles in advanced gastric cancer: a prospective randomized trial. World J. Surg. Oncol. 14 (1), 88–100.

Li, Y., Zheng, X., Zhang, X., Liu, S., Pei, Q., Zheng, M., Xie, Z., 2017. Porphyrin-based carbon dots for photodynamic therapy of hepatoma. Adv. Health Mater. 6, 1–6. 1600924.

Liu, S., Wei, L., Hao, L., Fang, N., Chang, M.W., Xu, R., Yang, Y., Chen, Y., 2009. Sharper and faster nano darts kill more bacteria: a study of antibacterial activity of individually dispersed pristine single-walled carbon nanotube. ACS Nano 3 (12), 3891–3902.

Liu, Y., Zhao, B., Sun, B., Chen, C., 2013. Understanding the toxicity of carbon nanotubes. Acc. Chem. Res. 46, 702–713.

Liu, J., Wang, T., Wang, H., Gu, H., Xu, Y., Tang, H., Jia, G., Liu, Y., 2015. Biocompatibility of graphene oxide intravenously administrated in mice-effects of dose, size and exposure protocols. Toxicol. Res. 4 (1), 83–91.

Lu, J., Bowles, M., 2013. How will nanotechnology affect agricultural supply chains. Int. Food Agribus. Manag. Rev. 16, 21–42.

Luo, P.G., Yang, F., Yang, S.T., Sonkar, S.K., Yang, L., Broglie, J.J., Sun, Y.P., 2014. Carbon-based quantum dots for fluorescence imaging of cells and tissues. RSC Adv. 4 (21), 10791–10807.

Magrez, A., Kasas, S., Salicio, D., Pasquier, N., Seo, J.W., Celio, M., Forró, L., 2006. Cellular toxicity of carbon-based nanomaterials. Nano Lett. 6 (6), 1121–1125.

Ma-Hock, L., Treumann, S., Strauss, V., Brill, S., Luizi, F., Mertler, M., Wiench, K., Gamer, A.O., Ravenzwaay, B.V., Landsiedel, R., 2009. Inhalation toxicity of multiwall carbon nanotubes in rats exposed for 3 months. Toxicol. Sci. 112 (2), 468–481.

McComb, T., Bischoff, K., Thompson, B., Smith, M.C., Mohammed, H.O., Ebel, J., Hillebrandt, J., 2010. An investigation of blood selenium concentrations of goats in New York State. J. Vet. Diagn. Investig. 22, 696–701.

Meena, N.S., Sahni, Y.P., Dinesh, T., Singh, R.P., 2018. Applications of nanotechnology in veterinary therapeutics. J. Entomol. Zool. Stud. 6 (2), 167–175.

Merkoçi, A., 2006. Review: carbon nanotubes in analytical sciences. Microchim. Acta 152, 157. https://doi.org/10.1007/s00604-005-0439-z.

Mitchell, L.A., GAO, J., Vander Wal, R., Gigliotti, A., Burchiel, S.W., McDonaold, S.W., 2007. Pulmonary and systemic immune response to inhaled multi-walled carbon nanotubes. Toxicol. Sci. 100, 203–214.

Monerris, M.J., Arévalo, F.J., Fernández, H., Zon, M.A., Molina, P.G., 2012. Integrated electrochemical immunosensor with gold nanoparticles for the determination of progesterone. Sensors Actuators B Chem. 166, 586–592.

Mubarak, N.M., Yusof, F., Alkhatib, M.F., 2011. The production of carbon nanotubes using two stage chemical vapor deposition and their potential use in protein purification. Chem. Eng. J. (Amsterdam, Neth.) 168, 461–469.

Muktar, Y., Bikila, T., Keffale, M., 2015. Application of nanotechnology for animal health and production improvement: a review. World Appl. Sci. J. 33, 1588–1596.

Muller, J., Huaux, F., Moreau, N., Misson, P., Heilier, J.F., Delos, M., Arras, M., Fonseca, A., Nagy, J.B., Lison, D.B., 2005. Respiratory toxicity of multi-wall carbon nanotubes. Toxicol. Appl. Pharmacol. 207 (3), 221–231.

Murr, L.E., Bang, J.J., Esquivel, E.V., Guerrero, P.A., Lopez, D.A., 2004. Carbon nanotubes, nanocrystal forms, and complex nanoparticle aggregate in common fuel-gas combustion sources and the ambient air. J. Nanopart. Res. 6, 241–251.

Neethirajan, S., Tuteja, S.K., Huang, S.T., Kelton, D., 2017. Recent advancement in biosensors technology for animal and livestock health management. Biosens. Bioelectron. 98, 398–407.

Novoselov, K.S., 2004. Development and Applications of Mesoscopic Hall Microprobes. (Ph.D. thesis) Radboud University Nijmegen, Nijmegen, the Netherlands.

Osawa, E., Ho, D., Huang, H., Korobov, M.V., Rozhkova, N.N., 2009. Consequences of strong and diverse electrostatic potential fields on the surface of detonation nanodiamond particles. Diam. Relat. Mater. 18, 904–909.

Ouyang, M., Huang, J.L., Cheung, C.L., Lieber, C.M., 2001. Atomically resolved single-walled carbon nanotube intramolecular junctions. Science 291 (5501), 97–100.

Parvongnukul, K., Lumb, W.V., 1978. Evaluation of polytetrafluoroethylene-graphite-coated total hip prostheses in goats. Am. J. Vet. Res. 39, 221–228.

Patolsky, F., Zheng, G., Hayden, O., Lakadamyali, M., Zhuang, X., Lieber, C.M., 2004. Electrical detection of single viruses. Proc. Natl. Acad. Sci. U. S. A. 101, 14017–14022.

Peer, D., Karp, J.M., Hong, S., Farokhzad, O.C., Margalit, R., Langer, R., 2007. Nanocarriers as an emerging platform for cancer therapy. Nat. Nanotechnol. 2, 751–760.

Pietroiusti, A., Massimiani, M., Fenoglio, I., Colonna, M., Valentini, F., Palleschi, G., Sgambato, A., 2011. Low doses of pristine and oxidized single-wall carbon nanotubes affect mammalian embryonic development. ACS Nano 5 (6), 4624–4633.

Plata, D.L., Gschwend, P.M., Reddy, C.M., 2008. Industrially synthesized single walled carbon nanotubes: compositional data for users, environmental risk assessments and source apportionment. Nanotechnology 19 (18), 185706–185715.

Rahimi, E., 2013. Lead and cadmium concentrations in goat, cow, sheep, and buffalo milks from different regions of Iran. Food Chem. 136, 389–391.

Rakesh, P.S., Abdalla, A.M., Abdel Fattah, A., Ghosh, S., Puri, I.K., 2018. Synthesis, characterization, and applications of carbon nanotubes functionalized with magnetic nanoparticles. In: Balasubramanian, G. (Ed.), Advances in Nanomaterials (Chapter 2). Springer International Publishing AG, pp. 37. https://doi.org/10.1007/978-3-319-64717-3_2. https://www.springer.com/cda/content/document/cda_downloaddocument/9783319647159-c1.pdf?SGWID=0-0-45-1623598-p181071311.

Refai, M.K., Hassan, A.A., 2013. Monograph on Mycotoxigenic Fungi and Mycotoxins in Foods and Feeds with Synopsis on studies done by authors on mycotoxigenic fungi and mycotoxins in foods and feeds. https://www.academia.edu.eg./7688721-AtefHassan. https://www.researchgate.net/publication/AtefHassan.

Reina, G., González-Dominguez, J.M., Criado, A., Vázquez, E., Bianco, A., Prato, M., 2017. Promises, facts and challenges for graphene in biomedical applications. Chem. Soc. Rev. 46, 4400–4416.

Rolfe, P., 2012. Micro- and Nano sensors for medical and biological measurement. Sens. Mater. 24, 275–302.

Sagadevan, S., Periasamy, M., 2014. Recent trends in nanobiosensors and their applications—a review. Rev. Adv. Mater. Sci. 36, 62–69.

Sakamoto, Y., Nakae, D., Fukumori, N., Tayama, K., Maekawa, A., Imai, K., Ogata, A., 2009. Induction of mesothelioma by a single intrascrotal administration of multi-wall carbon nanotube in intact male Fischer 344 rats. J. Toxicol. Sci. 34 (1), 65–76.

Sanchez, V.C., Jachak, A., Hurt, R.H., Kane, A.B., 2012. Biological interactions of graphene-family nanomaterials: an interdisciplinary review. Chem. Res. Toxicol. 25 (1), 15–34.

Saragusty, J., Arav, A., 2011. Current progress in oocyte and embryo cryopreservation by slow freezing and vitrification. Reproduction 141, 1–19.

Sayes, C.M., Liang, F., Hudson, J.L., Mendez, J., Guo, G., Beach, J.M., Colvin, V.L., 2006. Functionalization density dependence of single-walled carbon nanotubes cytotoxicity in vitro. Toxicol. Lett. 161 (2), 135–142.

Schipper, M.L., Nakayama, R.N., Davis, G.R., Kam, N.W., Chu, P., Liu, Z., Sun, X., Dai, H., Gambhir, S.S., 2008. A pilot toxicology study of single-walled carbon nanotubes in a small sample of mice. Nat. Nanotechnol. 3 (4), 216–221.

Sharma, C.S., Sarkar, S., Periyakaruppan, A., Barr, J., Wise, K., Thomas, R., Wilson, B.L., Ramesh, G.T., 2007. Single walled carbon nanotubes induce oxidative stress in rat lung epithelial cells. J. Nanosci. Nanotechnol. 7 (7), 2466–2472.

Shvedova, A.A., Kisin, E.R., Mercer, R.T., Murray, A.R., Johnson, V.G., Potapovich, A.L., Tyurina, L.W., Gorelik, O., Arepalli, S., Schwegler, D.B., Baron, P., 2005. Unusual inflammatory and fibrogenic pulmonary responses to single-walled carbon nanotubes in mice. Am. J. Phys. Lung Cell. Mol. Phys. 289 (5), L698–L708.

Sijbesma, R., Srdanov, G., Wudl, F., 1993. Synthesis of a fullerene derivative for the inhibition of HIV enzymes J. Am. Chem. Soc. 115, 6510–6512.

Snyder-Talkington, B.N., Dong, C., Porter, D.W., Ducatman, B., Wolfarth, M.G., Andrew, M., Battelli, L., Raese, R., Castranova, V., Guo, N.L., 2016. Multiwall carbon nanotube-induced pulmonary inflammatory and fibrotic responses and genomic changes following aspiration exposure in mice: a 1-year post-exposure study. J. Toxicol. Environ. Health A 79 (8), 352–366.

Son, K.H., Hong, J.H., Lee, J.W., 2016. Carbon nanotubes as cancer therapeutic carriers and mediators. Int. J. Nanomedicine 11, 5163–5172.

Sun, X., Liu, Z., Welsher, K., Robinson, J.T., Goodwin, A., Zaric, S., Dai, H., 2008. Nano-graphene oxide for cellular imaging and drug delivery. Nano Res. 1, 203–212.

Surrendering, A., Sandhiya, S., Pradhan, S.C., Adithan, C., 2009. Novel applications of nanotechnology in medicine. Indian J. Med. Res. 130, 689–701.

Thakare, V.S., Das, M., Jain, A.K., Patil, S., Jain, S., 2010. Carbon nanotubes in cancer theragnosis. Nanomedicine 5 (8), 1277–1301.

Thess, A., Lee, R., Nikolaev, P., Dai, H., Petit, P., Robert, J., Xu, C., Lee, Y.H., Kim, S.G., Rinzler, A.G., 1996. Crystalline ropes of metallic carbon nanotubes. Science 273 (5274), 483–487.

Tu, X., Manohar, S., Jagota, A., Zheng, M., 2009. DNA sequence motifs for structure specific recognition and separation of carbon nanotubes. Nature 460 (7252), 250–253.

Vecitis, C.D., Zodrow, K.R., Kang, S., Elimelech, M., 2010. Electronic structure dependent bacterial cytotoxicity of single walled carbon nanotubes. ACS Nano 4 (9), 5471–5479.

WanderWal, R.L., 2003. Carbon nanotube synthesis in a flame using laser ablation for in situ catalyst generation. Appl. Phys. A Mater. Sci. Process. 77 (7), 885–889.

Wang, C., Waje, M., Wang, X., Tang, J.M., Haddon, R.C., Yan, Y., 2004. Proton exchange membrane fuel cells with carbon nanotube based electrodes. Nano Lett. 4 (2), 345–348.

Wang, H., Gu, L., Lin, Y., Lu, F., Meziani, M.J., Luo, P.G., Wang, G., Cao, L., Sun, Y.P., 2008. Unique aggregation of anthrax (*Bacillus anthracis*) spores by sugar coated single walled carbon nanotubes. J. Am. Chem. Soc. 128 (41), 13364–13365.

Wang, H., Yang, S.T., Cao, A., Liu, Y., 2013. Quantification of carbon nanomaterials in vivo. Acc. Chem. Res. 46, 750–760.

Wang, P., Liu, J., GAO, H., Hu, Y., Hou, X., Le Croy, G.E., Bunker, C.E., Liu, Y., Sun, Y.P., 2017. Host–guest carbon dots as high-performance fluorescence probes. J. Mater. Chem. 5, 6328–6335.

Warheit, D.B., Laurence, P.R., Reed, K.L., Roach, D.H., Reynolds, G.A.M., Webb, T.R., 2004. Comparative pulmonary toxicity assessment of single-wall carbon nanotubes in rats. Toxicol. Lett. 77 (1), 117–125.

Wei, Q., Bi, J., Zhang, X., Wang, J., Wang, J., Liu, P., Li, Z., Wu, W., 2014. Damaging effects of multi-walled carbon nanotubes on pregnant mice with different pregnancy times. Sci. Rep. 4, 4352–4360.

Wu, W., Wieckoski, S., Pastorin, G., Benincasa, M., Klumpp, C., Briand, J.P., Gennaro, R., Prato, M., Bianco, A., 2004. Targeted delivery of amphotericin B to cells by using functionalized carbon nanotubes. Angew. Chem. 117 (39), 6516–6520.

Wu, P., Chen, X., Hu, N., Tam, U.C., Blixt, O., Zettl, A., Bertozzi, C.R., 2008. Biocompatible carbon nanotubes generated by functionalization with glycodendrimers. Angew. Chem. 120 (27), 5100–5103.

Wu, X., Lin, Q., Chen, G., Lu, J., Zeng, Y., Chen, X., Yan, Y., 2015. Sentinel lymph node detection using carbon nanoparticles in patients with early breast cancer. PLoS One 10 (8), e0135714.

Xie, P., Tang, X., Li, L., Qian, Z., Ran, M., Zhang, X., Xin, X., Luo, H., 2016. Drug-loaded carbon nanoparticle suspension injection, drug selection, releasing behavior, in vitro cytotoxicity and in vivo lymph node targeting. J. Nanosci. Nanotechnol. 16, 6910–6918.

Xie, P., Yang, S.T., He, T., Yang, S., Tang, X.H., 2017a. Bioaccumulation and toxicity of carbon nanoparticles suspension injection in intravenously exposed mice. Int. J. Mol. Sci. 18 (12), 2562. 1–14.

Xie, P., Yang, S.T., He, T., Yang, S., Tang, X.H., 2017b. Bioaccumulation and toxicity of carbon nanoparticles suspension injection in intravenously exposed mice. Int. J. Mol. Sci. 18 (12), 2562–2570.

Xie, P., Xin, Q., Yang, S.T., He, T., Huang, Y., Zeng, G., Ran, M., Tang, X.H., 2017c. Skeleton labeled ^{13}C-carbon nanoparticles for the imaging and quantification in tumor drainage lymph nodes. Int. J. Nanomedicine 12, 4891–4899.

Yamashita, K., Yoshioka, Y., Higashisaka, K., Mimura, K., Morishita, Y., Nozaki, M., Abe, Y., 2011. Silica and titanium dioxide nanoparticles cause pregnancy complications in mice. Nat. Nanotechnol. 6 (5), 321–330.

Yang, S.T., Wang, X., Jia, G., Gu, Y., Wang, T., Nie, H., Ge, C., Wang, H., Liu, Y., 2008. Long-term accumulation and low toxicity of single-walled carbon nanotubes in intravenously exposed mice. Toxicol. Lett. 181, 182–189.

Yang, S.T., Wang, X., Wang, H., Lu, F., Luo, P.G., Cao, L., Meziani, M.J., Liu, J.H., Liu, Y., Chen, M., 2009. Carbon dots as nontoxic and high performance fluorescence imaging agents. J. Phys. Chem. 113, 18110–18114.

Yang, C., Mamouni, J., Tang, Y., Yang, L., 2010. Antimicrobial activity of single walled carbon nanotubes length effect. Langmuir 26 (20), 16013–16019.

Yang, H., Sun, C., Fan, Z., Tian, X., Yan, L., Du, L., Keelan, J.A., 2012a. Effects of gestational age and surface modification on materno-fetal transfer of nanoparticles in murine pregnancy. Sci. Rep. 2, 847–855.

Yang, Q., Wang, X., Chen, J., Tian, C., Li, T., Chen, Y., Lv, C., 2012b. A clinical study on regional lymphatic chemotherapy using an activated carbon nanoparticle–epirubicin in patients with breast cancer. Tumor Biol. 33, 2341–2348.

Ye, S., Shao, K., Li, Z., Guo, N., Zuo, Y., Li, Q., Lu, Z., Chen, L., He, Q., Han, H., 2015. Antiviral activity of grapheme oxide: how sharp edged structure and charge matter. ACS Appl. Mater. Interfaces 7, 21571–21579.

Zhang, L., Liu, W., Lin, L., Chen, D., Stenzel, M.H., 2008. Degradable disulfide core- cross-linked micelles as a drug delivery system prepared from vinyl functionalized nucleosides via the RAFT process. Biomacromolecules 9, 3321–3331.

Zhang, D., Deng, X., Ji, Z., Shen, L., Dong, M., Wu, G., Liu, Y., 2010. Long-term hepatotoxicity of polyethylene-glycol functionalized multi-walled carbon nanotubes in mice. Nanotech. 21 (17), 175101.

Zhang, J.Q., Sun, Q., Bo, J., Huang, R., Zhang, M., Xia, Z., Ju, L., Xiang, G., 2014. Single-walled carbon nanohorn (SWNH) aggregates inhibited proliferation of human liver cell lines and promoted apoptosis, especially for hepatoma cell lines. Int. J. Nanomedicine 9, 759–773.

Zhang, B., Wang, Y., Liu, J., Zhai, J., 2017. Recent developments of phototherapy based on graphene family nanomaterials. Curr. Med. Chem. 24, 268–291.

Zhu, Y., Chen, X., Zhang, H., Chen, L., Zhou, S., Wu, K., Wang, Z., Kong, L., Zhuang, H., 2016. Carbon nanoparticle–guided central lymph node dissection in clinically node-negative patients with papillary thyroid carcinoma. Head Neck 38, 840–845.

Carbon nanomaterials (CNTs) phytotoxicity: Quo vadis?

24

Kamel A. Abd-Elsalam[a], Khaled Kasem[b], Hassan Almoammar[c,d]
[a]Plant Pathology Research Institute, Agricultural Research Center (ARC), Giza, Egypt, [b]Department of Plant Protection, Faculty of Agriculture, Hama University, Hama, Syria, [c]National Centre for Biotechnology, King Abdulaziz City for Science and Technology (KACST), Riyadh, Saudi Arabia, [d]ETH Zürich, Department of Biology, Institute of Microbiology, Zürich, Switzerland

1 Introduction

Several nanomaterials (NMs) have received popularity in scientific innovations due to their adjustable physical, chemical, and biological characteristics with improved functionality through their particular bulk counterparts. NMs are classified based on their size, structure, shape, and origin. The capability to forecast the exclusive properties of NMs raises the importance of each group. Due to higher progress in the development of NMs and their economic applications, concerns associated with toxicity are unavoidable (Jeevanandam et al., 2018; Rekulapally et al., 2019). The increased consumption of constructed nanoparticles increases the hazards linked with their release into the environment. For example, studies have shown that nanomaterials may be absorbed and translocated within plants. Because plants represent a main part of the ecosystem, the accumulation of nanomaterials in plants is a danger to plants and the food chain (Prakash and Nair, 2018). In recent years, nanomaterial applications in agriculture have risen significantly, which usually enforced the threat of crops by harming xylem vessels and photosynthetic systems. Plastoglobule lipid bodies appeared as an essential function in the chloroplast lipid metabolism, activating as a thylakoid microdomain for metabolite synthesis, repair, and removal. The cross-sectional shape of spring barley leaves showed that the copper oxide nanoparticles (CuO-NPs) reduced the counts of plastoglobule and starch granule in the chloroplast. The current findings suggest that chloroplast could be among the essential features in the analysis of the NP level of toxicity in plants (Rajput et al., 2019). The phytotoxicity of various nanoparticles is mainly affected by the shape, size, chemical structure, and covering material formation. Often, the phytotoxicity of the nanoparticle cloud is a response to the toxicity of chemicals that have been used for its preparation. Additionally, phytotoxicity might be based upon the ecosystem and on the physical and chemical characteristics of some plant species. The nanoparticles may have a repressive impact on plant growth in diverse growing phases. A few nanoparticles are absorbed by plant roots and transferred to the vegetative units of the plant by using the vascular system, according to the structure, shape, size of the nanoparticle, and plant physiology. In the reviewed reports, nanoparticles have not been correctly described and/or their structure versus their particular shapes has not

Carbon Nanomaterials for Agri-food and Environmental Applications. https://doi.org/10.1016/B978-0-12-819786-8.00024-4

been examined, which may be one of the most significant challenges that will have to be overcome for additional thinking about the application of nanoparticles in plant science research (Krug, 2014; Nau et al., 2016; Ruttkay-Nedecky et al., 2017). To date, work on CNM phytotoxicity has been conflicting. Multiple observations showed that various carbon nanomaterials may have interaction with the soil rhizosphere, microbes, and plant species. Therefore, they influence plant growth and production by gathering in plant tissues, specifically in the consumable portion, which may consequently trigger agricultural economic deficits in terms of crop yield and quality and cause a probable risk to human health. Therefore, new investigations on CNM activated toxicity for crop plants are justified (Hao et al., 2019; Oyelami and Semple, 2015; Verma et al., 2019; Wagay et al., 2019). Few reports to date have targeted the phytotoxicity of carbon-based nanomaterials (Afreen et al., 2018; Wagay et al., 2019). Examining the toxicity, uptake, transport, dismissions, and deterioration of GFNs within plants is still effortlessly identified and recognized (Huang et al., 2018). We need to give more interest to the phytotoxicity of CNMs and its affecting factors, along with its probable toxicity mechanisms. Recently, an extensive review discussing the phytotoxicity of graphene-family nanomaterials and its mechanism outlined existing information gaps in CNM toxicity to numerous plant species that were explored (Verma et al., 2019; Wagay et al., 2019; Wang et al., 2019). For further knowledge of the phytotoxicity mode of action of CNMs in plants, the main aim of the current chapter is to seriously review the existing literature on the phytotoxicity, the toxic influencing factors, and the toxicological mechanisms of CNMs, in addition to discovering the probable hazards presented through the toxicity of various CNMs in the microbial community.

2 Negative effects of nanomaterials

Nanotoxicity is extremely resulting from the small size and large surface area of nanomaterials (NMs), which usually enables fast distribution and attack of physiological barriers in the human body. Exclusive physicochemical characteristics of nanoparticles make the examination of their particular toxic effects complex and complicated. This makes it essential to obtain specific data on distinct mechanisms engaged in nanomaterial action and toxicity (Jain et al., 2018). Natural NMs are generally present in the environment for seasons, and they have a few mechanisms to cause less harmful effects among living organisms. However, research advancements have found some acute toxic impacts of nano-sized particles in biosystems. Besides that, growing NPs, such as viral NPs and nanozymes, must be subjected to strict cytotoxicity assays to determine harmless mechanisms of application and dose levels to be able to reduce or prevent the probable dangers of manufactured NMs in consumer products; policies are generally applied in some countries. Intensive studies in nanotoxicology and tight regulations by federal agencies are important to determine and prevent toxic NPs (Jeevanandam et al., 2018). The phytotoxicology of manufactured nanomaterials is a fairly unique discipline of research, which as stated by Nowack and Bucheli (2007) wasn't present prior to 2007. Appropriately, the probable toxicity of nanoparticles on plants has not quite been broadly examined, and the recorded results are regularly detailed and contrary to the modest information on the fundamental mechanism (Juganson et al., 2015). The majority of phytotoxicity investigations have evaluated germination, cell cultures, and genetic changes (Kaphle et al., 2017). Although carbon nanomaterials have been proven to penetrate seed coats, enter the plant cells, and translocate into several plant parts more effectively, many studies suggested that exposure to carbon nanomaterials reduces seed germination and root growth and changes the root structures (Begum and Fugetsu, 2012; Kaphle et al., 2017). Likewise, it was stated that carbon nanotubes and fullerene could potentially cause destruction to plants (Begum and Fugetsu, 2012; Liu et al., 2010; Shen et al., 2010). Carbon nanomaterials may be suppress seedling growth and effects on morphological, physiological, biochemical, molecular, nutritional, and genetic levels of various plants (Nair, 2018). Different types of nanomaterials have a small size, they stick to air flows very easily than microparticles and simply accumulated in commonly aired enclosures (Al-Halafi, 2014; Baughman et al., 2002; Lang et al., 2011; Safiuddin et al., 2014; Zhou et al., 2014). Accordingly, a special challenge is the capability of the nanoparticle immediately absorbed by individual cells and cell nuclei, specifically throughout the respiratory system.

Nanoparticle toxicity in plants has been described multiple times (Arruda et al., 2015; Husen and Siddiqi, 2014). The findings demonstrated that only some plants processed with nanoparticles showed toxic issues; more extensive studies revealed positive or no resultant effects on plants. Phytotoxicity mechanisms continue to be undiscovered. Nevertheless, they might be expected to be linked to the chemical formula, size, and surface area of nanoparticles. The existence of nanoparticles on the plant root surface can modify the surface chemistry of the roots and therefore affect the uptake of nutrients into the plant root (Lee et al., 2012; Mirzajani et al., 2013); thus, these must be considered. A number of the produced NPs consist of some heavy metals. Hence, soil and water contaminants with metallic NPs are becoming an essential environmental concern. Nanoparticles are able to interact with plants, leading to the uptake and deposition that influence their fate and transportation in the ecosystem. Furthermore, NPs could continue to be attached to the plant surface and impart physical and chemical damage to the plant organs. Usually, NPs enter the plant root system through the lateral root junctions and spread to the xylem during the cortex and the pericycle (Dietz and Herth, 2011). Metallic nanoparticles release ionic salts, and the nanoparticles and ionic salts have an identical toxicity impact; this prevents us from discriminating whether metallic nanoparticles exert particle-specific toxicity. Ag nanoparticle toxicity in Arabidopsis is size- and concentration-dependent. Ag nanoparticles may be apoplastically transferred into the cell wall and accumulate at the plasmodesmata (Geisler-Lee et al., 2013). The phytotoxicity of Ag NPs on barley that was exposed to Ag^+ as $AgNO_3$ and Ag NPs was investigated, seed germination and seedling growth of barley were reduced at 1 mM Ag^+. Low concentration of Ag NPs (0.1 mM) could be beneficial for barley grain germination and seedling growth. Nevertheless, the excessive concentrations of Ag NPs (0.5 and 1 mM) minimized grain germination and showed a better decrease in the root length. A drop in the photosynthetic pigments and disruption of chloroplast grana thylakoids in Ag NP-treated plants verified the leaf chlorosis (Fayez et al., 2017). There is certainly restricted literature presented on NP toxicity on the chloroplast ultrastructure and rarely on plastoglobules and starch granules. Ma et al. (2015) found the enlarged and ruptured outer surface of chloroplast in Bt-transgenic cotton leaves by CeO2 NPs. Cvjetko et al. (2018) also detected some variations in the size of the chloroplast in Ag NP-treated and $AgNO_3$-treated tobacco leaves. Fluctuations in chloroplast size would affect the light path inside leaf cells, and consequently photosynthesis (Zhang et al., 2010) Fig. 1.

2.1 Negative effects of CNMs

Various natural carbon-based nanoparticles are present only in minimal portions, and the overwhelming majority are engineered or synthetically produced. Consequently, their availability does not depend on natural reserves (such as diamonds), and hypothetical production may be performed in unrestricted amounts as long as raw materials for synthesis are available. According to the most recent estimations, a regular boost in production volumes is predicted within the next few years (Zaytseva and Neumann, 2016). On the other hand, in spite of the fact that carbon nanomaterials can enhance commercial improvement, there are issues regarding a probable release into the environment, interactions of introduced nanomaterials with different organisms, and incorporation into food chains with yet mysterious implications. The harmful impact of carbon nanomaterials in several plants was generally observed on their particular influence on seed germination, plant growth, and production. Only a couple of investigations were generally published on the toxicity of carbon nanomaterials in hybrids with several other chemicals and on the mechanism of toxicity in plants. According to the presented studies, the responses of carbon nanomaterials in plants varied based on the plant species, the concentration tested, and the stage of development. Based on up-to-date reviews, carbon nanotubes and fullerenes may cause harm to plants (Begum and Fugetsu, 2012; Chen et al., 2010; Liu et al., 2010; Shen et al., 2010; Verma et al., 2019; Wagay et al., 2019; Wang et al., 2019).

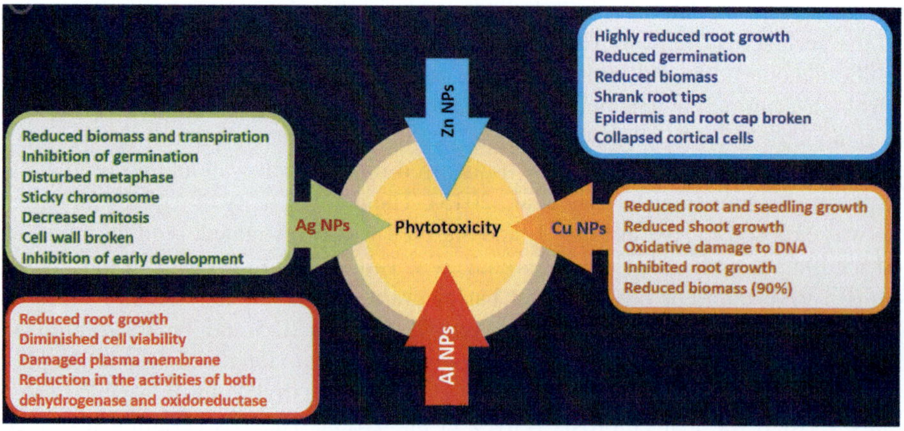

Fig. 1 The negative effect of some nanomaterials on plants.

2.1.1 Phytotoxic effects of graphene nanomaterials

Graphene and its nanocomposites are considered one of the most growing dynamic carbon nanomaterials their broad scope in diverse areas of scientific research, furthermore, it poses toxic effects on distinct plant species. Arvidsson et al. (2013) suggested that the fate of graphene in the ecosystem and its toxicity needed additional study. A few reviews have been published on the toxic effects of GNMs in many organisms and cell models (Ou et al., 2016). Moreover, toxicity, uptake, and translocation of designed nanomaterials in vascular plants was observed (Miralles et al., 2012). Begum and his team reported that an excessive concentration of graphene between 500 and 2000 mg/L considerably reduced plant growth and the biomass of cabbage, tomato, and red spinach, except for lettuce (Begum et al., 2011). The interaction mechanism of graphene and its derivatives on root and shoot growth, biomass, shape, reduction of the antioxidant enzyme, cell death, and reactive oxygen species formation in cabbage, tomato, red spinach, and lettuce was investigated (Begum et al., 2011). Various morphological and physiological parameters mentioned that exposure to graphene drastically inhibited plant growth and decreased biomass. A dose-dependent reduction in the percentage and width of leaves was detected in graphene-treated plants. A specific concentration boosts reactive oxygen species production and leads to cell death in cabbage, tomato, and red spinach, implying that the harmful impact was mediated through oxidative stress. A comparable toxicity outcome was also exhibited on red spinach treated with 0–1000 mg/L of MWCNTs after 15 days of hydroponic culture, presenting a growth decrease and cell death (Begum et al., 2012). The phytotoxic issues significantly correlated with graphene induced-oxidative stress suggested by improved lipid peroxidation and increased antioxidant enzyme activity. Permanent exposure to grapheme triggered a decrease in shoot biomass, chlorophyll content, PSII activity, and levels of nutrient elements such as N, K, Ca, Mg, Fe, Zn, and Cu. The results showed that exposure to graphene decreased plant growth and photosynthesis and induced a difference in nutrient homeostasis. Researchers determined that graphene has a negative effect on plant growth, incorporating root hair reduction, oxidative burst, inhibition of photosynthesis, and nutritional disorder (Zhang et al., 2016). The prevalent toxicity symptoms viewed in plants treated with GFNs are serious damage to morphology and reduced growth in variables, consisting of root and shoot length, root number, root diameter, and biomass production (Table 1). To illustrate, the morphology of rice seedlings was significantly reduced if the graphene concentration reached 100 mg/L (Liu et al., 2015). However, GO in the amount of 10 μg/L could hardly stimulate important modifications in shoot and root development of Arabidopsis seedlings, or flowering time (Zhao et al., 2015). GO also negatively influenced biomass accumulation and stem elongation in wheat seedlings (Chen et al., 2017a,b). After exposure to 50 and 500 mg/kg rGO, several harmful issues on the shoot height and root length of rice seedlings were discovered by Hao et al. (2018); also, the root diameter and the number of cells in the transverse section considerably decreased. Unfunctionalized graphene oxide (GO) and amine-functionalized graphene oxide (G-NH2) on wheat demonstrated phytotoxicity in the concentration range from 125 to 2000 μg/mL after 9 days of hydroponic culture; GO revealed unfavorable results on the development of seedlings, including root cell death (Chen et al., 2018).

Table 1 Negative effects of graphene on the germination and growth of plant species.

Plant species	GFNs	Exposure concentration	Exposure time	Toxic symptoms of growth	References
Rice (*Oryza sativa*)	Graphene	100 and 200 mg/L	16 days	Inhibition of root and stem length, adventitious root number, and root and stem fresh weight	Liu et al. (2015)
	rGO	50, 500 mg/kg	30 days	Reduced shoot height and root length, decreased root diameter and number of cells in transverse section	Hao et al. (2018)
Wheat (*Triticum aestivum*)	GO	>1000 μg/mL	9 days	Decrease in root length, shoot length, and relative biomass; obvious damage to plant tissue structures	Chen et al. (2018)
	Graphene	250, 500, 1000, and 1500 mg/L	30 days	Root hair reduction, oxidative burst, photosynthesis inhibition, and nutritional disorder	Zhang et al. (2016)
		200 mg/L	5 days	Inhibition in the number of wheat roots	Hu et al. (2014)
Maize (*Zea mays*)	GO	100, 500, 1500 mg/L	15 days	Significant decrease of shoot and root weight	Yin et al. (2018)
Brassica napus	GO	25–100 mg/L			
50-100 mg/L	15 days	Shorter seminal root length Decrease in fresh root weight	Cheng et al. (2016)		
Cabbage (*Brassica oleracea*), tomato (*Lycopersicon esculentum*), red spinach (*Amaranthus tricolor*and *A. lividus*)	Graphene	500–2000 mg/L	20 days	Significant inhibition of plant growth and biomass. Decrease in the number and size of leaves	Begum et al. (2011)

2.1.2 Phytotoxic effects of fullerenes

A small number of investigations evaluated the impact of fullerene C60 on seed germination and plant growth in various crops. A huge percentage of nanoscience research highlighting the effects of C60 on marine and terrestrial plants recorded no negative or positive effects of C60 on plant growth (Tao et al., 2015). On the other hand, not all functionalized fullerenes display a stimulus impact on plant growth. Growing rice plants treated with C70 fullerene were postponed by at least 1 month and their seed-setting rate decreased by 4.6% in comparison with the controls (Lin et al., 2009). Liu and his coworkers studied the inhibitive effects of water-soluble fullerene C70 (C(COOH)2) 4-8 at the cellular level by using transgenic seedling lines providing fluorescent DNA makers (Liu et al., 2010). Root growth retardation and reduction of root geotropism were detected while fluorescence imaging discovered the abnormalities of root tips in hormone dismissions, cell division, microtubule organization, and mitochondrial activity. Mushtaq (2011) examined the reduction results of carbon nanoparticles (size 30–50 nm; 0–5000 μg/mL) on seed germination and root elongation of cucumber. The inhibitory effect was more apparent in root growth compared to the seed germination rate. Root growth reduction was also observed in *C. pepo* by the application of fullerene NP (Kelsey and White, 2013). Liu et al. (2013) offered the most important direct proof for the difference of the plant cell wall structure below the fullerene application using atomic force microscopy ligand-receptor binding force data to the living plant cell. They analyzed the variations of the tobacco plant cell wall (*Nicotiana tobacum*) after treatment with water-soluble carboxy fullerenes (C70(C(COOH)2) (2-4). It was detected that the C70(C(COOH)2) (2-4) were adsorbed to the cell wall, which resulted in the interruption of the cell wall and membrane resulting in cell growth prohibition. A time- and dose-dependent increase of glycosyl residue on the cell wall accompanied by improved reactive oxygen species generation was detected.

Jiang et al. (2014) recognized that the existence of 150 μg/mL of carbon nanotubes reduced the root length, root activity, and stem length of the germinated seedlings of rice. The current findings demonstrated the phytotoxicity of excessive concentrations of carbon nanotubes on seed germination and root growth. Kumar et al. (2015) examined the influence of nano-ZnO, nano-Fe_2O_3, and fullerene NPs on seed germination in diverse plant species. Among NPs, ZnO NPs triggered significant inhibition in seed germination in the tested species, thus cucumber was determined to be highly susceptible and rice slightly susceptible to NP phytotoxicity. However, nano-Fe_2O_3 and fullerene NPs revealed there was simply no impact on seed germination. Wang et al. (2016) explored the bioaccumulation of water-soluble fullerene nanoparticles on wheat seedlings using 13 C-labeling techniques. Soil processed with fullerene nanoproducts such as C60 or nC60 provided a modest effect on the composition and action of the soil microorganisms similar to Eubacteria and Kinetoplastida (protozoans). The population diversity was investigated using denaturing gradient gel electrophoresis profiles and showed 70% similarity level (Johansen et al., 2008; Tong et al., 2007). While nC60 accumulated in smaller sizes could have adverse effects on soil microbiota, soil organic matter may possibly perform an important role in modulating the environmental influence of applied nanomaterials (Tong et al., 2016). Fullerol of a suitable size and concentration will probably consequently be applied to boost the yield of cereal and fruit crops; however, the negative effect on the ecosystem should be evaluated prior to proposing its application in the open field (Husen and Siddiqi, 2014). These results indicate that the plant supply of fullerenes and/or organic pollutants is dependent not simply on substrate characteristics, but even on variations among plant species. There is a need for an extra investigation to explore the fate of fullerenes in diverse plant systems.

2.1.3 Phytotoxic effects of carbon nanotubes

Plant scientists have proven that the treatment of carbon nanotubes to plant seeds may boost the germination ratio and increase the growth of seedlings; also, some slight negative effects were observed. Single-walled carbon nanotubes (SWCNTs) of lengths less than 500 nm labeled with fluorescein isothiocyanate penetrated the cell wall of the living plants by endocytosis (Samaj et al., 2004). The current findings revealed that though fluorescein isothiocyanate alone is certainly not easily absorbed by the plants, both of them collectively helped with the absorption and penetration of carbon nanomaterials. Nevertheless, Stampoulis et al. (2009) recorded that multiwalled carbon nanotubes (MWCNTs) (13–16 nm) in zucchini seedlings at a concentration of 1000 mg/L confirmed decreased biomass, but their seed germination continued to be unaffected. Shen et al. (2010) determined that specific quantities of SWCNTs are able to stimulate the formation of reactive oxygen species (ROS), which usually contributes to programmed cell death, in Arabidopsis leaves and protoplasts. Some plant species sensitive to the toxicity of SWCNTs were analyzed by Canas et al. (2008). The treatment of a SWCNT was investigated for a considerably damaged root elongation of tomato, cabbage, carrot, and lettuce. In addition, tomato plants revealed the greatest level of sensitivity to SWCNTs. The functionalized SWCNTs may have a diverse toxic actions and are much less poisonous when compared to nonfunctionalized SWCNTs. The report outlined the benefit of examining the surface characteristics of carbon nanotubes in identifying their phytotoxicity. Pourkhaloee et al. (2011) mentioned that in *S. macrosiphon, C. annuum,* and tall fescue *F. arundinacea,* exposure to excessive concentrations of SWCNTs has influenced seedling growth. Smirnova et al. (2011) have proven the existence of MWCNTs from the Taunit interior of the cells and seedling root tissues. The data highlighted the capability of MWCNTs to enter and aggregate inside the roots as well as their particular potential to be moved to the seedling leaves. In red spinach (*Amaranthus tricolor* L), the phytotoxicity of MWCNTs triggers growth prohibition and cell death; MWCNTs likewise induce ROS production and hypersensitive response-type necrotic lesions of leaf cells and tissues (Begum and Fugetsu, 2012).

Maize seedlings treated with SWCNTs activated a reduction in root hair growth. A gene transcription assay revealed that seedling exposure to SWCNTs decreased the root hair based on the associated gene (RTH1, RTH3) expression (Yan et al., 2013). Also, treatment with SWCNTs upregulated the expression of epigenetic modification enzyme genes, leading to the global deacetylation of histone H3. It was determined that exposure to SWCNTs improved the histone deacetylation as a result of amassing the nanoparticles in the root apex. The interaction between MWCNTs and the vegetable/commercial crops such as red spinach, lettuce, rice, cucumbers, chili, lady's fingers, and soybeans was evaluated, root, shoot growth, and plant cell death was recognized at the seedling stage by Begum et al. (2012), while red spinach, lettuce, rice, and cucumber experience great toxicity from MWNTs at the seedling stage (Begum et al., 2014). Both positive and harmful impacts activated by the same concentrations and types of MWCNTs are generally recognized, according to the plant developmental phase and the growing conditions (Zaytseva and Neumann, 2016). Haghighi and da Silva (2014) confirmed that SWCNTs from 8 to 15 nm with a concentration of 10–40 mg/L led to phytotoxicity by reducing the percentage of germination in *Raphanus sativus* and *Brassica rapa*. Additional SWCNTs (50–800 μg/mL) investigated a promise to improve drought stress up to modest levels and caused inhibition in germination and growth features of *Hyoscya musniger* (Hatami et al., 2017). Valdimir et al. (2019) offered new findings of long-term phytotoxicity evaluations of SWCNTs, MWCNTs, and nanodiamonds (ND), utilizing the cactus species *Ferocactus latispinus, Melocactus matanzanus,* and *Parodia ayopayana*. They found small and poor growing plants that are simple for long-term analysis utilizing substrate quantities as small as a few milliliters, in contrast to the generally recorded tests with various crop species. After the first three months of exposure, Valdimir and coworkers determined that nanomaterials may generate a relatively beneficial impact on seedling growth, which may change soon after extra prolonged exposure. Consequently, cautious analysis of the possible phytotoxic effects needs to involve assays for many months while shorter tests may possibly generate inaccurate results.

2.2 CNMs toxicity for soil-microbes

Soil microorganism populations have an immediate effect on soil health and quality throughout procedures including beneficial nutrient cycling, degradation of organic and natural matter, and symbiotic interactions with various plant species (Kennedy and Smith, 1995). As a result, safeguarding the soil microbial biomass and variability is the main obstacle in the agriculture sector. Presently, controlled information is obtainable on the relationship between CNTs and soil microorganisms. Because of the reduced size and modified physical, chemical, and structural characteristics of carbon nanomaterials, the absorption and translocation of diverse forms of carbon nanomaterials increase severe issues regarding their toxic impact on plants and the microbial community in soil. Minor data is presently observed about plant uptake of carbon nanotubes, or about toxicity mechanisms on different plant species. CNMs can be specifically toxic to soil microbes, may modify the bioavailability of nutrients, or can boost or decrease the toxicity of natural and organic substances and/or toxins (Dinesh et al., 2012). Furthermore, phytotoxicity might ultimately have an effect on microbial populations. CNTs have a special effect on soil microbial activity, which is contentious and only a little analyzed. Nevertheless, most of these reports appear to determine that CNTs reduced soil microbial activity (Chung et al., 2011; Ge et al., 2016; Jin et al., 2013, 2014). Boonyanitipong and his team discovered that about 10 g of MWCNTs per kg of soil is likely to boost the degraders of difficult pollutants such as PAH. Researchers ascribed this unique effect to the acidic nature of the applied MWCNT, which triggered a critical decline in soil pH at larger exposure concentrations and therefore the soil microbial population and its chemistry were entirely altered (Boonyanitipong et al., 2011). Enzymatic actions of soil bacteria were suppressed by both MWCNTs and SWCNTs because MWCNTs reduced the enzymatic activities of two types of natural soils, including sandy loam and loamy sand soils at 500 mg/kg (Chung et al., 2011). Jin et al. (2013) identified that SWCNTs decreased considerably the enzyme activities of natural sandy loam soil at concentrations ranging from 30–300 mg/kg. Much more interestingly and by using a metagenomic relative examination for the bacterial communities and their diversity, Khodakovskaya et al. detected that the bacterial diversity was not considerably influenced by MWCNTs while a notable change in the bacterial composition and their species was recorded (Jin et al., 2014; Khodakovskaya et al., 2013). In harmony with the Rodrigues review, the bacterial soil community was influenced by the existence of SWCNTs with a significant effect after 3 days, but bacteria restored totally after two weeks (Rodrigues et al., 2013). In contrast, Du et al. (2015) investigated that graphene oxide stimulated a particular improvement in a few nitrogen-fixing and dissimilatory iron-reducing bacteria, with following the compositional analysis in the whole soil bacterial population. Gurunathan (2015) investigated the harmful impact of graphene oxide nanoparticles (GO NPs) on collecting Rhizobacteria from the soil (five Bacillus species: *B. megaterium, B. cereus, B. subtilis, B. mycoides*, and *B. marisflavi*). Two weeks later, they documented that the soil's bacterial group variations were damaged by MWCNTs at maximum concentrations. Shrestha et al. (2013) detected a considerable change in the microbial populations in soils modified with MWCNTs and isolated from the alfalfa rhizosphere.

Shan et al. (2015) explored the negative effects of biochar, activated carbon (AC), and SWCNTs and MWCNTs in different concentrations (0, 0.2, 20, and 2000 $mg\,kg^{-1}$ dry soil) on the fate of 14C-catechol and microbial populations in soil. They recorded that SWCNTs at 2000 $mg\,kg^{-1}$ considerably decreased the mineralization of agricultural soil by bacteria. Ge et al. (2016) generated an interesting result regarding the impact of MWCNTs in comparison to organic or commercial carbonaceous materials on soil microorganism populations employing long-term reports on dry soil. They observed that MWCNTs minimized soil DNA diversity and modified bacterial communities soon after 1 year of treatment. Tong et al. (2016) recommended that nC60 accumulated in smaller sizes could have an adverse effect on soil microbiota and that soil organic matter might play an important function in modulating the environmental influence of various nanomaterials. The mode of action of CNM toxicity on microorganisms is related to animals. The main toxicity mechanism is due to oxidative stress (Zhu et al., 2017). However, studies applied to understand the toxicity mechanism of CNMs are significantly less than expected. CNTs can cause conformational improvements in microbial enzymes and impact the enzymatic catalytic procedure of microbes, which will, even more, bother the microbial metabolism (Chen et al., 2017b). To date, there are not enough investigations presented to determine a probable diverse effect between functionalized and unfunctionalized CNTs on soil microbial populations and activities.

2.3 CNM phytotoxicity mechanism

In carbon toxicology, extra reports are presented and researchers confirmed that short CNTs (30–100 nm) move more quickly because of their size and are less toxic because they are eradicated easily. Longer CNTs were compared to asbestos. This carbon form can be characterized as the most toxic because CNTs may enter various organisms, but they cannot be removed from the ecosystem and thus may possess toxic effects (Donaldson et al., 2013; Jaurand et al., 2009). Due to CNT formation and uses that are still rising, their distribution into the environment can keep growing. Therefore, it is very important to better examine CNT behavior and its effects on the agricultural environment. More reports are urgently required to recognize mechanistic pathways of puncture and biodistribution of CNTs in plants and microorganisms to provide practical and secure CNT use. Additionally, it is necessary to examine the influence of physicochemical variables of CNTs on their effects. Understanding the effects of these details will allow constructing CNTs to be "safer by design" (Liné et al., 2017). The phytotoxicity mechanism is still unclear and needs to be overestimated. Recent developments in the use of carbon nanotubes in plant systems have brought renewed interest in advanced applications, but at the same time, toxicity issues remain under serious discussion. The use of carbon nanotubes in plant systems can eventually increase human and animal exposure through the simplest approaches in the food chain. Recently, several investigations discussed the environmental effects and the security profile of carbon nanomaterials (Kaphle et al., 2017; Shen et al., 2010; Stampoulis et al., 2009). Some possible factors responsible for the toxicity of carbon nanomaterials in plant systems (Fig. 2). The toxicity of carbon nanomaterials is mostly due to the increase in oxidative stress and ROS generation in the formation of aggregates and clusters, causing the blocking of pores while interacting with plant cell walls (Bhati et al., 2018). When NPs' entrance into the plant cell may be degraded the cell wall, the specific characteristics of the cell wall permitting NP transfer could be attributed to the pore size of the cell wall (Fleischer et al., 1999). The NPs that are in the size range within the cell wall pore size could cross the cell wall and reach the plasma membrane (Navarro et al., 2008). The rate of entry relies on the size and surface areas of NPs. In fact, the smaller NPs can enter plant cells immediately. On the other hand, bigger NPs, as they can't enter cells, simply cannot influence the cell metabolic pathways (Verano-Braga et al., 2014). Larger NPs can only permeate through the hydathodes,

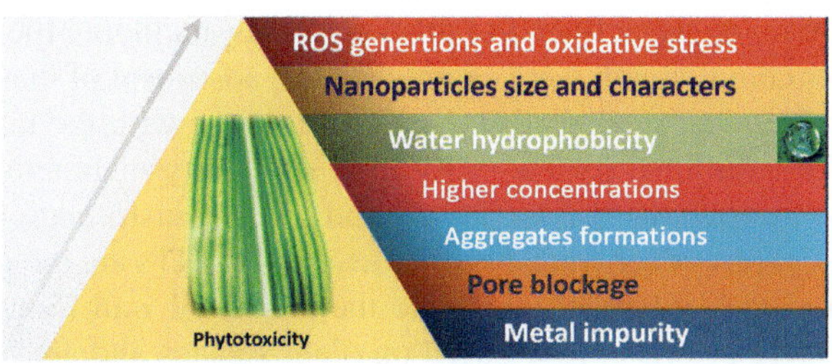

Fig. 2 A schematic representation of the factors responsible for the toxicity of nanocarbons in plant systems.

flower stigmas, and stomata. The mode of action between NPs and plants is based on chemical or physical properties. Chemical interactions include the formation of reactive oxygen species (Nel et al., 2006), disruption of ion cell membrane transport activity (Auffan et al., 2008), oxidative damage (Foley et al., 2002), and lipid peroxidation (Kamat et al., 2000).

Dietz and Herth (2011) clearly explained five models for the interaction between nanomaterials and various organisms: (1) metal ions introduced by nanoparticles in solution generate a chemical effect; (2) the hard spherical particles develop mechanical effects; (3) the nanoparticle surface delivers catalytic effects; (4) the nanoparticle surface binds proteins, either by noncovalent or covalent mechanisms, or causes oxidative effects; and (5) the nanoparticles change the chemical environment, especially the pH. The mechanism of phytotoxicity may include elements of the five models mentioned above and other effects. These adverse effects may be caused by auxin disruption, abnormal cell division, and microtubule disorganization. Furthermore, the fullerene effect on fruit and crop production in edible plants and vegetables, may increase the water retaining capacity, biomass, and fruit yield in plants up to ~118%. The internalization and accumulation of these compounds in plant roots and seedlings and their phytotoxicity have been investigated by Liu et al. (2013).

Ghosh and his team discovered that the internalization of MWCNTs interior the plant cells leading to chromosomal deviation, DNA splitting, and apoptosis in Allium root cells (Ghosh et al., 2015). Shen et al. (2010) observed that different concentrations of SWCNTs can stimulate the formation of reactive oxygen species (ROS), which usually causes programmed cell death in Arabidopsis leaves and protoplasts. In red spinach (*Amaranthus tricolor* L), the MWCNT phytotoxicity triggers growing prohibition and cell death; MWCNTs additionally cause ROS formation and hypersensitive response-type necrotic lesions of leaf cells and tissues (Begum and Fugetsu, 2012). While plant cells and mammalian cells have specific structures, which usually consist of solid and restricted plant cell walls, chloroplasts, and large central vacuoles, they demonstrate identical reactions to fullerene (Chen et al., 2010). The bioaccumulation of dichloro-diphenyl-dichloro-ethylene, a persistent and estrogenic pollutant, in some food crops such as zucchinis, soybeans, and tomatoes, increased in the presence of the C60 fullerene (Avanasi et al., 2014; De La Torre-Roche et al., 2013). Avanasi and his coworkers have assessed plant consumption of C60 fullerene and determined that 14C-labeled C60 may be gradually consumed by plants and can probably persevere in the soil for a prolonged time (Avanasi et al., 2014). The modification activated by carboxy-fullerenes to the cell wall of tobacco plants, including the mode of action and survival under containment, were examined. The adsorbent of these nanoparticles generated the dysfunction of the cell wall and membrane with the major prohibition of cell growth. An additional review proved the capability of functionalized carbon nanotubes to pass through the plant cell wall and cause transformations, in particular organelles. Also, the physicochemical properties of SWCNTs are the main preventing features for their successful penetration into the plant cell wall (Serag et al., 2013). The growth and yield profiles differ based on plant species and are based mostly on the number of CNMs used. Scientific reports observed an improvement in vegetative growth and yield of fruits/seeds at a lower concentration of CNMs, but a decline in these plant parameters was also documented once bigger concentrations of CNMs were employed (Verma et al., 2019). CNMs can improve nutrients and water consumption in plants, and CNTs increase seed germination in multiple plant species. MWCNTs assist in water movement and boost nitrogenase activity; also, carbon dots stimulate the plant photosystem (Fig. 3). In summary, diverse studies on CNT phytotoxicity and safety were performed and it was shown that functionalization minimizes the toxic impact while pristine CNTs induce different phytotoxic special effects. However, additional studies are required. A few examples of the mechanism of phytotoxic effects of different carbon nanomaterials on diverse plant species are shown in Table 2.

2.3.1 Molecular mechanism

In order to recognize the practical benefits and phytotoxicity of applying nanomaterials in the agriculture sector, the primary step must be to evaluate the transmission and adsorption of nanoparticles into various plant species (Gao et al., 2006). When researchers treated *Spinacia oleracea* with nano-anatase TiO2, the Rubisco carboxylase activity increased 2.67 times compared to the control. The molecular mechanism of carbon reaction offered by nano-anatase TiO2 during photosynthesis was still not definitely recognized. Molecular markers such as reverse transcription PCR and Northern blotting experiments proved that Rubisco small subunits and Rubisco large subunit messenger RNAs were elevated in the nano-treated plants (Xuming et al., 2008). Correctly, the protein expression of Rubisco from the nano-anatase-treated spinach was raised by 40% in contrast to the control. Evaluation of improvements in gene expression can be obtained by using great performance methods incorporating cDNA microarrays or quantitative real-time PCR. This shows a practical technique for the detection of the mechanisms of toxicity and molecular reactions in cells subjected to nanomaterials (Xu et al., 2011). The function of toxicogenomic tools in regulating environmental toxicology and risk analysis was examined (Ankley et al., 2006; Boverhof and Zacharewski, 2005; Magdolenova et al., 2014). Gene expression studies are commonly carried out to enhance morphological and/or physiological inspections and present exclusive data on specific toxicity pathways and mechanisms that simply cannot be acquired specifically from additional strategies (Ankley et al., 2006; Dietz and Herth, 2011). Furthermore, specific gene expression behavior may perhaps support the advancement and approval of offering biomarkers appropriate for high-efficiency verification tools, and for enhanced awareness of the phytotoxicity of nanoparticles (Merrick and Bruno, 2004; Thomas et al., 2011). Changes in gene expression have been shown to be induced by very low doses of contaminants and could support the analysis of the cellular effects of long-term phytotoxicity produced by some nanoparticles (Poma and Di Giorgio, 2008; Poynton and Vulpe, 2009). Khodakovskaya and his team in 2012 studied that CNTs discovered an upregulation of many genes, consisting of an aquaporin (NtPIP1) and two genes, namely, CycB and NtLRX, included in water transportation, cell wall development, and cell division. CNTs apparently control tobacco water channels (aquaporins) and have a strong effect on plant transcriptome and different necessary genes. MWCNTs are generally discovered to express the water channel gene (NtPIP1) and the creation of the related NtPIP1 protein in tobacco cells (Khodakovskaya et al., 2012).

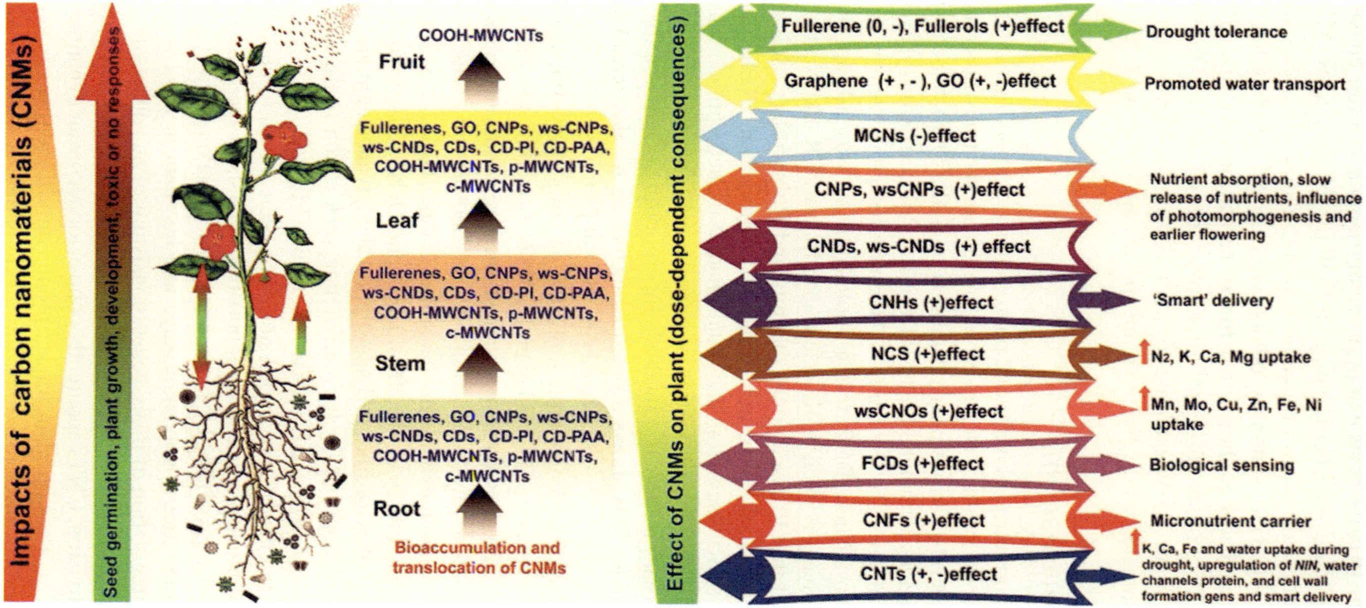

Fig. 3 Uptake, transmission, translocation, and accumulation of CNMs (fullerenes, fullerols, CNPs, and CNTs) CNMs' in different parts of the plants are still in infancy; especially uptake, translocation and accumulation of the CNMs in the plant body and their role in affecting the plant growth and plant's response to imparted toxicity.

Reprinted from Verma, S.K., Das, A.K., Gantait, S., Kumar, V., Gurel, E., 2019. Applications of carbon nanomaterials in the plant system: a perspective view on the pros and cons. Sci. Total Environ. 667, 485–499 with permission from Elsevier.

Table 2 The mechanism of phytotoxic effects of different carbon nanomaterials on different plant species (updated from Mohamed et al., 2018 with permission from Springer).

Carbon NMs	Plant	Mechanism	References
Graphene oxide (GO)	Arabidopsis	Fragmented nuclei, membrane damage, and mitochondrial dysfunction	Begum and Fugetsu (2012)
	Faba bean	Concentration-dependent decrease in oxidative enzyme activity	Anjum et al. (2014)
Water-soluble graphene oxide (ws-GO)	Lettuce, cabbage, red spinach, tomato	Reduced plant growth, biomass, the number and size of leaves, increased ROS along with necrotic symptoms	Begum et al. (2011)
CNTs	Red spinach	Growth inhibition, changes to tissue structure	Begum and Fugetsu (2012)
MWCNTs	Rice cells	Increased ROS generation and decreased cell viability	Tan et al. (2009)
	Rice	Delayed flowering and seed setting. Reduced seed weight	Lin et al. (2009)
	Rice	Chromatin condensed inside the cytoplasm and caused cell death, plasma membrane detachment from cell wall, and cell shrinkage	Tan et al. (2009)
	Zucchini	Negatively affected biomass production and transpiration	Stampoulis et al. (2009)
	Wheat	Enhanced the uptake of phenantrene to the living cells	Wild and Jones (2009)
	Onion	Chromosomal aberrations, DNA fragmentation, and apoptosis in root cells	Ghosh et al. (2015)
	Onobrychis	Enhanced POD activity	Smirnova et al. (2011)
Fullerene C60	Arabidopsis	Stress response of gene expression	Landa et al. (2012)
C60(OH)20	Allium	Plant cell damage	Chen et al. (2010)
ws-C70	Tobacco BY-2 cells	Cell boundary disruption and growth inhibition	Liu et al. (2013)

In a study with *A. thaliana*, CNP treatments have stimulated quick flowering in both long-day and short-day growth conditions, which signifies a photoperiod-dependent effect (Kumar et al., 2018). The researchers identified the upregulation of the repressor of gibberellic acid 1 (RGA1), the early flowering that is most likely gibberellic acid-independent. They recommend that the quick flowering *A. thaliana* seedlings treated with CNP are phytochrome B and photoperiod-dependent.

2.3.2 Tools for detection of phytotoxicity mechanism

Integrated omics techniques may present a more detailed review of the complexities involved with nanomaterial phytotoxicity, in contrast to a single omics way. Therefore, these modern advances play a role in extending the safety analysis of nanotoxicity and offer correct detection of plant diseases (Fröhlich, 2017; Shin et al., 2018). Most reports conducted within the last 10 years were devoted to the effects of NPs on plants at the morphological and physiological levels; however, the serious impacts of NPs at the molecular level has not attracted enough interest. "PlantNanOmics" is a recently rising theme in agricultural research that examines the impact of nanomaterials on plant growth (Kole et al., 2016). Positive effects from the development of systems biology and multiple omics methodologies, such as transcriptomics, proteomics, and metabonomics, are usually utilized in the upcoming investigation to adequately evaluate the phytotoxicity mechanism of NPs and resistance mechanisms in plants (Ruotolo et al., 2018; Yan and Chen, 2019). Phytotoxicity mechanism detection methods including genomics, epigenomics, transcriptomics, proteomics, metabolomics, integrated multiomics tools, physiological methods (physionomics), and a phenomics approach to determine toxicity measurements in plants, nanotoxicogenomics, and nanotoxicoproteomics are described below (Fig. 4).

Nanotoxicogenomics method

Toxicogenomics terminology identifies the application of genomics methods to determine toxicology mechanisms (Nuwaysir et al., 1999). Examination of variations in gene expression throughout molecular techniques, including cDNA microarrays or quantitative real-time PCR (qRT-PCR), indicates a strong strategy for recognizing the mechanisms of toxicity and molecular results in cells subjected to nanomaterial treatment (Xu et al., 2011). The role of toxicogenomics in regulating environmental toxicology and risk examination was discussed (Ankley et al., 2006; Boverhof and Zacharewski,

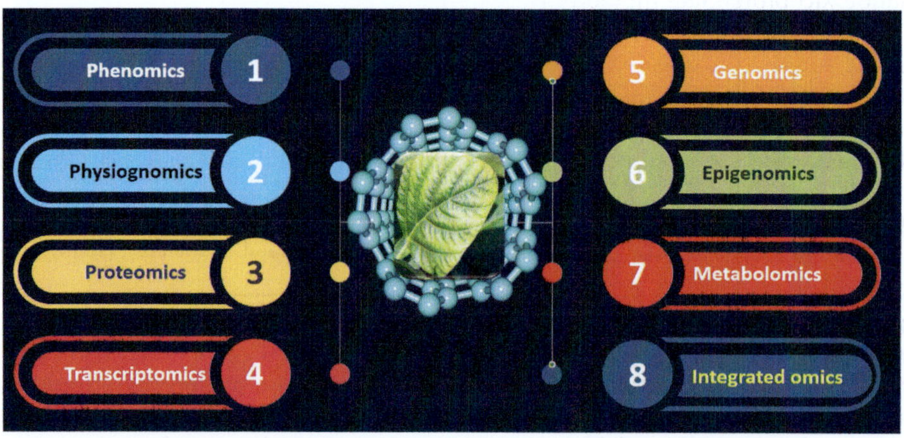

Fig. 4 Morphological, physiological, and molecular methods used for detection phytotoxicity mechanisms such as genomics, epigenomics, transcriptomics, proteomics, metabolomics, *integrated* multi*omics,* physiological methods (physionomics), and phenomics techniques.

2005; Magdolenova et al., 2014). Gene expression studies are commonly carried out to evaluate morphological and/or physiological inspections and present specific details on certain toxicity pathways and mechanisms that could not be provided from several other methods (Ankley et al., 2006; Dietz and Herth, 2011). Furthermore, specific gene expression profiles can assist in the design and approval of promising biomarkers appropriate for high-efficiency testing techniques, and for greater understanding of the toxicity of various nanoparticles (Merrick and Bruno, 2004; Thomas et al., 2011). Variations in gene expression are generally demonstrated to be activated by minimal doses of pollutants and could support evaluation of the cellular effect of chronic toxicity linked with nanoparticles (Poma and Di Giorgio, 2008; Poynton and Vulpe, 2009). Nevertheless transcriptional examinations had been broadly applied to analyze the molecular mechanism of nanoparticle toxicity in diverse organisms, including microbes, humans, mammalian cell lines, and other types of organisms (Asharani et al., 2009). A few reports have been carried out to determine the molecular mechanism of nanoparticle-plant interactions and nanoparticle-mediated phytotoxicity (Aken, 2015; Remédios et al., 2012). These results could present a clear relationship between the plant phenotype and the connected genes influenced by nanoparticles, and clarify the molecular genetic mechanisms behind the toxic effects of nanoparticles on plants. SWCNTs have been proven to have negative issues with Arabidopsis and rice leaf protoplasts by using oxidative stress, producing a particular quantity of programmed cell death (PCD)/apoptosis, DNA destruction, and chromatin fusion, as exposed by terminal deoxynucleotidyl transferase dUTP nick end-labeling (TUNEL)-positive reaction (Shen et al., 2010). Alternatively, treat maize seedlings with SWCNTs that are able to stimulate the growth of seminal roots in *Hemarthria compressa* (Poaceae) (Yan et al., 2013). These results are associated with differential expression and upregulation of the affiliated genes encoding epigenetic amendment enzymes, resulting in global deacetylation of histone H3, identical to various other abiotic stress response mechanisms.

Nanotoxicoproteomics method

The term toxicoproteomics is a particular part of toxicology that includes the proteomic research of toxicity induced by external stress inducers, such as produced chemicals, engineered nanomaterials, or any additional toxic material that may trigger significant cellular reactions (Gao et al., 2009; Wetmore and Merrick, 2004). Nanotoxicoproteomics strategy combines concepts and techniques of toxicology and proteomics to explain the molecular mechanism of nanoparticle toxicity. Practical proteomics is usually applied as an effective approach for abiotic stress-related analysis in different crops through the detection of new stress-responsive proteins (Barkla et al., 2013). Recognizing the mechanics of expression of targeted proteins could present special observations into their function and interactions. Alteration in protein accumulation under nanoparticle stress is closely interrelated to plant phenotypic response, as variations at the transcript level usually are not usually in agreement with modification at the protein level. Consequently, the examination of alterations in the plant proteome is extremely crucial because proteins are special effectors of the plant stress response. Evolution of cellular reactions in protein quantity after treating plants with nanomaterials may considerably lead to clarifying our awareness of physiological mechanisms fundamental to plant-nanoparticle interaction. Differential protein expression patterns signify the core of proteomic methods, and many approaches are presented for the differential examination of protein expression. Two-dimensional polyacrylamide gel electrophoresis (2-DE) is still a major technique for this, and densitometric analysis of 2-DE gels along with mass spectrometry (MS) has allowed plant experts to determine differentially expressed protein bands. Protein chips and microarrays also appear to be a powerful device for the detection of protein interaction with other biomolecules. Thus, proteomic analyses may possibly result in recognition and description of protein markers under nanoparticle exposure, which could be applied as promising biomarkers for carbon nanomaterial phytotoxicity.

3 Conclusion and future trends

Recently, several research reports have confirmed the beneficial and also negative impact on various plants, according to the CNM groups, concentration, time of exposure, plant species, and growing conditions on which these are employed. Nevertheless, multiple excellent functions of carbon nanotubes were examined in the majority of the reports, and only a few carbon nanomaterials at particular concentrations have harmful or no critical effects on plant cells based on their concentration, size, shape, ultrasonication, etc. Engineered carbon nanomaterials exhibit unusual properties and diverse capabilities on numerous plants. CNMs can easily trigger adverse effects on plants by creating reactive oxygen species in plant cells, thus ultimately causing cell death. Microbial soil communities have a highly valuable role in preserving the soil environment, soil health, and plant yields. These types of microbial soil and plant roots may be impacted, positively as well as negatively, by manufactured NPs, but the available data on this particular feature is still restricted. In the current chapter, we made an effort to clarify the positive and negative effects of diverse carbon nanomaterials on plants and rhizospheric microorganisms. There is still a significant knowledge gap that may be required to cover the relationship between CNMs and phytotoxicity. However, mechanism knowledge concerning the dynamics of CNM interactions with plants or the microbial community and their effect on plant growth and development is still ambiguous. Nanoscience researchers need to target different techniques, including the incorporation of plant-omics techniques and physiological phenotyping, to clarify the carbon nanomaterial phytotoxicity and tolerance mechanisms in plants. Also, careful evaluation of the possible phytotoxic effects must include investigations for several months while shorter assays might generate deceptive findings.

References

Afreen, S., Omar, R.A., Talreja, N., Chauhan, D., Ashfaq, M., 2018. Carbon-based nanostructured materials for energy and environmental remediation applications. In: Prasad, R., Aranda, E. (Eds.), Approaches in Bioremediation, Nanotechnology in the Life Sciences. Springer, Cham.

Aken, B.V., 2015. Gene expression changes in plants and microorganisms exposed to nanomaterials. Curr. Opin. Biotechnol. 33, 206–219.

Al-Halafi, A.M., 2014. Nanocarriers of nanotechnology in retinal diseases. Saudi J. Ophthalmol.

Anjum, N.A., Singh, N., Singh, M.K., Sayeed, I., Duarte, A.C., Pereira, E., Ahmad, I., 2014. Single-bilayer graphene oxide sheet impacts and underlying potential mechanism assessment in germinating faba bean (Vicia faba L.). Sci. Total Environ. 472, 834–841.

Ankley, G.T., Daston, G.P., Degitz, S.J., Denslow, N.D., Hoke, R.A., Kennedy, S.W., Miracle, A.L., Perkins, E.J., Snape, J., Tillitt, D.E., 2006. Toxicogenomics in regulatory ecotoxicology. Environ. Sci. Technol. 40, 4055–4065.

Arruda, S.C.C., Silvabc, A.L.D., Galazzi, R.M., Azevedo, R.A., Arruda, M.A.Z., 2015. Nanoparticles applied to plant science: a review. Talanta 131, 693–705.

Arvidsson, R., Molander, S., Sandén, B.A., 2013. Review of potential environmental and health risks of the nanomaterial graphene. Hum. Ecol. Risk. Assess. Int. J. 19 (4), 873–887.

Asharani, P.V., Mun, G.L.K., Hande, M.P., Valiyaveettil, S., 2009. Cytotoxicity and genotoxicity of silver nanoparticles in human cells. ACS Nano 3, 279–290.

Auffan, M., Achouak, W., Rose, J., Roncato, M., Chanéac, C., Waite, D.T., Masion, A., Woicik, J.C., Wiesner, M.R., Bottero, J., 2008. Relation between the redox state of iron-based nanoparticles and their cytotoxicity toward *Escherichia coli*. Environ. Sci. Technol. 42, 6730–6735.

Avanasi, R., Jackson, W.A., Sherwin, B., Mudge, J.F., Anderson, T.A., 2014. C60 fullerene soil sorption, biodegradation, and plant uptake. Environ. Sci. Technol. 48 (5), 2792–2797.

Barkla, B.J., Vera-Estrella, R., Pantoja, O., 2013. Progress and challenges for abiotic stress proteomics of crop plants. Proteomics 13, 1801–1815.

Baughman, R.H., Zakhidov, A.A., De Heer, W.A., 2002. Carbon nanotubes the route toward applications. Science 297 (5582), 787–792.

Begum, P., Fugetsu, B., 2012. Phytotoxicity of multi-walled carbon nanotubes on red spinach (*Amaranthus tricolor* L.) and the role of ascorbic acid as an antioxidant. J. Hazard Mater. 243, 212–222.

Begum, P., Ikhtiari, R., Fugetsu, B., 2011. Graphene phytotoxicity in the seedling stage of cabbage, tomato, red spinach, and lettuce. Carbon 49 (12), 3907–3919.

Begum, P., Ikhtiari, R., Fugetsu, B., 2014. Potential impact of multi-walled carbon nanotubes exposure to the seedling stage of selected plant species. Nanomaterials 4 (2), 203–221.

Bhati, A., Gunture, K., Tripathi, K.M., Singh, A., Sarkar, S., Sonkar, S.K., 2018. Exploration of nano carbons in relevance to plant systems. New J. Chem. 42, 16411–16427.

Boonyanitipong, P., Kositsup, B., Kumar, P., Baruah, S., Dutta, J., 2011. Toxicity of ZnO and TiO_2 nanoparticles on germinating rice seed *Oryza sativa* L. Int. J. Biosci. Biochem. Bioinform. 1, 282–285.

Boverhof, D.R., Zacharewski, T.R., 2005. Toxicogenomics in risk assessment: applications and needs. Toxicol. Sci. 89, 352–360.

Canas, J.E., Long, M., Nations, S., Vadan, R., Dai, L., Luo, M., Ambikapathi, R., Lee, E.H., Olszyk, D., 2008. Effects of functionalized and nonfunctionalized single-walled carbon nanotubes on root elongation of select crop species. Environ. Toxicol. Chem. 27, 1922–1931.

Chen, R., Ratnikova, T., Stone, M., Lin, S., Lard, M., Huang, G., Hudson, J.S., 2010. Differential uptake of carbon nanoparticles by plant and mammalian cells. Small 6 (5), 612–617.

Chen, L., Wang, C., Li, H., Qu, X., Yang, S.T., Chang, X.L., 2017a. Bioaccumulation and toxicity of 13C-skeleton labeled graphene oxide in wheat. Environ. Sci. Technol. 51, 10146–10153.

Chen, M., Zeng, G., Xu, P., Yan, M., Xiong, W., Zhou, S., 2017b. Interaction of carbon nanotubes with microbial enzymes: conformational transitions and potential toxicity. Environ. Sci. Nano 4, 1954–1960.

Chen, J., Yang, L., Li, S., Ding, W., 2018. Various physiological response to graphene oxide and amine-functionalized graphene oxide in wheat (*Triticum aestivum*). Molecules 23, 1104.

Cheng, F., Liu, Y.F., Lu, G.Y., Zhang, X.K., Xie, L.L., Yuan, C.F., Xu, B.B., 2016. Graphene oxide modulates root growth of *Brassica napus* L., regulates ABA and IAA concentration. J. Plant Physiol. 193, 57–63.

Chung, H., Son, Y., Yoon, T.K., Kim, S., Kim, W., 2011. The effect of multi-walled carbon nanotubes on soil microbial activity. Ecotoxicol. Environ. Saf. 74, 569–575.

Cvjetko, P., Zovko, M., Štefanić, P.P., Biba, R., Tkalec, M., Domijan, A.M., Vrček, I.V., 2018. Phytotoxic effects of silver nanoparticles in tobacco plants. Environ. Sci. Pollut. Res. 25, 5590–5602.

De La Torre-Roche, R., Hawthorne, J., Deng, Y., Xing, B., Cai, W., Newman, L.A., Wang, Q., Ma, X., Hamdi, H., White, J.C., 2013. Multiwalled carbon nanotubes and C60 fullerenes differentially impact the accumulation of weathered pesticides in four agricultural plants. Environ. Sci. Technol. 47, 12539–12547.

Dietz, K.J., Herth, S., 2011. Plant nanotoxicology. Trends Plant Sci. 16, 582–589.

Dinesh, R., Anandaraj, M., Srinivasan, V., Hamza, S., 2012. Engineered nanoparticles in the soil and their potential implications to microbial activity. Geoderma 173–174, 19–27.

Donaldson, K., Poland, C.A., Murphy, F.A., Farlane, M.M., Chernova, T., Schinwald, A., 2013. Pulmonary toxicity of carbon nanotubes and asbestos similarities and differences. Adv. Drug Deliv. Rev. 65, 2078–2086.

Du, J., Hu, X., Zhou, Q., 2015. Graphene oxide regulates the bacterial community and exhibits property changes in soil. RSC Adv. 5, 27009–27017.

Fayez, K.A., El-Deeb, B.A., Mostafa, N.Y., 2017. Toxicity of biosynthetic silver nanoparticles on the growth, cell ultrastructure, and physiological activities of barley plant. Acta Physiol. Plantarum 39 (7). 155 pp.

Fleischer, A., Fleischer, A., O'Neill, M.A., Ehwald, R., 1999. The pore size of non-graminaceous plant cell walls is rapidly decreased by borate ester cross-linking of the pectic polysaccharide rhamnogalacturonan II. Plant Physiol. 121, 829–838.

Foley, S., Crowley, C., Smaihi, M., Bonfils, C., Erlanger, B.F., Seta, P., Larroque, C., 2002. Cellular localisation of a water-soluble fullerene derivative. Biochem. Biophys. Res. Commun. 294, 116–119.

Fröhlich, E., 2017. Role of omics techniques in the toxicity testing of nanoparticles. J. Nanobiotechnol. 15, 84.

Gao, F., Hong, F., Liu, C., Zheng, L., Su, M., Wu, X., Yang, F., Wu, C., Yang, P., 2006. Mechanism of nano-anatase TiO2 on promoting photosynthetic carbon reaction of spinach. Biol. Trace Elem. Res. 111, 239–253.

Gao, Y., Holland, R.D., Yu, L.R., 2009. Quantitative proteomics for drug toxicity. Brief. Funct. Genomic. Proteomic. 8, 158–166.

Ge, Y., Priester, J.H., Mortimer, M., Chang, C.H., Ji, Z., Schimel, J.P., Holden, P.A., 2016. Holden, Long-term effects of multiwalled carbon nanotubes and graphene on microbial communities in dry soil. Environ. Sci. Technol. 50, 3965–3974.

Geisler-Lee, J., Wang, Q., Yao, Y., Zhang, W., Geisler, M., Li, K., Huang, Y., Chen, Y., Kolmakov, A., Ma, X., 2013. Phytotoxicity, accumulation and transport of silver nanoparticles by *Arabidopsis thaliana*. Nanotoxicology 7, 323–337.

Ghosh, M., Bhadra, S., Adegoke, A., Bandyopadhyay, M., Mukherjee, A., 2015. MWCNT uptake in Allium cepa root cells induces cytotoxic and genotoxic responses and results in DNA hyper methylation. Mutat. Res. 774, 49–58.

Gurunathan, S., 2015. Cytotoxicity of graphene oxide nanoparticles on plant growth promoting rhizobacteria. J. Ind. Eng. Chem. 32, 282–291.

Haghighi, M., da Silva, J.A.T., 2014. The effect of carbon nanotubes on the seed germination and seedling growth of four vegetable species. J. Crop. Sci. Biotechnol. 17 (4), 201–208.

Hao, Y., Ma, C., Zhang, Z., Song, Y., Cao, W., Guo, J., Zhou, G., Rui, Y., Liu, L., Xing, B., 2018. nanomaterials alter plant physiology and soil bacterial community composition in a rice–soil–bacterial ecosystem. Environ. Pollut. 232, 123–136.

Hao, Y., Xu, B., Ma, C., Shang, J., Gu, W., Li, W., Hou, T., Xiang, Y., Cao, W., Xing, B., Rui, Y., 2019. Synthesis of novel mesoporous carbon nanoparticles and their phytotoxicity to rice (*Oryza sativa* L.). J. Saudi Chem. Soc. 23, 75–82.

Hatami, M., Hadian, J., Ghorbanpour, M., 2017. Mechanisms underlying toxicity and stimulatory role of single–walled carbon nanotubes in *Hyoscyamus niger* during drought stress simulated by polyethylene glycol. J. Hazard. Mater. 324, 306–320.

Hu, X., Mu, L., Kang, J., Lu, K., Zhou, R., Zhou, Q., 2014. Humic acid acts as a natural antidote of graphene by regulating nanomaterial translocation and metabolic fluxes *in vivo*. Environ. Sci. Technol. 48, 6919–6927.

Huang, C., Xia, T., Niu, J., Yang, Y., Lin, S., Wang, X., Yang, G., Mao, L., Xing, B., 2018. Transformation of 14C–labeled graphene to 14CO2 in the shoots of a rice plant. Angew. Chem. Int. Ed. 57, 9759–9763.

Husen, A., Siddiqi, K.S., 2014. Carbon and fullerene nanomaterials in plant system. J. Nanobiotech. 12, 16–25.

Jain, A., Ranjan, S., Dasgupta, N., Ramalingam, C., 2018. Nanomaterials in food and agriculture: An overview on their safety concerns and regulatory issues. Crit. Rev. Food Sci. Nutr. 58 (2), 297–317.

Jaurand, M.C.F., Renier, A., Daubriac, J., 2009. Mesothelioma: do asbestos and carbon nanotubes pose the same health risk? Part. Fibre Toxicol. 6, 16.

Jeevanandam, J., Barhoum, A., Chan, Y.S., Dufresne, A., Danquah, M.K., 2018. Review on nanoparticles and nanostructured materials: history, sources, toxicity and regulations. Beilstein J. Nanotechnol. 9, 1050–1074.

Jiang, Y., Hua, Z., Zhao, Y., Liu, Q., Wang, F., Zhang, Q., 2014. The effect of carbon nanotubes on rice seed germination and root growth. In: Zhang, T.C., Ouyang, P., Kaplan, S., Skarnes, B. (Eds.), Proceedings of the 2012 International Conference on Applied Biotechnology (ICAB 2012). Lecture Notes Electrical Eng. 250, pp. 1207–1212.

Jin, L., Son, Y., Yoon, T.K., Kang, Y.J., Kim, W., Chung, H., 2013. High concentrations of single-walled carbon nanotubes lower soil enzyme activity and microbial biomass. Ecotoxicol. Environ. Saf. 88, 9–15.

Jin, L., Son, Y., DeForest, J.L., Kang, Y.J., Kim, W., Chung, H., 2014. Single-walled carbon nanotubes alter soil microbial community composition. Sci. Total Environ., 466, 533–538.

Johansen, A., Pedersen, A.L., Jensen, K.A., Karlson, U., Hansen, B.M., Scott-Fordsmand, J.J., Winding, A., 2008. Effects of C60 fullerene nanoparticles on soil bacteria and protozoans. Environ. Toxicol. Chem. 27, 1895–1903.

Juganson, K., Ivask, A., Blinova, I., Mortimer, M., Kahru, A., 2015. NanoE–Tox: New and in-depth database concerning ecotoxicity of nanomaterials. Beilstein J. Nanotechnol. 6, 1788–1804.

Kamat, J.P., Devasagayam, T.P.A., Priyadarsini, K.L., Mohan, H., 2000. Reactive oxygen species mediated membrane damage induced by fullerene derivatives and its possible biological implications. Toxicology 155, 55–61.

Kaphle, A., Navya, N., Umapathi, A., Chopra, M., Daima, H.K., 2017. Nanomaterial impact, toxicity and regulation in agriculture, food and environment. Nanoscience in Food and Agriculture 5. In: Sustainable Agriculture Reviews. 26. Springer International, Publishing, AG, pp. 205–242.

Kelsey, J.W., White, J.C., 2013. Effect of C60 fullerenes on the accumulation of weathered p,p′–DDE by plant and earthworm species under single and multispecies conditions. Environ. Toxicol. Chem. 32, 1117–1123.

Kennedy, A.C., Smith, K.L., 1995. Soil microbial diversity and the sustainability of agricultural soils. Plant and Soil 170, 75–86.

Khodakovskaya, M.V., De Silva, K., Biris, A.S., Dervishi, E., Villagarcia, H., 2012. Carbon nanotubes induce growth enhancement of tobacco cells. ACS Nano 6, 2128–2135.

Khodakovskaya, M.V., Kim, B.S., Kim, J.N., Alimohammadi, M., Dervishi, E., Mustafa, T., Cernigla, C.E., 2013. Carbon nanotubes as plant growth regulators: effects on tomato growth, reproductive system, and soil microbial community. Small 9, 115–123.

Kole, C., Kumar, D.S., Khodakovskaya, M.V. (Eds.), 2016. Plant Nanotechnology: Principles and Practices. Springer.

Krug, H.F., 2014. Nanosafety research-are we on the right track? Angew. Chem. Int. Ed. 53, 12304–12319.

Kumar, S., Patra, A.K., Datta, S.C., Rosin, K.G., Purakayastha, T.J., 2015. Phytotoxicity of nanoparticles to seed germination of plants. Int. J. Adv. Res. 3, 854–865.

Kumar, A., Singh, A., Panigrahy, M., Sahoo, K., Panigrahi, K.C., 2018. Carbon nanoparticles influence photomorphogenesis and flowering time in Arabidopsis thaliana. Plant Cell Rep. 37, 901–912.

Landa, P., Vankova, R., Andrlova, J., Hodek, J., Marsik, P., Storchova, H., White, J.C., Vanek, T., 2012. Nanoparticle-specific changes in *Arabidopsis thaliana* gene expression after exposure to ZnO, TiO_2, and fullerene soot. J. Hazard. Mater. 241–242, 55–62.

Lang, X., Hirat, A., Fujita, T., Chen, M., 2011. Nanoporous metal/oxide hybrid electrodes for electrochemical super capacitors. Nat. Nanotechnol. 6 (4), 232–236.

Lee, S., Kim, S., Lee, I., 2012. Effects of soil-plant interactive system on response to exposure to ZnO nanoparticles. J. Microbiol. Biotechnol. 22, 1264–1270.

Lin, S., Reppert, J., Hu, Q., Hudson, J.S., Reid, M.L., Ratnikova, T.A., Rao, A.M., Luo, H., Ke, P.C., 2009. Uptake, translocation, and transmission of carbon nanomaterials in rice plants. Small 5, 1128–1132.

Liné, C., Larue, C., Flahaut, E., 2017. Carbon nanotubes: Impacts and behaviour in the terrestrial ecosystem—a review. Carbon 123, 767–785.

Liu, Q., Zhao, Y., Wan, Y., Zheng, J., Zhang, X., Wang, C., Fang, X., Lin, J., 2010. Study of the inhibitory effect of water-soluble fullerenes on plant growth at the cellular level. ACS Nano 4 (10), 5743–5748.

Liu, Q., Zhang, X., Zhao, Y., Lin, J., Shu, C., Wang, C., Fang, X., 2013. Fullerene-induced increase of glycosyl residue on living plant cell wall. Environ. Sci. Technol. 47 (13), 7490–7498.

Liu, S., Wei, H., Li, Z., Li, S., Yan, H., He, Y., Tian, Z., 2015. Effects of graphene on germination and seedling morphology in rice. J. Nanosci. Nanotechnol. 15, 2695–2701.

Ma, C., Rui, Y., Liu, S., Li, X., Xing, B., Liu, L., 2015. Phytotoxic mechanism of nanoparticles: destruction of chloroplasts and vascular bundles and alteration of nutrient absorption. Sci. Rep. 5, 11618.

Magdolenova, Z., Collins, A., Kumar, A., Dhawan, A., Stone, V., Dusinska, M., 2014. Mechanisms of genotoxicity. A review of in vitro and in vivo studies with engineered nanoparticles. Nanotoxicology 8, 233–278.

Merrick, B.A., Bruno, M.E., 2004. Genomic and proteomic profiling for biomarkers and signature profiles of toxicity. Curr. Opin. Mol. Ther. 6, 600–607.

Miralles, P., Church, T.L., Harris, A.T., 2012. Toxicity, uptake, and translocation of engineered nanomaterials in vascular plants. Environ. Sci. Technol. 46 (17), 9224–9239.

Mirzajani, F., Ghassempour, A., Farzaneh, M., Hamzelou, S., Askar, H., 2013. Effect of silver nanoparticles on *Oryza sativa* L. and its rhizosphere bacteria. Ecotoxcol. Environ. Saf. 88, 48–54.

Mohamed, M.A., Hashim, A.F., Alghuthaymi, M.A., Abd-Elsalam, K.A., 2018. Nano-carbon: Plant Growth Promotion and Protection. In: Nanobiotechnology Applications in Plant Protection. Springer, Cham, pp. 155–188.

Mushtaq, Y.K., 2011. Effect of nanoscale Fe_3O_4, TiO_2 and carbon particles on cucumber seed germination. J. Environ. Sci. Health, Part A: Tox. Hazard. Subst. Environ. Eng. 46, 1732–1735.

Nair, M.G., 2018. Toxicological impact of carbon nanomaterials on plants. In: Nanotechnology, Food Security and Water Treatment. Springer, Cham, pp. 163–183.

Nau, K., Bohmer, N., Kühnel, D., Marquardt, C., Paul, F., Steinbach, C., Krug, H.F., 2016. The DaNa(2.0) knowledge base on nanomaterials-communicating current nanosafety research based on evaluated literature data. J. Mater. Educ. 38, 93–109.

Navarro, E., Baun, A., Behra, R., Hartmann, N.B., Filser, J., Miao, A., Quigg, A., Santschi, P.H., Sigg, L., 2008. Environmental behavior and ecotoxicity of engineered nanoparticles to algae, plants, and fungi. Ecotoxicology 17, 372–386.

Nel, A., Xia, T., Mädler, L., Li, N., 2006. Toxic potential of materials at the nanolevel. Science 311, 622–627.

Nowack, B., Bucheli, T.D., 2007. Occurrence, behavior and effects of nanoparticles in the environment. Environ. Pollut. 150, 5–22.

Nuwaysir, E.F., Bittner, M., Trent, J., Barrett, J.C., Afshari, C.A., 1999. Microarrays and toxicology: the advent of toxicogenomics. Mol. Carcinog. 24, 153–159.

Ou, L., Song, B., Liang, H., Liu, J., Feng, X., Deng, B., Sun, T., Shao, L., 2016. Toxicity of graphene-family nanoparticles: a general review of the origins and mechanisms. Part. Fiber Toxicol 13 (1), 57.

Oyelami, A.O., Semple, K.T., 2015. Impact of carbon nanomaterials on microbial activity in soil. Soil Biol. Biochem. 86, 172–180.

Poma, A., Di Giorgio, M.L., 2008. Toxicogenomics to improve comprehension of the mechanisms underlying responses of in vitro and in vivo systems to nanomaterials: a review. Curr. Genomics 9, 571–585.

Pourkhaloee, A., Haghighi, M., Saharkhiz, M.J., Jouzi, H., Doroodmand, M.M., 2011. Carbon nano tubes can promote seed germination via seed coat penetration. Seed Technol. 33 (2), 155–169.

Poynton, H., Vulpe, C., 2009. Ecotoxicogenomics: emerging technologies for emerging contaminants. J. Am. Water Resour. Assoc. 45, 83–96.

Prakash, M., Nair, G., 2018. Toxicological impact of carbon nanomaterials on plants. In: Gothandam, K.M., et al. (Eds.), Nanotechnology, Food Security and Water Treatment. Springer, Cham, pp. 163–184.

Rajput, V.D., Minkina, T., Fedorenko, A., Mandzhieva, S., Sushkova, S., Lysenko, V., Duplii, N., Azarov, A., Aleksandrovich, V., 2019. Destructive effect of copper oxide nanoparticles on ultrastructure of chloroplast, plastoglobules and starch grains in spring barley (*Hordeum sativum*). Int. J. Agric. Biol. 21, 171–174.

Rekulapally, R., Murthy Chavali, L.N., Idris, M.M., Singh, S., 2019. Toxicity of TiO_2, SiO_2, ZnO, CuO, Au and Ag engineered nanoparticles on hatching and early nauplii of *Artemia* sp. Peer J. e6138.

Remédios, C., Rosário, F., Bastos, V., 2012. Environmental nanoparticles interactions with plants: morphological, physiological, and genotoxic aspects. J. Bot. 1–8, 751686. https://doi.org/10.1155/2012/751686.

Rodrigues, D.F., Jaisi, D.P., Elimelech, M., 2013. Toxicity of functionalized single- walled carbon nanotubes on soil microbial communities: implications for nutrient cycling in soil. Environ. Sci. Technol. 47, 625–633.

Ruotolo, R., Maestri, E., Pagano, L., Marmiroli, M., White, J.C., Marmiroli, N., 2018. Plant response to metal-containing engineered nanomaterials: an omics-based perspective. Environ. Sci. Technol. 2018 (52), 2451–2467.

Ruttkay-Nedecky, B., Krystofova, O., Nejdl, L., Adam, V., 2017. Nanoparticles based on essential metals and their phytotoxicity. J. Nanobiotechnol. 15–33.

Safiuddin, M., Gonzalez, M., Cao, J., Tighe, S.L., 2014. State of the art report on use of nano-materials in concrete. Int. J. Pavement Eng. 15 (10), 940–949.

Samaj, J., Baluška, F., Voigt, B., Schlicht, M., Volkmann, D., Menzel, D., 2004. Endocytosis, actin cytoskeleton, and signaling. Plant Physiol. 135, 1150–1161.

Serag, M.F., Kaji, N., Habuchi, S., Bianco, A., Baba, Y., 2013. Nanobiotechnology meets plant cell biology: carbon nanotubes as organelle targeting nanocarriers. RSC Adv. 3, 4856–4862.

Shan, J., Ji, R., Yu, Y., Xie, Z., Yan, X., 2015. Biochar, activated carbon, and carbon nanotubes have different effects on fate of 14C-catechol and microbial community in soil. Sci. Rep. 5, 16000.

Shen, C.X., Zhang, Q.F., Li, J., Bi, F.C., Yao, N., 2010. Induction of programmed cell death in Arabidopsis and rice by single-wall carbon nanotubes. Am. J. Bot. 97 (10), 1602–1609.

Shin, T.H., Da Yeon Lee, H.S.L., Park, H.J., Jin, M.S., Paik, M.J., Manavalan, B., Mo, J.S., Lee, G., 2018. Integration of metabolomics and transcriptomics in nanotoxicity studies. BMB Rep. 51, 14.

Shrestha, B., Acosta-Martinez, V., Cox, S.B., Green, M.J., Li, S., Cañas-Carrell, J.E., 2013. An evaluation of the impact of multiwalled carbon nanotubes on soil microbial community structure and functioning. J. Hazard. Mater. 261, 188–197.

Smirnova, E.A., Gusev, A.A., Zaitseva, O.N., Lazareva, E.M., Onishchenko, G.E., Kuznetsova, E.V., Tkachev, A.G., Feofanov, A.V., Kirpichnikov, M.P., 2011. Multiwalled carbon nanotubes penetrate in to plant cells and affects the growth of on *Brychis arenaria* seedlings. Acta Nat. 3 (1), 99–106.

Stampoulis, D., Sinha, S.K., White, J.C., 2009. Assay-dependent phytotoxicity of nanoparticles to plants. Environ. Sci. Technol. 43, 9473–9479.

Tan, X.-M., Lin, C., Fugetsu, B., 2009. Studies on toxicity of multi-walled carbon nanotubes on suspension rice cells. Carbon 47, 3479–3487.

Tao, X., Yu, Y., Fortner, J.D., He, Y., Chen, Y., Hughes, J.B., 2015. Effects of aqueous stable fullerene nanocrystal (nC60) on *Scenedesmus obliquus*: evaluation of the sub-lethal photosynthetic responses and inhibition mechanism. Chemosphere 122, 162–167.

Thomas, C.R., George, S., Horst, A.M., Ji, Z.X., Miller, R.J., Peralta-Videa, J.R., Xia, T.A., Pokhrel, S., Madler, L., Gardea-Torresdey, J.L., 2011. Nanomaterials in the environment: from materials to high-throughput screening to organisms. ACS Nano 5, 13–20.

Tong, Z., Bischoff, M., Nies, L., Applegate, B., Turco, R.F., 2007. Impact of fullerene (C60) on a soil microbial community. Environ. Sci. Technol. 41 (8), 2985–2991.

Tong, Z.H., Bischoff, M., Nies, L.F., Carroll, N.J., Applegate, B., Turco, R.F., 2016. Influence of fullerene (C60) on soil bacterial communities: aqueous aggregate size and solvent co-introduction effects. Sci. Rep. 6, 28069.

Valdimir, A.B., Terrazas, T., Luna-Martínez, N., Basiuk, E.V., 2019. Phytotoxicity of carbon nanotubes and nanodiamond in long-term assays with Cactaceae plant seedlings. Fuller. Nanotubes Carbon Nanostruct. 27 (2), 141–149.

Verano-Braga, T., Miethling-Graff, R., Wojdyla, K., Rogowska-Wrzesinska, A., Brewer, J.R., Erdmann, H., Kjeldsen, F., 2014. Insights into the cellular response triggered by silver nanoparticles using quantitative proteomics. ACS Nano 8, 2161–2175.

Verma, S.K., Das, A.K., Gantait, S., Kumar, V., Gurel, E., 2019. Applications of carbon nanomaterials in the plant system: a perspective view on the pros and cons. Sci. Total Environ. 667, 485–499.

Wagay, J.A., Singh, S., Raffi, M., Rahman, Q.I., Husen, A., 2019. Effect of carbon-based nanomaterials on rhizosphere and plant functioning. In: Nanomaterials and Plant Potential. Springer, Cham, pp. 553–575.

Wang, C., Zhang, H., Ruan, L., Chen, L., Li, H., Chang, X.L., Zhang, X., Yang, S.T., 2016. Bioaccumulation of 13 C–fullerenol nanomaterials in wheat. Environ. Sci. Nano 3, 799–805.

Wang, Q., Li, C., Wang, Y., Que, X., 2019. Phytotoxicity of graphene family nanomaterials and its mechanisms: a review. Front. Chem. 7, 292.

Wetmore, B.A., Merrick, B.A., 2004. Toxicoproteomics: proteomics applied to toxicology and pathology. Toxicol. Pathol. 32 (6), 619–642.

Wild, E., Jones, K.C., 2009. Novel method for the direct visualization of in vivo nanomaterials and chemical interactions in plants. Environ. Sci. Technol. 43, 5290–5294.

Xu, L.M., Takemura, T., Xu, M., Hanagata, N., 2011. Toxicity of silver nanoparticles as assessed by global gene expression analysis. Mater. Express 1, 74–79.

Xuming, W., Fengqing, G., Linglan, M., Jie, L., Sitao, Y., Ping, Y., Fashui, H., 2008. Effects of nano-anatase on Ribulose-1,5-Bisphophate carboxylase/oxygenase mRNA expression in Spinach. Biol. Trace Elem. Res. 126, 280–289.

Yan, A., Chen, Z., 2019. Impacts of silver nanoparticles on plants: a focus on the phytotoxicity and underlying mechanism. Int. J. Mol. Sci. 20 (5), 1003.

Yan, S., Zhao, L., Li, H., Zhang, Q., Tan, J., Huang, M., He, S., Li, L., 2013. Single-walled carbon nanotubes selectively influence maize root tissue development accompanied by the change in the related gene expression. J. Hazard. Mater. 246–247, 110–118.

Yin, L., Wang, Z., Wang, S., Xu, W., Bao, H., 2018. Effects of graphene oxide and/or Cd^{2+} on seed germination, seedling growth, and uptake to Cd^{2+} in solution culture. Water Air Soil Pollut. 229, 1–12.

Zhang, R., Wise, R.R., Struck, K.R., Sharkey, T.D., 2010. Moderate heat stress of *Arabidopsis thaliana* leaves causes chloroplast swelling and plastoglobule formation. Photosynth. Res. 105, 123.

Zaytseva, O., Neumann, G., 2016. Carbon nanomaterials: production, impact on plant development, agricultural and environmental applications. Chem. Biol. Technol. Agric. 3 (1), 17.

Zhang, P., Zhang, R., Fang, X., Song, T., Cai, X., Liu, H., Du, S., 2016. Toxic effects of graphene on the growth and nutritional levels of wheat (*Triticum aestivum* L.): short- and long-term exposure studies. J. Hazard. Mater. 317, 543–551.

Zhao, S., Wang, Q., Zhao, Y., Rui, Q., Wang, D., 2015. Toxicity and translocation of graphene oxide in *Arabidopsis thaliana*. Environ. Toxicol. Phar. 39, 145–156.

Zhou, X., Torabi, M., Lu, J., Shen, R., Zhang, K., 2014. Nanostructured energetic composites: synthesis, ignition/combustion modeling, and applications. ACS Appl. Mater. Interfaces 6 (5), 3058–3074.

Zhu, S., Luo, F., Chen, W., Zhu, B., Wang, G., 2017. Toxicity evaluation of graphene oxide on cysts and three larval stages of Artemia salina. Sci. Total Environ. 595, 101–109.

Further reading

Kerfahi, D., Tripathi, B.M., Singh, D., Kim, H., Lee, S., Lee, J., Adams, J.M., 2015. Effects of functionalized and raw multi-walled carbon nanotubes on soil bacterial community composition. PLoS One 10 (3), e0123042.

Li, K., Chang, Z., Shen, C., Yao, N., 2015. Toxicity of nanomaterials to plants. In: Siddiqui, M.H., Al-Whaibi, M.H., Mohammad, F. (Eds.), Nanotechnology and Plant Sciences. Nanoparticles and Their Impact on Plants. Springer International Publishing, Switzerland.

Application of Carbon-Based Nanomaterials in Food Preservation Area

25

Abuelmagd M. Abdelmonem[a,b]
[a]Special Food and Nutrition Research, Food Technology Research Institute, Agricultural Research Center, Oula, Egypt, [b]Functional Nanostructures, Institute of Physical Chemistry and Electrochemistry, Leibniz Universität Hannover, Hannover, Germany

1 Introduction

Due to the unique and fascinating properties (physical, chemical, mechanical, biological, etc.) of nanomaterials and as a result of fast-growing development in nanomaterial synthesis, surface modification, and design, nanotechnology applications are expected to contribute and greatly transform many areas of research/disciplines from electronic and energy applications to environmental and biological disciplines. Food science is a multidisciplinary field of special importance for life. Food science is an applied science that deals with food, and it is a highly multidisciplinary field of chemistry, physics, engineering, and biology. Food science and technology include a wide range of disciplines such as food processing, industries, food packaging, food chemistry, analysis, nutrition, etc. The food science and food industry and related disciplines are among the fastest-growing and most-promising areas of nanomaterial/nanotechnology applications. The application of nanomaterials and nanotechnologies in the food sector and related areas (food nanotechnology) offers great potential features in many foods and food sector-related disciplines, including food processing, active and intelligent packaging, preservation, food analysis, safety, and delivery (Bajpai et al., 2018; He and Min Hwang, 2016).

This chapter focuses on recent investigations/developments in carbon-based nanomaterials (e.g., carbon nanoparticles or carbon quantum dots, graphene and graphene oxide (GO) nanomaterials, single-walled and multiwalled carbon nanotubes (SWCNTs and MWCNTs)) and nanotechnologies in food science, industries, and related areas. We are especially focused on food packaging and smart packaging, food analysis and safety (for quality control, environmental and health purposes, and public concerns), and risk assessment of the possible environmental, toxicological, and health issues (biosafety) of carbon-based nanomaterial applications in food disciplines.

2 Carbon nanotube polymer nanocomposites

Due to their unique physicochemical and mechanical properties, carbon nanotubes (CNTs), either single-walled nanotubes (SWCNTs) or multiwalled nanotubes

Carbon Nanomaterials for Agri-food and Environmental Applications. https://doi.org/10.1016/B978-0-12-819786-8.00025-6

(MCWNTs), are widely investigated to improve the tensile strength/modulus of several polymers such as polyethylene naphtalate, polyvinyl alcohol, polyamide, and polypropylene. CNTs have unique mechanical properties (e.g., elastic modulus of up to 1 TPa and tensile strength of 200 GPa) and additionally exhibit proven antibacterial properties against many microorganisms, which could be attributed to their direct penetration through microbial cells. The unique physicochemical and mechanical properties of the nanoparticle as well as their antimicrobial properties make CNTs promising materials in the food sector in many aspects, especially in food packaging, food analysis, and safety (Sharma et al., 2017).

3 Food packaging

Food packaging is a main and vital part of the food industry, the food supply chain (all overall stages), safety, and public health. Food packaging plays an important role in for example, protection/preservation, waste reduction, traceability, and marketing. The mechanical, barrier, and antimicrobial properties are some of the main or most important properties in the development and evaluation of food packaging. Recent advances focus on and make use of certain directions such as being novel/innovative, cost-effective, active and smart, natural, ecofriendly, and resourceful materials in food-packaging systems. The unique properties of nanomaterials (e.g., mechanical properties of graphene) make them promising for food packaging. A combination of cinnamaldehyde (an aromatic unsaturated aldehyde from a natural source with well-proven antimicrobial properties against a wide range of microorganisms, including bacteria, yeasts, and molds), chitosan (a linear natural polycationic polysaccharide polymer obtained by deacetylation of chitin), and graphene stacks has been developed and used in food-packaging systems. In this study, chitosan was functionalized with antimicrobial cinnamaldehyde using the Schiff base formation. Under suitable conditions, the Schiff base products are reversible and thus a controlled release of the functional bioactive materials (cinnamaldehyde) from chitosan (used as the carrier) is achieved. The investigated packaging material (film) showed superior properties such as antimicrobial effects and enhanced mechanical properties (Demitri et al., 2016). Naskar et al. (2018) have been able to incorporate silver-zinc oxide nanoparticles (Ag-ZnO), reduced graphene oxide (rGO), and polyethylene glycol (PEG) (AZGP) at room temperature to design and develop multifunctional nanocomposites with antibiofilm activities as a food packaging. The silver and zinc oxide nanoparticles are distributed uniformly over rGO nanosheets (characterized and confirmed by different techniques such as X-ray diffraction and TEM, FTIR, Raman, UV–visible and X-ray photoelectron spectroscopy). The antibiofilm activity of the investigated silver-zinc oxide nanoparticles, rGO, and PEG nanocomposite (AZGPN) has been studied and confirmed on two types of bacteria (*Staphylococcus aureus* and *Pseudomonas aeruginosa*), which resulted in significant inhibition activity toward biofilm formation as well as eradication of preformed biofilm. These silver-zinc oxide nanoparticles, rGO, and AZGPNs have also been incorporated into agar film (AZGP-based agar film), which has been fabricated and characterized by XRD as a possible material for food packaging. The proposed AZGP-based agar film showed an enhanced textural and tensile strength as well as excellent antimicrobial activity, even after a long period (about 90 days).

In a similar work in the same direction and making use of graphene in food packaging, a nanosheet of polyactide/graphene contains clove essential oil as a potential food packaging. GO improved the complex viscosity and glass transition temperature while lowering the oxygen permeability of the plasticized polyactide. The matrix transition temperature (T_g) and the reinforcement of GO prevented the porosity of the plasticized polylactide (PLA)/clove essential oil (CLO) film surface. The addition of clove essential oil contributed to the mechanical properties enhancing the flexibility of films by lowering the tensile stress, the complex viscosity, and the glass transition temperature as well as the excellent antimicrobial properties (against *Staphylococcus aureus* and *Escherichia coli*) of composite film (Arfat et al., 2018).

In a recent work by Gouvêa et al. (2018) for applications of graphene in food packaging, a reduced graphene oxide-zinc oxide (rGO-ZnO) composite was synthesized by the simultaneous reduction of zinc acetate and GO in a single reaction step. The obtained rGO-ZnO composite was incorporated into glycerol-plasticized poly(3-hydroxybutyrate-*co*-3-hydroxy valerate) (PHBV) by the melted extrusion technique. The incorporation of the rGO-ZnO composite into the plasticized PHBV matrix contributed to maintaining the modulus and glass transition temperature (Tg). The antimicrobial activity of the film was investigated on *Escherichia coli*. The migration tests to food simulants revealed that rGO had no migration and the ZnO migration was below the upper level recommended for Zn^{2+}. Multifunctional, multicomponent, and bio-based films of silver (Ag) silica-carbon nanoparticles incorporated into biopolymer thin films were fabricated using three-dimensional (3D) printing techniques and investigated for food-packaging applications. Ag nanoparticles were prepared from $AgNO_3$ by a top-down approach through one-step ball milling in the presence of silica-carbon hybrids obtained from the pyrolysis of agrowaste (rice husks). Different techniques such as XRD, TEM, XPS, FE-SEM, Raman microanalysis, TGA, DSC, and tensile testing were employed for characterization of the as-synthesized nanoparticles, the printed films, and the influence of nanoparticles on the thermal and mechanical properties of the films. The antimicrobial activity of the films (performed on *Salmonella enteritidis*) showed no bacterial growth and the exerted bacteriostatic effect relied on direct contact of the films with bacteria (Biswas et al., 2019).

Smart packaging through printed functional labels on packaging materials using inkjet printing and inks with changeable photoluminescence properties is of great importance for food packaging and safety purposes. Carbon nanoparticles are a promising ink candidate for such a purpose. Carbon nanoparticle-based compositions have been developed and investigated for such inkjet printing on paper packaging materials for smart packaging labels (Hrytsenko et al., 2018). Due to their unique properties (e.g., elastic modulus and strength), CNTs have been successfully applied to improve electrical, thermal, mechanical, and gas barrier properties. In this regard, the bionanocomposites of PHBV reinforced with PHBV-grafted multiwalled carbon nanotubes (PHBV-g-MWCNTs) were fabricated by a simple solution casting method. The obtained film (as food packaging material) was transparent and the PHBV-g-MWCNTs uniformly distributed throughout the PHBV matrix significantly improved the thermal stability and mechanical, barrier, and migration properties of PHBV. PHBV-g-MWCNT nanocomposites showed a wider melt-processing window, low water uptake, and water vapor permeability while the migration levels in food simulants were below the upper migration limits (Yu et al., 2014). Torres-Giner et al. (2018) reported a composite of poly(ethylene-*co*-vinyl alcohol) and graphene nanoplatelets for intelligent food-packaging applications. In this work, graphene nanoplatelets prepared from graphene powder were embedded in a polymeric fiber of poly(ethylene-*co*-vinyl alcohol) (EVOH) by electrospinning. The electrospun nanocomposite was investigated and characterized (chemically, thermally, and morphologically). The graphene nanoparticles (GNPs) were efficiently dispersed in the electrospun nanocomposite fiber mats and rolled along the EVOH fibrillar matrix up to the contents of 0.5 wt%. The dielectric behavior of the nanocomposite fibers (expressed as a function of the frequency range and nanoparticle content) showed that the dielectric constant rapidly decreased with a frequency increase. The dielectric constant only increased at a low concentration of GNPs and the nanocomposite fiber mats became electrically conductive with the maximum at 0.5 wt% of the nanoparticle content. The results of this study indicate the importance of such electrospun nanocomposites and electrospinning as a possible technology in intelligent food-packaging applications such as smart labels or tags.

4 Analytical tools and biosensors for food analysis and safety

Carbon-based nanomaterials have been widely investigated as analytic tools and biosensors in food analysis, chemistry, and food safety (Jiao et al., 2014; Sukanya et al., 2018; Vasilescu et al., 2018). Graphene is one of the most reported and promising carbon-based nanomaterials investigated in food analysis, biosensing, and safety (Jiao et al., 2014). Aflatoxin is a class of life-threatening toxins that has received great interest. Basu et al. (2015) reported the rGO and graphene field-effect capacitive as a biosensor (immunosensor) for detection of Aflatoxin B1 (AF1) with subfemtomolar detection food capacity, which is one of the magnitudes.

Polychlorinated biphenyls (PCBs), which are a series of lipid-soluble organic pollutants, are a group of food and environmental pollutants. PCBs are known for their carcinogenic activity at ultralow concentrations. The detection of RCB traces in food has high importance for safety and health (Fangying et al., 2017). A composite of molybdenum disulfide/reduced graphene oxide (MoS_2/rGO) in the form of a fiber coating was developed and employed in the microextraction of PCBs. The developed microextraction system (MoS_2/rGO deposited on stainless steel wire) coupled with gas chromatography spectrometry was successfully employed for the quantification of PCBs. The investigated developed coating (MoS_2/rGO) was able to absorb the analytes (PCBs) 1.1–2.9 times higher in comparison to the commercially available SPME fiber (coated with divinyl-benzene/carboxen/polydimethylsiloxane) and 1.5–2.7 times that of the prepared RGO modified fiber. This ecofriendly, inexpensive, and reusable system with high affinity toward PCBs, good sensitivity and detection limits ($<0.1\,ng\,mL^{-1}$), linearity (from 0.25 to $100\,ng\,mL^{-1}$), and reproducibility is of great value in the detection of PCBs in food. A similar technique was developed for the microextraction and detection of heterocyclic amines in food samples. Heterocyclic amines are cyclic compounds formed in cooked food due to high-temperature cooking methods, especially such as pan frying or grilling directly over an open flame. Heterocyclic amines are a public concern for human health because they are potential mutagens and carcinogens. Acrylamide-modified graphene has been developed and successively used as an adsorbent for micro solid-phase extraction and determination of heterocyclic amine traces in food samples (spicy salted duck, baked fish, and fried chicken) by using (linked to) high-performance chromatography with detection limits ranging from 0.70 to 2.5 ng/L (Zhang et al., 2015).

The color of the food is one of the most important physical properties in quality and attraction and it is given a high priority in the food industry. Therefore, processing the food color to stabilize, homogenize, enhance, or even change it by using external compound "colorants" is a normal practice in the food industry to make it more attractive for customers. The colorants used in the food industry are natural or synthetic. As is the case with other food additives, the use of food colorants–especially synthetic colorants as a very commonly used important class of food additives–is strictly controlled by legislation and health agencies all over the world due to the possible risk to human health associated with some of the colorants. In this direction and for food safety and human health, it is of great importance to develop accurate and reliable techniques for the detection/determination of synthetic colorants in foodstuffs. A nanocomposite of graphene and silver in combination with surface-enhanced Raman scattering was fabricated and employed as a sensor for simple and rapid detection of prohibited colorants in food (Xie et al., 2012).

The applications of carbon-based nanomaterials in food chemistry, analysis, and food safety purposes are fast growing in promising areas of research (Fig. 1). Table 1 lists additional publications in this regard.

Fig. 1 summarizes the possible applications of nanomaterials in food science.

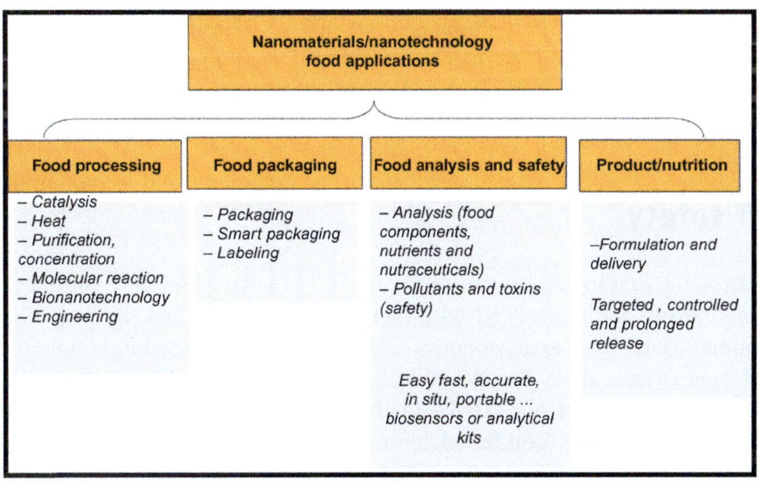

Fig. 1 The possible applications of nanomaterials in the food sector.

Table 1 Applications of carbon-based nanomaterials in food analysis and safety.

Nanomaterials/ nanocomposite	Analyte	Reference
MWCNTs	L-tyrosine	Shrestha et al. (2016)
GO-wrapped tin oxide@carbon nanospheres	Maltol	Gan et al. (2017)
Graphene-cobalt hexacyanoferrate nanocomposite	Hydrazine and nitrite	Luo et al. (2015)
MWCNTs	Cadmium	Behbahani et al. (2013)
Carbon aerogel/palladium-ruthenium nanoparticles	Food dyes	Thirumalraj et al. (2018)
MWCNTs	Fe^{+3}, Cu^{+2},Mn^{+2} Pb^{+2},	Ozcan et al. (2010)
SWCNTs	Chromium and iron	Soylak and Unsal (2011)
MWCNTs	Illegal dyes	Ya-hui and Jing (2018)
CNTs	Ascorbic acid	Ping et al. (2012)
GO	Carbamate pesticide	da Silva et al. (2018)
CNTs	Copper (II)	Fan et al. (2017)
MWCNTs	Mutagenic amines	Bueno et al. (2016)
GO	Carbofuran	Wong et al. (2014)
Doped graphene/Cu nanocomposite	Glucose	Shabnam et al. (2017)
Sulfur-doped carbon quantum dots	Caffeine	Nemati et al. (2018)
Fluorescent carbon nanoparticles	Glucose	Wang et al. (2018)
Magnetic graphene	Fluoroquinolones	He et al. (2017)
Graphene oxide	Polycyclic aromatic hydrocarbons	Hou et al. (2016)
Magnetic graphene sol–gel hybrid	Acrylamide	Nodeh et al. (2018)
3D graphene	Caffeine	Rahimi et al. (2018)
Fluorescent nitrogen-doped carbon nanoparticles	Hg^{+2}	Liu et al. (2016)
Carbon dots	Nitrite	Xiang et al. (2018)

5 CNT safety

The increased interest in carbon nanotubes (SWCNTs and MWCNTs) and their widespread applications are associated with increased concern about the possible risk of these nanomaterials. Risk assessment of CNTs and carbon-based nanomaterials is an important area of research. The possible in vivo multiparametric toxicity of CNTs has been studied (Fang et al., 2018). In this study, the in vivo toxicity of pristine MWCNTs and MWCNTs combined with nonylphenol (NP) were studied in male mice. The investigated materials were administrated to male mice by gavage for five days and some toxicity-related parameters and this included the (a) in vivo oxidative stress such as intracellular superoxide dismutase and glutathione peroxidase activity, and malondialdehyde and glutathione levels in tissues were determined to as indicators for the and toxicity and (b) genotoxicity (DNA damage in mouse liver and sperm via the comet assay). Additionally, the mitochondrial damage in the liver cells was investigated by direct visual observations using transmission electron microscopy (TEM). The MWCNTs and NP at high doses induced oxidative lesions in the liver and caused DNA damage in mouse sperm, as was revealed from the results of this in vivo multiparametric in vivo toxicity (oxidative damage and DNA damage experiments).

In another study in this regard, the health risks and derived occupational exposure limit of CNTs via inhalation was reported. In this study, two methods were developed: the integration of the intratracheal instillation data with the inhalation data, and the "biaxial approach." Results revealed that toxicity levels were correlated with a specific surface area of the tested CNTs. On the genotoxicity level, the study indicated that CNTs do not directly interact with genetic materials.

6 Conclusion

Carbon-based nanomaterials are promising materials for the food sector, especially food analysis and food packaging applications, through improving the mechanical and barrier properties and functionality of biodegradable polymers used in this sector. However, as is the case with nanomaterials in general, the fast-growing application of nanotechnology and nanomaterials resulted in an increased interest in research related to the risk assessment, biosafety, and environmental impact of carbon-based nanomaterials. Many questions have been raised about the increased probability of consumption and exposure of humans and the environment to such nanomaterials, for example about the possible migration, release, and interaction (end of life) of such nanomaterials. The release of graphene nanoplatelets and multiwalled CNTs impeded in polylactic-based film at the end of life of waste treatment has been studied (Kotsilkov et al., 2018) during degradation and combustion/burning to address questions about (i) biodegradation of matrix nanocomposite film at waste disposal; and (ii) combustion and fire of nanocomposite wastes. Different techniques such as thermogravimetric analysis, electron microscopy (TEM and SEM), and AFM were employed to verify the release of nanoparticles from nanocomposite films. The results indicated that single graphene nanoplatelets of nanosized thickness and length from 100 to 1000 nm as well as their micron-sized aggregates were indeed released in relatively large amounts from the polylactic nanocomposite film to high-temperature treatment, possibly due to partial degradation of the PLA polymer. On the contrary, the release of the entangled MWSNTs was possible only after full degradation (hydrolysis) of the film. While the combustion at 500°C and 650°C resulted in residue ash containing mainly single airborne GNPs and MWCNTs, the combustion at 850°C resulted in a small amount of ash (~0.07%) containing only GNPs and amorphous carbon soot. This means that at 850°C, the MWCNTs fully degrade the GNP remaining in the nanoparticles.

In conclusion, despite the unique properties of carbon-based nanomaterials and increased investigations in the food sector, applications of nanomaterials in food-related disciplines (e.g., food processing/industry, packaging, analysis, safety, and nutrition) are poorly developed compared to other applications such as biomedical, energy, electronics, information technology, and other industries. This could be attributed to the possible toxic effect and the need for in-depth research on risk assessment and possible health and environmental problems associated with applications of such nanomaterials in the agrifood sector.

References

Arfat, A.Y., Ahmed, J., Ejaz, M., Mullah, M., 2018. Polylactide/graphene oxide nanosheets/clove essential oil composite films for potential food packaging applications. Int. J. Biol. Macromol. 107, 194–203.

Bajpai, V.K., Kamle, M., Shukla, S., Mahato, D.K., Chandra, P., Hwang, S.K., Kumar, P., Huh, Y.S., Han, Y., 2018. Prospects of using nanotechnology for food preservation, safety, and security. J. Food Drug Anal. 26, 1201–1214.

Basu, J., Datta, S., Roy Chaudhuri, C., 2015. A graphene field-effect capacitive immunosensor for sub-femtomolar food toxin detection. Biosens. Bioelectron. 68, 544–549.

Behbahani, M., Bagheri, A., Amini, M.M., Sadeghi, O., Salarian, M., Najafi, F., Taghizadeh, M., 2013. Application of multiwalled carbon nanotubes modified by diphenylcarbazide for selective solid phase extraction of ultra traces Cd(II) in water samples and food products. Food Chem. 141, 48–53.

Biswas, M.C., Tiimob, B.J., Abdela, W., Jeelani, S., Rangari, V.K., 2019. Nano silica-carbon-silver ternary hybrid induced antimicrobial composite films for food packaging application. Food Packag. Shelf Life 19, 04–113.

Bueno, A.M., Marín, M.A., Contento, A.M., Ríos, Á., 2016. Determination of mutagenic amines in water and food samples by high-pressure liquid chromatography with amperometric detection using a multiwall carbon nanotubes-glassy carbon electrode. Food Chem. 192, 343–350.

da Silva, M.K.L., Vanzela, H.C., Defavari, L.M., Cesarino, I., 2018. Determination of carbamate pesticide in food using a biosensor based on reduced graphene oxide and acetylcholinesterase enzyme. Sens. Actuators B 277, 555–561.

Demitri, C., De Benedictis, V.M., Madaghiele, M., Corcione, C.E., Maffezzoli, A., 2016. Nanostructured active chitosan-based films for food packaging applications: effect of graphene stacks on mechanical properties. Measurement 90, 418–423.

Fan, Y., Xu, C., Wang, R., Hu, G., Miao, J., Hai, M., Lin, L., 2017. Determination of copper(II) ion in food using an ionic liquids-carbon nanotubes-based ion-selective electrode. J. Food Compos. Anal. 62, 63–68.

Fang, H., Cui, Y., Wang, Z., Wang, S., 2018. Toxicological assessment of multi-walled carbon nanotubes combined with nonylphenol in male mice. PLoS One 13 (7): e0200238. https://doi.org/10.1371/journal.pone.0200238.

Fangying, L.F., Gan, N., Cao, Y., Zhou, Y., Zuo, R., Dong, Y., 2017. A molybdenum disulfide/reduced graphene oxide fiber coating coupled with gas chromatography-mass spectrometry for the saponification-headspace solid-phase microextraction of polychlorinated biphenyls in food. J. Chromatogr. A 1525, 42–50.

Gan, T., Sun, J., Yu, M., Wang, K., Lv, Z., Liu, Y., 2017. Amplified electrochemical determination of maltol in food based on graphene oxide-wrapped tin oxide@carbon nanospheres. Food Chem. 214, 82–89.

Gouvêa, R.F., Del Aguila, E.M., Paschoalin, V.M.F., Andradea, C.T., 2018. Extruded hybrids based on poly(3-hydroxybutyrate-co-3-hydroxyvalerate) and reduced graphene oxide composite for active food packaging. Food Packag. Shelf Life 16, 77–85.

He, X., Min Hwang, H., 2016. Nanotechnology in food science: functionality, applicability, and safety assessment. J. Food Drug Anal. 24, 671–681.

He, X., Wang, J.N., Yang, K., Liu, H.Z., Wu, X.J., Wang, J.P., 2017. Magnetic graphene dispersive solid phase extraction combining high-performance liquid chromatography for determination of fluoroquinolones in foods. Food Chem. 221, 1226–1231.

Hou, X., Guo, Y., Liang, X., Wang, X., Wang, L., Wang, L., Liu, X., 2016. Bis(trifluoromethanesulfonyl)imide-based ionic liquids grafted on graphene oxide-coated solid-phase microextraction fiber for extraction and enrichment of polycyclic aromatic hydrocarbons in potatoes and phthalate esters in food-wrap. Talanta 153, 392–400.

Hrytsenko, O., Hrytsenko, D., Shvalagin, V., Grodziuk, G., Kompanets, M., 2018. The use of carbon nanoparticles for inkjet-printed functional labels for smart packaging. J. Nanomater. https://doi.org/10.1155/2018/6485654.

Jiao, S., Jin, J., Wang, L., 2014. Tannic acid functionalized N-doped graphene modified glassy carbon electrode for the determination of bisphenol A in food package. Talanta 122, 140–144.

Kotsilkov, S., Ivanov, E., Vitanov, N.K., 2018. Release of graphene and carbon nanotubes from biodegradable poly(lactic acid) films during degradation and combustion: risk associated with the end-of-life of nanocomposite food packaging materials. Materials 11, 2346.

Liu, J., Chen, Y., Wang, W., Feng, J., Peng, S., Ma, S., Chen, X., 2016. Effective synthesis of highly fluorescent nitrogen doped carbon nanoparticles for selective sensing of Hg^{+2} in food and cosmetics samples. RSC Adv. 6, 89916.

Luo, X., Pan, J., Pan, K., Yu, Y., Zhong, A., Wei, S., Li, J., Shi, J., Li, S., 2015. An electrochemical sensor for hydrazine and nitrite based on graphene–cobalt hexacyanoferrate nanocomposite: toward environment and food detection. J. Electroanal. Chem. 745, 80–87.

Naskar, A., Khan, H., Sarkar, R., Kumar, S., Halder, D., Jana, S., 2018. Anti-biofilm activity and food packaging application of room temperature solution process based polyethylene glycol capped Ag-ZnO-graphene nanocomposite. Mater. Sci. Eng. C 91, 743–753.

Nemati, F., Hosseini, M., Zare-Dorabei, R., Salehnia, F., Ganjali, M.R., 2018. Fluorescent turn on sensing of caffeine in food sample based on sulfurdoped carbon quantum dots and optimization of process parameters through response surface methodology. Sens. Actuators B 273, 25–34.

Nodeh, H.R., Ibrahim, W.A.W., Kamboh, M.A., Sanagi, M.M., 2018. Magnetic graphene sol-gel hybrid as clean-up adsorbent for acrylamide analysis in food samples prior to GC-MS. Food Chem. 239, 208–216.

Ozcan, S.G., Satiroglu, G., Soylak, M., 2010. Column solid phase extraction of iron(III), copper(II), manganese(II) and lead(II) ions food and water samples on multi-walled carbon nanotubes. Food Chem. Toxicol. 48, 2401–2406.

Ping, J., Wang, Y., Wu, J., Ying, Y., Ji, F., 2012. Determination of ascorbic acid levels in food samples by using an ionic liquid–carbon nanotube composite electrode. Food Chem. 135, 362–367.

Rahimi, A., Zanjanchi, M.A., Bakhtiari, S., Dehsaraei, M., 2018. Selective determination of caffeine in foods with 3D-graphene based ultrasound-assisted magnetic solid phase extraction. Food Chem. 262, 206–214.

Shabnam, L., Faisal, S.M., Roy, A.K., Haque, E., Minett, A.I., Gomes, V.G., 2017. Doped graphene/Cu nanocomposite: A high sensitivity non-enzymatic glucose sensor for food. Food Chem. 221, 751–759.

Sharma, C., Dhiman, R., Rokana, N., Panwar, H., 2017. Nanotechnology: an untapped resource for food packaging. Front. Microbiol. 8, 1735. http://dx.doi.org/ 10.3389/fmicb.2017.01735.

Shrestha, S., Mascarenhas, R.J., D'Souza, O.J., Satpati, A.K., Mekhalif, Z., Dhason, A., Marti, P., 2016. Amperometric sensor based on multi-walled carbon nanotube and poly (bromocresol purple) modified carbon paste electrode for the sensitive determination of L-tyrosine in food and biological samples. J. Electroanal. Chem. 778, 32–40.

Soylak, M., Unsal, Y.E., 2011. Chromium and iron determinations in food and herbal plant samples by atomic absorption spectrometry after solid phase extraction on single-walled carbon nanotubes (SWCNTs) disk. Food Chem. Toxicol. 48, 511–1515.

Sukanya, R., Sakthivel, M., Chen, S., Chen, T., 2018. A new type of terbium diselenide nano octagon integrated oxidized carbon nanofiber: An efficient electrode material for electrochemical detection of morin in the food sample. Sens. Actuators B 269, 354–367.

Thirumalraj, B., Rajkumar, C., Chen, S., Veerakumar, P., Perumal, P., Liu, S., 2018. Carbon aerogel supported palladium-ruthenium nanoparticles for electrochemical sensing and catalytic reduction of food dye. Sens. Actuators B 257, 48–59.

Torres-Giner, S., Echegoyen, Y., Teruel-Juanes, R., Badia, J.D., Ribes-Greus, A., Lagaron, J.M., 2018. Electrospun poly(ethylene-co-vinyl alcohol)/graphene nanoplatelets composites of interest in intelligent food packaging applications. Nanomaterials 8, 745. http://dx.doi.org/10.3390/nano8100745.

Vasilescu, A., Hayat, A., Gáspár, S., Marty, J., 2018. Advantages of carbon nanomaterials in electrochemical aptasensors for food analysis. Electroanalysis 30, 2–19.

Wang, M., Cui, M., Zhao, M., Cao, H., 2018. Sensitive determination of Amaranth in foods using graphene nanomeshes. J. Electroanal. Chem. 809, 117–124.

Wong, A., Materon, E.M., Sotomayor, M.D.T., 2014. Development of a biomimetic sensor modified with hemin and graphene oxide for monitoring of carbofuran in food. Elcctrochim. Acta 146, 830–837.

Xiang, G., Wang, Y., Zhang, H., Fan, H., Fan, L., He, L., Jiang, X., Zhao, L., 2018. Carbon dots based dual-emission silica nanoparticles as ratiometric fluorescent probe for nitrite determination in food samples. Food Chem. 260, 13–18.

Xie, F., Li, Y., Niu, L., Wang, H., Qian, H., Yao, W., 2012. A novel surface-enhanced Raman scattering sensor to detect prohibited colorants in food by graphene/silver nanocomposite. Talanta 100, 32–37.

Ya-hui, H.E., Jing, W., 2018. Detection of illegal dyes in foods using a polyethersulfone/multi-walled carbon nanotubes composite membrane as a cleanup method. J. Integr. Agric. 17 (3), 716–722.

Yu, H., Qin, Z., Sun, B., Yang, X., Yao, J., 2014. Reinforcement of transparent poly(3-hydroxybutyrate-co-3-hydroxyvalerate) by incorporation of functionalized carbon nanotubes as a novel bionanocomposite for food packaging. Compos. Sci. Technol. 94, 96–104.

Zhang, Q., Li, G., Xiao, X., 2015. Acrylamide-modified graphene for online micro-solid-phase extraction coupled to high-performance liquid chromatography for sensitive analysis of heterocyclic amines in food samples. Talanta 31, 127–135.

Further reading

Nakanishi, J., Morimoto, Y., Ogura, I., Kobayashi, N., Naya, M., Ema, M., Endoh, S., Shimada, S., Ogami, A., Myojyo, T., Oyabu, T., Gamo, M., Kishimoto, A., Igarashi, T., Hanai, S., 2015. Risk assessment of the carbon nanotube group. Risk Anal. 35 (10), 1940–1956.

Wang, H., Xie, Y., Liu, S., Cong, S., Song, Y., Xu, X., Tan, T., 2017. Presence of fluorescent carbon nanoparticles in baked lamb: their properties and potential application for sensors. J. Agric. Food Chem. 65, 7553–7559.

Risk management and regulatory aspects of carbon nanomaterials

26

Khalid Parwez[a], *Suman V. Budihal*[b]
[a]Department of Clinical Laboratory Science, Faculty of Applied Medical Science, Al-Dawadmi, Shaqra University, Al-Dawadmi, Saudi Arabia, [b]Department of Physiology, Kasturba Medical College, Mangalore, Manipal Academy of Higher Education, Manipal, India

1 Introduction

Studying the toxic effects of carbon nanotubes is the latest concern in the scientific community. Many investigations have been made to elucidate these effects, but there are variations in the elucidations of these reports. A diverse group of scientists (Dumortier et al., 2006; Kam et al., 2004; Schipper et al., 2008; Wu et al., 2008) proved that suitably functionalized carbon nanotubes are nontoxic to animals, whereas raw carbon nanotubes were shown to be toxic to mice lungs in in vivo studies (Lam et al., 2004; Muller et al., 2005; Shvedova et al., 2005; Warheit et al., 2004). Long MWNTs might be carcinogenic to mice (Poland et al., 2008) whereas pristine nanotubes cause oxidative stress and decrease cell viability (Cui et al., 2005; Manna et al., 2005). However, it is also known that leftover catalyst particles also contribute to this effect (Kagan et al., 2006). The covalently attached polar functional group can decrease the toxicity of the nanotubes (Sayes et al., 2006), whereas the toxicity of noncovalently functionalized carbon nanotubes depends on the noncovalently attached group. For example, upon internalization of encapsulated DNA-wrapped SWNT complexes, the animal cells were viable (Kam et al., 2005). Therefore, the toxicity of the carbon nanotube depends on the following criteria: (a) type of functionalization (b) aggregation behavior, and (c) presence of a metal catalyst.

This chapter aims to discuss the applications, qualitative and quantitative toxicological assessments, and regulatory aspects of carbon nanomaterials. In this chapter, we provide information about the current and near future applications of carbon nanotubes in sensor development, CNT-reinforced nanocomposite materials, energy production, and storage. CNTs sensors have a wide range of applications from environment gas sensing to biomedical sensors to packaging sensors for food and beverages. This chapter also outlines the scheme of functionalization of carbon nanotubes, which is a very important modification that needs to be accomplished before real-time application of the material. Furthermore, we have included the in vivo, in vitro, and eco (toxicological) data of the CNTs, which can be helpful in framing safety rules for this modern engineered nanoparticle. This data can be a reference the regulatory organizations and government bodies.

Carbon Nanomaterials for Agri-food and Environmental Applications. https://doi.org/10.1016/B978-0-12-819786-8.00026-8

2 An overview of carbon nanotube research

Manufacturing fundamental elements with a high strength-to-weight ratio using a carbon nanotube composite is the contemporary focus of researchers. A possible utilization of the polymer nanocomposite is the CNT-augmented ultrafine fiber via electrospinning (Formhals, 1934; Rajendran et al., 2012), which we have known since the 1930s. With the help of electrospinning technology, we can inexpensively produce polymer fibers with a nanometer diameter (less than 100 nm). These fibers are useful for drug delivery, energy storage, and improved functional garments (Kenawy et al., 2002; Laxminarayana and Jalili, 2005; Reneker and Chun, 1996). Incorporating carbon nanotubes (CNTs) in electrospinning can improve these three properties of the fibers (Dror et al., 2003; Ko et al., 2003; Lim et al., 2006): (i) strength, (ii) thermal conductivity, and (iii) electrical conductivity. Enhancing the dispersion and alignment of CNTs in the matrix is the real issue with electrospinning technology. Better dispersion of CNTs in the polymer matrix may lead to applications and developments of polymer/CNT nanocomposites (Xie et al., 2005). The dispersions mainly depend on the type of nanomaterials and methods of functionalization (Fig. 1). Electrochemical or electronic detection techniques are successful in developing biosensors due to good sensitivity, high specificity, and low cost. These techniques comprise voltammetric techniques (cyclic voltammetry and differential pulse voltammetry), chronocoulometry, electrochemical impedance spectroscopy, and electronic detection based on an electric field (Cai et al., 2003). CNT sensors (developed from CNTs) can detect many analytes such as particular DNA sequences (Tu et al., 2009), cancer biomarkers (Thakare et al., 2010), and larger entities such as viruses (Patolsky et al., 2004). Moreover, these sensors can monitor enzymatic activities and study the behavior of potential drug molecules in the drug development process (Prato et al., 2008). Both SWCNTs and MWCNTs can be altered and conjugated to a bioactive unit for various biological applications. Those biological applications are plausible only because the carbon nanotubes own some unusual properties such as a one-dimensional arrangement, large aspect ratio, outstanding mechanical characteristics, and chemical inertness (Endo et al., 2008). The carbohydrate-functionalized carbon nanotubes were

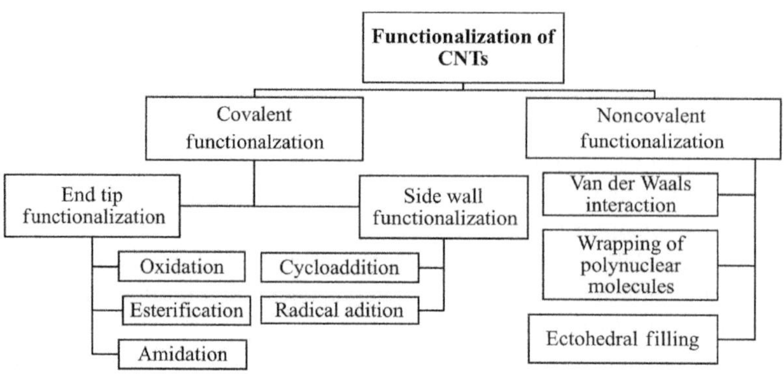

Fig. 1 Outline of the functionalization of carbon nanotubes.

previously used for identification of pathogenic microorganisms, namely *Escherichia coli* (Elkin et al., 2005) and *Bacillus anthracis* (Wang et al., 2008).

In advancement of energy production and storage, nanotubes exhibit exceptional potential in supercapacitors (Kim et al., 2006), Li-ion batteries (Leroux et al., 1999), solar cells (Landi et al., 2005), and fuel cells (Wang et al., 2004). Energy applications could become the broadest realm in the fat utilization of carbon nanotubes. For advancement of Li-ion battery performance, MnO_2 and $LiFePO_4$ are being used as cathodes while MWCNTs and graphene are being used as anodes. CNTs have been widely studied for the preparation of the fuel cell and proton exchange membrane fuel cells (PEMFCs) (Kamavaram et al., 2009). Catalyst-mediated electrochemical reaction is responsible for conversion of chemical energy to electrical energy in a PEMFC. PEMFC efficiency is subjected to the catalysts used (Wang et al., 2005). The catalysts should have high endurance, low cost, and more top activities in oxygen reduction and/or fuel oxidation reaction (Kundu et al., 2009). The most regularly used catalysts in PEMFCs are metal NPs, mainly Pt and/or Pt-based alloys, due to a better Fermi level and a high surface-to-volume ratio. (Li et al., 2002). Nonetheless, metal NPs are generally unstable and suffer loss of their catalytic activity due to their irreversible aggregation during electrochemical processes. Consequently, appropriate methods are obliged to fix and restrict these metal NPs from aggregation, for example, carbon nanotubes (CNTs) are the most extensively adopted provision in modern development. Though the evolution of PEMFCs is under commercialization, obstructions continue, including how CNTs influence the catalytic action of metal/CNTs and high material costs. Development of numerous profoundly dynamic catalysts with an economical price for fuel cell commercialization would be one of the important studies in this domain. Because of the enhanced production and intended use of CNTs in consumer commodities, there is a necessity for evaluation of the implied toxicity of these nanoparticles.

3 Toxicity studies of carbon nanotubes in vivo

In vivo toxicity knowledge plays a vital role in risk evaluation. Those techniques can be applied to determine acute toxicity, chronic toxicity, developmental toxicity, genotoxicity, and reproductive toxicity. In vivo studies are indispensable in the fields of medicine, including cancer therapy. Several animal trials are performed to highlight the possibly dangerous impressions of newly formed drugs and chemical substances on humans. In some experiments, researchers attempt to simulate situations concerning humans (e.g., arthritis, cystic fibrosis, and cancer) in animals to assess the capabilities of new medicines in treatment. Animal models are the perfect selection to investigate the potential toxic effect of CNTs on human health and the environment. Nonfunctionalized CNTs were instilled intratracheally (IT) into animals, exhibited as pulmonary toxicity including inflammation and fibrotic responses due to the collection of raw CNTs in lung airways (Lam et al., 2004). These outcomes suggest that aerosol vulnerability of untreated CNTs in the workplace should be shunned to preserve human health. Notwithstanding, intratracheal instillation of functionalized soluble CNTs has little inference on the toxicology profile. Asbestos-like pathogenicity was observed when the mice were exposed to large MWCNTs (80–160 nm diameter and 10–50 nm length) (Poland et al., 2008). The assumption of this finding for probable adverse effects of CNTs on human health is inadequate. It should be heeded that the MWCNT materials utilized in this research were just sonicated in bovine serum albumin (BSA) without surface functionalization.

Furthermore, no noticeable toxic result was observed for smaller and tinier MWCNTs of 1–20 nm length and 10–14 nm diameter, appreciating that the toxicology characterizations of CNTs may vary between CNTs of different sizes. Functionalized SWCNTs utilized in biomedical research have a length of 50–300 nm and diameter of 1–2 nm, which is entirely distinct from the geometry of MWCNTs adopted by Poland et al. (2008). Gambhir and group applied covalently and noncovalently PEGylated SWCNTs to investigate in vivo toxicity (Schipper et al., 2008). The PEGylated SWCNTs (−3 mg kg^{-1}) were intravenously infused into mice and inspected for four consecutive months. All the biochemical tests in blood, systolic blood pressure, and total blood counts are considered as the parameters to evaluate toxicity. Necropsy and tissue histology analyses were executed after four months. The blood chemistry and histological investigations were standard. Those experiments insinuate that functionalized biocompatible SWCNTs may be secured for in vivo biological reinforcements. An added investigation revealed related outcomes, confirming that PEGylated SWCNTs are gradually eliminated from the body after systemic administration in mouse models, without manifesting apparent toxicity in the system (Liu et al., 2008). Yang et al. (2008a, b) acknowledged that SWCNTs dangled in Tween-80 exhibited lesser toxicities to the experimented mice at a high dose of 40 mg kg^{-1}, following intravenous inoculation for three months. The reason for toxicity may be due to the oxidative stress engendered by SWCNTs assembled in the liver and lungs of mice (Yang et al., 2008a, b). The toxicity published was dose-dependent and appeared to be less acceptable at lower doses. A current article by the same group reveals that covalently PEGylated SWCNTs displayed an ultralong blood dissemination half-life in rodents. The long-term toxicology of altered SWCNTs remains to be investigated. No critical toxicity has been recorded at a higher dose of 24 mg kg^{-1}.

4 Respiratory toxicity

A guinea pig was inoculated intratracheally with the soot of CNT. Breathing rate, tidal capacity, pulmonary obstruction, broncho alveolar fluid, and protein content were estimated. The authors admitted that working with soot carrying CNT probably did not result in jeopardy to health, but they did not present their pathological investigation (Huczko et al., 2001). Research in mice was conducted by Lam et al. (2004), and they authenticated that SWCNT could be toxic if entering the lungs. Warheit et al. (2004) conveyed a related investigation in rats, reporting granuloma development apparently due to the collection of CNT. Muller et al. analyzed carbon black, MWCNT, and asbestos influences implanted in the trachea of rodents. Scholars demonstrated dose-dependent inflammation, and granuloma production increased considerably with

MWCNT compare to carbon black and asbestos. Early granulomatous reaction, abnormal acute inflammatory response, and progressive fibrosis were observed upon exposure of SWCNT in mice. Pharyngeal aspiration was used alternately of the intratracheal installation used in the earlier investigations and rendered aerosolization of fine SWNCT particles. Another contemporary study insinuated shifts in deposition prototype and pulmonary response when SWCNTs are uniformly dispersed in the suspension antecedent to pharyngeal aspiration (Mercer et al., 2008). Research insinuates MWCNT immigration to the subpleural and associated pleural mononuclear cells, and subpleural fibrosis in mice upon inhalation (Ryman-Rasmussen et al., 2009), further admonition and decent security models are prescribed when manipulating CNT. Research by Wang et al. (2010), confirmed the earlier reports, as they characterized in vitro and in vivo stimulation of collagen deposition, lung fibroblast propagation, and metalloproteinase intensified expression without inflammation when dispersed SWCNTs were applied. Following inhalation, the different variety of nanoparticles may enter the central nervous system (CNS) (Elder et al., 2010) by a method called transcytosis (Zensi et al., 2009). The investigation revealed that sniffed gold nanoparticles aggregate in the olfactory tubercle of rats and enter the cerebral cortex, lungs, and distinct organs such as the tongue, esophagus, kidney, spleen, aorta, septum, heart, and blood (Yu et al., 2007). Those remarks vindicate that nanoparticles can infiltrate the CNS via olfactory venation if they are in high doses in the air. Those nanoparticles may impact not only on the respiratory tract and neighboring organs, but can also be disseminated to distant organs.

5 Biodistribution of carbon nanotubes

Knowledge of biodistribution of CNTs following systemic inoculation inside animals is a pretty serious concern. Numerous investigators studied in vivo biodistribution and pharmacokinetic investigations in recent years. Scientists adopted various CNT materials, different surface functionalization methods, and various tracking methodologics. The result was unsteady and ambiguous. Singh et al. (2006) and Lacerda et al. (2008) utilized radio-labeled (1n-DTPA) SWCNTs and MWCNTs to describe biodistribution (Singh et al., 2006). Exceptionally, following intravenous inoculation of CNTs into mice, no uptake in the reticuloendothelial system (RES) such as the liver and spleen was witnessed. However, quick urinal removal of CNTs was witnessed. More than 95% of CNTs were removed within 3 h. Those results are comparable to the in vivo response of minute particles, yet distinct from that prognosticated of maximum nanoparticles with sizes exceeding the glomerular filtration threshold. The researchers justified their research by stating that the short CNT diameters were eliminated in urine, notwithstanding that they were significant in length. However, this theory is unsettled. For example, in the protein biodistribution and elimination function of quantum dots (QDs) published by Choi et al., it is observed that the maximum 6 nm size of spherical QDs, including coatings, was obliged for fast urinal elimination.

Nevertheless, the QDs are much shorter than the diameter of SWCNT bundles (10–40 nm) or MWCNTs (20–80 nm). Lacerda et al. (2008) used in those biodistribution investigations. Therefore, the inscribed fast urinal excretion of CNTs requires validation. Various other labs have also assessed the biodistribution of radio-labeled CNTs in rodents. Wang et al. noticed delayed urinal elimination and weak RES uptake in their primary research. However, the following articles by the same association utilizing^{14} C-taurine-functionalized CNTs recorded steadfast liver accumulation of CNTs following intravenous inoculation (Deng et al., 2008). Research done by McDevitt et al. (2007) utilizing antibody-conjugated radio-labeled CNTs functionalized by 1,3-dipolar cycloaddition also confirmed delayed urinal excretion and high CNT uptake in the liver and spleen. The biodistribution investigations of radio-labeled PEGylated SWNTs unveiled uptake of SWNT in RES organs without active clearance (Liu et al., 2007). A substantial quantity of CNTs persisted in the body for another 15 days. The radio-label system is a proper technique to identify the biodistribution of material, but may reach inaccurate outcomes, if excess free radioisotopes in the radio-labeled CNT specimens are not separated effectively. The free radioisotopes are tiny particles that could be quickly excreted in urine following intravenous inoculation.

Moreover, radio labels could be undeviatedly released from CNTs in vivo, and be regularly eliminated in the free form. Consequently, radio-labeling is not an excellent approach to investigate the elimination and long-term predestination of CNTs. The experts have discovered that photoluminescence is the inherent characteristic of CNTs. Cherukuri et al. (2006) used single semiconducting SWNTs that show NIR photoluminescence to trace nanotubes in rabbits. Without obtaining complete biodistribution data, the expert could not testify to SWNT photoluminescence signals in every organ besides the liver. Yang et al. (2007) researched the biodistribution of 13^{c} fortified unfunctionalized SWNTs over a month utilizing isotope ratio mass spectroscopy. The event conferred significant nanotube uptake in the liver, lung, and spleen without notable elimination within 28 days. Raman spectroscopy has been applied to analyze the long-term predestination of PEGylated nanotubes in rodents. It was reported that most of the PEGylated SWNTs were assembled in the liver and spleen following intravenous inoculation, but gradually eliminated through the biliary pathway toward the feces within months. A low SWNT Raman signal was also identified in the mouse kidney and bladder. It was shown that a small portion of SWNTs with short lengths was eliminated into the urine.

6 Toxicity of carbon nanotubes in vitro

In vitro, toxicological investigations are a highly significant means for nanotoxicology, corresponding to in vivo investigations because of moderate expense, lessening ethical anxieties, and a diminishing number of laboratory animals needed for trial.

7 CNT toxicity investigations in animal cell lines

The subject of carbon nanotube toxicity is still unresolved even in cell culture experiments. Inhibition of HEK 293 and cell proliferation following exposure to SWCNTs were reported by Cui et al. (2005). Cell cycle arrest and increased necrosis of human skin fibroblasts following MWCNT exposure were examined by different research groups (Ding et al., 2005). Nevertheless, it is worth stating that functionalized CNTs were not used in those investigations. Bottini et al. (2006) observed T lymphocyte apoptosis evoked by oxidized MWCNTs. It is pertinent to mention that functionalization via simple oxidation cannot disperse carbon nanotubes in saline and cell culture media. Sayes et al. (2006) indicated that CNT toxicity was also dependent on the density of functionalization. Inconsiderable toxicity was observed for those functionalized with a high frequency of phenyl-SO3X groups. Reasonably, these findings make us understand that unfunctionalized CNTs carry a hydrophobic surface and get aggregated in culture media or any solvent. These aggregated CNTs allows several biological species to interact via hydrophobic interactions and evoke cell toxicity. Khalid et al. (2014) reported no toxicity of functionalized MWCNTs to Saos cell lines up to the tested concentration of 1000 μg/mL. Surfactant, used for CNT dispersion in culture media and the metal catalyst, used during CNT synthesis may also add toxicity to the cells. So, these factors must also be analyzed when the toxicity of CNTs is evaluated (Dong et al., 2008; Plata et al., 2008).

Furthermore, proper analytical methods must be used in toxicity analysis to prevent interference of carbon nanotubes with the test reagents (Casey et al., 2007). Davoren et al. (2007) reported concentration-dependent cytotoxicity of SWCNTs on a lung carcinoma cell line (A549). Another study, led by Sharma, revealed that SWCNTs induced oxidative stress in rat lung cells (Sharma et al., 2007). Toxicity studies on primary bronchial epithelial cells and A549 cells confirmed oxidative stress, and the dispersion medium is the reason for cell death (Herzog et al., 2007). Furthermore, oxidative stress is considered the reason for toxicity in human A549 and rat macrophage NR8383 cell lines. However, when purified SWCNTs correspond with commercial CNT, it was revealed that all the biological consequences are associated with the metal traces. There is a complicated result between WST 2-(4- Iodopheny1)-3-(4-nitropheny1)-5-(2,4-disulfopheny1)- 2H-tetrazolium, monosodium salt) and MTT (3-(4-5-Dimethylthiazol-2-yl)-2, 5-diphenyltetrazolium bromide) viability assays. These dyes depend on mitochondrial dehydrogenases activity (Pulskamp et al., 2007). The modifications can only be described based on CNT associations with nonsoluble formazan crystals in MTT. That is why suitable assay methods and well-characterized materials are the most important requirements for in vitro toxicity assays of carbon nanotubes.

8 CNT toxicity investigations in bacteria and yeast cells

As an option to animal cell lines, bacteria and yeast can be relevant models for studying how single-celled microorganisms react to environmental stressors such as CNTs (Boor, 2006). Numerous toxicity mechanisms have been suggested (Kang et al., 2007) for CNTs, including: (a) penetration of the cell envelope, (b) oxidation of cell ingredients, (c) arrest of transmembrane electron transfer, and (d) reactive oxygen species (ROS).

CNT toxicity CNT depends on its structure along with its geometry and surface functionalization. Various studies have shown that adequately functionalized, serum-stable CNTs are innoxious to animal cells, whereas CNTs without functionalization seemed critically toxic to human or animal cell lines at a moderate dosage (Dumortier et al., 2006). SWCNTs display a potent antimicrobial response for both suspended and deposited bacteria and interrupt the accumulation of bacterial films. The immediate interaction among the SWCNTs and bacteria is the central cause to induce cell death (Kang et al., 2007).

The individual CNT can physically puncture the cell membrane; hence this becomes more toxic to the cells (Liu et al., 2009). The CNT-bacteria interplay is determined by surface functionalization and CNT length. A functionalized SWCNT bearing -OH or -COOH groups aggregates more efficiently with bacteria and diminishes bacteria viability as contrasted to the positively charged SWCNTs functionalized with $-NH_2$ (Arias and Yang, 2009). Likewise, longer SWCNTs exhibited concentration- and time-dependent toxicity to bacteria, whereas short SWCNTs were limited in toxicity as they aggregated themselves (Yang et al., 2010). SWCNT purity may also influence bacterial toxicity. Pure SWCNTs were observed to be less toxic than SWCNTs with higher metal content due to glutathione oxidation following contact (Vecitis et al., 2010).

Moreover, greater ionic strength suspensions, such as phosphate-buffered saline (PBS) or brain heart infusion broth, decreased the intensity of interactions between SWCNTs and cells, hence toxicity decreased. Whereas, low ionic strength suspensions such as deionized water or saline increased the intensity of interactions between SWCNTs and cells, resulting in an increase in toxicity. Likewise, a film with natural organic matter (NOM) limits SWCNT toxicity, notwithstanding diminished aggregation (Kang et al., 2009). Other studies revealed that SWCNTs reduce enzyme activity and microbial biomass at concentrations of $300\,mg\,kg^{-1}$ and above (Jin et al., 2013). As it is clear that SWCNTs provoke bacterial death, a surface coating with SWCNTs would decrease biofilm expansion in both real and industrial settings (Rodrigues and Elimelech, 2010). MWCNTs appear to be runty toxic to bacteria as contrasted to SWCNTs (Kang et al., 2008a, b). The decreased toxicity may be due to minor interactions among bacteria and MWCNTs. The limited interaction might be due to the greater rigidity and presumably inferior van der Waals forces at the MWCNT surface. Thin MWCNTs with less diameter exhibit greater toxicity to bacteria compared to larger ones (Zheng et al., 2010). When the consequence of MWCNT length was estimated, shorter MWCNTs were extratoxic to Pseudomonas fluorescence compared to long MWCNTs (Riding et al., 2012).

Toxicity to bacteria raised, when uncapped, debundled MWCNTs are dispersed in the solution (Kang et al., 2008a, b). CNT purity has also been vindicated to influence toxicity in microorganisms. Furthermore, when the toxicity within pristine and purified MWCNTs was studied in two bacterial strains (*E. coli* and *Cupriavidus metallidurans*), no variation in MWCNT toxicity was perceived between the two forms (Simon-Deckers et al., 2009). The heating refinement of CNTs presumably has inadequate ability to modify the surface corresponding to acid processing, consequently sustaining the toxicity of the raw form. However, in both investigations, Gum Arabic (GA, 0.25 wt%) was used to suspend CNTs, which might have altered the surface and influenced the toxicity. Meanwhile, in a soil toxicity assay, MWCNTs reduced microbial biomass and enzyme activity at concentration of 5000 mg kg^{-1} (Chung et al., 2011). Mycelium growth on a solid medium was observed upon incubation of the conidia of the fungi *Paecilomyces fumosoroseus* for 865 h with 0.2 mg L^{-1} raw and/or carboxylated MWCNTs. Associations among the fungi and CNTs had no notable effect on germination and biomass production, but the loss of biomass was witnessed following exposure to raw MWCNTs for 865 h (Gorczyca et al., 2009). The mechanical impacts of CNTs, as observed in bacteria, might have caused the effects.

9 Ecotoxicity of carbon nanotubes

As the production and widespread application of CNTs in industrial and customer products progresses, the release of this nanomaterial into the environment will also scale up. Many scientific reviews have evaluated the sources, behavior, fate, and mechanisms of toxicity of carbon nanomaterials. Most of these assessments state that additional research is obligatory in the field of nanoecotoxicology (Table 1).

10 Regulation of carbon nanomaterials

Scientists, regulatory agencies, and governments are taking several measures to institute standard guidelines for handling of nanomaterials around the world (NIOSH, 2006, 2009; Paik et al., 2008). Because of inadequate toxicological/ecotoxicological data of carbon nanomaterials as well as qualitative and quantitative characterization data, public authorities and regulatory agencies face difficulties in framing new laws or adapting incremental approaches to handling the risk management of carbon nanomaterials. The Commission of the European Community has adopted the incremental approach, which emphasizes existing laws and amends them to regulate nanotechnology and nanomaterials. The Commission also acknowledged that if a prevalent material that is already present on the market as an existing entity and is reintroduced in the nano form, that is, less than 100 nm in diameter (nanomaterial), the data sheet of the material shall be updated. The supplementary information, including classification of the nanomaterial, risk assessment, and management techniques must be incorporated in the registration profile and shall be communicated to the buyers (Commission of the European Communities, 2008). Both the Commission (2008) and the Scientific Committee on Emerging and Newly Identified Health Risks (SCENIHR, 2007, 2009) also recommended that existing assessment guidelines for evaluating risks are not suitable for assessing risks associated with nanomaterials.

Previously, carbon and graphite materials were spared from registration under REACH (Registration, Evaluation, and Authorization of Chemicals) in European countries. However, this exemption stand-off after the safety issue of this material surfaced by the scientist (Führ et al., 2007). Currently, the company has to register

Table 1 Summary of the studies related to ecotoxicity of CNTs on different organisms.

Organism tested	Types of CNTs	LOEC	EC 50	Mechanism of toxicity	References
Chlorella vulgaris	Pristine CNT	$0.053\,mg\,L^{-1}$	$1.8\,mg\,L^{-1}$	Oxidative stress, agglomeration, and physical interactions	Schwab et al. (2011)
	MWCNT of diameter 10, 20–40, and 60–100 lllll	NA	41.0, 12.7, and $12.4\,mg\,L^{-1}$, respectively	Oxidative stress, agglomeration, and physical interactions	Long et al. (2012)
Pseudokirchn eriella subcapitata	Pristine CNT	$0.053\,mg\,L^{-1}$	$2.5\,mg\,L^{-1}$	Oxidative stress, agglomeration, and physical interactions	Schwab et al. (2011)
	SWCNT	0.25 mg L	NA	Oxidative stress, agglomeration, and physical interactions	Youn et al. (2011)
Thalassiosira pseudonanas	DWCNTs	$0.1\,mg\,L^{-1}$	$1.86\,mg\,L^{-1}$	Oxidative stress, agglomeration, and physical interactions	Kwok et al. (2010)
Dunaliella tertiolecta	MWCNT	NA	$0.8\,mg\,L^{-1}$	Oxidative stress and photosynthesis inhibition	Wei et al. (2010)
Tetrahymena thermophila	SWCNT	$1.6\,mg\,L^{-1}$	NA	Physical interactions	Ghafari et al. (2008)
Stylonychia mytilus	Functionalized MWCNT	$1\,mg\,L^{-1}$	NA	Physical interactions	Zhu et al. (2006)
Hydra attenuata	SWCNT	$1.10\,mg\,L^{-1}$	*Hydra* NA	Physical interactions	Calabrese (2005)
Daphnia magna	SWCNT (60% pure)	NA	$1.3\,mg\,L^{-1}$	Physical interactions	Zhu et al. (2009)
	MWCNT resuspended in NOM	NOEC $20\,mg\,L^{-1}$	NA	No toxicity	Kim et al. (2009)
	MWCNT grafted with polyethylenimine	NA	$25\,mg\,L^{-1}$	Increased size of surface coating	Petersen et al. (2011)

Ceriodaphnia. dubia	MWCNT resuspended in NOM	0.25 mg L^{-1}	NA	Agglomeration	Li and Huang (2011)
Danio rerio embryo	SWCNT	120 mg L^{-1}	NA	Agglomeration	Cheng et al. (2007)
Oryzias melastigma	DWCNT	10 mg L^{-1}	NA	Agglomeration	Kwok et al. (2010)
Xenopus leavis larvae	DWCNT	10 mg L^{-1}	NA	Physical interactions	Bourdiol et al. (2013)
Drosophila melanogaster	SWCNT and MWCNT in 1 g kg^{-1} food			No toxicity	Leeuw et al. (2007)
Female Fisher rats	Oral gavage of 0.64 mg ka^{1} SWCNT	NA	NA	Increased levels of oxidative damage to DNA in liver and lung tissue	Aschberger et al. (2010)
Sprague–Dawley rat	1000 mg kg^{-1} of SWCNT from gestation day 6 to 19	NA	NA	No teratogenicity	Lim et al. (2011)

Abbreviations: *EC 50*: effective concentration 50; *LOEC*: least observable effect concentration; *NOEC*: No observed effect concentration; *NOM*: natural organic matter.

under REACH if they are producing these nanomaterials in quantities above one ton per year. Furthermore, the Chemical Safety Assessment must be produced if the production is more than 10 tons per year. If the material falls under PBT (Persistent, Bioaccumulative, and Toxic) or vPvB (very Persistent and very Accumulative), the company must develop an exposure scenario and risk assessment procedure to ensure the safety of these engineered nanomaterials.

11 Conclusions

Toxicity of carbon nanomaterials is a critical concern in the modern world for the scientific community, environmentalists and governments. The chances of exposure to the environment are greater as the application of carbon nanomaterial is increasing every day. Some research shows different toxicity patterns for the materials when exposed to living cells in vitro or in vivo, whereas other studies say that the adequately functionalized bearing carboxylic or hydroxyl group and serum-stable CNTs are safe for living cells. I want to conclude my discussion by highlighting the factors involved in toxicity and the toxicity mechanism. The toxicity of the nanomaterials depends on many factors, including functionalization, catalyst, size, shape, dimension, dispersion, and methods used for detecting toxicity. The pristine carbon nanotubes are more damaging to cells compared to functionalized ones. The covalently functionalized CNTs are more compatible for cells compared to noncovalent functionalization. The catalysts used during the production of the nanotubes such as platinum or iron also contribute to the toxicity of the cells. Hence, it is imperative to differentiate the toxicity of carbon nanotubes and catalysts. Dispersion in a high ionic strength solvent such as PBS makes CNTs more compatible with living cells compared to the less ionic strength solvent such as deionized water. Hence, it is always recommended to prepare the solution in PBS or other high ionic strength solvents for better compatibility and less toxicity. The short and broken CNTs with small diameters are observed to be damaging to bacterial cells because of physical puncturing. In vivo studies help us to understand the acute toxicity, chronic toxicity, developmental toxicity, genotoxicity, and reproductive toxicity of CNTs in laboratory animals. No critical acute toxicity, chronic toxicity, developmental toxicity, genotoxicity, or reproductive toxicity is observed following intravenous or intratracheal instillation of CNTs. Adequately functionalized CNTs are biocompatible and promptly eliminated through urine or the biliary pathway following intravenous inoculation. Pharmacokinetics studies of CNTs show less or no uptake of CNTs to the reticuloendothelial systems, including the liver, lungs, and spleen. Various mechanisms are also listed to study CNT toxicity to living cells, including oxidation of cell components, arrest of electron transport chain, reactive oxygen species, and physical puncturing of the cell. Further studies need to be conducted in the field of ecotoxicity of CNTs and validation of the toxicological data for the safety of aquatic and aerial animals. These studies shall help public regulatory organizations to frame rules to ensure the safety of this modern engineered nanoparticle.

References

Arias, L.R., Yang, L., 2009. Inactivation of bacterial pathogens by carbon nanotubes in suspensions. Langmuir 25 (5), 3003–3012. https://doi.org/10.1021/1a802769m.

Aschberger, K., Johnston, H.J., Stone, V., Aitken, R.J., Hankin, S.M., Peters, S.A., Tran, C.L., Christensen, F.M., 2010. Review of carbon nanotubes toxicity and exposure appraisal of human health risk assessment based on open literature. Grit. Rev. Toxicol. 40 (9), 759–790. https://doi.org/10.3109/10408444.2010.506638.

Boor, K.J., 2006. Bacterial stress responses: what doesn't kill them can make them stronger. PLoS Biol. 4 (1), e23. https://doi.org/10.1371/journal.pbio.0040023.

Bottini, M., Bruckner, S., Nika, K., Bottini, N., Bellucci, S., Magrini, A., Bergamaschi, A., Mustelin, T., 2006. Multi- walled carbon nanotubes induce T lymphocyte apoptosis. Toxicol. Lett. 160 (2), 121–126. https://doi.org/10.1016/j.toxlet.2005.06.020.

Bourdiol, F., Mouchet, F., Perrault, A., Fourquaux, I., Datas, L., Gancet, C., Boutonnet, J.C., Pinelli, E., Gauthier, L., Flahaut, E., 2013. Biocompatible polymer-assisted dispersion of multi walled carbon nanotubes in water, application to the investigation of their ecotoxicity using *Xenopus laevis* amphibian larvae. Carbon 54, 175–191. https://doi.org/10.1016/j.carbon.2012.11.024.

Cai, H., Cao, X., Jiang, Y., He, P., Fang, Y., 2003. Carbon nanotube-enhanced electrochemical DNA biosensor for DNA hybridization detection. Anal. Bioanal. Chem. 375 (2), 287–293. https://doi.org/10.1016/j.envpol.2004.10.001.

Calabrese, E.J., 2005. Paradigm lost, paradigm found: the reemergence of hormesis as a fundamental dose response model in the toxicological sciences. Environ. Pollut. 138 (3), 378–411. https://doi.org/10.1016/j.envpol.2004.10.001.

Casey, A., Herzog, E., Davoren, M., Lyng, F.M., Byrne, H.J., Chambers, G., 2007. Spectroscopic analysis confirms the interactions between single walled carbon nanotubes and various dyes commonly used to assess cytotoxicity. Carbon 45 (7), 1425–1432. https://doi.org/10.1016/j.carbon.2007.03.033.

Cheng, J., Flahaut, E., Cheng, S.H., 2007. Effect of carbon nanotubes on developing zebrafish (*Danio rerio*) embryos. Environ. Toxicol. Chem. 26 (4), 708–716. https://doi.org/10.1897/06-272R. 1.

Cherukuri, P., Gannon, C.J., Leeuw, T.K., Schmidt, H.K., Smalley, R.E., Curley, S.A., Weisman, R.B., 2006. Mammalian pharmacokinetics of carbon nanotubes using intrinsic near infrared fluorescence. Proc. Natl. Acad. Sci. 103 (50), 18882–18886. https://doi.org/10.1073/pnas.0609265103.

Chung, H., Son, Y., Yoon, T.K., Kim, S., Kim, W., 2011. The effect of multi walled carbon nanotubes on soil microbial activity. Ecotoxicol. Environ. Saf. 74 (4), 569–575. https://doi.org/10.1016/j.ecoenv.2011.01.004.

Cui, D.X., Tian, F.R., Ozkan, C.S., Wang, M., Gao, H.J., 2005. Effect of single wall carbon nanotubes on human HEK293 cells. Toxicol. Lett. 155 (1), 73–85. https://doi.org/10.1016/j.toxlet.2004.08.015.

Commission of the European Communities, 2008. Communication From The Commission To The European Parliament, The Council And The European Economic And Social Committee Regulatory Aspects Of Nanomaterials [Sec(2008) 2036] Com(2008) 366 final. Commission of the European Communities, Brussels.

Davoren, M., Herzog, E., Casey, A., Cottineau, B., Chambers, G., Byrne, H.J., Lyng, F.M., 2007. In vitro toxicity evaluation of single walled carbon nanotubes on human A549 lung cells. Toxicol. In Vitro 21 (3), 438–448. https://doi.org/10.1016/j.tiv.2006.10.007.

Deng, X.Y., Yang, S.T., Nie, H.Y., Wang, H.F., Liu, Y.F., 2008. A generally adoptable radiotracing method for tracking carbon nanotubes in animals. Nanotechnology 19 (7), 75101. https://doi.org/10.1088/0957-4484/19/7/075101.

Ding, L.H., Stilwell, J., Zhang, T.T., Elboudwarej, O., Jiang, H.J., Selegue, J.P., Cooke, P.A., Gray, J.W., Chen, F.Q.F., 2005. Molecular characterization of the cytotoxic mechanism of multiwall carbon nanotubes and nano-onions on human skin fibroblast. Nano Lett. 5 (12), 2448–2464. https://doi.org/10.1021/nl051748o.

Dong, L., Joseph, K.L., Witkowski, C.M., Craig, M.M., 2008. Cytotoxicity of single-walled carbon nanotubes suspended in various surfactants. Nanotechnology 19 (25), 255702. https://doi.org/10.1088/0957-4484/19/25/255702.

Dror, Y., Salalha, W., Khalfin, R.L., Cohen, Y., Yarin, A.L., Zussman, E., 2003. Carbon nanotubes embedded in oriented polymer nanofibers by electrospinning. Langmuir 19 (17), 7012–7020. https://doi.org/10.1021/la034234i.

Dumortier, H., Lacotte, S., Pastorin, G., Marega, R., Wu, W., Bonifazi, D., Briand, J.P., Prato, M., Muller, S., Bianc, A., 2006. Functionalized carbon nanotubes are noncytotoxic and preserve the functionality of primary immune cells. Nano Lett. 6 (7), 1522–1528. https://doi.org/10.1021/nl061160x.

Elder, A., Gelein, R., Silva, V., Feikert, T., Opanashuk, L., Carter, J., Potter, R., Maynard, A., Ito, Y., Finkelstein, J., Oberdorster, G., 2010. Translocation of inhaled ultrafine manganese oxide particles to the central nervous system. Environ. Health Perspect. 114 (8), 1172–1178. https://doi.org/10.1289/ehp. 9030.

Elkin, T., Jiang, X., Taylor, S., Lin, Y., Gu, L., Yang, H., Brown, J., Collins, S., Sun, Y.P., 2005. Immuno carbon nanotubes and recognition of pathogens. Chem. Biochem. 6 (4), 640–643. https://doi.org/10.1002/cbic.200400337.

Endo, M., Strano, M.S., Ajayan, P.M., 2008. Potential applications of carbon nanotubes. Appl. Phys. 111, 13–61. https://doi.org/10.1007/978-3-540-72865-8.

Formhals, A., Process and apparatus for preparing artificial threads, U.S. Patent 1,975,504. (1934).

Führ, M., Hermann, A., Merenyi, S., Moch, K., Möller, M., 2007. Legal appraisal of nanotechnologies. Available: http://www.umweltdaten.de/publikationen/fpdf-l/3198.pdf (Accessed 26-03-2019). Umwelt Bundes Amt UBA-FB 000996. ISSN 1862–4804.

Ghafari, P., St-Denis, C.H., Power, M.E., Jin, X., Tsou, V., Mandal, H.S., Bols, N.C., Shirley, X., 2008. Impact of carbon nanotubes on the ingestion and digestion of bacteria by ciliated protozoa. Nat. Nanotechnol. 3 (6), 347–351. https://doi.org/10.1038/nnano.2008.109.

Gorczyca, A., Kasprowicz, M.J., Lemek, T., 2009. Physiological effect of multi walled carbon nanotubes (MWCNTs) on conidia of the entomopathogenic fungus, Paecilomyces fumosoroseus (Deuteromycotina: Hyphomycetes). J. Environ. Sci. Health A 44 (14), 1592–1597. https://doi.org/10.1080/10934520903263603.

Herzog, E., Casey, A., Lyng, F.M., Chambers, G., Byrne, H.J., Davoren, M., 2007. A new approach to the toxicity testing of carbon-based nanomaterials—the clonogenic assay. Toxicol. Lett. 174 (1–3), 49–60. https://doi.org/10.1016/j.toxlet.2007.08.009.

Huczko, A., Lange, H., Calstrokko, E., Grubek-Jaworska, H., Droszcz, P., 2001. Physiological test of carbon nanotubes: are they asbestos like? Fuller. Sci. Technol. 9 (2), 251–254. https://doi.org/10.1081/FST-100102973.

Jin, L., Son, Y., Yoon, T.K., Kang, Y.J., Kim, W., Chung, H., 2013. High concentrations of single walled carbon nanotubes lower soil enzyme activity and microbial biomass. Ecotoxicol. Environ. Saf. 88, 9–15. https://doi.org/10.1016/j.ecoenv.2012.10.031.

Kagan, V.E., Tyurina, Y.Y., Tyurin, V.A., Konduru, N.V., Potapovich, A.I., Osipov, A.N., Shvedova, A.A., 2006. Direct and indirect effects of single walled carbon nanotubes on RAW 264.7 macrophages: role of iron. Toxicol. Lett. 165 (1), 88–100. https://doi.org/10.1016/j.toxlet.2006.02.001.

Kam, N.W.S., Jessop, T.C., Wender, P.A., Dai, H.J., 2004. Nanotube molecular transporters: Internalization of carbon nanotube-protein conjugates into mammalian cells. J. Am. Chem. Soc. 126 (22), 6850–6851. https://doi.org/10.1021/j a0486059.

Kam, N.W.S., O'Connell, M., Wisdom, J.A., Dai, H., 2005. Carbon nanotubes as multifunctional biological transporters and near-infrared agents for selective cancer cell destruction. Proc. Natl. Acad. Sci. 102 (33), 11600–11605. https://doi.org/10.1073/pnas.0502680102.

Kamavaram, V., Veedu, V., Kannan, A.M., 2009. Synthesis and characterization of platinum nanoparticles on in situ grown carbon nanotubes based carbon paper for proton exchange membrane fuel cell cathode. J. Power Sources 188 (1), 51–56. https://doi.org/10.1016/j jpowsour.2008.11.084.

Kang, S., Herzberg, M., Rodrigues, D.F., Elimelech, M., 2008a. Antibacterial effects of carbon nanotubes: Size does matter. Langmuir 24 (13), 6409–6413. https://doi.org/10.1021/1a800951v.

Kang, S., Mauter, M.S., Elimelech, M., 2008b. Physicochemical determinants of multiwalled carbon nanotube bacterial cytotoxicity. Environ. Sci. Technol. 42 (19), 7528–7534. https://doi.org/10.1021/es8010173.

Kang, S., Mauter, M.S., Elimelech, M., 2009. Microbial cytotoxicity of carbon based nanomaterials: Implications for river water and wastewater effluent. Environ. Sci. Technol. 43 (7), 2648–2653. https://doi.org/10.1021/es8031506.

Kang, S., Pinault, M., Pfefferle, L.D., Elimelech, M., 2007. Single walled carbon nanotubes exhibit strong antimicrobial activity. Langmuir 23 (17), 8670–8673. https://doi.org/10.1021/1a701067r.

Kenawy, E.R., Bowlin, G.L., Mansfield, K., Layman, J., Simpson, D.G., Sanders, E.H., Wnek, G.E., 2002. Release of tetracycline hydrochloride from electrospun poly(ethylene-co-vinylacetate), poly(lactic acid) and a blend. J. Control Release 81 (1), 57–64. https://doi.org/10.1016/S0168-3659(02)00041-X.

Khalid, P., Hussain, M.A., Rekha, P.D., Arun, A.B., 2014. Carbon nanotube-reinforced hydroxyapatite composite and their interaction with human osteoblast in vitro. Hum. Exp. Toxicol. 33 (9), 897–990.

Kim, K.T., Edgington, A.J., Klaine, S.J., Cho, J.W., Kim, S.D., 2009. Influence of multi-walled carbon nanotubes dispersed in natural organic matter on speciation and bioavailability of copper. Environ. Sci. Technol. 43 (23), 8979–8984.

Kim, Y.J., Kim, Y.A., Chino, T., Suezaki, H., Endo, M., Dresselhaus, M.S., 2006. Chemically modified multi walled carbon nanotubes as an additive for Supercapacitors. Small 2 (3), 339–345. https://doi.org/10.1002/sm11.200500327.

Ko, F., Gogotsi, Y., Ali, A., Naguib, N., Ye, H., Yang, G.L., Cristopher, L., Willis, P., 2003. Electrospinning of continuous carbon nanotube filled nanofiber yarns. Adv. Mater. 15 (14), 1161–1165. https://doi.org/10.1002/adma.200304955.

Kundu, S., Nagaiah, T.C., Xia, W., Wang, Y., Dommele, S.V., Bitter, J.H., Santa, M., Grundmeier, G., Bron, M., Schuhmann, W., Muhler, M., 2009. Electrocatalytic activity and stability of nitrogen-containing carbon nanotubes in the oxygen reduction reaction. J. Phys. Chem. C 113 (32), 14302–14310. https://doi.org/10.1021/jp811320d.

Kwok, K.W., Leung, K.M., Flahaut, E., Cheng, J., Cheng, S.H., 2010. Chronic toxicity of double walled carbon nanotubes to three marine organisms: Influence of different dispersion methods. Nanomedicine 5 (6), 951–961. https://doi.org/10.2217/nnm.10.59.

Lacerda, L., Soundararajan, A., Singh, R., Pastorin, G., Al-Jamal, K.T., Turton, J., Frederik, P., Herrero, M.A., Bao, S.L.A., Emfietzoglou, D., Kostarelos, K., 2008. Dynamic imaging of functionalized multi-walled carbon nanotube systemic circulation and urinary excretion. Adv. Mater. 20 (2), 225–230. https://doi.org/10.1002/adma.200890002.

Lam, C.W., James, J.T., McCluskey, R., Hunter, R.L., 2004. Pulmonary toxicity of single wall carbon nanotubes in mice 7 and 90 days after intratracheal instillation. Toxicol. Lett. 77 (1), 126–134. https://doi.org/10.1093/toxsci/kfg243.

Landi, B.J., Castro, S.L., Ruf, H.J., Evans, C.M., Bailey, S.G., Raffaelle, R.P., 2005. CdSe quantum dot single wall carbon nanotube complexes for polymeric solar cells. Sol. Energy Mater. Sol. C 87 (1), 733–746. https://doi.org/10.1016/j.solmat.2004.07.047.

Laxminarayana, K., Jalili, N., 2005. Functional nanotube-based textiles: Pathway to next generation fabrics with enhanced sensing capabilities. Text. Res. J. 75 (9), 670–680. https://doi.org/10.1177/0040517505059330.

Leeuw, T.K., Reith, R.M., Simonette, R.A., Harden, M.E., Cherukuri, P., Tsyboulski, D.A., Beckmingham, K.M., Weisman, R.B., 2007. Single-walled carbon nanotubes in the intact organism: near-TR imaging and biocompatibility studies in Drosophila. Nano Lett. 7 (9), 2650–2654. https://doi.org/10.1021/n10710452.

Leroux, F., Metenier, K., Gautier, S., Frackowiak, E., Bonnamy, S., Beguin, F., 1999. Electrochemical insertion of lithium in catalytic multi walled carbon nanotubes. J. Power Sources 81, 317–322. https://doi.org/10.1016/S0378-7753(99)00130-5.

Li, M., Huang, C.P., 2011. The responses of *Ceriodaphnia dubia* toward multi-walled carbon nanotubes: effect of physical–chemical treatment. Carbon 49 (5), 1672–1679.

Li, W., Liang, C., Qiu, J., Zhou, W., Han, H., Wei, Z., Sun, G., Xin, Q., 2002. Carbon nanotubes as support for cathode catalyst of a direct methanol fuel cell. Carbon 40 (5), 791–794. https://doi.org/10.1016/S0008-6223. (02)00039-8.

Lim, J.H., Kim, S.H., Lee, I.C., Moon, C., Kim, S.H., Shin, D.H., Kim, H.C., Kim, J.C., 2011. Evaluation of maternal toxicity in rats exposed to multi-wall carbon nanotubes during pregnancy. Environ. Health Toxicol. 26, e2011006. https://doi.org/10.5620/eht.2011.26.e2011006.

Lim, J.Y., Lee, C.K., Kim, S.J., Kim, I.Y., Kim, S.I., 2006. Controlled Nanofiber Composed of Multinwall carbon nanotube/poly (ethylene oxide). J. Macromol. Sci. A 43 (4–5), 785–796. https://doi.org/10.1080/10601320600598936.

Liu, S., Wei, L., Hao, L., Fang, N., Chang, M.W., Xu, R., Yang, Y., Chen, Y., 2009. Sharper and faster nano darts kill more bacteria: A study of antibacterial activity of individually dispersed pristine single-walled carbon nanotube. ACS Nano 3 (12), 3891–3902. https://doi.org/10.1021/nn901252r.

Liu, Z., Cai, W.B., He, L.N., Nakayama, N., Chen, K., Sun, X.M., Chen, X.Y., Dai, H.J., 2007. In vivo biodistribution and highly efficient tumour targeting of carbon nanotubes in mice. Nat. Nanotechnol. 2 (1), 47–52. https://doi.org/10.1038/nnano.2006.170.

Liu, Z., Davis, C., Cai, W., He, L., Chen, X., Dai, H., 2008. Circulation and long-term fate of functionalized, biocompatible single-walled carbon nanotubes in mice probed by Raman spectroscopy. Proc. Natl. Acad. Sci. 105 (5), 1410–1415. https://doi.org/10.1073/pnas.0707654105.

Long, Z., Ji, J., Yang, K., Lin, D., Wu, F., 2012. Systematic and quantitative investigation of the mechanism of carbon nanotubes' toxicity toward algae. Environ. Sci. Technol. 46 (15), 8458–8466. https://doi.org/10.1021/es301802g.

Manna, S.K., Sarkar, S., Barr, J., Wise, K., Barrera, E.V., Jejelowo, O., Ramesh, G.T., 2005. Single walled carbon nanotube induces oxidative stress and activates nuclear transcription factor-KB in human keratinocytes. Nano Lett. 5 (9), 1676–1684. https://doi.org/10.1021/n10507966.

McDevitt, M.R., Chattopadhyay, D., Kappel, B.J., Jaggi, J.S., Schiffman, S.R., Antczak, C., Njardarson, J.T., Brentjens, R., Scheinberg, D.A., 2007. Tumor targeting with antibody functionalized, radiolabeled carbon nanotubes. J. Nucl. Med. 48, 1180–1189. https://doi.org/10.2967/jnumed.106.039131.

Mercer, R.R., Scabilloni, J., Wang, L., Kisin, E., Murray, A.R., Schwegler-Berry, D., Shvedova, A.A., Castranova, V., 2008. Alteration of deposition pattern and pulmonary response as a result of improved dispersion of aspirated single-walled carbon nanotubes in a mouse model. Am. J. Physiol. Lung Cell Mol. Physiol. 294 (1), L87–L97. https://doi.org/10.1152/ajplung.00186.2007.

Muller, J., Huaux, F., Moreau, N., Misson, P., Heilier, J.F., Delos, M., Arras, M., Fonseca, A., Nagy, J.B., Lison, D., 2005. Respiratory toxicity of multi-wall carbon nanotubes. Toxicol. Appl. Pharmacol. 207 (3), 221–231. https://doi.org/10.1016/j.taap.2005.01.008.

NIOSH, 2006. Approaches to Safe Nanotechnology: An Information Exchange with NIOSH. Department Of Health And Human Services Centers for Disease Control and Prevention National Institute for Occupational Safety and Health, Washington, D.C.

NIOSH, 2009. Approaches to safe nanotechnology: Managing the health and safety concerns associated with engineered nanomaterials. Department Of Health And Human Services Centers for Disease Control and Prevention National Institute for Occupational Safety and Health, Washington, D.C. Available: http://cdc.gov/niosh/docs/2009-125/. Accessed 26 March 2019.

Patolsky, F., Zheng, G., Hayden, O., Lakadamyali, M., Zhuang, X., Lieber, C.M., 2004. Electrical detection of single viruses. Proc. Natl. Acad. Sci. 101, 14017–14022. https://doi.org/10.1073/pnas.0406159101.

Petersen, E.J., Pinto, R.A., Mai, D.J., Landrum, P.F., Weber Jr., W.J., 2011. Influence of polyethyleneimine graftings of multi walled carbon nanotubes on their accumulation and elimination by and toxicity to *Daphnia magna*. Environ. Sci. Technol. 45 (3), 1133–1138. https://doi.org/10.1021/es1030239.

Plata, D.L., Gschwend, P.M., Reddy, C.M., 2008. Industrially synthesized single walled carbon nanotubes: Compositional data for users, environmental risk assessments and source apportionment. Nanotechnology 19 (18), 185706. https://doi.org/10.1088/0957-4484/19/18/185706.

Poland, C.A., Duffin, R., Kinloch, I., Maynard, A., Wallace, W.A.H., Seaton, A., Stone, V., Brown, S., MacNee, W., Donaldson, K., 2008. Carbon nanotubes introduced into the abdominal cavity of mice show asbestos like pathogenicity in a pilot study. Nat. Nanotechnol. 3 (7), 423–428. https://doi.org/10.1038/nnano.2008.111.

Prato, M., Kostarelos, K., Bianco, A., 2008. Functionalized carbon nanotubes in drug design and discovery. Acc. Chem. Res. 41 (1), 60–68. https://doi.org/10.1021/ar700089b.

Pulskamp, K., Diabate, S., Krug, H.F., 2007. Carbon nanotubes show no sign of acute toxicity but induce intracellular reactive oxygen species in dependence on contaminants. Toxicol. Lett. 168 (1), 58–74. https://doi.org/10.1016/j.toxlet.2006.11.001.

Paik, S.Y., Zalk, D.M., Swuste, P., 2008. Application of a pilot control banding tool for risk level assessment and control of nanoparticle exposures. Ann. Occup. Hyg. 52 (6), 419–428.

Rajendran, V., Selvamurthy, W., Saminathan, K., 2012. Compendium on Indian Capability on Nano Science and Technology. Macmillan Publishers, India. ISBN 978–93–5059-030-0.

Reneker, D.H., Chun, I., 1996. Nanometre diameter fibres of polymer, produced by electrospinning. Nanotechnology 7 (3), 216. https://doi.org/10.1088/0957-4484/7/3/009.

Riding, M.J., Martin, F.L., Trevisan, J., Llabjani, V., Patel, I.I., Jones, K.C., Semple, K.T., 2012. Concentration dependent effects of carbon nanoparticles in gram negative bacteria determined by infrared spectroscopy with multivariate analysis. Environ. Pollut. 163, 226–234. https://doi.org/10.1016/j.envpol.2011.12.027.

Rodrigues, D.F., Elimelech, M., 2010. Toxic effects of single walled carbon nanotubes in the development of *E. coli* biofilm. Environ. Sci. Technol. 44 (12), 4583–4589. https://doi.org/10.1021/es1005785.

Ryman-Rasmussen, J.P., Cesta, M.F., Brody, A.R., Shipley-Phillips, J.K., Everitt, J.I., Tewksbury, E.W., Moss, O.R., Wong, B.A., Dodd, D.E., Andersen, M.E., Bonner, J.C., 2009. Inhaled carbon nanotubes reach the subpleural tissue in mice. Nat. Nanotechnol. 4 (11), 695–747. https://doi.org/10.1038/nnano.2009.305.

Sayes, C.M., Liang, F., Hudson, J.L., Mendez, J., Guo, W., Beach, J.M., Colvin, V.L., 2006. Functionalization density dependence of single-walled carbon nanotubes cytotoxicity in vitro. Toxicol. Lett. 161 (2), 135–142. https://doi.org/10.1016/j.toxlet.2005.08.011.

Schipper, M.L., Nakayama-Ratchford, N., Davis, C.R., Kam, N.W.S., Chu, P., Liu, Z., Sun, X., Dai, H., Gambhir, S.S., 2008. A pilot toxicology study of single-walled carbon nanotubes in a small sample of mice. Nat. Nanotechnol. 3 (4), 216–221. https://doi.org/10.1038/nnano.2008.68.

Schwab, F., Bucheli, T.D., Lukhele, L.P., Magrez, A., Nowack, B., Sigg, L., Knauer, K., 2011. Are carbon nanotube effects on green algae caused by shading and agglomeration? Environ. Sci. Technol. 45 (14), 6136–6144. https://doi.org/10.1021/es200506b.

Sharma, C.S., Sarkar, S., Periyakaruppan, A., Barr, J., Wise, K., Thomas, R., Wilson, B.L., Ramesh, G.T., 2007. Single walled carbon nanotubes induces oxidative stress in rat lung epithelial cells. J. Nanosci. Nanotechnol. 7 (7), 2466–2472. https://doi.org/10.1166/jnn.2007.431.

Shvedova, A.A., Kisin, E.R., Mercer, R., Murray, A.R., Johnson, V.J., Potapovich, A.I., Tyurina, Y.Y., Gorelik, O., Arepalli, S., Schwegler-Berry, D., Baron, P., 2005. Unusual inflammatory and fibrogenic pulmonary responses to single-walled carbon nanotubes in mice. Am. J. Phys. Lung Cell Mol. Physiol. 289 (5), L698–L708. https://doi.org/10.1152/ajplung.00084.2005.

Simon-Deckers, A., Loo, S., Mayne-L'hermite, M., Herlin-Boime, N., Menguy, N., Reynaud, C., Gouget, B., Carriere, M., 2009. Size composition and shape dependent toxicological impact of metal oxide nanoparticles and carbon nanotubes toward bacteria. Environ. Sci. Technol. 43 (21), 8423–8429. https://doi.org/10.1021/es9016975.

Singh, R., Pantarotto, D., Lacerda, L., Pastorin, G., Klumpp, C., Prato, M., Bianco, A., Kostarelos, K., 2006. Tissue biodistribution and blood clearance rates of intravenously administered carbon nanotube radiotracers. Proc. Natl. Acad. Sci. 103 (9), 3357–3362. https://doi.org/10.1073/pnas.0509009103.

Scientific Committee for Emerging and Newly-Identified Health Risks (SCENIHR), 2007. The Appropriateness of the Risk Assessment Methodology in Accordance With the Technical Guidance Documents for New and Existing Substances for Assessing the Risks of Nanomaterials. European Commission Health & Consumer Protection Directorate-General, Brussels.

Scientific Committee on Emerging and Newly Identified Health Risks SCENIHR, 2009. Risk Assessment of Products of Nanotechnologies. European Commission Health & Consumer Protection Directorate-General, Brussels.

Thakare, V.S., Das, M., Jain, A.K., Patil, S., Jain, S., 2010. Carbon nanotubes in cancer theragnosis. Nanomedicine 5 (8), 1277–1301. https://doi.org/10.2217/nnm.10.95.

Tu, X., Manohar, S., Jagota, A., Zheng, M., 2009. DNA sequence motifs for structure specific recognition and separation of carbon nanotubes. Nature 460 (7252), 250–253. https://doi.org/10.1038/nature08116.

Vecitis, C.D., Zodrow, K.R., Kang, S., Elimelech, M., 2010. Electronic structure dependent bacterial cytotoxicity of single walled carbon nanotubes. ACS Nano 4 (9), 5471–5479. https://doi.org/10.1021/nn101558x.

Wang, C., Waje, M., Wang, X., Tang, J.M., Haddon, R.C., Yan, Y., 2004. Proton exchange membrane fuel cells with carbon nanotube based electrodes. Nano Lett. 4 (2), 345–348. https://doi.org/10.1021/n1034952p.

Wang, H., Gu, L., Lin, Y., Lu, F., Meziani, M.J., Luo, P.G., Wang, W., Cao, L., Sun, Y.P., 2008. Unique aggregation of anthrax (*Bacillus anthracis*) spores by sugar coated single walled carbon nanotubes. J. Am. Chem. Soc. 128 (41), 13364–13365. https://doi.org/10.1021/ja065455o.

Wang, L., Mercer, R.R., Rojanasakul, Y., Qiu, A., Lu, Y., Scabilloni, J.F., Wu, N., Castranova, V., 2010. Direct fibrogenic effects of dispersed single walled carbon nanotubes on human lung fibroblasts. J. Toxicol. Environ. Health A 73 (5–6), 410–422. https://doi.org/10.1080/15287390903486550.

Wang, X., Waje, M., Yan, Y., 2005. CNT based electrodes with high efficiency for PEMFCs. Electrochem. Solid-State Lett. 8 (1), A42–A44. https://doi.org/10.1149/1.1830397.

Warheit, D.B., Laurence, B.R., Reed, K.L., Roach, D.H., Reynolds, G.A.M., Webb, T.R., 2004. Comparative pulmonary toxicity assessment of single-wall carbon nanotubes in rats. Toxicol. Lett. 77 (1), 117–125. https://doi.org/10.1093/toxsci/kfg228.

Wei, L.P., Thakkar, M., Chen, Y.H., Ntim, S.A., Mitra, S., Zheng, X.Y., 2010. Cytotoxicity effects of water dispersible oxidized multiwalled carbon nanotubes on marine alga, *Dunaliella tertiolecta*. Aquat. Toxicol. 100 (2), 194–201. https://doi.org/10.1016/j.aquatox.2010.07.001.

Wu, P., Chen, X., Hu, N., Tam, U.C., Blixt, O., Zettl, A., Bertozzi, C.R., 2008. Biocompatible carbon nanotubes generated by functionalization with glycodendrimers. Angew. Chem. 120 (27), 5100–5103. https://doi.org/10.1002/ange.200705363.

Xie, X.L., Mai, Y.W., Zhou, X.P., 2005. Dispersion and alignment of carbon nanotubes in polymer matrix: a review. Mater. Sci. Eng. R 49 (4), 89–112. https://doi.org/10.1016/j.mser.2005.04.002.

Yang, C., Mamouni, J., Tang, Y., Yang, L., 2010. Antimicrobial activity of sigle walled carbon nanotubes lenght effect. Langmuir 26 (20), 16013–16019. https://doi.org/10.1021/la103110g.

Yang, S.T., Fernando, K.A., Liu, J.H., Wang, J., Sun, H.F., Liu, Y., Chen, M., Huang, Y., Wang, X., Wang, H., 2008a. Covalently PEGylated carbon nanotubes with stealth character in vivo. Small 4 (7), 940–944. https://doi.org/10.1002/sml1.200700714.

Yang, S.T., Guo, W., Lin, Y., Deng, X.Y., Wang, H.F., Sun, H.F., Liu, Y.F., Wang, X., Wang, W., Chen, M., Sun, Y.P., 2007. Biodistribution of pristine single-walled carbon nanotubes in vivo. J. Phys. Chem. C 111 (48), 17761–17764. https://doi.org/10.1021/jp070712c.

Yang, S.T., Wang, X., Jia, G., Gu, Y., Wang, T., Nie, H., Ge, C., Wang, H., Liu, Y., 2008b. Long term accumulation and low toxicity of single walled carbon nanotubes in intravenously exposed mice. Toxicol. Lett. 181 (3), 182–189. https://doi.org/10.1016/j.toxlet.2008.07.020.

Youn, S., Wang, R., Gao, J., Hovespyan, A., Ziegler, K.J., Bonzongo, J.C., Bitton, G., 2011. Mitigation of the impact of single-walled carbon nanotubes on a freshwater green algae: Pseudokirchneriella subcapitata. Nanotoxicology 6 (2), 161–172. https://doi.org/10.3109/17435390.2011.562329.

Yu, L.E., Lanry, Y.L.Y., Ong, C.N., Tan, Y.L., Balasubramaniam, K.S., Hartono, D., Shui, G., Wenk, M.R., Ong, W.Y., 2007. Translocation and effects of gold nanoparticles after inhalation exposure in rats. Nanotoxicology 1 (3), 235–242. https://doi.org/10.1080/17435390701763108.

Zensi, A., Begley, D., Pontikis, C., Legros, C., Mihoreanu, L., Wagner, S., Buchel, C., von Briesen, H., Kreuter, J., 2009. Albumin nanoparticles targeted with Apo E enter the CNS by transcytosis and are delivered to neurons. J. Control Release 137 (1), 78–86. https://doi.org/10.1016/j.jconrel.2009.03.002.

Zheng, H., Liu, L., Lu, Y., Long, Y., Wang, L., Ho, K.P., 2010. Rapid determination of nanotoxicity using luminous bacteria. Anal. Sci. 26 (1), 125–128. https://doi.org/10.2116/analsci.26.125.

Zhu, X., Zhu, L., Chen, Y., Tian, S., 2009. Acute toxicities of six manufactured nanomaterial suspensions to Daphnia manga. J. Nanopart. Res. 11 (1), 67–75. https://doi.org/10.1007/s11051-008-9426-8.

Zhu, Y., Zhao, Q., Li, Y., Cai, X., Li, W., 2006. The interaction and toxicity of multi walled carbon nanotubes with *Stylonychia mytilus*. J. Nanosci. Nanotechnol. 6 (5), 1357–1364. https://doi.org/10.1166/jnn.2006.194.

Index

Note: Page numbers followed by *f* indicate figures and *t* indicate tables.

C

D

E

F

G

H

O

Q

R

S

T

U

V

W

X

Y

Z

CPI Antony Rowe
Eastbourne, UK
November 23, 2019